Optogenetics
From Neuronal Function to Mapping and Disease Biology

Discovered little more than a decade ago, optogenetics – a revolutionary technique combining optical and genetic methods to observe and control the function of neurons – is now a widely used research tool. Optogenetics-driven research has led to insights into Parkinson's disease and other neurological and psychiatric disorders. With contributions from leaders and innovators from both academia and industry, this volume explores the discovery and application of optogenetics, from the basic science to its potential clinical use. Chapters cover a range of optogenetics applications, including for brain circuits, plasticity, memory, learning, sleep, vision and neurodegenerative and neuropsychiatric diseases. Providing authoritative coverage of the huge potential that optogenetics research carries, this is an ideal resource for researchers and graduate students and those working in the biotechnology and pharmaceutical industries and in a clinical setting.

Krishnarao Appasani is the Founder and Chief Executive Officer of GeneExpression Systems, a global conference-producing organization focusing on biomedical and physical sciences. He is an award-winning scientist, Fellow of the Royal Society of Biology and the editor of *Genome-Wide Association Studies: From Polymorphism to Personalized Medicine* (2016), *Epigenomics: From Chromatin Biology to Therapeutics* (2012), *MicroRNAs: From Basic Science to Disease Biology* (2007) and *RNA Interference: From Basic Science to Drug Development* (2005), all published by Cambridge University Press.

Optogenetics

FROM NEURONAL FUNCTION TO MAPPING AND DISEASE BIOLOGY

Edited by

Krishnarao Appasani

GeneExpression Systems, Inc., Waltham, MA, USA

Foreword by

Georg Nagel

University of Würzburg, Würzburg, Germany

CAMBRIDGE
UNIVERSITY PRESS

CAMBRIDGE
UNIVERSITY PRESS

University Printing House, Cambridge CB2 8BS, United Kingdom

One Liberty Plaza, 20th Floor, New York, NY 10006, USA

477 Williamstown Road, Port Melbourne, VIC 3207, Australia

4843/24, 2nd Floor, Ansari Road, Daryaganj, Delhi – 110002, India

79 Anson Road, #06–04/06, Singapore 079906

Cambridge University Press is part of the University of Cambridge.

It furthers the University's mission by disseminating knowledge in the pursuit of education, learning, and research at the highest international levels of excellence.

www.cambridge.org
Information on this title: www.cambridge.org/9781107053014
10.1017/9781107281875

First published 2017

Printed in the United Kingdom by TJ International Ltd. Padstow Cornwall in March 2017

A catalogue record for this publication is available from the British Library.

Library of Congress Cataloging-in-Publication Data
Names: Appasani, Krishnarao, 1959– editor.
Title: Optogenetics : from neuronal function to mapping and disease biology / edited by Krishnarao Appasani, GeneExpression Systems, Inc., Waltham, MA, USA.
Description: Cambridge : Cambridge University Press, 2017. | Includes bibliographical references.
Identifiers: LCCN 2016058565 | ISBN 9781107053014
Subjects: LCSH: Light – Physiological effect. | Cells. | Tissues – Optical properties. | Biotechnology.
Classification: LCC QH642 .O68 2017 | DDC 571.4/55–dc23
LC record available at https://lccn.loc.gov/2016058565

ISBN 978-1-107-05301-4 Hardback

Dedicated to:

The late ***Francis H. C. Crick*** *(1916–2004) 1962 Nobel Laureate Medicine or Physiology*

British-born legendary scientist, atheist, co-discoverer of DNA double helix, and a futurist, who foresaw (a new field) in which light might have the properties to serve as a tool to control one type of cell in the brain.

Contents

The color plates are to be found between pages 324 and 325

Contributors

Dany S. Adams
Department of Biology &
Tufts Center for Regenerative and Developmental Biology
Tufts University
Medford, Massachusetts, USA

Krishnarao Appasani
GeneExpression Systems, Inc.
Waltham, Massachusetts, USA

Raghu K. Appasani
University of Massachusetts Medical School
Worcester, Massachusetts, USA &
MINDS Foundation USA & India

Benjamin R. Arenkiel
Department of Molecular and Human Genetics &
Department of Neuroscience
Baylor College of Medicine
Jan and Dan Duncan Neurological Research Institute
Houston, Texas, USA

Taras Balandin
Institute of Complex Systems
Structural Biochemistry (ICS-6),
Jülich, Germany

Valentin Borshchevskiy
Institute of Complex Systems
Structural Biochemistry (ICS-6),
Jülich, Germany &

Laboratory for Advanced Studies of Membrane Proteins
Moscow Institute of Physics and Technology
Moscow, Russia

M. Christian Brown
Department of Otology and Laryngology
Massachusetts Eye and Ear Infirmary
Harvard Medical School
Boston, Massachusetts, USA

J. Barney Bryson
Institute of Neurology
University College London
London, UK

Karl Emanuel Busch
Centre for Integrative Physiology
University of Edinburgh
Edinburgh, UK

William N. Butler
Center for Social Brain Sciences
Department of Psychological and Brain Sciences
Dartmouth College
Hanover, New Hampshire, USA

Antoine Chaffiol
European Vision Institute
INSERM University Pierre and Marie Curie
Paris, France

Stephen E. Chang
Center for Social Brain Sciences
Department of Psychological and Brain Sciences
Dartmouth College
Hanover, New Hampshire, USA

Alexandre Charlet
Centre National de la Recherche Scientifique &
Institute of Cellular and Integrative Neurosciences &
University of Strasbourg Institute for Advanced Study
University of Strasbourg
Strasbourg, France

Jenny X. Chen
Department of Otology and Laryngology
Massachusetts Eye and Ear Infirmary
Harvard Medical School
Boston, Massachusetts, USA

Adam C. G. Crego
Center for Social Brain Sciences
Department of Psychological and Brain Sciences
Dartmouth College
Hanover, New Hampshire, USA

Jens Duebel
European Vision Institute
INSERM University Pierre and Marie Curie
Paris, France

Federico Esposti
Università Vita-Salute San Raffaele
Milan, Italy

Joan Fallon
Curemark, Inc.
Rye, New York, USA

Allyson K. Friedman
Department of Pharmacology and Systems Therapeutics
Institute for Systems Biomedicine
Icahn School of Medicine at Mount Sinai &
Department of Biological Sciences
Hunter College of City University of New York
New York, New York, USA

Marc Folcher
Department of Biosystems Science and Engineering
ETH Zürich, Zürich Switzerland

Grégory Gauvain
European Vision Institute
INSERM University Pierre and Marie Curie
Paris, France

Inna Gefen
Department of Medical Engineering
School of Engineering

Ruppin Academic Center
Netanya, Israel

Valentin Gordeliy
Institute of Structural Biology
University of Grenoble Alps
Grenoble, France &
Institute of Complex Systems
Structural Biochemistry (ICS-6),
Jülich, Germany &
Laboratory for Advanced Studies of Membrane Proteins
Moscow Institute of Physics and Technology
Moscow, Russia

Linda Greensmith
Institute of Neurology
University College London
London, UK

Valery Grinevich
German Cancer Research Center DKFZ
University of Heidelberg
Heidelberg, Germany &
Central Institute of Mental Health
Mannheim, Germany

Ivan Gushchin
Institute of Complex Systems
Structural Biochemistry (ICS-6),
Jülich, Germany

Andrea L. Gutman
Department of Psychology
University of Iowa
Iowa City, Iowa, USA

Ming-Hu Han
Department of Neuroscience
Friedman Brain Institute &
Department of Pharmacology and Systems Therapeutics
Icahn School of Medicine at Mount Sinai
New York, New York, USA

Alexander M. Herman
Department of Molecular and Human Genetics &
Department of Neuroscience
Baylor College of Medicine
Jan and Dan Duncan Neurological Research Institute
Houston, Texas, USA

Victor H. Hernandez Gonzalez
Department of Chemical, Electronic and Biomedical Engineering
Division of Sciences and Engineering
University of Guanajuato
Guanajuato, Mexico

Ute Hochgeschwender
Department of Neuroscience
College of Medicine
Central Michigan University
Mount Pleasant, Michigan, USA

Barbara Juarez
Department of Pharmacology and Systems Therapeutics
Institute for Systems Biomedicine
Icahn School of Medicine at Mount Sinai
New York, New York, USA

William F. Kaemmerer
Medtronic, Inc.
Minneapolis, Minnesota, USA

Sergey Kasparov
School of Physiology and Pharmacology
University of Bristol
Bristol, UK

Yonatan Katz
Department of Neurobiology
Weizmann Institute of Science
Rehovot, Israel

Dennis Kätzel
Institute of Applied Physiology
University of Ulm
Ulm, Germany &
Department of Experimental Psychology
University of Oxford
Oxford, UK

Koutarou D. Kimura
Laboratory of Neural Circuit Function
Department of Biological Sciences
Graduate School of Science
Osaka University
Osaka, Japan

Sonja Kleinlogel
Department of Physiology
University of Bern
Bern, Switzerland

Michael M. Kohl
Department of Physiology, Anatomy and Genetics
University of Oxford
Oxford, UK

Merab Kokaia
Epilepsy Center, Wallenberg Neuroscience Center
Faculty of Medicine
Lund University Hospital
Lund, Sweden

Kirill Kovalev
Institute of Complex Systems
Structural Biochemistry (ICS-6),
Jülich, Germany &
Laboratory for Advanced Studies of Membrane Proteins
Moscow Institute of Physics and Technology
Moscow, Russia

Elliott Kozin
Department of Otology and Laryngology
Massachusetts Eye and Ear Infirmary
Harvard Medical School
Boston, Massachusetts, USA

Ryan T. LaLumiere
Department of Psychology
University of Iowa
Iowa City, Iowa, USA

Ilan Lampl
Department of Neurobiology
Weizmann Institute of Science
Rehovot, Israel

Marco Ledri
Laboratory of Molecular Neurobiology
Department of Molecular and Developmental Neurobiology
Institute of Experimental Medicine &
Epilepsy Center, Wallenberg Neuroscience Center
Lund University Hospital
Lund, Sweden

Daniel J. Lee
Department of Otology and Laryngology
Massachusetts Eye and Ear Infirmary
Harvard Medical School
Boston, Massachusetts, USA

Michael Levin
Department of Biology &
Tufts Center for Regenerative and Developmental Biology
Tufts University
Medford, Massachusetts, USA

Olga Malyshevskaya
Division of Cellular and Molecular Neurobiology
Graduate School of Frontier Biosciences
Osaka University
Osaka, Japan

Meggane Melchior
Centre National de la Recherche Scientifique
University of Strasbourg
Institute of Cellular and Integrative Neurosciences

Tobias Moser
Institute for Auditory Neuroscience & Inner Ear Lab
Collaborative Research Center 889
University Medical Center Göttingen &
Synaptic Nanophysiology Group
Max-Planck-Institute for Biophysical Chemistry
Göttingen, Germany

Sabine Müller
Division of Mind and Brain Research
Department of Psychiatry and Psychotherapy
Charité Universitätsmedizin Berlin, Germany

Georg Nagel
Julius-von-Sachs Institute
University of Würzburg
Würzburg, Germany

Jatin Nagpal
Institute of Biochemistry &
Buchmann Institute for Molecular Life Sciences
Goethe University
Frankfurt, Germany

Dasha Nelidova
Friedrich Miescher Institute for Biomedical Research
Basel, Switzerland

Jay M. Patel
Department of Molecular and Human Genetics &
Department of Neuroscience
Baylor College of Medicine
Jan and Dan Duncan Neurological Research Institute
Houston, Texas, USA

Serge A. Picaud
European Vision Institute
INSERM University Pierre and Marie Curie
Paris, France

Vitaly Polovinkin
Institut de Biologie Structurale
Centre National de la Recherche Scientifique &
Institut de Biologie Structurale
Direction des Sciences du Vivant
Commissariat à l'Énergie Atomique
Grenoble, France &
Institute of Complex Systems
Structural Biochemistry (ICS-6),
Jülich, Germany &
Laboratory for Advanced Studies of Membrane Proteins
Moscow Institute of Physics and Technology
Moscow, Russia

Priyamvada Rajasethupathy
Stanford University School of Medicine
Standford, California, USA

Christian Renicke
Fachbereich Biologie – Genetik
Philipps-Universität Marburg
Marburg, Germany

Adi Schejter Bar-Noam
Faculty of Bioengineering
Israel Institute of Technology
Haifa, Israel

Vitaly Shevchenko
Institute of Complex Systems
Structural Biochemistry (ICS-6),
Jülich, Germany &
Laboratory for Advanced Studies of Membrane Proteins
Moscow Institute of Physics and Technology
Moscow, Russia

Hee-Sup Shin
Center for Cognition and Sociality
Institute for Basic Science
Daejeon, South Korea

Shy Shoham
Faculty of Bioengineering
Israel Institute of Technology
Haifa, Israel

Kyle S. Smith
Center for Social Brain Sciences
Department of Psychological and Brain Sciences
Dartmouth College
Hanover, New Hampshire, USA

Eriko Sugano
Laboratory of Visual Neuroscience
Department of Chemistry and Bioengineering
Iwate University
Morioka, Japan

Yan Tang
Schaller Research Group on Neuropeptides
German Cancer Research Center and CellNetworks Cluster of Excellence

University of Heidelberg
Heidelberg, Germany &
Key Laboratory of Brain Functional Genomics (Ministry of Education)
East China Normal University
Shanghai, China

Christof Taxis
Fachbereich Biologie – Genetik
Philipps-Universität Marburg
Marburg, Germany

Anja G. Teschemacher
School of Physiology and Pharmacology
University of Bristol
Bristol, UK

Hiroshi Tomita
Laboratory of Visual Neuroscience
Department of Chemistry and Bioengineering
Iwate University
Morioka, Japan

Jan Tønnesen
Achucarro Basque Center for Neuroscience,
Zamudio, Spain

Ai-Sun Tseng
School of Life Sciences
University of Nevada
Las Vegas, Nevada, USA

Charles-Francois Vincent Latchoumane
Center for Cognition and Sociality
Institute for Basic Science
Daejeon, South Korea

Jérôme Wahis
Centre National de la Recherche Scientifique
Orléans, France &
Institute of Cellular and Integrative Neurosciences
University of Strasbourg
Strasbourg, Germany

Henrik Walter
Division of Mind and Brain Research

Department of Psychiatry and Psychotherapy
Charité Universitätsmedizin Berlin, Germany

Nobuhiko Yamamoto
Division of Cellular and Molecular Neurobiology
Graduate School of Frontier Biosciences
Osaka University
Osaka, Japan

Boris V. Zemelman
Center for Learning and Memory
University of Texas
Austin, Texas, USA

Foreword

I am thankful to Krishnarao Appasani for making the phenomenal effort to prepare this volume on the rapidly expanding field of optogenetics, and for inviting me to offer a few of my own introductory statements. Optogenetics is an illuminating and a very simple, straight-forward concept: a heterologously expressed (i.e. **genetic**ally encoded) protein is activated by illumination (**opto**). The name "optogenetics" was first proposed for the light-induced depolarization of cells via Channelrhodopsin-2. The earlier, very successful experiments with genetically encoded GFP (and the whole family of derivatives and fusion constructs) could certainly also be included within optogenetics, in this case with *optogenetic sensors* whereas channelrhodopsins belong to *optogenetic actuators*. Even prior to gene technology, biochemists had the idea to use light (mostly UV) as a trigger for biochemical or cell biological reactions which might be called "optobiochemistry." Interestingly, some of the light-sensitive reagents from optobiochemistry resurfaced again in a kind of "chemical optogenetics." With chemical optogenetics I refer to optogenetics with synthesized light-sensitive reagents, e.g., caged ATP or caged neurotransmitters, and the very successful family of light-sensitive compounds based on azobenzene. These chemical optogenetic approaches have, so far, only been used by a few very active groups and consequently chemical optogenetics is only treated briefly in Chapters 2 and 23 of this book. Some important model animals such as *Drosophila* and zebra fish are missing from the book, but this just underlines the tremendous growth of the field of optogenetics in little more than a decade, necessarily leading to restrictions in the coverage possible in a single book.

In fact, it is impressive how much precious information is sampled in this newest collection, *Optogenetics: From Neuronal Function to Mapping and Disease Biology*, although optogenetic actuators took off only fifteen years ago. The book shows that optogenetics is not only a simple concept, but that optogenetics became widely accepted and successful due to a simple tool. This tool is Channelrhodopsin-2 (ChR2), nicely highlighted by its presence in the majority of the contributed chapters.

ChR2 was not "invented for optogenetics," its discovery resulted from basic research on light-sensitive membrane proteins and on phototaxis of green algae. Of course, we didn't invent the Channelrhodopsins, they were invented in microbes (over millions of years) by nature, whereas their use in animals is a recent human invention. Two other opsins (chlamyopsin-1 and 2 or cop1 & cop2) were already cloned before from the green alga *Chlamydomonas reinhardtii* by Peter Hegemann's group but nobody could confirm their proposed function as light-gated ion channels or channel regulators, although Peter sent the DNA to several laboratories (and unfortunately, their function is still unknown).

In 2002, in collaboration with Peter, who provided a cDNA for a presumable opsin from an EST bank in Kazusa, Japan, I could clearly show that this new opsin forms a directly light-gated ion channel with high permeability for protons. My collaborators Ernst Bamberg and Peter Hegemann, then agreed to the name channelopsin, instead of Peter's initial suggestion chlamyopsin-3 (a name which is still found in data banks!), as there is no homology to cop1 or 2. I studied this new opsin from the green alga *Chlamydomonas reinhardtii* by heterologous expression in vertebrate cells. After covalent binding of the chromophor retinal the protein became functional and, according to convention, was then called Channelrhodopsin (ChR). The publication of our discovery – channelopsin-1 or chop1 – therefore happened in the same year as the seminal work, describing light-activated neurons, by Zemelmann and Miesenböck in 2002 (see Chapter 2 of the present book). Only a few requests for chop1 DNA reached me in 2003. And in fact, chop1 from *Chlamydomonas reinhardtii* never made it to a useful optogenetic tool. Nowadays we know that the relatively small light-induced conductance change by ChR1 is mostly due to poor expression, caused by protein degradation. Nevertheless, after nailing its function down, we were confident enough that more such directly light-activated proteins would be found in nature and that they would become extremely useful – so we applied for patents on their use which got granted in the EU and US, and were, to my knowledge, the first ones.

We applied for the patent of channelopsin use because, already in 2002, we had the feeling that we'd discovered a protein of similar importance to GFP, although we did not expect its initial applications to be almost exclusively in the neurosciences. As you can see from the chapters in this book, channelrhodopsin-2 is the most-used optogenetic tool. Its application in the neurosciences took off after we published the characterization of this much-better expressing channelrhodopsin-1 homolog and explicitly showed in 2003 fast light-induced membrane depolarization and fast recovery in human embryonic kidney (HEK293) and other cells. In 2004 I got several requests for the DNA and provided it to Ishizuka/Yawo in Japan, to Stefan Herlitze (via my collaborator Hegemann), and to Karl Deisseroth, both at the time in the US, and others. Even before that, I convinced Alexander Gottschalk at the University of Frankfurt to work together to test light-effects in *C. elegans*, mediated by chop2 expression. After some initial difficulties we succeeded to light-manipulate *C. elegans*, already with the more efficient ChR2 mutant H134R (Nagel *et al.*, 2005). Another parallel effort, even without our knowledge and material transfer, started shortly after the 2003 ChR2 publication and should also be mentioned: Zhuo-Hua Pan trusted our publication and went on to synthesize the DNA for the active chop2 fragment, for expression in mouse retina. This is a very cheap and advisable approach nowadays but was, at a cost of 10,000 US$, quite risky and expensive for Zhuo-Hua Pan at the time (personal communication). His group's very early and successful vision restoration report was delayed in the publication process and appeared "only" at the beginning of 2006 (Bi *et al.*). But for some reason (my guess: "creative writing" in reviews) only our collaborative work with Karl Deisseroth (Boyden *et al.,* 2005) is considered by most as the starting point for chop2-mediated optogenetics. The four other publications; (Bi *et al.,* 2006, *Neuron*; Ishizuka *et al.,* 2006, *Neuroscience Research*; Li *et al.,* 2005, *Proceedings of the National Academy of Sciences*, USA; Nagel *et al.,* 2005, *Current Biology*) are often completely disregarded.

As evidenced by the Bi *et al.*, (2006) paper, vision restoration efforts started early and seem to belong to the most promising therapeutic applications of ChR2, summarized in several chapters which form Part V of this book. But what I had hoped for in the beginning, and what now seems surprising to some, is that we are also seeing increasing applications of microbial opsins in non-neuronal tissue, which is again covered in this up-to-date book. And we see increasing other "optogenetic tools," including the other microbial opsins which are light-activated ion pumps, the light-activated adenylyl and guanylyl cyclases, and very promising novel synthetic constructs on the basis of the LOV (light-oxygen-voltage) domain. After some refinement of these tools (especially to guarantee inactivity, if not illuminated with blue light), I expect that they will contribute to a further boost for optogenetics in all fields of biological research, not just in the neurosciences.

The light-activated chloride pumps – the halorhodopsins – and the light-activated proton pumps, like bacteriorhodopsin, Arch, and Mac, entered optogenetics well after channelrhodopsin-2. This might seem surprising, as bacteriorhodopsin from *Halobacterium salinarum* was shown to be functional in animal cells as early as 1995 (Nagel *et al., Federation of the European Biochemists Society Letters*). But "only" light-activated currents were shown and no light-induced hyper-polarization, which seemed to us trivial, as it is a necessary consequence of the light-activated outward currents.

In fact, halorhodopsin was introduced to nerve and muscle cells two years after the above-mentioned five channelrhodopsin papers, as the result of a discussion (and subsequent collaboration) about possible "optogenetic inhibitors" between Ernst Bamberg, Karl Deisseroth, Alexander Gottschalk, and myself in a typical Frankfurt cider eatery in 2006. This happened after a lecture of Karl's at the Max-Planck-Institute in Frankfurt. He was invited by me (already in nearby Würzburg, Bavaria) and Ernst Bamberg from the Max-Planck-Institute. His trip to Frankfurt was funded by a grant to me and Karl from BaCaTec, a Bavarian-Californian funding agency, and the Max-Planck-Institute in Frankfurt. The collaboration yielded a common paper (Zhang *et al.,* 2007) showing our ten- year-old halorhodopsin data in animal cells, the light-induced inhibition of action potentials (from Karl's lab) and light-inhibited swimming of *C. elegans* (from Alexander's lab). But, as you might guess, again different recounting exists about the beginning of the use of light-activated ion pumps as optogenetic tools: the importance of creative writing.

With these rather personal accounts I want to conclude by saying that I strongly recommend this carefully assembled book by Krishnarao Appasani to as many readers as possible. The book is a rich source for scientists who already use chop2 or related tools, as well as for scientists who are considering using optogenetic approaches in their research. Therefore, I wish this book great success and I wish chop2 and the other optogenetic tools a continuing bright future!

March 01, 2017

Georg Nagel
Julius-von-Sachs Institute
University of Würzburg
Würzburg, Germany

Preface

"You", your joys and your sorrows, your memories and your ambitions, your sense of personal identity and free will, are in fact no more than the behavior of a vast assembly of nerve cells ...

Francis H. C. Crick, *The Astonishing Hypothesis: The Scientific Search for the Soul*, New York, Scribner, 1994.

As the famous German scientist Max K.E.L. Planck (1858–1947), once said, "Knowledge precedes application." Later, the Austrian-British philosopher of science Karl R. Popper (1902–1994) believed that "before we can find the answers, we need the power to ask new questions, in other words, we need new technology." A key example of such a technological advance is the development of the new field of *optogenetics* by Karl Deisseroth, Edward Boyden, Ernst Bamberg Peter Hagemann, Georg Nagel and Gero Miesenböck. Gero Miesenböck defines optogenetics as "using genetic strategies for observing and controlling the function of neurons with light," whereas Deisseroth states that it is "the combination of genetic and optical methods to achieve gain or loss of function of well-defined events in specific cells of living tissue." The biochemistry of opsins helped to develop the new field of optogenetics and molecular engineering. Use of optogenetics as an important research tool continues to grow rapidly, and it is now used in more than 1,000 laboratories. Use of its associated technologies for delivering light deep into organisms as complex as freely moving mammals, for targeting light-sensitivity to cells of interest, and for assessing specific readouts, or effects, of this optical control are also expanding.

Although *optogenetics* was discovered little more than a decade ago, it has recently been applied to few representative neurological and cardiac diseases. The potential utility of these techniques to help us understand *"neuro-cognitive biology and genetics"* more broadly is just beginning. There is a growing interest in utilizing these techniques in academic research laboratories worldwide, to better understand their applications in neurobiology and neurology in order to develop new diagnostic tools and drug targets for neurological disorders. The emergence of these and other neuro-techniques led US President Barack Obama to prioritize brain research. In 2013, he announced the **B**rain **R**esearch through **A**dvancing

Innovative **N**eurotechnologies (BRAIN) Initiative, a collaborative, public-private research project for the development and application of innovative technologies that can create a dynamic understanding of brain function, and uncover the mysteries of brain disorders, such as Alzheimer's and Parkinson's diseases, depression and traumatic brain injury.

Optogenetics: From Neuronal Function to Mapping and Disease Biology is intended for those in the neuroscience, neurology, neural engineering, biotechnology, genetics, genomics, pharmaco-genomics and molecular medicine fields. There are a few books already available covering optogenetics. *(Miranda, E. (2010). Neural Engineering, BCI, Optogenetics, and Brain Implant. General Books, LLC; Surhone, L., Tennoe, M. and Henssonow, S. (2011). Optogenetics. Betascript Publishing Press; Scaglia, B. (2011). Optogenetics and Anxiety. Webster's Digital Services; Miesenböck, G. (2011). Optogenetics: Circuits, Genes, and Photons in Biological Systems. John Wiley & Sons; Knopfel, T. and Boyden, E. (2012). Optogenetics: Tools for Controlling and Monitoring Neuronal Activity. Elsevier Science Press; Hegemann, P. and Sigris, S. (2013). Optogenetics. Walter De Gruyter, Press; Ilango, A. and Lobo, M.K. (2014). Neural Circuits Underlying Emotion and Motivation: Insights from Optogenetics and Pharmacogenetics. Frontiers in Behavioral Neurosciences. Volume 8; and Yawo, H., Kandori, H. and Koizumi, A. (2015). Optogenetics: Light Sensing Proteins and their Application. Springer).*

For example, E. Miranda's *Neural Engineering, BCI, Optogenetics, and Brain Implant* highlighted the use of optogenetics for neural engineering and brain implant purposes, whereas L. Surhone, M. Tennoe and S. Henssonow's *Optogenetics* gave an overview of the field. Miesenböck (2011) covered the role of optogenetics in studying neural circuits, whereas Knopfel and Boyden (2012) emphasized the importance of optogenetic tools in the study of neural cell activity. Hegemann and Sigris (2013) and Ilango and Lobo (2014) summarized the use of opsin proteins in optogenetics and pharmacogenetics, respectively. A recent book by Yawo et al. (2015) focused on the light-sensing proteins. This present book differs, in that it is the first text completely devoted to combining optogenetics' use in model organisms, opsin biological tools and their applications in neurobiology and neurodegenerative diseases. Special emphasis is placed on the development of new-generation tools for in vivo imaging, and for the study of learning, neuro-psychiatric diseases and behavior. This book also focuses on the use of optogenetics in the context of vision re-storage, memory, plasticity and sleep.

The goal is for this book to serve as a reference for graduate students, post-doctoral researchers and instructors, and as an explanatory analysis for executives and scientists in molecular medicine, molecular engineering, biotechnology and pharmaceutical companies. Our hope is that this volume will serve as a prologue to the field for newcomers and as a reference for those already active in the field. We have carefully chosen the chapters, written by experts in the field from both academia and industry, and have divided the chapters into appropriate sections to support the theme expressed in the subtitle of this book: *From Neuronal Function to Mapping and Disease Biology*. Developing *"novel optogenetic tools"* is likely to become a prerequisite for *"deep-brain stimulations"* used for treating patients with epilepsy and other neurological disorders. The elegant and revolutionary optogenetic approaches that are covered in this book will undoubtedly have tremendous commercial promise for the development of innovative surgical procedures for human neurological diseases.

Many people have contributed to making our involvement in this project possible. We thank our teachers for their excellent teaching, guidance and mentorship, which have helped us to bring about this educational enterprise. We are extremely grateful to all of the contributors, without whose commitment this book would not have been possible. Many people have had a hand in the preparation of this book. Each chapter has been passed back and forth between the authors for criticism and revision; hence, each chapter represents a joint contribution. We thank our reviewers, who have made the hours spent putting this volume together worthwhile. We are indebted to the staff of Cambridge University Press, and, in particular, to Katrina Halliday for her generosity and efficiency throughout the editing of this book; she truly understands the urgency and need for this volume. We also extend our appreciation to Megan Waddington and Caroline Mowatt for their excellent cooperation during the development of this volume. We also want to thank Professor Georg Nagel of the University of Würzburg, Germany, (who is one of the discoverers of Channelrhodopsin, which paved the way for the field of optogenetics) for his thoughtfulness in writing a foreword to this book. Last, but not least, we thank Shyamala Appasani for her understanding and support during the development of this volume.

This book is the eighth in the series of *Gene Expression and Regulation*, and the fourth joint project of father and son. A portion of the royalties will be contributed to the Dr. Appasani Foundation (a non-profit organization devoted to bringing social change through the education of youth in developing nations) and The MINDS Foundation (**M**ental **I**llness and **N**eurological **D**isease**s**), which is committed to taking a grassroots approach to providing high-quality mental healthcare in rural India.

Krishnarao Appasani
Raghu K. Appasani

Part I

Optogenetics in Model Organisms

1 Introduction to Optogenetics: From Neuronal Function to Mapping and Disease Biology

Krishnarao Appasani and Raghu K. Appasani

In our quest to move the enterprise of science forward we should never forget that we do not know where this long march is taking us or what we will need to get there.

Karl Deisseroth, *Scientific American*, October 20, 2010.

The brain, a highly heterogeneous organ, consists of many individual cell types. Understanding the properties of these cell types is critically important in forming a collective appreciation of how the brain works, and how various diseases and disorders arise when the components that constitute the brain are impaired or their normal functions altered. Toward this end, it is vital to develop and employ methods for precisely manipulating specific cell types within the intact brain. For hundreds of years, humanity has tried to understand the nervous system. Since the pioneering electrophysiology studies of Italian scientist Luigi Galvani in the late eighteenth century (1791), in which he developed the theory of electrical excitation of neurons while studying the muscles in frogs, advances in techniques, methods and knowledge have brought us closer to understanding the mechanisms governing neuronal activity (Goldensohn, 1998). In 1979, Francis Crick suggested that the major challenge facing neuroscience was the need to control one type of cell in the brain while leaving others unaltered. He later speculated:

> One of the next requirements is to be able to turn the firing of one or more types of neuron on or off in the alert animal in a rapid manner. The ideal signal would be light, probably at an infrared wavelength to allow the light to penetrate far enough. This seems rather far-fetched but it is conceivable that molecular biologists could engineer a particular cell type to be sensitive to light in this way. (Crick, 1979, 1999)

He proposed that a light might have the necessary properties to serve as a control tool, but no one had experimentally demonstrated it.

It took almost three decades to obtain experimental evidence, which was provided for the first time in 2002 by Gero Miesenböck (then at Yale, now at Oxford), (Zemelman *et al.*, 2003), and in 2005 by Edward Boyden (then at Stanford, now at MIT) and Karl Deisseroth (of Stanford) using the microbial opsin proteins discovered by the German scientists Ernst Bamberg, Peter Hagemann and Georg Nagel

(Boyden *et al.*, 2005). In this context, it is very important to mention the origin and biology of these opsins, as they significantly contributed to the development of the new field of *optogenetics*. In the early 1970s, bacteriorhodopsin (a microbial equivalent of the vertebrate rhodopsin protein), a retinal-containing 7-helical transmembrane protein, was first discovered in the archaeal halophiles by Walther Stoeckenius and Dieter Oesterhelt, both then at the University of California, San Francisco (Oesterhelt and Stoeckenius, 1971). In the early 1980s, James Spudich and his colleagues identified these bacterial proteins as the light sensors (each tuned to a different wavelength) that produce complex photactic behavior in the extremophiles, which includes ionic homeostasis, energy storage, phototaxis, flagellar motor rotation and photosynthesis (Bogomolni and Spudich, 1982; Spudich, 2006).

Although not the focus here, it is worth mentioning that the biochemistry of both bacteriorhodopsin and bovine rhodopsin were significantly studied for a period of two decades using synthetic biology approaches by my post-doctoral mentor Nobel laureate H. Gobind Khorana's colleagues and several others (Sakmar, 2002; Kim *et al.*, 2005). During this time, in the early 2000s, scientists from Germany, including Ernst Bamberg, Peter Hegemann (then at the University of Regensburg, now at the Humboldt University of Berlin), Georg Nagel (then at University of Frankfurt) and their colleagues, discovered *channelrhodopsins* (*ChR1 and ChR2*), another class of microbial light-gated ion channel proteins, and demonstrated their use in studying rapid behavioral responses in excitable cells of *Caenorhabditis elegans* (Nagel *et al.*, 2002, 2003, 2005). These bacterial proteins act as a single-component ion pump that can be briefly activated by photons of green light – a remarkable all-in-one molecular machine (Deisseroth, 2010).

As mentioned to me by Edward Boyden in a private conversation (in the context of organizing a theme conference on the topic of optogenetics) in the summer of 2012, quoting his paper (Boyden, 2011), during 2004, Karl Deisseroth, a young physician-scientist in Robert Malenka's lab at Stanford, teamed up with his graduate student Feng Zhang and Edward Boyden (who was also a graduate student in Jennifer Raymond and Richard Tsien's lab at Stanford) and in collaboration with the previously described German team that discovered channelrhodopsins, introduced ChR2 into mammalian cell cultured neurons for the first time. Using pulses of visible light, they were able to attain reliable, millisecond-precision control over the patterns of firing of action potentials in the cells (Boyden *et al.*, 2005). In 2005, two seminal articles (Lima and Miesenböck, 2005; Boyden *et al.*, 2005) launched the era of optogenetics. This groundbreaking American–German collaborative work laid the foundations for the new field of optogenetics. Soon, several other groups had also demonstrated the same phenomenon in other model organisms. Most of these bacterial proteins were later proven to turn neurons on or off in response to light or photons. Since then, many groups have developed optogenetic tools, pushed opsin molecular engineering forward, and revolutionized our ability to study the processes in neuronal circuits, all with the promise of potential new approaches for the treatment of different diseases (Montogomery *et al.*, 2016; Vann and Xiong, 2016; Song and Knopfel, 2016; Hunter, 2016).

The term 'optogenetics' first appeared in a mini-review (Deisseroth, *et al.*, 2006), and several definitions of optogenetics exist: Ernst Bamberg, a German biochemist, defines optogenetics as "the use of genetically encoded light-activated proteins for manipulation of cells in an almost non-invasive way by light," (Adamantidis *et al.*, 2015) whereas Michael Häusser, a British neuroscientist, felt that "any approach that combines optical interrogation with genetic targeting qualifies as optogenetic, and that includes the use of genetically encoded activity sensors" (Adamantidis *et al.*, 2015). Richard Tsien, an American neuroscientist, considered optogenetics to be the "use of genetically encoded molecules to excite and inhibit neurons" (Adamantidis *et al.*, 2015). However the definitions vary, optogenetics is now more than ten years old, and it follows from the Nobel techniques such as Patch-Clamp (by German biophysicist Erwin Neher and German physiologist Bert Sakman) and the Nobel discovery of fluorescent proteins (by American biochemists Osamu Shimomura, Martin Chalfie, and Roger Tsien) that revolutionized the fields of neurophysiology and biological imaging. Although optogenetics arose from neuroscience, it addresses a much broader unmet need in the study of biological systems that extends beyond neuroscience to muscle, cardiac, and embryonic stem cells. Disease models have also been explored, including for Parkinson's disease, anxiety, retinal degeneration, respiration, addiction, and depression. It is hoped that a combination of viral optogenetic therapy and optical-fiber implants could provide treatments to combat specific neural abnormalities in neurodegenerative diseases. The present book details the historical perspectives of the field of optogenetics, its use in model organisms, the technology platform, applications in neurobiology for psychiatric and behavioral diseases, synaptic plasticity, and the restoration of vision, as well as its application for conditions relating to memory, learning and sleep.

Scope of This Book

This text consists of thirty-two chapters, grouped into six parts, starting with the biology and techniques, leading to descriptions of the applications as follows:

Part I: Optogenetics in Model Organisms

This section consists of five chapters. As Zemelman describes in the second chapter, directed manipulation of brain function can be traced back decades. Specifically, working with large mammals and primates, Delgado demonstrated that many properties attributed to the primitive brain – sleep, nurture, hunger, aggression – could be modified by stimulating narrowly circumscribed groups of neurons, including by "remote control" (Delgado 1964). Consequently, these studies motivated the performance of localized electrical stimulation to diagnose, condition, and treat human subjects suffering from behavioral and psychiatric disorders. Crick's prediction that "the ability to probe defined groups of neurons with [light] will hold the key to an understanding of neural systems" succinctly framed the revolution in experimental neuroscience that followed fifteen years later. Subsequently, Zemelman's initial and seminal experiments on studying ion currents in *Drosophila melanogaster* in the presence of retinal and visible light laid the foundation for what several years later became the field of optogenetics (Zemelman and Miesenböck, 2003).

Caenorhabditis elegans is one of the best-suited model organisms in which to use light stimulation to control gene expression. Most importantly, because of its transparent body, stimulatory light is readily delivered to the target cells, its "connectome biology" is well known, but the neuronal dynamics of its neural circuit signaling have not yet been mapped out. Therefore, in Chapter 3, Kimura and Busch describe the functional studies of chemical synapse function *in vivo*, using optical interrogation studies to understand the functional circuitry, and behavioral analyses. The complete "wiring diagram" of the nervous system in *C. elegans* is highly conserved, and exhibits a rich repertoire of quantifiable behavior and simple forms of associative learning. In fact, *C. elegans* was the first animal in which optogenetic tools based on microbial rhodopsins were successfully implemented and behavior was remotely controlled (Nagel *et al.*, 2005). In Chapter 4, Nagpal highlights the diversity of optogenetic tools, primarily the actuators that have been applied in *C. elegans* and the insight they have provided in the field of neuroscience.

Xenopus laevis (the African clawed frog) is an important model organism for the study of developmental biology due to its exceptional resistance to infection and the external development of its eggs in larger numbers. Importantly, its oocytes are used as expression systems in which to determine the characteristics of channels and pumps, thus techniques for successful expression of exogenous ion translocators are well established. In Chapter 5 Adams *et al.* show that optogenetics can be useful for studying ion-flux dependent developmental processes *in vivo*, and demonstrated the potential utility of optogenetics for regenerative medicine. They describe the use of the light-activated hydrogen pump Archaerhodopsin to initiate full regeneration of a complex vertebrate appendage, especially to study the role of V_{mem} during normal processes and the role of misregulation of V_{mem} in disease.

Part II: Opsin Biology, Tools, and Technology Platform

This section consists of six chapters. The proton-pumping and anion-pumping rhodopsins have been known for a long time. Therefore, it is natural that in addition to channelrhodopsin which is used to fire neurons, proton and chloride pumps were also used as optogenetics tools to silence neurons. For different reasons, sodium and, in particular, potassium light-driven pumps were long-sought tools of optogenetics. However, the cation (sodium) pumps were only described in 2013. In Chapter 6, Shevchenko *et al.* describe a family of sodium light-driven pumps, their functional, structural properties, and the applications of potassium-pumping proteins in optogenetic experiments. In combination with the light-activated channelrhodopsin 2, which is a well-known molecular off-switch, the KR2 potassium pump would form a perfect pair of tools for the precise control of nerve cell activity; those are described in detail by Shevchenko *et al.* in Chapter 6. Optogenetics was combined with electrophysiology in order to evaluate the effects on neuronal activity. Since the field of electrophysiology spans many different preparations, from tissue culture to live animals, different technical approaches had to be incorporated or invented when combined with optogenetics. Therefore, Katz and Lampl (Chapter 7) review the current methods for simultaneous electrophysiology and optogenetics *in vivo*. Depending on the electrophysiological requirements, different probes for both extracellular recordings, and intracellular recordings were developed.

During development, fundamental neuronal circuits are formed by a genetically defined program, and the fine connections are modified by neuronal activity (Katz and Shatz, 1996). Such refinement processes are essential to establish functional connectivity in the brain. A key issue is how neuronal activity influences axonal growth and branching. To address this issue, it is essential to manipulate the firing activity of the axons that are observed morphologically. Electrical stimulation with fine electrodes has been widely used for the manipulation. However, it is difficult to specify which cells or fibers are stimulated via the conventional method. The optogenetic method can be adopted to overcome this problem. This method is useful not only for stimulus-based physiological experiments, but also for morphological and developmental studies. An interesting question is how different patterns of spontaneous neuronal activity affect axonal growth, branching, and the refinement of neural circuits in general. The remote control of neuronal activity by light combined with morphological observation provides a new avenue for investigating the implication of activity in neuronal circuit formation. Thus, in Chapter 8, Malyshevskaya and Yamamoto demonstrate the methodological efficiency of the optogenetic technique for developmental studies of neuronal circuit formation, focusing on axonal growth of cortical neurons which was regulated by firing activity.

In recent years, many different tools have been developed that use light to influence protein activity by regulating synthesis, localization, activity, or stability, and these have been described in depth in several reviews. In Chapter 9, Renicke and Taxis discuss practical aspects of using the different psd module variants in yeast, and provide guidelines for experimental design and applications. Overall, photoactuators from the optogenetic toolbox offer ample opportunities to influence at the molecular level and to interfere with regulatory processes in a non-invasive way. Over the last decade, the use of seven-transmembrane opsin proteins as optogenetic tools has been extended to structurally similar proteins, including archaeal halorhodopsin. Folcher (Chapter 10) reviews the recent advances in the bioengineering of photo-activatable nucleotide cyclases, especially secondary messenger-based optogenetic actuators, in the context of their fields of application. It is possible to fine-tune secondary messenger signalling at the single cell level or in a whole animal experiment by expressing the optogenetic transgene under the control of a tissue-specific promoter. Moreover, as described elegantly by Folcher, the knowledge-based protein engineering strategies targeting enzyme activities or domain organization have further broadened the possible strategies for adapting optogenetic devices to specific application scenarios. Folcher is hoping to develop an implantable, wirelessly piloted machine-cell interface, based on a light communication channel, which will become an innovative strategy for tomorrow's therapeutics.

Comprehensive interrogation of neuronal circuits requires acute as well as chronic manipulations of spatially defined subpopulations as well as entire populations dispersed over the brain. Thus, there is a need to combine optogenetic and chemogenetic approaches to allow the use of both modes of interrogation in the same brain circuit, and ideally through the same actuator molecule, thereby facilitating the comprehensive study of neuronal systems. Although a number of groups have presented the combined toolbox of chemogenetic and optogenetics to address pertinent questions, in Chapter 11 Hochgeschwender introduces the basic building blocks for bioluminescence-driven optogenetics. These novel classes of tools can be improved and extended in numerous ways and find applications beyond photonic control of neurons in modifying many cell types and cellular processes.

Part III: Optogenetics in Neurobiology, Brain Circuits, and Plasticity

This section consists of five chapters. Compared to electrical stimulation, optical stimulation offers the potential for superior spatial control over what portion of brain tissue is affected by the stimulation. As with electrical stimulation, optical stimulation can be applied without the pharmacological side effects and dosing considerations that invariably accompany drug interventions. Current neuromodulation therapies for epilepsy may be a good first choice for the application of optogenetics to a central nervous system disorder. In Chapter 12, Kaemmerer describes his group's modified approaches to deep-brain stimulation. Over the past couple of decades, it has become increasingly clear that astroglia play a fundamental role in neurotransmission, not only by fuelling it and removing the waste, but also by integrating, amplifying, and modulating neuronal signals. Since astrocytes express a wide variety of receptors for neurotransmitters and can release glio-transmitters in response to stimulation, they are intimately linked with local neuronal activity. In Chapter 13, Teschemacher and Kasparov summarize the invaluable role of optogenetic approaches in the study of astrocytes and their natural signaling modes.

In recent years, viral techniques have been combined with optogenetics, which allowed study of the modifications of activity in different brain regions and circuits. Recently, several studies have focused on viral/optogenetic manipulation of "non-classical" neurotransmitters – neuromodulators and neuropeptides. In Chapter 14, Tang *et al.* provide an excellent overview of viral and transgenic techniques applied to specific targeting of hypothalamic neurons, expressing oxytocin–neuropeptide orchestrating social behavior. Using viral delivery methods in combination with optogenetics, Tang *et al.* have achieved the physiology of peptidergic neurons and peptidergic signaling mechanisms in the brain. To date, 5 out of 100 neuropeptidergic brain circuits have only been explored by optogenetics; that leaves the opportunity to explore the fundamental cell biology of "post-synaptic" signaling.

The ability to manipulate specified cell types in a targeted manner is essential for decoding the contribution of individual neuronal populations in various physiological states. Deciphering patterns of functional synaptic connectivity is the main focus of both basic and clinical neurosciences. The evolution of optogenetic methods to target neurons and their circuits for activity manipulations has provided unprecedented opportunities to investigate functional connectivity in previously unimaginable detail. In Chapter 15, Herman *et al.* review the current optogenetic "toolbox" that is available to investigate brain circuitry, and further highlight the various genetic methods used to express optogenetic reporters with cell-type specificity. The mapping of neural circuits allows the high-throughput determination of the anatomical rules that govern the connections between the different nerve cells in a neural network. Although the invention of optical mapping of neural circuits pre-dated the advent of optogenetics by a decade, the use of optogenetic tools for circuit mapping did not come easily to the field. This was primarily because of the difficulty in localizing optogenetic excitation to the somato-dendritic compartment as opposed to axons and synapses. Accordingly, in Chapter 16, Kohl and Kätzel discuss strategies surrounding how this feature has either been harnessed to map long-range connections, or was overcome technologically to map local circuits.

Part IV: Optogenetics in Learning, Neuropsychiatric Diseases, and Behavior

This section consists of five chapters. Traditional methods have provided a strong groundwork for understanding the neural correlates of drug addiction. However, the development of optogenetic approaches has significantly advanced the study of addiction and reward-related learning by affording the ability to more closely interrogate neural circuitry and pathway-specific synaptic adaptations with far greater temporal and spatial precision than was possible in the past. In Chapter 17, Gutman and LaLumiere summarize the use of optogenetic techniques in understanding the role of the mesocorticolimbic system in learning, and to investigate this in the context of drugs of abuse. The contributions of optogenetic studies to the field's understanding of the neural correlates of addiction will undoubtedly inform the identification and development of treatments for this disease. In the last sixty years, research into the biological basis of neuropsychiatric disorders has progressed tremendously because of technological precision in studying cellular neurobiology, advances in the scale of genome-wide studies, refined animal models, and the tuned dissection of neural circuits. However, with the success of deep-brain stimulation to treat major depressive disorder (MDD) and substance-use disorders – a technique that uses chronically implanted stimulating electrodes to specifically target brain regions in the reward system for electrical stimulation – there is an ever-evolving hypothesis that neuropsychiatric disorders are a neural circuit disorder (Volkow and Koob, 2015). Optogenetics has offered researchers the unique ability to selectively target genetically distinct cell populations or neural circuits for temporally precise stimulation or inhibition in awake and behaving animal models. This has shed light on the neural circuit mechanisms involved in depression and substance abuse. In Chapter 18, Juarez *et al.* describe the optogenetics, in conjunction with validated animal models for neuropsychiatric disorders, which have dramatically altered how researchers approached and interpreted the study of depression and substance-use disorders.

In recent years, there has been a rapid expansion of insights into brain function related to reward learning and goal-directed behavior through the use of optogenetics technologies. In addition, we also have a sense of the brain circuits and neural dynamics related to reward and goal-directed behavior in learning and motivational processes, but little understanding of how specific patterns of neural activity or specific circuits control these behaviors. However, optogenetics has begun to help researchers circumvent this gap by employing highly precise interventions in neural activity as applied to specific questions. In Chapter 19, Crego *et al.* outline several advantages and considerations for using optogenetics in answering behavioral questions, especially in understanding the neural basis of goal-directed learning and reward-related behavior.

Epilepsy is one of the most common neurological disorders, affecting up to 60 million people worldwide and potentially occurring at all ages. Importantly, epilepsy is not a single disease, but is considered to be a family of disorders with different etiologies (Shorvon, 2011). Epilepsy was among the first diseases to be explored using optogenetics approaches to intercept the neural activity underlying seizures. Tønnesen *et al.* introduce optogenetic strategies for the first time to treat epilepsy, as described in Chapter 20, but they are still facing obstacles prior to initiating clinical trials. The optogenetic therapy for epilepsy does not seem to provide either a cure or disease modification; rather, it offers a prosthetic approach by suppressing seizures when they are about to occur or once initiated.

The major function of the nervous system in all higher organisms is to receive and process sensory information from the environment and, in turn, to produce an appropriate response to adapt to changes in the environment. In most cases, the elicited response takes the form of movement or 'motor function'. Almost all human behavioral output, ranging from locomotion and articulated hand movement to speech and expression of emotions, is underpinned by motor functions. The neural circuits that govern motor function are extremely complex, and damage to almost any part of the motor system, either as a result of trauma or disease, can result in paralysis, i.e. an inability to control muscles. In Chapter 21, Bryson and Greensmith focus on two such insults that can critically affect the motor system: traumatic spinal cord injury and the progressive neurodegenerative disease, amyotrophic lateral sclerosis. Additionally, they emphasize the translational application of stem cell–based neural replacement techniques and the artificial control of motor function using optogenetics, as a therapeutic strategy to restore the lost motor function.

Part V: Optogenetics in Vision Restoration and Memory

This section consists of six chapters. The retina is our most complex sensory organ. Blindness often results from dysfunction of this complex neural network. At present, degenerative diseases of the retina remain incurable. Patients with these conditions suffer progressive visual decline resulting from the ongoing dysfunction or loss of retinal cells, with photoreceptors being particularly prone to degeneration.

One of the main goals of translational retinal research is to develop strategies to repair the retina. Given the great complexity of this tissue, any attempt to restore retinal function requires quantitative knowledge of retinal neural circuits as well as methods to target these circuits specifically. In Chapter 22, Nelidova and Esposti summarize the viral delivery of optogenetic tools to retinal cell types to restore natural vision. It is noteworthy that delivery of Adeno-associated virus to the retina of human patients has been shown to be safe in multiple clinical trials. All the pre-clinical reports thus far have demonstrated that optogenetic therapy is safe and is capable of inducing light-driven activity in mouse models of retina degeneration, irrespective of the cell type chosen for therapy. To date, the only therapy available for late-stage photoreceptor degeneration is based on electronic chips that are implanted in close proximity to the retina. The chips are costly and, due to their small size, recover vision only locally, with crude and non-physiological electrical activation. Because of this, many investigators have explored optogenetic therapies which collectively aim to render surviving inner retinal cells directly sensitive to light. Recently, optogenetic vision recovery has entered a new era, where light-sensing proteins are being engineered according to the need for a specific application. The potential advantages of optogenetic vision recovery treatment are that they are ambulant, long-lived, low-cost, and have the theoretical potential to recover high-resolution vision across the entire visual field. In Chapter 23, Sonja Kleinlogel discusses the use of such an optogenetic approach using opsin-based light sensors for potential vision restoration in blind patients suffering from photoreceptor degeneration.

Degeneration of photoreceptors (the cells responsible for the canonical image forming photo-transduction) is the cause of the two most common blinding diseases – retinitis pigmentosa and age-related macular degeneration – that affect millions of people worldwide. It would thus be highly beneficial to reintroduce light sensitivity, using optogenetic therapy in surviving cells, in order to restore vision. The use of genetically encoded light-activated proteins has revolutionized all fields of neuroscience by allowing the control of the excitation state of neurons with high spatial and temporal precision. These new powerful methods have had a major impact on every research field, and the optogenetic toolbox is still evolving as it becomes more highly optimized and diversified. In Chapter 24, Gauvain *et al.* describe pre-clinical tests in humans using optogenetics-based vision-restoration strategies. A retinal neuroprosthesis is a device aimed at providing an artificial sense of vision by translating visual scenes into appropriate spatio-temporal patterns of neuronal activity. A number of studies by several research groups have explored the fundamental feasibility of using ChR2 and other optogenetic probes in an optical retinal prosthesis, clearly suggesting that this technology may provide a viable path to vision restoration (Roska and Pepperberg, 2014). Combining optogenetics with advanced light projection technologies can potentially allow a relatively direct route toward non-contact restoration of high-acuity vision. In Chapter 25, Bar-Noam *et al.* describe how spatial light modulation technology can be harnessed as a powerful tool for large-scale, cellular-resolution, massively parallel artificial control of neuron population activity in the isolated blinded retinas through "holographic optical neural interfaces."

Even after a patient has become blind, the retinal neurons other than the photoreceptor cells are relatively well preserved, and techniques to use these surviving neurons to restore visual function are being investigated. In addition to its involvement in photoreception, channelrhodopsin-2 (ChR2), derived from the green algae *Chlamydomonas*, functions as a light-gated cation-selective channel that permits the intracellular passage of cations (Nagel *et al.*, 2003). In Chapter 26, Tomita and Sugano have described a gene therapy approach for vision regeneration by using ChR2 and a modified *Volvox*-derived channelrhodopsin-1, which was developed in their laboratory. Vision-regeneration research using optogenetics is not limited to the endowment of light sensitivity to ganglion cells (tertiary neurons), as presented in this chapter.

One of the great mysteries of the human mind is its ability to form memories. The ability to faithfully store and retrieve information is an important adaptive quality that allows humans to reflect on, engage with, and respond to an ever-changing environment. Disturbances of memory significantly diminish quality of life and also contribute to a number of disorders, including Alzheimer's, fronto-temporal dementias, drug addictions, and post-traumatic stress disorder. Pioneering studies to understand memory processes have been transformative in illuminating the molecular and physiological mechanisms that occur at the level of individual neurons and individual synapses (Kandel *et al.*, 2014). It is not well understood how individual neurons contribute to a "memory network," their contributions from diverse brain regions, or their precise functional connections in memory storage and retrieval. In Chapter 27, Rajasethupathy provides an overview of what has been learned about memory circuits using optogenetics in the mammalian brain. Optogenetics has been used to shed light on a wide variety of questions involving memory circuits, with much work already having focused on delineating the role of bottom-up circuits in memory allocation, processing, and retrieval. Specifically in the memory context, Rajasethupathy *et al.* have suggested that the adaptive emergence of contextual memory-specific hub neurons may facilitate efficient access to the engram, suitable for memories that are strongly encoded via repetition, reward, or saliency.

Part VI: Optogenetics in Sleep, Prosthetics, and Epigenetics of Neurodegenerative Diseases

This section consists of five chapters. Sleep is undeniably recognized as one of the most naturally beneficial forms of rest for corporal and mental health. Modern lifestyles entrain a forceful habit of maintenance of wakefulness for both professional and recreational reasons, and this has led to the increased propensity for sleep-related disorders and its associated socio-economical downfalls. Prolonged wakefulness and sleep deprivation are known, for example, to result in impaired cognition, a drunk-like state, hallucination, and hormonal and immune system deregulation, and could potentially cascade to an irreversible loss of overall homeostasis, eventually leading to death (Rechtschaffen *et al.*, 1983). Which brain centers are involved in sleep/wake regulation, and what is the nature of the interactions between those centers for the modulation of sleep architecture? These two questions require new tools and approaches to be elucidated. The decade-old revolutionary optogenetic techniques are now being used in this field. In Chapter 28, Latchoumane and Shin summarize some of the latest discoveries in the control of the sleep/wake architecture, focusing particularly on the thalamic control of brain oscillations and non-rapid eye movement sleep.

The World Health Organization estimates that 360 million people, representing nearly 5 percent of the world's population, suffer from disabling hearing loss, with functional, social, and economic consequences. For the treatment of severe to profound hearing loss, the development of cochlear implants and auditory brainstem implants enables patients to regain partial restoration of hearing. Presently, these hearing prostheses convert acoustic information into pulses of electricity that directly stimulate neurons along the auditory pathways to generate a meaningful percept. Recent research proposes applying light in the place of electricity for stimulation using optogenetic tools to create more sophisticated implants with greater numbers of independent channels of frequency information. Optogenetics can potentially harness the fine-grained spatial-temporal resolution of the auditory nervous system and reduce side effects from off-target activation of non-auditory pathways. Chen *et al.*, in Chapter 29, review the application of optogenetics-based auditory implants that were discovered in their laboratory and others. It is now clear that optogenetics provides a real possibility for the stimulation of the auditory pathway and the development of new otoprosthesis, either cochlear or in the brainstem. Chapter 30, by Hernandez Gonzalez and Moser, summarizes the scope of optogenetics in auditory research and prosthetics. The optical implants have to be probed in chronic implanted animal models first in order to assess their reliability, safety, auditory perception, frequency, and the intensity resolution of coding before taking them further to clinical trials in human beings.

Amino acids have long been described as the building blocks of proteins. Amino acid pool depletion and the lack of the building blocks for new proteins represents one way in which diminished numbers of amino acids, especially essential amino acids, could impact neurodegeneration in a number of well-known conditions such as autism, Parkinson's disease, and schizophrenia. Further recent literature points to the significant role that the essential amino acid phenylalanine plays in the pathophysiology of addiction, especially cocaine addiction (Nestler *et al.*, 2001). Essential amino acids as well as non-essential amino acids play important roles in neurological function, by directly affecting gene expression. The role of amino acids in gene expression opens up new doors for the role of epigenetics and disease, especially neurodegenerative ones (Appasani, 2012). In Chapter 31, Joan Fallon discusses the role of amino acids in neurodegenerative and addictive diseases. Although the regulation of amino acids in epigenetics is not a perfect fit for this volume, their importance in neurodegenerative diseases is well appreciated and thus needs some attention here.

Although optogenetic deep-brain stimulation (DBS) has not yet been tested in humans, it can be considered as the next logical step in the development of DBS. Deep-brain stimulation has been used to treat more than 100,000 patients suffering from advanced Parkinson's disease, and since 1999 it has been tested in several hundreds of psychiatric patients. It can be expected that optogenetics might improve DBS, allowing it to which can replace electrical DBS. Although this development is still in its infancy, a bioethical debate on the use optogenetics in DBS has already begun in several countries. In a review by Müller and Walter, the pros and cons of optogenetics-based DBS are detailed in Chapter 32.

Future Perspectives

The early success of applying optogenetic tools to the study of astrocytes will continue to drive these lines of research forward and may potentially lead to therapeutic applications targeting astrocytes. The continued development of more sophisticated optical control devices, which has rapidly evolved during the short history of optogenetics, could see the goal of restoring more complex motor functions come to fruition in the foreseeable future.

Optogenetic techniques are unlikely to provide a way to address the underlying pathology and progression of neurodegenerative diseases. The potential disadvantages of optogenetics over other therapeutic modalities are largely unknown, but some can be anticipated. The optogenetic therapy for epilepsy does not seem to provide a cure or disease modification, but, rather, offers a prosthetic approach suppressing seizures when they are about to occur or have been initiated. Since opsin is an ectopically expressed protein, there is the potential for the development of a limiting and possibly damaging immune response to the foreign protein. There is also the danger that the effects of opsins and optical stimulation in pathological tissue may be unpredictably different from their effects in normal tissue. The path to the clinic is unlikely to be a short one; however, the new therapy that optogenetic technology undoubtedly offers for epilepsy is a unique opportunity. The potential of such clinical applications of optogenetics may open the door to further therapeutic possibilities, providing hope for sufferers of many other neurological disorders.

Optogenetic technology also opens new doors to the design of a better auditory prosthesis. The optical implants first have to be probed in chronic implanted animal models, and then their reliability, safety, and effectiveness assessed before being taken to clinical trials in human beings. The use of optogenetic tools to restore the sight of blind patients has an extremely promising future. The combination of both genetic and optical tools offer the promise of enabling the dissection of neural circuits that contribute to the complex cognitive processing underlying molecular memory formation in the mammalian brain.

However, there are still major challenges to overcome before it will be possible to use optogenetics to treat retinal cell diseases and vision restoration. First and foremost, safety concerns need to be reduced by carrying out thorough studies in non-human primates. These pre-clinical tests will need to carefully assess the immunogenicity of the gene candidates, the correct trafficking of the proteins, and the stability of expression, as well as the bio-compatibility of the light stimulation if a device is required. The first clinical study sponsored by RetroSense Therapetics Inc. (Ann Arbor, Michigan) involved a Texas woman who is legally blind due to retinitis pigmentosa; her vision was treated using a combination of gene therapy and light to precisely control nerve cells (Bourzac, 2016). According to Antonelli Bonci, Director of the National Institute of Drug Abuse (NIH) "this first clinical trial is going to be a gold mine of information about doing optogenetics studies in humans."

Now, the conceptual and methodological advances associated with optogenetic approaches are providing fresh momentum to neuroscience drug discovery, particularly in areas that are stalled on the concept of "fixing the brain chemistry." Optogenetics is beginning to translate into drug discovery in several domains, including target discovery, high-throughput screening, and novel therapeutic approaches to disease states (Song and Knopfel, 2016). It has also been shown that one can incorporate nanomaterials with different properties into optogenetics, which brings new opportunities to solve the problems encountered in these techniques, from the delivery/expression of the optogene to the stimulation/inhibition and follow-up sensing of neural activity. Since optogenetics techniques are relatively new and constantly evolving, they will surely shed new "light" on how we smell, learn, sleep, and remember (Grimaud and Lledo, 2016). It is anticipated that an implantable wirelessly piloted machine-cell interface, based on a light communication channel, will be developed in the future, which will become an innovative strategy for tomorrow's therapeutics.

The optogenetics technology platform is already helping scientists make rapid progress in understanding what brain cells underlie movement, motivation, pain, and many other basic brain functions in animals. However, it should be mentioned with caution that before optogenetics can be used therapeutically in the brain, researchers will need more information about which cells to target. In order to test optogenetic deep-brain stimulation in clinical trials, more research is needed in order to have a sufficient level of safety and adherence to certain bioethical rules.

REFERENCES

Adamantidis, A., Arber, S., Bains, J.S., *et al.*, (2015). Optogenetics: 10 years after ChR2 in neurons-views from the community. *Nature Neuroscience*, **18**, 1202–1212.

Appasani, K. (2012). *Epigenomics: From chromatin biology to therapeutics*. Cambridge: Cambridge University Press.

Bogomolni, R.A. and Spudich, J.L. (1982). Identification of a third rhodopsin-like pigment in photactic *Halobacterium halobium*. *Proceedings of the National Academy of Sciences, United States of America*, **79**, 6250–6254.

Bourzac, K. (2016). In first human test of optogenetics, doctors aim to restore sight to the blind. *MIT-Technology Review*, **Feb 19**, 3.

Boyden, E.S. (2011). A history of optogenetics: the development of tools for controlling brain circuits with light. *F1000 Biology Reports*, **3**: 11 (doi:10.3410/B3-11), 1–12.

Boyden, E.S., Zhang, F., Bamberg, E., Nagel, G. and Deisseroth (2005). Millisecond-timescale, genetically targeted optical control of neural activity. *Nature Neuroscience*, **8**, 1263–1268.

Crick, F.H.C. (1979). Thinking about the brain. *Scientific American*, **241**, 219–232.

Crick, F.H.C. (1999). Impact of molecular biology on neuroscience. *Philosophical Transactions of the Royal Society, London, B. Biological Sciences*, **354**, 2021–2025.

Deisseroth, K. (2010). Optogenetics: Controlling the brain with light. *Scientific American, October* **20**, 1–12.

Deisseroth, K., Feng, G., Majewska, A.K., *et al.* (2006). Next-generation optical technologies for illuminating genetically targeted brain circuits. *Journal of Neuroscience*, **26**, 10380–10386.

Delgado, J.M.R. (1964). Free behavior and brain stimulation. *International Review of Neurobiology*, **6**, 349–449.

Goldensohn, E.S. (1998). Animal electricity from Bologna to Boston. *Electroencephalygraphy in Clinical Neurophysiology*, **106**, 94–100.

Grimaud, J. and Lledo, P.M. (2016). Illuminating odors: when optogenetics brings to light unexpected olfactory abilities. *Learning and Memory*, **23**, 249–254.

Hunter, P. (2016). Shining a light on optogenetics. *European Molecular Biology Organization Reports*, **17**, 634–637.

Kandel, E.R., Dudai, Y. and Mayford, M.R. (2014). The molecular and systems biology of memory. *Cell*, **157**, 163–186.

Katz, L.C. and Shatz, C.J. (1996). Synaptic activity and the construction of cortical circuits. *Science*, **274**, 1133–1138.

Kim, J.M., Hwa, J., Garriga, P.J., *et al.* (2005). Light-driven activation of beta 2-adrenergic receptor signaling by a chimeric rhodopsin containing the beta 2-adrenergic receptor cytoplasmic loops. *Biochemistry*, **44**, 2284–2292.

Montogomery, K.L., Iyer, S.M., Christensen, A.J., *et al.* (2016). Beyond the brain: Optogenetic control in the spinal cord of peripheral nervous system. *Science Translational Medicine*, **8**, 337rv5.

Nagel, G., Ollig, D., Fuhrmann, M. Kateriya, S., *et al.* (2002). Channelrhodopsin-1: a light-gated proton channel in green algae. *Science*, **296**, 2395–2398.

Nagel, G., Szellas, T., Huhn, W., *et al.* (2003). Channelrhodopsin-2: a directly light-gated cation-selective membrane channel. *Proceedings of the National Academy of Sciences, United States of America*, **100**, 13940–13945.

Nagel, G., Brauner, M., Liewald, J.F., *et al.* (2005). Light activation of channelrhodopsin-2 in excitable cells of *Caenorhabditis elegans* triggers rapid behavioral responses. *Current Biology*, **15**, 2279–2284.

Nestler, E., Barrot, M. and Self, D. (2001). ΔFosB: A sustained molecular switch for addiction. *Proceedings of the National Academy of Sciences, United States of America*, **98**, 11042–11046.

Oesterhelt, D. and Stoeckenius, W. (1971). Rhodopsin-like protein from the purple membrane of Halobacterium halobium. *Nature New Biology*, **233**, 149–152.

Rechtschaffen, A., Gilliland, M.A., Bergmann, B.M., *et al.* (1983). Physiological correlates of prolonged sleep deprivation in rats. *Science*, **221**, 182–184.

Roska, B. and Pepperberg, D. (2014). Restoring vision to the blind: Optogenetics. *Translational Visual Science and Technology*, **3**, 4–8.

Sakmar, T.P. (2002). Structure of rhodopsin and the superfamily of seven-helical receptors: the same and not the same. *Current Opinions in Cell Biology*, **14**, 189–195.

Shorvon, S.D. (2011). The etiologic classification of epilepsy. *Epilepsia*, **52**, 1052–1057.

Song, C. and Knopfel, T. (2016). Optogenetics enlightens neuroscience drug discovery. *Nature Reviews Drug Discovery*, **15**, 97–109.

Spudich, J.L, (2006). The multitalented microbial sensory rhodopsins. *Trends in Microbiology*, **14**, 480–487.

Vann, K.T. and Xiong, S. (2016). Optogenetics for neurodegenerative diseases. *International Journal of Pathophysiology and Pharmacology*, **8**, 1–8.

Volkow, N.D. and Koob, G. (2015). Brain disease model of addiction: Why is it so controversial? *Lancet Psychiatry*, **2**, 677–679.

Yizhar, O., Fenno, L.E., Davidson, T., *et al.* (2011). Optogenetics in neural systems. *Neuron*, **71**, 9–34.

Zemelman, B.V., Nesnas, N., Lee G.A., *et al.* (2003). Photochemical gating of heterologous ion channels: Remote control over genetically designated populations of neurons. *Proceedings of the National Academy of Sciences, United States of America*, **100**, 1352–1357.

Zhang, F., Vierock, J., Yizhar, O., *et al.* (2011). The microbial opsin family of optogenetic tools. *Cell*, **147**, 1446–1457.

2 Uncovering Key Neurons for Manipulation in Mammals

Boris V. Zemelman

2.1 Introduction

In one guise or another, directed manipulation of brain function can be traced back decades. While much has been made of Francis Crick's musings in the 1990s on the potential power of selective neuronal stimulation (Crick, 1999), remarkably effective – albeit controversial – *in vivo* experiments had been conducted by Jose Delgado nearly 30 years earlier. Working with large mammals and primates, Delgado demonstrated that many properties attributed to the primitive brain (MacLean, 1990; Panksepp, 2004) – sleep, nurture, hunger and aggression – could be modified by stimulating narrowly circumscribed groups of neurons, including by "remote control" (Delgado, 1964; Delgado, 1969). These studies motivated ever more troubling attempts at *psychosurgery* in the form of localized electrical stimulation in order to diagnose, condition and treat human subjects suffering from behavioral and psychiatric disorders (Heath, Monroe and Mickle, 1955; King, 1961; Sweet, Ervin and Mark, 1969).

Current selective activation techniques, however, stem from efforts to image, rather than perturb, brain function. The advent of fluorescent protein-based cellular markers and reporters in the late 1990s represented a critical advance in combining the convenience of light with the precision of genetic encoding. Such sensors helped catalyze our plan to design a method for stimulating neurons, as opposed to passively monitoring them. Our 2001 overview of optical sensors ends with this prescient hypothesis: "... schemes [that] localize the response to illumination could [feed] patterns of distributed activity ... directly into a genetically circumscribed population of neurons, irrespective of the anatomical location of its members or their connection to sensory input. Perhaps the ability to probe defined groups of neurons with [light] will hold the key to an understanding of neural systems" (Zemelman and Miesenböck, 2001). These sentences succinctly frame the revolution that would take place in experimental neuroscience over the following 15 years. While the tools for selective photoactivation of neurons have benefitted from numerous much-needed refinements, chief among them the cloning of channelrhodopsin (Nagel *et al.*,

2003), the prediction that light and a heterologous light receptor could be used to dissect neuronal mechanisms *in vivo* has been exhaustively validated across numerous systems and species.

Our hypothesis was especially relevant because the first experiments sensitizing mammalian neurons to light using fly rhodopsin had already succeeded, but had not yet been published. Early in 2000, when I joined his newly established laboratory to work on photostimulation, Gero Miesenböck had proposed that the functional elements for our nascent technique could be adopted from the invertebrate visual system, in which photoreceptor cells depolarize in response to illumination. I began by expressing nine proteins required for *Drosophila* phototransduction in a *Xenopus* oocyte, and soon detected ion currents in the presence of retinal and visible light. This seminal experiment laid the foundation for what several years later would become the field of *optogenetics*, as it involved both heterologous opsin and stimulating light. As with many such breakthroughs, the success of that initial experiment hinged on a timely hunch. Among the mRNA transcripts, I decided to include one for arrestin-2 – despite the fact that arrestin was considered necessary for termination of signaling – reasoning on the basis of studies by Kislev and Subramaniam (1997) on *Drosophila* metarhodopsin that, in an *in vitro* system, arrestin might aid the binding of retinal, added in trans, to opsin in order to form functional rhodopsin. Arrestin turned out to be a critical component of the cascade – in subsequent experiments, I identified three fly proteins out of the original nine that were necessary and sufficient for light-dependent ion conductance: arrestin-2, rhodopsin and Gqα (chARGe) (Zemelman *et al.*, 2002). While chARGe tapped into host cell pathways, including channels, the mechanistic role of arrestin in our scheme has never been fully elucidated. Its requirement was subsequently confirmed in a study that reconstituted the melanopsin pathway *in vitro* (Melyan *et al.*, 2005; Qiu *et al.*, 2005). Furthermore, our own attempts to circumvent arrestin suggested that the original reasoning may have been correct, as multi-hour retinal incubations partially alleviated the need for arrestin (unpublished observations). It is still unclear, however, whether arrestin itself participates in retinal loading or resets the invertebrate photocycle, indirectly aiding retinal–opsin binding. Recent experimental findings, for example, indicate that arrestins contribute to the activation and deactivation of G-protein-coupled receptor (GPCR) signaling (Lee *et al.*, 2016; Nuber *et al.*, 2016).

The photocurrents produced by chARGe proteins exceeded those generated by vertebrate rhodopsin by 10–100-fold, acting as a reliable stimulus for neurons (Khorana *et al.*, 1988). However, despite its effectiveness, this first activator exhibited significant limitations. Introducing four coding sequences into neurons (the fourth was GFP in order to tag transfected cells) was, at that time, a non-trivial enterprise. Because invertebrate opsin binds covalently to retinal, we had postulated that different wavelengths of light could be used to turn the receptor on and off, generating temporally constrained epochs of activity; however, perhaps because the spectral overlap was too great, light-based deactivation proved impossible. As a result, the selection of a metabotropic receptor to act solely as an activator meant that illumination and cell stimulation were not tightly coupled, reducing temporal precision. Finally, neurons had to be incubated with retinal for several minutes before stimulation could commence. Indeed, one of the truly surprising (and arguably underappreciated) features of channelrhodopsin is that it can adsorb bio-available retinal.

2.2 Goals and Reality

Lofty plans are built on compromises. The chARGe activator fulfilled our primary goal of a selective light sensitizer, but it was cumbersome and relatively slow to respond. In moving beyond chARGe, it was important to transform our wish list into a set of priorities. The main one was to perform experiments in an intact brain that would reveal key cellular substrates of behaviors and physiological states. Another was to feed naturalistic signals into the system, mimicking the information content of neuronal spike trains. In practice, however, even these two goals could not be easily reconciled. At first glance, light (as stimulus) is a good solution: it is non-destructive, able to illuminate brain regions containing sensitized neurons and the light pulse frequency can also be modulated. However, when target cells are broadly distributed (including in larger brains) or located deep in the brain and in situations where cell locations are not known in advance, light requirement is a hindrance. Moreover, our neurons could not reliably follow light pulses and, even if they could – channelrhodopsin and its many variants certainly can – beyond synchronizing cell responses (a potentially problematic outcome, as described below), what stimulus frequencies would we want to achieve? To this day, that question has not been adequately answered, and so, even with channelrhodopsin, light pulses are used *in vivo* primarily to minimize channel inactivation, rather than to generate a particular firing pattern.

In the context of neuronal manipulations that are designed to affect behavior, a principal requirement, therefore, is to address enough genetically sensitized cells to produce an observable change. This is more easily achieved with a systemically administered drug than with light. An additional element is stimulus duration – the magnitude of currents generated by chARGe caused neuron action potential frequencies to increase dramatically during the first few seconds of illumination, quickly reaching unsustainable levels. A system that could raise and maintain cell activity for significantly longer intervals became the goal. As an aside, additional underappreciated advantages of channelrhodopsin are that it supports far lower conductances and is prone to inactivation. These features reduce the likelihood of runaway cell excitation, albeit at the cost of overall effectiveness. Lastly, we anticipated the need to vary cell excitability, changing the likelihood that target neurons would respond to intrinsic signals, as opposed to stimulating or silencing them outright. After all, a unique drawback of opsin-based methods is that they synchronize the target cell population, essentially producing a localized seizure. Asynchrony is a hallmark of normal brain function, and thus opsins, which are so useful for demonstrating functional connections, are not able to mimic this aspect of endogenous circuit behavior. For all of these reasons, a systemic trigger and a robust, but tunable, actuator were needed.

2.3 Metabotropic and Ionotropic Actuators

As the chARGe story went to press, we began to address some of the shortcomings of this original three-protein system. To improve temporal control, I selected three ionotropic receptors: TRPV1 (Caterina *et al.*, 1997), TRPM8 (Peier *et al.*, 2002) and $P2X_2$ (Brake, Wagenbach and Julius, 1994). The first two were thought to be absent from the mammalian central nervous system and the last was present in mammals, but absent in *Drosophila*, yielding three independent activator–ligand pairs (Zemelman *et al.*, 2003). These activators had several characteristics that both surpassed and complemented chARGe. Neuronal depolarization and spiking frequency were now tunable, proportional to ligand concentration. Peak ion currents were robust, such that trace quantities of heterologous receptors were required. Because second messenger cascades were not directly engaged, cell depolarization was more naturalistic and easily reversible. While the requisite ligands could also be chemically inactivated (Zemelman *et al.*, 2003) or made photoswitchable (Stein *et al.*, 2013) in order to enable light-dependent regulation, the enumerated benefits were especially well suited to *in vivo* applications, where the light requirement could be cumbersome and its temporal precision less essential.

It is reasonable, therefore, to regard ionotropic receptor–ligand actuators as complementary to light-based systems. The latter are most effective when target neurons are relatively easy to access and timing is critical (Losonczy and Zemelman, 2016). On the other hand, ligand-gated channels are more appropriate for longer-duration behavioral manipulations, where target cells are not clustered and where actuation must be tuned – light stimulation, then and now, is largely an all-or-none affair. In fact, I think that the *opto* in *optogenetics* may lead many *opto* newcomers to focus on an arbitrary aspect of the method. The primary goal, for us then and others since, has been to activate (or silence) genetically sensitized neurons in an intact brain irrespective of their location. As long as the encoded receptor is normally absent from the central nervous system and can reliably and reversibly alter cell activity, the choice of light or ligand is best dictated by the task at hand.

The two-component ionotropic receptor–ligand pairs have indeed been directed to *in vivo* applications: $P2X_2$ and caged ATP were used to generate – for the first time – naturalistic behavior in an animal, causing headless flies to take to the air (Lima and Miesenböck, 2005). Exogenous TRPV1 has also been effectively applied *in vivo* to alter locomotion and feeding in rodents (Arenkiel *et al.*, 2008; Güler *et al.*, 2012; Wang *et al.*, 2014). While broader use of the ionotropic channels has been curtailed by their diverse roles in the brain (Edwards, 2014), I anticipate that, aided by protein engineering, analogous receptor–ligand actuators still have much to offer.

2.4 The Advent of Channelrhodopsin

Channelrhodopsin2, cloned in 2003 (Nagel *et al.*, 2003), rapidly transformed the activator landscape (Boyden *et al.*, 2005). Remarkably, it embodied key elements of both the metabotropic and the ionotropic systems. It was not a GPCR and, as such, supported very rapid cell responses. In addition, channelrhodopsin2 itself was the light receptor, one that even managed to load systemic retinal, so that none had to be added in trans. It was indeed a fully genetically encodable, fully autonomous light-based activator. One might quibble with its relatively low conductance (Nagel *et al.*, 2003), long-term toxicity and rapid inactivation – subsequent engineered iterations of the channel have addressed some of these shortcomings – but, in the key respects, it evolved to be quite suitable before it saw the first mammalian neuron, and its singular accomplishment was to make neuronal light activation broadly accessible, transforming experimental neuroscience. As many fine reviews of channelrhodopsin and its applications attest, Karl Deisseroth had literally brought fire to the masses (in a fashion Miesenböck and I were not prepared to do), but here, alas, his story and that of Prometheus poignantly diverge.

2.5 The Challenge of Neuronal Selection

As we proceed through the second decade of selectively manipulating neuronal function, we find that the *opto* field has expanded to include not only actuators, but also genetically encoded activity reporters and other light-sensing proteins that are able to alter aspects of cellular function (Cambridge *et al.*, 2009; Stierl *et al.*, 2011; Chen *et al.*, 2013; Wend *et al.*, 2014; Kyung *et al.*, 2015; Rost *et al.*, 2015). Yet, despite this plethora of tools, one prominent goal implicit in the original design remains unfulfilled.

How do we select and access the target cells? Arguably, this quandary has (so far) curtailed the number of genuine discoveries made possible by *optogenetics* (Adamantidis *et al.*, 2015). After all, the methods in question, be they light or ligand based, are not end points, but rather elegant ways of tackling a pressing scientific question. In offering the means for altering the properties or functions of precisely defined cell ensembles, the techniques place the onus on the experimenter to provide such ensembles. Instead, in the most rudimentary scenarios, *optogenetics* has been used to affect cells regionally, as would a lesion or an electrical stimulus, without regard for the variety of ensnared cell types. More refined studies have focused on known neuron classes, but most of these classes, expressing a specific neurochemical maker, are defined based on brain anatomy, not function, and also comprise multiple cell types. Here, it is important to distinguish *in vitro* applications of *optogenetics*, which benefit from imposed anatomical, spatial and genetic constraints – these studies have significantly advanced the understanding of neural circuit mechanisms. *In vivo*, however (from cortex to the amygdala), optogenetic tools have too often been used to confirm insights gleaned years ago using less flashy methods.

It is clear, therefore, that the amalgam of genetics-based tools will reach their full potential to illuminate brain mechanisms once we identify functionally relevant cell assemblies and learn how to access those cells in order to produce naturalistic activity patterns. Some promising directions, including our own ongoing efforts, are described below.

2.6 The Identification of Activated Neurons

Sensory signals take a convoluted path through multiple neuronal circuit elements as they are perceived, transformed and acted on by the organism. Initial breakthroughs in neural connectivity relied on the accessibility of the critical structures: a sensory epithelium to initiate a signal – often artificial – and the somatosensory cortex to read out and decipher the transformations of the signal by the intervening circuitry.

A persistent challenge has been to flesh out the cellular substrates of affective processes that originate or terminate deeper in the brain. Where the hedonic value of experience reinforces memory and motivates actions, stimuli can be multimodal and the experimental readout is often behavioral. The clinical impetus for exploring cellular mechanisms of such processes is clear: addiction, depression and fear affect numerous lives worldwide.

Optogenetic and pharmacogenetic tools have helped probe the roles of established neuron classes in a variety of such behaviors. However, these methods alone cannot uncover novel cell classes in order to complete the neuronal network puzzle. Such heterogeneous cell ensembles can only be identified on the basis of concurrent activity.

In setting out to develop a genetic activity sensor, we have adopted the following set of priorities: (1) universality – cell type and signal independence; (2) versatility – minimal species restrictions; (3) modularity – a single framework for tagging, imaging and manipulating active neurons; (4) ease of use – avoiding animal transgenics; (5) fine temporal control – to target cell populations that are active during behaviorally relevant epochs; and (6) resolution – the ability to label multiple independent cell populations in a single animal; these populations may represent the "before" and "after" brain states, synaptic partners or recruited orthogonal assemblies.

2.7 Evolution of Activity-based Cell Detection

Considerable effort has been expended on classifying neuronal populations based on morphology, location, active and passive properties, the complement of cellular proteins and other tangible (and relatively constant) features. It is now clear, however, that the neuronal assemblies supporting mental phenomena – memories, emotions and motivational states – are transient and not grouped according to these attributes. As a result, we know very little about the causal significance of activity patterns within and across sparse heterogeneous populations of neurons.

Existing techniques for identifying neurons on the basis of activity provide a foundation and a springboard for our efforts. Nearly 40 years ago, it was observed that growth factors and depolarization-dependent calcium influx rapidly increased *c-fos* transcription, linking neuronal activation with protein synthesis. (Hunt, Pini and Evan, 1987; Sagar, Sharp and Curran, 1988; Ceccatelli *et al.*, 1989; Morgan and Curran, 1991; Loebrich and Nedivi, 2009). Evidence also mounted that *c-fos*, along with several other so-called immediate–early genes (IEGs), was briefly induced in subsets of neurons *in vivo* in response to sensory and pharmacological stimuli. These observations fostered the widespread reliance on elevated c-fos protein expression as a generic marker of recent neuronal activation (Minatohara, Akiyoshi and Okuno, 2016).

Post hoc histochemical methods for identifying activated neurons on the basis of IEG expression help pair stimuli and behavioral states with anatomical domains, but offer limited cell-autonomous information and preclude functional analyses of labeled neurons. Transgenic animals in which IEG promoters had been linked to heterologous reporters provided a continuous readout of cellular activation and the means to identify activated neurons for *in vitro* studies (Yassin *et al.*, 2010). In fos-GFP and Arc-GFP mice (Barth, Gerkin and Dean, 2004; Wang *et al.*, 2006), a short-lived fluorescent protein appeared *in vivo* in stimulated neurons. In one elegant study, different stimuli were used to detect and characterize distinct, presumably orientation-tuned cells in the mouse visual cortex (Wang *et al.*, 2006).

The choice of a destabilized fluorophore in the previous study highlights a Goldilocks problem associated with *in vivo* cell detection: there is generally either too much or too little labeling. The brain is continuously active, so that, if a reporter accumulates in a cell, brief and ongoing neuronal activity become indistinguishable. On the other hand, short-lived reporters reduce sensitivity and impose significant strictures on the timing of downstream analyses. A recent strategy has been to couple IEG expression to that of conditional intermediates, such as the chimeric tetracycline transactivator (tTA) (Gossen and Bujard, 1992; Reijmers *et al.*, 2007) or inducible Cre recombinase (CreER) (Feil *et al.*, 1997; Guenthner *et al.*, 2013), in order to control the onset and duration of labeling. Here the bipartite recombinase approach appears to be superior to the tTA transactivator: the pulse of conditional intermediate activity is adjustable in both instances; however, the recombinase-dependent heterologous protein expression is permanent, while expression continues only as long as transactivator is present in the tetracycline system. Permanence of expression represents a milestone in the evolution of labeling techniques, as it provides the means to examine and manipulate (using actuators) neurons identified on the basis of elevated activity at any future experimentally suitable time point.

While these genetic approaches represent important technical advances over traditional IEG immunostaining, shortcomings abound. One aspect of the Goldilocks problem remains largely unresolved: the labeling epochs are still too long to reliably link identified neurons to brain processes – hours for the CreER system (targeted recombination of active populations [TRAP]) (Guenthner *et al.*, 2013) and days for tTA-based one (TetTag) (Cowansage *et al.*, 2014). Like others (Guenthner *et al.*, 2013; Denny *et al.*, 2014), we have generated animals in which *c-fos*, *Arc* and *zif268* promoters drive CreER expression, and have used them to label neurons activated by cognitive tasks (Figure 2.1). However, we have observed significant home cage labeling and dramatic differences in labeling among experimental cohorts (unpublished observations), tempering our enthusiasm for the technique. Consequently, absent meaningful within-animal controls (e.g. two

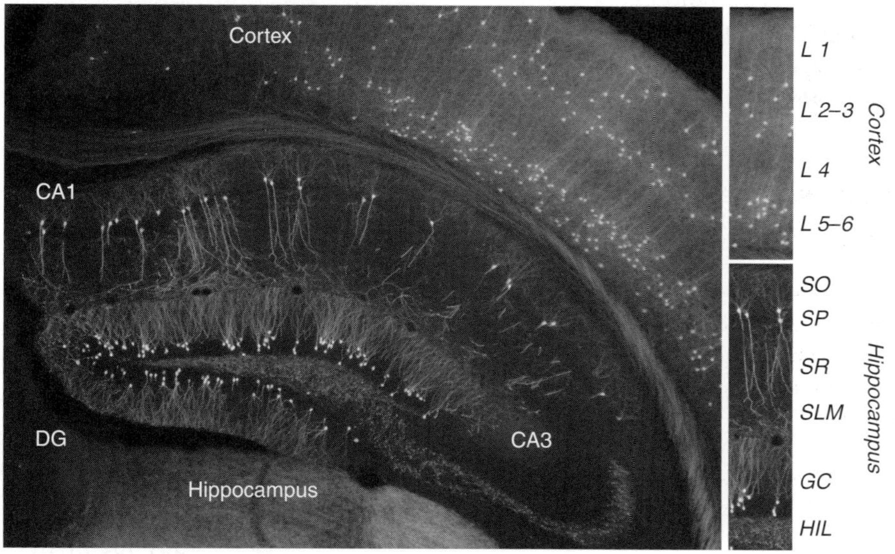

Figure 2.1 Activity-dependent cortical and dorsal hippocampal neuron labeling in an Arc-CreER mouse. Representative coronal brain section from the progeny of Arc-CreER and CAG-(tdTomato)^{Cre} reporter mice (Ai14) (Madisen *et al.*, 2009) following exposure to an enriched environment. Hydroxytamoxifen was administered 2 hours prior to the start of the experiment. Subsequently, animals were returned to home cages and sacrificed 2 weeks later. Note the absence of label in the CA3 region and cortical layer IV. In addition, inhibitory interneurons that are normally abundant in the SO and HIL are not labeled. *DG:* dentate gyrus; *SO:* stratum oriens; *SP:* stratum pyramidale; *SR:* stratum radiatum, *SLM:* stratum lacunosum-moleculare; *GC:* granule cell layer of the dentate gyrus; *HIL:* hilus. (A black-and-white version of this figure will appear in some formats. For the color version, please refer to the plate section.)

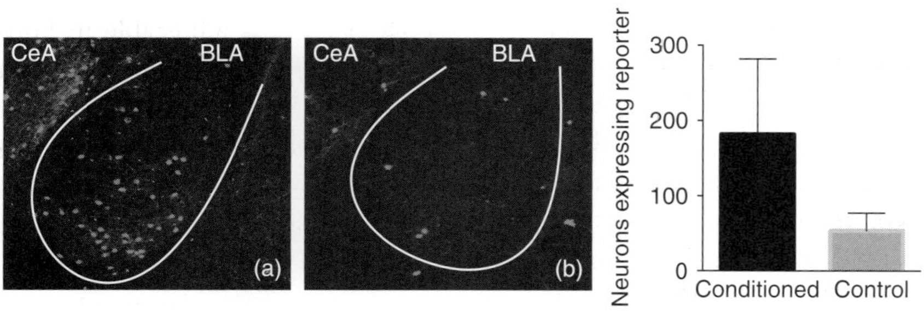

Figure 2.2 Activity-dependent amygdala labeling in Arc-CreER mice. (A) Animals were subjected to appetitive conditioning, with labeling initiated using hydroxytamoxifen at asymptotic task performance (n = 4). (B) Control animals were labeled in home cages (n = 3). Representative amygdala sections and aggregate cell counts are shown. The experimental preparation was similar to that described in Figure 2.1. *CeA:* central amygdala; *BLA:* basolateral amygdala (based on Maya *et al.*, 2016, in preparation). (A black-and-white version of this figure will appear in some formats. For the color version, please refer to the plate section.)

time interval labeling steps using identical methods), these techniques are best restricted to exploring brain regions where basal activity can be regulated, such as the visual cortex, or where it is normally low (Figure 2.2) and (Root *et al.*, 2014).

In addition, methods that track IEG expression are subject to complex stimulus-specific and cell-autonomous regulation that can bias outcomes. For example, the TetTag system appears to detect only CaMKIIa-positive excitatory neurons (Garner *et al.*, 2012). Likewise, Arc-CreER labels neither inhibitory interneurons nor neurons in key domains, such as the hippocampal CA3 region and cortical layer IV (Figure 2.1) and (Guenthner *et al.*, 2013). Therefore, putting aside for the moment the thorny question of reporter calibration (Minatohara, Akiyoshi and Okuno, 2016), the ensemble of neurons detected using these techniques is likely to be both too broad and incomplete, hindering comprehensive neuronal circuit analysis (nonetheless some *post hoc in vivo* manipulations have proven effective: Koya *et al.*, 2009; Bossert *et al.*, 2011; Garner *et al.*, 2012; Liu *et al.*, 2012; Ramirez *et al.*, 2013; Redondo *et al.*, 2014).

2.8 Distilling Cellular Assemblies

We have been developing a novel and timely technique for identifying and targeting neuronal assemblies on the basis of function, rather than intrinsic characteristics. Our approach is distinctly versatile because it is built around a virus-based system that reads out elevated neuronal activity in any brain region and species. It is designed for unprecedented temporal precision, matching the timescale of naturalistic behavior. Further, the functional role of each uncovered assembly element can be tested using direct molecular–genetic manipulations.

Our methods for identifying and targeting activated neurons rely on the ubiquitous cAMP response element-binding protein (CREB), which plays a critical role in activity-dependent protein synthesis. Neurons receive a multitude of signals through GPCRs, receptor kinases and differentially gated ion channels. Correct processing of those signals depends on CREB and is essential for proper cell and circuit operations. In the neuron, CREB binds to the palindromic TGACGTCA sequence – the CREB response element (CRE) – found within the regulatory regions of thousands of genes (Impey *et al.*, 2004; Zhang *et al.*, 2005). Depending on the route of neuronal activation, MAPK-activated ribosomal S6 kinase (RSK), calcium/calmodulin-dependent kinase IV (CaMKIV) or cAMP-dependent protein kinase A (PKA) phosphorylate CREB at Ser133 (Gonzalez and Montminy, 1989; Sheng, Thompson and Greenberg, 1991). Phosphorylated CREB (pCREB) is a transcriptional activator (Gonzalez and Montminy, 1989; Sheng, Thompson and Greenberg, 1991), but normally recruits transcriptional cofactors, such as the CREB-binding protein (CBP) (Kwok *et al.*, 1994). Since kinase activity is modulated by intracellular calcium, CREB may be considered to be a calcium integrator. In the neuronal activation hierarchy, it is upstream of IEGs such as *c-fos*, whose expression is regulated by pCREB and additional transcription factors in a stimulus-specific fashion (Sheng, McFadden and Greenberg, 1990; Bading, Ginty and Greenberg, 1993; Bonni *et al.*, 1995). CREB is therefore a more universal sensor of cellular activation than IEGs.

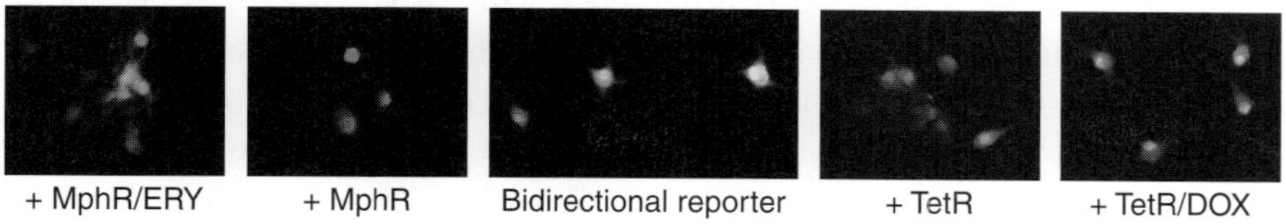

Figure 2.3 A bidirectional virus-based neuronal activity reporter. *Middle panel:* Cultured hippocampal neurons stimulated with forskolin express red and green fluorophores, producing yellow–orange labeling. *+MphR/ +TetR panels:* The indicated bacterial repressor blocks expression of green or red fluorophores. *ERY/DOX panels:* The addition of respective ligands restores two-color expression in activated neurons. (A black-and-white version of this figure will appear in some formats. For the color version, please refer to the plate section.)

In our reporter system, endogenous CREB binds to a synthetic CRE enhancer that is tunable through multimerization. All other transcription factor binding sequences are absent, consistent with reporter universality. As a result, the CREB reporter compares favorably in terms of sensitivity and universality to recently described IEG-based viral systems (Kawashima *et al.*, 2009; Kawashima *et al.*, 2013). We have also taken advantage of the palindromic nature of the CRE sequence to create a bidirectional promoter that is able to express two transgenes.

In this system, CREB-dependent transcription is robust, but tightly regulated. Short, minimal promoters flank the core enhancer. Each minimal promoter is, in turn, interrupted by a unique bacterial operator for differential control over transgene expression. We use native versions of bacterial repressors, rather than the better-known VP16 and KRAB chimeras (Gossen and Bujard, 1992; Freundlieb, Schirra-Müller and Bujard, 1999), to block RNA polymerase while preserving the activity-dependent promoter (Yao and Eriksson, 1999). Expression of each transgene can therefore take place only when a neuron has been activated and the appropriate repressor ligand is present, enabling the identification of neuronal ensembles that have been active during the two experimentally defined time intervals (Figure 2.3).

2.9 Labeling Relevant Neurons with Light

Since our goal is to identify cellular ensembles supporting mental and behavioral states, the neuronal labeling technique must match the timescale of naturalistic behavior. To focus on memory formation as an illustrative example, the dentate gyrus contains more than 1 million essentially indistinguishable granule cells, but less than five percent of these seemingly identical neurons are active during a single memory event (Barnes *et al.*, 1990; Kee *et al.*, 2007; Stone *et al.*, 2011). If we are to identify the cellular ensemble representing a specific contextual experience – say, memory formation or recall – temporal precision is needed. Using behavioral cues, we deduce that contextual memories form and are extinguished over the course of minutes, not hours or days. Our solution is to identify activated neurons on the basis of elevated intracellular calcium, but to restrict the activity-based labeling interval using light.

Light application is particularly attractive because it yields easily to temporal control. We reasoned that, just as other optogenetic methods have aided neuronal circuit analysis by approximating the timescale of cell activity, so too can a light-dependent labeling technique illuminate transient cell assemblies. This advance is made possible by a novel way of controlling gene expression using light. While descriptions of light-dependent gene expression can be found in the literature (Cambridge *et al.*, 2009), our approach is mechanistically quite different. When bacterial repressors are used to block transcription, orthogonal ligands can selectively displace individual repressors. If the ligands are initially inactivated (caged) using a photo-labile chemical modification, akin to how we had restricted ionotropic actuators, light application can specify periods during which transcription can take place. In our prior experience with photo-labile caging groups, there was a tradeoff between the stability of the caged reagent – free ligand should not be generated in the absence of light – and the efficiency of light-dependent de-protection (Zemelman *et al.*, 2003). Choosing the more stable caging group is, in fact, a benefit, since the small amount of liberated ligand is rapidly replaced by excess caged variant once illumination ceases, terminating labeling.

According to the outlined scheme, activity-dependent labeling can occur when neurons are activated and when light is present. To take advantage of the bidirectional reporter, we are focusing on chemical modifications that enable differential uncaging (i.e. the use of distinct wavelengths of visible light for each ligand). The experimenter can therefore specify the gene product and the time interval for its expression. In addition to enhancing the temporal resolution of activity reporters, the delineated system also extends the range of options for transcription regulation using more traditional exogenous, as well as endogenous promoters.

2.10 Cellular Targeting Using Iterative Bioinformatics

As a complementary strategy to the identification of neurons based on activity, we are developing a set of broadly applicable tools for targeting functional cell classes across species, with an initial focus on the neocortex of non-human primates. Cortical neurons vary according to laminar location, morphology, synaptic targets and molecular identity. To decipher cortical information processing, we would like to identify, monitor and manipulate functional subsets of these neurons using genetically encoded markers, activity reporters and actuators. High-specificity neuronal targeting has been achieved in laboratory rodents with the help of transgenic techniques. Additionally, pan-neuronal, excitatory and inhibitory cell targeting has been reported using viruses encoding upstream regions of the mouse *synapsin, CaMKIIα* and *GAD2* genes, respectively (Schoch, Cibelli and Thiel, 1996; Pinal, Cortessis and Tobin, 1997; Dittgen *et al.*, 2004). In the primate, we and others have used recombinant adeno-associated viruses and lentiviruses with *chicken β-actin, synapsin* and *CamKIIα* promoter elements to obtain pan-neuronal and excitatory neuron expression of fluorescent proteins and actuators (Han *et al.*, 2009; Diester *et al.*, 2011; Galvan *et al.*, 2012; Gerits *et al.*, 2012; Jazayeri, Lindbloom-Brown and Horwitz, 2012; Ruiz *et al.*, 2013; Yaguchi *et al.*,

2013; Seidemann *et al.*, 2016), as well as genetically encoded calcium indicators (Seidemann *et al.*, 2016), but successful expression of heterologous proteins in other cell types has not yet been achieved. A principal limitation of the viral approach is that short promoters for targeting the most prominent neuron classes have not been mapped (Nathanson *et al.*, 2009), and the traditional method for testing large upstream regions of cognate neurochemical markers for expression selectivity is hampered by the limited genome sizes of adeno-associated viruses and lentiviruses. Consequently, the promoters listed above represent the few known cases of virus-based, cell type-specific transgene expression. Since targeting neuron classes through genomic manipulations is impractical in primates (including humans), viruses represent the one viable alternative.

In designing promoters for viral use, one option is to identify domains within untranslated regions of genes of interest that are conserved across species (Ovcharenko *et al.*, 2004) and fusing these regions into a single chimeric promoter. We have used this approach successfully to selectively target amacrine cells in the mouse retina (Borghuis *et al.*, 2011) and homogeneous cell classes in the gerbil inferior colliculus (Figure 2.4A & B). However, these are exceptional cases, and a cross-species, genome-wide rational approach is needed.

The basic premise is that genes that are co-expressed in specific neurons share regulatory programs. Therefore, the first task is to extend recent efforts (Sugino *et al.*, 2006; Rossier *et al.*, 2014; Tasic *et al.*, 2016) in order to identify such genes in cell classes of interest (e.g. the parvalbumin interneurons). We are also comparing gene expression patterns of primate and rodent neurons of the same class in order to restrict subsequent analyses to evolutionarily conserved gene sets. Since gene regulatory elements are also highly conserved (www.encodeproject.org), we can identify regulatory domains of candidate genes by screening the genomes of distantly related species.

A subsequent coarse-grained sequence analysis integrates transcription factor mapping (Blanco *et al.*, 2006) with metrics of sequence properties, such as GC content, in order to produce a set of features for supervised modeling algorithms. After assembling combinations of candidate motifs at their preferred positions from a transcription start site, we compare the resulting synthetic promoters with prospective regulatory domains of known neuron class-specific genes.

The endogenous non-coding domains most closely matching the synthetic template are tested for the ability to recapitulate the original gene expression pattern. This is most easily accomplished with viral vectors encoding a fluorescent reporter. The lengths of candidate domains are variable, ranging from a few hundred to two thousand base pairs and determined primarily by the lengths of applicable templates. Templates are subsequently refined based on the performance of candidate regulatory regions.

However, few proteins are found exclusively in specific cell classes (Sharma *et al.*, 2015). Consequently, promoters designed on the basis of co-expressed markers are unlikely to target a single class of neurons. Indeed, this prediction is borne out by experimental evidence (Nathanson *et al.*, 2009; unpublished observations). Our solution is to use viruses intersectionally, stacking promoter

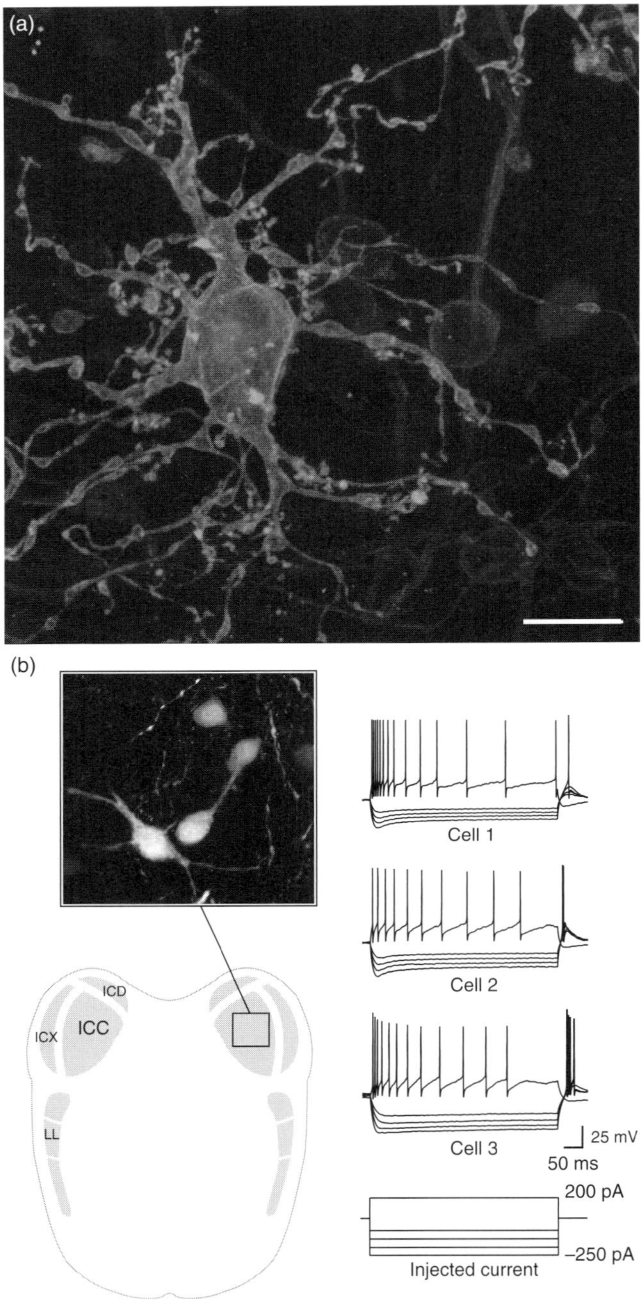

Figure 2.4 Cell type-specific promoters drive the expression of virus-encoded transgenes in rodent neurons. (A) A promoter composed of conserved non-coding regions of the *MGluR1* gene exclusively labels retinal amacrine AII cells (based on Borghuis *et al.*, 2011). Scale bar: 10 μm. (B) Interdependent viruses comprising truncated calbindin and calcium/calmodulin-dependent protein kinase II (CaMKII) non-coding regions restrict fluorophore expression to a morphologically and electrically homogeneous population of neurons within the gerbil inferior colliculus. Injected current steps are indicated (based on Kreeger *et al.*, 2016, in preparation). (A black-and-white version of this figure will appear in some formats. For the color version, please refer to the plate section.)

specificity. We postulate that, while expression from a single candidate promoter may not be limited to a single class of neurons, the concurrent use of several promoters associated with markers co-expressed in target cells can achieve cell type specificity (Figure 2.4B).

We rely on site-specific recombinases (Cre, Dre and Flp) in order to create a *nested doll* arrangement of up to four unique promoters encoded by high-titer viruses. Flanking cleavage sites allow the expression of one recombinase to be predicated from the activity of another. The final construct in the hierarchy expresses the gene of interest in a fashion that is defined by overlapping promoter specificity.

An additional and likely more onerous limitation of current schemes for cellular categorization and related promoter selection stems from prevalent definitions of cellular classes. As already proposed, groupings based on neurochemical markers are reasonably effective at distinguishing excitatory from inhibitory neurons (although even these seemingly rudimentary categories break down in the amygdala and other variegated brain regions). Within each class, moreover, neurons differ widely according to functions and properties. Consequently, striving to target all parvalbumin neurons, for example, is harder to rationalize than, say, targeting a functionally homogeneous subset of parvalbumin neurons.

Importantly, the single and intersectional promoter schemes outlined above demonstrably help achieve this very goal. Our empirical findings demonstrate that cells that are able to support transgene expression from one or more truncated promoters are highly likely to share other characteristics, such as firing properties and connectivity. As with any biological screen, it is therefore critical to evaluate promoter efficacy based not on an ability to support expression in all parvalbumin, somatostatin or cholecystokinin neurons, but, rather, based on the ultimate goal of the targeting approach – to select functionally related neuronal ensembles. It is convenient to evaluate such ensembles in the rodent and primate visual cortices by combining *in vivo* functional imaging and fluorescent ensemble tagging with presentations of diverse visual stimuli.

2.11 Concluding Remarks

Effective examination of the cellular basis of behavior rests on three experimental pillars: (1) precise identification; (2) comprehensive targeting; and (3) temporally restricted, yet reliable, activity manipulation of neurons. In the years since the first genetic methods for selective neuronal activation were described, progress has been made toward meeting each of these requirements. However, as the adoption of various *chemogenetic* and *optogenetic* methods by the neuroscience community has grown, the remaining challenges have become more pressing. The transition away from transgenic rodents to larger species with more complex behavioral phenotypes, and the goal of identifying and examining the interactions among multiple elements of previously uncharted neuronal circuits *in vivo* using both functional imaging and activity perturbations, have exposed the limitations of existing techniques. It is quite likely, therefore, that the story that began when the first neuron was sensitized to respond to a pulse of light has many chapters yet to be written.

ACKNOWLEDGMENTS

I am indebted to Michael Drew, Attila Losonczy, Daniel Johnston, Richard Aldrich and Robert Messing for critical comments on this manuscript. The Zemelman laboratory is grateful for the support of the Human Frontiers Science Program, the Brain Research Foundation, the Winkler Family Foundation, the National Institute of Mental Health, NIH BRAIN Initiative grants from the National Eye Institute and the National Institute of Neurological Disorders and Stroke, as well as grants from the University of Texas System BRAIN Initiative Seed Grant program.

REFERENCES

Adamantidis, A., Arber, S., Jaideep, S., *et al.* (2015). Optogenetics: 10 years after ChR2 in neurons – views from the community. *Nature Neuroscience*, **18**, 1202–1212.

Arenkiel, R., Marguerita E. K., Davison, I. G., *et al.* (2008). Genetic control of neuronal activity in mice conditionally expressing TRPV1. *Nature Methods*, **5**, 299–302.

Bading, H., Ginty, D. D. and Greenberg, M. E. (1993). Regulation of gene expression in hippocampal neurons by distinct calcium signaling pathways. *Science*, **260**, 181–186.

Barnes, C. A., McNaughton, B. L., Mizumori, S. J., *et al.* (1990). Comparison of spatial and temporal characteristics of neuronal activity in sequential stages of hippocampal processing. *Progress in Brain Research*, **83**, 287–300.

Barth, A. L., Gerkin, R. C. and Dean, K. L. (2004). Alteration of neuronal firing properties after *in vivo* experience in a FosGFP transgenic mouse. *The Journal of Neuroscience*, **24**, 6466–6475.

Blanco, E., Messeguer, X., Smith T. F., *et al.* (2006). Transcription factor map alignment of promoter regions. *PLoS Computational Biology*, **2**, e49.

Bonni, A., Ginty, D. D., Dudek, H., *et al.* (1995). Serine 133-phosphorylated CREB induces transcription via a cooperative mechanism that may confer specificity to neurotrophin signals. *Molecular and Cellular Neurosciences*, **6**, 168–183.

Borghuis, B. G., Tian, L., Xu, Y., *et al.* (2011). Imaging light responses of targeted neuron populations in the rodent retina. *The Journal of Neuroscience*, **31**, 2855–2867.

Bossert, J. M., Stern, A. L., Theberge, F. R. M., *et al.* (2011). Ventral medial prefrontal cortex neuronal ensembles mediate context-induced relapse to heroin. *Nature Neuroscience*, **14**, 420–422.

Boyden, E. S., Zhang F., Bamberg E., *et al.* (2005). Millisecond-timescale, genetically targeted optical control of neural activity. *Nature Neuroscience*, **8**, 1263–1268.

Brake, A. J., Wagenbach, M. J. and Julius, D. (1994). New structural motif for ligand-gated ion channels defined by an ionotropic ATP receptor. *Nature*, **371**, 519–523.

Cambridge, S. B., Geissler, D., Calegari, F., *et al.* (2009). Doxycycline-dependent photoactivated gene expression in eukaryotic systems. *Nature Methods*, **6**, 527–531.

Caterina, M. J., Schumacher, M. A., Tominaga, M., *et al.* (1997). The capsaicin receptor: a heat-activated ion channel in the pain pathway. *Nature*, **389**, 816–824.

Ceccatelli, S., Villar, M. J., Goldstein, M., *et al.* (1989). Expression of c-Fos immunoreactivity in transmitter-characterized neurons after stress. *Proceedings of the National Academy of Sciences USA*, **86**, 9569–9573.

Chen, T.-W., Wardill, T. J., Sun, Y., *et al.* (2013). Ultrasensitive fluorescent proteins for imaging neuronal activity. *Nature*, **499**, 295–300.

Cowansage, K. K., Shuman, T., Dillingham, B. C., *et al.* (2014). Direct reactivation of a coherent neocortical memory of context. *Neuron*, **84**, 432–441.

Crick, F. (1999). The impact of molecular biology on neuroscience. *Philosophical Transactions of the Royal Society of London*, **354**, 2021–2025.

Delgado, J. M. R. (1964). Free behavior and brain stimulation. *International Review of Neurobiology*, **6**, 349–449.

Delgado, J. M. R. (1969). *Physical Control of the Mind: Toward a Psychocivilized Society*. New York, NY: Harper & Row.

Denny, C. A., Kheirbek, M. A., Alba, E. L., *et al.* (2014). Hippocampal memory traces are differentially modulated by experience, time, and adult neurogenesis. *Neuron*, **83**, 189–201.

Diester, I., Kaufman, M. T., Mogri, M., *et al.* (2011). An optogenetic toolbox designed for primates. *Nature Neuroscience*, **14**, 387–397.

Dittgen, T., Nimmerjahn, A., Komai, S., *et al.* (2004). Lentivirus-based genetic manipulations of cortical neurons and their optical and electrophysiological monitoring *in vivo*. *Proceedings of the National Academy of Sciences USA*, **101**, 18206–18211.

Edwards, J. G. (2014). TRPV1 in the central nervous system: synaptic plasticity, function, and pharmacological implications. *Progress in Drug Research*, **68**, 77–104.

Feil, R., Wagner, J., Metzger, D., *et al.* (1997). Regulation of Cre recombinase activity by mutated estrogen receptor ligand-binding domains. *Biochemical and Biophysical Research Communications*, **237**, 752–757.

Freundlieb, S., Schirra-Müller, C. and Bujard, H. (1999). A tetracycline controlled activation/repression system with increased potential for gene transfer into mammalian cells. *The Journal of Gene Medicine*, **1**, 4–12.

Galvan, A., Hu, X., Smith, Y., *et al.* (2012). *In vivo* optogenetic control of striatal and thalamic neurons in non-human primates. *PLoS ONE*, **7**, e50808.

Garner, A. R., Rowland, D. C., Hwang, S. Y., *et al.* (2012). Generation of a synthetic memory trace. *Science*, **335**, 1513–1516.

Gerits, A., Farivar, R., Rosen, B. R., *et al.* (2012). Optogenetically induced behavioral and functional network changes in primates. *Current Biology*, **22**, 1722–1726.

Gonzalez, G. A. and Montminy, M. R. (1989). Cyclic AMP stimulates somatostatin gene transcription by phosphorylation of CREB at serine 133. *Cell*, **59**, 675–680.

Gossen, M. and Bujard, H. (1992). Tight control of gene expression in mammalian cells by tetracycline-responsive promoters. *Proceedings of the National Academy of Sciences USA*, **89**, 5547–5551.

Guenthner, C. J., Miyamichi, K., Yang, H. H., *et al.* (2013). Permanent genetic access to transiently active neurons via TRAP: targeted recombination in active populations. *Neuron*, **78**, 773–784.

Güler, A. D., Rainwater, A., Parker, J. G., *et al.* (2012). Transient activation of specific neurons in mice by selective expression of the capsaicin receptor. *Nature Communications*, **3**, 746–756.

Han, X., Qian, X., Bernstein, J. G., *et al.* (2009). Millisecond-timescale optical control of neural dynamics in the nonhuman primate brain. *Neuron*, **62**, 191–198.

Heath, R. G., Monroe, R. R. and Mickle, W. A. (1955). Stimulation of the amygdaloid nucleus in a schizophrenic patient. *American Journal of Psychiatry*, **111**, 862–863.

Hunt, S. P., Pini, A. and Evan, G. (1987). Induction of c-Fos-like protein in spinal cord neurons following sensory stimulation. *Nature*, **328**, 632–634.

Impey, S., McCorkle, S. R., Cha-Molstad, H., *et al.* (2004). Defining the CREB regulon: a genome-wide analysis of transcription factor regulatory regions. *Cell*, **119**, 1041–1054.

Jazayeri, M., Lindbloom-Brown, Z. and Horwitz, G. D. (2012). Saccadic eye movements evoked by optogenetic activation of primate V1. *Nature Neuroscience*, **15**, 1368–1370.

Kawashima, T., Okuno, H., Nonaka, M., *et al.* (2009). Synaptic activity–responsive element in the Arc/Arg3.1 promoter essential for synapse-to-nucleus signaling in activated neurons. *Proceedings of the National Academy of Sciences USA*, **106**, 316–321.

Kawashima, T., Kitamura, K., Suzuki, K., *et al.* (2013). Functional labeling of neurons and their projections using the synthetic activity-dependent promoter E-SARE. *Nature Methods*, **10**, 889–895.

Kee, N., Teixeira, C. M., Wang, A. H., *et al.* (2007). Preferential incorporation of adult-generated granule cells into spatial memory networks in the dentate gyrus. *Nature Neuroscience*, **10**, 355–362.

Khorana, H. G., Knox, B. E., Nasi, E., *et al.* (1988). Expression of a bovine rhodopsin gene in *Xenopus* oocytes: demonstration of light-dependent ionic currents. *Proceedings of the National Academy of Sciences USA*, **85**, 7917–7921.

King, H. E. (1961). Psychological effects of excitation in the limbic system. In: *Electrical Stimulation of the Brain: An Interdisciplinary Survey of Neurobehavioral Integrative Systems*. D. E. Sheer (ed.), pp. 477–486. Austin, TX: University of Texas Press.

Kiselev, A. and Subramaniam, S. (1997). Studies of Rh1 metarhodopsin stabilization in wild-type *Drosophila* and in mutants lacking one or both arrestins. *Biochemistry*, **36**, 2188–2196.

Koya, E., Golden, S. A., Harvey, B. K., *et al.* (2009). Targeted disruption of cocaine-activated nucleus accumbens neurons prevents context-specific sensitization. *Nature Neuroscience*, **12**, 1069–1073.

Kwok, R. P., Lundblad, J. R., Chrivia, J. C., *et al.* (1994). Nuclear protein CBP is a coactivator for the transcription factor CREB. *Nature*, **370**, 223–226.

Kyung, T., Lee, S., Kim, J. E., *et al.* (2015). Optogenetic control of endogenous Ca^{2+} channels *in vivo*. *Nature Biotechnology*, **33**, 1092–1096.

Lee, M.-H., Appleton, K. M., Strungs, E. G., *et al.* (2016). The conformational signature of B-arrestin2 predicts its trafficking and signalling functions. *Nature*, **531**, 665–668.

Lima, S. Q. and Miesenböck, G. (2005). Remote control of behavior through genetically targeted photostimulation of neurons. *Cell*, **121**, 141–152.

Liu, X., Ramirez, S., Pang, P. T., *et al.* (2012). Optogenetic stimulation of a hippocampal engram activates fear memory recall. *Nature*, **484**, 381–385.

Loebrich, S., and Nedivi, E., (2009). The function of activity-regulated genes in the nervous system. *Physiological Reviews*, **89 (4)**, 1079–1103. doi:10.1152/physrev.00013.2009.

Losonczy, A. and Zemelman, B. V. (2016). Illuminating memory circuit dynamics. *Learning & Memory* (In preparation).

MacLean, P. D. (1990). *The Triune Brain in Evolution: Role in Paleocerebral Functions*. New York, NY: Plenum Press.

Madisen, L., Zwingman, T. A., Sunkin, S. M., *et al.* (2009). A robust and high-throughput Cre reporting and characterization system for the whole mouse brain. *Nature Neuroscience*, **13**, 133–140.

Melyan, Z., Tarttelin, E. E., Bellingham, J., *et al.* (2005). Addition of human melanopsin renders mammalian cells photoresponsive. *Nature*, **433**, 741–745.

Minatohara, K., Akiyoshi, M. and Okuno, H. (2016). Role of immediate–early genes in synaptic plasticity and neuronal ensembles underlying the memory trace. *Frontiers in Molecular Neuroscience*, **8**, 78.

Morgan, J. I. and Curran, T. (1991). Stimulus–transcription coupling in the nervous system: involvement of the inducible proto-oncogenes *Fos* and *Jun*. *Annual Review of Neuroscience*, **14**, 421–451.

Nagel, G., Szellas, T., Huhn, W., *et al.* (2003). Channelrhodopsin-2, a directly light-gated cation-selective membrane channel. *Proceedings of the National Academy of Sciences USA*, **100**, 13940–13945.

Nathanson, J. L., Jappelli, R., Scheeff, E. D., *et al.*, (2009). Short promoters in viral vectors drive selective expression in mammalian inhibitory neurons, but do not restrict activity to specific inhibitory cell-types. *Frontiers in Neural Circuits*, **3**, 19.

Nuber, S., Zabel, U., Lorenz, K., *et al.* (2016). B-arrestin biosensors reveal a rapid, receptor-dependent activation/deactivation cycle. *Nature*, **531**, 661–664.

Ovcharenko, I., Nobrega, M. A., Loots, G. G., *et al.* (2004). ECR Browser: a tool for visualizing and accessing data from comparisons of multiple vertebrate genomes. *Nucleic Acids Research*, **32**, W280–W286.

Panksepp, J. (2004). *Affective Neuroscience: the Foundations of Human and Animal Emotions*. Oxford: Oxford University Press.

Peier, A. M., Moqrich, A., Hergarden, A. C., *et al.* (2002). A TRP channel that senses cold stimuli and menthol. *Cell*, **108**, 705–715.

Pinal, C. S., Cortessis, V. and Tobin, A. J. (1997). Multiple elements regulate GAD65 transcription. *Developmental Neuroscience*, **19**, 465–475.

Qiu, X., Kumbalasiri, T., Carlson, S. M., *et al.* (2005). Induction of photosensitivity by heterologous expression of melanopsin. *Nature*, **433**, 745–749.

Ramirez, S., Liu, X., Lin, P.-A., *et al.* (2013). Creating a false memory in the hippocampus. *Science*, **341**, 387–391.

Redondo, R. L., Kim, J., Arons, A. L., *et al.* (2014). Bidirectional switch of the valence associated with a hippocampal contextual memory engram. *Nature*, **513**, 426–430.

Reijmers, L. G., Perkins, B. L., Matsuo, N., *et al.* (2007). Localization of a stable neural correlate of associative memory. *Science*, **317**, 1230–1233.

Root, C. M., Denny, C. A., Hen, R., *et al.* (2014). The participation of cortical amygdala in innate, odour-driven behaviour. *Nature*, **515**, 269–273.

Rossier, J., Bernard, A., Cabungcal, J.-H., *et al.* (2014). Cortical fast-spiking parvalbumin interneurons enwrapped in the perineuronal net express the metallopeptidases Adamts8, Adamts15 and neprilysin. *Molecular Psychiatry*, **20**, 154–161.

Rost, B. R., Schneider, F., Grauel, M. K., *et al.* (2015). Optogenetic acidification of synaptic vesicles and lysosomes. *Nature Neuroscience*, **18**, 1845–1852.

Ruiz, O., Lustig, B. R., Nassi, J. J., *et al.* (2013). Optogenetics through windows on the brain in the nonhuman primate. *Journal of Neurophysiology*, **110**, 1455–1467.

Sagar, S. M., Sharp, F. R. and Curran, T. (1988). Expression of c-Fos protein in brain: metabolic mapping at the cellular level. *Science*, **240**, 1328–1331.

Schoch, S., Cibelli, G. and Thiel., G. (1996). Neuron-specific gene expression of synapsin I: major role of a negative regulatory mechanism. *The Journal of Biological Chemistry*, **271**, 3317–3323.

Seidemann, E., Chen, Y., Bai, Y., *et al.* (2016). Calcium imaging with genetically encoded indicators in behaving primates. *eLife*, **5**, 3771.

Sharma, K., Schmitt, S., Bergner, C. G., *et al.* (2015). Cell type- and brain region-resolved mouse brain proteome. *Nature Neuroscience*, **18**, 1819–1831.

Sheng, M, McFadden, G. and Greenberg, M. E. (1990). Membrane depolarization and calcium induce c-Fos transcription via phosphorylation of transcription factor CREB. *Neuron*, **4**, 571–582.

Sheng, M., Thompson, M. A. and Greenberg, M. E. (1991). CREB: a Ca^{2+}-regulated transcription factor phosphorylated by calmodulin-dependent kinases. *Science*, **252**, 1427–1430.

Stein, M., Breit, A., Fehrentz, T., *et al.* (2013). Optical control of TRPV1 channels. *Angewandte Chemie*, **52**, 9845–9848.

Stierl, M., Stumpf, P., Udwari, D., *et al.* (2011). Light modulation of cellular cAMP by a small bacterial photoactivated adenylyl cyclase, bPAC, of the soil bacterium *Beggiatoa*. *The Journal of Biological Chemistry*, **286**, 1181–1188.

Stone, S. S., Teixeira, C. M., Zaslavsky, K., *et al.* (2011). Functional convergence of developmentally and adult-generated granule cells in dentate gyrus circuits supporting hippocampus-dependent memory. *Hippocampus*, **21**, 1348–1362.

Sugino, K., Hempel, C. M., Miller, M. N., *et al.* (2006). Molecular taxonomy of major neuronal classes in the adult mouse forebrain. *Nature Neuroscience*, **9**, 99–107.

Sweet, W. H., Ervin, F. and Mark, V. H. (1969). The relationship of violent behaviour to focal cerebral disease. In: *Aggressive Behaviour*. S. Garattini and E. B. Sigg (Eds.), pp. 336–352. Amsterdam: Excerpta Medica Foundation.

Tasic, B., Menon, V., Nguyen, T. N., *et al.*, (2016). Adult mouse cortical cell taxonomy revealed by single cell transcriptomics. *Nature Neuroscience*, **19**, 335–346.

Wang, K. H., Majewska, A., Schummers, J., *et al.* (2006). *In vivo* two-photon imaging reveals a role of Arc in enhancing orientation specificity in visual cortex. *Cell*, **126**, 389–402.

Wang, M., Perova, Z., Benjamin R. A., *et al.* (2014). Synaptic modifications in the medial prefrontal cortex in susceptibility and resilience to stress. *The Journal of Neuroscience*, **34**, 7485–7492.

Wend, S., Wagner, H. J., Konrad Müller, K., *et al.* (2014). Optogenetic control of protein kinase activity in mammalian cells. *ACS Synthetic Biology*, **3**, 280–285.

Yaguchi, M., Ohashi, Y., Tsubota, T., *et al.* (2013). Characterization of the properties of seven promoters in the motor cortex of rats and monkeys after lentiviral vector-mediated gene transfer. *Human Gene Therapy Methods*, **24**, 333–344.

Yao, F. and Eriksson, E. (1999). A novel tetracycline-inducible viral replication switch. *Human Gene Therapy*, **10**, 419–427.

Yassin, L., Benedetti, B. L., Jouhanneau, J.-S., *et al.* (2010). An embedded subnetwork of highly active neurons in the neocortex. *Neuron*, **68**, 1043–1050.

Zemelman, B. V. and Miesenböck, G. (2001). Genetic schemes and schemata in neurophysiology. *Current Opinion in Neurobiology*, **11**, 409–414.

Zemelman, B. V., Lee G. A., Ng, M., *et al.* (2002). Selective photostimulation of Genetically chARGed neurons. *Neuron*, **33**, 15–22.

Zemelman, B. V., Nesnas, N., Lee G. A., *et al.* (2003). Photochemical gating of heterologous ion channels: remote control over genetically designated populations of neurons. *Proceedings of the National Academy of Sciences USA*, **100**, 1352–1357.

Zhang, X., Odom D. T., Koo, S.-H., *et al.* (2005). Genome-wide analysis of cAMP-response element binding protein occupancy, phosphorylation, and target gene activation in human tissues. *Proceedings of the National Academy of Sciences USA*, **102**, 4459–4464.

3 From Connectome to Function: Using Optogenetics to Shed Light on the *Caenorhabditis elegans* Nervous System

Koutarou D. Kimura and Karl Emanuel Busch

3.1 Introduction

Optogenetics emerged from the discovery of light-activated ion channels and ion pumps (Nagel *et al.*, 2002; Nagel *et al.*, 2003; Zhang *et al.*, 2007), which have revolutionized our ways of controlling neural activities non-invasively and with high spatiotemporal precision. For optogenetics, as its name indicates, the combination of two techniques is essential: light stimulation and the control of gene expression. Namely, the light-activated ion pumps and channels should be expressed in the target neurons, and light should be delivered to those neurons.

The nematode *Caenorhabditis elegans* is one of the best-suited model animals for meeting these two requirements. Because of its transparent body, stimulatory light is readily delivered to the target cells. In addition, cell- or tissue-specific expression of ion channel/pump genes can be easily achieved by combining them with specific promoters and by creating transgenic animals with these cloned constructs: a few kilobases of the upstream sequence of a gene typically confers cell- or tissue-specific expression. This is further aided by the very small size of the entire *C. elegans* nervous system. It consists of 302 neurons, which show invariant cell lineage and can be uniquely identified by their cell shapes and positions, and/or by cell-specific promoters (Figure 3.1). Of all of these neurons, 20 are pharynx specific, 60–70 are sensory and ~100 are motor neurons. Therefore, only roughly 100 neurons can be regarded as constituting the "central nervous system" of these worms. Due to these advantages, *C. elegans* was used in two foundational studies of optogenetics: the first behavioral regulation by channelrhodopsin (ChR) (Nagel *et al.*, 2005) and the first bidirectional behavioral regulation by ChR and halorhodopsin (NpHR) (Zhang *et al.*, 2007). Moreover, despite its small size, the *C. elegans* nervous system exhibits simple but sophisticated "brain" functions, such as learning and memory, navigation and decision making (Bendesky *et al.*, 2011; Lockery, 2011; Faumont *et al.*, 2012; Sasakura and Mori, 2013). Thus, through optogenetic analyses, we can conceivably interrogate how basic brain functions can be generated from a small-scale neural circuit.

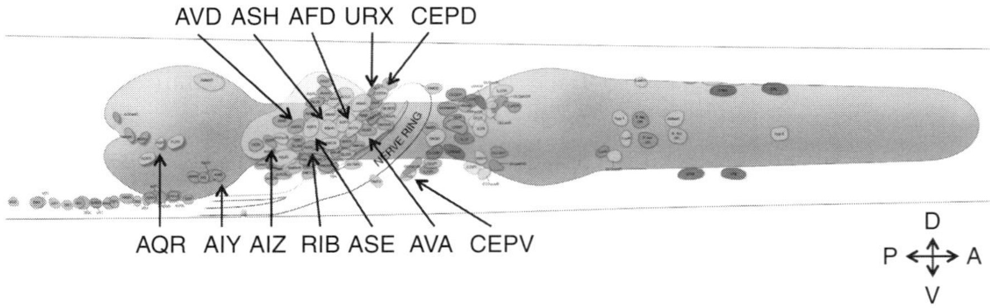

Figure 3.1 Positions of neuronal cell bodies in the head of *C. elegans*. The cell bodies of the neurons (small circles) mentioned in this chapter are indicated. The large, gourd-like structure is the pharyngeal musculature. Reproduced and modified with permission from Altun and Hall (2016) www.wormatlas.org.

To achieve an understanding of how neural circuits function, we first need to understand the connectivity of the neurons that form these networks. *C. elegans* has a unique advantage in this respect, as complete information regarding the chemical synaptic connections and gap junctions between these neurons – the first "connectome" information – has been available since 1986 (White *et al.*, 1986). By reconstructing synaptic contacts from serial electron micrograph sections by hand, 5000 synaptic connections and 600 gap junctions between neurons have been identified.

Although it is an invaluable resource for *C. elegans* neurobiology, the anatomical connectivity map of the *C. elegans* nervous system has proved insufficient for obtaining a full understanding of how the worm processes information; knowing whether and how many synapses connect neuron A with neuron B does not reveal if the connection is excitatory or inhibitory, or whether it is relevant for specific behaviors. Therefore, the *functional* relations between neurons and the dynamics of neural circuit signaling need to be mapped and are still unknown for most *C. elegans* neurons.

To this end, we have to consider the input, information flow and output of neuronal networks. Light-gated channels can provide the input and allow the bypassing or simulating of sensory stimulation. Behavior is the ultimate output function, but is rarely shaped exclusively by a single circuit; to more specifically probe the consequence of a specific optogenetic manipulation, one also needs to record the activity of the neurons that receive input from the stimulated neuron(s).

This review aims to provide readers with a brief overview as to how we analyze basic brain functions in *C. elegans* with optogenetics. Following this introduction, we will summarize in Section 3.2 how optogenetics is used in *C. elegans* to study chemical synapse function *in vivo*. In Section 3.3, we will describe all-optical interrogation (i.e. optogenetic stimulation and optical monitoring of neural activities) for studying functional circuitry. In Section 3.4, we will recapitulate optogenetic behavioral analyses performed in worms, especially in terms of the advantages and disadvantages of particular methods or setups used for the assays – this is to help readers to choose the proper experimental setups for their purposes.

In Section 3.5, we will describe our own recent results of optogenetic analyses of *C. elegans* neural function: neural multitasking in O_2 sensation and functional asymmetry of anatomically symmetric dopaminergic neurons. Finally, in Section 3.6, we will describe the future directions of optogenetic analyses of *C. elegans*. Descriptions of the history of optogenetics and of the diverse genes harnessed for optogenetic analyses in *C. elegans* are beyond the scope of our review, and we recommend that interested readers consult other recent reviews (Husson *et al.*, 2013; Fang-Yen *et al.*, 2015).

3.2 Using Optogenetics to Study Synaptic Functions

Neurons communicate directly with each other via synaptic transmission, and the molecular mechanisms that govern information transfer at synapses are of intense interest in neuroscience. *C. elegans* has greatly contributed to this field through the discovery and characterization of mutations of evolutionarily conserved genes that disrupt synaptic function, such as *unc-13* (Richmond, 2005). Optogenetics has greatly improved our technical arsenal for elucidating synaptic function in *C. elegans* and has advanced our understanding of information transfer at synapses, synaptic vesicle biology and synaptic plasticity.

3.2.1 Synaptic Neurotransmission

Neuromuscular junctions (NMJs) are the most experimentally accessible synapses in *C. elegans*, especially through electrophysiological means. They are formed by interconnected excitatory cholinergic and inhibitory GABAergic motoneurons that synapse with body wall muscles on opposite sides of the body. Together, they create a balanced system that contracts muscles on one side of the body while inhibiting them on the other. Muscle contraction propagates along the body through proprioception in the preceding body segment in order to produce locomotion (Wen *et al.*, 2012). Neurotransmission at *C. elegans* NMJs can be studied by voltage or current clamp techniques, but presynaptic stimulation by electric shock is difficult, damaging when done repeatedly, and cannot selectively induce cholinergic or GABAergic transmission. Therefore, optogenetic presynaptic stimulation is superior and was used in several studies in combination with electrophysiological recordings from body wall muscle cells (Liu *et al.*, 2009). These "opto-electrophysiological" studies were complemented by behavioral readouts (muscle relaxation or contraction) and were used to shed light on the general properties of synaptic transmission, synaptic vesicle biology and especially synaptic plasticity. They showed that transmission at *C. elegans* NMJs is graded; these synapses can transmit stimulation intensity precisely and in a broad dynamic range. Recordings in a range of mutants with pre- or post-synaptic defects, including in ACh or GABA transmission, largely confirmed the expected phenotypes (Liewald *et al.*, 2008).

To functionally map how information propagates in neural circuits, one needs to examine the transfer properties of synapses between neurons, and opto-electrophysiology has also been applied to this task, as paired electrophysiological recordings of neurons are not feasible in *C. elegans* due to the neurons' small sizes. Narayan *et al.* used the technique to probe synaptic information transfer (Narayan *et al.*, 2011). They showed that an anatomically defined synaptic connection between the thermosensory neuron pair AFD and the AIY interneuron pair had evolved in order to faithfully transmit temperature information. The synapse is peptidergic and the transfer function is excitatory, graded and tonic, with sustained release upon depolarization; it shows no depression or facilitation within short timespans (Narayan *et al.*, 2011). One concern highlighted in the study is that the expression levels of optogenetic channels in neurons vary extensively between individual animals. Therefore, the quantitative measurement of the synaptic transfer function requires calibrating with channel expression (Narayan *et al.*, 2011).

A second study accomplished opto-electrophysiological recordings from a neuronal synapse by using a less invasive dissection procedure (Lindsay *et al.*, 2011). Animals were glued into grooves on an agar surface imprinted with a microfabricated stamp, which position them such that a tiny slit in the worm adjacent to the target neuron allowed access to it. With this preparation, the researchers studied the synaptic connection between the polymodal sensory neuron pair ASH and the AVA command interneuron pair. ASH is the main nociceptor of worms and confers avoidance of chemicals, touch, high osmolarity, etc. It synapses directly with AVA, which is thought to coordinate reversal behavior, a transient escape movement. ASH was photostimulated by ChR2. The signal received in AVA was excitatory, graded and used glutamate as the neurotransmitter.

3.2.2 Synaptic Vesicle Biology

As optogenetics allows the manipulation of neurons *in vivo* with high temporal and spatial precision, it could also be used to shed light on the dynamics of synaptic vesicle release and recycling. In an ingenious study, Watanabe and colleagues coupled optogenetically induced vesicle release with rapid high-pressure freezing of whole *C. elegans* animals in order to shed light on this process (Watanabe *et al.*, 2013a). They modified the specimen carrier of a high-pressure freezing system to contain a window for illumination and coordinated it with freezing with millisecond precision. They expressed the ChR2 variant ChIEF in motoneurons, gave a single 3-ms stimulus, froze the specimen within 20 ms after the light pulse and at several time points thereafter and examined synaptic densities with transmission electron microscopy. This technique enabled the studying of the ultrastructure of synaptic vesicle release *in vivo* in response to a single physiological stimulus.

They discovered that endocytic invaginations appeared after 20 ms and turned into large vesicles within less than 50 ms. This ultrafast endocytosis process occurred at two sites – at dense projections in the center of the synapse and later at adherens junctions flanking the synapse – and depended on dynamin. Ultrafast endocytosis appears to represent a novel model of synaptic vesicle endocytosis, distinct from clathrin-mediated or kiss-and-run endocytosis. It also appears to be evolutionarily conserved, as it was subsequently demonstrated to occur in cultured mouse hippocampal neurons (Watanabe *et al.*, 2013b). This novel endocytosis process may serve to rapidly restore the surface area of the membrane.

In sharp contrast, strong and prolonged neuronal activity, such as during epileptic seizures, exhausts the capacity of synapses to regenerate synaptic vesicles and causes clathrin-independent bulk endocytosis. Kittelmann and colleagues studied this process by continuously photoactivating ChR2 in the cholinergic motoneurons for 30 s before freezing them at high pressure. Synapses took up to 20 s to regain full release capability, and locomotory activity took up to 60 s to recover; again, dynamin was required for synaptic vesicle regeneration (Kittelmann *et al.*, 2013).

3.2.3 Synaptic Plasticity

Transmission at chemical synapses is highly plastic, and because this plasticity is a key element of learning and memory, it is intensively studied; a vast number of molecules and pathways have been shown to regulate this process. In *C. elegans*, optogenetic channels enabled the possibility of conducting electrophysiological studies of synaptic plasticity at NMJs, because they made possible the repeated and prolonged stimulation of presynaptic motoneurons. High-frequency stimulation of the excitatory cholinergic synapses at NMJs were found to depress transmission; this depended on the nicotine-sensitive ACh receptor complex. Inhibitory GABAergic synapses, in contrast, showed short-term facilitation of transmission before depression set in (Liu *et al.*, 2009).

Two studies from the Maricq laboratory shed further light on the molecular mechanisms of synaptic plasticity by using chronic stimulation with light-gated channels. Jensen *et al.* showed that Wnt signaling regulates experience-dependent synaptic plasticity at *C. elegans* NMJs by changing the localization of the nicotinic ACh receptor towards or away from the cell surface (Jensen *et al.*, 2012). ChR2 was expressed in the cholinergic motoneurons, and animals were photostimulated for up to an hour. This chronic stimulation induced the activity-dependent secretion of the Wnt ligand from the presynaptic cells, a relocation of the ACh receptor to the cell surface in the muscle cells and an increase in the AChR-mediated current. These changes were abolished in Wnt mutants.

Hoerndli and colleagues investigated the activity-dependent plasticity of α-amino-3-hydroxy-5-methyl-4-isoxazolepropionic acid (AMPA) receptors at glutamatergic synapses (Hoerndli *et al.*, 2015). To induce activity-dependent changes at specific synapses *in vivo*, a ChR variant was expressed in the mechanosensory neurons, and adult worms underwent optogenetic "training" for 1 hour with repetitive illumination. This long-lasting stimulation increased AMPA receptor localization in the AVA interneurons, specifically at the synapses they receive from the mechanosensors. This relocation depended on UNC-43, the homolog of Ca^{2+}/calmodulin-dependent protein kinase II, acting via microtubule-dependent transport.

3.3 All-optical Interrogation to Study Functional Circuitry

Optogenetics is a vital tool for achieving an understanding of how information flows in the nervous system of *C. elegans*, as mapping synaptic information transfer requires a defined signal in the upstream neuron and must be accompanied by functional recording in other circuit neurons. Several studies have demonstrated the simultaneous manipulation and recording of *C. elegans* neurons. The opto-electrophysiological studies described above are still technically challenging and invasive; they are also limited by the recording taking place in the soma and thus are unable to capture neural activity that is confined to the processes of neurons.

All-optical interrogation, in which both neuronal manipulation and the monitoring of neural activity are performed by using optogenetic tools and genetically encoded calcium indicators simultaneously, is seen as the ideal solution in order to probe neural circuit function in the living animal, as it is less invasive and enables the study of many neurons in parallel (Emiliani *et al.*, 2015). However, all-optical approaches encounter the obstacle that the excitation spectra of many optogenetic tools and calcium sensors overlap; either the wavelengths or the illumination fields for manipulation and recording must be kept separate. Secondly, limiting optogenetic stimulation to the target neurons is a challenge, as promoter expression in most cases is simply not specific enough and will have off-target effects. For this, targeted illumination can be the solution, or the use of combinations of promoters.

The first study to take an all-optical approach to functionally mapping *C. elegans* neural circuits by Guo and colleagues took advantage of the fact that ChR2 requires strong photostimulation in order to efficiently depolarize neurons, whereas the genetically encoded calcium sensor GCaMP can be recorded at considerably weaker light intensity (Guo *et al.*, 2009). These researchers thus used two light sources of different intensities: a 488-nm laser at ~0.1 mW mm^{-2} to record GCaMP fluorescence and a xenon light source at 8.0 mW mm^{-2} to stimulate ChR2. The bright illumination light was passed through a digital light processing (DLP) mirror array to target the illumination to specific neurons. Optical sectioning with a spinning disk unit and a sensitive electron multiplying charge coupled device camera helped with monitoring the weakly excited GCaMP signal. The system was used to examine the functional coupling of the polymodal nociceptor neuron pair ASH with the AVA and AVD command interneurons.

This approach, however, is not optimal, in that the weak illumination used for GCaMP recording may still gate the optogenetic channels in many cases. Indeed, other studies irradiated neuronally expressed ChR2 at 0.1 mW mm^{-2} or even lower intensities and elicited behavioral responses (Piggott *et al.*, 2011; Schild and Glauser, 2015). It also means accepting a lower signal-to-noise ratio in the calcium imaging traces. Animals were paralyzed with a GABA agonist in order to aid neuron-specific targeting and recording, which precluded correlating neural activity with behavioral responses.

Shipley and colleagues addressed this limitation by using targeted illumination for both the optogenetic stimulation and the calcium imaging of selected neurons (Shipley *et al.*, 2014). They extended their previously described setup for the optogenetic stimulation of freely moving *C. elegans*, consisting of a DLP system, a high-speed camera and real-time tracking ("Colbert" in Leifer *et al.*, 2011; see also Section 3.4.4), in order to selectively illuminate both mechanosensory neurons with blue laser light for stimulating ChR, and downstream command interneurons, termed AVA, with the same wavelength in order to excite the calcium sensor GCaMP3 and record its fluorescence signal with an additional complementary metal oxide semiconductor camera (Shipley *et al.*, 2014). Since the worms here were freely moving and tracked using near-infrared light, the system even allowed the recording of behavioral readout at the same time.

This setup, in which both stimulation and recording uses the same wavelength, can only be used for neurons that are spatially separated. Another solution would be to use either red-shifted, light-gated channels or calcium sensors for probing neighboring neurons. A proof of principle of this approach was first demonstrated by stimulating cholinergic motoneurons with ChR2 and visualizing the resulting Ca^{2+} increase in the body wall muscles with the red calcium sensor RCaMP1 in animals immobilized on agar pads (Akerboom *et al.*, 2013). A more recent study used the improved red calcium sensor R-CaMP2 to demonstrate the coupling of the presynaptic polymodal nociceptor neuron pair ASH with the postsynaptic AVA interneurons in freely moving animals. Like the majority of neurons, the cell bodies of ASH and AVA are very close to each other around the nerve ring in the animal's head, precluding the use of targeted illumination approaches. ChR2 stimulation in ASH will cause a response in AVA expressing R-CaMP2, and induces transient reversal movement (Inoue *et al.*, 2015). The method is limited by requiring sufficiently specific promoters, and novel methods to limit expression in any given neuron will be needed in order to extend these approaches to probing all synaptic connections.

Using a similar tracking setup with whole-field blue and yellow light stimulation, respectively, of the stimulation-recording gene pairs ChR2-RCaMP and NpHR-GCaMP3, an extensive dissection of a novel circuit motif in *C. elegans* revealed that the first layer interneuron AIY encodes two different types of behavioral outputs via two segregated outputs to different second layer interneurons: it excites RIB in order to control locomotion speed and it inhibits AIZ in order to induce a transient switch in direction of movement, termed "reversal". Both outputs use acetylcholine, but act through different receptors on the two postsynaptic neurons, an ACh-gated chloride channel in AIZ and a nicotine-type excitatory nAChR in RIB (Li *et al.*, 2014).

3.4 Optogenetic Systems for Freely Behaving *C. elegans*

Behavior is the major output of an animal's nervous system. Therefore, to understand how optogenetic control of neural activity affects the dynamic functioning of a neural circuit, it is quite important to quantitatively assess the changes in animal behavior during optogenetic analysis. However, in order to monitor a worm's behavior for sufficiently long periods of time, we need to track the worm (~1 mm in length) while it is moving and behaving. In this section, we will summarize the advantages and disadvantages of different solutions in terms of ease of setting up the system and resolution of behavioral analyses, among other factors (Table 3.1), of:

Table 3.1 Typical properties of optogenetic microscope systems.

	i	ii	iii	iv	v
Microscope	None (a high-resolution camera plus lens)	Stereo	Upright or inverted	Upright or inverted	Upright or inverted
Objective lens	None	1× objective lens with ≤5–6× zoom	4–10×	4–10×	15–40×
Motorized stage	None	None	Yes	Yes	Yes
Behavioral arena	Large (can be the standard 9-cm diameter plate)	Small (ø3.5 cm with 5× zoom)	Not limited (because of the tracking stage)	Not limited (because of the tracking stage)	Not limited (because of the tracking stage)
Light stimulation	Strong LED	Mercury, xenon or LED[a]	Mercury, xenon or LED[a]	LCD or laser + DLP	LCD or laser + DLP
Maximum light intensity	0.5 mW mm^{-2} (blue)	1 mW mm^{-2} (blue) 5–10 mW mm^{-2} (green or yellow)	5–10 mW mm^{-2} (blue) 50 mW mm^{-2} (green or yellow)	5–10 mW mm^{-2} (blue, LCD or laser + DLP)	5–10 mW mm^{-2} (blue, LCD or laser + DLP)
Spatial resolution of target illumination	Entire view field (whole body)	Entire view field (whole body)	Entire view field (whole body)	13–30 μm (a part of the body)	≤5 μm (single neuron position)
Application	All-or-none activation of neuron(s) or muscles expressing a light-activated ion channel or pump	All-or-none activation of neuron(s) or muscles expressing a light-activated ion channel or pump	All-or-none activation of neuron(s) or muscles with stronger light; behavior can be analyzed in detail	Selected activation of neuron (s) or muscles that are separate by more than 15–30 μm from the others	Selected activation of neuron from other neurons in the head ganglia
References	Kawazoe *et al.*, 2013; Husson *et al.*, 2012 (for OptoTracker)	Schild and Glauser, 2015; Okazaki and Takagi, 2013; Busch *et al.*, 2012	Satoh *et al.*, 2014	Stirman *et al.*, 2011; Leifer *et al.*, 2011	Kocabas *et al.*, 2012; Tanimoto *et al.*, 2016[b]

[a] These systems can also integrate a LCD or laser + DMP system to selectively illuminate a part of the view field.

[b] Based on the criteria described in Stirman *et al.* (2011), Tanimoto *et al.*'s system has an imaging resolution of 1.6 mm, an illumination resolution of 2 mm and a tracking accuracy of 1.3 mm (= 250 mm/200 Hz).

(i) A non-microscope system with illumination of the entire field of view
(ii) A stereo microscope system with illumination of the entire field of view
(iii) A compound microscope system with illumination of the entire field of view
(iv) A microscope system with targeted illumination of a part of the worm's body
(v) A microscope system with high-speed tracking and illumination of a single neuron

For the systems mentioned above, the following hardware and/or software are required:

- A camera for capturing the worm's image (required for all systems)
- A lighting system for optogenetic stimulation, which illuminates the entire field of view or specific region of the worm with ≥0.5 mW mm^{-2} intensity (required for all systems)
- A machine vision program and a motorized stage to maintain the worm's position in the view field (required for systems iii–v)
- A machine vision program to recognize and illuminate a specific part of the worm's body (required for systems iv and v)

3.4.1 A Non-microscope System with Illumination of the Entire Field of View

This type of system integrates a fixed, high-resolution camera system for monitoring the behavior of multiple worms, such as the Parallel Worm Tracker or the Multi-Worm Tracker (Ramot *et al.*, 2008; Swierczek *et al.*, 2011) with a strong light source for optogenetic stimulation.

Its advantages are: (1) the behavior of multiple worms can be analyzed simultaneously; and (2) this system can be set up easily and at comparatively low cost because it does not require a motorized stage controlled by machine vision software.

Its disadvantages are: (1) delivering stronger light stimuli (≥0.5 mW mm^{-2}) evenly to the relatively large area is technically difficult. Therefore, optogenetic ion channels/pumps with high light sensitivity may be required; and (2) the spatiotemporal resolution of behavioral monitoring of this type of system is generally sufficient to record speed, large angle turns and reversals (short backward movements), but not for detailed analyses to measure head swing or body curvature.

Its applications are: Kawazoe *et al.* combined an inexpensive, high-resolution USB camera (~2000 × 2000 pixels) and a high-power ring LED (~0. 5mW mm^{-2}). As a proof-of-principle experiment using the standard 9-cm Petri dish behavioral arena, these authors optogenetically stimulated the body wall muscles of worms that expressed ChRGR, an improved version of ChR2 (Kawazoe *et al.*, 2013).

3.4.2 A Stereo Microscope System with Illumination of the Entire Field of View

This type of system uses a stereo (i.e. dissecting) microscope with a fluorescence light source for optogenetic stimulation, coupled with a camera for behavioral observation.

Its advantages are: (1) similarly to system i, this setup does not require the controlling of a motorized stage; and (2) with the standard fluorescence stereo microscope system, one can utilize the bright emission lines at 546 and 577 nm for green–yellow light in the mercury spectrum, which allows illumination at 5–10 mW mm^{-2} or more with regular zoom power (e.g. ≥5× with a 1× objective lens).

Its disadvantages are: (1) because of the absence of a bright emission line in the mercury spectrum, the maximum power of blue light is approximately 1 mW mm^{-2} in the conditions described above. Xenon or metal halide lamps or LED light sources may produce much higher intensities; and (2) when a high-power zoom is used, it gives a higher power density of the fluorescence light, but restricts the area for behavioral observation. With 6× zoom and a 1× objective lens, an area of 3.5 mm in diameter can be observed. Because a worm generally moves ~0.2 mm s^{-1} in the absence of food, less than 20 s of movement can be observed continuously. Nevertheless, this system is useful when a worm's speed is slower (e.g. in the presence of food bacteria), when short-period responses are observed and/or when multiple worms are observed simultaneously.

Its applications are: this system was used to study the effect of GABA signaling on body muscle contraction, the effect of mechanosensory stimulation on reversal responses and how the O_2-sensing neurons control locomotory speed (Busch *et al.*, 2012; Han *et al.*, 2015; Schild and Glauser, 2015).

3.4.3 A Compound Microscope System with Illumination of the Entire Field of View

This type of system integrates an upright or inverted microscope system with a motorized stage to maintain a freely moving worm in the field of view. It also requires an illumination system, which can be a regular mercury lamp, as well as a camera.

Its advantages are: (1) even with the standard mercury lamp, the blue light intensity reaches a sufficient level for optogenetic regulation (5–10 mW mm^{-2} with a 10× objective lens); (2) because of the motorized stage, one can track a freely behaving worm for extended periods, typically for more than several minutes; (3) multiple methods for online (i.e. real-time) regulation of motorized stages based on a worm's image are available today (Husson, 2012); (4) compared to system iv (see below), this method does not require precise control of the illuminated area; and (5) because of the higher magnification compared to systems i and ii, the images obtained from this system enable detailed analyses of a worm's posture and movement.

Its disadvantages are: (1) although relatively simple, the system still requires a machine vision system that regulates a motorized stage; and (2) similarly to systems i and ii, due to the illumination of the whole field of view, expression of the light-activated ion channels/pumps needs to be restricted to the target cells.

Its applications are: Satoh *et al.* combined this type of system with online posture/movement analysis software and revealed that the response of the ASER sensory neuron to subtle changes in sensed salt concentration (caused by the animal's own movement) plays a significant role in the regulation of the animal's migratory course (Satoh *et al.*, 2014).

3.4.4 A Compound Microscope System with Targeted Illumination of a Part of the Worm's Body

This type of system integrates an upright or inverted microscope system with a motorized stage to maintain a freely moving worm in the view field. In addition, a machine vision program enables the restriction of illumination to only a part of a moving worm (e.g. to its head). Hang Lu and Aravinthan Samuel's laboratories have published similar but independent systems of this type (Leifer *et al.*, 2011; Stirman *et al.*, 2011).

Its advantages is: by using excellent machine vision technologies, both systems can target a specific area of a moving worm with a spatial resolution of 15–30 μm, which can be updated every 20–40 ms. Given that the average worm's size is ~1 mm in the anterior–posterior axis and 40–50 μm in width, this resolution allows us to target only the dorsal or ventral side of a worm's body, or a part of the body along the anterior–posterior axis, for example. Its other advantages are the same as for system iii.

Its disadvantages are: (1) although the machine vision programs used in these systems are highly sophisticated, they do not provide a spatial resolution that is sufficient for specifically illuminating one of multiple neurons in the worm's brain (head ganglion), where most neurons' cell bodies (2–3 μm in diameter) are tightly packed within a 30–40 μm^2 area; and (2) the two systems use different light sources, each of which has their own advantages and disadvantages. The system from Lu's laboratory uses a regular LCD projector, which provides a low-cost and low-complexity system setup and enables multicolor stimulation. However, in principle, LCD projectors always emit certain levels of light, even with a command of "complete black." This may lead to unintended activation of neuron(s). The system from Samuel's laboratory used a DLP and a laser. With the DLP system, "black" is completely black and there will be no background signal. However, its handling is more difficult and its cost is higher than for the projector system (Husson, 2012).

Its applications are: both systems published by Lu and Samuel's laboratories have been used widely by multiple groups in order to optogenetically stimulate mechanosensory neurons (whose cell bodies are well separated along the worm's anterior–posterior axis), motor neurons (either ventral or dorsal motor neurons were selectively stimulated) or body wall muscles (either the anterior or posterior part of the body was stimulated) (Leifer *et al.*, 2011; Schultheis *et al.*, 2011; Stirman *et al.*, 2011; Husson *et al.*, 2012a; Husson *et al.*, 2012b; Schmitt *et al.*, 2012; Wen *et al.*, 2012; Donnelly *et al.*, 2013; Williams *et al.*, 2013; Cohen *et al.*, 2014; Krieg *et al.*, 2014; Luo *et al.*, 2014; Shipley *et al.*, 2014; Trojanowski *et al.*, 2014).

3.4.5 A Microscope System with High-speed Tracking and with Illumination of a Single Neuron Position

To illuminate a single neuron (in reality, a pair of left–right neurons aligned vertically, as *C. elegans* worms lie on their side and move either left- or right-side down), a much faster tracking subsystem and higher power of objective lens are required to obtain a 2–4-μm spatial precision. For these systems, in order to control the motorized stage effectively, a machine vision subsystem for tracking the worm's position is separated from the subsystem for the optogenetic illumination. Simply put, a 200-Hz cycle from image acquisition to stage control is required for a 1-μm spatial resolution since worms move with an average locomotory speed of up to ~200 μm s^{-1}. As for the lens, because objective lenses with a magnification greater than 10× do not enable the capturing of the entire body of a worm, a lower-power lens should be used in parallel (Kocabas *et al.*, 2012), or a part of the worm (e.g. neuronal fluorescence or an image of the worm's head) should be recognized by the tracking software (Faumont *et al.*, 2011; Tanimoto *et al.*, 2016).

Its advantage is: such systems should enable the targeted stimulation of individual neurons, even when a cell-specific promoter is not available.

Its disadvantage is: this method is technically difficult. Only a few works using this method have been published (Faumont *et al.*, 2011; Kocabas *et al.*, 2012; Tanimoto *et al.*, 2016).

Its applications are: Kocabas *et al.* selectively targeted the AIY or AIZ neurons in the worm's brain (the head ganglion) and observed different behavioral responses depending on whether the head was bent dorsally or ventrally (Kocabas *et al.*, 2012). One of the author's groups selectively targeted either dorsal or ventral pairs of dopaminergic neurons located in the head and revealed functional asymmetry between the dopaminergic neurons (see Section 3.5) (Tanimoto *et al.*, 2016).

3.5 Recent Optogenetic Probing of Neural Functions that Regulate Behavior

3.5.1 Neuronal Multitasking in Oxygen O_2 Responses

Many neurons in *C. elegans* "multitask" to sense multiple modalities and/or to direct different behaviors, and we can tease apart such different functional outputs by applying optogenetic stimulation in the right way. Busch and colleagues studied the oxygen-sensing neurons of the worm with optogenetics, calcium imaging and behavioral assays in order to uncover how they mediate hyperoxia avoidance (Busch *et al.*, 2012). They showed that O_2 evokes two distinct behavioral programs: the concentration of ambient O_2 sets locomotory speed, whereas rising or falling O_2 levels induce or suppress escape movements, termed reversals and omega turns. These two behavioral outputs are distinct: the control of speed lasts for as long as a given $[O_2]$ persists, allowing the animals to migrate to or away from environments with uniform levels of O_2. Reversals and omega turns are transient and controlled by the change of $[O_2]$ (dc/dt) instead, allowing the animal to navigate local O_2 gradients. Remarkably, both behaviors are generated by the same set of sensory neurons, URX, AQR and PQR, which show tonic calcium responses to elevated O_2 (Figure 3.2A). Taking advantage of the fact that the transgene array expressing ChR in these neurons frequently causes mosaic expression, Busch *et al.* showed that each of the neurons contributes cumulatively in order to control steady-state speed. Stimulating URX, AQR and PQR with ChR2 for 15 minutes caused an enduring rise in locomotion, and conversely, NpHR illumination inhibited it, demonstrating that the tonic activity of the sensory neurons themselves is necessary and sufficient for generating the behavior. URX and AQR are located in the head and PQR in the tail of the animal, suggesting that their positions could allow a head-to-tail comparison of $[O_2]$ and orientation within local O_2 gradients. Targeted illumination of ChR2 in the head indeed induced a reversal, whereas tail illumination of PQR only produced forward movement. Both behavioral programs act together for the efficient avoidance of hyperoxia.

The same O_2-sensing neurons have also been implicated in leaving of bacterial food sources that are close to depletion (Figure 3.2A). Activating AQR, PQR and URX with ChR2 induced strong food leaving, again illustrating that multiple different behaviors can be recapitulated by the optogenetic manipulation of the same neurons (Milward *et al.*, 2011).

3.5.2 Asymmetric Activation of Dopaminergic Neurons

Dopamine is a neuromodulator that affects locomotor activity, cognition, emotion and learning and memory in mammals (Dayan and Balleine, 2002; Schultz, 2007), but how external and/or internal conditions of an animal are integrated to regulate dopamine release from the dopaminergic neurons and how dopamine signaling modulates the activity of target neurons are still not well understood.

In *C. elegans*, dopamine signaling is also known to modulate various neural functions, such as locomotion, stimulus sensitivity and associative and non-associative learning through different types of neurons (Chase and Koelle, 2007). In the worms, only four left–right pairs of neurons are dopaminergic (Sulston *et al.*, 1975), and they are suggested to sense the presence of food bacteria not chemically, but mechanically (Sawin *et al.*, 2000; Kindt *et al.*, 2007). However, the actual dynamics of the food-dependent activation of dopaminergic neurons has not been revealed.

By using a microscope system with high-speed tracking and illumination of a single neuron position, the Kimura laboratory functionally dissected the *C. elegans* dopaminergic neurons. They found that the head dorsal pair of dopaminergic neurons (CEPDs) is more strongly activated than the ventral pair (CEPVs) by the mechanical stimulus that a bacterial lawn provides (Figure 3.2B). Furthermore, they showed that the selective optogenetic activation of CEPD, but not CEPV, mimicked the bacterial food-induced behavioral response, indicating that CEPD plays a more primary role than CEPV, despite their strong similarity in terms of anatomy and gene expression patterns revealed so far (Tanimoto *et al.*, 2016). Further analysis may lead to an understanding of the mechanisms as to how functional differences are generated among neurons that are anatomically and genetically homologous to each other. Moreover, their system will enable detailed

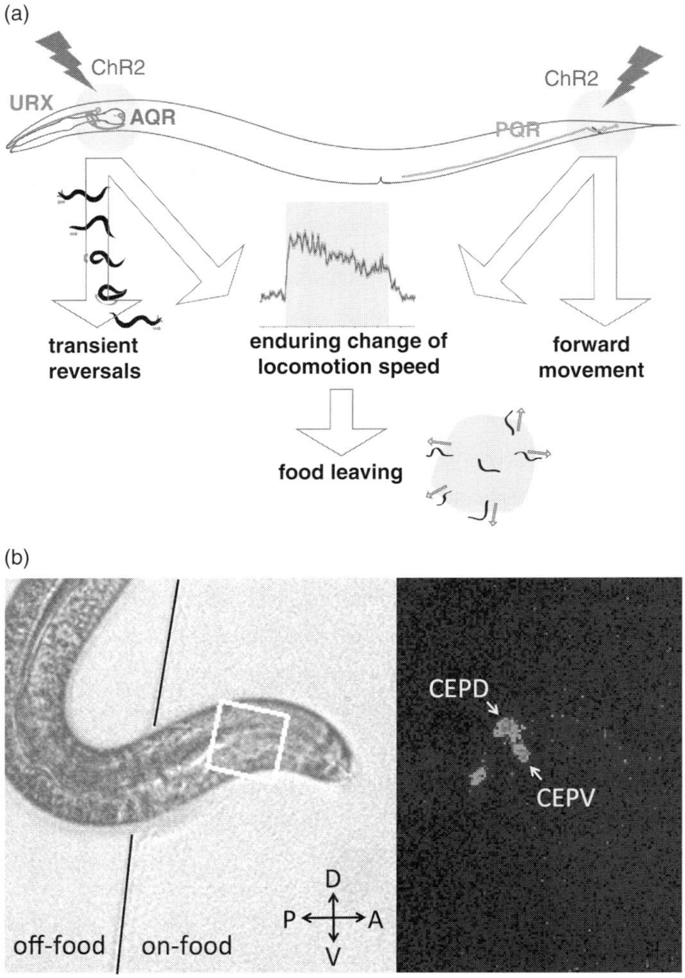

Figure 3.2 Neuronal targets for optogenetics in *C. elegans*. (A) Selective optogenetic activation of the URX, AQR (head) and PQR (tail) O_2-sensing neurons showed that the neurons contribute differently to multiple behavioral outputs. (B) Asymmetric activation of CEP neuron pairs. A bright-field image of *C. elegans* upon entering a lawn of bacteria (left) and fluorescent calcium signals from CEPD and CEPV (right) are shown. (A black-and-white version of this figure will appear in some formats. For the color version, please refer to the plate section.)

optogenetic interrogation of neural circuits in the worm's brain during behavior with single-neuron resolution.

3.6 Future Prospects

3.6.1 Gradual Light Stimulation

Most of the optogenetic light stimulation so far has been "all-or-none." It is suitable for a qualitative understanding of neuronal activity or a quantitative understanding of the upper limits and/or delays of neuronal outputs to constant inputs. However, actual neural activities vary depending on external stimuli and internal states. In particular, the membrane potentials of *C. elegans* neurons change gradually without action potentials, similarly to many sensory neurons and non-spiking neurons, such as retinal bipolar cells in mammals. Thus, to understand the actual neural computations that regulate worm behavior, it may be necessary to first understand the gradually changing temporal pattern of the target neuron's activity during behavior through calcium imaging or voltage imaging, and then to mimic the pattern with optogenetics.

3.6.2 Feedback ("Closed-loop") Regulation of Light Stimulation Based on Behavioral Output

To adapt themselves to the environment, animals respond to sensory inputs and change their behavior. The behavioral response in turn modulates the absolute value of the input sensed, altering the rate of change of the value and/or the direction of the sensory input, which further affects the behavior. Therefore, to understand an animal's sensory behavior, it is essential to regulate optogenetic neural stimulation dynamically based on this behavior.

Such optogenetic control requires a machine vision-based analysis and classification of an animal's behavior in several categories and a subsequent illumination based on this classification, which should be completed in real time. So far, Ramanathan and Iino's groups have revealed such dynamic neuronal activities in freely behaving worms during chemotaxis using their optogenetic systems (Kocabas *et al.*, 2012; Satoh *et al.*, 2014). Ramanathan's group in particular established an excellent system with closed-loop control and gradual illumination – the timing of light stimulation is dependent on the worm's posture (whether its head is bending dorsally or ventrally), and the strength of the light stimulation depends on the animal's position, making the worm migrate on a "virtual" odor gradient. This study has set a new standard for the optogenetic analysis of *C. elegans* behavior.

3.7 Concluding Remarks

As described in this chapter, *C. elegans* has proven to be one of the best-suited models for the optogenetic analyses of neural function, and various microscope systems have been developed for this purpose. Nevertheless, even if we were able to modulate any neural activity we want, it may be meaningless if the neural activities induced are physiologically unrealistic. Thus, it is important for optogenetic analyses to first understand real-life neural activities. Whole-brain imaging of *C. elegans* has recently become available (Schrödel *et al.*, 2013; Prevedel *et al.*, 2014; Tokunaga *et al.*, 2014). Whole-brain imaging will provide a complete overview of information processing, which then needs to be interrogated by optogenetic analyses. By taking advantage of *C. elegans* for optogenetics, the worms may become the first animal in which we can grasp the "information process-ome."

REFERENCES

Akerboom, J. *et al.* (2013). Genetically encoded calcium indicators for multi-color neural activity imaging and combination with optogenetics. *Frontiers in Molecular Neuroscience*, **6**, 2.

Bendesky, A. *et al.* (2011). Catecholamine receptor polymorphisms affect decision-making in *C. elegans*. *Nature*, **472**(7343), 313–318.

Busch, K.E. *et al.* (2012). Tonic signaling from O_2 sensors sets neural circuit activity and behavioral state. *Nature Neuroscience*, **15**(4), 581–591.

Chase, D.L. and Koelle, M.R. (2007). Biogenic amine neurotransmitters in C. elegans. *WormBook*, 1–15.

Cohen, E. *et al.* (2014). *Caenorhabditis elegans* nicotinic acetylcholine receptors are required for nociception. *Molecular and Cellular Neurosciences*, **59**, 85–96.

Dayan, P. and Balleine, B.W. (2002). Reward, motivation, and reinforcement learning. *Neuron*, **36**(2), 285–298.

Donnelly, J.L. *et al.* (2013). Monoaminergic orchestration of motor programs in a complex *C. elegans* behavior. *PLoS Biology*, **11**(4), e1001529.

Emiliani, V. *et al.* (2015). All-optical interrogation of neural circuits. *The Journal of Neuroscience*, **35**(41), 13917–13926.

Fang-Yen, C., Alkema, M.J. and Samuel, A.D.T. (2015). Illuminating neural circuits and behaviour in *Caenorhabditis elegans* with optogenetics. *Philosophical Transactions of the Royal Society B: Biological Sciences*, **370**(1677), 20140212.

Faumont, S. *et al.* (2011). An image-free opto-mechanical system for creating virtual environments and imaging neuronal activity in freely moving *Caenorhabditis elegans*. *PLoS ONE*, **6**(9), e24666.

Faumont, S., Lindsay, T.H. and Lockery, S.R. (2012). Neuronal microcircuits for decision making in *C. elegans*. *Current Opinion in Neurobiology*, **22**(4), 580–591.

Guo, Z.V., Hart, A.C. and Ramanathan, S. (2009). Optical interrogation of neural circuits in *Caenorhabditis elegans*. *Nature Methods*, **6**(12), 891–896.

Han, B., Bellemer, A. and Koelle, M.R. (2015). An evolutionarily conserved switch in response to GABA affects development and behavior of the locomotor circuit of *Caenorhabditis elegans*. *Genetics*, **199**(4), 1159–1172.

Hoerndli, F.J. *et al.* (2015). Neuronal activity and CaMKII regulate kinesin-mediated transport of synaptic AMPARs. *Neuron*, **86**(2), 457–474.

Husson, S.J. (2012). Keeping track of worm trackers. *WormBook*, 1–17.

Husson, S.J. *et al.* (2012a). Optogenetic analysis of a nociceptor neuron and network reveals ion channels acting downstream of primary sensors. *Current Biology*, **22**(9), 743–752.

Husson, S.J. *et al.* (2012b). Microbial light-activatable proton pumps as neuronal inhibitors to functionally dissect neuronal networks in *C. elegans*. *PLoS ONE*, **7**(7), e40937

Husson, S.J., Gottschalk, A. and Leifer, A.M. (2013). Optogenetic manipulation of neural activity in *C. elegans*: from synapse to circuits and behaviour. *Biology of the Cell*, **105** (6), 235–250.

Inoue, M. *et al.* (2015). Rational design of a high-affinity, fast, red calcium indicator R-CaMP2. *Nature Methods*, **12**(1), 64–70.

Jensen, M. *et al.* (2012). Wnt signaling regulates acetylcholine receptor translocation and synaptic plasticity in the adult nervous system. *Cell*, **149**(1), 173–187.

Kawazoe, Y., Yawo, H. and Kimura, K.D. (2013). A simple optogenetic system for behavioral analysis of freely moving small animals. *Neuroscience Research*, **75**(1), 65–68.

Kindt, K.S. *et al.* (2007). Dopamine mediates context-dependent modulation of sensory plasticity in *C. elegans*. *Neuron*, **55**(4), 662–676.

Kittelmann, M. *et al.* (2013). *In vivo* synaptic recovery following optogenetic hyperstimulation. *Proceedings of the National Academy of Sciences*, **110**(32), E3007–E3016.

Kocabas, A. *et al.* (2012). Controlling interneuron activity in *Caenorhabditis elegans* to evoke chemotactic behaviour. *Nature*, **490**(7419), 273–277.

Krieg, M., Dunn, A.R. and Goodman, M.B. (2014). Mechanical control of the sense of touch by β-spectrin. *Nature Cell Biology*, **16**(3), 224–233.

Leifer, A.M. *et al.* (2011). Optogenetic manipulation of neural activity in freely moving *Caenorhabditis elegans*. *Nature Methods*, **8**(2), 147–152.

Li, Z. *et al.* (2014). Encoding of both analog- and digital-like behavioral outputs by one *C. elegans* interneuron. *Cell*, **159**(4), 751–765.

Liewald, J.F. *et al.* (2008). Optogenetic analysis of synaptic function. *Nature Methods*, **5**(10), 895–902.

Lindsay, T.H., Thiele, T.R. and Lockery, S.R. (2011). Optogenetic analysis of synaptic transmission in the central nervous system of the nematode *Caenorhabditis elegans*. *Nature Communications*, **2**, 306–309.

Liu, Q., Hollopeter, G. and Jorgensen, E.M. (2009). Graded synaptic transmission at the *Caenorhabditis elegans* neuromuscular junction. *Proceedings of the National Academy of Sciences of the United States of America*, **106**(26), 10823–10828.

Lockery, S.R. (2011). The computational worm: spatial orientation and its neuronal basis in *C. elegans*. *Current Opinion in Neurobiology*, **21**(5), 782–790.

Luo, L. *et al.* (2014). Dynamic encoding of perception, memory, and movement in a *C. elegans* chemotaxis circuit. *Neuron*, **82**(5), 1115–1128.

Milward, K. *et al.* (2011). Neuronal and molecular substrates for optimal foraging in *Caenorhabditis elegans*. *Proceedings of the National Academy of Sciences*, **108**(51), 20672–20677.

Nagel, G. *et al.* (2002). Channelrhodopsin-1: a light-gated proton channel in green algae. *Science*, **296**(5577), 2395–2398.

Nagel, G. *et al.* (2003). Channelrhodopsin-2, a directly light-gated cation-selective membrane channel. *Proceedings of the National Academy of Sciences of the United States of America*, **100**(24), 13940–13945.

Nagel, G. *et al.* (2005). Light activation of channelrhodopsin-2 in excitable cells of *Caenorhabditis elegans* triggers rapid behavioral responses. *Current Biology*, **15**(24), 2279–2284.

Narayan, A., Laurent, G. and Sternberg, P.W. (2011). Transfer characteristics of a thermosensory synapse in *Caenorhabditis elegans*. *Proceedings of the National Academy of Sciences*, **108**(23), 9667–9672.

Piggott, B.J. *et al.* (2011). The neural circuits and synaptic mechanisms underlying motor initiation in *C. elegans*. *Cell*, **147**(4), 922–933.

Prevedel, R. *et al.* (2014). Simultaneous whole-animal 3D imaging of neuronal activity using light-field microscopy. *Nature Methods*, **11**(7), 727–730.

Ramot, D. *et al.* (2008). The Parallel Worm Tracker: a platform for measuring average speed and drug-induced paralysis in nematodes. *PLoS ONE*, **3**(5), e2208.

Richmond, J. (2005). Synaptic function. *WormBook*, 1–14.

Sasakura, H. and Mori, I. (2013). Behavioral plasticity, learning, and memory in *C. elegans*. *Current Opinion in Neurobiology*, **23**(1), 92–99.

Satoh, Y. *et al.* (2014). Regulation of experience-dependent bidirectional chemotaxis by a neural circuit switch in *Caenorhabditis elegans*. *Journal of Neuroscience*, **34**(47), 15631–15637.

Sawin, E.R., Ranganathan, R. and Horvitz, H.R. (2000). *C. elegans* locomotory rate is modulated by the environment through a dopaminergic pathway and by experience through a serotonergic pathway. *Neuron*, **26**(3), 619–631.

Schild, L.C. and Glauser, D.A. (2015). Dual color neural activation and behavior control with Chrimson and CoChR in Caenorhabditis elegans. *Genetics*, **200**(4), 1029–1034.

Schmitt, C. *et al.* (2012). Specific expression of channelrhodopsin-2 in single neurons of *Caenorhabditis elegans*. *PLoS ONE*, **7**(8), e43164.

Schrödel, T. *et al.* (2013). Brain-wide 3D imaging of neuronal activity in *Caenorhabditis elegans* with sculpted light. *Nature Methods*, **10**(10), 1013–1020.

Schultheis, C. *et al.* (2011). Optogenetic analysis of GABAB receptor signaling in *Caenorhabditis elegans* motor neurons. *Journal of Neurophysiology*, **106**(2), 817–827.

Schultz, W. (2007). Multiple dopamine functions at different time courses. *Annual Review of Neuroscience*, **30**, 259–288.

Shipley, F.B. *et al.* (2014). Simultaneous optogenetic manipulation and calcium imaging in freely moving *C. elegans*. *Frontiers in Neural Circuits*, **8**, 28.

Stirman, J.N. *et al.* (2011). Real-time multimodal optical control of neurons and muscles in freely behaving *Caenorhabditis elegans*. *Nature Methods*, **8**(2), 153–158.

Sulston, J., Dew, M. and Brenner, S. (1975). Dopaminergic neurons in the nematode *Caenorhabditis elegans*. *The Journal of Comparative Neurology*, **163**(2), 215–226.

Swierczek, N.A. *et al.* (2011). High-throughput behavioral analysis in *C. elegans*. *Nature Methods*, **8**(7), 592–598.

Tanimoto, Y. *et al.* (2016). *In actio* optophysiological analyses reveal functional diversification of dopaminergic neurons in the nematode *C. elegans*. *Scientific Reports*, **6**, 26297.

Tokunaga, T. *et al.* (2014). Automated detection and tracking of many cells by using 4D live-cell imaging data. *Bioinformatics (Oxford, England)*, **30**(12), i43–i51.

Trojanowski, N.F. *et al.* (2014). Neural and genetic degeneracy underlies *Caenorhabditis elegans* feeding behavior. *Journal of Neurophysiology*, **112**(4), 951–961.

Watanabe, S. *et al.* (2013a). Ultrafast endocytosis at *Caenorhabditis elegans* neuromuscular junctions. *eLife*, **2**, e00723.

Watanabe, S. *et al.* (2013b). Ultrafast endocytosis at mouse hippocampal synapses. *Nature*, **504**(7479), 242–247.

Wen, Q. *et al.* (2012). Proprioceptive coupling within motor neurons drives *C. elegans* forward locomotion. *Neuron*, **76**(4), 750–761.

White, J.G. *et al.* (1986). The structure of the nervous system of the nematode *Caenorhabditis elegans*. *Philosophical Transactions of the Royal Society B: Biological Sciences*, **314**(1165), 1–340.

Williams, D.C. *et al.* (2013). Rapid and permanent neuronal inactivation *in vivo* via subcellular generation of reactive oxygen with the use of KillerRed. *Cell Reports*, **5**(2), 553–563.

Zhang, F. *et al.* (2007). Multimodal fast optical interrogation of neural circuitry. *Nature*, **446** (7136), 633–639.

4 From Synapse to Behavior: Optogenetic Tools for the Investigation of the *Caenorhabditis elegans* Nervous System

Jatin Nagpal

4.1 Introduction

Optogenetics, though still only a decade old as a field, has revolutionized research in neuroscience and cell biology. It allows for the non-invasive, spatiotemporally precise and genetically targeted light control of neural activity. The optogenetic actuators or the genetically encoded light-addressable elements mediate the light-driven manipulation of membrane potential, intracellular signaling, neuronal network activity and behavior (Fenno, Yizhar and Deisseroth, 2011; Dugué, Akemann and Knöpfel, 2012; Husson, Gottschalk and Leifer, 2013; Fang-Yen *et al.*, 2015).

The transparent and genetically amenable nematode *Caenorhabditis elegans* is a versatile model organism that has been used extensively in the fields of molecular, cellular and systems neuroscience. It possesses a nervous system comprising 302 neurons (in adult hermaphrodites), which has been anatomically mapped to the synapse resolution (White *et al.*, 1986). This complete "wiring diagram" of the nervous system is highly conserved between individual *C. elegans* worms, which are eutelic organisms (having a fixed number of somatic cells). Despite a seemingly simple nervous system comprising 7000 synapses in between 302 neurons and muscle cells, *C. elegans* exhibits a rich repertoire of quantifiable behavior (de Bono and Maricq, 2005; Sengupta and Samuel, 2009; Yemini *et al.*, 2013). It can respond to a variety of sensory stimuli, such as chemical, mechanical, thermal, gaseous and magnetic ones. It also exhibits habituation and simple forms of associative learning (Ardiel and Rankin, 2010). Furthermore, there is a remarkable level of conservation in the genes and neurochemistry of *C. elegans* and mammals.

Its optical transparency and genetic tractability make *C. elegans* a highly attractive model system for the use of optogenetics when performing neuroscientific investigations (Xu and Kim, 2011). In fact, it was the first animal in which optogenetic tools based on microbial rhodopsins were successfully implemented and behavior was remotely controlled using them (Nagel *et al.*, 2005; Zhang *et al.*, 2007). Since then, *C. elegans* has been a model of choice for the development and application of novel optogenetic tools and techniques. The combination of optogenetic actuators with optogenetic reporters of neural activity (genetically encoded calcium and voltage indicators) are providing hitherto unprecedented access to the dissection of neural circuits and behavior in *C. elegans* (Husson *et al.*, 2012b; Akerboom *et al.*, 2013; Flytzanis *et al.*, 2014; Wabnig *et al.*, 2015). In this chapter, I will highlight the diversity of optogenetic tools that have been applied in the worms, primarily the actuators, and the insights they have provided in the field of neuroscience (Figure 4.1).

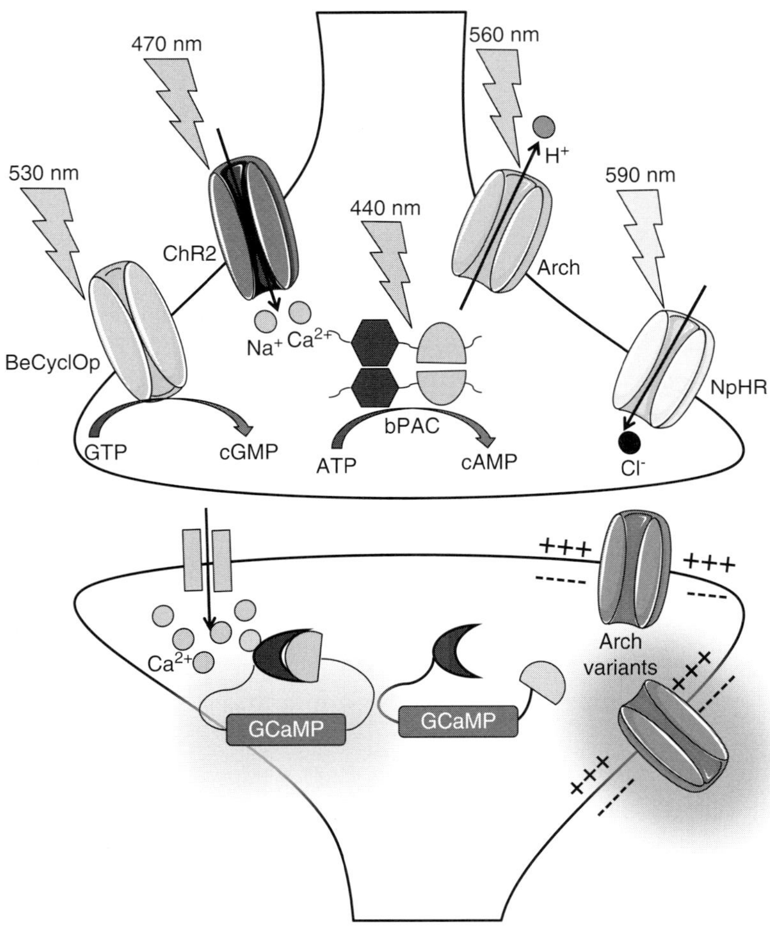

Figure 4.1 An illustration of the synapse featuring the prominent optogenetic actuators and sensors discussed in the review. The blue light-activated cation channel, channelrhodopsin-2 (ChR2), is the primary optogenetic depolarizer activating the excitable cell, while yellow light-activated chloride pump, halorhodopsin (NpHR), and green–yellow light-activated proton pump, archaerhodopsin (Arch), have been used as optogenetic hyperpolarizers, thereby inhibiting the cell. *Beggiatoa*-derived photoactivated adenylyl cyclase (bPAC) is a blue light sensing using flavin (BLUF) domain-containing enzyme catalyzing the synthesis of cAMP upon blue light illumination. Blue–green light control of cGMP synthesis has been achieved with a guanylyl cyclase opsin from *Blastocladiella emersonii* (BeCyclOp). The genetically encoded calcium indicator GCaMP and the voltage sensor Arch, which are the optogenetic sensors reporting on neural activation, have been found in the postsynaptic compartment. (A black-and-white version of this figure will appear in some formats. For the color version, please refer to the plate section.)

4.2 Microbial Rhodopsin-based Optogenetic Tools Affecting the Membrane Potential

4.2.1 Photoactivation of Excitable Cells

The earliest attempts at controlling neural activity with light were made by the Miesenböck laboratory when they employed the components of the *Drosophila* photoreceptor signaling cascade to activate cultured hippocampal neurons (Zemelman *et al.*, 2002). Later, they light activated heterologously expressed, ligand-gated ionotropic receptors ($P2X_2$ or TRPV1) by uncaging orthogonal ligands in order to trigger escape behavior in *Drosophila* (Zemelman *et al.*, 2003; Lima and Miesenböck, 2005). At approximately the same time, opto-chemical genetic approaches were developed wherein synthetic photoswitchable agonists were used to gate the cognate receptors with light (Banghart *et al.*, 2004; Volgraf *et al.*, 2006). However, these approaches are subject to certain limitations. The responses of protein cascades are relatively slow and difficult to control; caged agonists and synthesized organic photoswitches have to be additionally supplemented (e.g. by microinjection), which may complicate targeting or tissue penetration. The application of single-component optogenetic tools overcame these difficulties.

The first member of the single-component optogenetic tool family to be used for neuroscience-based applications was channelrhodopsin-2 (ChR2). ChR2 is a blue light-gated cation channel found in the single-cell alga *Chlamydomonas reinhardtii*, in which it mediates phototaxis (Nagel *et al.*, 2003). ChR2 is a type-1 rhodopsin consisting of the 7-transmembrane spanning apoprotein channelopsin-2 (chop-2) bound to the cofactor retinal, a vitamin A-related organic molecule, which isomerizes from all-*trans* to 13-*cis* upon 470-nm blue light illumination. Apart from being a chromophore, retinal also stabilizes the ChR2 with respect to degradation (Ullrich *et al.*, 2013). Upon photo-isomerization of retinal, a conformational change of the protein causes the opening of the channel. It induces an inward rectifying current carried primarily by monovalent cations (e.g. Na^+, K^+ and H^+) and some divalent cations (including small amounts of Ca^{2+}), which depolarizes the plasma membrane of a cell (or a neuron) (Bamann *et al.*, 2008; Feldbauer *et al.*, 2009). Heterologous expression of ChR2 in cultured neurons conferred on them blue light sensitivity, wherein ChR2 reliably evoked temporally precise action potentials (Boyden *et al.*, 2005). ChR2, when expressed in specific neurons or muscle cells of *C. elegans*, could evoke predictable behaviors, making *C. elegans* the first animal in which microbial rhodopsin-based optogenetics was used to evoke coordinated behavior (Nagel *et al.*, 2005). Unlike mammals, *C. elegans* does not synthesize retinal and hence it must be supplemented in its bacterial diet for the ChR2 to be functional. This provides an ideal internal control for any optogenetic experiment with ChR2 or any retinal-based rhodopsin in *C. elegans*.

ChR2, upon illumination with blue light (450–470 nm), activates on a sub-millisecond timescale and deactivates within approximately 20 ms, having a relatively small single-channel conductance (Feldbauer *et al.*, 2009). Over the years, several variants of ChR2 have been engineered with altered photocycle kinetics, conductance, ion selectivity and optical spectra in order to allow for specific optogenetic applications (Nagel *et al.*, 2005; Berndt *et al.*, 2009; Lin *et al.*, 2009; Berndt *et al.*, 2011; Lin, 2011). ChR2(H134R) shows reduced inactivation and an extended open-state lifetime, therefore exhibiting higher steady-state currents as compared to ChR2(wild-type) (Nagel *et al.*, 2005), and thus has been the most commonly used ChR2 variant in *C. elegans*. Combining H134R with T159C further improves the function in *C. elegans* due to higher conductance and better expression (Berndt *et al.*, 2011; Erbguth *et al.*, 2012). ChIEF (a chimera of ChR1 and ChR2 with an I170V point mutation) exhibits significantly faster rates of channel closure, as well as reduced inactivation (Lin *et al.*, 2009), and thus has been used to evoke ultrafast depolarization of neurons at the neuromuscular junction of *C. elegans* for studying synaptic vesicle recycling using high-pressure freeze-electron microscopy (Watanabe *et al.*, 2013). ChR2(E123T) also shows fast off-kinetics and has been used to evoke spiking in mammalian neurons at 200 Hz (Gunaydin *et al.*, 2010). For applications in which a longer period of activation is desired, slower-kinetic mutants – ChR2(C128X) – have also been engineered that have deactivation time constants ranging from 2 to 106 s, depending on the specific mutation (Berndt *et al.*, 2009). These ChR2(C128X) variants (or the step function opsins) have been expressed in worms, enabling long-term depolarization of neurons, and so having implications even at the developmental time scale (Schultheis *et al.*, 2011). Attempts have also been made to obtain ChR2 variants that have altered excitation spectral peaks (Zhang *et al.*, 2008; Yizhar *et al.*, 2011; Prigge *et al.*, 2012). The green light-activated variant C1V1-ETET (a chimeric protein composed of *C. reinhardtii* ChR1 and *Volvox carteri* ChR1, with two additional mutations) has been expressed in *C. elegans* (Yizhar *et al.*, 2011; Erbguth *et al.*, 2012) and used in conjunction with the highly efficient ChR2(H134R, T159C) in order to achieve independent control of two distinct neuron populations. However, as their excitation spectra are not widely segregated, there is a requirement for fine-tuning expression levels and/or illumination parameters in order to avoid the risk of cross-activation. In a recent study (Schild and Glauser, 2015), two newly discovered channelrhodopsins (i.e. the red light-sensitive Chrimson from *Chlamydomonas noctigama* and the blue light-sensitive CoChR from *Chloromonas oogama*) (Klapoetke *et al.*, 2014) have been successfully employed in order to achieve efficient blue/red dual color optogenetic excitation of different mechanosensory neurons in *C. elegans*. Though Chrimson exhibits a shoulder toward the blue side of the activation spectrum, low blue light intensities could be used for efficient bimodal activation without cross-activation of Chrimson, since the blue light (470 nm) sensitivity of CoChR is five-times greater than that of ChR2 (Klapoetke *et al.*, 2014).

4.2.2 Photoinhibition of Excitable Cells

Two years after the first use of ChR2 in *C. elegans*, optical inhibition of neural activity was made possible using the yellow light-activated chloride pump halorhodopsin (NpHR) from the archaebacterium *Natronomonas pharaonis* (Zhang *et al.*, 2007). The peak of its excitation spectrum lies at 590 nm. The activation of NpHR in the *C. elegans* body wall muscle cells led to hyperpolarization and relaxation of the muscle cells, resulting in the extension of the worm's body. Due to the inefficient trafficking of the opsin to the plasma membrane, improved versions were made with mammalian endoplasmic reticulum export and trafficking motifs (Gradinaru *et al.*, 2008). However, these did not fare better compared to NpHR in worms due to the non-conservation of these signals in *C. elegans* (Husson *et al.*, 2012b).

Chow *et al.* performed an extensive screen of microbial opsins and identified light-activated proton pumps that can act as neuronal inhibitors (Chow *et al.*, 2010). These include archaerhodopsin-3 (Arch) from *Halorubrum sodomense* and Mac from the fungus *Leptosphaeria maculans*, which are activated by yellow–green and blue–green wavelengths of light, respectively, and pump protons out of the cell, leading to hyperpolarization of excitable cells. They express well on the neural plasma membrane and have higher photocurrents as compared to NpHR, making them more potent photo-inhibitors. Both Arch and Mac were successfully expressed in *C. elegans* muscle cells and neurons (Husson *et al.*, 2012b) and could be combined with ChR2 for studying the causal relationships of single neurons in the worm nociceptive behavior circuit. ArchT, another proton pump reported to be three-times more light sensitive than Arch (Han *et al.*, 2011), was also expressed in *C. elegans* muscle cells and serotonergic neurons (Okazaki and Takagi, 2013) but, unlike in mouse cells, was found to have similar light sensitivity as Arch in *C. elegans*. Recently, a spectrally red-shifted cruxhalorhodopsin, Jaws, derived from the species *Haloarcula (Halobacterium) salinarum* (strain Shark) and engineered to produce red light-induced photocurrents three times larger than ArchT and NpHR, has also been described (Chuong *et al.*, 2014). There is currently no report describing its expression in *C. elegans*. In an alternative strategy, synthetic retinal analogs have been combined with microbial rhodopsins such as ChR2(H134R), Mac and Arch to alter their photocycle kinetics and spectral tuning in order to expand the applications of the existing optogenetic toolkit (AzimiHashemi *et al.*, 2014).

The hyperpolarizing chloride or proton pumps such as NpHR, Mac and Arch are less efficient than channels such as ChR2, transporting only one ion per photocycle instead of the many ions that channels can allow per photon absorption. Rational structure-guided molecular engineering of channelrhodopsins resulted in the generation of inhibitory chloride channels (inhibitory C1C2 and SloChloC) (Berndt *et al.*, 2014; Wietek *et al.*, 2014). Recently, two naturally occurring blue and green light-gated, chloride-conducting channelrhodopsin from the algae *Guillardia theta* have been reported and have been shown to have six- to eight-times larger stationary-state photocurrents than ChR2 and much higher light sensitivities than any previously known optical inhibitors (Govorunova *et al.*, 2015). However, none of these chloride channels have been implemented in *C. elegans* yet.

4.3 Optogenetic Tools Targeting Intracellular Signaling Pathways

Optogenetic tools that affect membrane potential, though allow control over the output of the cell, bypass the intracellular signal transduction and thereby completely override the intrinsic activity of the cell. The cyclic nucleotides cAMP and cGMP are universal secondary messengers that regulate metabolic and behavioral responses in diverse organisms (Lucas *et al.*, 2000; Beavo and Brunton, 2002; Gancedo, 2013). The ability to turn on and off cyclic nucleotide-mediated signal transduction pathways *in vivo* in specific cell types, and at desired times, could lead to valuable mechanistic insights into cellular regulatory networks. Such precise spatiotemporal control of cellular cAMP and cGMP has been achieved by photo-activatable cyclases.

Photo-activatable adenylyl cyclase (PAC) obtained from *Euglena* (euPAC) has been shown to control cAMP levels in *Xenopus* oocytes and the neurons of *Drosophila* and *C. elegans* (Schröder-Lang *et al.*, 2007; Weissenberger *et al.*, 2011). euPAC photostimulation in the cholinergic neurons of *C. elegans* resulted in increased locomotive activity that was coordinated, as opposed to the spastic paralysis observed when photostimulating these neurons via ChR2. Another PAC obtained from *Beggiatoa* species, bPAC, has been shown to exhibit less dark activity than euPAC (Ryu *et al.*, 2010) and to trigger light-induced cAMP production in cultured neurons, neurons of *Drosophila* (Stierl *et al.*, 2011) and neurons of *C. elegans* (Flavell *et al.*, 2013). These cyclases are blue light sensing using flavin (BLUF) domain-containing enzymes that use flavin adenine dinucleotide (FAD) as the cofactor and catalyze the synthesis of cAMP from ATP upon activation by blue light (Stierl *et al.*, 2011). Since the worms endogenously produce and use FAD, these cyclases do not require any external addition of the cofactor.

Recently, Gao *et al.* employed a novel green light-activated, rhodopsin-based guanylyl cyclase 'BeCyclOp' derived from the fungus *Blastocladiella emersonii* (Avelar *et al.*, 2014; Scheib *et al.*, 2015) to optogenetically manipulate cGMP levels in heterologous cells (*Xenopus* oocytes, HEK293T cells) and in the muscle cells of *C. elegans* (Gao *et al.*, 2015). Specifically, in the O_2/CO_2 sensory neurons of *C. elegans*, BeCyclOp activation evoked behavioral responses that were consistent with their normal sensory functions.

Optogenetic tools have also been developed and used for selective, spatiotemporally precise and acute ablation of cells and proteins within a cell in order to dissect the function of the corresponding cell and protein within a complex network such as the nervous system. Mitochondrially targeted miniSOG, a monomeric fluorescent protein of 106 amino acids that generates singlet oxygen upon blue light illumination, was transgenically expressed in *C. elegans* neurons. Upon blue light illumination, mito-miniSOG causes fast and effective death of neurons due to reactive oxygen species (ROS)-mediated cellular oxidative stress without causing detectable damage to surrounding tissues (Qi *et al.*, 2012). KillerRed, a dimeric red fluorescent protein, upon illumination with green light (540–580 nm) produces high levels of ROS, primarily the superoxide anion radicals instead of singlet oxygen. A plasma membrane-targeted tandem-dimer version of KillerRed was transgenically expressed in the sensory neurons, interneurons and motor neurons of *C. elegans* worms and could rapidly and efficiently ablate the cells upon green light illumination without affecting the neighboring neurons (Kobayashi *et al.*, 2013; Williams *et al.*, 2013). When fused to proteins of interest, miniSOG has also been used for acute, fast, light-induced protein destruction (e.g. the destruction of the presynaptic proteins involved in neurotransmission in worms) (Lin *et al.*, 2013). However, the non-localized effects of the generated ROS affecting bystander proteins has prompted another approach (i.e. with a photosensitive degron [psd]) in order to acutely degrade only the proteins of interest (Hermann *et al.*, 2015). In this recent study, the authors demonstrated that psd, comprising a light-switchable LOV2 domain with a carboxy terminal degradation sequence from mouse ornithine decarboxylase (Renicke *et al.*, 2013), quickly and selectively targets the protein it is fused to through a ubiquitin-independent proteasomal recognition-based process in a light-dependent fashion in the nervous system of *C. elegans*.

4.4 Optogenetic Sensors to Detect Neural Activity

Calcium dynamics are a proxy for the activity state of excitable cells such as neurons and muscles. Genetically encoded Ca^{2+} indicators (GECIs) (single green channel GCaMP and cyan fluorescent protein–yellow fluorescent protein Förster resonance energy transfer-based cameleon) have been extensively used to report on the activity of the primary sensory neurons, interneurons and muscle cells of *C. elegans* in response to external stimuli such as mechanical touch, specific chemicals, temperature and an electric field (Kerr, 2006; Chung *et al.*, 2013). The most widely used GECI – GCaMP – consists of circularly permuted green fluorescent protein (cpEGFP) inserted between calmodulin (CaM) and an M13 peptide. Upon calcium binding, conformational changes in the CaM–M13 complex induce fluorescence changes in cpEGFP that relate to the activity state of the cell. GCaMPs have undergone iterative improvements in their intrinsic properties such as signal-to-noise ratio, brightness, kinetics and dynamic range, with the latest in the series being GCaMP6 (Tian *et al.*, 2009; Akerboom *et al.*, 2012; Chen *et al.*, 2013; Broussard *et al.*, 2014).

The use of the combination of optogenetic actuators (primarily with blue–green excitation spectra) with GCaMPs to probe functional connectivity in the nervous system has been limited due to the substantial spectral overlap of the actuators and the sensors. To overcome this problem and to enable independent and simultaneous use of both tools, red-shifted calcium indicators have been developed. RCaMP, which is engineered from circular permutation of the red fluorescent protein mRuby, has been used to report on the calcium responses of ChR2-evoked muscle and neuron activity in *C. elegans* (Husson *et al.*, 2012a; Akerboom *et al.*, 2013; Wabnig *et al.*, 2015).

Another way to measure neural activity is to directly observe the voltage changes across the cell. Genetically encoded voltage indicators enable direct optical imaging of neuronal circuit operations (Mutoh *et al.*, 2012). The optogenetic inhibitor proton pump Arch has been shown to emit voltage-sensitive fluorescence (Kralj *et al.*, 2012; Gong *et al.*, 2013; Maclaurin *et al.*, 2013), and two variants of the wild-type Arch have been used for fluorescence-based voltage sensing in behaving *C. elegans* (Flytzanis *et al.*, 2014).

The combination of optogenetic manipulation with calcium imaging in freely moving worms has been recently achieved (Shipley *et al.*, 2014). With these advances, closed-loop optogenetics with online tracking of behavior becomes conceivable in the near future, which will advance our knowledge on how the relatively small nervous system of *C. elegans* is able to accomplish complex behavioral tasks.

REFERENCES

Akerboom J, Carreras Calderón N, Tian L, Wabnig S, Prigge M, Tolö J, Gordus A, Orger MB, Severi KE, Macklin JJ, Patel R, Pulver SR, Wardill TJ, Fischer E, Schüler C, Chen T-W, Sarkisyan KS, Marvin JS, Bargmann CI, Kim DS, Kügler S, Lagnado L, Hegemann P, Gottschalk A, Schreiter ER, Looger LL (2013) Genetically encoded calcium indicators for multi-color neural activity imaging and combination with optogenetics. *Front Mol Neurosci*, **6**:2.

Akerboom J, Chen T-W, Wardill TJ, Tian L, Marvin JS, Mutlu S, Calderón NC, Esposti F, Borghuis BG, Sun XR, Gordus A, Orger MB, Portugues R, Engert F, Macklin JJ, Filosa A, Aggarwal A, Kerr RA, Takagi R, Kracun S, Shigetomi E, Khakh BS, Baier H, Lagnado L, Wang SS-H, Bargmann CI, Kimmel BE, Jayaraman V, Svoboda K, Kim DS, Schreiter ER, Looger LL (2012) Optimization of a GCaMP calcium indicator for neural activity imaging. *J Neurosci*, **32**:13819–13840.

Ardiel EL, Rankin CH (2010) An elegant mind: learning and memory in Caenorhabditis elegans. *Learn Mem*, **17**:191–201.

Avelar GM, Schumacher RI, Zaini PA, Leonard G, Richards TA, Gomes SL (2014) A Rhodopsin-Guanylyl cyclase gene fusion functions in visual perception in a fungus. *Curr Biol*, **24**:1234–1240.

AzimiHashemi N, Erbguth K, Vogt A, Riemensperger T, Rauch E, Woodmansee D, Nagpal J, Brauner M, Sheves M, Fiala A, Kattner L, Trauner D, Hegemann P, Gottschalk A, Liewald JF (2014) Synthetic retinal analogues modify the spectral and kinetic characteristics of microbial rhodopsin optogenetic tools. *Nat Commun*, **5**:5810.

Bamann C, Kirsch T, Nagel G, Bamberg E (2008) Spectral characteristics of the photocycle of channelrhodopsin-2 and its implication for channel function. *J Mol Biol*, **375**:686–694.

Banghart M, Borges K, Isacoff E, Trauner D, Kramer RH (2004) Light-activated ion channels for remote control of neuronal firing. *Nat Neurosci*, **7**:1381–1386.

Beavo JA, Brunton LL (2002) Cyclic nucleotide research – still expanding after half a century. *Nat Rev Mol Cell Biol*, **3**:710–718.

Berndt A, Lee SY, Ramakrishnan C, Deisseroth K (2014) Structure-guided transformation of channelrhodopsin into a light-activated chloride channel. *Science*, **344**:420–424.

Berndt A, Schoenenberger P, Mattis J, Tye KM, Deisseroth K, Hegemann P, Oertner TG (2011) High-efficiency channelrhodopsins for fast neuronal stimulation at low light levels. *Proc Natl Acad Sci U S A*, **108**:7595–7600.

Berndt A, Yizhar O, Gunaydin LA, Hegemann P, Deisseroth K (2009) Bi-stable neural state switches. *Nat Neurosci*, **12**:229–234.

Boyden ES, Zhang F, Bamberg E, Nagel G, Deisseroth K (2005) Millisecond-timescale, genetically targeted optical control of neural activity. *Nat Neurosci*, **8**:1263–1268.

Broussard GJ, Liang R, Tian L (2014) Monitoring activity in neural circuits with genetically encoded indicators. *Front Mol Neurosci*, **7**:97.

Chen T-W, Wardill TJ, Sun Y, Pulver SR, Renninger SL, Baohan A, Schreiter ER, Kerr RA, Orger MB, Jayaraman V, Looger LL, Svoboda K, Kim DS (2013) Ultrasensitive fluorescent proteins for imaging neuronal activity. *Nature*, **499**:295–300.

Chow BY, Han X, Dobry AS, Qian X, Chuong AS, Li M, Henninger M a, Belfort GM, Lin Y, Monahan PE, Boyden ES (2010) High-performance genetically targetable optical neural silencing by light-driven proton pumps. *Nature*, **463**:98–102.

Chung SH, Sun L, Gabel CV (2013) *In vivo* neuronal calcium imaging in *C. elegans*. *J Vis Exp*, **74**:50357.

Chuong AS, Miri ML, Busskamp V, Matthews GAC, Acker LC, Sørensen AT, Young A, Klapoetke NC, Henninger MA, Kodandaramaiah SB, Ogawa M, Ramanlal SB, Bandler RC, Allen BD, Forest CR, Chow BY, Han X, Lin Y, Tye KM, Roska B, Cardin JA, Boyden ES (2014) Noninvasive optical inhibition with a red-shifted microbial rhodopsin. *Nat Neurosci*, **17**:1123–1129.

de Bono M, Maricq AV (2005) Neuronal substrates of complex behaviors in *C. elegans*. *Annu Rev Neurosci*, **28**:451–501.

Dugué GP, Akemann W, Knöpfel T (2012) A comprehensive concept of optogenetics. *Prog Brain Res*, **196**:1–28.

Erbguth K, Prigge M, Schneider F, Hegemann P, Gottschalk A (2012) Bimodal activation of different neuron classes with the spectrally red-shifted channelrhodopsin chimera C1V1 in *Caenorhabditis elegans*. *PLoS One*, **7**:e46827.

Fang-Yen C, Alkema MJ, Samuel ADT, Fang-yen C (2015) Illuminating neural circuits and behaviour in Caenorhabditis elegans with optogenetics. *Philos Trans R Soc Lond B Biol Sci*, **370**:20140212.

Feldbauer K, Zimmermann D, Pintschovius V, Spitz J, Bamann C, Bamberg E (2009) Channelrhodopsin-2 is a leaky proton pump. *Proc Natl Acad Sci U S A*, **106**:12317–12322.

Fenno L, Yizhar O, Deisseroth K (2011) The development and application of optogenetics. *Annu Rev Neurosci*, **34**:389–412.

Flavell SW, Pokala N, Macosko EZ, Albrecht DR, Larsch J, Bargmann CI (2013) Serotonin and the neuropeptide PDF initiate and extend opposing behavioral states in *C. elegans*. *Cell*, **154**:1023–1035.

Flytzanis NC, Bedbrook CN, Chiu H, Engqvist MKM, Xiao C, Chan KY, Sternberg PW, Arnold FH, Gradinaru V (2014) Archaerhodopsin variants with enhanced voltage-sensitive fluorescence in mammalian and *Caenorhabditis elegans* neurons. *Nat Commun*, **5**:4894.

Gancedo JM (2013) Biological roles of cAMP: variations on a theme in the different kingdoms of life. *Biol Rev Camb Philos Soc*, **88**:645–68.

Gao S, Nagpal J, Schneider MW, Kozjak-pavlovic V, Nagel G, Gottschalk A (2015) Optogenetic manipulation of cGMP in cells and animals by the tightly light-regulated guanylyl-cyclase opsin CyclOp. *Nat Commun*, **6**:1–12.

Gong Y, Li JZ, Schnitzer MJ (2013) Enhanced archaerhodopsin fluorescent protein voltage indicators. *PLoS One*, **8**:e66959.

Govorunova EG, Sineshchekov OA, Janz R, Liu X, Spudich JL (2015) Natural light-gated anion channels: a family of microbial rhodopsins for advanced optogenetics. *Science*, **349**:647–650.

Gradinaru V, Thompson KR, Deisseroth K (2008) eNpHR: a *Natronomonas* halorhodopsin enhanced for optogenetic applications. *Brain Cell Biol*, **36**:129–139.

Gunaydin LA, Yizhar O, Berndt A, Sohal VS, Deisseroth K, Hegemann P (2010) Ultrafast optogenetic control. *Nat Neurosci*, **13**:387–392.

Han X, Chow BY, Zhou H, Klapoetke NC, Chuong A, Rajimehr R, Yang A, Baratta MV, Winkle J, Desimone R, Boyden ES (2011) A high-light sensitivity optical neural silencer: development and application to optogenetic control of non-human primate cortex. *Front Syst Neurosci*, **5**:18.

Hermann A, Liewald JF, Gottschalk A (2015) A photosensitive degron enables acute light-induced protein degradation in the nervous system. *Curr Biol*, **25**: R749–R750.

Husson SJ, Costa WS, Wabnig S, Stirman JN, Watson JD, Spencer WC, Akerboom J, Looger LL, Treinin M, Miller DM, Lu H, Gottschalk A (2012a) Optogenetic analysis of a nociceptor neuron and network reveals ion channels acting downstream of primary sensors. *Curr Biol*, **22**:743–752.

Husson SJ, Gottschalk A, Leifer AM (2013) Optogenetic manipulation of neural activity in *C. elegans*: from synapse to circuits and behaviour. *Biol Cell*, **105**:235–250.

Husson SJ, Liewald JF, Schultheis C, Stirman JN, Lu H, Gottschalk A (2012b) Microbial light-activatable proton pumps as neuronal inhibitors to functionally dissect neuronal networks in *C. elegans*. *PLoS One*, **7**:e40937.

Kerr RA (2006) Imaging the activity of neurons and muscles. *WormBook*, 1–13.

Klapoetke NC, Murata Y, Kim SS, Pulver SR, Birdsey-Benson A, Cho YK, Morimoto TK, Chuong AS, Carpenter EJ, Tian Z, Wang J, Xie Y, Yan Z, Zhang Y, Chow BY, Surek B, Melkonian M, Jayaraman V, Constantine-Paton M, Wong GK-S, Boyden ES (2014) Independent optical excitation of distinct neural populations. *Nat Methods*, **11**: 338–346.

Kobayashi J, Shidara H, Morisawa Y, Kawakami M, Tanahashi Y, Hotta K, Oka K (2013) A method for selective ablation of neurons in C. elegans using the phototoxic fluorescent protein, KillerRed. *Neurosci Lett*, **548**:261–264.

Kralj JM, Douglass AD, Hochbaum DR, Maclaurin D, Cohen AE (2012) Optical recording of action potentials in mammalian neurons using a microbial rhodopsin. *Nat Methods*, **9**:90–95.

Lima SQ, Miesenböck G (2005) Remote control of behavior through genetically targeted photostimulation of neurons. *Cell*, **121**:141–152.

Lin JY (2011) A user's guide to channelrhodopsin variants: features, limitations and future developments. *Exp Physiol*, **96**:19–25.

Lin JY, Lin MZ, Steinbach P, Tsien RY (2009) Characterization of engineered channelrhodopsin variants with improved properties and kinetics. *Biophys J*, **96**:1803–1814.

Lin JY, Sann SB, Zhou K, Nabavi S, Proulx CD, Malinow R, Jin Y, Tsien RY (2013) Optogenetic inhibition of synaptic release with chromophore-assisted light inactivation (CALI). *Neuron*, **79**:241–253.

Lucas KA, Pitari GM, Kazerounian S, Ruiz-Stewart I, Park J, Schulz S, Chepenik KP, Waldman SA (2000) Guanylyl cyclases and signaling by cyclic GMP. *Pharmacol Rev*, **52**:375–414.

Maclaurin D, Venkatachalam V, Lee H, Cohen AE (2013) Mechanism of voltage-sensitive fluorescence in a microbial rhodopsin. *Proc Natl Acad Sci U S A*, **110**:5939–5944.

Mutoh H, Akemann W, Knöpfel T (2012) Genetically engineered fluorescent voltage reporters. *ACS Chem Neurosci*, **3**:585–592.

Nagel G, Brauner M, Liewald JF, Adeishvili N, Bamberg E, Gottschalk A (2005) Light activation of channelrhodopsin-2 in excitable cells of caenorhabditis elegans triggers rapid behavioral responses. *Curr Biol*, **15**:2279–2284.

Nagel G, Szellas T, Huhn W, Kateriya S, Adeishvili N, Berthold P, Ollig D, Hegemann P, Bamberg E (2003) Channelrhodopsin-2, a directly light-gated cation-selective membrane channel. *Proc Natl Acad Sci U S A*, **100**:13940–13945.

Okazaki A, Takagi S (2013) An optogenetic application of proton pump ArchT to C. elegans cells. *Neurosci Res*, **75**:29–34.

Prigge M, Schneider F, Tsunoda SP, Shilyansky C, Wietek J, Deisseroth K, Hegemann P (2012) Color-tuned channelrhodopsins for multiwavelength optogenetics. *J Biol Chem*, **287**:31804–31812.

Qi YB, Garren EJ, Shu X, Tsien RY, Jin Y (2012) Photo-inducible cell ablation in *Caenorhabditis elegans* using the genetically encoded singlet oxygen generating protein miniSOG. *Proc Natl Acad Sci U S A*, **109**:7499–7504.

Renicke C, Schuster D, Usherenko S, Essen L-O, Taxis C (2013) A LOV2 domain-based optogenetic tool to control protein degradation and cellular function. *Chem Biol*, **20**:619–626.

Ryu M-H, Moskvin OV, Siltberg-Liberles J, Gomelsky M (2010) Natural and engineered photoactivated nucleotidyl cyclases for optogenetic applications. *J Biol Chem*, **285**:41501–41508.

Scheib U, Stehfest K, Gee CE, Korschen HG, Fudim R, Oertner TG, Hegemann P (2015) The rhodopsin-guanylyl cyclase of the aquatic fungus *Blastocladiella emersonii* enables fast optical control of cGMP signaling. *Sci Signal*, **8**:1–8.

Schild LC, Glauser DA (2015) Dual color neural activation and behavior control with Chrimson and CoChR in *C. elegans*. *Genetics*, **200**:1029–1034.

Schröder-Lang S, Schwärzel M, Seifert R, Strünker T, Kateriya S, Looser J, Watanabe M, Kaupp UB, Hegemann P, Nagel G (2007) Fast manipulation of cellular cAMP level by light *in vivo*. *Nat Methods*, **4**:39–42.

Schultheis C, Liewald JF, Bamberg E, Nagel G, Gottschalk A (2011) Optogenetic long-term manipulation of behavior and animal development. *PLoS One*, **6**:e18766.

Sengupta P, Samuel ADT (2009) *Caenorhabditis elegans*: a model system for systems neuroscience. *Curr Opin Neurobiol*, **19**:637–643.

Shipley FB, Clark CM, Alkema MJ, Leifer AM (2014) Simultaneous optogenetic manipulation and calcium imaging in freely moving *C. elegans*. *Front Neural Circuits*, **8**:28.

Stierl M, Stumpf P, Udwari D, Gueta R, Hagedorn R, Losi A, Gärtner W, Petereit L, Efetova M, Schwarzel M, Oertner TG, Nagel G, Hegemann P (2011) Light modulation of cellular cAMP by a small bacterial photoactivated adenylyl cyclase, bPAC, of the soil bacterium *Beggiatoa*. *J Biol Chem*, **286**:1181–1188.

Tian L, Hires SA, Mao T, Huber D, Chiappe ME, Chalasani SH, Petreanu L, Akerboom J, McKinney SA, Schreiter ER, Bargmann CI, Jayaraman V, Svoboda K, Looger LL (2009) Imaging neural activity in worms, flies and mice with improved GCaMP calcium indicators. *Nat Methods*, **6**:875–881.

Ullrich S, Gueta R, Nagel G (2013) Degradation of channelopsin-2 in the absence of retinal and degradation resistance in certain mutants. *Biol Chem*, **394**:271–280.

Volgraf M, Gorostiza P, Numano R, Kramer RH, Isacoff EY, Trauner D (2006) Allosteric control of an ionotropic glutamate receptor with an optical switch. *Nat Chem Biol*, **2**:47–52.

Wabnig S, Liewald JF, Yu S-C, Gottschalk A (2015) High-throughput all-optical analysis of synaptic transmission and synaptic vesicle recycling in *Caenorhabditis elegans*. *PLoS One*, **10**:e0135584.

Watanabe S, Liu Q, Davis MW, Hollopeter G, Thomas N, Jorgensen NB, Jorgensen EM (2013) Ultrafast endocytosis at *Caenorhabditis elegans* neuromuscular junctions. *Elife*, **2**:e00723.

Weissenberger S, Schultheis C, Liewald JF, Erbguth K, Nagel G, Gottschalk A (2011) PACα – an optogenetic tool for *in vivo* manipulation of cellular cAMP levels, neurotransmitter release, and behavior in *Caenorhabditis elegans*. *J Neurochem*, **116**:616–625.

White JG, Southgate E, Thomson JN, Brenner S (1986) The structure of the nervous system of the nematode *Caenorhabditis elegans*. *Philos Trans R Soc B Biol Sci*, **314**:1–340.

Wietek J, Wiegert JS, Adeishvili N, Schneider F, Watanabe H, Tsunoda SP, Vogt A, Elstner M, Oertner TG, Hegemann P (2014) Conversion of channelrhodopsin into a light-gated chloride channel. *Science*, **344**:409–412.

Williams DC, Bejjani RE, Ramirez PM, Coakley S, Kim SA, Lee H, Wen Q, Samuel A, Lu H, Hilliard MA, Hammarlund M (2013) Rapid and permanent neuronal inactivation *in vivo* via subcellular generation of reactive oxygen with the use of KillerRed. *Cell Rep*, **5**:553–563.

Xu X, Kim SK (2011) The early bird catches the worm: new technologies for the Caenorhabditis elegans toolkit. *Nat Rev Genet*, **12**:793–801.

Yemini E, Jucikas T, Grundy LJ, Brown AEX, Schafer WR (2013) A database of *Caenorhabditis elegans* behavioral phenotypes. *Nat Methods*, **10**:877–879.

Yizhar O, Fenno LE, Prigge M, Schneider F, Davidson TJ, O'Shea DJ, Sohal VS, Goshen I, Finkelstein J, Paz JT, Stehfest K, Fudim R, Ramakrishnan C, Huguenard JR, Hegemann P, Deisseroth K (2011) Neocortical excitation/inhibition balance in information processing and social dysfunction. *Nature*, **477**:171–178.

Zemelman BV, Nesnas N, Lee GA, Miesenbock G (2003) Photochemical gating of heterologous ion channels: remote control over genetically designated populations of neurons. *Proc Natl Acad Sci U S A*, **100**:1352–1357.

Zemelman BV, Lee GA, Ng M, Miesenböck G (2002) Selective photostimulation of genetically chARGed neurons. *Neuron*, **33**:15–22.

Zhang F, Prigge M, Beyrière F, Tsunoda SP, Mattis J, Yizhar O, Hegemann P, Deisseroth K (2008) Red-shifted optogenetic excitation: a tool for fast neural control derived from *Volvox carteri*. *Nat Neurosci*, **11**:631–633.

Zhang F, Wang L-P, Brauner M, Liewald JF, Kay K, Watzke N, Wood PG, Bamberg E, Nagel G, Gottschalk A, Deisseroth K (2007) Multimodal fast optical interrogation of neural circuitry. *Nature*, **446**:633–639.

5 Using Optogenetics *In Vivo* to Stimulate Regeneration in *Xenopus laevis*

Dany S. Adams, Ai-Sun Tseng, and Michael Levin

5.1 Introduction

Optogenetics has revolutionized neuroscience by giving researchers the power to stimulate or prevent action potentials using light. This enables an unprecedented level of resolution, both in space and time (Knopfel *et al.*, 2010). Another important use of optogenetics is now being recognized, which is to use optogenetics to probe the activity and roles of ion flux in non-excitable cells. This chapter is about our use of the light-activated hydrogen pump archaerhodopsin to initiate full regeneration of a complex vertebrate appendage – the tail of the tadpole of *Xenopus laevis*, the African clawed frog.

Xenopus laevis is an important organism for the study of developmental biology due to its exceptional resistance to infection, the external development of its large eggs (1.2 mm in diameter) and the large number of eggs produced when the female is stimulated to ovulate by injection of human chorionic gonadotropin (indeed, before over-the-counter pregnancy tests, and before rabbits were used, a woman's urine was injected into a female *Xenopus* and, if the frog produced eggs, this indicated that the woman was pregnant). A single female can produce hundreds of eggs in one day, making it possible to collect sample sizes large enough for statistical analysis. The long history of *Xenopus* as a model organism means that there is a tremendous store of knowledge about a great many aspects of its development. Finally, and uniquely to *Xenopus*, its oocytes are used as expression systems in which to determine the characteristics of channels and pumps; thus, techniques for the successful expression of exogenous ion translocators are well established.

The life cycle of *Xenopus* is typical of amphibians. Females and males both spawn into the water so fertilization and subsequent development are external; a couple of hours after fertilization the zygote begins cleavage, a series of rapid mitoses that lack gap phases and are unaccompanied by growth. Following gastrulation (the formation of the three germ layers), neurulation (the formation of the neural tube that will develop into the spinal cord and brain) and a period of organogenesis during which the different tissues are formed, the embryo enters

Figure 5.1 *Xenopus laevis* tadpoles. (A) Position of amputation on stage 40 tadpoles. The tail is defined as the distance from the posterior edge of the gut to the tip of the tail. The cut is made between the halfway point and a third of the way from the tip. Healing occurs within 1 hour. After 8 days at 22°C, the tail has regenerated. (B) A stage 47 tadpole. At this stage, an appreciable proportion of tadpole tails are refractory to regeneration. (C) A stage 47 tadpole that has healed following amputation. While the tadpole will continue to develop, the tail will not regrow. All figures were adapted from images originally published in *Biology Open* (Adams *et al.*, 2013) and are freely available under a Creative Commons License (http://creativecommons.org/licenses/by/3.0/legal code); copyright belongs to the authors. Images have been adjusted for color and print size only.

the larval or tadpole phase. It is during this time that the yolk supplied by the oocyte is used up and the tadpole begins to eat and grow, undergoing metamorphosis and reaching sexual maturity in approximately 1 year. Under laboratory conditions at a constant temperature of 18–23°C, the larval phase is reached in approximately 1 week (Nieuwkoop and Faber, 1994). This phase is divided into about 15 stages, numbered 40–55. Up to about stage 46, the tadpole tail will fully regenerate if it is amputated. Likewise, after stage 48, the tadpole tail can regenerate. Between stages 46 and 48, however, the tadpole tail will not regenerate if amputated; this has been termed the refractory stage (Figure 5.1) (Beck *et al.*, 2003). The ultimate goal of this work was to induce regeneration of refractory-stage tails using optogenetics, both to show that optogenetics can be useful for studying ion flux-dependent developmental processes *in vivo* and to demonstrate the potential utility of optogenetics for regenerative medicine.

Controlled ion flux is the basis of an action potential and thus is the hallmark of excitable cells; however, all cells translocate ions across their membranes, using energy to maintain and modulate a particular resting membrane voltage (V_{mem}). V_{mem}, a stable, long-lasting characteristic of a cell, has now been shown to be much more than a side effect of other processes. V_{mem} regulates proliferation, cell cycle progression, apoptosis, cell migration, cell orientation, differentiation and de-differentiation, all of which are necessary for regeneration (Adams *et al.*, 2007; Tseng *et al.*, 2010; Beane *et al.*, 2011; Levin and Stevenson, 2012; Pai *et al.*, 2012).

We have published evidence that endogenous V_{mem}, controlled in part by the proton pump H^+-V-ATPase, plays a fundamental role in the control of *Xenopus* tadpole tail regeneration (Adams *et al.*, 2007). Therefore, in order to test optogenetics *in vivo*, we used refractory tail regeneration so as to assay for light-controlled proton efflux and subsequent V_{mem} hyperpolarization. The full description of this work can be found in Adams *et al.* (2013).

5.2 Methodology

5.2.1 Husbandry of *Xenopus laevis*

All care and treatment of *Xenopus* were performed in strict accordance with the recommendations in the Guide for the Care and Use of Laboratory Animals of the National Institutes of Health; the protocol was approved by Tufts University's Institutional Animal Care and Use Committee (#M2011-70). Maintenance of adult *Xenopus* and collection of gametes were as described previously (Sive *et al.*, 2000; Adams *et al.*, 2007). Tadpoles were staged according to Nieuwkoop and Faber (1994). Amputation of a third to half of the tail was performed at refractory stage 47 under MS-222 (tricaine) anesthesia (Figure 5.1). For molecular studies, tadpoles were given an overdose of tricaine then fixed in 3.7% paraformaldehyde (Sive *et al.*, 2000).

5.2.2 Molecular Biology

60 ± 10 pg of Arch-GFP or Arch-tdTom mRNA were injected into fertilized eggs (i.e. the one-cell stage) or into one of two cells at the two-cell stage; this dose minimized lethality still leading to observable expression as judged by imaging of the fluorescent protein. For rescue experiments, 4 ± 1 ng of mRNA for a dominant-negative E subunit of H^+-V-ATPase (YCHE78) (Adams *et al.*, 2007) was injected at the one-cell stage, either alone as a positive control for YCHE78 activity, or co-injected with the Arch mRNA.

In situ hybridization for markers of regeneration was performed according to a standard protocol (Harland, 1991) using probes for Notch1 (Coffman *et al.*, 1990) and Msx1 (Feledy *et al.*, 1999). Antibodies against acetylated α-tubulin (Sigma #T6793) and phospho-histone-3 (H3P; Upstate #05-598) were used at 1:1000 dilution.

5.2.3 Illumination of Arch-expressing Embryos

Two light delivery systems were employed. Petri dishes of tadpoles were either placed under an array of six LEDs (471 nm) at a distance resulting in irradiance of 0.5 ± 0.1 mW mm^{-2} or into a light-tight box containing two fiber optic cables, each connected to SugarCube LEDs (463 nm; Boston Engineering, Waltham, MA) delivering a total of 0.8 ± 0.1 mW mm^{-2}. Because exposure lasted hours to days, thermo- and photo-toxicity were concerns; therefore, dishes were kept on cooling stages set to 18°C and lights were turned on for 0.5 s then off for 2 s. Prior to measuring V_{mem} of Arch-expressing cells using microelectrodes, lights were turned on for 5 s then off for 10 s.

5.2.4 Electrophysiology and Measurement of pH

V_{mem} was measured directly using standard techniques on a Warner Instruments Oocyte Clamp Amplifier OC-725C with oocyte patch-clamp model 7251.I (Harvard Apparatus, Hampden, CT) as described previously (Adams *et al.*, 2007). To confirm the activity of Arch, pH measurements were made on tails of lightly anesthetized tadpoles using two different pH reporting dyes: 5 μM BCECF-AM and 5 μM SNARF-5F (Invitrogen, Waltham, MA) (Adams and Levin, 2012). Two dyes were necessary, one that was imaged with wavelengths that activated Arch, and the other imaged using wavelengths that did not activate Arch; a difference between the measurements made with these two reagents would therefore confirm that the activating wavelength was indeed inducing H^+ pumping. Tadpoles were soaked in dye for 2 hours, then the dye was rinsed away. For BCECF, a dual excitation dye, filter sets were EX 450/20, D 460, EM 535/30 and EX 500/20, D 515, EM 535/30. For SNARF-5F, a dual emission dye, filter sets were EX 540/25, D 565, EM 580/25 and EX 540/25, D 610, EM 640/25. Two regions of interest – an area with clear signal from the fluorescent dye and an area clearly lacking such a signal – were chosen on images of the fluorescent dye then transferred to darkfield- and flatfield-corrected images of dye ratios. The difference in brightness (mean pixel value) between the two regions was calculated and converted to a measure of pH difference using published calibrations of the two dyes (Sasaki *et al.*, 1992; Morley *et al.*, 1996).

5.2.5 Imaging and Scoring of Regeneration

Arch-GFP-expressing tadpoles were imaged using filters EX 470/20, D 485, EM 517/23. Arch-tdTom-expressing tadpoles were imaged using filters EX 545/20, D 565, EM 595/50.

Tail regeneration was scored as described in Adams *et al.* (2007). Briefly, tails from a given dish were sorted into four categories: perfect regeneration, good regeneration (minor imperfections), bad regeneration (missing or grossly malformed parts) and no regeneration. Regenerative ability was then summarized as a score for that dish (sample), called the regeneration index, with 0 signifying no regeneration and 300 signifying perfect regeneration of all tadpoles in the dish.

5.2.6 Statistical Analysis

The effect of light on non-injected animals was examined and we found no evidence for a difference in regeneration index of non-injected animals exposed to light versus those kept in the dark. The effect size considered meaningful was 10% in all cases and results were considered statistically significant at $p \leq 0.01$. To analyze Arch activity, the number of H3P-positive cells and Arch + YCHE78 activity, we defined the sample sizes as the number of tadpoles; for comparing regeneration indices after refractory regeneration, sample sizes were the numbers of dishes. T-tests were used to compare V_{mem} values measured using a microelectrode, Arch experiments and numbers of H3P-positive cells. A Mann–Whitney U-test was used to compare the distributions of wild-type, abnormal and dead (or otherwise unscorable) tadpoles resulting from YCHE78 injections.

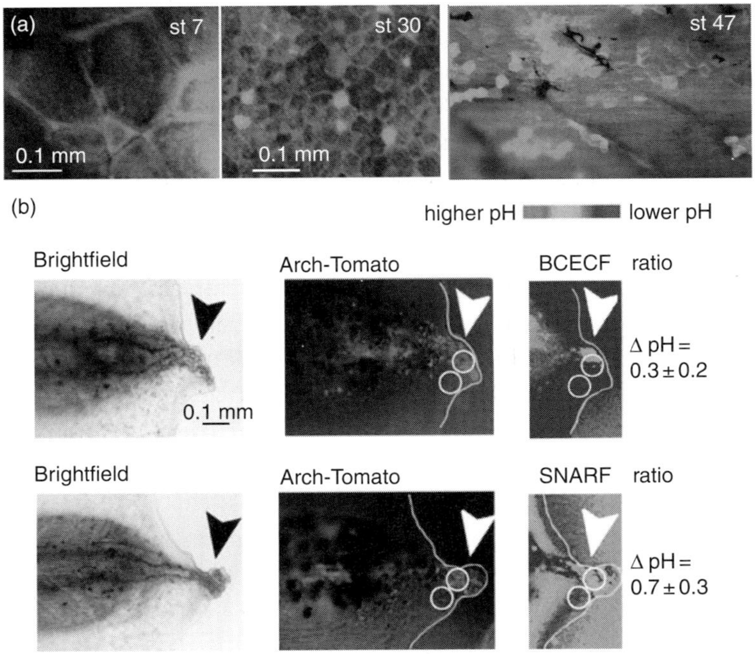

Figure 5.2 GFP-Arch mRNA was expressed and Arch is active in *Xenopus* cells. (A) By stage 7, expression is visible in some cells. Expression in cell membranes is clear by stage 30, and still present at stage 47, the middle of the refractory stage. (B) The change of pH in Arch-expressing cells is consistent with the presence of Arch activity. Two dyes were used to measure pH: one that is only weakly activated by the wavelengths that activate arch, BCECF, and another that is strongly activated by the light that activates Arch. Thus, BCECF reveals the pH of dark-maintained Arch and SNARF-5F reveals the pH of light-activated Arch. As predicted, pH is higher when Arch is activated by light. All figures were adapted from images originally published in *Biology Open* (Adams *et al.*, 2013) and are freely available under a Creative Commons License (http://creativecommons.org/licenses/by/3.0/legalcode); copyright belongs to the authors. Images have been adjusted for color and print size only. (A black-and-white version of this figure will appear in some formats. For the color version, please refer to the plate section.)

5.3 Results and Discussion

5.3.1 Exogenous Arch is Expressed in *Xenopus* Embryonic Cells and is Still Present During Tadpole Stages

To confirm the expression of the injected mRNA, we monitored the fluorescence of either GFP or tdTom. Fluorescence was detectable in most embryos, and only embryos with detectable fluorescence were used in experiments (Figure 5.2A).

5.3.2 Arch is Active in *Xenopus* Blastomeres and Cells of the Tadpole Tail

The cells of the tadpole tail are too small to be accurately measured using microelectrodes; therefore, we looked for Arch activity in tails by looking for differences in pH, predicting that when Arch is active, pH will be higher due to pumping of H^+ out of the cell. We measured the difference in pH between cells expressing Arch and cells not expressing Arch under lighting conditions when Arch is not active, using BCECF, and under lighting conditions when it is active, using SNARF-5F. We found that when

Arch is not being activated, the pH difference between cells that are expressing Arch and those that are not expressing Arch was 0.3 ± 0.2. In contrast, in cells being activated by light during measurement, the difference conferred by Arch expression more than doubled to 0.7 ± 0.3 (Figure 5.2B). Thus, we conclude that in the presence of 540-nm light, Arch pumps H^+ out of *Xenopus* tail cells, thus hyperpolarizing them.

The classical electrophysiology technique of impalement with a microelectrode was used to measure the V_{mem} of early embryonic cells, called blastomeres (Figure 5.3A). Early in cleavage, blastomeres are large enough to accurately measure V_{mem}, and non-injected embryos kept in either light or dark conditions measured −29 ± 2 mV. In contrast, we detected a meaningful and statistically significant difference between light- and dark-exposed injected embryos, with light-exposed embryos averaging −40 ± 7 mV and dark-exposed embryos averaging −22 mV ± 8 mV ($n = 12$, $t = 3.541$, $p = 0.002$). Thus, as predicted, activation of Arch by light hyperpolarizes cells.

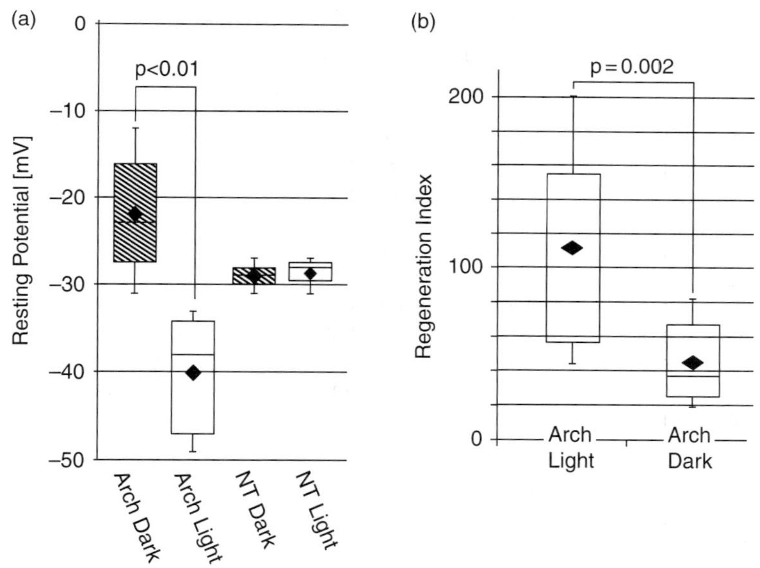

Figure 5.3 V_{mem} and regeneration index in dark- versus light-exposed cells. (A) Controls were (i) un-injected and maintained in the normal way (i.e. no treatment) and (ii) un-injected and exposed to the same light as the experimentally injected embryos (i.e. light only). Both types of control were found to have V_{mem} values of −29 ± 2 mV, which is consistent with other studies. Arch-injected, dark-maintained embryos were somewhat depolarized relative to non-injected embryos, measuring −22 ± 8 mV (see text for statistics). These data are consistent with other data indicating that, in the dark, when expressed in *Xenopus* embryos, Arch pumps H^+ into the cell (data not shown). When exposed to light, Arch-injected cells are hyperpolarized to −40 ± 7 mV. (B) The effect of 48 hours of Arch activity on the regeneration of refractory amputated tails. Arch-expressing tadpoles were divided into two groups: one was maintained in the dark, while the other was exposed to blinking blue light for 48 hours then returned to the dark for 6 days. Light-activated, Arch-expressing tadpoles have a significantly higher regeneration index than dark-maintained tadpoles. This is consistent with our hypothesis that light-activated Arch can trigger regeneration. All figures were adapted from images originally published in *Biology Open* (Adams *et al.*, 2013) and are freely available under a Creative Commons License (http://creativecommons.org/licenses/by/3.0/legal code); copyright belongs to the authors. Images have been adjusted for color and print size only.

Interestingly, the hyperpolarized cells from light-exposed embryos were not impaled and measured while being exposed to activating light. Because this was technically not possible, these embryos were removed from the light up to 15 minutes prior to being impaled, and kept in dishes under ambient light conditions. Thus, the hyperpolarized state induced by 471-nm light did not revert to normal once the activating light was removed. This could be because once the V_{mem} was "set" by the Arch activity, endogenous ion translocators maintained that polarity. Another possibility is that, as we found with many of the optogenetics reagents we tried (Adams *et al.*, 2014), in *Xenopus* cells, Arch does not behave as predicted from studies in mammalian neurons.

5.3.3 Arch Proton Pumping Can Rescue Developmental Defects Caused by Inhibition of the Endogenous Proton Pump

Another positive control confirmed that Arch-dependent H^+ efflux hyperpolarizes cells. We co-injected mRNA for Arch and a dominant-negative subunit of the endogenous complex that hyperpolarizes cells under normal conditions. On its own, injection of this mRNA, YCHE78, blocks the normal, endogenous H^+ efflux mediated by the H^+-V-ATPase, causing mispatterning of the left–right axis, as well as craniofacial defects that are characteristic of V_{mem} changes in ectodermal cells (Vandenberg *et al.*, 2011). We therefore hypothesized that Arch activity could counteract the effect of H^+-V-ATPase inhibition, thus rescuing the normal left–right patterning and preventing craniofacial anomalies. As predicted, co-injection of Arch abrogated the effect of YCHE78, confirming that Arch activity can functionally replace the normal signaling pathway ($n = 149$, $t = 3.423$, $p < 0.001$). We found the same result when we blocked H^+-V-ATPase using a pharmacological inhibitor.

5.3.4 Arch Activity Can Trigger the Endogenous Regeneration Pathway in Tails Amputated During the Regeneration-inhibited Refractory Stage

We amputated the tails of non-injected controls and Arch-expressing tadpoles at stage 47, incubating the controls and half of the Arch-expressing tadpoles in the dark, while exposing the other Arch-expressing tadpoles to light in order to activate H^+ efflux for 48 hours. A meaningful and statistically significant increase in regeneration was observed in the light-exposed tadpoles, with the regeneration index almost doubling ($n = 16$ dishes representing 428 tadpoles, $t = 3.709$, $p = 0.002$; Figure 5.3B). We consider these data to be solid evidence that optogenetics can be used to induce regeneration in otherwise non-regenerative tissue. These results are summarized in Figure 5.4A.

We next compared induced regeneration to the normal process seen in tails undergoing regeneration during stages before or after the refractory stage. Three facets of regeneration were examined: expression of regeneration-specific markers, number of H3P-positive and therefore proliferating cells and the growth of neurons into the growing regeneration bud. The induced tails matched normally regenerating tails in all three of these measures, while the tails of dark-maintained tadpoles looked like refractory tails (Figure 5.1). Both *Notch1* and *Msx1* were

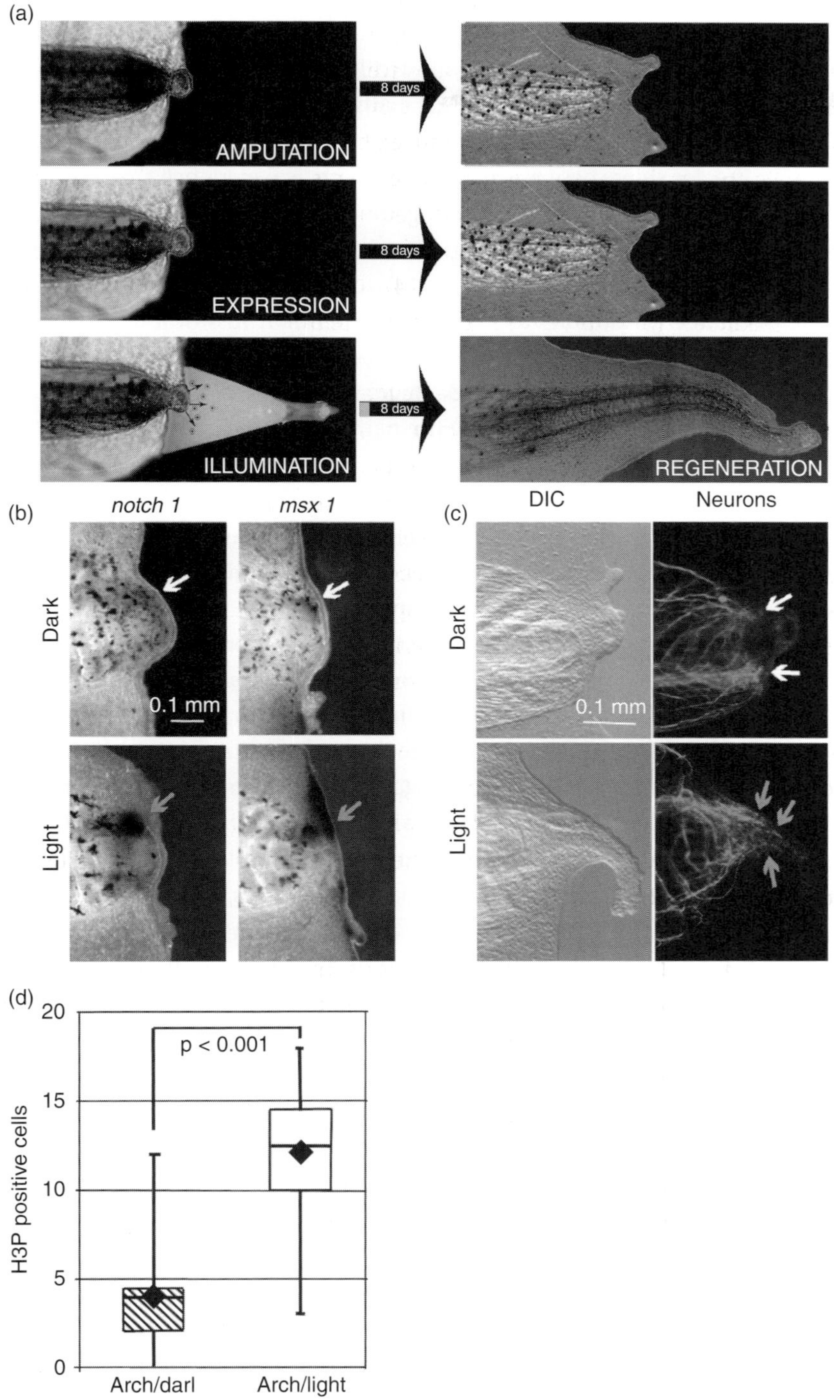

Figure 5.4 Light-activated arch triggers regeneration along the normal pathways. (A) A summary of the main results. Tails amputated at stage 47 do not grow back. Expression of Arch alone does not trigger regeneration, indicating that the presence of the exogenous protein does not, on its own, affect regeneration. However, when Arch-expressing tadpoles are exposed to light for 48 hours, many refractory tails regenerate in the normal 8day time period. (B) Two markers of standard regeneration, notch1 and msx1, are both expressed in the right place – the tip of the tail – only when Arch is light

present at the tip of Arch-expressing tails that were exposed to light, but were absent in Arch-expressing tails maintained in the dark (Figure 5.4B). Neurons were found to be growing into light-exposed tails, while those in dark-maintained tails were not activated to grow (Figure 5.4C). Finally, light-exposed tails had more than double the number of proliferating cells ($n = 22$, $t = 5.273$, $p < 0.001$; Figure 5.4D). We consider these results to provide critical evidence that induced H^+ efflux triggers regeneration by restarting the endogenous regeneration program that is active at other times.

This finding is more relevant to human medicine than is obvious at first. The refractory phase of tadpole tails is an age-dependent loss of a regenerative ability that is present in younger animals. Humans also have an age-dependent loss of regenerative ability in an appendage: up to the age of about 7 years, an entire fingertip can regenerate, including the nail (Illingworth and Barker, 1980). Moreover, other human tissues show some regenerative ability (e.g. the liver). Thus, it is entirely possible that human cells retain the regeneration pathway, like the refractory tadpole tail, and that manipulation of V_{mem} could be used to trigger the regeneration of normal tissues, and even complex appendages, in humans via endogenous, and therefore regulated, pathways.

Importantly, normal regeneration, without overproliferation or any of the other side effects seen with gene therapy, was induced by a very short period of hyperpolarization; that is, only 48 hours of light exposure induced complete regeneration of a normal tail. We see these results as evidence that modulation of V_{mem} by short-term expression and illumination of a light-gated ion translocator is sufficient to trigger the entire endogenous regeneration program, obviating the need for detailed micromanagement of the regeneration process. In other words, it allows us to trigger both the process and the regulation of the process, including cessation when the appendage has reached the correct size and differentiation state.

5.4 Concluding Remarks

We are extremely optimistic about the potential use of optogenetics in regenerative medicine. As we learn more about the role of V_{mem} during normal processes and the role of misregulation of V_{mem} in disease, more targets for minimally

Caption for Figure 5.4 (cont.)

activated. (C) Arch-expressing embryos maintained in the dark have neurons that terminate at the amputation plane, like refractory tails. The neurons in light-activated, Arch-expressing tails, however, grow into the tip of the tail, just as they do in normally regenerating tails. (D) The number of proliferating cells, assayed by the number of phospho-histone 3-positive cells, is much higher in the tips of light-activated, Arch-expressing tails than in those maintained in the dark, which is consistent with the need for cell proliferation during regeneration. All figures were adapted from images originally published in *Biology Open* (Adams *et al.*, 2013) and are freely available under a Creative Commons License (http://creativecommons.org/licenses/by/3.0/legalcode); copyright belongs to the authors. Images have been adjusted for color and print size only. (A black-and-white version of this figure will appear in some formats. For the color version, please refer to the plate section.)

invasive biomedical intervention will be revealed. We believe that the work described here is an important proof of principle. Moreover, we have presented evidence that V_{mem} has an unexpected yet essential role in initiating and regulating complex processes such as regeneration.

ACKNOWLEDGMENTS

The authors would like to thank Punita Koustubhan and Amber Brand for *Xenopus* care; Ed Boyden for Arch-GFP and advice on optogenetics; Karl Deisseroth, Richard Kramer and John Lin for reagents and for sharing their knowledge; Ravshan Sabirov for the NHE3 plasmid; Joan Lemire and Claire Stevenson for cloning of the plasmids; and Tal Shomrat and Brook Chernet for help with electrophysiology. M.L. gratefully acknowledges support from NIH (GM078484, AR055993), NSF (DBI-1152279) and the G. Harold and Leila Y. Mathers Charitable Foundation. This work was also supported by the Telemedicine and Advanced Technology Research Center (TATRC) at the US Army Medical Research and Materiel Command (USAMRMC) through award W81XWH-10 2-0058. D.S.A. prepared this manuscript with support from NIH (5R01-HD081326).

REFERENCES

Adams, D.S., Lemire, J.M., Kramer, R.H. and Levin, M. (2014). Optogenetics in developmental biology: using light to control ion flux-dependent signals in *Xenopus* embryos. *The International Journal of Developmental Biology*, **58**, 851–861.

Adams, D.S. and Levin, M. (2012). General principles for measuring resting membrane potential and ion concentration using fluorescent bioelectricity reporters. *Cold Spring Harbor Protocols*, 2012, 385–397.

Adams, D.S., Masi, A. and Levin, M. (2007). H^+-pump-dependent changes in membrane voltage are an early mechanism necessary and sufficient to induce *Xenopus* tail regeneration. *Development*, **134**, 1323–1335.

Adams, D.S., Tseng, A.S. and Levin, M. (2013). Light-activation of the archaerhodopsin H^+ -pump reverses age-dependent loss of vertebrate regeneration: sparking system-level controls *in vivo*. *Biology Open*, **2**, 306–313.

Beane, W.S., Morokuma, J., Adams, D.S. and Levin, M. (2011). A chemical genetics approach reveals H,K-ATPase-mediated membrane voltage is required for planarian head regeneration. *Chemistry & Biology*, **18**, 77–89.

Beck, C.W., Christen, B. and Slack, J.M.W. (2003). Molecular pathways needed for regeneration of spinal cord and muscle in a vertebrate. *Developmental Cell*, **5**, 429–439.

Coffman, C., Harris, W. and Kintner, C. (1990). Xotch, the *Xenopus* homolog of *Drosophila* notch. *Science*, **249**, 1438–1441.

Feledy, J.A., Beanan, M.J., Sandoval, J.J. *et al.* (1999). Inhibitory patterning of the anterior neural plate in *Xenopus* by homeodomain factors Dlx3 and Msx1. *Developmental Biology*, **212**, 455–464.

Harland, R.M. (1991). *In situ* hybridization: an improved whole mount method for *Xenopus* embryos. In: B.K. Kay and H.B. Peng, eds., Xenopus laevis: *Practical Uses in Cell and Molecular Biology*. San Diego, CA: Academic Press, pp. 685–695.

Illingworth, C.M. and Barker, A.T. (1980). Measurement of electrical currents emerging during the regeneration of amputated fingertips in children. *Clinical Physics and Physiological Measurement*, **1**, 87–89

Knopfel, T., Lin, M.Z., Levskaya, A. *et al.* (2010). Toward the second generation of optogenetic tools. *The Journal of Neuroscience*, **30**, 14998–15004.

Levin, M. and Stevenson, C.G. (2012). Regulation of cell behavior and tissue patterning by bioelectrical signals: challenges and opportunities for biomedical engineering. *Annual Review of Biomedical Engineering*, **14**, 295–323.

Morley, G.E., Taffet, S.M. and Delmar, M. (1996). Intramolecular interactions mediate pH regulation of connexin43 channels. *Biophysical Journal*, **70**, 1294–1302.

Nieuwkoop, P.D. and Faber, J. (1994). *Normal Table of* Xenopus laevis *(Daudin): A Systematical and Chronological Survey of the Development from the Fertilized Egg till the End of Metamorphosis*. New York, NY: Garland Publishing.

Pai, V.P., Aw, S., Shomrat, T., Lemire, J.M. and Levin, M. (2012). Transmembrane voltage potential controls embryonic eye patterning in *Xenopus laevis*. *Development*, **139**, 313–323.

Sasaki, S., Ishibashi, K., Nagai, T. and Marumo, F., (1992). Regulation mechanisms of intracellular pH of *Xenopus laevis* oocyte. *Biochimica et Biophysica Acta*, **1137**, 45–51.

Sive, H., Grainger, R.M. and Harland, R. (2000). *Early Development of* Xenopus laevis. New York, NY: Cold Spring Harbor Laboratory Press.

Tseng, A.-S., Beane, W.S., Lemire, J.M., Masi, A. and Levin, M., (2010). Induction of vertebrate regeneration by a transient sodium current. *Journal of Neuroscience*, **30**, 13192–13200.

Vandenberg, L.N., Morrie, R.D. and Adams, D.S. (2011). V-ATPase-dependent ectodermal voltage and pH regionalization are required for craniofacial morphogenesis. *Developmental Dynamics*, **240**, 1889–1904.

Part II

Opsin Biology, Tools, and Technology Platform

6 Sodium and Engineered Potassium Light-Driven Pumps

Vitaly Shevchenko, Ivan Gushchin, Vitaly Polovinkin, Kirill Kovalev, Taras Balandin, Valentin Borshchevskiy, and Valentin Gordeliy

6.1 Introduction

Optogenetics is one of the most important recent technological advances in biology. It revolutionized our ability to study the processes in neuronal circuits, promising new approaches to the treatment of different diseases and disabilities. The core tools of optogenetics are light-driven retinal membrane proteins. Retinal-containing membrane proteins are present in all domains of life. They employ light energy for a wide range of different functions such as ion transport, photosensing and channel activity (Gushchin *et al.*, 2013). All of these proteins contain seven transmembrane α-helices and a chromophore-retinal molecule, which is covalently bound via the Schiff base to the side-chain of a lysine amino acid (Oesterhelt and Stoeckenius, 1971; Stoeckenius *et al.*, 1979).

Microbial rhodopsins (MRs) were first discovered in the early 1970s in the archaeal halophiles (salt-loving microbes that live in saturated brines). These extremophiles have a remarkable variety of rhodopsins. The first known was called bacteriorhodopsin (BR) because of its structural similarities with vertebrate eye rhodopsins (Oesterhelt and Stoeckenius, 1971; Stoeckenius *et al.*, 1979). However, BRs that are widely distributed in different genera of archaeal halophiles are proton pumps used to generate light-driven proton gradients as sources of energy for the cell. Then, in 1979, again in the same microbe, a second pump was found; however, it was erroneously concluded that this pump functions as an outward sodium pump (Greene and Lanyi, 1979; Lindley and MacDonald, 1979; MacDonald *et al.*, 1979). Only later was it recognized that, in fact, this rhodopsin is an inward-directed chloride pump that helps maintain the proper electrochemical balance (Schobert and Lanyi, 1982). Soon after halorhodopsin, two sensory rhodopsins (also known as slow rhodopsins) were found in the same microbe (*Halobacterium salinarum*) (Bogomolni and Spudich, 1982). These are light sensors (each tuned to a different wavelength) that produce complex photactic behavior in these microbes. All of these rhodopsins were found in different representatives of this archaeal clade.

It was only after the 2000s that, thanks to metagenomics, rhodopsins were also found in bacteria. An uncultivated marine gammaproteobacterium of the 16S rRNA-defined clade SAR86 was confirmed to have a functional proton pump that was named proteorhodopsin (Béjà *et al.*, 2000). The rapid increase in sequenced bacterial genomes of the first decade of the twenty-first century enabled the discovery of several examples of rhodopsin genes in multiple branches of bacteria and archaea (de la Torre *et al.*, 2003; DeLong and Karl, 2005; O'Malley, 2007). A clear illustration of the widespread transmission of these genes in the prokaryotic domain was found in the extremely halophilic bacterium *Salinibacter*, the genomes of which produced four rhodopsins, one proton pump, one chloride pump and two sensory rhodopsins, representing exactly the same combination as the environment-sharing archaeon *Halobacterium* (Mongodin *et al.*, 2005). But nature plays tricks on us. *Salinibacter*'s proton pump, later called xanthorhodopsin, is different from the *Halobacterium* one in both structure and function, being the first discovered rhodopsin to use a carotenoid pigment antenna in the same manner as chlorophylls to increase the wavelength functional range (Balashov *et al.*, 2005). Since then, rhodopsins have been found in most clades of microbes exposed to light, including firmicutes, actinobacteria, bacteroidetes, most proteobacteria and marine archaea.

The most studied protein of this family is BR from a halophilic archaea *H. salinarum*, which is a proton pump that provides the first key universal step of transformation of energy in cells: the generation of a proton electrochemical gradient across the cell membrane. Beginning with electron microscopy (Henderson *et al.*, 1990) and then X-ray high-resolution studies, the analysis of halorhodopsin and sensory rhodopsins provided a basis for the fundamental hypothesis of transport mechanisms (Luecke *et al.*, 1999; Kolbe *et al.*, 2000; Gordeliy *et al.*, 2002; Moukhametzianov *et al.*, 2006) and motivated the development of new experimental technologies (Gushchin *et al.*, 2013).

Among the most important applications of rhodopsins is optogenetics. Unfortunately, few retinal proteins are available for optogenetics. Specifically, the most used are the non-selective channel rhodopsin, ChR2, which is used to depolarize and thus activate neurons (Nagel *et al.*, 2003), the Cl^- pump halorhodopsin, NpHR (Zhao *et al.*, 2008), and the H^+ pump archaerhodopsin, Arch3 (Han *et al.*, 2011). The latter two are used to hyperpolarize and thus silence neurons. A crucial aspect for future progress in optogenetics is to identify and/or engineer new light-driven proteins with novel properties, such as channels and transporters with high selectivity and conductivity for a particular ion. The ions Na^+, K^+ and Ca^{2+}, which are relevant for neuronal function, have not been addressable by MR pumps so far (the first experiments with Na^+ were conducted very recently). In particular, an outward K^+ pump would be highly desirable, as K^+ is the main ion used for neuronal re- or hyper-polarization. In our review, we describe very recent work on the structure–function characterization and engineering of sodium and potassium light-driven pumps.

Interestingly, the known structures led to a widely accepted belief that sodium pumps cannot exist. This erroneous conclusion was based on the assumption that the known rhodopsins had to have quite the same structures; for instance, similar to that of BR. If this were true, it would not be clear how a cation can be transferred from extracellular to cytoplasmic parts of the protein by the positively charged Schiff base of the chromophore upon its isomerization. However, the recently elucidated structure of the KR2 sodium pump, as is discussed in this chapter, is remarkably different from that of BR, and this difference explains why a cation pumps exist. Indeed, while investigating the marine bacterium *Dokdonia eikasta*, Inoue and co-workers discovered that *D. eikasta* cells are able to pump out sodium ions during early periods of growth under illumination (Inoue *et al.*, 2013). It turned out that the corresponding ion transporter, dubbed KR2, belongs to a new type of MR in which some of the most important ionizable amino acids are replaced with polar ones, and vice versa. Even when expressed in another commonly employed bacterium, *Escherichia coli*, KR2 could still transport sodium ions. Yet it appeared that potassium ions, despite being very similar to sodium ions, were not transported by KR2.

This and other curiosities displayed by KR2 attracted our interest because we had already carried out several structural studies of other MRs in order to investigate the molecular mechanisms behind their actions. However, membrane proteins are notoriously difficult to handle, and producing them is usually expensive and time consuming, with KR2 being no exception. Since the protein could be produced only in batches of several milligrams despite our new efficient approaches to membrane protein crystallization, while thousands of crystallization conditions should have been tested, we had to miniaturize and automate the process using robotic and imaging systems. The diffraction signal from the first crystals that were obtained, which measured less than 20 μm across, was too weak to be measured using in-house X-ray sources. Recent advances at the European Synchrotron Radiation Facility's macromolecular crystallography beamlines – such as microfocus X-ray beams, automatic sample changers and convenient software protocols – greatly helped with the search for the best crystals. After several months of trials at beamlines ID23-1 and ID29, we were able to improve the resolution of the diffraction data collected from around 20 Å to better than 1.5 Å. Meanwhile, we were able to extract information about KR2's spectroscopic characteristics using optical absorption spectra at the cryobench laboratory at beamline ID29S. Before we describe Gordeliy and Kandori's groups' structures (Gushchin *et al.*, 2015b; Kato H.E. *et al.*, 2015), we shall describe the family of sodium pumps and illustrate their general functional characterization.

6.2 The Family of Sodium-pumping Rhodopsins

The first light-driven sodium pump KR2, from *D. eikasta*, was discovered and characterized in 2013 (Inoue *et al.*, 2013). Other representatives rapidly populated this class of rhodopsins. The gene sequences had already been acquired during whole-genome sequencing of various marine, sedimentation and more exotic species (Figure 6.1) (Ivanova *et al.*, 2011; Riedel *et al.*, 2012; Singh *et al.*, 2013;

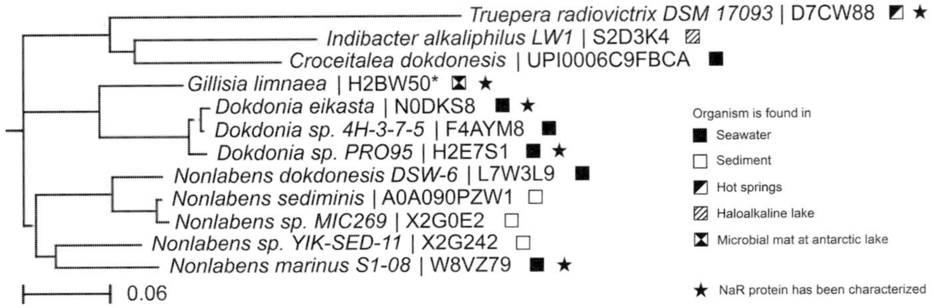

Figure 6.1 Phylogenetic tree of rhodopsins with the NDQ motif, and putative sodium-pumping rhodopsins. The name of the organism and corresponding UniProtID of the protein is shown. The media in which the species have been found are indicated by squares with different fills. The star indicates that the NaR protein has been demonstrated to pump sodium ions. * *Gillisia limnaea* GLR protein is represented in its truncated version (61 amino acids are truncated from the N-terminus).

Nakanishi *et al.*, 2014; Kwon *et al.*, 2016). The proteins are distinguished by their characteristic "NDQ motif" of the active site residues (N112, D116 and Q123 in KR2), replacing the "DTD motif" (D85, T89 and D96) in BR or DTE/DTK in other proton pumps (Balashov *et al.*, 2013; Ernst *et al.*, 2014) and NTQ/TSA in chloride pumps.

Sequence alignment of all available members of putative sodium-pumping rhodopsins shows high sequence similarity among them (Figure 6.2), with the transmembrane regions being more conserved than extra- and intra-cellular loops. In order to analyze the conservation of the functional units in the protein, we provide a brief description of the structure of KR2 in the following section.

6.3 Structures of the Wild-type Sodium Pump KR2

Several crystallographic structures of the light-driven sodium pump KR2 have been determined (Gushchin *et al.*, 2015b; Kato H.E. *et al.*, 2015). The structures have recently been reviewed by us (Gushchin *et al.*, 2015a) and here we will provide a brief synopsis.

Overall, similarly to other MRs, KR2 consists of a seven-transmembrane helical framework, with the light-absorbing cofactor retinal bound in the middle. Somewhat unexpectedly, the extracellular side of the protein is capped with the N-terminal α-helix, which is not observed in other MRs. The elongated loop connecting helices B and C is similar to xanthorhodopsin's B–C loop in structure (Luecke *et al.*, 2008) and makes β-sheet interactions with the linker joining the N-terminal α-helix and helix A.

Inside the protein, the crystallographic structures reveal the ion translocation pathway. Structural and mutational analysis enabled functional assignment of three major segments: the ion uptake cavity, the Schiff base cavity and the putative ion release cavity (Figure 6.3) (Gushchin *et al.*, 2015b; Kato H.E. *et al.*, 2015).

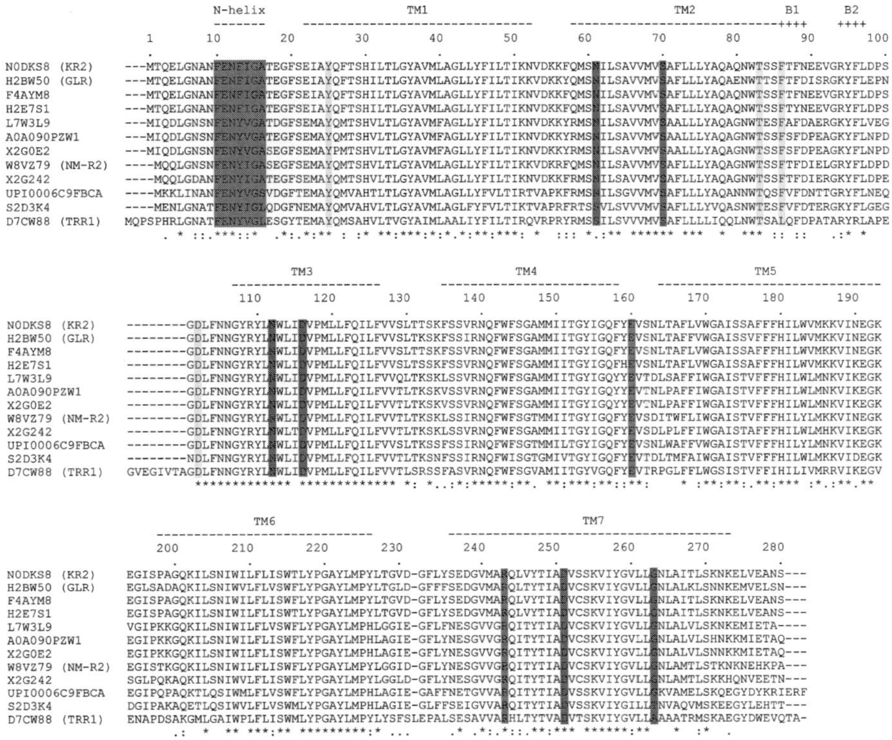

Figure 6.2 Structure-based sequence alignment of putative sodium-pumping rhodopsins. In order to precisely specify the proteins, we indicate the UniProtIDs. However, some of the proteins have "established" common names in the literature, so we cite them as well. TM indicates transmembrane regions, B shows the location on beta-sheets. The ion uptake cavity (cyan) is represented by an N61–G263 pair, the two latter members have an S61–T/A263 pair. The cavity in proximity to retinal (magenta) is completely conserved in the class (i.e. S70, N112, D116 and D251). The ion release cavity contains a cluster of three conserved ionizable residues: E11, E160 and R243 (green). All members possess an N-terminal α-helix that caps the putative ion release cavity. The oligomerization interface is also well conserved (yellow) among the members of sodium-pumping rhodopsins (Y25, T83, F/L86 and D102). (A black-and-white version of this figure will appear in some formats. For the color version, please refer to the plate section.)

The ion uptake cavity has a large opening into the cytoplasm. Obstructing this cavity through various mutations either inhibits ion pumping or imparts the protein with novel selectivities toward monovalent cations (Gushchin *et al.*, 2015b; Kato H. E. *et al.*, 2015; Konno *et al.*, 2016). Indeed, while the wild-type KR2 can only pump H^+, Li^+ and Na^+, the mutations G263F and G263W result in potassium-pumping variants, whereas the double-mutant N61L/G263F can pump ions as large as Cs^+. Overall, it has been found that raising the side-chain volumes at positions 61 and 263 leads to stronger conductivity of larger cations (Konno *et al.*, 2015). At the bottom of the KR2's ion uptake cavity is a polar residue Glu-123, the analog of the ionizable proton donor residues of the proton pumps. Mutations of Glu-123 to either alanine or valine result in decreased efficiency of the pump (Inoue *et al.*, 2015).

The retinal-binding pocket contains all-*trans* retinal, with its Schiff base embedded in a polar cavity, lined with Ser-70, Asn-112, Asp-116 and Asp-251.

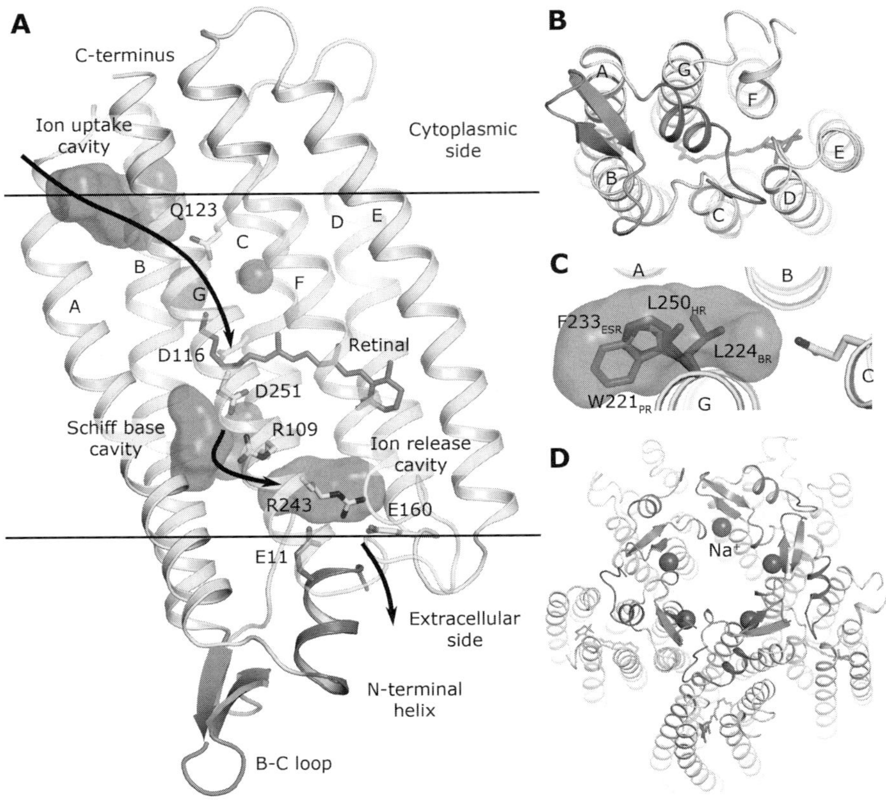

Figure 6.3 Structure of KR2. The N-terminal helix is shown in blue and the B–C loop is shown in orange. Water-accessible cavities are shown as red surfaces. (A) Side view. The ion path is shown with arrows. (B) View from the extracellular side. (C) KR2 ion uptake cavity. Hypothetical effects of the G263L, G263F and G263W mutations are shown using the conformations of the homologous residues L224 of bacteriorhodopsin (BR), L250 of [20], F233 of *Exiguobacterium sibiricum* rhodopsin (ESR) and W221 of blue proteorhodopsin (PR). (D) Pentameric assembly of KR2. Sodium ions are shown in violet. (A black-and-white version of this figure will appear in some formats. For the color version, please refer to the plate section.)

The Schiff base cavity is separated from the extracellular side of the protein by an Arg-109 side-chain. Asn-112 and Asp-116 are among the residues defining the sodium pumping ability of KR2, and all of the polar and ionizable residues of the Schiff base cavity are crucial for the pump's function (Inoue *et al.*, 2013).

There are several conformations of the Schiff base cavity that have been observed in the crystallographic structures (reviewed in Gushchin *et al.*, [2015a]), most notably a compact one and an expanded one. This is similar to the case of the ion-free and ion-bound chloride pump halorhodopsin (Kouyama *et al.*, 2010; Kanada *et al.*, 2011). Consequently, we hypothesized this ability of the Schiff base cavity to expand to be a requisite for a MR to be able to pump an ion that is not a proton (Gushchin *et al.*, 2015a).

Finally, the ion release cavity of KR2 is centered on three ion-bridged residues Glu-11, Glu-160 and Arg-243. Since mutations of any of these residues do not result in qualitative changes in the pump's function, it must be concluded that these residues have a mostly structural function (Inoue *et al.*, 2013; Gushchin *et al.*, 2015b; Kato H.E. *et al.*, 2015).

While KR2 crystallizes both as a monomer and a pentamer, we believe that the physiological state is pentameric (Gushchin *et al.*, 2015a, 2015b), similar to pentameric proteorhodopsins (Ran *et al.*, 2013). The sodium ion that binds to KR2 in the ground state is found both in monomer and at the pentamerization interface of a pentamer (Gushchin *et al.*, 2015a) and most probably plays a structural role and is not the ion that is being transported.

6.4 The Mechanism of Sodium Translocation

A number of representatives of putative sodium-pumping rhodopsins have been functionally characterized (Figure 6.1), and it has been demonstrated that they can pump sodium ions (Inoue *et al.*, 2013; Balashov *et al.*, 2014; Yoshizawa *et al.*, 2014; Bertsova *et al.*, 2015). Many of the studied proteins were found to be hybrid Na^+/H^+ pumps. The proteins transport sodium ions if they are present in sufficiently large amounts, but transport protons if the opposite is the case. KR2 from *D. eikasta*, GLR from *G. limnaea* and TRR1 from *Truepera radiovictrix* (Figure 6.4) are hybrid Na^+/H^+ pumps. NaR from *Dokdonia* sp. PRO95 and NM-R2 from *Nonlabens marinus* S1-08 display less pronounced H^+ transport in the absence of sodium ions (Yoshizawa *et al.*, 2014; Bertsova *et al.*, 2015).

All rhodopsins that have been characterized in this class have a common absorption maximum in the ground state of 523 nm. The presence or absence of sodium ions barely affected this value.

Time-resolved visible spectroscopy has been done for KR2 and GLR, sharing the primary result that photocycle turnover is faster in the presence of sodium ions. During the photocycle, a red-shifted K intermediate is rapidly formed; blue-shifted L/M states then appear, and then a red-shifted O state appears. The rate of formation of the O state is affected by the sodium concentration and it is increased with an increase of sodium concentration, implying that Na^+ uptake

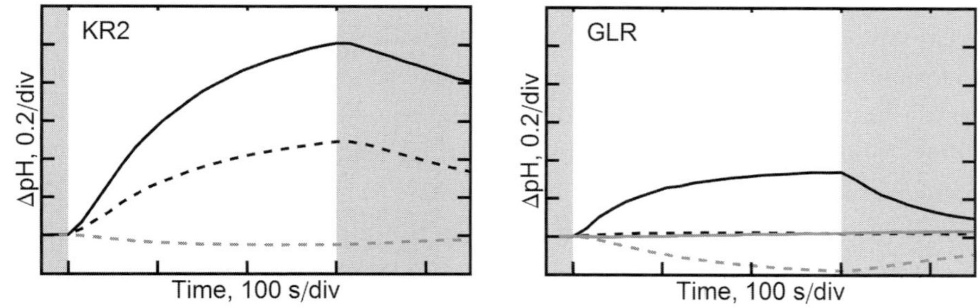

Figure 6.4 *Escherichia coli* activity tests of KR2 and GLR. The pH changes upon illumination in the media containing KR2 or GLR-expressing *E. coli* cells. The solutions contain 100 mM NaCl (black, dashed), 100 mM NaCl and 30 μM carbonylcyanide m-chlorophenylhydrazone (CCCP) (black, solid), 100 mM KCl (gray, dashed) and 100 mM KCl and 30 μM CCCP (gray, solid). The cells were illuminated for 300 s (light areas on the plots).

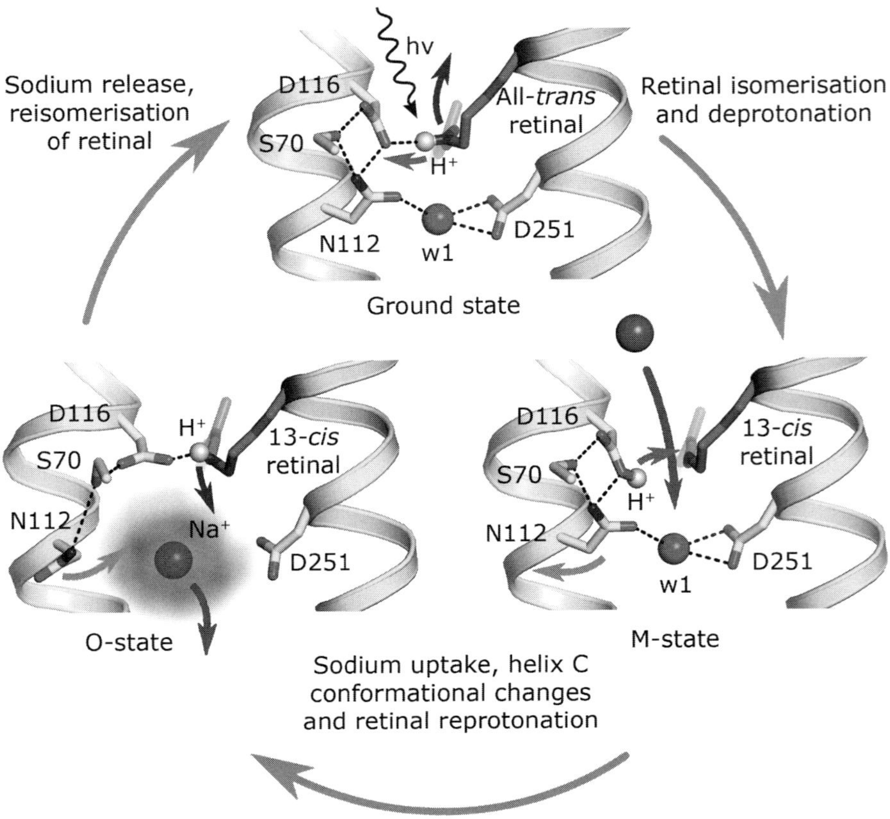

Figure 6.5 Proposed model of the structural changes during sodium translocation. (A black-and-white version of this figure will appear in some formats. For the color version, please refer to the plate section.)

occurs in the M-to-O transition. A natural competitive model was recently proposed for the Na^+/H^+ uptake for KR2. According to the presented kinetic analysis, the H^+/Na^+ selectivity is determined by the ion uptake rate constants k_H/K_{Na} and is approximately 7000–9000 at a neutral pH. This means that under physiological conditions KR2 will pump sodium much more efficiently than protons (Kato Y. *et al.*, 2015).

Crystallographic structures of KR2 together with spectroscopic data allowed us to formulate a hypothesis on the mechanism of the sodium translocation in the family of sodium-pumping rhodopsins (Figure 6.5) (Gushchin *et al.*, 2015a). Briefly, absorption of a photon by the retinal leads to its isomerization and consequent deprotonation. Then, in the M state to O state transition, the Schiff base cavity expands and the sodium ion enters the pump. During the relaxation of the O state to the ground state, the ion is released into the extracellular space and the Schiff base contracts to the resting conformation. The directionality of the sodium translocation is achieved via the proton residing at the Asp-116 side-chain that prevents the sodium from exiting into the cytoplasm.

The proposed mechanism is still under debate, and the experience obtained with BR may be valuable in this respect. BR is the most heavily studied rhodopsin protein and is regarded as a reference model for the family. The enormous scientific efforts and critical analysis applied to BR resulted in a unified view on the mechanism of unidirectional proton pumping (Wickstrand *et al.*, 2015), which is not consistent with all of the experimental data. The unified view gave answers to two major questions: (1) what forces a proton to move uphill against a transmembrane potential? And (2) what ensures the unidirectionality of the proton transfer? The state-of-the-art answer to these questions is that structural changes caused by retinal isomerization affect the local molecular surrounding of the proton translocation pathway groups (mostly the Schiff base and Asp-85), which changes their pKa, and the change of proton affinity drives proton translocation. At the same time, proton accessibility to the Schiff base switches from the extracellular to the cytoplasmic side and back during the photocycle, which ensures the unidirectionality of the proton transfer.

The KR2 crystal structure shows a putative ion translocation pathway in fine detail. However, the exact force that drives the ions and the mechanism for preventing backward ion leakage are still to be elucidated. Clear answers, just like those in the BR case, are important for rational manipulations of KR2 optogenetic properties.

To completely characterize KR2 and understand the possible mechanism of its function, we discuss in the next section an interesting phenomenon – oligomerization of KR2, which seems to be coupled with the switch from H^+ to Na^+ pumping.

6.5 Oligomeric States of KR2

KR2 can form both a monomeric state at low pH (around 4.3) and a pentameric state at higher pH while crystallizing *in meso*. It also appears as a pentamer in detergent when solubilized and purified by size-exclusion chromatography (SEC) at higher pH values, but comes in a variety of forms at low pH values (Figure 6.6). Because no sodium pumping activity has been observed at pH values under 5.0, we tried to determine whether the oligomeric state directly affects ion transport.

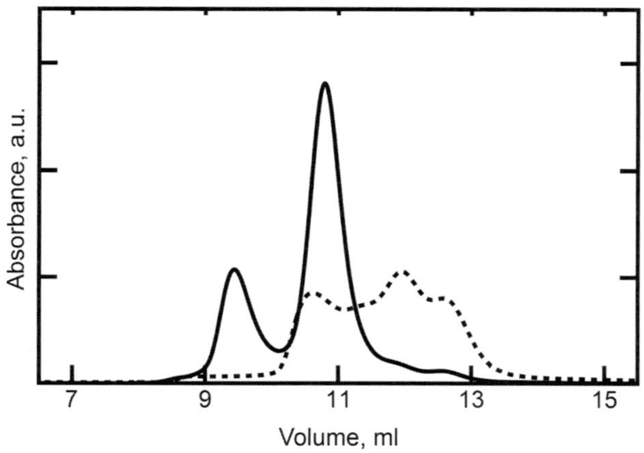

Figure 6.6 Size-exclusion chromatography of wild-type KR2 at pH 7.5 (solid) and pH 4.6 (dashed).

Crystal structures of KR2 show that in the red pentameric state, the His-30 residue of helix A can form hydrogen bonds with the Tyr-154 residue of helix D of a nearby protomer. To break this bond, H30A, H30L and Y154F mutants were expressed in *E. coli*. The H30L mutant showed no sodium pumping activity in cell suspensions at neutral pH values, nor did it form pentamers when purified. Pentamer formation is completely abolished by introducing amino acids with large side-chains and removing charge. The Y154F and H30A mutants displayed observable sodium pumping activity, although this was reduced in comparison to that of the wild-type protein (Figure 6.7). In SEC, these mutants partially appear as pentamers, thereby explaining the lower transport activity.

As histidine residues have relatively low pKa values and can undergo changes in their biochemical properties even if pH changes are small, we prepared a H30K mutant in order to make the protein less sensitive to proton concentrations at pH values of approximately 7. The H30K mutant showed the same sodium pumping activity as the Y154F mutant (Figure 6.7). SEC at physiological pH showed that the pentameric state fraction is even lower. The protein is found only in a monomeric state when crystallized *in meso* at pH 8.0 or lower (crystal packing is similar to one of the wild type in the blue monomeric form (PDB ID 4XTL). Its structure is almost the similar as that of wild-type KR2 in the blue state. This indicates that the "switch" between oligomeric states happens at higher pH values as compared to the wild-type protein, as was expected.

6.6 Optogenetic Applications of the Sodium Pump and Structure-based Engineering of the Light-driven Potassium Pump

Gordeliy and Kandori's groups suggested that KR2 could work as an inhibitory optogenetics tool in heterologous systems. Indeed, Kato *et al.* have demonstrated successful application of a eubacterial rhodopsin (KR2) in optogenetics (Kato, H.E. *et al.*, 2015). In contrast to halorhodopsin and archaearhodopsin-3, KR2 is not expected to induce unintentional side effects, such as the activation of pH-sensitive channels. It has been argued that the flow of Na^+ would be more physiological than that of proton or chloride ions. It is also natural that the voltage-independent photocurrent of KR2 would enable efficient sodium flux, even under highly negative membrane potential conditions. KR2 may have the potential to become an ideal tool in many optogenetics experiments.

As we mentioned previously, an interesting feature of KR2 is the unusual structure of the inward-facing ion uptake cavity, which is unexpectedly large and protrudes from the protein surface, suggesting that this structure could act as a filter that allows KR2 to be selective for sodium ions. We swapped specific amino acids at the relevant site using targeted mutations and found that KR2 indeed loses its sodium-pumping ability under such conditions. However, we also found that one of the mutations (G263F) transforms KR2 into a light-driven potassium pump – the first of its kind. This result is especially interesting in terms of potential optogenetic applications because transporting potassium ions from the cell is a natural neuron deactivation

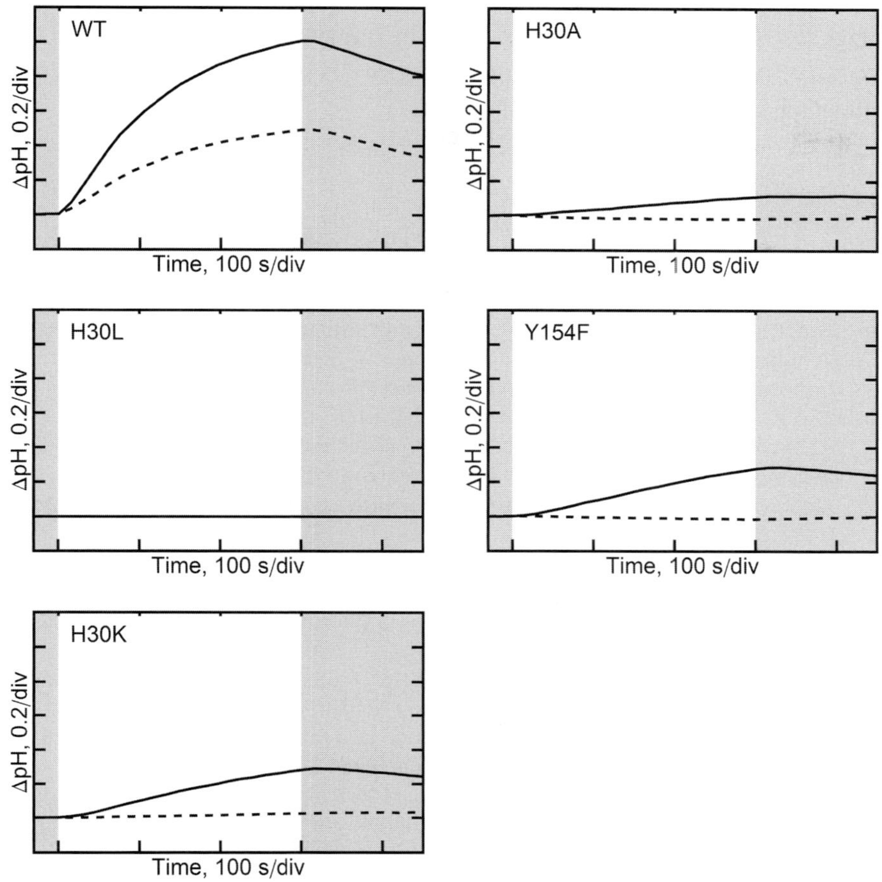

Figure 6.7 *Escherichia coli* activity tests of KR2 mutants related to pentamerization. pH changes upon illumination in the media containing KR2 or GLR-expressing *E. coli* cells. The solutions contain 100 mM NaCl (black, dashed) and 100 mM NaCl and 30 μM CCCP (black, solid). The cells were illuminated for 300 s (light areas on the plots).

mechanism. Normally, activated neurons release the ions through passive potassium channels in the membrane, but a light-activated potassium pump would enable this process to be precisely controlled; therefore, KR2 provides a highly effective off switch for neurons (Kato, H.E. *et al.*, 2015). The next step is to find ways of integrating the pump into different types of cells. The light-activated channelrhodopsin 2, which is a well-known molecular off switch, in combination with the KR2 potassium pump would form a perfect pair of tools for the precise control of nerve cell activity.

ACKNOWLEDGMENTS

The work was supported by the CEA(IBS)–HGF(FZJ) STC 5.1 specific agreement, the 5top100-program of the Ministry for Science and Education of the Russian Federation. The work was also supported by FRISBI (ANR-10-INSB-05-02) and

GRAL (ANR-10-LABX-49-01) within the Grenoble Partnership for Structural Biology (PSB). The work was performed in the framework of ERA.Net RUS PLUS grant (ID 323, Russian Federal Target Program "Research and Development" contract 14.587.21.0011, RFMEFI58715X0011).

REFERENCES

Balashov, S.P., Imasheva, E.S., Boichenko, V.A., Antón, J., Wang, J.M., Lanyi, J.K., (2005). Xanthorhodopsin: a proton pump with a light-harvesting carotenoid antenna. *Science* **309**, 2061–2064.

Balashov, S.P., Imasheva, E.S., Dioumaev, A.K., Wang, J.M., Jung, K.-H., Lanyi, J.K., (2014). Light-driven Na^+ pump from *Gillisia limnaea*: a high-affinity Na^+ binding site is formed transiently in the photocycle. *Biochem. (Mosc.)* **53**, 7549–7561.

Balashov, S.P., Petrovskaya, L.E., Imasheva, E.S., Lukashev, E.P., Dioumaev, A.K., Wang, J. M., Sychev, S.V., Dolgikh, D.A., Rubin, A.B., Kirpichnikov, M.P., Lanyi, J.K., (2013). Breaking the carboxyl rule lysine 96 facilitates reprotonation of the Schiff base in the photocycle of a retinal protein from *Exiguobacterium sibiricum*. *J. Biol. Chem.* **288**, 21254–21265.

Béjà, O., Aravind, L., Koonin, E.V., Suzuki, M.T., Hadd, A., Nguyen, L.P., Jovanovich, S.B., Gates, C.M., Feldman, R.A., Spudich, J.L., Spudich, E.N., DeLong, E.F., (2000). Bacterial rhodopsin: evidence for a new type of phototrophy in the sea. *Science* **289**, 1902–1906.

Bertsova, Y.V., Bogachev, A.V., Skulachev, V.P., (2015). Proteorhodopsin from *Dokdonia* sp. PRO95 is a light-driven Na^+-pump. *Biochem. Mosc.* **80**, 449–454.

Bogomolni, R.A., Spudich, J.L., (1982). Identification of a third rhodopsin-like pigment in phototactic *Halobacterium halobium*. *Proc. Natl. Acad. Sci.* **79**, 6250–6254.

de la Torre, J.R. Christianson, L.M., Béjà, O., Suzuki, M.T., Karl, D.M., Heidelberg, J., DeLong, E.F., (2003). Proteorhodopsin genes are distributed among divergent marine bacterial taxa. *Proc. Natl. Acad. Sci.* **100**, 12830–12835.

DeLong, E.F., Karl, D.M., (2005). Genomic perspectives in microbial oceanography. *Nature* **437**, 336–342.

Ernst, O.P., Lodowski, D.T., Elstner, M., Hegemann, P., Brown, L.S., Kandori, H., (2014). Microbial and animal rhodopsins: structures, functions, and molecular mechanisms. *Chem. Rev.* **114**, 126–163.

Gordeliy, V.I., Labahn, J., Moukhametzianov, R., Efremov, R., Granzin, J., Schlesinger, R., Büldt, G., Savopol, T., Scheidig, A.J., Klare, J.P., Engelhard, M., (2002). Molecular basis of transmembrane signalling by sensory rhodopsin II–transducer complex. *Nature* **419**, 484–487.

Greene, R.V., Lanyi, J.K., (1979). Proton movements in response to a light-driven electrogenic pump for sodium ions in *Halobacterium halobium* membranes. *J. Biol. Chem.* **254**, 10986–10994.

Gushchin, I., Chervakov, P., Kuzmichev, P., Popov, A.N., Round, E., Borshchevskiy, V., Ishchenko, A., Petrovskaya, L., Chupin, V., Dolgikh, D.A., Arseniev, A.S., Kirpichnikov, M., Gordeliy, V., (2013). Structural insights into the proton pumping by unusual proteorhodopsin from nonmarine bacteria. *Proc. Natl. Acad. Sci.* **110**, 12631–12636.

Gushchin, I., Shevchenko, V., Polovinkin, V., Borshchevskiy, V., Buslaev, P., Bamberg, E., Gordeliy, V., (2015a). Structure of the light-driven sodium pump KR2 and its implications for optogenetics. *FEBS J.* **283**, 1232–1238.

Gushchin, I., Shevchenko, V., Polovinkin, V., Kovalev, K., Alekseev, A., Round, E., Borshchevskiy, V., Balandin, T., Popov, A., Gensch, T., Fahlke, C., Bamann, C., Willbold, D., Büldt, G., Bamberg, E., Gordeliy, V., (2015b). Crystal structure of a light-driven sodium pump. *Nat. Struct. Mol. Biol.* **22**, 390–395.

Han, X., Chow, B.Y., Zhou, H., Klapoetke, N.C., Chuong, A., Rajimehr, R., Yang, A., Baratta, M.V., Winkle, J., Desimone, R., Boyden, E.S., (2011). A high-light sensitivity optical neural silencer: development and application to optogenetic control of non-human primate cortex. *Front. Syst. Neurosci.* **5**, 18.

Henderson, R., Baldwin, J.M., Ceska, T.A., Zemlin, F., Beckmann, E., Downing, K.H., (1990). Model for the structure of bacteriorhodopsin based on high-resolution electron cryo-microscopy. *J. Mol. Biol.* **213**, 899–929.

Inoue, K., Konno, M., Abe-Yoshizumi, R., Kandori, H., (2015). The role of the NDQ motif in sodium-pumping rhodopsins. *Angew. Chem.* **127**, 11698–11701.

Inoue, K., Ono, H., Abe-Yoshizumi, R., Yoshizawa, S., Ito, H., Kogure, K., Kandori, H., (2013). A light-driven sodium ion pump in marine bacteria. *Nat. Commun.* **4**, 1678.

Ivanova, N., Rohde, C., Munk, C., Nolan, M., Lucas, S., Del Rio, T.G., Tice, H., Deshpande, S., Cheng, J.-F., Tapia, R., Han, C., Goodwin, L., Pitluck, S., Liolios, K., Mavromatis, K., Mikhailova, N., Pati, A., Chen, A., Palaniappan, K., Land, M., Hauser, L., Chang, Y.-J., Jeffries, C.D., Brambilla, E., Rohde, M., Göker, M., Tindall, B.J., Woyke, T., Bristow, J., Eisen, J.A., Markowitz, V., Hugenholtz, P., Kyrpides, N.C., Klenk, H.-P., Lapidus, A., (2011). Complete genome sequence of *Truepera radiovictrix* type strain (RQ-24T). *Stand. Genomic Sci.* **4**, 91–99.

Kanada, S., Takeguchi, Y., Murakami, M., Ihara, K., Kouyama, T., (2011). Crystal structures of an O-like blue form and an anion-free yellow form of *pharaonis* halorhodopsin. *J. Mol. Biol.* **413**, 162–176.

Kato, H.E., Inoue, K., Abe-Yoshizumi, R., Kato, Y., Ono, H., Konno, M., Hososhima, S., Ishizuka, T., Hoque, M.R., Kunitomo, H., Ito, J., Yoshizawa, S., Yamashita, K., Takemoto, M., Nishizawa, T., Taniguchi, R., Kogure, K., Maturana, A.D., Iino, Y., Yawo, H., Ishitani, R., Kandori, H., Nureki, O., (2015). Structural basis for Na^+ transport mechanism by a light-driven Na^+ pump. *Nature* **521**, 48–53.

Kato, Y., Inoue, K., Kandori, H., (2015). Kinetic analysis of H^+–Na^+ selectivity in a light-driven Na^+-pumping rhodopsin. *J. Phys. Chem. Lett.* **6**, 5111–5115.

Kolbe, M., Besir, H., Essen, L.-O., Oesterhelt, D., (2000). Structure of the light-driven chloride pump halorhodopsin at 1.8 Å resolution. *Science* **288**, 1390–1396.

Konno, M., Kato, Y., Kato, H.E., Inoue, K., Nureki, O., Kandori, H., (2016). Mutant of a light-driven sodium ion pump can transport cesium ions. *J. Phys. Chem. Lett.* **7**, 51–55.

Kouyama, T., Kanada, S., Takeguchi, Y., Narusawa, A., Murakami, M., Ihara, K., (2010). Crystal structure of the light-driven chloride pump halorhodopsin from *Natronomonas pharaonis*. *J. Mol. Biol.* **396**, 564–579.

Kwon, Y.M., Kim, S.-Y., Jung, K.-H., Kim, S.-J., (2016). Diversity and functional analysis of light-driven pumping rhodopsins in marine *Flavobacteria*. *Microbiologyopen* **5**, 212–223.

Lindley, E.V., MacDonald, R.E., (1979). A second mechanism for sodium extrusion in *Halobacterium halobium*: a light-driven sodium pump. *Biochem. Biophys. Res. Commun.* **88**, 491–499.

Luecke, H., Schobert, B., Richter, H.-T., Cartailler, J.-P., Lanyi, J.K., (1999). Structure of bacteriorhodopsin at 1.55 Å resolution 1. *J. Mol. Biol.* **291**, 899–911.

Luecke, H., Schobert, B., Stagno, J., Imasheva, E.S., Wang, J.M., Balashov, S.P., Lanyi, J.K., (2008). Crystallographic structure of xanthorhodopsin, the light-driven proton pump with a dual chromophore. *Proc. Natl. Acad. Sci.* **105**, 16561–16565.

MacDonald, R.E., Greene, R.V., Clark, R.D., Lindley, E.V., (1979). Characterization of the light-driven sodium pump of *Halobacterium halobium*. *J Biol Chem* **254**, 11831–11838.

Mongodin, E.F., Nelson, K.E., Daugherty, S., DeBoy, R.T., Wister, J., Khouri, H., Weidman, J., Walsh, D.A., Papke, R.T., Perez, G.S., Sharma, A.K., Nesbø, C.L., MacLeod, D., Bapteste, E., Doolittle, W.F., Charlebois, R.L., Legault, B., Rodriguez-Valera, F., (2005). The genome of *Salinibacter ruber*: convergence and gene exchange among hyperhalophilic bacteria and archaea. *Proc. Natl. Acad. Sci.* **102**, 18147–18152.

Moukhametzianov, R., Klare, J.P., Efremov, R., Baeken, C., Göppner, A., Labahn, J., Engelhard, M., Büldt, G., Gordeliy, V.I., (2006). Development of the signal in sensory rhodopsin and its transfer to the cognate transducer. *Nature* **440**, 115–119.

Nagel, G., Szellas, T., Huhn, W., Kateriya, S., Adeishvili, N., Berthold, P., Ollig, D., Hegemann, P., Bamberg, E., (2003). Channelrhodopsin-2, a directly light-gated cation-selective membrane channel. *Proc. Natl. Acad. Sci.* **100**, 13940–13945.

Nakanishi, M., Meirelles, P., Suzuki, R., Takatani, N., Mino, S., Suda, W., Oshima, K., Hattori, M., Ohkuma, M., Hosokawa, M., Miyashita, K., Thompson, F.L., Niwa, A., Sawabe, T., Sawabe, T., (2014). Draft genome sequences of marine flavobacterium nonlabens strains NR17, NR24, NR27, NR32, NR33, and Ara13. *Genome Announc.* **2**, e01165-14.

Oesterhelt, D., Stoeckenius, W., (1971). Rhodopsin-like protein from the purple membrane of *Halobacterium halobium. Nature* **233**, 149–152.

O'Malley, M.A., (2007). Exploratory experimentation and scientific practice: metagenomics and the proteorhodopsin case. *Hist. Philos. Life Sci.* **29**, 337–360.

Riedel, T., Held, B., Nolan, M., Lucas, S., Lapidus, A., Tice, H., Del Rio, T.G., Cheng, J.-F., Han, C., Tapia, R., Goodwin, L.A., Pitluck, S., Liolios, K., Mavromatis, K., Pagani, I., Ivanova, N., Mikhailova, N., Pati, A., Chen, A., Palaniappan, K., Land, M., Rohde, M., Tindall, B.J., Detter, J.C., Göker, M., Bristow, J., Eisen, J.A., Markowitz, V., Hugenholtz, P., Kyrpides, N.C., Klenk, H.-P., Woyke, T., (2012). Genome sequence of the Antarctic rhodopsins-containing flavobacterium *Gillisia limnaea* type strain (R-8282(T)). *Stand. Genomic Sci.* **7**, 107–119.

Schobert, B., Lanyi, J.K., (1982). Halorhodopsin is a light-driven chloride pump. *J. Biol. Chem.* **257**, 10306–10313.

Singh, A., Kumar Jangir, P., Sharma, R., Singh, A., Kumar Pinnaka, A., Shivaji, S., (2013). Draft genome sequence of *Indibacter alkaliphilus* strain LW1T, isolated from Lonar Lake, a haloalkaline lake in the Buldana District of Maharashtra, India. *Genome Announc.* **1**, e00515-13.

Stoeckenius, W., Lozier, R.H., Bogomolni, R.A., (1979). Bacteriorhodopsin and the purple membrane of halobacteria. *Biochim. Biophys. Acta BBA Rev. Bioenerg.* **505**, 215–278.

Wickstrand, C., Dods, R., Royant, A., Neutze, R., (2015). Bacteriorhodopsin: would the real structural intermediates please stand up? *Biochim. Biophys. Acta* **1850**, 536–553.

Yoshizawa, S., Kumagai, Y., Kim, H., Ogura, Y., Hayashi, T., Iwasaki, W., DeLong, E.F., Kogure, K., (2014). Functional characterization of *Flavobacteria* rhodopsins reveals a unique class of light-driven chloride pump in bacteria. *Proc. Natl. Acad. Sci.* **111**, 6732–6737.

Zhao, S., Cunha, C., Zhang, F., Liu, Q., Gloss, B., Deisseroth, K., Augustine, G.J., Feng, G., (2008). Improved expression of halorhodopsin for light-induced silencing of neuronal activity. *Brain Cell Biol.* **36**, 141–154.

7 Methods for Simultaneous Electrophysiology and Optogenetics *In Vivo*

Yonatan Katz and Ilan Lampl

7.1 Introduction

For hundreds of years, humanity has tried to understand the nervous system. Since the pioneering studies of Galvani (Galvani *et al.*, 1791) in electrophysiology, in which he developed the theory of electrical excitation of neurons while studying the frog muscles, great advancement in techniques, methods and knowledge have brought us closer to understanding the mechanisms governing neuronal activity. Optogenetics (OG), a method to control neuronal activity by light, was revolutionized a decade ago in Karl Deisseroth's laboratory (Deisseroth *et al.*, 2006). This approach had a huge impact on neuroscience by enabling the manipulation of specific types of neurons in space and time. Since it has been introduced, thousands of papers using OG have been published. At the core of OG, a light-sensitive molecule, opsin, is coupled to an ion channel or pump and, upon exposure to light, ion flow through the channel or pump changes the membrane potential. Many new optogenetic tools have been developed and include various types of light-activated channels and ion pumps that are sensitive to different wavelengths across the whole visible spectrum (Hegemann and Möglich, 2011). Before OG, neurons were activated mainly by the use of electrical stimulation or pharmacology. The use of electrical stimulation is non-specific and affects many neurons and other cells close to the stimulating electrode, while pharmacology is more specific but has very low time resolution. When using OG, only the neurons expressing the light-sensitive proteins are activated with millisecond resolution (Boyden *et al.*, 2005). Since the expression can be regulated by a certain promotor or enhancer of choice, only the neurons in which the gene of choice is expressed will respond to the light. In this way, OG enables us to selectively manipulate subpopulations of cells based on genetic markers.

Since it has been introduced, OG has been combined with electrophysiology in order to evaluate the effects on neuronal activity. Since the field of electrophysiology spans many different preparations, from tissue culture to behaving animals, different technical approaches had to be incorporated or invented when combined with OG. This chapter reviews the current methods for simultaneous

OG and electrophysiology (SOGEP) *in vivo*. Depending on the electrophysiological requirements, different probes were developed, mainly for extracellular, but also for intracellular, recordings.

The initial probes were made of a single optical fiber glued to a metal electrode in anesthetized animals (Nakamura *et al.*, 2012). Later, multi-contact silicon probes were developed and integrated with optical fibers. Since one of the experimenter's goals is to collect data and to minimally interfere with animal behavior, the preparation of choice in order to obtain high-quality data is freely behaving animals. Indeed, the growing use of electrophysiological experiments in behaving animals increased the need to tether light delivery. Such technology for integrating the light source within the brain was recently developed and enabled tethered experiments in behaving animals. The technology for performing such experiments is demanding, but it is developing fast and is already available on the shelf.

7.2 *In Vivo* Electrophysiology

Neuronal activity can be monitored in various techniques such as functional magnetic resonance imaging, positron emission tomography, intrinsic signal imaging, voltage-sensitive dyes, calcium indicators and others. Yet electrophysiology remains one of the common and deep-rooted methods used in neuroscience. It enables single-cell resolution, but also the probing of network activity. The electrodes are abundant and relatively cheap. Signal recording is immediate and does not require long procedures such as waiting for viral expression or loading tissues with indicators. Most importantly, this is the most direct method for probing neuronal activity. Today, *in vivo* electrophysiology is mainly governed by extracellular recordings, which are easier to achieve than intracellular recording, particularly in freely moving animals. Extracellular recording probes come in different flavors, from single metal electrodes to twisted wire tetrodes and up to multi-electrode arrays (MEAs) with multiple contacts (many channels). All of them contain a conducting contact to receive the electrical signals in their surroundings. Since, in most instances, action potential signals, rather than local field potentials, are required for the question in hand, the recording electrode has to be very close to the recorded neuron. Due to the dense arrangement of neurons within the brain, reliable identification of more than one neuron requires the use of more than one electrode. Recording from several sites in very close proximity to each other enables us to differentiate between spikes generated by different neurons. This method is based on the fact that spikes of certain neurons will have a different shape in each recording site due to changes in the signal's amplitude as a function of distance and direction (Drake *et al.*, 1988). Although, theoretically, wire tetrodes can receive signals from a radius larger than 100 μm, which can contain 1000 neurons (Henze *et al.*, 2000), they can differentiate reliably only about ten neurons due to the low firing rates and small amplitudes of the signals (Tolhurst *et al.*, 2009; Clancy *et al.*, 2015). New developments of miniature silicon probes reduce tissue damage and have more recording contacts per unit area (Kipke *et al.*, 2008). They are formed by lithographic patterning, which enables us to produce the probe in almost any shape to specifically suit the experimental design and recorded tissue properties.

Intracellular recording *in vivo* occupies a smaller subfield of electrophysiology and recently it has gained increasing popularity. This method gives a direct insight into the subthreshold activity of neurons. In some neuronal tissues, such as the cortex, neurons receive inputs from a large number of nearby neurons. Therefore, this method statistically probes the activity of a large population of functional neurons whose activity shapes the recorded synaptic inputs. Another important property of intracellular recordings is their specificity. There are many types of neurons with different functional specializations, potentially having opposite roles (e.g. excitatory vs. inhibitory); one must take extreme care when averaging data from heterogeneous populations. Due to activity specificity, collecting data from thousands of neurons without the ability to sort them into functional groups might bias the conclusions. Hence, a trade-off between quality and quantity should be taken into consideration. Histological identification of the neuron using intracellular recording enables functional cellular identification and the ability to group similar neuronal populations. Intracellular recording is done using sharp or patch pipettes, but intracellular recordings combined with OG are dominated by the patch clamp method (Jouhanneau *et al.*, 2014; Pala and Petersen, 2015; Schiemann *et al.*, 2015) due to its greater stability of recording, especially in behaving animals where neuron–electrode contact should be maintained during the animal's movements.

7.3 OG: Controlling the Brain by Light

In 2005, two seminal articles (Boyden *et al.*, 2005; Lima and Miesenbock, 2005) launched the era of OG. Miesenbock and colleagues have been striving for years to remotely control behavior through genetically targeted photostimulation of neurons (Miesenböck and Rothman, 1997; Miesenböck *et al.*, 1998; Zemelman *et al.*, 2002; Lima and Miesenböck, 2005); however, the main catalyst for OG came from Deisseroth and colleagues, who successfully modified and functionally expressed channelrhodopsin-2 (ChR2), a single light-sensitive, non-specific cation channel from algae, in neurons. Upon activation by light, the cation channel opens and the neuron is depolarized (Boyden *et al.*, 2005). Since then, researchers have developed and modified many light-activated proteins in order to control neuronal activity. Different properties were modified according to necessities such as conductance, spectral sensitivity, expression level, ion selectivity, membrane trafficking and kinetics (Zhang *et al.*, 2008; Berndt *et al.*, 2009; Lin *et al.*, 2009; Chow *et al.*, 2010; Han *et al.*, 2011; Lin *et al.*, 2013; Sudo *et al.*, 2013; Hochbaum *et al.*, 2014; Klapoetke *et al.*, 2014). In order to activate the light-driven proteins, a strong light source is required.

7.3.1 Optogenetic Light Sources

The two main light sources used for OG are lasers and LEDs, but other sources such as arc lamps are also used. Optical fibers can potentially be connected to any light source, but one should carefully control the light power. On the one hand, we prefer to expose the tissue to as little light power as possible in order to reduce damage by heating, but on the other hand, we should supply sufficient light power to reliably activate the opsins. Standard scientific white light sources (halogen, xenon and mercury) can easily drive expressed opsins, but since light should be band-passed in order to reduce the total power, the right filter should be mounted. Furthermore, a shutter is needed in order to control exposure time and the light should be coupled into the optical fiber, creating complex and expensive systems that are not common to *in vivo* OG. LEDs and lasers are widely used due to their simplicity, effectiveness and cost. Their disadvantage is having a narrow spectral bandwidth, which necessitates two different devices in order to independently activate two spectrally different opsins. Laser power is much higher than that of LEDs, which are usually too weak when coupled with small fibers (<100 μm). Thus, when optical fiber illumination is used in anesthetized, restrained or tethered animal, lasers will give more experimental options than LEDs (Wang *et al.*, 2015).

There are two main methods of introducing light into the tissue: (1) using a distant light source that requires the delivery of light to the tissue, usually through an optical fiber; or (2) using an in-site light source (such as an LED), which can be implanted (Kim *et al.*, 2013) or can directly illuminate the cortical surface (Huber *et al.*, 2008; Ruiz *et al.*, 2013). In the first method, the optical fiber from the light source should be linked to the recording probe. The regular coupling is via a second optical fiber integrated or glued to the recording probe. Commercial optical fibers for introducing the light into the tissue are already available, but their variety is still limited, necessitating specifically fabricated fibers. An example of a specifically fabricated probe is the multi-waveguide probe (Zorzos *et al.*, 2010). This probe can accept light from a set of sources and guides the light to a tiny mirror that reflects the light into the tissue (Figure 7.1A).

An array of such probes is used for 3D waveguide arrays that use a scanning mirror or digital micro-mirror device in order to couple light in a wide brain volume (~4 × 4mm) with any desired temporal pattern (Figure 7.1B) (Zorzos *et al.*, 2012).

In the second method, a combined light source with a recording electrode is implanted into the desired optostimulation location. One example is the micro-LED (μLED), a tiny LED that can be inserted into the brain. This is an evolving technology with high light intensity, which can be applied in all LED materials (colors). The use of μLED for OG was recently established (Cao *et al.*, 2013) and this promising technology will be widely used in freely behaving animals. In this method, light stimulation is coupled to the recording electrode, hence photostimulating neurons in the vicinity of the recording. Unlike standard OG recordings where the light comes from a separate, distant probe, in this case, light penetration does not depend on light scattering and absorption, since it is located within the tissue. Initially, the μLED probe width was huge (~900 μm) and surely deformed the surrounding tissue; however, a recently fabricated probe (McAlinden *et al.*, 2015) integrated with extracellular recording contacts has comparable size to commercially used extracellular probes (150 μm wide).

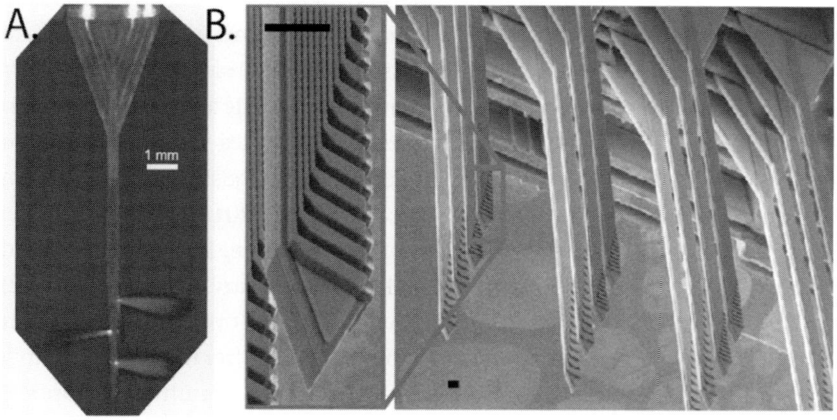

Figure 7.1 (A) Waveguide probe containing 12 channels. Each channel guides the light along the shank through the bends down to the corner mirror that deflects the light into the tissue. In this case, different light sources are coupled to the probe, showing emissions of 473nm (right) and 632nm (left) light out of three separate ports. Scale bar: 1 mm. (B) Scanning electron micrograph of an assembled 3D waveguide array (similar to [A]) with a zoomed-in view of the output apertures. Output apertures shown here are 9 μm × 30 μm. Scale bar: 100 μm. Adapted from Zorzos *et al.*, 2010, 2012, with permission.

7.4 Combined Simultaneous OG and Electrophysiology

The success of OG drove the development of different tools for introducing light into the tissue using different recording probes. SOGEP recordings *in vivo* (Gradinaru *et al.*, 2007) were started shortly after ChR2 was expressed in neurons (Boyden *et al.*, 2005). SOGEP can be used to identify specific genetically tagged neurons and to characterize them physiologically.

Generally, there are few options when recording and stimulating in the same location. The first basic optrode consisted of a single optical fiber glued to a single metal electrode; other options consist of any configuration between the basic optrode and MEA (silicon probes) with single or multiple light guides. During SOGEP, the light should activate neurons proximate to the electrode. For a few years now, the basic optrodes (and others SOGEP probes) have been commercially available (e.g. FHC and Thomas RECORDING). If one wishes to reduce expense, an even simpler configuration is possible during superficial cortical recordings of up to 1 mm by using a bare optical fiber directly illuminating the area of recording (the correct power can be roughly calculated at http://web.stanford.edu/group/dlab/cgi-bin/graph/chart.php). However, photostimulating deeper brain structures requires a reliable light delivery method. Light can be delivered via coupled optical fibers or via a waveguide deposited onto the recording probe using various techniques (e.g. photoresist deposition) (Cao *et al.*, 2013; Rubehn *et al.*, 2013). Both recording and optical stimulation probes require space, and a compromise should be made when designing a combined probe in order to minimize tissue damage.

When testing the effects of OG using a MEA probe, light diffraction and scattering by the tissue should be taken into consideration (Favre-Bulle *et al.*, 2015). If a single light guide is used within a square MEA (Wang *et al.*, 2012), the light does not evenly modulate neuronal activity across the probe and will strongly affect electrodes close to the light source (Figure 7.2B). Such a probe can be useful for following the propagation of neuronal waves (e.g. in epileptic activity). In order to achieve better control of light stimulation, the light guides can be attached to planar electrode arrays (these probes are now commercially available from Neuronexus) (Figure 7.2A) (Royer *et al.*, 2010; Stark *et al.*, 2012); in these cases, a few obstacles had to be overcome when shaping the probe. Initially, an optical cannula that has a single diode-fiber assembly was necessary. Stark *et al.* (2012) used two manipulators in order to position the LED and the optical fiber such that light output is optimal in close proximity to the electrode; after that, the assembly was glued together, grounded and shielded. They used small multi-mode optical fibers (50 μm), which were glued on top of each electrode shaft in order to reduce tissue damage. In this way, not only did each recording shank have its own light guide, but also each shank could have a different light source.

Light stimulation in proximity to the recording electrode was also solved (for extracellular recordings) in another "reversed" approach; in this case, the core of the optrode was composed of an optical fiber that tappers down to a few microns at the tips. The fiber is coated with a conductive material that is electrically isolated by epoxy (Figure 7.3B, left) (Zhang *et al.*, 2009). Since the conductor surrounds the optical fiber, the tip size dictates the sizes of both the light source and the metal electrode. In this case, a very small aperture (1 μm) was fabricated in order to achieve point source electrical recording.

7.4.1 Intracellular OG

Many *in vivo* patch clamp recording studies use multi-photon imaging to specifically target subtypes of neuronal populations in behaving mice. Usually, OG is combined with targeted recordings using full-field illumination to activate all opsin-expressing neurons in the recorded area, but sometimes a more specific patterned photostimulation is required. The simplest approach is to target a light beam using a pair of galvanometers. This enables a serial activation of different points in space. A more complex illumination pattern can be induced using a digital light processing (DLP) mirror array (Guo *et al.*, 2009), in which each mirror is independently controlled. Another alternative is to use holographic projection with a spatial light modulator (Reutsky-Gefen *et al.*, 2013), which is more power efficient than DLP. These photostimulations are confined to superficial layers due to the poor penetration of light through tissue. In order to achieve intracellular recording in deep structures while optogenetically stimulating the vicinity of the recorded site, light should be introduced into the tissue and to the exact recording location.

Classic *in vivo* blind patch clamp recordings are widely used in awake mice today, and this method is mostly used in superficial brain structures, such as the cerebral cortex, rather than deep structures, due to the simplicity of light delivery. Superficial SOGEP can be done using an articulated arm to position an optical fiber, but deeper brain recordings are more demanding and the light source must

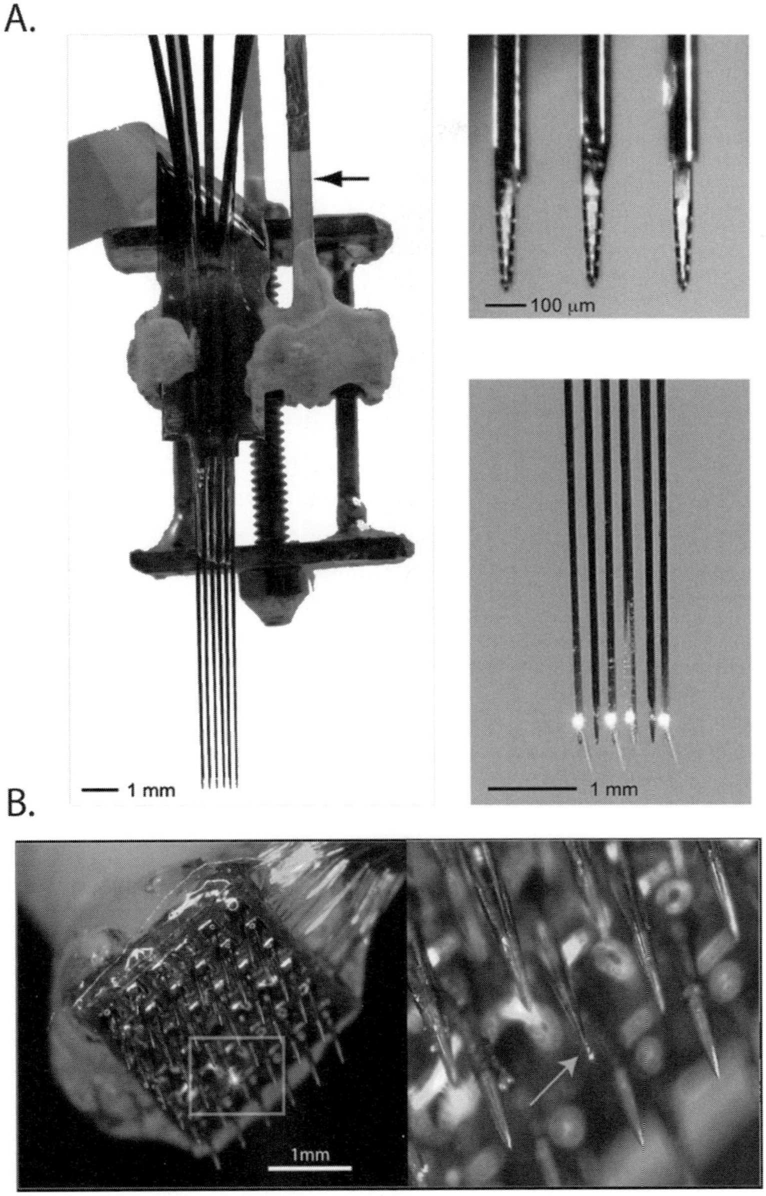

Figure 7.2 OG multi-electrode arrays. (A) Diode probe example. Six diode-fiber assemblies (four blue and two red) attached to separate shanks of a six-shank silicon probe. Left: probe on a movable drive; arrow: metal rod used for connecting the diode ends of the fibers. Top right: magnified frontal view of three shanks. Bottom right: four shanks illuminated with blue light. Adapted from Stark *et al.* (2012), with permission. (B) Optrode MEA. Image of the 6 × 6 multi-electrode array device with one element being replaced by an optrode (square). The spacing between neighbor electrodes is 400 μm and the electrode shank's length is 1 mm. Right: a close-up view of the optrode shows the laser light being emitted from the tip of the optrode (arrow). Adapted from Wang *et al.* (2012), with permission.

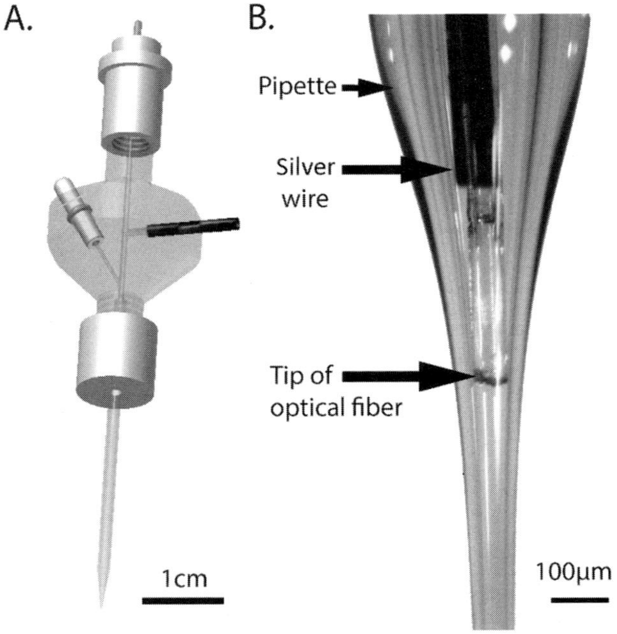

Figure 7.3 Optopatcher. (A) Drawing of the electrode holder. Additional to the regular structure of the intracellular pipette holder, the optopatcher has a port that enables the insertion of an optical fiber into the pipette. (B) The optical fiber is inserted up to the tip of the glass pipette, below the silver wire, in order to reduce photoelectric artifacts. Adapted from Kwon *et al.* (2015), with permission.

be accurately positioned in proximity to the recording pipette. The growing use of OG raised the need for light stimulation of deep structures, which requires a second manipulator and high precision. This issue was solved by inserting the optical fiber into the recording glass pipette (Katz *et al.*, 2013). This configuration ensures that light stimulation will always cover the recording site without the complexity of positioning a second manipulator (Figure 7.4).

Another benefit of this configuration when using OG is its ability to specifically identify depolarizing opsin-expressing neurons before patching (Muñoz *et al.*, 2014). The latency of light-evoked spikes is very short (~3 ms) and it can withstand high-frequency stimulation with reliable responses and little jitter, making it easy to identify an opsin-expressing neuron.

7.4.2 Awake OG

In recent years, there has been an increase in studies performed on awake animals due to technological improvements, but mostly due to the comprehensive understanding that many responses related to the activity of brain circuits are different under anesthesia. When anesthetized, the core neuronal systems that control behavior are out of function, and thus the study of activity related to decision making, for example, has been opted to be performed in awake animals.

There are three main approached to recording activity in awake animals: head restrained, tethered and freely behaving (wireless recordings). The use of each of

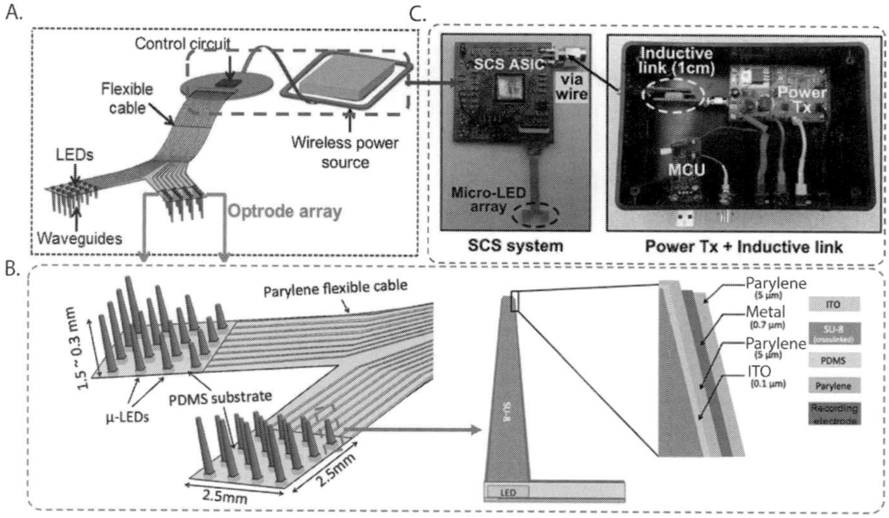

Figure 7.4 Wireless powered optrode array. (A) Conceptual illustration of the wireless optrode neural interface. (B) Overall design of the LED-coupled optrode array with 32 μLEDs, micro-waveguides and microelectrodes integrated on a Parylene-C substrate. Adapted from Katz *et al.* (2013), with permission. (A black-and-white version of this figure will appear in some formats. For the color version, please refer to the plate section.)

these techniques is determined by the research question, which will dictate the experimental platform. The use of head-restrained animals is widespread in monkeys, but since only a few studies of OG in monkeys have been published, this line of experiments will not be described here. In head-restrained animals, a head bar is strongly attached to the skull and a small access window is opened over the cranium for recording. During experiments, the animal is set within a stereotaxic apparatus and the head bar is fastened to it. Head restraining stabilizes the brain and enables intracellular recordings; thus, it is the most common technique for intracellular SOGEP in mice and rats (Gentet *et al.*, 2012; Eggermann *et al.*, 2014; Lee *et al.*, 2014). The main problem is that the animal may be stressed and therefore the range of behavioral tasks might be limited. Currently, head-restrained mice manage to perform a demanding task only under a strict water restriction protocol (Guo *et al.*, 2014), and the most complex training they can learn involves choosing between right or left leak ports (Musall *et al.*, 2014; Ollerenshaw *et al.*, 2014).

Tethered recordings, in which the animals are moving relatively freely in the cage with an implant on their head connected via cables to the recording and stimulating devices, currently dominate the behavioral SOGEP field due to the simplicity of transferring light, electricity and data via cables. This approach enables very complex recording configurations with multiple probes. One example is the "hyperdrive", which consists of an optical fiber surrounded by six mobile tetrodes (Siegle *et al.*, 2011). The system can be mounted for many months and over different brain areas. On the one hand, the advantage of tethered systems is that the controls are away from the animal and hence there are no constraints on the light source or acquisition system. On the other hand, the animal can become entangled in the wires or deliberately disconnect them, as they put strain on the animal's head during movements. Other constraints of tethered devices are the small size of the experimental arena and the limited movement of the animal as it cannot go under any object.

Though animal behavior is more naturalistic in tethered rather than in head-restrained animals, tethering the animal still alters behavior, disrupts movements and potentially confines behavioral experiments. Wireless SOGEP can potentially overcome these problems. Currently, only a few instruments are able for wirelessly recording and delivering photostimulation.

7.4.3 Probes for Freely Behaving Animals

When dealing with freely behaving animals (no tethered cables), all electronic elements and attached hardware (i.e. photostimulation and recording devices) should be minimized in order to have the smallest possible effect on the animal's behavior. Since most of the animals that are currently used for OG are mice, this poses a technical constraint, as the head-mounted apparatus weight should not exceed 10% of the animal's weight.

Due to the miniature size of wireless systems, the light source in all cases will be an LED, which can be mounted on the cortical surface, coupled to an optical fiber or even inserted into the brain. In cortical surface stimulation, the LED is mounted over the area of interest, and often it will be placed over a hole in the skull, though newly developed red-shifted opsins enable transcranial optical activation of neurons in deep brain structures without the need to surgically thin the skull (but other problems are created, such as the broad spread of light and light reduction) (Lin *et al.*, 2013). For deep brain photostimulation, LEDs mounted on top of an optical cannula are widely used (Manita *et al.*, 2015; Sato *et al.*, 2015).

Wireless photostimulation can be powered by battery (Iwai *et al.*, 2011) or wirelessly via magnetic field (Wentz *et al.*, 2011). Battery-powered wireless systems are commercially available from various suppliers, but they are relatively easy to assemble and can be built in the laboratory. The advantage of magnetic field-powered systems is that the LED can be activated repeatedly without the need for replacing the power source. Due to the small size of such devices, which contain coil, circuit and μLED, they can even be implanted subcutaneously in mice (Montgomery *et al.*, 2015).

Immense effort is being invested in building a standalone system that will integrate both OG and recording in a freely behaving animal. Such a system not only needs to withstand the power demands dictated by the OG stimulation power, but also must integrate a high-bandpass telemetry system. Recently, Kim *et al.* (2013) fabricated a wireless system containing a radiofrequency (RF) power module in which they printed μLEDs on thin, flexible, plastic strips that can be positioned at any desired location within a soft tissue. Those cellular-sized optostimulation devices enable minimally invasive operation and localized stimulation. This method enables us to print multiple sensors on to the film, including microelectrodes, thermal sensors and photodetectors. The addition of such probes to the arsenal of the researcher will enable us to explore new fields not yet integrated in biological research using behaving animals. Though only a single probe was introduced in this study, the flexibility of such probes will eventually enable us to introduce many probes and to record and manipulate simultaneously the activity in different brain regions in behaving animals.

Another wireless-powered approach uses an LED array that has multiple optrodes, each surrounded by a recording electrode (Kwon *et al.*, 2015). The wireless interface system is composed of three different units: the μLED array coupled to a waveguide array with conducting electrodes; an inductive link for power transfer to the LEDs; and a wireless switched capacitor-based stimulator that harvests the RF power (Figure 7.3).

A custom-designed computer interface enables the user to control each LED and a commercial neural recording system is used for signal detection. The recording array can be used only on the cortical surface and the depth of the recording is dictated by the length of the waveguides. Use of such a system allows us to obtain valuable data from many probes during behavior, together with OG control. Such research can take place within the animal's home cage and could possibly be used in natural habitats.

7.5 Future Directions

In order to have the means to understand the neuronal code in behaving animals, we need to have the wiring diagram of neuronal connections, the inputs to the neurons (membrane voltages) and the outputs (action potentials). Obtaining such information would require simultaneous intracellular recording from many neurons in different brain areas, together with targeted stimulation of neuronal subpopulations. Single recording electrodes have been used for almost a century, but technological developments have brought us closer to reaching the goal of multi-site cellular-resolution recordings. Since brain activity is sparse and dispersed, the probes should have the ability to be independently positioned in a 3D space. Current methods for intracellular recordings (sharp or patch clamp recordings) are invasive and enable only short recording epochs; different strategies might solve these problems. A novel approach uses nano-devices that are "swallowed" by the neurons, enabling multi-site intracellular recording (Fendyur and Spira, 2012). These devices are plates with protruding gold-plated mushroom electrodes, and neurons engulf the electrodes and strongly adhere to them. This enables reliable extracellular recording. By using an electroporating current pulse, it is possible to obtain temporal intracellular configuration data. Currently, this method has only been applied *in vitro*. Other evolving technologies for intracellular recording are carbon nanotubes. The small sizes of the tubes permit them to invade into the neuron with minimal damage and record electrical signals with sub-millivolt precision (Yoon *et al.*, 2013). Currently, difficulties in assembly and interface limit nano-probes to single electrodes, which constrains the wide use of these tubes. With further improvements and refinements, this technology could promote easy intracellular recording of many neurons. Both of these emerging techniques suffer from technical constraints and have not yet been applied to *in vivo* experiments.

Intracellular probing of multiple neurons can be achieved by imaging. Using voltage-sensitive dyes (Grinvald *et al.*, 1981), one can simultaneously monitor intracellular voltages in many neurons, but due to difficulties in dye loading and bleaching, along with a lack of single-cell resolution *in vivo*, this technology is not widely used in behaving animals. A different technique that is widely used in monitoring intracellular neuronal activity is calcium imaging. Calcium imaging depends upon a fluorescent calcium indicator. The most widely used genetically encoded indicator is the GCaMP (Nakai *et al.*, 2001). Though calcium imaging cannot monitor intracellular membrane voltages, it reliably differentiates between spikes (during low firing rates) at cellular resolution, and it can also simultaneously probe many neurons in space and time (Ahrens *et al.*, 2013). By using red-shifted opsin, which has a separate excitation spectrum, GCaMP can be used simultaneously with OG in order to record neuronal activity *in vivo* in an all-optical strategy (Packer *et al.*, 2015). Calcium imaging can be used *in vivo* to monitor deep brain structures such as the hippocampus (Ghosh *et al.*, 2011). Though hundreds of neurons can be simultaneously imaged through such a tiny microscope, this probe causes extensive damage to all tissues above the hippocampus as they are removed, and no more than one probe can be mounted over a mouse head due to size limitations. There are different methods for monitoring voltage other than GCaMP (FlaSh5, SPARC, VSFP1/2, DPA–diO hybrid, etc.), but the main obstacle is the imaging systems that are needed to gather the information, which are currently too large to be mounted on a mouse head. Currently, the use of combined OG and imaging techniques is steadily increasing due to the ability to collect data from many tagged neurons simultaneously. Since most of the imaging systems are currently too complex for experiments in freely behaving animals, these experiments are mostly done using electrophysiology by exploiting state-of-the-art multi-electrode arrays.

Presently, simultaneous OG and electrophysiology technologies are the most advanced tools for manipulating and collecting neuronal data from any location in the brain. Even though the methods in use are invasive and incorporate tissue damage, no other alternative technique for single-cell resolution activity in freely behaving animals has been developed. State-of-the-art fabrication techniques and electronics enable us to manufacture micron-scale integrated probes that can carry sophisticated sensors in order to collect light, temperature and electrical signals, and they also include actuators in order to deliver light and electrical stimulation. In the near future, SOGEP will flourish and contribute to our understanding of the neuronal activity underlying behavior.

REFERENCES

Ahrens, M.B., Orger, M.B., Robson, D.N., Li, J.M., Keller, P.J., 2013. Whole-brain functional imaging at cellular resolution using light-sheet microscopy. *Nat. Methods* **10**, 413–420.

Berndt, A., Yizhar, O., Gunaydin, L.A., Hegemann, P., Deisseroth, K., 2009. Bi-stable neural state switches. *Nat. Neurosci.* **12**, 229–234.

Boyden, E.S., Zhang, F., Bamberg, E., Nagel, G., Deisseroth, K., 2005. Millisecond-timescale, genetically targeted optical control of neural activity. *Nat. Neurosci.* **8**, 1263–1268.

Cao, H., Gu, L., Mohanty, S.K., Chiao, J.-C., 2013. An integrated μLED optrode for optogenetic stimulation and electrical recording. *IEEE Trans. Biomed. Eng.* **60**, 225–229.

Chow, B.Y., Han, X., Dobry, A.S., Qian, X., Chuong, A.S., Li, M., Henninger, M.A., Belfort, G. M., Lin, Y., Monahan, P.E., Boyden, E.S., 2010. High-performance genetically targetable optical neural silencing by light-driven proton pumps. *Nature* **463**, 98–102.

Clancy, K.B., Schnepel, P., Rao, A.T., Feldman, D.E., 2015. Structure of a single whisker representation in layer 2 of mouse somatosensory cortex. *J. Neurosci.* **35**, 3946–3958.

Deisseroth, K., Feng, G., Majewska, A.K., Miesenböck, G., Ting, A., Schnitzer, M.J., 2006. Next-generation optical technologies for illuminating genetically targeted brain circuits. *J. Neurosci.* **26**, 10380–10386.

Drake, K.L., Wise, K.D., Farraye, J., Anderson, D.J., BeMent, S.L., 1988. Performance of planar multisite microprobes in recording extracellular single-unit intracortical activity. *IEEE Trans. Biomed. Eng.* **35**, 719–732.

Eggermann, E., Kremer, Y., Crochet, S., Petersen, C.C.H., 2014. Cholinergic signals in mouse barrel cortex during active whisker sensing. *Cell Rep.* **9**, 1654–1660.

Favre-Bulle, I.A., Preece, D., Nieminen, T.A., Heap, L.A., Scott, E.K., Rubinsztein-Dunlop, H., 2015. Scattering of sculpted light in intact brain tissue, with implications for optogenetics. *Sci. Rep.* **5**, 11501.

Fendyur, A., Spira, M.E., 2012. Toward on-chip, in-cell recordings from cultured cardiomyocytes by arrays of gold mushroom-shaped microelectrodes. *Front. Neuroengineering* **5**, 21.

Galvani, L., Volta, A., Zambelli, J., 1791. *Aloysii Galvani De Viribus Electricitatis in Motu Musculari Commentarius*. Bononiae: Ex Typographia Instituti Scientiarium.

Gentet, L.J., Kremer, Y., Taniguchi, H., Huang, Z.J., Staiger, J.F., Petersen, C.C.H., 2012. Unique functional properties of somatostatin-expressing GABAergic neurons in mouse barrel cortex. *Nat. Neurosci.* **15**, 607–612.

Ghosh, K.K., Burns, L.D., Cocker, E.D., Nimmerjahn, A., Ziv, Y., Gamal, A.E., Schnitzer, M.J., 2011. Miniaturized integration of a fluorescence microscope. *Nat. Methods* **8**, 871–878.

Gradinaru, V., Thompson, K.R., Zhang, F., Mogri, M., Kay, K., Schneider, M.B., Deisseroth, K., 2007. Targeting and readout strategies for fast optical neural control *in vitro* and *in vivo*. *J. Neurosci.* **27**, 14231–14238.

Grinvald, A., Cohen, L.B., Lesher, S., Boyle, M.B., 1981. Simultaneous optical monitoring of activity of many neurons in invertebrate ganglia using a 124-element photodiode array. *J. Neurophysiol.* **45**, 829–840.

Guo, Z.V., Hart, A.C., Ramanathan, S., 2009. Optical interrogation of neural circuits in Caenorhabditis elegans. *Nat. Methods* **6**, 891–896.

Guo, Z.V., Hires, S.A., Li, N., O'Connor, D.H., Komiyama, T., Ophir, E., Huber, D., Bonardi, C., Morandell, K., Gutnisky, D., Peron, S., Xu, N., Cox, J., Svoboda, K., 2014. Procedures for behavioral experiments in head-fixed mice. *PLoS One* **9**, e88678.

Han, X., Chow, B.Y., Zhou, H., Klapoetke, N.C., Chuong, A., Rajimehr, R., Yang, A., Baratta, M.V., Winkle, J., Desimone, R., Boyden, E.S., 2011. A high-light sensitivity optical neural silencer: development and application to optogenetic control of non-human primate cortex. *Front. Syst. Neurosci.* **5**, 18.

Hegemann, P., Möglich, A., 2011. Channelrhodopsin engineering and exploration of new optogenetic tools. *Nat. Methods* **8**, 39–42.

Henze, D.A., Borhegyi, Z., Csicsvari, J., Mamiya, A., Harris, K.D., Buzsáki, G., 2000. Intracellular features predicted by extracellular recordings in the hippocampus *in vivo*. *J. Neurophysiol.* **84**, 390–400.

Hochbaum, D.R., Zhao, Y., Farhi, S.L., Klapoetke, N., Werley, C.A., Kapoor, V., Zou, P., Kralj, J. M., Maclaurin, D., Smedemark-Margulies, N., Saulnier, J.L., Boulting, G.L., Straub, C., Cho, Y.K., Melkonian, M., Wong, G.K.-S., Harrison, D.J., Murthy, V.N., Sabatini, B.L., Boyden, E.S., Campbell, R.E., Cohen, A.E., 2014. All-optical electrophysiology in mammalian neurons using engineered microbial rhodopsins. *Nat. Methods* **11**, 825–833.

Huber, D., Petreanu, L., Ghitani, N., Ranade, S., Hromádka, T., Mainen, Z., Svoboda, K., 2008. Sparse optical microstimulation in barrel cortex drives learned behaviour in freely moving mice. *Nature* **451**, 61–64.

Iwai, Y., Honda, S., Ozeki, H., Hashimoto, M., Hirase, H., 2011. A simple head-mountable LED device for chronic stimulation of optogenetic molecules in freely moving mice. *Neurosci. Res.* **70**, 124–127.

Jouhanneau, J.-S., Ferrarese, L., Estebanez, L., Audette, N.J., Brecht, M., Barth, A.L., Poulet, J. F.A., 2014. Cortical fosGFP expression reveals broad receptive field excitatory neurons targeted by POm. *Neuron* **84**, 1065–1078.

Katz, Y., Yizhar, O., Staiger, J., Lampl, I., 2013. Optopatcher – an electrode holder for simultaneous intracellular patch-clamp recording and optical manipulation. *J. Neurosci. Methods* **214**, 113–117.

Kim, T., McCall, J.G., Jung, Y.H., Huang, X., Siuda, E.R., Li, Y., Song, J., Song, Y.M., Pao, H.A., Kim, R.-H., Lu, C., Lee, S.D., Song, I.-S., Shin, G., Al-Hasani, R., Kim, S., Tan, M.P., Huang, Y., Omenetto, F.G., Rogers, J.A., Bruchas, M.R., 2013. Injectable, cellular-scale optoelectronics with applications for wireless optogenetics. *Science* **340**, 211–216.

Kipke, D.R., Shain, W., Buzsáki, G., Fetz, E., Henderson, J.M., Hetke, J.F., Schalk, G., 2008. Advanced neurotechnologies for chronic neural interfaces: new horizons and clinical opportunities. *J. Neurosci.* **28**, 11830–11838.

Klapoetke, N.C., Murata, Y., Kim, S.S., Pulver, S.R., Birdsey-Benson, A., Cho, Y.K., Morimoto, T.K., Chuong, A.S., Carpenter, E.J., Tian, Z., Wang, J., Xie, Y., Yan, Z., Zhang, Y., Chow, B.Y., Surek, B., Melkonian, M., Jayaraman, V., Constantine-Paton, M., Wong, G.K.-S., Boyden, E.S., 2014. Independent optical excitation of distinct neural populations. *Nat. Methods* **11**, 338–346.

Kwon, K.Y., Lee, H.-M., Ghovanloo, M., Weber, A., Li, W., 2015. Design, fabrication, and packaging of an integrated, wirelessly-powered optrode array for optogenetics application. *Front. Syst. Neurosci.* **9**, 69.

Lee, D., Shtengel, G., Osborne, J.E., Lee, A.K., 2014. Anesthetized- and awake-patched whole-cell recordings in freely moving rats using UV-cured collar-based electrode stabilization. *Nat. Protoc.* **9**, 2784–2795.

Lima, S.Q., Miesenböck, G., 2005. Remote control of behavior through genetically targeted photostimulation of neurons. *Cell* **121**, 141–152.

Lin, J.Y., Knutsen, P.M., Muller, A., Kleinfeld, D., Tsien, R.Y., 2013. ReaChR: a red-shifted variant of channelrhodopsin enables deep transcranial optogenetic excitation. *Nat. Neurosci.* **16**, 1499–1508.

Lin, J.Y., Lin, M.Z., Steinbach, P., Tsien, R.Y., 2009. Characterization of engineered channelrhodopsin variants with improved properties and kinetics. *Biophys. J.* **96**, 1803–1814.

Manita, S., Suzuki, T., Homma, C., Matsumoto, T., Odagawa, M., Yamada, K., Ota, K., Matsubara, C., Inutsuka, A., Sato, M., Ohkura, M., Yamanaka, A., Yanagawa, Y., Nakai, J., Hayashi, Y., Larkum, M.E., Murayama, M., 2015. A top-down cortical circuit for accurate sensory perception. *Neuron* **86**, 1304–1316.

McAlinden, N., Gu, E., Dawson, M.D., Sakata, S., Mathieson, K., 2015. Optogenetic activation of neocortical neurons in vivo with a sapphire-based micro-scale LED probe. *Front. Neural Circuits* **9**, 25.

Miesenböck, G., De Angelis, D.A., Rothman, J.E., 1998. Visualizing secretion and synaptic transmission with pH-sensitive green fluorescent proteins. *Nature* **394**, 192–195.

Miesenböck, G., Rothman, J.E., 1997. Patterns of synaptic activity in neural networks recorded by light emission from synaptolucins. *Proc. Natl. Acad. Sci. U. S. A.* **94**, 3402–3407.

Montgomery, K.L., Yeh, A.J., Ho, J.S., Tsao, V., Mohan Iyer, S., Grosenick, L., Ferenczi, E.A., Tanabe, Y., Deisseroth, K., Delp, S.L., Poon, A.S.Y., 2015. Wirelessly powered, fully internal optogenetics for brain, spinal and peripheral circuits in mice. *Nat. Methods* **12**, 969–974.

Muñoz, W., Tremblay, R., Rudy, B., 2014. Channelrhodopsin-assisted patching: *in vivo* recording of genetically and morphologically identified neurons throughout the brain. *Cell Rep.* **9**, 2304–2316.

Musall, S., von der Behrens, W., Mayrhofer, J.M., Weber, B., Helmchen, F., Haiss, F., 2014. Tactile frequency discrimination is enhanced by circumventing neocortical adaptation. *Nat. Neurosci.* **17**, 1567–1573.

Nakai, J., Ohkura, M., Imoto, K., 2001. A high signal-to-noise Ca^{2+} probe composed of a single green fluorescent protein. *Nat. Biotechnol.* **19**, 137–141.

Nakamura, S., Baratta, M.V., Pomrenze, M.B., Dolzani, S.D., Cooper, D.C., 2012. High fidelity optogenetic control of individual prefrontal cortical pyramidal neurons in vivo. *F1000Research* **1**, 7.

Ollerenshaw, D.R., Zheng, H.J.V., Millard, D.C., Wang, Q., Stanley, G.B., 2014. The adaptive trade-off between detection and discrimination in cortical representations and behavior. *Neuron* **81**, 1152–1164.

Packer, A.M., Russell, L.E., Dalgleish, H.W.P., Häusser, M., 2015. Simultaneous all-optical manipulation and recording of neural circuit activity with cellular resolution in vivo. *Nat. Methods* **12**, 140–146.

Pala, A., Petersen, C.C.H., 2015. *In vivo* measurement of cell-type-specific synaptic connectivity and synaptic transmission in layer 2/3 mouse barrel cortex. *Neuron* **85**, 68–75.

Reutsky-Gefen, I., Golan, L., Farah, N., Schejter, A., Tsur, L., Brosh, I., Shoham, S., 2013. Holographic optogenetic stimulation of patterned neuronal activity for vision restoration. *Nat. Commun.* **4**, 1509.

Royer, S., Zemelman, B.V., Barbic, M., Losonczy, A., Buzsáki, G., Magee, J.C., 2010. Multi-array silicon probes with integrated optical fibers: light-assisted perturbation and recording of local neural circuits in the behaving animal. *Eur. J. Neurosci.* **31**, 2279–2291.

Rubehn, B., Wolff, S.B.E., Tovote, P., Lüthi, A., Stieglitz, T., 2013. A polymer-based neural microimplant for optogenetic applications: design and first in vivo study. *Lab. Chip* **13**, 579–588.

Ruiz, O., Lustig, B.R., Nassi, J.J., Cetin, A., Reynolds, J.H., Albright, T.D., Callaway, E.M., Stoner, G.R., Roe, A.W., 2013. Optogenetics through windows on the brain in the nonhuman primate. *J. Neurophysiol.* **110**, 1455–1467.

Sato, M., Ito, M., Nagase, M., Sugimura, Y.K., Takahashi, Y., Watabe, A.M., Kato, F., 2015. The lateral parabrachial nucleus is actively involved in the acquisition of fear memory in mice. *Mol. Brain* **8**, 22.

Schiemann, J., Puggioni, P., Dacre, J., Pelko, M., Domanski, A., van Rossum, M.C.W., Duguid, I., 2015. Cellular mechanisms underlying behavioral state-dependent bidirectional modulation of motor cortex output. *Cell Rep.* **11**, 1319–1330.

Siegle, J.H., Carlen, M., Meletis, K., Tsai, L.-H., Moore, C.I., Ritt, J., 2011. Chronically implanted hyperdrive for cortical recording and optogenetic control in behaving mice. *Conf. Proc. IEEE Eng. Med. Biol. Soc.* **2011**, 7529–7532.

Stark, E., Koos, T., Buzsáki, G., 2012. Diode probes for spatiotemporal optical control of multiple neurons in freely moving animals. *J. Neurophysiol.* **108**, 349–363.

Sudo, Y., Okazaki, A., Ono, H., Yagasaki, J., Sugo, S., Kamiya, M., Reissig, L., Inoue, K., Ihara, K., Kandori, H., Takagi, S., Hayashi, S., 2013. A blue-shifted light-driven proton pump for neural silencing. *J. Biol. Chem.* **288**, 20624–20632.

Tolhurst, D.J., Smyth, D., Thompson, I.D., 2009. The sparseness of neuronal responses in ferret primary visual cortex. *J. Neurosci.* **29**, 2355–2370.

Wang, J., Wagner, F., Borton, D.A., Zhang, J., Ozden, I., Burwell, R.D., Nurmikko, A.V., van Wagenen, R., Diester, I., Deisseroth, K., 2012. Integrated device for combined optical neuromodulation and electrical recording for chronic *in vivo* applications. *J. Neural Eng.* **9**, 016001.

Wang, Y., Gong, Q., Li, Y.Y., Li, A.Z., Zhang, Y.G., Cao, C.F., Xu, H.X., Cui, J., Gao, J.J., 2015. A wireless remote high-power laser device for optogenetic experiments. *Laser Phys.* **25**, 045601.

Wentz, C.T., Bernstein, J.G., Monahan, P., Guerra, A., Rodriguez, A., Boyden, E.S., 2011. A wirelessly powered and controlled device for optical neural control of freely-behaving animals. *J. Neural Eng.* **8**, 046021.

Yoon, I., Hamaguchi, K., Borzenets, I.V., Finkelstein, G., Mooney, R., Donald, B.R., 2013. Intracellular neural recording with pure carbon nanotube probes. *PLoS One* **8**, e65715.

Zemelman, B.V., Lee, G.A., Ng, M., Miesenböck, G., 2002. Selective photostimulation of genetically chARGed neurons. *Neuron* **33**, 15–22.

Zhang, F., Prigge, M., Beyrière, F., Tsunoda, S.P., Mattis, J., Yizhar, O., Hegemann, P., Deisseroth, K., 2008. Red-shifted optogenetic excitation: a tool for fast neural control derived from *Volvox carteri*. *Nat. Neurosci.* **11**, 631–633.

Zhang, J., Laiwalla, F., Kim, J.A., Urabe, H., Van Wagenen, R., Song, Y.-K., Connors, B.W., Zhang, F., Deisseroth, K., Nurmikko, A.V., 2009. Integrated device for optical stimulation and spatiotemporal electrical recording of neural activity in light-sensitized brain tissue. *J. Neural Eng.* **6**, 055007.

Zorzos, A.N., Boyden, E.S., Fonstad, C.G., 2010. Multiwaveguide implantable probe for light delivery to sets of distributed brain targets. *Opt. Lett.* **35**, 4133–4135.

Zorzos, A.N., Scholvin, J., Boyden, E.S., Fonstad, C.G., 2012. Three-dimensional multiwaveguide probe array for light delivery to distributed brain circuits. *Opt. Lett.* **37**, 4841–4843.

8 Application of an Optogenetic Technique to a Study of Activity-dependent Axon Growth in the Developing Cortex

Olga Malyshevskaya and Nobuhiko Yamamoto

8.1 Introduction

During development, fundamental neuronal circuits are formed by a genetically defined program, and the fine connections are modified by neuronal activity (Katz and Shatz, 1996). Such refinement processes are essential to establishing functional connectivity in the brain. A key issue is how neuronal activity influences axonal growth and branching. To address this issue, it is essential to manipulate the firing activity of the axons that are observed morphologically. Electrical stimulation with fine electrodes has been widely used for such manipulation. However, it is difficult to specify which cells or fibers are stimulated by the conventional method. The optogenetic method can allow us to overcome this problem (Boyden *et al.*, 2005). This method is useful not only for stimulus-based physiological experiments, but also for morphological and developmental studies.

In this article, we demonstrate the methodological efficiency of the optogenetic technique for developmental studies of neuronal circuit formation, focusing on axonal growth of cortical neurons. Cortical neurons form not only extrinsic connections with subcortical regions, but also intrinsic connections with adjacent cortical regions. Horizontal connections are the intrinsic circuits that connect upper layer neurons over distance (Callaway and Katz, 1990; Durack and Katz, 1996) and are thought to be involved in the information processing of specific visual inputs (Gilbert and Wiesel, 1989; Lowel and Singer, 1992). It is known that neuronal activity is necessary to form these horizontal connections (Callaway and Katz, 1991; Ruthazer and Stryker, 1996), but the underlying mechanism is still vague. Our experimental results using the optogenetic method and *in vitro* technique showed that axonal growth of cortical neurons was regulated by firing activity (Malyshevskaya *et al.*, 2013).

8.2 Methodology

The outline of the experiment is shown in Figure 8.1.

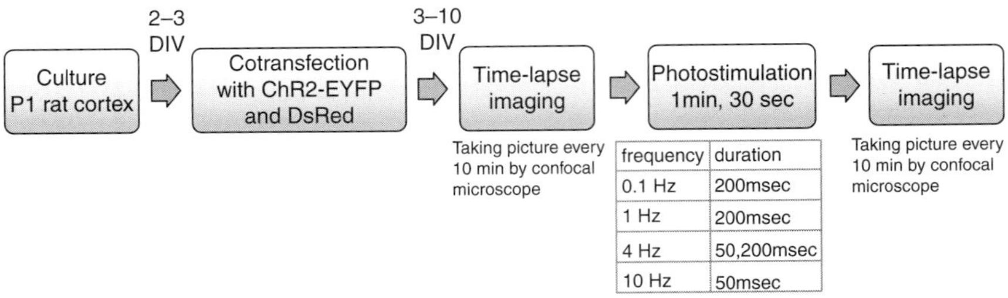

Figure 8.1 Time schedule for the experiment, which consists of slice culture, electroporation, time-lapse observation and optogenetic stimulation.

8.2.1 Organotypic Slice Culture

An organotypic slice culture technique was used for the observation of horizontal axons (Uesaka *et al.*, 2005). In this culture system, cortical circuits, including horizontal connections, are recapitulated with the cortical laminar structure preserved (Yamamoto *et al.*, 1989; Yamamoto *et al.*, 1992). The following procedures were carried out under sterilized conditions. After taking the whole brain from postnatal day (P) 0 or P1 rats, coronal slices (250–300 μm thick) were cut with small scissors in ice-cold Hanks' solution. These slices were immediately placed on a collagen-coated membrane filter (MilliCell-CM Low Height PICMORG-50, Millipore; see below), which had been immersed in the culture medium. The culture medium consisted of a 1:1 mixture of Dulbecco's modified essential medium and Ham's F12 with several supplements containing insulin and transferrin. The volume of the medium was adjusted to slightly below the surface of the cortical slices so that they could receive a sufficient supply of the medium and the mixed gas. The cultures were maintained at 37°C in an environment of humidified 95% air and 5% CO_2. The medium was exchanged with the serum-free medium every 3–4 days.

For the collagen coating, collagen fibers were collected from an adult rat tail under sterilized conditions and were solubilized with approximately 100 ml of 0.1% acetic acid solution at 4°C overnight. After centrifugation at 15 000 *g* for 30 minutes, the supernatant was collected. The membrane filter was inserted into a 35-mm Petri dish. The collagen solution (70–100 μl) was spread on the central part of the membrane filter. Once completely dry, approximately 1.5 ml of the culture medium was added to the culture dish.

Multi-electrode dishes (MEDs; Alpha MED Sciences) were used for the extracellular recording of spontaneous and ChR2-evoked firing activity (Uesaka *et al.*, 2005; Uesaka *et al.*, 2007). The collagen solution (40 μl) was spread on the MEDs. Once completely dry, a small amount of the culture medium was added to the culture dish. Then, a cortical slice was plated on to the collagen-coated MEDs. After plating, the medium level was adjusted to slightly below the surface of the cortical slices.

8.2.2 Plasmid Preparation and Electroporation

The coding region of hChR2 (H134R)-EYFP was placed in the pCAGGS vector (pCAGGS-ChR2-EYFP) (the original plasmid map is available at

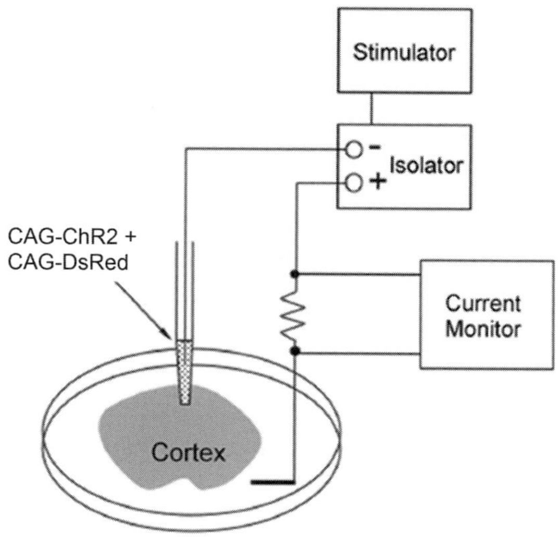

Figure 8.2 Experimental setup for electroporation. Electrical pulses were generated in the stimulator (pulse generator) and were applied to cortical cells through a glass microelectrode with the isolator. The current was monitored with the resistance in the electrical circuit.

www.stanford.edu/group/dlab/optogenetics/sequence_info.html). The plasmid was amplified and purified using a MaxiPrep kit (PureLink, Life Technologies) according to the manufacturer's protocol, and was dissolved in Hanks' solution at appropriate concentrations. Similarly, pCAGGS-DsRed plasmids were prepared.

To transfect the plasmid DNA into upper layer cortical cells, electroporation with glass microelectrodes was carried out under sterilized conditions (Figure 8.2) (Uesaka *et al.*, 2005). First, an electrode glass pipette (200–250 μm diameter) and an ejection glass pipette (70 μm diameter) were attached to micro-manipulators. These pipettes were filled with a mixture of pCAGGS-ChR2-EYFP (final concentration: 3 μg μl^{-1}) and pCAGGS-DsRed (final concentration: 2 μg μl^{-1}). The culture dish was placed on a stage with micro-manipulators. A silver wire (anode) was inserted into the culture medium. The plasmid solution (approximately 0.5 μl) was ejected from the glass pipette on to the surface of the cultured cortical slice at 1–4 days *in vitro* (DIV). Subsequently, the electrode pipette (cathode) was softly placed on the upper layer location. Then, electrical pulses (10 trains of 200 square pulses of 1-ms duration, 200 Hz, 400–500 μA) were delivered between the electrode pipette and silver wire using a pulse generator (Master 8, AMPI) and an isolator (BSI-2, Bak Electronics). The current passing through the micropipette was monitored across a resistor (100 Ω) in series with an oscilloscope.

8.2.3 Time-lapse Study

To observe axonal growth, a culture dish containing cortical slices was sealed with a coverslip and transferred to a microscopic stage. The slices were maintained at 37°C using a heating apparatus (Kokensha Engineering) and were supplied with a 95% air and 5% CO_2 humidified gas mixture. The objective lens was coiled with a band-heater (Thermoplate, Tokai Hit) in order to prevent condensation.

Cortical axons labeled with DsRed and ChR2-EYFP were observed by confocal microscopy (Nikon C2, Nikon Instech) with a 10× objective lens (Nikon Plan Fluor, numerical aperture 0.3) at 560-nm excitation, which never activates ChR2. After 1–2 hours of incubation, images were acquired at 10–20 minute intervals before and after optogenetic stimulation (see below). For each imaging session, a total of three to four optical sections were sampled at different depths (5-μm intervals).

8.2.4 Optogenetic Stimulation

A solid-state illuminator (475-nm peak wavelength; maximal power: 20 mW; Lumencor SPECTRA, Lumencor) was used for optogenetic stimulation through the objective lens. The beam position was controlled manually through the microscopic stage. The intensity of the light stimulus was adjusted using a Lumencor Remote Control Accessory. The light intensity was occasionally measured with a light power meter. The duration and the frequency of the light stimulus were controlled by the pulse generator.

To stimulate observed axons, excitation light (475 nm) with the following pulse frequencies and durations was delivered to the observed axons: 0.1 Hz (duration: 200 ms, 60 pulses), 1 Hz (1 ms, 60 pulses), 4 Hz (200 ms, 240 pulses) and 10 Hz (1 ms, 600 pulses). To confirm whether the optogenetic stimulation efficiently generates action potentials in ChR2-expressing cells, whole-cell patch clamp recording was performed.

8.2.5 Electrophysiological Experiments

To test whether the optogenetic stimulation induces action potentials in ChR2-expressing cells, extracellular recording was carried out for cortical slices on MEDs. For this experiment, *exo utero* electroporation was performed on the embryonic cortex in order to transfect a large number of cortical cells with ChR2 plasmids. Briefly, pregnant mice at embryonic day 14.5 were deeply anesthetized with Nembutal (pentobarbital, 50 mg kg^{-1}). The abdomen was surgically opened, and then embryos were taken. The plasmid solution of pCAGGS-ChR2-EYFP (3 μg μl^{-1}) was injected into the lateral ventricle of each embryo with a glass pipette (tip diameter: 20–30 μm). Paddle-like electrodes were positioned on the forebrain region outside the skin, and square pulses (30 V, 50 ms) were delivered five times to the embryonic head through the electrodes by an electroporator. After electroporation, the brains were taken out immediately. Then, the cortical slices containing the transfection region were dissected and placed on the collagen-coated MED. All cultures were maintained at 37°C in an environment of humidified 95% air and 5% CO_2.

The extracellular voltages at the electrodes were amplified by a factor of ten using an eight-channel head amplifier (SH-MED8; Alpha MED Sciences), further amplified by an eight-channel main amplifier (bandwidth filter, 0.1–20 kHz; SU-MED8; Alpha MED Sciences) and digitized at 10 kHz (DigiData 1200; Axon

Instruments). The digitized data were analyzed using AxoScope software (Axon Instruments).

The noise level on each electrode was within the range of 10–20 μV. Negative potentials with amplitudes above a set threshold were counted as one spike. The threshold of 1.5-times the maximal amplitude of the baseline noise (20 μV) was used. After completion of the recording, the MED probe was disconnected from the connector and returned to the incubator. The frequency of spontaneous firing activity was quantified with AxoGraph software.

Whole-cell patch recording was also performed in order to determine the efficiency of the optogenetic stimulation. At 1–3 days after transfection, whole-cell patch clamp recording was conducted from ChR2-EYFP-expressing cells in cultured rat cortical slices. To do this, an epifluorescence microscope (Olympus) equipped with a 40× water immersion objective (Olympus, numerical aperture 0.8) was used. Excitation light was introduced through the objective lens. Whole-cell recording was carried out at room temperature (27–28°C). Resistances of patch pipettes were 7–8 mΩ when filled with an intracellular solution containing (in mM): K methane-sulfonate 130, Hepes 10, KCl 10, ethylene glycol tetraacetic acid 0.5, spermine 0.1 and phosphocreatine 10 (pH 7.3, adjusted with KOH). The composition of the standard bathing solution was (in mM): 124 NaCl, 3 KCl, 2 $CaCl_2$, 1.3 $MgSO_4$, 1.2 KH_2PO_4, 26 $NaHCO_3$ and 10 glucose, bubbled with 95% O_2 and 5% CO_2 (pH 7.4). The signals were amplified with Axoclamp 2A (Axon Instruments) and analyzed by AxoScope 10.2 software.

8.3 Results

8.3.1 Observation of the Cell Morphology and Axon Growth of Upper Layer Neurons

Horizontal axons, collaterals of upper layer cell axons (layer 2/3 cells), were observed in cortical slice cultures by electroporating DsRed and ChR2-EYFP plasmids into a small number of these cells (Figure 8.3). Horizontally elongating, labeled axons were found to originate from pyramidal cells in the upper layers and run along the same layer by confocal microscopy with an excitation wavelength of 560 nm. Such clearly labeled axons were selected for the time-lapse study. Labeled axons with growth cones at their tips grew continuously during the first week *in vitro*. The growth rates varied among axons.

8.3.2 Spontaneous Firing Activity in Cultured Cortical Slices and Evoked Activity by Optogenetic Stimulation

Based on our previous study, the mean firing rate of spontaneous firing activity is quite low in the early stages (5–8 DIV), but increases strikingly by 11 DIV, with a further increase at later stages (9–14 DIV) (Uesaka *et al.*, 2005). We confirmed the time course of the occurrence of neuronal activity. To reduce the effect of spontaneous firing activity, the time-lapse study with optogenetic stimulation was performed before 11 DIV (mostly at 5–9 DIV).

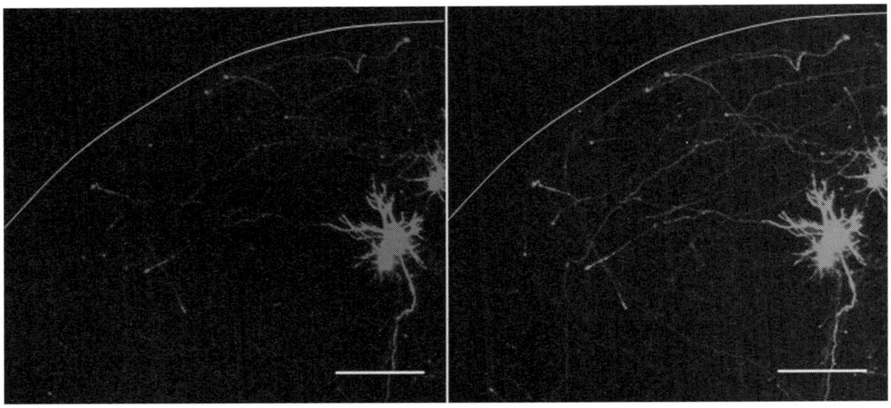

Figure 8.3 Cellular morphology of cortical upper layer neurons and horizontal axons in slice culture. A small number of upper layer neurons were co-transfected with DsRed and ChR2 plasmids. Expression of the fluorescent protein shows a pyramidal-shaped cell with a primary axon extending into the deeper layers and horizontally elongating collaterals extending into the upper layers (left). Scale bar: 100 μm. ChR2 is expressed in soma and axons (right). (A black-and-white version of this figure will appear in some formats. For the color version, please refer to the plate section.)

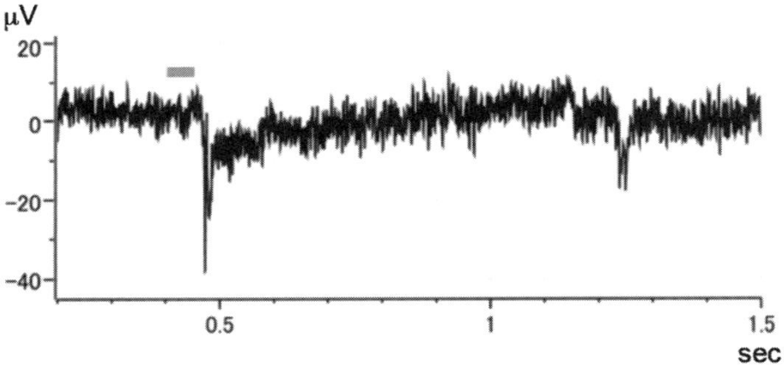

Figure 8.4 Evoked potentials on MEDs by optogenetic stimulation. Excitation light (50 ms, gray bar) was applied to the ChR2-expressing region of a cortical explant on an MED, and the evoked responses were recorded extracellularly from an electrode. Note that field potentials were elicited with latencies via the optogenetic stimulation.

To test whether the present optogenetic stimulation effectively induces action potentials in cortical neurons, extracellular recording using MEDs was performed in ChR2-expressing cortical slices. In this experiment, cortical slices that contained a large number of ChR2-expressing cells were plated on MEDs. After several days in culture, brief pulses of blue laser light were applied to the ChR2-transfected region, and extracellular recording was performed in the area 600–1000 μm away from the transfected area. We found that field potentials with clear latencies were elicited by optogenetic stimulation (Figure 8.4).

The efficiency of the optogenetic stimulation was also examined in the electrophysiological experiment with whole-cell patch recording under the current clamp mode. Several days after transfection with ChR2 plasmid, ChR2-positive cells were found in the upper layers and were patch clamped. Next, depolarizing step pulses were introduced to the patched cells, which had stable resting potentials of –60 mV. Action potentials were elicited by blue light photostimulation (475 nm) projected on to the cell area. Stimuli with different durations – 50, 100, 200 ms (0.5 Hz) – were introduced to the same cells. Interestingly, even pulses with long durations (200 ms) also resulted in single action potentials. Therefore, at least to this extent, the duration of the pulse was not very influential (significant), and even pulses with short durations (50 ms) were enough to cause the same response in transfected cells. No potentials were recorded in non-transfected neurons, indicating that the irradiance alone does not elicit any response (Trachtenberg *et al.*, 2002; Nagel *et al.*, 2005; Nirenberg and Pandarinath, 2012).

8.3.3 The Influence of Firing Activity Due to Optogenetic Stimulation on Horizontal Axon Growth

The influence of firing activity on horizontal axon growth was studied using optogenetic stimulation. After confirming that horizontal axons transfected with DsRed and ChR2-EYFP grew at constant growth rates, optogenetic stimulation, which consisted of a 1-minute stimulation period at different frequencies (0.1, 1, 4 and 10 Hz), was applied to these axons or cell bodies.

The observed axons grew at a constant rate (on average: 35 μm h^{-1}) before the optogenetic stimulation. Most of them transiently stopped growing immediately after high-frequency stimulation (4 and 10 Hz). Some axons exhibited obvious retraction with growth cone collapse (Figure 8.5). Quantitative analysis further demonstrated that the growth rates were significantly reduced after the stimulation. In contrast, axons did not stop growing after low-frequency stimulation (0.1 Hz). After moderate-frequency stimulation (1Hz), half of the observed axons paused in their growth, while the remaining half did not show any pause behavior.

8.4 Summary and Discussion

Our previous study in cortical slice cultures has demonstrated that horizontal axon branching is inhibited by the blockade of firing activity (Uesaka *et al.*, 2005).

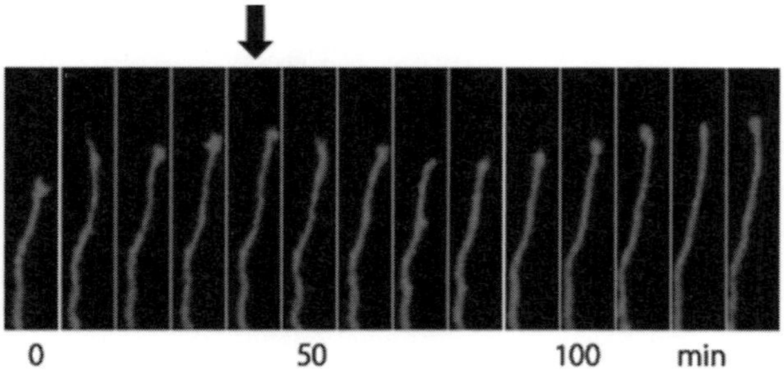

Figure 8.5 Stop behavior of horizontal axons after optogenetic stimulation. The horizontal axons that expressed ChR2 and DsRed were followed every 10 minutes under excitation wavelengths of 540 nm. The axons paused growing for >20 minutes after repetitive photostimulation (arrow) applied to the soma. The axons paused in their growth and slightly retracted. After 50 minutes, the axons restarted growing and recovered their previous growth speed.

The result we obtained using the optogenetic stimulation technique demonstrated that horizontal axons paused in their growth after high-frequency stimulation, but not low-frequency stimulation. Taken together, it is likely that horizontal axons stop growing and begin to form branches as high-frequency neuronal activity occurs during development.

In vitro experiments in developing rat neocortical slices have demonstrated the presence of spatiotemporally organized patterns of spontaneous cortical activity (Yuste *et al.*, 1992). Different firing patterns of activity are likely to have different effects on neural circuitry (Itoh *et al.*, 1997; Huberman *et al.*, 2008). Therefore, another interesting question relates to how different patterns of spontaneous neuronal activity affect axonal growth, branching and the refinement of neural circuits in general. Optogenetic stimulation could be a powerful tool for addressing this issue.

Previous studies have shown that axon growth from both mouse dorsal root ganglion neurons and *Helisoma* neurons is arrested when the cells are electrically stimulated (Cohan and Kater, 1986; Fields *et al.*, 1990). Furthermore, in the case of *Helisoma* neurons, this arrest has been attributed to a rise in the calcium concentration in the growth cones (Cohan *et al.*, 1987). It has also been demonstrated that growth cone stalling induced by electrical stimulation is mediated by calcium entry from the extracellular medium, as well as by calcium release from intracellular stores, which define spatially restricted microdomains directly affecting cytoskeletal dynamics (Gomez and Spitzer, 1999; Ibarretxe *et al.*, 2007). Similar mechanisms may be applied to the stop behavior of horizontal axons. Calcium imaging combined with optogenetic stimulation would determine the optimal calcium ion concentration for the stop behavior. At the same time, the effective frequency of firing activity could be demonstrated.

Thus, remote control of neuronal activity by light combined with morphological observation provides a new avenue for investigating the implications of activity in neuronal circuit formation.

REFERENCES

Boyden ES, Zhang F, Bamberg E, Nagel G, Deisseroth K (2005) Millisecond-timescale, genetically targeted optical control of neural activity. *Nat Neurosci* **8**:1263–1268.

Callaway EM, Katz LC (1990) Emergence and refinement of clustered horizontal connections in cat striate cortex. *J Neurosci* **10**:1134–1153.

Callaway EM, Katz LC (1991) Effects of binocular deprivation on the development of clustered horizontal connections in cat striate cortex. *Proc Natl Acad Sci U S A* **88**:745–749.

Cohan CS, Kater SB (1986) Suppression of neurite elongation and growth cone motility by electrical activity. *Science* **232**:1638–1640.

Cohan CS, Connor JA, Kater SB (1987) Electrically and chemically mediated increases in intracellular calcium in neuronal growth cones. *J Neurosci* **7**:3588–3599.

Durack JC, Katz LC (1996) Development of horizontal projections in layer 2/3 of ferret visual cortex. *Cereb Cortex* **6**:178–183.

Fields RD, Neale EA, Nelson PG (1990) Effects of patterned electrical activity on neurite outgrowth from mouse sensory neurons. *J Neurosci* **10**:2950–2964.

Gilbert CD, Wiesel TN (1989) Columnar specificity of intrinsic horizontal and corticocortical connections in cat visual cortex. *J Neurosci* **9**:2432–2442.

Gomez TM, Spitzer NC (1999) *In vivo* regulation of axon extension and pathfinding by growth-cone calcium transients. *Nature* **397**:350–355.

Huberman AD, Feller MB, Chapman B (2008) Mechanisms underlying development of visual maps and receptive fields. *Annu Rev Neurosci* **31**:479–509.

Ibarretxe G, Perrais D, Jaskolski F, Vimeney A, Mulle C (2007) Fast regulation of axonal growth cone motility by electrical activity. *J Neurosci* **27**:7684–7695.

Itoh K, Ozaki M, Stevens B, Fields RD (1997) Activity-dependent regulation of N-cadherin in DRG neurons: differential regulation of N-cadherin, NCAM, and L1 by distinct patterns of action potentials. *J Neurobiol* **33**:735–748.

Katz LC, Shatz CJ (1996) Synaptic activity and the construction of cortical circuits. *Science* **274**:1133–1138.

Lowel S, Singer W (1992) Selection of intrinsic horizontal connections in the visual cortex by correlated neuronal activity. *Science* **255**:209–212.

Malyshevskaya O, Shiraishi Y, Kimura F, Yamamoto N (2013) Role of electrical activity in horizontal axon growth in the developing cortex: a time-lapse study using optogenetic stimulation. *PLoS One* **8**:e82954.

Nagel G, Brauner M, Liewald JF, Adeishvili N, Bamberg E, Gottschalk A (2005) Light activation of channelrhodopsin-2 in excitable cells of *Caenorhabditis elegans* triggers rapid behavioral responses. *Curr Biol* **15**:2279–2284.

Nirenberg S, Pandarinath C (2012) Retinal prosthetic strategy with the capacity to restore normal vision. *Proc Natl Acad Sci U S A* **109**:15012–15017.

Ruthazer ES, Stryker MP (1996) The role of activity in the development of long-range horizontal connections in area 17 of the ferret. *J Neurosci* **16**:7253–7269.

Trachtenberg JT, Chen BE, Knott GW, Feng G, Sanes JR, Welker E, Svoboda K (2002) Long-term *in vivo* imaging of experience-dependent synaptic plasticity in adult cortex. *Nature* **420**:788–794.

Uesaka N, Hayano Y, Yamada A, Yamamoto N (2007) Interplay between laminar specificity and activity-dependent mechanisms of thalamocortical axon branching. *J Neurosci* **27**:5215–5223.

Uesaka N, Hirai S, Maruyama T, Ruthazer ES, Yamamoto N (2005) Activity dependence of cortical axon branch formation: a morphological and electrophysiological study using organotypic slice cultures. *J Neurosci* **25**:1–9.

Yamamoto N, Kurotani T, Toyama K (1989) Neural connections between the lateral geniculate nucleus and visual cortex in vitro. *Science* **245**:192–194.

Yamamoto N, Yamada K, Kurotani T, Toyama K (1992) Laminar specificity of extrinsic cortical connections studied in coculture preparations. *Neuron* **9**:217–228.

Yuste R, Peinado A, Katz LC (1992) Neuronal domains in developing neocortex. *Science* **257**:665–669.

9 Development of an Optogenetic Tool to Regulate Protein Stability *In Vivo*

Christian Renicke and Christof Taxis

9.1 Introduction

The introduction of the green fluorescent protein as a genetically encoded tool for the observation of physiological events in living organisms revolutionized life sciences (Tsien, 1998; Chudakov *et al.*, 2010). Recently, another branch evolved that used light to precisely manipulate cellular functions via genetically expressed photoactuators. Both approaches, aiming at either observation or regulation of cellular processes, are referred to as optogenetics (Deisseroth *et al.*, 2006; Miesenbock, 2009). The second strategy requires a bifunctional protein that comprises a sensor domain for photo-reception and an effector domain facilitating a specific output. Light as a signal has certain advantages compared to, for example, small-molecule approaches to regulate cell behavior or protein activities. Mainly, these relate to unmatched temporal and spatial control, as well as precise regulation of the quantity and quality of light. However, many biological systems rely on and react to light as an important environmental cue. This has to be considered in the experimental design of an optogenetic approach.

In recent years, many different tools have been developed using light to influence protein activity by regulating synthesis, localization, activity or stability, which has been described, in depth, by several reviews (Gautier *et al.*, 2014; Zhang and Cui, 2015; Ziegler and Moglich, 2015). These tools fall into two broad classes: the first comprises naturally occurring photoactuators that needed only minimal adjustments for usage in heterologous systems. One example is the channelrhodopsins, which revolutionized neuronal studies at all levels, from single-cell measurements in isolated neurons up to behavioral studies in whole animals (Hausser, 2014). Although such photoactuators might be directly transferable into the experimental organism of choice, the generation of improved variants by knowledge-based, site-directed mutagenesis or directed evolution is a way to improve the applicability of these tools. This offers researchers the opportunity to use an experimental setup that is optimized for their needs (Lin, 2011). The second class is synthetic, modular photoactuators that provide a novel cellular function by controlling the activity of an effector domain with a photoreceptor.

In general, an in-depth understanding of the light-induced changes in the photoreceptor and the regulation mechanism of the effector domain is necessary for this approach. This has been very successfully utilized in order to achieve site-specific regulation of protein activities, such as a small GTPase or actin filament formation (Wu *et al.*, 2009; Baarlink *et al.*, 2013).

Some generalizable trends became evident that have been employed effectively for several approaches. One example is the development of tools for the light regulation of transcription, which took advantage of the characterization of proteins that change their association with a photoreceptor depending on its signaling state (Shimizu-Sato *et al.*, 2002; Levskaya *et al.*, 2005; Ohlendorf *et al.*, 2012; Polstein and Gersbach, 2012; Wang *et al.*, 2012; Kim and Lin, 2013; Konermann *et al.*, 2013). Another way to control protein abundance generically is to regulate the degradation of a target protein. Light control of protein stability has been established by regulating the activity of a degradation sequence with a photoreceptor domain (Renicke *et al.*, 2013; Bonger *et al.*, 2014) or by the uncaging of a small chemical compound (Delacour *et al.*, 2015). The ubiquitin–proteasome system, which is the main proteolytic machinery in eukaryotes (Goldberg, 2007), is used for proteolysis of the target protein in all three methods. Thus, these methods are restricted to target proteins that reside in the nucleus or the cytosol of a eukaryotic cell.

The LOV2 domains of *Arabidopsis thaliana* and *Avena sativa* phototropin 1 have been used to regulate the function of an effector domain in several optogenetic tools (Pudasaini *et al.*, 2015; Ziegler and Moglich, 2015). The LOV domains are used in these constructs in a similar way to their original role in plant phototropins, in which they regulate the activity of an adjacent kinase domain (Harper *et al.*, 2004). The photoreceptors of the LOV2 domain family are well studied; they use a flavin mononucleotide (FMN) cofactor as the primary light-sensing molecule. After excitation of FMN by blue light, a covalent bond is formed between the carbon atom at position 4a of FMN and the sulfur of a cysteine residue. This adduct formation induces conformational rearrangements in the LOV2 domain that lead to the unfolding of a C-terminal helix – the so-called Jα helix (Pudasaini *et al.*, 2015). The structural change at the C-terminus of the LOV2 domain upon excitation with blue light has been used to regulate the accessibility and activity of synthetic degrons. The LOV2 domain of *A. sativa* phototropin 1 and a degradation sequence consisting of four amino acids were used to generate the so-called blue light-inducible degradation construct, which has been shown to mediate the light control of protein stability in mammalian cell culture and zebrafish embryos (Bonger *et al.*, 2014). In another implementation, the degron is a synthetic variant of the C-terminal degron of murine ornithine decarboxylase (ODC), which is called cODC1 (Jungbluth *et al.*, 2010), linked to the LOV2 domain of *A. thaliana* phototropin 1 (Figure 9.1A) (Renicke *et al.*, 2013). The ODC degron belongs to the few known degrons that induce proteasomal degradation independent of ubiquitylation, a common prerequisite for proteins to be degraded by the proteasome. It has been shown to be useful for the *in vivo* destabilization of proteins in budding yeast, tobacco plants

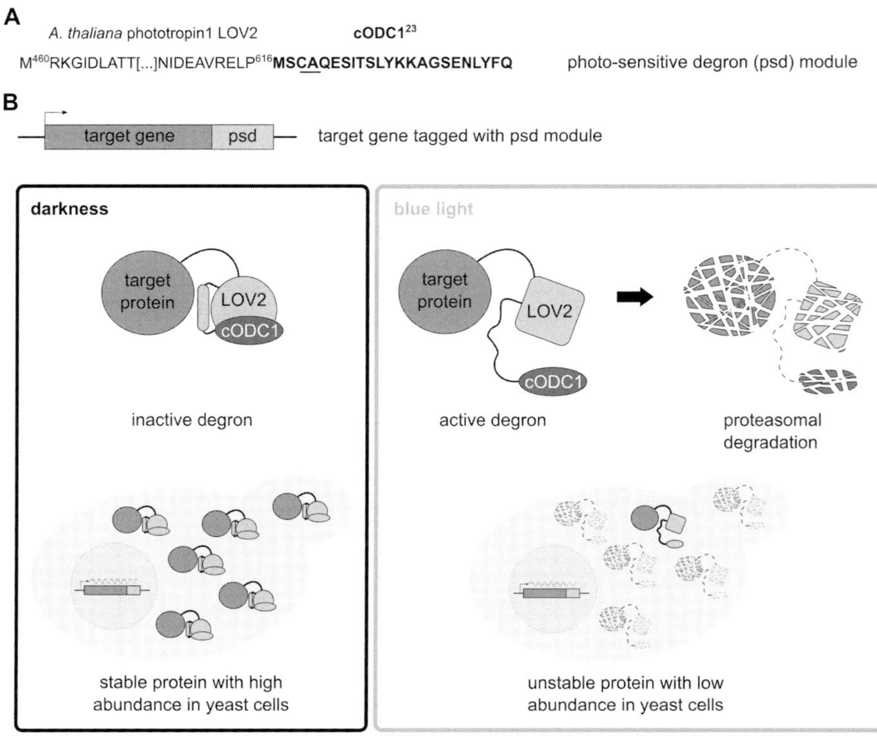

Figure 9.1 The photosensitive degron (psd) module. (A) The psd module consists of the *Arabidopsis thaliana* phototropin 1 LOV2 domain (amino acids 460–616) with a C-terminal extension of 23 amino acids from the synthetic degron cODC1 (letters in bold) containing the crucial CA motif (underlined). This degron was derived from the murine ornithine decarboxylase C-terminal degradation sequence. (B) Activation mechanism of the psd module exemplified in the yeast *Saccharomyces cerevisiae*. The gene encoding the psd module can be used to extend the target gene at the 3´-end. This leads to the formation of a fusion protein that is stable in darkness. Exposure of the yeast cells to blue light leads to a structural change in the LOV2 core domain and the unfolding of a C-terminal α-helix, which exposes and activates the cODC1 degron. This induces proteasomal degradation of the target. Subsequently, target protein levels drop until a new equilibrium between ongoing biosynthesis and degradation has been reached.

and mammalian cell culture, or for the study of protein degradation *in vitro* (Ghoda *et al.*, 1989; Loetscher *et al.*, 1991; DeScenzo and Minocha, 1993; Hoyt *et al.*, 2003; Matsuzawa *et al.*, 2005).

To induce degradation, two requirements have to be met for ODC-derived degrons: a stretch of 37 amino acids without a secondary structure at the very C-terminus of the protein; and a cysteine–alanine motif, which has to be present roughly in the center of the unfolded sequence stretch. The cysteine residue has been shown to be required for proteasome association, and the lack of secondary structure is important for the induction of proteolysis (Takeuchi *et al.*, 2007; Ravid and Hochstrasser, 2008; Takeuchi *et al.*, 2008). Thus, it can be assumed that light-induced unfolding of the Jα helix is necessary not only for exposure, but also for the activation of the degron within the context of the photosensitive degron (psd) module (Figure 9.1B).

The psd module has been developed and characterized in the yeast *Saccharomyces cerevisiae*. There, it was applied successfully in order to regulate growth, secretion, cell cycle events and enzymatic activity with blue light; recently, it was shown to work in the nematode *Caenorhabditis elegans* as well (Renicke *et al.*, 2013; Hermann *et al.*, 2015).

In this chapter, we discuss the practical aspects of using the psd module in yeast, point out problems that might arise and highlight strategies to circumvent these obstacles. Furthermore, we provide guidelines for the usage of the different psd module variants and we describe typical applications and experimental designs (e.g. target protein inactivation by light during the developmental program of sporulation and the following of target protein depletion by live-cell imaging).

9.2 Methodology

9.2.1 Yeast Strains, Plasmids and Growth Conditions

No specialized yeast strain is necessary to control protein stability by light. The psd module can be inserted at a chromosomal locus at the 3′ end of a gene by standard procedures or the desired fusion gene may be created on a plasmid (Janke *et al.*, 2004; Renicke *et al.*, 2013). The latter approach might require a yeast strain lacking the target gene, but this in turn provides the opportunity to uncouple target gene expression from intrinsic regulation by substitution of the original promoter. No specialized requirements are necessary for medium composition. We used all kinds of standard solid media for plate assays (Guthrie *et al.*, 1991). Clear plastic cell culture flasks with a ventilated cap were used to grow cells in liquid cultures. In this case, coloring of a medium might be an issue due to absorption of relevant wavelengths, resulting in decreased light penetration. Therefore, low-fluorescence medium was used to grow yeast cells in shaking cultures (Usherenko *et al.*, 2014). It might be noteworthy that low-fluorescence medium is quite similar to synthetic complete medium; however, the latter contains riboflavins. The riboflavin derivative FMN is used as an essential cofactor by the LOV2 domain. In yeast strains or under growth conditions, in which riboflavins are limiting due to reduced biosynthesis, the addition of FMN to the medium might be a way of ensuring the availability of the cofactor.

9.2.2 Illumination Conditions

Common LEDs, either high-power blue light LEDs or RGB LEDs, were used to illuminate yeast cells. In general, the light intensity was adjusted with a dimmer connected to the LEDs to a photon flux of 30 μmol m^{-2} s^{-1} at the level of the cells. The light flux was monitored with an optometer (e.g. P2000, equipped with light detector D-9306-2, Gigahertz-Optik, Türkenfeld, Germany). Light-proof boxes with LEDs mounted to the lids were used to expose the cells to specific illumination conditions. The interiors of the boxes were lined with reflective film to increase light yield. For live-cell and time-lapse microscopy experiments, a single blue light LED was placed in the transmitted light path to allow specific illumination of the sample. During recording of fluorescence images, the LED was switched off. A light flux of up to 30 μmol m^{-2} s^{-1} does not influence the growth rate of wild-type cells – even very light-sensitive *yap1Δ* mutant cells were still proliferating under these conditions. Moreover, the presence of a psd module construct did not result in a decreased growth rate under these conditions (Renicke *et al.*, 2013; Usherenko *et al.*, 2014).

9.2.3 Detection of Target Proteins in Yeast Cells

Standard techniques were used to quantify the target protein content in yeast cells (Ausubel *et al.*, 1995). Immunoblotting was used to follow target protein abundance in yeast cultures. Whole-cell extracts were prepared by alkaline lysis and trichloroacetic acid precipitation (Yaffe and Schatz, 1984). SDS-PAGE and blotting were performed following standard procedures (Laemmli, 1970; Towbin *et al.*, 1979). Fluorescence microscopy was used for single-cell measurements of target protein levels modified with a fluorescent protein and the psd module (Renicke *et al.*, 2013), whereas quantification at a population scale was done by fluorometer measurements (Usherenko *et al.*, 2014).

9.3 Results and Discussion

9.3.1 Regulation of Protein Abundance by Light

Control of protein levels by light requires the fusion of the psd module to the carboxy-terminus of the target, which has to reside in the nucleus or the cytosol in order to be accessible for proteasomal degradation. In addition, successful regulation depends on the intrinsic stability of the target protein. Regulation of protein abundance with the psd module works best for stable proteins with long half-lives. If a protein has a very short half-life, it may not be further destabilized by the psd module. For such proteins, it might be more advisable to control the gene expression of the target by light (Pathak *et al.*, 2014). The half-lives of the diverse psd module variants range from 6 to 20 minutes at destabilizing conditions (Table 9.1) (Renicke *et al.*, 2013; Usherenko *et al.*, 2014). The half-lives of most yeast proteins have been measured (Belle *et al.*, 2006), which might help in finding the most promising method for initial experiments.

Table 9.1 Characteristics of psd module variants.[a]

Name	Half-life in darkness[b] (minutes)	Half-life in 30 μmol m^{-2} s^{-1} blue light[b] (minutes)	Dark/light switching ratio
Wild-type psd module	123 ± 21	20 ± 1	11
K92R E132A E155G	102 ± 41	12 ± 0.4	22
K121M N128Y G138A	92 ± 28	13 ± 1	22
K92R E132A E139N N148D E155G	66 ± 10	10 ± 0.4	16
K121M N128Y	44 ± 8	9 ± 0.3	14

[a] The measurements have been published previously (Usherenko *et al.*, 2014).
[b] Mean +/– SD.

In yeast, a target gene can be expanded conveniently at the 3′-end by homologous recombination with polymerase chain reaction (PCR)-generated cassettes (Janke *et al.*, 2004), a technique that is also suitable for the generation of chromosomally integrated gene fusions with the psd module (Renicke *et al.*, 2013). Furthermore, this does not interfere with the naturally evolved regulation of target gene expression, which preserves the normal response to changes in the environment. An alternative is the usage of a strain with a deletion of the target gene complemented by a plasmid bearing the psd module-modified target gene. In such a construct, it might be worth exchanging the promoter of the target gene with a regulatable promoter like P_{GAL1} or the variants P_{GALL} or P_{GALS}, which have reduced expression strengths (Janke *et al.*, 2004). Such dual control enables transcriptional repression concomitantly with protein destabilization; however, it requires a switch in growth medium if a *GAL1* promoter variant is used to regulate the expression of the target gene.

Light exhibits unique advantages compared to other signals that may be applied to switching conditional mutants to the inactive state. One advantage is spatial control of illumination, which can be used to regulate the abundance of a target very precisely (Figure 9.2).

Such spatial control has also been used for the light regulation of enzymatic activity in order to conduct yeast photography (Renicke *et al.*, 2013). It convincingly shows that enzymatic activity can be regulated with high spatial control. This spatial control is not only achievable on a macroscopic scale, but also experiments on the microscopic level could benefit from selective illumination approaches: a specific protein might be inactivated in a fraction of the cells, whereas both light-exposed and undisturbed cells can be observed at the same time. The proof of principle for such an experimental setup has been achieved with a chemical biology method for the regulation of protein stability (Delacour *et al.*, 2015).

9.3.2 Application of the psd Module to Generating Conditional Mutants

Regulation of protein abundance by light is a powerful approach for the creation of conditional mutants. It has been shown that proteins with diverse cellular functions can be regulated by light using the psd module. Among others, the abundance of the Polo-like kinase Cdc5, the phosphatase Cdc14, the AAA-ATPase Cdc48, the iron–sulfur cluster biogenesis factor Yae1, and Sec62, a subunit of the translocon that is necessary for import into the endoplasmic reticulum, have been regulated by light (Renicke *et al.*, 2013; Paul *et al.*, 2015). Indirectly, reversed regulation has been achieved with constitutively active variants of cyclin-dependent kinase regulators modified with the psd module in an otherwise wild-type background. In this case, accumulation of the modified regulators in darkness leads to a block in cell cycle progression at a distinct step in each case (Renicke *et al.*, 2013). Thus, the psd module can be employed to create conditional mutants by inactivating proteins during vegetative growth. Moreover, it is also a useful tool for investigating a developmental process. In yeast, the differentiation program of sporulation is coupled to meiotic cell divisions; it is induced by starvation conditions and leads to

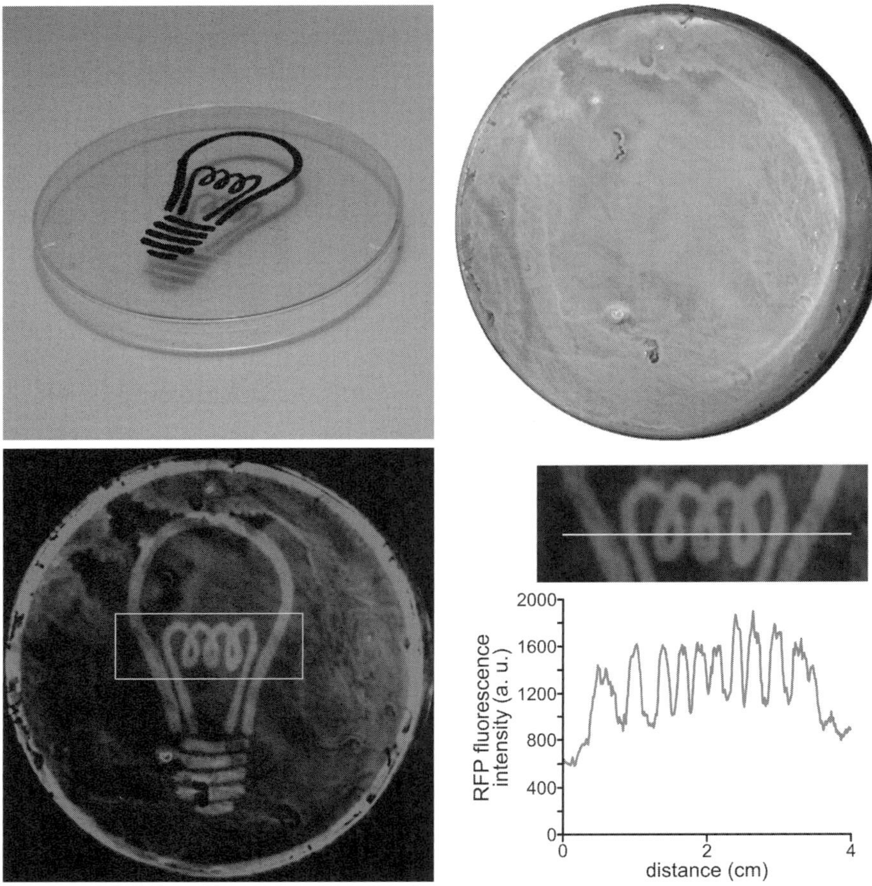

Figure 9.2 Construction of a light sensor using the psd module. Yeast cells expressing P_{TDH3}-*RFP-psd* (plasmid based) were spread on solid medium and incubated at 30°C for 24 hours. Blue light (465 nm, 10 μmol m^{-2} s^{-1}) was applied on parts of the plate using a mask (upper left panel). Please note that the rim of the plate was not illuminated due to the usage of a hood to block lateral light. RFP fluorescence (lower left image) and yeast cell growth (upper right image) were imaged with a fluorescence image analyzer and a digital camera, respectively. The graph (lower right panel) shows the line plot of fluorescence (measured along the yellow line) after background subtraction. The magnified area is indicated in the fluorescence image. Background fluorescence was obtained from cell-free areas. (A black-and-white version of this figure will appear in some formats. For the color version, please refer to the plate section.)

the generation of up to four spores containing haploid genomes from a diploid mother cell (Neiman, 2005). Usage of a conditional mutant to inactivate a protein during meiosis has the advantage that perturbation of the vegetative growth phase is minimized and inactivation is reversible. With the psd module, inactivation of the target protein is induced by blue light exposure after entry into the meiotic cell divisions has been initiated in most cells (Figure 9.3).

Light-induced depletion of Cdc5-psd, Cdc14-psd and Sec62-psd during meiosis suggested that target protein activity is lost quickly after induction of the sporulation program, resulting in specific defects in progression through different phases of the meiotic divisions or spore formation. Inactivation of the Polo-like kinase

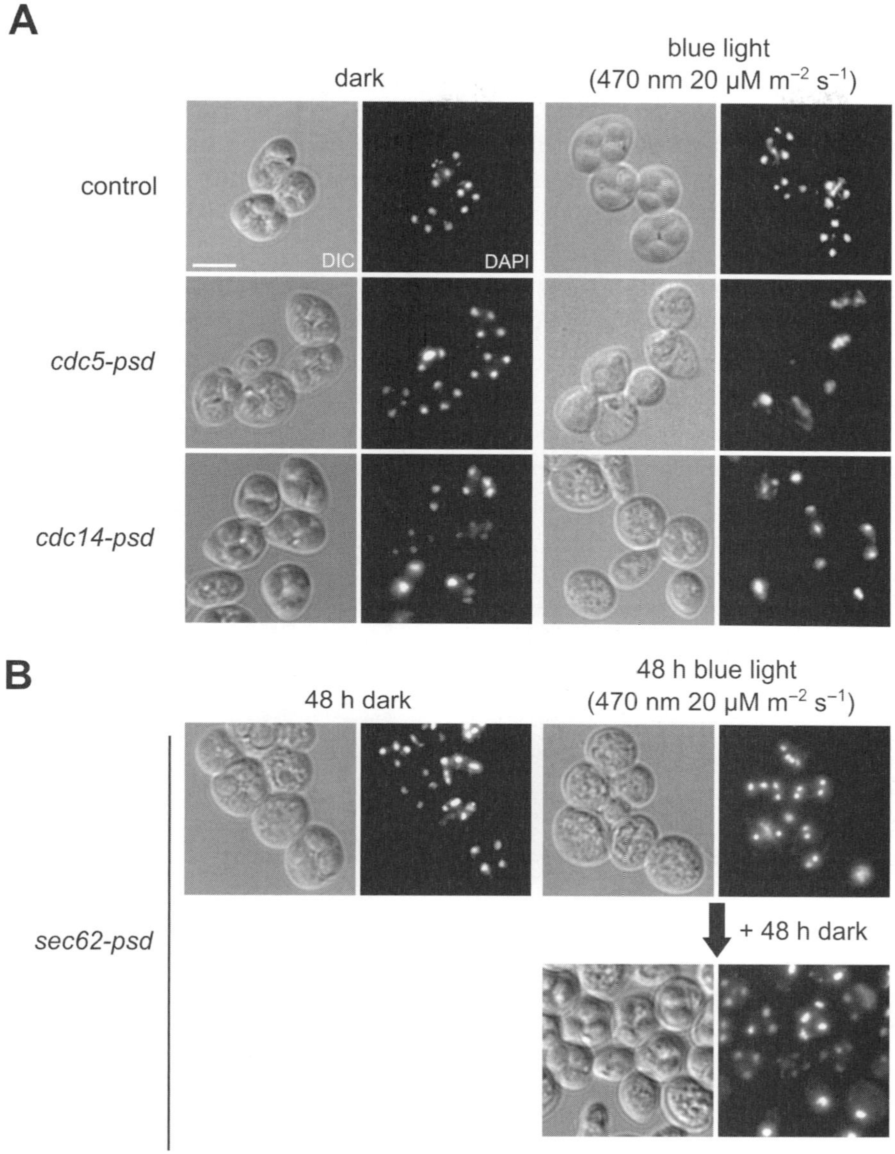

Figure 9.3 Generation of conditional mutants with the psd module. (A) Depletion of Cdc5-RFP-psd and Cdc14-RFP-psd by blue light after induction of sporulation leads to the blocking of spore formation. Diploid yeast cells (genotypes indicated in the figure) were subjected to sporulation conditions at 25°C for 2 days in the absence or presence of blue light (470 nm, 20 μmol m^{-2} s^{-1}). The images show cells in differential interference contrast and Hoechst 33342-stained DNA in the 4′,6-diamidino-2-phenylindole (DAPI) channel. Scale bar: 5 μm. (B) Depletion of Sec62-RFP-psd leads to a reversible block in sporulation. Cells were exposed for 48 hours to sporulation conditions as described in (A) and then subjected to an additional 48 hours of incubation in darkness.

Cdc5 resulted in cells that were mostly arrested during early phases of meiosis I, inactivation of Cdc14 blocked cells at the metaphase to anaphase transition of meiosis I, whereas cells with reduced Sec62 levels most likely finished both meiotic divisions, but were unable to form refractive spores (Figure 9.3A and 9.3B)

(Renicke and Taxis, unpublished observations). Interestingly, the cells that were arrested in sporulation due to Sec62 depletion completed the developmental program when placed into darkness after 48 hours of blue light incubation (Figure 9.3B). This behavior neatly demonstrated the reversibility of target protein depletion by protein biosynthesis. Similarly, the reappearance of a target has been observed previously by following a fluorescent protein modified with the psd module (Renicke *et al.*, 2013).

9.3.3 Detection of Target Proteins Modified with the psd Module

During the regulation of protein stability with the psd, it is important to determine the abundance of the target protein before and after the transition from permissive to restrictive conditions. Commonly available methods can be used to record the changes over time, such as fluorescence microscopy (Figure 9.4), fluorometer measurements and immunoblotting (Renicke *et al.*, 2013; Usherenko

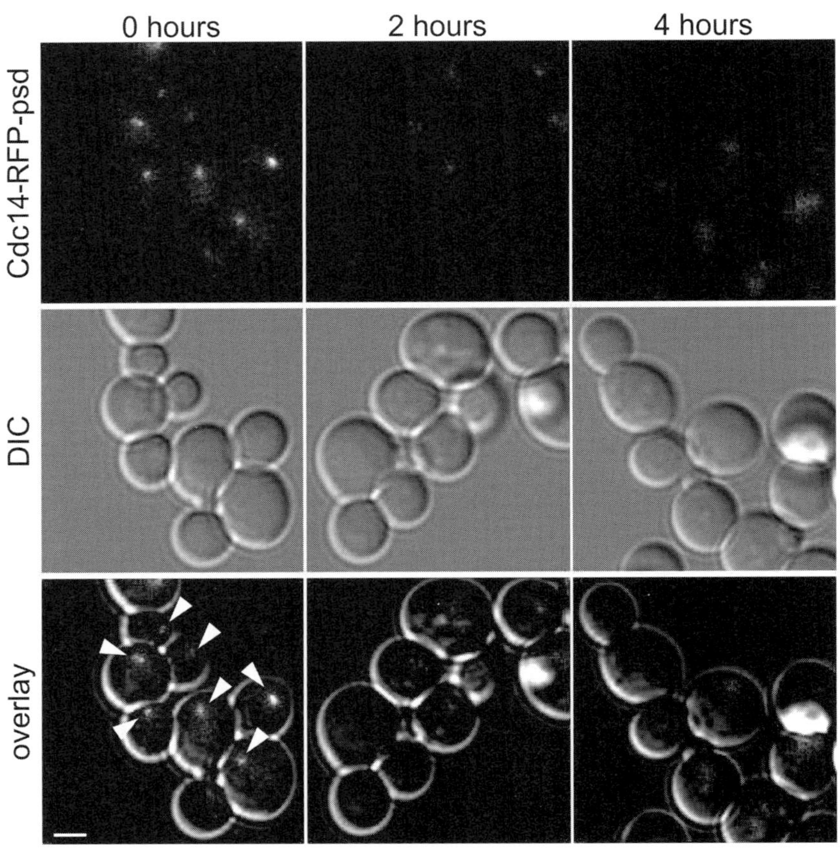

Figure 9.4 Live-cell microscopy of Cdc14-RFP-psd. Images of cells containing Cdc14-RFP-psd were recorded before and after illuminating the cells with blue light (465 nm; 30 μmol m^{-2} s^{-1}) for the times indicated in the figure (scale bar: 2 μm). The fluorescence signals show the typical nucleolar localization of Cdc14 at the beginning of the experiment (marked by arrowheads in the overlay of fluorescence and the differential interference contrast channel), which disappear after 2 hours of blue light exposure.

et al., 2014). Although not tested by us, methods such as flow cytometry or mass spectrometry are also applicable for this purpose (Newman *et al.*, 2006; Renicke *et al.*, 2013; Usherenko *et al.*, 2014; Selevsek *et al.*, 2015). For fluorescence-based approaches, the target protein has to be modified with a fluorescent tag in addition to the psd module. In yeast, such a gene extension requires only an appropriate cassette plasmid that contains the gene for the fluorescent protein of choice N-terminally fused to the psd module for PCR-based tagging of the target gene (Renicke *et al.*, 2013). Modification of a target with such an approach has been used to observe the disappearance of the target protein *in vivo* after illuminating the cells with blue light (Figure 9.4).

Fluorimeter measurements or immunoblotting may require more time and several steps of sample preparation, but offer advantages in the form of signal-to-noise ratio, specificity and sensitivity. When immunoblotting, the target protein is modified with a smaller tag. The decision as to which of these methods to use can be based largely on the available equipment and the specific experimental requirements.

9.3.4 Variants of the psd Module

A mutagenesis-based approach was undertaken in order to create improved variants of the psd module (Usherenko *et al.*, 2014). Firstly, mutations were tested in the setting of the psd module, which have been described in the literature to improve the switching behavior of homologous LOV2 domains. Secondly, we made mutants by random mutagenesis and selected variants with interesting characteristics. Finally, we combined both approaches and merged promising mutations into one construct (Figure 9.5). Our efforts resulted in variants with improved switching behavior, which, in case of the psd module, is the abundance of the target protein in cells kept in darkness divided by its abundance in restrictive conditions. In addition, we identified variants with a threefold decrease in half-life in blue light-illuminated cells. However, the latter variants showed decreased stability in darkness as well. An overview of selected psd module variants is given in Table 9.1.

The diversification of psd module characteristics has consequences for its usage: the choice of psd module variants depends on the characteristics of the target protein. It is probably best to modify a target protein that is highly abundant and is required in high levels in order to sustain its specific cellular function with the original psd module or one of the variants with a high switching ratio (K121M N128Y G138A and K92R E132A E155G). These two variants showed reduced half-lives in restrictive conditions (Usherenko *et al.*, 2014), which was crucial to obtaining a yeast strain in which the cyclin-dependent kinase Cdc28 could be inactivated by light (Figure 9.6).

Target proteins that are required only in low amounts need a psd module variant that leads to minimal levels at restrictive conditions. Examples are variants with half-lives of 10 minutes and below (K92R E132A E139N N148D E155G and K121M N128Y) in blue light-illuminated cells (Usherenko *et al.*, 2014). The abundance of a specific yeast protein has been measured by different

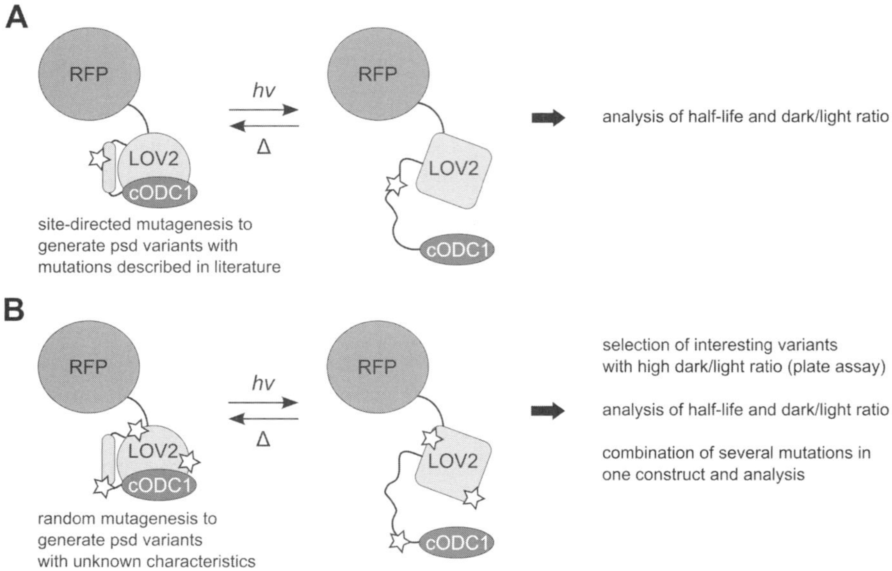

Figure 9.5 Mutagenesis strategies to improve the psd module. (A) Site-directed mutagenesis was used to generate variants of the psd module containing mutations that have been shown to change the signaling characteristics of homologous LOV2 domains. These variants were then tested for their dark/light switching behavior and their half-lives in cells kept in darkness or exposed to blue light. (B) Random mutagenesis was used to generate a library of psd module variants. Fluorescence measurements were used to quantify the abundance of the red fluorescent protein that is fused to the psd module in cells grown in darkness and in cells exposed to blue light. Interesting clones were analyzed as described in (A). Site-directed mutagenesis was used to combine promising mutations in one construct, followed by further characterization.

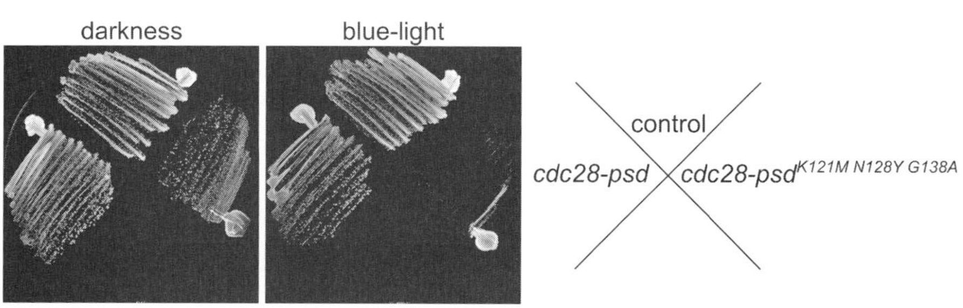

Figure 9.6 Inactivation of the yeast cyclin-dependent kinase Cdc28 by blue light. Yeast strains with Cdc28-psd variants (as indicated in the figure) and an isogenic control strain were streaked on solid rich medium and incubated for 2 days in darkness or exposed to blue light (465 nm, 30 μmol m^{-2} s^{-1}) at 30°C.

approaches (Ghaemmaghami *et al.*, 2003; Newman *et al.*, 2006; Kulak *et al.*, 2014; Chong *et al.*, 2015; Selevsek *et al.*, 2015), and the results of these efforts might be helpful in deciding which psd module variant is appropriate for successfully creating the desired mutant; alternatively, several psd module variants can be tested for a specific target.

9.4 Concluding remarks

Light regulation of protein stability has strong potential as an easy-to-implement tool that can be used to great effect. Light as a signal provides outstanding features, especially in terms of spatial precision and modulation of illumination strength. It has been shown that this can be used to gradually tune the activity of an enzyme in yeast (Renicke *et al.*, 2013). Such tunability suggests novel experimental approaches to, for example, gaining information about network behavior after reducing the amounts of single nodes or determining cellular thresholds for a protein's function. Importantly, the psd module induces changes in protein levels fairly rapidly, which allows us to record the dynamic behavior of a system in a reasonable timeframe. One way to improve the method would be to combine it with light-repressable transcription (Sorokina *et al.*, 2009) or light-activated gene silencing (Konermann *et al.*, 2013) in order to control protein abundance through both biosynthesis and stability. A variant of the psd module that is excited by different wavelengths would also be of advantage. However, there are currently no known photoreceptors with a similar activation mechanism that are excited by longer wavelengths. Such variants could be used in analogy to the multi-colored family of fluorescent proteins; target proteins modified with psd modules that react to different excitation wavelengths would enable the destabilizing of several proteins in the same cell at distinct time points.

Overall, photoactuators from the optogenetic toolbox offer ample opportunities for influencing cells on a molecular level and interfering with regulatory processes in a non-invasive way. Regulation of protein abundance by light with the psd is a valuable addition to this package that complements existing tools. While light has been used for centuries in order to observe biological systems using microscopy, recent years have shown that it is also a versatile signal for controlling cellular events in a precise way.

REFERENCES

Ausubel, F.M., Kingston, R.E., Seidman, F.G., *et al.* (ed.) (1995). *Current Protocols in Molecular Biology*, New York, NY: John Wiley and Sons.

Baarlink, C., Wang, H. and Grosse, R. (2013). Nuclear actin network assembly by formins regulates the SRF coactivator MAL. *Science*, **340**, 864–867.

Belle, A., Tanay, A., Bitincka, L., *et al.* (2006). Quantification of protein half-lives in the budding yeast proteome. *Proc Natl Acad Sci U S A*, **103**, 13004–13009.

Bonger, K.M., Rakhit, R., Payumo, A.Y., *et al.* (2014). General method for regulating protein stability with light. *ACS Chem Biol*, **9**, 111–115.

Chong, Y.T., Koh, J.L., Friesen, H., *et al.* (2015). Yeast proteome dynamics from single cell imaging and automated analysis. *Cell*, **161**, 1413–1424.

Chudakov, D.M., Matz, M.V., Lukyanov, S., *et al.* (2010). Fluorescent proteins and their applications in imaging living cells and tissues. *Physiol Rev*, **90**, 1103–1163.

Deisseroth, K., Feng, G., Majewska, A.K., *et al.* (2006). Next-generation optical technologies for illuminating genetically targeted brain circuits. *J Neurosci*, **26**, 10380–10386.

Delacour, Q., Li, C., Plamont, M.A., *et al.* (2015). Light-activated proteolysis for the spatiotemporal control of proteins. *ACS Chem Biol*, **10**, 1643–1647.

Descenzo, R.A. and Minocha, S.C. (1993). Modulation of cellular polyamines in tobacco by transfer and expression of mouse ornithine decarboxylase cDNA. *Plant Mol Biol*, **22**, 113–127.

Gautier, A., Gauron, C., Volovitch, M., *et al.* (2014). How to control proteins with light in living systems. *Nat Chem Biol*, **10**, 533–541.

Ghaemmaghami, S., Huh, W.K., Bower, K., *et al.* (2003). Global analysis of protein expression in yeast. *Nature*, **425**, 737–741.

Ghoda, L., Van Daalen Wetters, T., Macrae, M., *et al.* (1989). Prevention of rapid intracellular degradation of ODC by a carboxyl-terminal truncation. *Science*, **243**, 1493–1495.

Goldberg, A.L. (2007). Functions of the proteasome: from protein degradation and immune surveillance to cancer therapy. *Biochem Soc Trans*, **35**, 12–17.

Guthrie, C., Fink, G., Abelson, J., *et al.* (1991). Guide to yeast genetics and molecular biology. *Methods Enzymol*, **194**, 1–863.

Harper, S.M., Christie, J.M. and Gardner, K.H. (2004). Disruption of the LOV-Jalpha helix interaction activates phototropin kinase activity. *Biochemistry*, **43**, 16184–16192.

Hausser, M. (2014). Optogenetics: the age of light. *Nat Methods*, **11**, 1012–1014.

Hermann, A., Liewald, J.F. and Gottschalk, A. (2015). A photosensitive degron enables acute light-induced protein degradation in the nervous system. *Curr Biol*, **25**, R749–R750.

Hoyt, M.A., Zhang, M. and Coffino, P. (2003). Ubiquitin-independent mechanisms of mouse ornithine decarboxylase degradation are conserved between mammalian and fungal cells. *J Biol Chem*, **278**, 12135–12143.

Janke, C., Magiera, M.M., Rathfelder, N., *et al.* (2004). A versatile toolbox for PCR-based tagging of yeast genes: new fluorescent proteins, more markers and promoter substitution cassettes. *Yeast*, **21**, 947–962.

Jungbluth, M., Renicke, C. and Taxis, C. (2010). Targeted protein depletion in *Saccharomyces cerevisiae* by activation of a bidirectional degron. *BMC Syst Biol*, **4**, 176.

Kim, B. and Lin, M.Z. (2013). Optobiology: optical control of biological processes via protein engineering. *Biochem Soc Trans*, **41**, 1183–1188.

Konermann, S., Brigham, M.D., Trevino, A.E., *et al.* (2013). Optical control of mammalian endogenous transcription and epigenetic states. *Nature*, **500**, 472–476.

Kulak, N.A., Pichler, G., Paron, I., *et al.* (2014). Minimal, encapsulated proteomic-sample processing applied to copy-number estimation in eukaryotic cells. *Nat Methods*, **11**, 319–324.

Laemmli, U.K. (1970). Cleavage of structural proteins during the assembly of the head of bacteriophage T4. *Nature*, **227**, 680–685.

Levskaya, A., Chevalier, A.A., Tabor, J.J., *et al.* (2005). Synthetic biology: engineering Escherichia coli to see light. *Nature*, **438**, 441–442.

Lin, J.Y. (2011). A user's guide to channelrhodopsin variants: features, limitations and future developments. *Exp Physiol*, **96**, 19–25.

Loetscher, P., Pratt, G. and Rechsteiner, M. (1991). The C terminus of mouse ornithine decarboxylase confers rapid degradation on dihydrofolate reductase. Support for the pest hypothesis. *J Biol Chem*, **266**, 11213–11220.

Matsuzawa, S., Cuddy, M., Fukushima, T., *et al.* (2005). Method for targeting protein destruction by using a ubiquitin-independent, proteasome-mediated degradation pathway. *Proc Natl Acad Sci U S A*, **102**, 14982–14987.

Miesenbock, G. (2009). The optogenetic catechism. *Science*, **326**, 395–399.

Neiman, A.M. (2005). Ascospore formation in the yeast *Saccharomyces cerevisiae*. *Microbiol Mol Biol Rev*, **69**, 565–584.

Newman, J.R., Ghaemmaghami, S., Ihmels, J., *et al.* (2006). Single-cell proteomic analysis of *S. cerevisiae* reveals the architecture of biological noise. *Nature*, **441**, 840–846.

Ohlendorf, R., Vidavski, R.R., Eldar, A., *et al.* (2012). From dusk till dawn: one-plasmid systems for light-regulated gene expression. *J Mol Biol*, **416**, 534–542.

Pathak, G.P., Strickland, D., Vrana, J.D., *et al.* (2014). Benchmarking of optical dimerizer systems. *ACS Synth Biol*, **3**, 832–838.

Paul, V.D., Muhlenhoff, U., Stumpfig, M., *et al.* (2015). The deca-GX3 proteins Yae1-Lto1 function as adaptors recruiting the ABC protein Rli1 for iron–sulfur cluster insertion. *Elife*, **4**, e08231.

Polstein, L.R. and Gersbach, C.A. (2012). Light-inducible spatiotemporal control of gene activation by customizable zinc finger transcription factors. *J Am Chem Soc*, **134**, 16480–16483.

Pudasaini, A., El-Arab, K.K. and Zoltowski, B.D. (2015). LOV-based optogenetic devices: light-driven modules to impart photoregulated control of cellular signaling. *Front Mol Biosci*, **2**, 18.

Ravid, T. and Hochstrasser, M. (2008). Diversity of degradation signals in the ubiquitin-proteasome system. *Nat Rev Mol Cell Biol*, **9**, 679–690.

Renicke, C., Schuster, D., Usherenko, S., *et al.* (2013). A LOV2 domain-based optogenetic tool to control protein degradation and cellular function. *Chem Biol*, **20**, 619–626.

Selevsek, N., Chang, C.Y., Gillet, L.C., *et al.* (2015). Reproducible and consistent quantification of the *Saccharomyces cerevisiae* proteome by SWATH-mass spectrometry. *Mol Cell Proteomics*, **14**, 739–749.

Shimizu-Sato, S., Huq, E., Tepperman, J.M., *et al.* (2002). A light-switchable gene promoter system. *Nat Biotechnol*, **20**, 1041–1044.

Sorokina, O., Kapus, A., Terecskei, K., *et al.* (2009). A switchable light-input, light-output system modelled and constructed in yeast. *J Biol Eng*, **3**, 15.

Takeuchi, J., Chen, H. and Coffino, P. (2007). Proteasome substrate degradation requires association plus extended peptide. *EMBO J*, **26**, 123–131.

Takeuchi, J., Chen, H., Hoyt, M.A., *et al.* (2008). Structural elements of the ubiquitin-independent proteasome degron of ornithine decarboxylase. *Biochem J*, **410**, 401–407.

Towbin, H., Staehelin, T. and Gordon, J. (1979). Electrophoretic transfer of proteins from polyacrylamide gels to nitrocellulose sheets: procedure and some applications. *Proc Natl Acad Sci U S A*, **76**, 4350–4354.

Tsien, R.Y. (1998). The green fluorescent protein. *Annu Rev Biochem*, **67**, 509–544.

Usherenko, S., Stibbe, H., Musco, M., *et al.* (2014). Photo-sensitive degron variants for tuning protein stability by light. *BMC Syst Biol*, **8**, 128.

Wang, X., Chen, X. and Yang, Y. (2012). Spatiotemporal control of gene expression by a light-switchable transgene system. *Nat Methods*, **9**, 266–269.

Wu, Y.I., Frey, D., Lungu, O.I., *et al.* (2009). A genetically encoded photoactivatable Rac controls the motility of living cells. *Nature*, **461**, 104–108.

Yaffe, M.P. and Schatz, G. (1984). Two nuclear mutations that block mitochondrial protein import in yeast. *Proc Natl Acad Sci U S A*, **81**, 4819–4823.

Zhang, K. and Cui, B. (2015). Optogenetic control of intracellular signaling pathways. *Trends Biotechnol*, **33**, 92–100.

Ziegler, T. and Moglich, A. (2015). Photoreceptor engineering. *Front Mol Biosci*, **2**, 30.

10 Photoactivatable Nucleotide Cyclases for Synthetic Photobiology Applications

Marc Folcher

10.1 Introduction

The light signal is one of the most important environmental cues regulating microbial life, from bacterial photosynthesis to algal phototaxis. Natural photoreceptors have versatile characteristics related to the ecology of their original host organisms. *Halobacterium* is a waterborne microbe that uses the environmental light sensor protein bacteriorhodopsin to drive proton transport actively through the cell membrane in order to generate energy. Neuroscientists capitalized on this single-component light sensor to trigger an action potential in mammalian neurons by using light as an inducer (Yizhar *et al.*, 2011).

Proteins from the seven-transmembrane opsin gene family function as light-gated ion pumps. Over the last decade, their use as optogenetic tools has been extended to structurally similar proteins, including archaeal halorhodopsin. The channelrhodopsins (ChRs) from unicellular algae and vertebrate rhodopsins (the photoreceptors that sense light in the retina) are associated with a retinal cofactor that changes conformation upon photon absorption to induce a precise signal transduction cascade in the cell within milliseconds (Boyden *et al.*, 2005; Yizhar *et al.*, 2011).

In contrast to microbial opsins that directly control the membrane potential, the vertebrate rhodopsin phototransduction pathway recruits a specific G-protein-coupled receptor (GPCR; transducin) to control the degradation of the cyclic guanosine 3′5′-monophosphate (cGMP) secondary messenger molecules by cognate phosphodiesterase (PDE) (Yizhar *et al.*, 2011). Ultimately, the sudden drop in the secondary messenger concentration controls closure of the cGMP-gated cation channel and affects membrane voltage. The cGMP-dependent channels are not the only possible targets; retina proteins can be engineered to harness other secondary messenger transduction pathways. Chimeric proteins (Optoβ_2-AR) composed of the transmembrane light-harvesting domain of mammalian rhodopsin and the cytoplasmic loop of a conventional ligand-gated GPCR enable pathway rewiring to control Gα-dependent signaling. In this configuration, the light input induces a subsequent cyclic adenosine 3′,5′-monophosphate (cAMP) secondary messenger surge (Kim *et al.*, 2005; Airan *et al.*, 2009; Bailes *et al.*, 2012) in the cell (Figure 10.1A). The portfolio of natural photosensors available for optogenetic applications extends far beyond retinylidene proteins. A large variety of non-opsin photoactivatable proteins often consist of a chromophore harboring an N-terminal photosensor domain associated with a C-terminal effector domain. These domains include the blue light-induced light oxygen voltage sensing (LOV) photoreceptor domain (Guenther *et al.*, 2012), the blue light-using FAD (BLUF) domain, the blue light-sensing cryptochrome 2 (CRY2) domain, the red light-sensing phytochrome B protein (PHYB) domain (Figure 10.2) and the drompa photoactivatable fluorescent proteins. The identification of natural non-opsin composite photoreceptor genes in available genome sequences reveals the broad biological diversity of effector domains associated with light sensor domains (Losi and Gartner, 2008). Recent biochemical characterization of some of these light protein switches brings photobiologists new optogenetic tools for harnessing specific pathways in non-excitable cells (Losi and Gartner, 2008; Mandalari *et al.*, 2013).

Furthermore, the modular architecture of photoreceptors allows their use as a template for protein engineering approaches. It was thus possible to place several output functions – for example, transcription control (Motta-Mena *et al.*, 2014), inducible protein–protein association (Motta-Mena *et al.*, 2014), aggregation-based inhibition (Bugaj *et al.*, 2013), oligomerization (Leung *et al.*, 2008) and secondary messenger signaling – under light control.

Cyclic nucleotide secondary messenger molecules play an essential role in the amplification of signals in pathway transduction mechanisms. cAMP is ubiquitous in all life kingdoms, in contrast to cGMP, which was thought for a long time to be unique to eukaryotes. Recent discoveries in the secondary messenger signaling field shed light on a cGMP-based prokaryotic signaling pathway (Gomelsky, 2011), but also on new emerging classes of cyclic dinucleotide (CDN) secondary messengers that comprise cyclic guanosine monophosphate–adenosine monophosphate (cGAMP), cyclic diadenylate monophosphate (c-di-AMP) and cyclic diguanylate monophosphate (c-di-GMP) (Gomelsky, 2011; Abdul-Sater *et al.*, 2012; Sun *et al.*, 2013). The CDNs are important secondary messengers in prokaryotes, where they orchestrate lifestyle decisions. In prokaryotic pathogens, the CDNs are key players in the infection mechanism by acting as a triggering signal of the eukaryotic host innate immune response. Nucleotide secondary messengers and nucleoside analogs represent an important source of pharmacological agents for cancer treatments and antiviral therapies (Jordheim *et al.*, 2013). Pioneering studies in photochemistry identified key photosensitive residues that function as a "photocaging" group associated with nucleotides for light-controlled nucleotide release (Tang *et al.*, 2013; Samanta *et al.*, 2014). In the cell, cyclic nucleotides and CDN secondary messenger concentrations are regulated through the activity of nucleotide cyclases, which are responsible for their synthesis, and of PDEs, which control their degradation. The secondary messenger regulatory enzymes and, in particular, the PDEs are the subject of large screening approaches in order to identify potential therapeutic molecules for many types of

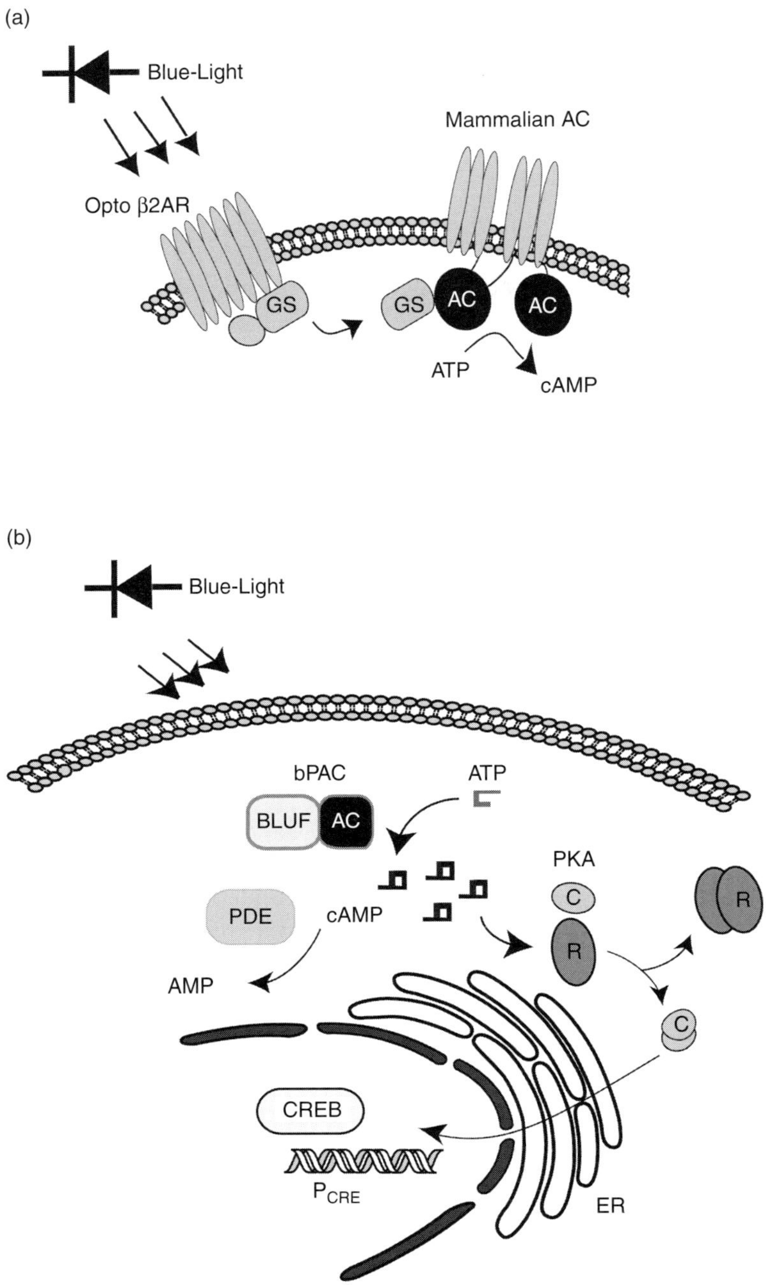

Figure 10.1 (A) Chimeric Optoβ_2-AR-dependent pathway; blue light-induced Gs-α subunit signaling activates adenylyl cyclase (AC). (B) Bacterial photoactivatable adenylyl cyclase (bPAC) cAMP-dependent pathway; blue light-induced AC activity results in cAMP surges, protein kinase A activation and dissociation of the catalytic subunit (c) and its translocation to the nucleus, where it phosphorylates cAMP response element binding protein (CREB) to mediate activation of P_{cre} promoter elements.

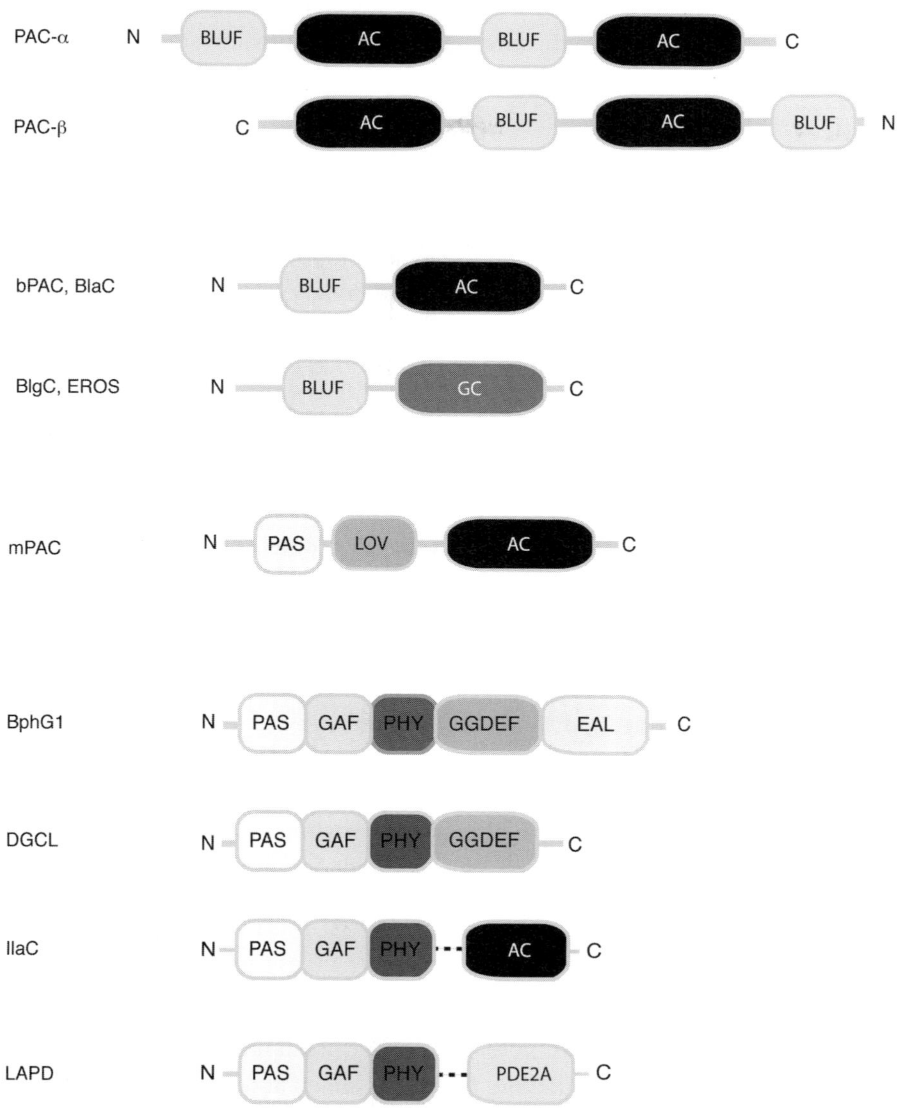

Figure 10.2 Domain organization of light-activatable nucleotide cyclases. AC, adenylyl cyclase; BLUF, blue light-using FAD; EAL, cyclic di-GMP-specific phosphodiesterase; GC, guanylate cyclase; GGDEF, diguanylate cyclase domain; LOV, light oxygen voltage; PAS, Per-ARNT-Sim repeats; PDE, phosphodiesterase; PHY, phytochrome domain.

disease. Moreover, a new class of optogenetic actuator has been gaining momentum: the natural photoactivatable nucleotide cyclases (PACs) are versatile modular enzymes composed of a light-sensing domain linked to a nucleotide cyclase domain (Figure 10.2). Their mechanism of action relies on conformation changes induced by light on the sensor domain control that affects the activity of the nucleotide cyclase. In addition, the recent identification of light-dependent PDEs complements the secondary messenger optogenetic tool family that is available for precise fine-tuning of secondary messenger signaling. The validation of optogenetic techniques to control secondary messenger signaling in electrically non-excitable cells opens new therapeutic avenues. The goal of this chapter is to review the recent advances in the bioengineering of photoactivatable nucleotide cyclases in the context of their fields of application.

10.2 BLUF Domain and Blue Light-activatable Adenylyl Cyclase

The blue light photoreceptor complex that controls photo-avoidance in the freshwater unicellular eukaryote *Euglena gracilis* is composed of PAC-α and PAC-β, which form hetero-tetrameric light-sensitive adenylyl cyclase (AC) (Figure 10.2). Most prokaryotic AC catalytic domains function as homodimers. The cyclization reaction takes place at the dimer interface of the prokaryotic enzymes, where both monomers contribute to the reaction that converts ATP into cAMP (Looser *et al.*, 2009). In contrast, the more complex metazoan AC structure is often composed of two homologous contiguous AC domains that form heterodimers (Sinha and Sprang, 2006). The *E. gracilis* photoreceptor connects metazoan AC domain architecture to tandem repeats of the BLUF domain photoreceptors (Gomelsky and Klug, 2002). Expression of the PAC-α domain is sufficient to form an active blue light-regulated AC enzyme and has been further used for diverse photobiology applications (Schroder-Lang *et al.*, 2007). Functional validation of PAC-α in a heterologous host was accomplished by co-transfecting the PAC-α with a cyclic nucleotide-gated channel and successfully measuring the light-mediated membrane depolarization in *Xenopus laevis* oocytes and HEK293 cells (Schroder-Lang *et al.*, 2007). Furthermore, transgenic expression of PAC-α in *Drosophila* brain neurons, *Caenorhabditis elegans* cholinergic neurons and *Aplysia* sensory neurons confirmed that the photoactivatable enzyme mediates rapid control of cAMP in freely moving animals (Nagahama *et al.*, 2007; Schroder-Lang *et al.*, 2007; Bucher and Buchner, 2009; Weissenberger *et al.*, 2011).

The *Drosophila* neuronal network that controls olfactory stimulus links a unique olfactory receptor to each olfactory neuron. Placing an optogenetic allele under a particular olfactory receptor promoter enables the possibility of light-triggering a specific olfactory neuron population. In a study aimed at deciphering *Drosophila* olfactory neuronal network activation, two optogenetic alleles were compared. Driving expression of either PAC-α or the ChR-2 under the control of a particular olfactory receptor promoter permitted the light-triggering of olfactory neuron activation, mimicking an odor stimulus. It also offered the opportunity to compare the phenotypes resulting from the different pathway activation routes, either by using a PAC-α-controlled intracellular cAMP surge or ChR-2-mediated membrane depolarization (Bellmann *et al.*, 2010; Stortkuhl and Fiala, 2011).

One drawback associated with recombinant PAC-α expression is the observed high AC dark activity that raises basal cAMP levels in host cells. One alternative consists of using the well-characterized bacterial blue light-activated AC (bPAC) from the sulfur-oxidizing bacteria *Beggiatoa* or one of its orthologues (Figure 10.2) (Yasukawa *et al.*, 2013; Stierl *et al.*, 2014).

In comparison to the eukaryotic PAC-α, the bPAC photosensor domain organization is simpler, associating a homodimer formed by a BLUF photosensor domain with a single prokaryotic AC effector domain (Figure 10.2). In addition, functional expression of the recombinant bPAC photosensor shows lower basal AC activity in the dark and higher activity in the light than does the eukaryotic PAC-α (Stierl *et al.*, 2014). In combination with tissue-specific promoters, bPAC represents a unique, non-invasive tool for deciphering the cAMP-dependent pathway (Figure 10.1B). It is possible to observe the physiological impacts associated with cAMP levels in cell culture and tissue. The *Drosophila* renal tubule can be reprogrammed by the recombinant expression of bPAC in order to study cAMP-dependent signaling (Efetova *et al.*, 2013). Furthermore, the optogenetic actuator allows blue light control of the endogenous cortisol level in zebrafish (Gutierrez-Triana *et al.*, 2015).

PAC offers an alternative method to assaying the influence of cAMP on the different stages of host–pathogen interactions that are not compromised by the addition of exogenous AC agonists such as forskolin. Optogenetic modulation of adenylate cyclase in *Toxoplasma gondii* demonstrates the requirement of parasite cAMP-dependent signaling for host cell invasion and stage differentiation (Hartmann *et al.*, 2013). The fusion of a shield1-regulatable destabilization domain to bPAC provides an additional regulatory input for fine-tuning parasite intracellular cAMP levels during host cell invasion by this intracellular protozoan parasite.

10.2.1 Bacterial PAC Applications for Studying Fertility

A complex cyclic nucleotide-dependent signaling pathway orchestrates the sexual reproduction mechanism in mammals. For example, in male fertility, the entrance of Ca^{2+} and HCO_3^- into sperm is a critical step in sperm "capacitation" that results in an intracellular increase of cAMP and plays a significant role in sperm motility.

During the transit of sperm in the female genital tract, environmental sensing by bicarbonate-inducible soluble AC (SACY) activity controls flagellar beating. It is not possible to directly engineer germinal cells, but it is possible to collect the sperm from a Prm1-bPAC transgenic mouse with recombinant bPAC expression driven by a testis-specific promoter (Jansen *et al.*, 2015). In sperm collected from an infertile transgenic mouse devoid of SACY activity, the bPAC allele fully complemented the infertility phenotype by restoring sperm motility under blue light illumination (Jansen *et al.*, 2015). In the presence of blue light irradiation, bPAC-recombinant sperm cells were twice as successful as wild-type cells in passing a functional test for sperm capacitation. Moreover, a red-shifted calcium fluorescent indicator compatible with the bPAC activation wavelength can be used to monitor Ca^{2+} influx (Arrenberg *et al.*, 2010; Jansen *et al.*, 2015). The study offers compelling evidence that a cAMP boost does not induce Ca^{2+} influx in mouse sperm.

10.2.2 Optogenetic Applications to Control Vascular Tone

Therapeutic control of tissue vasodilatation is a major field of pharmacological research. Optogenetic approaches supported a breakthrough in the analysis of cardiovascular regulatory functions (Arrenberg *et al.*, 2010). The transgenic expression of light-gated ion channels in cardiac tissue provided the basis for locating pacemaker cells in zebrafish (Arrenberg *et al.*, 2010). Optogenetic control can also mediate synchronized contraction of cardiomyocytes (Bruegmann *et al.*, 2010). Reprogramming of vascular smooth muscle cells by ChR allowed light to control vasoconstriction (Wu *et al.*, 2015). Moreover, native melanopsin (opsin 4) expression was detected in specific vascular tissues and was shown to play a role in light-mediated vasorelaxation associated with vascular hyperpolarization (Sikka *et al.*, 2014).

Photo-manipulation of vascular tone with blue light is not restricted to opsin photosensors. A classic phototherapy approach relies on the sole use of blue light to treat skin conditions. Blue light at a wavelength of 420–453 nm was shown to induce the non-enzymatic release of nitric oxide (NO) from photolabile NO derivatives (Oplander *et al.*, 2013).

NO is a gaseous secondary messenger that controls vasodilation by activating a soluble guanylate cyclase (GC). The subsequent increase of cGMP intracellular concentration results in the closing of voltage-gated calcium channels, which decreases intracellular calcium, causing relaxation of vascular tissues. The design of photo-triggered, caged NO compounds allows for spatiotemporal control of NO release in order to control vasodilatation (Diring *et al.*, 2013; Ieda *et al.*, 2014). Light-mediated NO delivery could find its place in future localized pharmacological treatments. NO plays a critical role in the mechanism of erection. In the penis erectile tissues, the nerve cells and the endothelial cells enmeshing the corpus cavernosum smooth muscle cells produce NO to stimulate cGMP synthesis. The increase in cGMP levels affects the calcium channel and relaxation of the corpus cavernosum smooth muscle, blood influx and, ultimately, the erection. In mammalian cells, intracellular concentrations of cGMP are controlled by the activity of GC, which synthesizes cGMP from GTP, and by PDE, which hydrolyzes it into GMP. Small-molecule compounds that affect cGMP levels, in particular inhibitors of cGMP-specific PDE, are front-line drugs in the treatment of cardiovascular diseases and erectile dysfunction.

The recent identification of a cGMP-dependent transcription pathway in prokaryotes placed a cGMP biosensor in the synthetic biology toolbox (Marden *et al.*, 2011; Ryu *et al.*, 2015). As a blueprint for eukaryotic biosensor design, a prokaryotic transcription regulator domain is fused to a viral activation domain. By linking the *Rhodospirillum centenum* cGMP receptor protein to the herpes simplex-derived VP16 transactivation domain, it is possible to convert a bacterial transcriptional regulator into a mammalian transcription factor (GTA) in order to determine intracellular cGMP concentration (Figure 10.3). The cGMP reporter is an essential asset in engineering photoactivatable GC and screening for biopharmaceuticals (Kim *et al.*, 2015a; Kim *et al.*, 2015b; Ryu *et al.*, 2015).

The catalytic domain of GC is similar to the catalytic domain of AC; by exchanging only three amino acid residues, it is possible to convert an AC into a GC enzyme (Figure 10.2) (Kasahara *et al.*, 2001). This knowledge-driven, directed mutagenesis approach was successfully chosen by the Gomelsky group in order

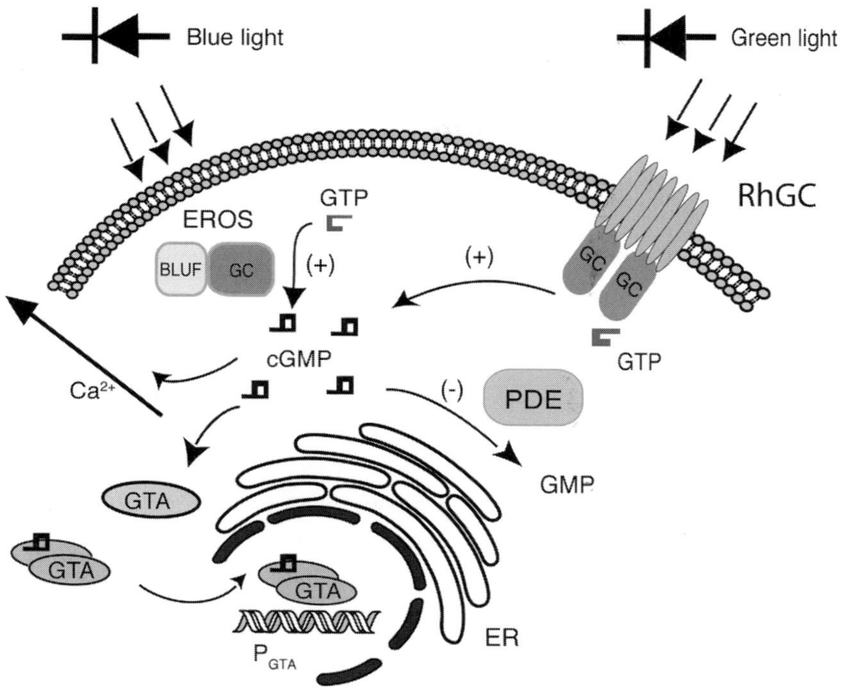

Figure 10.3 Light-activatable GC pathway. Erectile optogenetic stimulator protein (EROS) activity can be scored by the cGMP biosensor GTA, activating transcription of the chimeric promoter P_{GTA} that harbors cognate GTA operators. In penis erectile tissue, the cGMP concentration is controlled by the opposing activity of GC and PDE; a surge of cGMP triggering Ca^{2+} efflux ultimately results in relaxation of the corpus cavernosum. Alternatively, green light activation of rhodopsin guanylate cyclase (RhGC) can promote cGMP biosynthesis.

to reprogram a photoactivatable nucleotidyl cyclase. Using bPAC (also named BlaC) as a model template for protein engineering, it was possible to generate one triple-mutant BlgC that shows blue light-dependent GC activity and residual AC activity *in vitro* (Figure 10.2) (Ryu *et al.*, 2010). A similar site-directed mutagenesis approach performed with the human codon-optimized, BlgC-derived GC as a template generated erectile optogenetic stimulator (EROS) alleles. A combination of both *R. centenum* cGMP-derived biosensor and cAMP response element binding protein reporter was used to score the AC and GC light-dependent activity of the different clones in mammalian cells (Figure 10.3) (Kim *et al.*, 2015b). Interestingly, even though the approach produced alleles with an observed increased dark basal activity, it also gave enzymes showing mixed AC and GC light-dependent activity.

A photoactivatable GC offers a unique opportunity to monitor the influence that cGMP exercises on the mechanism of erection in male rats. In an *in vivo* study, a vector harboring the EROS gene under a constitutive promoter was injected directly in corpora cavernosa tissues (Kim *et al.*, 2015b). Some of the EROS-treated rats that were placed in a metabolic cage with a blue light retro-illumination source showed photo-stimulated erectile responses and ejaculation within a minute of light exposure. The precise signaling pathways that tie the GC-dependent erectile pathway to ejaculation remain elusive. Interestingly, in these experimental conditions, optogenetic intervention may have had an influence on both erection and ejaculation (Kim *et al.*, 2015b).

The recent characterization and validation in the *Xenopus* oocyte, Chinese hamster ovary cells and mammalian neurons of a light-activatable GC from phototactic fungi, offers photobiologists new perspectives in fine-tuning cGMP-dependent pathways. The newly identified protein presents an unusual domain organization composed of an input microbial rhodopsin domain linked to an output GC domain (Scheib *et al.*, 2015).

The LOV protein family associates a LOV flavin-containing photoreceptor input domain with a diverse signaling output domain that shares homology with His-kinase, CDN turnover enzymes (Cao *et al.*, 2010) and, in rare cases, ACs (Figure 10.2) (Herrou and Crosson, 2011). On blue-light irradiation, the ubiquitous flavin chromophore forms a covalent bond with a cysteine from the active site. The LOV domain proteins are building templates for biologically inspired protein engineering and synthetic biology applications (Moglich and Moffat, 2010). *Microcoleus* photoactivated AC (mPAC) domain organization resembles bPAC organization; it connects a LOV sensor domain to a microbial AC domain (Figure 10.2). mPAC induction improves fruiting body formation, but only partially complements a *Dictyostelium* AC null mutant (Chen *et al.*, 2014).

10.3 Near-infrared Light-responsive Bacteriophytochromes

The use of blue light for therapeutic applications is limited by its low penetration and associated cytotoxic effects. In contrast, the near-infrared (NIR) window in biological tissue combines the power of high tissue penetration with low phototoxicity and appears to be safe for clinical applications (Myakishev-Rempel *et al.*, 2012). Bacteriophytochromes (BphPs) are a family of red to NIR light-photosensory modules composed of PAS-GAF-PHY domains that are often coupled to a biliverdin chromophore. The biliverdin chromophore gives CRY2s many advantages because it is naturally present in animal tissue; the biliverdin NIR absorption spectrum also makes it an attractive photosensor for therapeutic applications. The chromophore undergoes reversible transformation from a red-absorbing form to a far red-absorbing form. The engineering of NIR-fluorescent BphP derivatives has brought forward new classes of fluorescent proteins that are compatible with whole-animal imaging involving murine models (Piatkevich *et al.*, 2013).

BphPs show a modular domain organization in which light-induced conformational changes transmit from the input light sensor domain to the output effector domain through an α-helical linker. The family of output domains associated with the BphPs are similar to that associated with the LOV photoreceptors (Losi and Gartner, 2008). As with the LOV domains, one important class of output domains associated with BphPs belongs to the GGDEF and EAL domain family (Figure 10.2). These protein domains relate to the turnover of the ubiquitous prokaryotic secondary messenger c-di-GMP. This secondary messenger

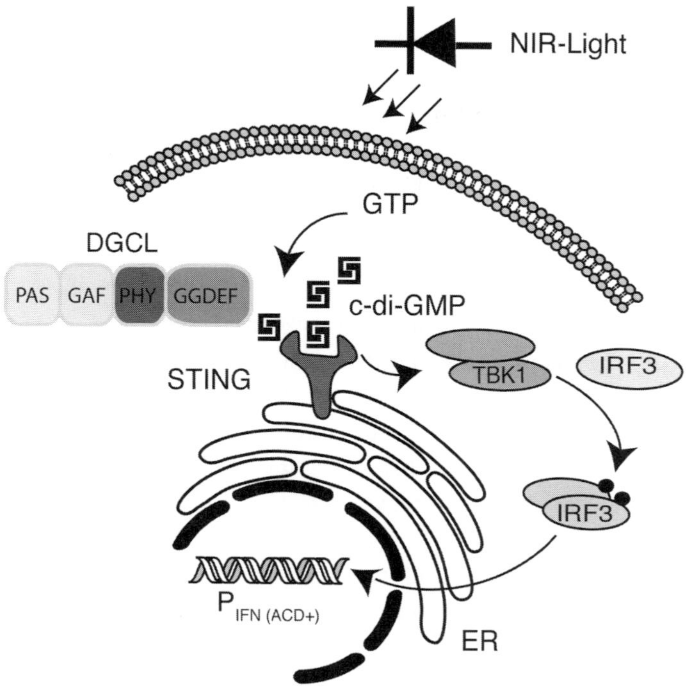

Figure 10.4 NIR light-phytochrome-associated c-di-GMP-dependent pathway. c-di-GMP activates stimulator of interferon genes (STING) at the endoplasmic reticulum (ER) and triggers tank-binding kinase 1 (TBK1)-mediated phosphorylation of interferon regulatory factor 3 (IRF3). Phosphorylated IRF3 translocates to induce the optimized type 1 interferon promoters ($P_{IFN[ACD+]}$).

orchestrates the environmental light-triggered transition from motile cells to biofilm-forming microbial communities (Jenal and Malone, 2006). The GGDEF domain has a homologous structure with type III AC. Diguanylate cyclases (DGCs) harboring the GGDEF domain synthetize c-di-GMP from two GTP molecules (Figure 10.4). c-di-GMP-specific PDEs harboring an EAL domain degrade the secondary messenger into GMP. The use of c-di-GMP optogenetic actuators is limited in prokaryotes by the abundance of naturally occurring enzymes that can interfere in order to control c-di-GMP-dependent pathways. In contrast, the absence of DGC activity in most eukaryotic cells makes the bacterial secondary messenger an attractive module for building an artificial secondary messenger signaling pathway.

Recent reports identified stimulator of interferon genes (STING) as a human innate immunity receptor for CDNs (cGAMP, c-di-AMP and c-di-GMP). CDNs are STING agonist-inducing interferon regulatory factor 3 (IRF3)-dependent transcriptional activators of type I interferon (IFN) promoters (Tang *et al.*, 2013). Placing the human secreted embryonic alkaline phosphatase (SEAP) under the control of an IRF3-optimized IFN-β promoter offers a sensitive method for detecting type I IFN pathway activation (Escalante *et al.*, 2007) and scoring STING signaling (Figure 10.4) (Folcher *et al.*, 2014). *Rhodobacter sphaeroides* BphG1 associates a tandem GGDEF with a PAS-GAF-PHY, and an EAL domain is putatively a bifunctional enzyme capable of both synthesis (DGC) and degradation (PDE) of c-di-GMP. Light-controlled diguanylate cyclase (DGCL), a BphG1-truncated allele lacking the EAL domain, reveals interesting NIR light-controlled DGC enzymatic properties that qualify it for state-of-the-art optogenetic applications (Figures 10.2 and 10.4) (Ryu *et al.*, 2014).

In contrast to the fast-acting dynamics associated with optogenetic applications in electrically excitable cells, optogenetic transcription control is a much slower process. Nonetheless, in non-excitable cells, a synthetic DGCL optogenetic pathway that harnesses an optimized IFNβ promoter offers fast light responsiveness in comparison to previously studied melanopsin-based systems (Folcher *et al.*, 2014).

A method to quantify the signaling molecules is required for the development of an artificial secondary messenger pathway. A c-di-GMP fluorescent-based biosensor can be used to detect secondary messenger dynamics *in vivo* and *in vitro* at the single-cell level (Christen *et al.*, 2010; Roembke *et al.*, 2014). NIR light illumination in the range of minutes of exposure is enough to induce an immediate c-di-GMP surge in mammalian cells expressing DGCL. This transient increase in the secondary messenger leads to activation of a consequential STING-dependent type I IFN pathway in mammalian cells.

10.3.1 Engineering BphPs

c-di-GMP signaling is not the only possible application for BphPs. A synthetic biology approach was used to convert the NIR-dependent DGC photo-switch in order to engineer an NIR-conditioned AC (Ryu *et al.*, 2014). The structural information available on a *Deinococcus radiodurans* phytochrome photosensory core in activated and resting states shed light on the conformational changes associated with the signal transduction of the BphP (Takala *et al.*, 2014). The fold of the photosensor input domain controls bending of the phytochrome monomer, which can affect realignment of the output domain. The blueprint of the phytochrome signal transduction mechanism can be followed in order to design a NIR light-controlled AC. This can be done by fusing the *R. sphaeroides* BphG1 sensory core with a new αhelical linker region connected to a domain sharing a similar type III AC fold (Figure 10.2) (Werner *et al.*, 1993). The systematic changes made during the protein fusion in the domain linker structure affect the performance of the light-induced AC. Again, the use of a cAMP reporter is an essential step in the selection of the newly engineered enzymes. Further engineering by rounds of mutagenesis is necessary in order to improve the photodynamic range of the best-performing clones. This study confirms that the BphPs represent a key engineering template for the development of future optogenetic tools (Ryu *et al.*, 2014).

10.4 Prosthetic Networks and Optogenetics

One limitation that hinders the therapeutic applications of optogenetics in general is related to the risks of invasive molecular biology techniques associated with tissue reprogramming. The risks are evident when it comes to reprogramming brain neural tissue with viral vectors. In contrast, the photoactivatable nucleotide cyclase field of application offers new opportunities for the design of safe optogenetic applications on the basis of *in vitro* reprogramming of non-excitable cells. A solution to minimize the risks associated with future cell-based therapeutics is to confine the xenotransplanted or allotransplanted reprogrammed cells into a semi-permeable membrane (Auslander *et al.*, 2012). The confinement protects the engineered cells from the immune system and protects the body from genetically foreign material. At the forefront of clinical applications to cure neurodegenerative disease are cell-based therapeutic devices used to deliver protein-based therapeutics behind the blood–brain barrier (Karami *et al.*, 2015).

With the help of synthetic biology, designer cells can be genetically programmed for light-adjustable expression of a protein-based therapy and encapsulated in an implantable micro-container. As an example, melanopsin was used to reprogram mammalian cells to respond to blue light illumination and to coordinate the biosynthesis and release of diabetes therapeutics in a type 2 diabetes mouse model (Ye *et al.*, 2011). Designer cells encapsulated in alginate beads were placed subcutaneously so that blood glucose homeostasis was controlled by transdermal blue light illumination. In an alternative implantation scenario, the designer cells reside in an optogenetic device to release therapeutic proteins in the peritoneal cavity. A hollow-fiber micro-container coupled to an optical fiber transmitting the blue light from an external source hosts the designer cells (Ye *et al.*, 2011). Optical fiber-coupled devices restrict animal movements and limit possible implantation strategies. The use of an extracorporeal portable light source could partially circumvent these problems. Last-generation optogenetic stimulators engineered to deliver light into the rodent brain often rely on miniaturized flexible electronics associated with an externally mounted micro-LED (Bagheri *et al.*, 2013; Kim *et al.*, 2013; Kale *et al.*, 2015; Montgomery *et al.*, 2015).

The implantation of battery-powered electronics in the body has led to numerous medical advances (e.g. pacemakers, cochlear implants and deep brain stimulation) (Joung, 2013). However, one technical challenge remains: how to wirelessly power wearable body electronic devices. Energy-harvesting devices or transmission of energy from an external source (Basar *et al.*, 2014) can now be used as battery substitutes to power a light source. Wireless transmission of power is possible with an inductive link (Figure 10.5). The principle relies on the coupling of an electromagnetic field between an emitter antenna and a receiver coil to provide constant power for an implanted or wearable device (Figure 10.5) (Basar *et al.*, 2014).

10.4.1 Wireless-powered Optogenetic Implants

Medical implants are made either of metal hardware that can host electronics, as in limb prostheses and pacemakers, or of biological soft matter, as in cell-based implants (Auslander *et al.*, 2012). A new class of prosthetic device associates, in a modular implant, an electronic module powered by an inductive link with a miniaturized bioreactor module hosting the designer cells (Figure 10.2) (Folcher *et al.*, 2014). Three orthogonally oriented coils of energy-harvesting antennae offer freedom of implantation orientation. The electronic module pilots a light

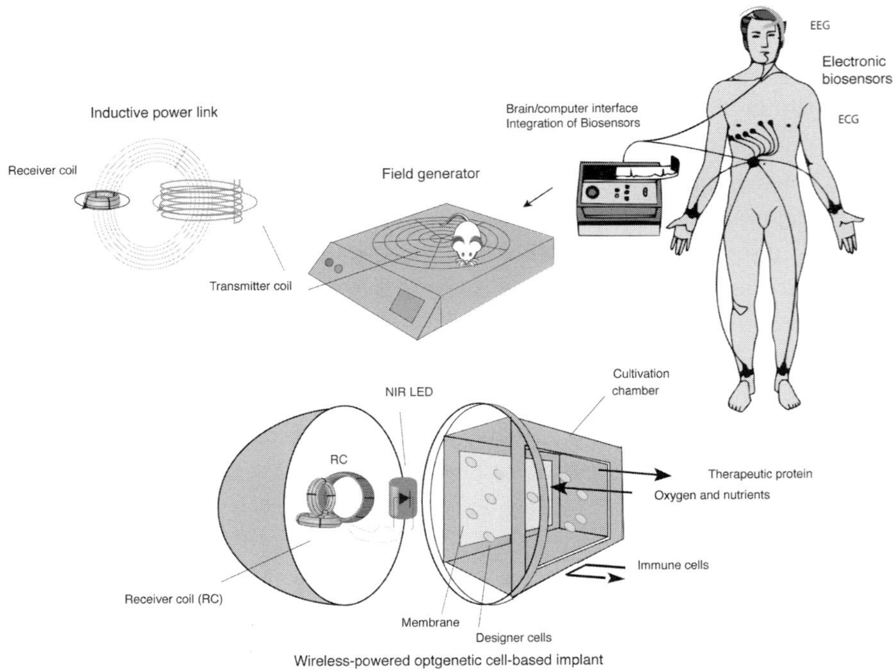

Figure 10.5 In mind-controlled transgene expression, the wireless-powered optogenetic cell implant is controlled by an electroencephalography (EEG) headset that captures brain-wave activities. The EEG (or any other electronic biosensor) is coupled to a computer interface that programs the wireless power of the implant. This power is transmitted by an inductive power link from a field generator coil to a receiver coil in the implant. The wireless-powered optogenetic implant is composed of an electronic module that hosts the receiver antenna coupled to a NIR LED and a cell chamber that hosts designer cells. Mice with a subcutaneous implant can move freely on the field generator.

source at the interface of the bioreactor module to control a synthetic NIR light optogenetic pathway. The electronic module establishes communication between a wirelessly transmitted input signal and a NIR light source that irradiates the cell chamber hosting the designer cells engineered for NIR light-adjustable expression of a transgene (Folcher *et al.*, 2014). The application field of the device can be extended by choosing the excitation wavelength of the LED as another value that is compatible with an alternative optogenetic actuator. Other than cells, the device could also host drug-release systems that are based on light-controlled polymeric capsules (He *et al.*, 2013; Baumann *et al.*, 2014). A new principal application of the wireless-powered optogenetic device is as a remote-controlled biopharmaceutical syringe.

10.4.2 Mind-controlled Transgene Expression by Brain–Computer Interface

Any electronic-based biosensor giving an electrical input signal can potentially pilot a wireless-powered, optogenetic cell-based device. Breakthroughs in cybernetics research have paved the way for prosthetic limb control by brain–computer interfaces (BCIs) (Daly and Wolpaw, 2008; Galan *et al.*, 2008). Mind-triggered electrophysiological signals captured by an electroencephalography (EEG)-based

BCIs can control a robotic arm (Clement *et al.*, 2011) or a simple light switch. As a proof of concept, a BCI was set to pilot a wireless-powered, optogenetic cell-based implant and to link brain-wave patterns to transgene expression (Folcher *et al.*, 2014). The BCI identified the mental state-specific electrical patterns resulting from concentration (computer gaming) or meditation (relaxation) in order to define an illumination time of the designer cells. To test the device, human subjects wearing an EEG headset BCI were trained to obtain different mental states by visualizing a biofeedback indicator on a computer screen. The BCI identified the task and associated it with the illumination time of the implant placed subcutaneously in mice. The subjects were thus able to program transgene expression in a subcutaneous implant (Figure 10.4). The information originating from the electrical activity recorded on the scalp was integrated by the electronics to pilot transgene expression in the implant. The interface controlled the release of the SEAP reporter protein in the animal's blood circulatory system (Folcher *et al.*, 2014). Combining the device and electronic miniaturization in order to develop the BCI algorithm could provide the basis of future implant prototypes controlling therapeutic delivery behind the blood–brain barrier. The BCI would score an electrical pattern associated with the onset of a neurological disorder, such as a seizure, and trigger the local delivery of a protein-based biopharmaceutical.

10.5 Conclusions and Perspectives

The aim of this chapter was to update and summarize the applications associated with the development of secondary messenger-based optogenetic actuators.

The biochemical characterization of natural photoactivatable nucleotide cyclase has established a universal strategy for controlling secondary messenger signaling across kingdoms. It has also contributed to the expansion of the optogenetic field of applications to non-excitable cells. The pleiotropic actions of cyclic nucleotides that orchestrate many fundamental processes, including cell differentiation, proliferation and motility, can now be assessed by non-invasive means. It is possible to fine-tune secondary messenger signaling at the single-cell level or in a whole-animal experiment by expressing the optogenetic transgene under the control of a tissue-specific promoter. The tailor-made optogenetic actuator proteins have moved beyond the direct control of the membrane potential of versatile protein–protein interactions. The portfolio of light-responsive proteins is now so extended that it offers the possibility to unravel signal transduction events by tapping into successive steps of a cyclic nucleotide-dependent transduction pathway. Moreover, knowledge-based protein engineering strategies targeting enzyme activities or domain organization have further broadened the possible strategies for adapting optogenetic devices to specific application scenarios. By capitalizing on the conformational changes transmitted by NIR light illumination of BphPs, it will be possible to engineer new optogenetic tools. Interestingly, the recent characterization and engineering of light-controlled PDE (Figure 10.2) (Gasser *et al.*, 2014; Enomoto *et al.*, 2015) complements the possibility of harnessing secondary messenger signaling. The developments of wireless-powered devices, flexible electronic circuits and state-of-the-art biosensors associated with optogenetic applications have modified the therapeutic perspectives of what was primarily designed as an investigation technique. An implantable, wirelessly piloted machine–cell interface, based on a light communication channel, represents a new option for tomorrow's therapeutics. Combining a therapeutic cell confinement strategy with electronic-based control will support the safe design of implants in clinical settings.

REFERENCES

Abdul-Sater, A.A., A. Grajkowski, H. Erdjument-Bromage, C. Plumlee, A. Levi, M.T. Schreiber, C. Lee, H. Shuman, S.L. Beaucage, and C. Schindler (2012). The overlapping host responses to bacterial cyclic dinucleotides. *Microbes Infect*, **14**, 188–97.

Airan, R.D., K.R. Thompson, L.E. Fenno, H. Bernstein, and K. Deisseroth (2009). Temporally precise in vivo control of intracellular signalling. *Nature*, **458**, 1025–9.

Arrenberg, A.B., D.Y. Stainier, H. Baier, and J. Huisken (2010). Optogenetic control of cardiac function. *Science*, **330**, 971–4.

Auslander, S., M. Wieland, and M. Fussenegger (2012). Smart medication through combination of synthetic biology and cell microencapsulation. *Metab Eng*, **14**, 252–60.

Bagheri, A., S.R. Gabran, M.T. Salam, J.L. Perez Velazquez, R.R. Mansour, M.M. Salama, and R. Genov (2013). Massively-parallel neuromonitoring and neurostimulation rodent headset with nanotextured flexible microelectrodes. *IEEE Trans Biomed Circuits Syst*, **7**, 601–9.

Bailes, H.J., L.Y. Zhuang, and R.J. Lucas (2012). Reproducible and sustained regulation of Galphas signalling using a metazoan opsin as an optogenetic tool. *PLoS One*, **7**, e30774.

Basar, M.R., M.Y. Ahmad, J. Cho, and F. Ibrahim (2014). Application of wireless power transmission systems in wireless capsule endoscopy: an overview. *Sensors (Basel)*, **14**, 10929–51.

Baumann, P., M. Spulber, I.A. Dinu, and C.G. Palivan (2014). Cellular Trojan horse based polymer nanoreactors with light-sensitive activity. *J. Phys. Chem. B*, **118**, 9361–70.

Bellmann, D., A. Richardt, R. Freyberger, N. Nuwal, M. Schwarzel, A. Fiala, and K.F. Stortkuhl (2010). Optogenetically induced olfactory stimulation in *Drosophila* larvae reveals the neuronal basis of odor-aversion behavior. *Front. Behav. Neurosci.*, **4**, 27.

Boyden, E.S., F. Zhang, E. Bamberg, G. Nagel, and K. Deisseroth (2005). Millisecond-timescale, genetically targeted optical control of neural activity. *Nat. Neurosci.*, **8**, 1263–8.

Bruegmann, T., D. Malan, M. Hesse, T. Beiert, C.J. Fuegemann, B.K. Fleischmann, and P. Sasse (2010). Optogenetic control of heart muscle in vitro and in vivo. *Nat Methods*, **7**, 897–900.

Bucher, D. and E. Buchner (2009). Stimulating PACalpha increases miniature excitatory junction potential frequency at the *Drosophila* neuromuscular junction. *J. Neurogenet.*, **23**, 220–4.

Bugaj, L.J., A.T. Choksi, C.K. Mesuda, R.S. Kane, and D.V. Schaffer (2013). Optogenetic protein clustering and signaling activation in mammalian cells. *Nat Methods*, **10**, 249–52.

Cao, Z., E. Livoti, A. Losi, and W. Gartner (2010). A blue light-inducible phosphodiesterase activity in the cyanobacterium *Synechococcus elongatus*. *Photochem. Photobiol.*, **86**, 606–11.

Chen, Z.H., S. Raffelberg, A. Losi, P. Schaap, and W. Gartner (2014). A cyanobacterial light activated adenylyl cyclase partially restores development of a *Dictyostelium discoideum*, adenylyl cyclase a null mutant. *J. Biotechnol.*, **191**, 246–9.

Christen, M., H.D. Kulasekara, B. Christen, B.R. Kulasekara, L.R. Hoffman, and S.I. Miller (2010). Asymmetrical distribution of the second messenger c-di-GMP upon bacterial cell division. *Science*, **328**, 1295–7.

Clement, R.G., K.E. Bugler, and C.W. Oliver (2011). Bionic prosthetic hands: a review of present technology and future aspirations. *Surgeon*, **9**, 336–40.

Daly, J.J. and J.R. Wolpaw (2008). Brain–computer interfaces in neurological rehabilitation. *Lancet Neurol*, **7**, 1032–43.

Diring, S., D.O. Wang, C. Kim, M. Kondo, Y. Chen, S. Kitagawa, K. Kamei, and S. Furukawa (2013). Localized cell stimulation by nitric oxide using a photoactive porous coordination polymer platform. *Nat Commun*, **4**, 2684.

Efetova, M., L. Petereit, K. Rosiewicz, G. Overend, F. Haussig, B.T. Hovemann, P. Cabrero, J.A. Dow, and M. Schwarzel (2013). Separate roles of PKA and EPAC in renal function unraveled by the optogenetic control of cAMP levels in vivo. *J. Cell Sci.*, **126**, 778–88.

Enomoto, G., Ni-Ni-Win, R. Narikawa, and M. Ikeuchi (2015). Three cyanobacteriochromes work together to form a light color-sensitive input system for c-di-GMP signaling of cell aggregation. *Proc. Natl. Acad. Sci. U. S. A.*, **112**, 8082–7.

Escalante, C.R., E. Nistal-Villan, L. Shen, A. Garcia-Sastre, and A.K. Aggarwal (2007). Structure of IRF-3 bound to the PRDIII-I regulatory element of the human interferon-beta enhancer. *Mol Cell*, **26**, 703–16.

Folcher, M., S. Oesterle, K. Zwicky, T. Thekkottil, J. Heymoz, M. Hohmann, M. Christen, M. Daoud El-Baba, P. Buchmann, and M. Fussenegger (2014). Mind-controlled transgene expression by a wireless-powered optogenetic designer cell implant. *Nat Commun*, **5**, 5392.

Galan, F., M. Nuttin, E. Lew, P.W. Ferrez, G. Vanacker, J. Philips, and R. Millan Jdel (2008). A brain-actuated wheelchair: asynchronous and non-invasive brain–computer interfaces for continuous control of robots. *Clin Neurophysiol*, **119**, 2159–69.

Gasser, C., S. Taiber, C.M. Yeh, C.H. Wittig, P. Hegemann, S. Ryu, F. Wunder, and A. Moglich (2014). Engineering of a red-light-activated human cAMP/cGMP-specific phosphodiesterase. *Proc. Natl. Acad. Sci. U. S. A.*, **111**, 8803–8.

Gomelsky, M. (2011). cAMP, c-di-GMP, c-di-AMP and now cGMP: bacteria use them all! *Mol. Microbiol.*, **79**, 562–5.

Gomelsky, M. and G. Klug (2002). BLUF: a novel FAD-binding domain involved in sensory transduction in microorganisms. *Trends Biochem. Sci.*, **27**, 497–500.

Guenther, T., N.H. Lovell, and G.J. Suaning (2012). Bionic vision: system architectures: a review. *Expert Rev. Med. Devices*, **9**, 33–48.

Gutierrez-Triana, J.A., U. Herget, L.A. Castillo-Ramirez, M. Lutz, C.M. Yeh, R.J. De Marco, and S. Ryu (2015). Manipulation of interrenal cell function in developing zebrafish using genetically targeted ablation and an optogenetic tool. *Endocrinology*, **156**, 3394–401.

Hartmann, A., R.D. Arroyo-Olarte, K. Imkeller, P. Hegemann, R. Lucius, and N. Gupta (2013). Optogenetic modulation of an adenylate cyclase in *Toxoplasma gondii* demonstrates a requirement of the parasite cAMP for host-cell invasion and stage differentiation. *J. Biol. Chem.*, **288**, 13705–17.

He, J., P. Zhang, T. Babu, Y. Liu, J. Gong, and Z. Nie (2013). Near-infrared light-responsive vesicles of Au nanoflowers. *Chem. Commun. (Camb.)*, **49**, 576–8.

Herrou, J. and S. Crosson (2011). Function, structure and mechanism of bacterial photosensory LOV proteins. *Nat. Rev. Microbiol.*, **9**, 713–23.

Ieda, N., Y. Hotta, N. Miyata, K. Kimura, and H. Nakagawa (2014). Photomanipulation of vasodilation with a blue-light-controllable nitric oxide releaser. *J. Am. Chem. Soc.*, **136**, 7085–91.

Jansen, V., L. Alvarez, M. Balbach, T. Strunker, P. Hegemann, U.B. Kaupp, and D. Wachten (2015). Controlling fertilization and cAMP signaling in sperm by optogenetics. *Elife*, **4**, e05161.

Jenal, U. and J. Malone (2006). Mechanisms of cyclic-di-GMP signaling in bacteria. *Annu. Rev. Genet.*, **40**, 385–407.

Jordheim, L.P., D. Durantel, F. Zoulim, and C. Dumontet (2013). Advances in the development of nucleoside and nucleotide analogues for cancer and viral diseases. *Nat. Rev. Drug Discov.*, **12**, 447–64.

Joung, Y.H. (2013). Development of implantable medical devices: from an engineering perspective. *Int. Neurourol. J.*, **17**, 98–106.

Kale, R.P., A.Z. Kouzani, K. Walder, M. Berk, and S.J. Tye (2015). Evolution of optogenetic microdevices. *Neurophotonics*, **2**, 031206.

Karami, A., H. Eyjolfsdottir, S. Vijayaraghavan, G. Lind, P. Almqvist, A. Kadir, B. Linderoth, N. Andreasen, K. Blennow, A. Wall, E. Westman, D. Ferreira, M. Kristoffersen Wiberg, L.O. Wahlund, A. Seiger, A. Nordberg, L. Wahlberg, T. Darreh-Shori, and M. Eriksdotter (2015). Changes in CSF cholinergic biomarkers in response to cell therapy with NGF in patients with Alzheimer's disease. *Alzheimers Dement.*

Kasahara, M., T. Unno, K. Yashiro, and M. Ohmori (2001). CyaG, a novel cyanobacterial adenylyl cyclase and a possible ancestor of mammalian guanylyl cyclases. *J. Biol. Chem.*, **276**, 10564–9.

Kim, J.M., J. Hwa, P. Garriga, P.J. Reeves, U.L. RajBhandary, and H.G. Khorana (2005). Light-driven activation of beta 2-adrenergic receptor signaling by a chimeric rhodopsin containing the beta 2-adrenergic receptor cytoplasmic loops. *Biochemistry*, **44**, 2284–92.

Kim, T., M. Folcher, G. Charpin-El Hamri, and M. Fussenegger (2015a). A synthetic cGMP-sensitive gene switch providing Viagra®-controlled gene expression in mammalian cells and mice. *Metab. Eng*, **29**, 169–79.

Kim, T., M. Folcher, M. Doaud-El Baba, and M. Fussenegger (2015b). A synthetic erectile optogenetic stimulator enabling blue-light-inducible penile erection. *Angew. Chem. Int. Ed. Engl.*, **54**, 5933–8.

Kim, T.I., J.G. McCall, Y.H. Jung, X. Huang, E.R. Siuda, Y. Li, J. Song, Y.M. Song, H.A. Pao, R. H. Kim, C. Lu, S.D. Lee, I.S. Song, G. Shin, R. Al-Hasani, S. Kim, M.P. Tan, Y. Huang, F. G. Omenetto, J.A. Rogers, and M.R. Bruchas (2013). Injectable, cellular-scale optoelectronics with applications for wireless optogenetics. *Science*, **340**, 211–6.

Leung, D.W., C. Otomo, J. Chory, and M.K. Rosen (2008). Genetically encoded photoswitching of actin assembly through the Cdc42–WASP–Arp2/3 complex pathway. *Proc. Natl. Acad. Sci. U. S. A.*, **105**, 12797–802.

Looser, J., S. Schroder-Lang, P. Hegemann, and G. Nagel (2009). Mechanistic insights in light-induced cAMP production by photoactivated adenylyl cyclase alpha (PACalpha). *Biol. Chem.*, **390**, 1105–11.

Losi, A. and W. Gartner (2008). Bacterial bilin- and flavin-binding photoreceptors. *Photochem. Photobiol. Sci.*, **7**, 1168–78.

Mandalari, C., A. Losi, and W. Gartner (2013). Distance-tree analysis, distribution and co-presence of bilin- and flavin-binding prokaryotic photoreceptors for visible light. *Photochem Photobiol Sci*, **12**, 1144–57.

Marden, J.N., Q. Dong, S. Roychowdhury, J.E. Berleman, and C.E. Bauer (2011). Cyclic GMP controls *Rhodospirillum centenum* cyst development. *Mol. Microbiol.*, **79**, 600–15.

Moglich, A. and K. Moffat (2010). Engineered photoreceptors as novel optogenetic tools. *Photochem. Photobiol. Sci.*, **9**, 1286–300.

Montgomery, K.L., A.J. Yeh, J.S. Ho, V. Tsao, S. Mohan Iyer, L. Grosenick, E.A. Ferenczi, Y. Tanabe, K. Deisseroth, S.L. Delp, and A.S. Poon (2015). Wirelessly powered, fully internal optogenetics for brain, spinal and peripheral circuits in mice. *Nat Methods*, **12**, 969–74.

Motta-Mena, L.B., A. Reade, M.J. Mallory, S. Glantz, O.D. Weiner, K.W. Lynch, and K.H. Gardner (2014). An optogenetic gene expression system with rapid activation and deactivation kinetics. *Nat. Chem. Biol.*, **10**, 196–202.

Myakishev-Rempel, M., I. Stadler, P. Brondon, D.R. Axe, M. Friedman, F.B. Nardia, and R. Lanzafame (2012). A preliminary study of the safety of red light phototherapy of tissues harboring cancer. *Photomed. Laser Surg.*, **30**, 551–8.

Nagahama, T., T. Suzuki, S. Yoshikawa, and M. Iseki (2007). Functional transplant of photoactivated adenylyl cyclase (PAC) into *Aplysia* sensory neurons. *Neurosci. Res.*, **59**, 81–8.

Oplander, C., A. Deck, C.M. Volkmar, M. Kirsch, J. Liebmann, M. Born, F. van Abeelen, E.E. van Faassen, K.D. Kroncke, J. Windolf, and C.V. Suschek (2013). Mechanism and biological relevance of blue-light (420–453 nm)-induced nonenzymatic nitric oxide generation from photolabile nitric oxide derivates in human skin in vitro and in vivo. *Free Radic. Biol. Med.*, **65**, 1363–77.

Piatkevich, K.D., F.V. Subach, and V.V. Verkhusha (2013). Far-red light photoactivatable near-infrared fluorescent proteins engineered from a bacterial phytochrome. *Nat Commun*, **4**, 2153.

Roembke, B.T., J. Zhou, Y. Zheng, D. Sayre, A. Lizardo, L. Bernard, and H.O. Sintim (2014). A cyclic dinucleotide containing 2-aminopurine is a general fluorescent sensor for c-di-GMP and 3′,3′-cGAMP. *Mol. Biosyst.*, **10**, 1568–75.

Ryu, M.H., I.H. Kang, M.D. Nelson, T.M. Jensen, A.I. Lyuksyutova, J. Siltberg-Liberles, D.M. Raizen, and M. Gomelsky (2014). Engineering adenylate cyclases regulated by near-infrared window light. *Proc Natl Acad Sci U S A*, **111**, 10167–72.

Ryu, M.H., O.V. Moskvin, J. Siltberg-Liberles, and M. Gomelsky (2010). Natural and engineered photoactivated nucleotidyl cyclases for optogenetic applications. *J. Biol. Chem.*, **285**, 41501–8.

Ryu, M.H., H. Youn, I.H. Kang, and M. Gomelsky (2015). Identification of bacterial guanylate cyclases. *Proteins*, **83**, 799–804.

Samanta, A., M. Thunemann, R. Feil, and T. Stafforst (2014). Upon the photostability of 8-nitro-cGMP and its caging as a 7-dimethylaminocoumarinyl ester. *Chem. Commun. (Camb.)*, **50**, 7120–3.

Scheib, U., K. Stehfest, C.E. Gee, H.G. Korschen, R. Fudim, T.G. Oertner, and P. Hegemann (2015). The rhodopsin-guanylyl cyclase of the aquatic fungus *Blastocladiella emersonii* enables fast optical control of cGMP signaling. *Sci Signal*, **8**, rs8.

Schroder-Lang, S., M. Schwarzel, R. Seifert, T. Strunker, S. Kateriya, J. Looser, M. Watanabe, U.B. Kaupp, P. Hegemann, and G. Nagel (2007). Fast manipulation of cellular cAMP level by light *in vivo*. *Nat Methods*, **4**, 39–42.

Sikka, G., G.P. Hussmann, D. Pandey, S. Cao, D. Hori, J.T. Park, J. Steppan, J.H. Kim, V. Barodka, A.C. Myers, L. Santhanam, D. Nyhan, M.K. Halushka, R.C. Koehler, S.H. Snyder, L.A. Shimoda, and D.E. Berkowitz (2014). Melanopsin mediates light-dependent relaxation in blood vessels. *Proc. Natl. Acad. Sci. U. S. A.*, **111**, 17977–82.

Sinha, S.C. and S.R. Sprang (2006). Structures, mechanism, regulation and evolution of class III nucleotidyl cyclases. *Rev. Physiol. Biochem. Pharmacol.*, **157**, 105–40.

Stierl, M., A. Penzkofer, J.T. Kennis, P. Hegemann, and T. Mathes (2014). Key residues for the light regulation of the blue light-activated adenylyl cyclase from *Beggiatoa* sp. *Biochemistry*, **53**, 5121–30.

Stortkuhl, K.F. and A. Fiala (2011). The smell of blue light: a new approach toward understanding an olfactory neuronal network. *Front. Neurosci.*, **5**, 72.

Sun, L., J. Wu, F. Du, X. Chen, and Z.J. Chen (2013). Cyclic GMP–AMP synthase is a cytosolic DNA sensor that activates the type I interferon pathway. *Science*, **339**, 786–91.

Takala, H., A. Bjorling, O. Berntsson, H. Lehtivuori, S. Niebling, M. Hoernke, I. Kosheleva, R. Henning, A. Menzel, J.A. Ihalainen, and S. Westenhoff (2014). Signal amplification and transduction in phytochrome photosensors. *Nature*, **509**, 245–8.

Tang, X., J. Zhang, J. Sun, Y. Wang, J. Wu, and L. Zhang (2013). Caged nucleotides/nucleosides and their photochemical biology. *Org. Biomol. Chem.*, **11**, 7814–24.

Weissenberger, S., C. Schultheis, J.F. Liewald, K. Erbguth, G. Nagel, and A. Gottschalk (2011). PACalpha – an optogenetic tool for *in vivo* manipulation of cellular cAMP levels, neurotransmitter release, and behavior in *Caenorhabditis elegans*. *J. Neurochem.*, **116**, 616–25.

Werner, G.S., C. Schaefer, R. Dirks, H.R. Figulla, and H. Kreuzer (1993). Doppler echocardiographic assessment of left ventricular filling in idiopathic dilated cardiomyopathy during a one-year follow-up: relation to the clinical course of disease. *Am. Heart J.*, **126**, 1408–16.

Wu, Y., S.S. Li, X. Jin, N. Cui, S. Zhang, and C. Jiang (2015). Optogenetic approach for functional assays of the cardiovascular system by light activation of the vascular smooth muscle. *Vascul. Pharmacol.*, **71**, 192–200.

Yasukawa, H., A. Sato, A. Kita, K. Kodaira, M. Iseki, T. Takahashi, M. Shibusawa, M. Watanabe, and K. Yagita (2013). Identification of photoactivated adenylyl cyclases in *Naegleria australiensis* and BLUF-containing protein in *Naegleria fowleri*. *J. Gen. Appl. Microbiol.*, **59**, 361–9.

Ye, H., M. Daoud-El Baba, R.W. Peng, and M. Fussenegger (2011). A synthetic optogenetic transcription device enhances blood-glucose homeostasis in mice. *Science*, **332**, 1565–8.

Yizhar, O., L.E. Fenno, T.J. Davidson, M. Mogri, and K. Deisseroth (2011). Optogenetics in neural systems. *Neuron*, **71**, 9–34.

11 Bioluminescence Activation of Light-sensing Molecules

Ute Hochgeschwender

11.1 Introduction

Optogenetic approaches provide control of neuronal activity in rapid, millisecond timescales achieved by physical light activation (Boyden *et al.*, 2005; Li *et al.*, 2005). However, many applications, specifically *in vivo*, often do not require the highest temporal resolution; instead, the usefulness of opsins for *in vivo* studies within the brain has been limited by the need for optical fibers for light delivery, specifically regarding the number and the location(s) of neurons that can be photostimulated. Because of the small dimensions of the optical fibers commonly used for optogenetic photostimulation, as well as the attenuation of light in brain tissue due to light scattering, the volume over which neurons can be excited is much smaller than the volume of most major brain structures; this means that only a small fraction of the relevant neurons will be activated. Simultaneous photostimulation of multiple locations can increase this number, but requires multiple light sources, which is often impractical. Finally, insertion of light guides or optic fibers in order to access deep brain structures is technically demanding and causes damage to brain tissue.

To allow manipulation of the activity of dispersed neuronal populations using optogenetic probes without fiber-optic implants, we developed an approach whereby bioluminescence – biological light produced by enzymatic reaction between a protein, luciferase and its diffusible substrate, luciferin – activates the opsin, which is tethered to the luciferase, creating a luminescent opsin, or luminopsin (LMO) (Figure 11.1) (Berglund *et al.*, 2013). After injection into the peripheral bloodstream, luciferin reaches a target in the brain because it crosses the blood–brain barrier (Birkner *et al.*, 2014). Light is generated by the luciferase and then activates the opsin, resulting in activation (in case of channelrhodopsins) or inhibition (in case of proton or chloride pumps or chloride channels) of the target neurons. This strategy takes full advantage of optogenetic elements – direct translation of light activation into neural effects, effective in any neuron or cell type and an ever-expanding repertoire of functionality – by using them as the common denominator, but switching out the light source from an invasive,

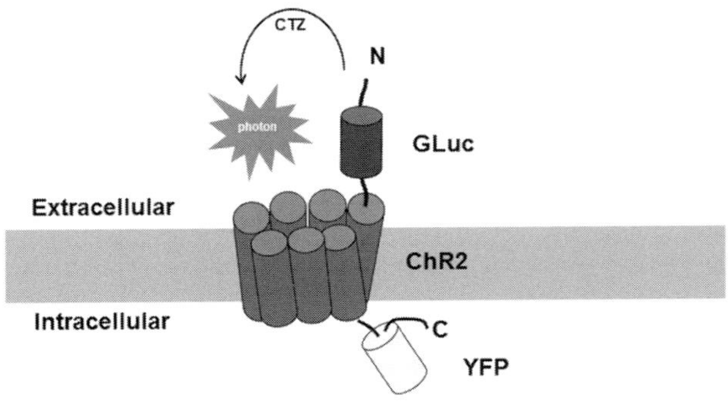

Figure 11.1 Design of luminopsins (LMOs). Here, we fused channelrhodopsin-2 from *Chlamydomonas* (ChR) (Nagel *et al.*, 2003) to a luciferase from the marine copepod *Gaussia princeps* (GLuc) (Verhaegen and Christopoulos, 2002), generating LMO1 (Berglund *et al.*, 2013). The luciferase is fused to the Nterminus of ChR2, while a fluorescent reporter (YFP) is fused to the Cterminus. The small-molecule substrate coelenterazine (CTZ) is oxidized by GLuc to generate bioluminescence with an emission peak at 470 nm, which is near the excitation peak of ChR2. ChR2 can be activated by physical light or by application of CTZ. This generic design can be modified by replacing GLuc with a different luciferase, or replacing ChR2 with a different channelrhodopsin or proton pump, as long as the emission wavelength of the luciferase matches the activation wavelength of the optogenetic element.

physical one to a non-invasive, biological one (i.e. a light-producing protein, or a luciferase).

This approach shares non-invasive neuronal modulation with other approaches, referred to as "chemogenetics" (Sternson and Roth, 2014), in which a genetically targeted actuator interacts with a systemically administered small molecule. However, our approach is unique in its integration of opto- and chemo-genetic technology. Capitalizing on the major advantage of opsins as powerful generators of electrical current, bioluminescence-driven optogenetics integrates opto- and chemo-genetic methods by preserving the conventional photoactivation of opsins in order to provide optimal time resolution in defined spaces, where desired, while at the same time providing chemogenetic access to the same molecules so as to permit chronic and non-invasive control of entire populations. This allows manipulation of neuronal activity over a range of spatial and temporal scales in the same experimental animal. Comprehensive interrogation of neuronal circuits requires acute and chronic manipulations of spatially defined subpopulations, as well as entire populations dispersed over the brain. Thus, there is a need to combine optogenetic and chemogenetic approaches to allow the use of both modes of interrogation in the same brain circuit, and, ideally, through the same actuator molecule, thereby facilitating the comprehensive study of neuronal systems.

Several groups, including our own, are continuously complementing the growing optogenetic toolset along this chemogenetic dimension by engineering fusions of light-generating luciferases to light-activated molecules, and these groups are starting to apply these versatile tools in order to address pertinent questions (Berglund *et al.*, 2013; Berglund *et al.*, 2015; Higashikubo *et al.*, 2015;

Kopparaju *et al.*, 2015; Tung, Gutekunst and Gross, 2015; Wen *et al.*, 2015). This chapter will introduce the basic building blocks for bioluminescence-driven optogenetics, expand on ways of combining them and discuss possible applications.

11.2 Methodology

Generating and testing the tools for the bioluminescent activation of optogenetic elements involves standard methods of molecular biology, the production of viral vectors, tissue culture methods and electrophysiology approaches (e.g. patch clamps and multi-electrode arrays [MEAs]). Tools are implemented *in vivo* (e.g. in behavioral experiments, in multi-array electrophysiology recordings in awake animals and in acute brain slice recordings).

11.2.1 Construction of LMOs

Starting materials for our LMO constructs were plasmids obtained from individual laboratories, from Addgene or from commercial vendors. We generated fusion constructs of luciferases and optogenetic elements by using a combination of DNA synthesis (Genscript), polymerase chain reaction, recombination cloning and standard cloning. In all of our current LMO constructs, the luciferase is separated from the opsin by a 15-amino acid linker. Furthermore, all constructs carry the signal peptide from *Gaussia* luciferase (Berglund *et al.*, 2013). The coding sequence was cloned in either a 5′ to 3′ direction after the promoter or in an inverted orientation and flanked by *loxP* sites (Schnütgen *et al.*, 2003; Atasoy *et al.*, 2008) for use in Cre recombinase-expressing animals. Cloning vectors were mammalian expression plasmids, such as pcDNA3.1, for use in tissue culture cells, or viral vectors (lentivirus and adeno-associated virus [AAV]) when high transduction efficiencies were needed (e.g. in primary neurons) and for *in vivo* applications.

11.2.2 Testing LMOs *In Vitro*

Constructs (expression plasmids with ubiquitous promoters) were first transfected into HEK cells; HEK cells provide a convenient heterologous expression system for testing and verifying new LMO constructs. Fluorescence microscopy was used to determine proper membrane expression of the LMOs. Bioluminescence emission was measured using a luminometer, including response kinetics, time of light onset and duration in response to the application of different concentrations of luciferase substrate. In addition to giving us basic parameters for the comparison of constructs, this step served as a checkpoint for the proper function and expression of constructs before going forward with functional tests.

The ability of the bioluminescence emitted by the luciferase to activate optogenetic elements was tested by single-cell patch clamp recordings in HEK cells and/or – in the case of neuron-specific promoters – in primary neurons. Bioluminescence emission from LMO-expressing cells was monitored concurrently with electrophysiological recordings. Neurons were also plated on MEAs and transduced with various LMO constructs in order to study the effects of LMOs in neuronal populations. Changes in spontaneous activity after adding the substrate were recorded.

11.2.3 Applying LMOs *In Vivo*

While the non-invasive manipulation of neural circuits in the behaving animal is the objective, new constructs need to be validated *in vivo* first. To this end, we measured, in mice stereotactically injected with LMO-expressing AAV, the effects of peripheral (intravenous and intraperitoneal) substrate application on neuronal response as assessed by *in vivo* multi-electrode recordings, and on stereotypic motor behavior as assessed by ipsi- and contra-versive circling (Clissold *et al.*, 2014). Expression of LMO in the brain was also confirmed by *in vivo* bioluminescence imaging (IVIS, Perkin Elmer) (Birkner *et al.*, 2014).

The same synaptic circuit analyzed by LMOs in behaving animals *in vivo* can also subsequently be interrogated, using tissue from the same animal, for higher-resolution analysis *ex vivo*. We used current clamp recordings to define the ability of the luciferase substrate to excite LMO-expressing neurons in acute brain slices. Furthermore, as the optogenetic elements in LMOs are as accessible to physical light stimulation as the original individual channelrhodopsins or pumps, we used LMOs to optically map the circuit between presynaptic LMO-expressing cells and their postsynaptic partners (Wen *et al.*, 2015).

Full details of all experimental methods can be found in recent publications from our laboratory and our colleagues' laboratories (Berglund *et al.* 2013; Tung *et al.* 2015; Berglund *et al.*, in press).

11.3 Results and Discussion

11.3.1 Light Emission

There is a wealth of luciferases available, the majority provided by marine organisms (Shimomura, 1985; Haddock, Moline and Case, 2010). Parameters for choosing a biological light source among different luciferases are the wavelength of emitted light, the overall luminescence, the kinetics of light emission, the substrate for the luciferase and the need for possible cofactors. Selected examples are listed in Table 11.1. The color of emitted light has to match the wavelength for activation of the targeted light sensor. The overall luminescence has to be intense enough to activate the target molecule. Different kinetics of light emission will allow for choosing between flash and glow kinetics, and between different half-lives, depending on the experiments planned. Size is a consideration when using the luciferase in fusion proteins that are later packaged into viral particles. All luciferases that are chosen should function at physiological pH and temperature, and should be codon-optimized for use in different host organisms. Depending on the application, the luciferase substrate needs to fulfill specific requirements, such as stability or the ability to cross the blood–brain barrier. Lastly, again depending on the specific application, the need for a cofactor, such as ATP in the case of firefly luciferase FLuc (Deluca, 1976), might interfere with, for example,

Table 11.1 Matching luciferases with optogenetic elements. Key characteristics are listed for four examples of luciferases with emission spectra extending from blue to red: NanoLuc, which is based on the smaller subunit from the luciferase of the deep sea shrimp *Oplophorus gracilirostris* (Shimomura *et al.*, 1978; Hall *et al.*, 2012); GLuc; NanoLantern, a fusion of the bright fluorescent protein Venus and an enhanced *Renilla* luciferase variant, RLuc8 (Saito *et al.*, 2012); and FLuc. These can be partnered with optogenetic elements falling in two broad classes distinguished by activation wavelength into blue- and red-shifted elements. Within each class there are two luciferases that can be partnered with activating and silencing opsins; in addition, the blue-shifted spectrum has activating and silencing step function opsins. Examples given here are VChR1 (Ernst *et al.*, 2008; Zhang *et al.*, 2008), ReaChR (Lin *et al.*, 2013), Mac (Chow *et al.*, 2010; Han *et al.*, 2011) and Halo (Lanyi *et al.*, 1990; Han and Boyden, 2007; Zhang *et al.*, 2007), as well as ChR2(C128S) (Berndt *et al.*, 2009) and iChloC (Wietek *et al.*, 2015).

	Blue-shifted		Red-shifted	
	400 nm ←	Wavelength		→ 600 nm
Luciferase	NLuc	GLuc	RLuc	FLuc
Organism	*Oplophorus gracilirostris*	*Gaussia princeps*	*Renilia reniformis*	*Photinus pyralis*
Variant	NanoLuc	GLuc	NanoLantern	Lic2
Size	19 kDa	20 kDa	36 kDa	61 kDa
Substrate	Furimazine	Coelenterazine	2-deoxycoelenterazine	D-luciferin
Cofactor	–	–	–	ATP
Emission peak	460 nm	470 nm	540 nm	565 nm
		Matching opsins		
Activating	VChR1, ChR2(C128S)		ReaChR	
Silencing	Mac, iChloC		Halo	

modulating neuronal activity (Le Masson, Przedborski and Abbott, 2014; Rangaraju, Calloway and Ryan, 2014).

So far, we have used *Gaussia* luciferase (GLuc) as a source of blue light, as it is the luciferase with the highest light emission (Verhaegen and Christopoulos, 2002; Tannous *et al.*, 2005). It is also one of the smallest luciferases available. Its substrate, coelenterazine (CTZ), reaches the brain after peripheral (intravenous or intraperitoneal) administration. While native GLuc generates strong bioluminescence, this might not be sufficient for all applications. Increasing the brightness of luciferases can be achieved in different ways. For example, mutations of GLuc have yielded superluminescent variants with a bioluminescence that is tenfold higher than the native form (Welsh *et al.*, 2009; Kim *et al.*, 2011). Sequence alignment of copepod luciferases inspired the creation of artificial luciferases (ALuc) with unique optical properties (Kim, Torimura and Tao, 2013). In the case of luminous organisms, the problem of low brightness has been solved in nature by the phenomenon of intramolecular bioluminescence resonance energy transfer (BRET), in which the excited energy of a luminescent substrate, bound to a luciferase or photoprotein, is efficiently transferred to the acceptor fluorescent protein by a Förster resonance energy transfer mechanism, thereby increasing the emitted photon number. Examples of such natural BRET pairs can be found in the sea pansy *Renilla reniformis* (RLuc and *Renilla* GFP) (Ward and Cormier, 1976) or the jellyfish *Aequorea victoria* (aequorin and *A. victoria* GFP) (Morise *et al.*, 1974). In an attempt to generate luciferase-based probes that can be used for high-resolution cell imaging, Hoshino *et al.* developed RLuc–EYFP fusion proteins, which enhanced the integrated luminescence intensity of the luciferase, and also induced a red shift in the emission peak (Hoshino, Nakajima and Ohmiya, 2007). By using a mutagenized brighter RLuc in combination with a circularly permutated Venus (a fluorescent protein with high BRET efficiency), luminescence was increased to a level that allowed real-time imaging of intracellular structures in living cells, as well as sensitive tumor detection in freely moving mice (Saito *et al.*, 2012). The high quantum efficiency of the Förster mechanism allows the overall light emission from the fluorescent protein acceptor in these intramolecular BRET pairs to be higher than that of the luciferase alone. In addition, these fusion proteins can be tuned to emit different wavelengths depending on the fluorescent protein they are partnered with, thus expanding the use of the original luciferase (Takai *et al.*, 2015).

Other sources of light are photoproteins, such as aequorin or obelin (Shimomura, Johnson and Saiga, 1962; Campbell, 1974). A photoprotein is a luciferase that forms a complex with its substrate first; light is produced upon the binding of calcium to the pre-complexed protein. Aequorin emits blue light in the presence of CTZ and calcium and thus has been used as a calcium indicator (Kendall *et al.*, 1994; Baubet *et al.*, 2000; Rogers *et al.*, 2005; Naumann *et al.*, 2010), including in transgenic mice (Rogers *et al.*, 2007). Employing calcium-sensing photoproteins as the light-emitting component will activate downstream light-sensing molecules in an activity-dependent manner. However, light emission from photoproteins is relatively weak. Brighter alternatives are split luciferases, where calcium sensing modules such as calmodulin-M13 are inserted into an otherwise not calcium-sensitive luciferase; only upon calcium binding are the two halves brought close enough together to allow normal luciferase function and light emission (Saito *et al.*, 2012).

For the substrate, we have used native CTZ in our studies. It is possible that analogs of this substrate could achieve even higher light emissions (Teranishi and Shimomura, 1997; Zhao *et al.*, 2004). Luciferase substrates are being utilized in a range of organisms in nature, and are without negative effects in the non-luminous organisms consuming them (e.g. in humans when eating seafood). Luciferases and their substrates have been widely used for whole-animal imaging studies; they are non-toxic and well tolerated over repeated applications. Specifically, CTZ has been used as a substrate for luciferases and photoproteins *in vivo* in flies, fish and mice, with no side effects reported (Tannous *et al.*, 2005; Martin *et al.*, 2007; Rogers *et al.*, 2007; Naumann *et al.*, 2010). For example, after incubating transgenic zebrafish expressing the calcium-sensing photoprotein aequorin in CTZ-containing medium, freely behaving fish have been monitored for neural activity (Naumann *et al.*, 2010). Expanding the pool of well-tolerated substrates is desirable, as the use of luciferases that act on different, non-overlapping substrates would allow the use of two distinct luciferase systems in the same experimental animal (see Table 11.1; CTZ versus furimazine versus D-luciferin).

11.3.2 Light Sensing

Combining light-producing and light-sensing molecules is limited by the rate of photons needed to activate the light sensor and by the rate of photon production by the luciferase. Our initial experiments with LMO1 and LMO2 demonstrated that the photons produced by GLuc-catalyzed CTZ oxidation are sufficient to activate channelrhodopsin. Using these designs as a baseline, there are several ways in which the coupling between the light-activated molecule and the light emitted by the luciferase can be improved. One is to improve on the side of light emission by the luciferase, as discussed in the previous paragraph. Another way is to partner the luciferases with optogenetic elements of increased light sensitivity. For example, the VChR1-containing LMO2 showed significantly improved coupling efficiency compared to the ChR2-containing LMO1, as VChR1 outperformed ChR2 due to its superior sensitivity to light (Berglund *et al.*, 2013). Recently, new optogenetic elements with higher light sensitivity have been reported, which should be ideal for combining with biological light emission (Chuong *et al.*, 2014; Klapoetke *et al.*, 2014; Govorunova *et al.*, 2015). In addition to newly discovered light-sensing molecules, variants of existing channelrhodopsins have been generated by directed mutagenesis. Several ChR2 mutants have been engineered that remain open for longer periods of time after light activation and show increased sensitivity to low light levels while generating large photocurrents (Berndt *et al.*, 2009; Berndt *et al.*, 2011; Berndt *et al.*, 2014; Wietek *et al.*, 2014; Wietek *et al.*, 2015). LMOs employing these ChR2 mutants will likely increase the coupling efficiency and thus will increase the potency of the LMOs for current modulation (Berglund *et al.*, 2015).

Higher light sensitivity resulting from the new optogenetic elements being identified, or the current ones being modified, will also allow us to extend the choice of partnering luciferases to those with lower light emission levels. This will enable the matching of existing (and future) optogenetic elements activated by light across the visible spectrum with luciferases emitting light along these wavelengths.

11.3.3 Combining Light Emitters with Light Sensors

Bringing the light-emitting luciferase in close enough proximity to the light-sensing molecule can, in principle, be achieved in several ways; from expressing both moieties in one fusion protein to simple co-transfection of the two partners.

Previously, we generated fusion proteins with the following basic design: luciferase–15-aa linker–opsin–fluorescent reporter (Figure 11.2A) (Berglund *et al.*, 2013). The arguments for this configuration are that it allows for the luciferase to be outside the cell, affording faster access to the substrate; it ensures the shortest distance from the luciferase to the opsin chromophore is achieved; and it guarantees equal amounts of both partners, minimizing inter-experimental variability. The luciferase could also be tethered to the C-terminus of the opsin, putting it inside the cell (Figure 11.2B); however, in this case, the fluorescent protein would be released from close proximity to the luciferase by introducing a self-cleaving peptide (2A) in order to avoid losing photons through BRET from the luciferase to the fluorescent reporter. Expressing the luciferase inside the cell potentially slows down light onset and dims light output compared to extracellular expression because the substrate has to get across the cell membrane. While extra- and intra-cellular configurations are both options for the naturally secreted luciferases, such as GLuc, other luciferases, such as the ATP-dependent FLuc or

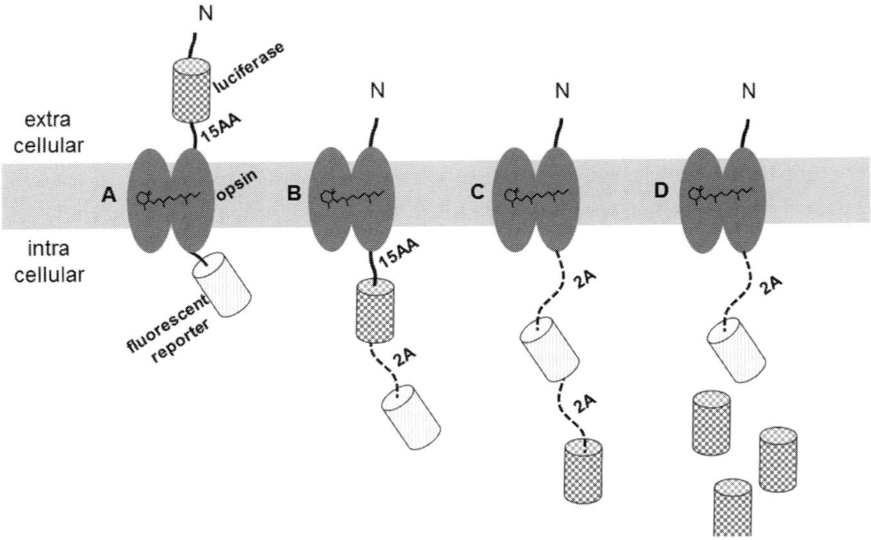

Figure 11.2 Configurations of luciferase–opsin combinations. (A) Initial design of LMO, which places the luciferase at the extracellular, N-terminal end of the opsin, and the fluorescent reporter at the intracellular, C-terminal end. (B) The luciferase can be placed inside the cell, still tethered to the opsin by a linker. In order not to divert light from the luciferase to the fluorescent reporter, a self-cleaving sequence (2A) releases the fluorescent molecule into the cell away from the luciferase. (C) Introducing into the cell a fusion protein in which all moieties are separated by self-cleaving sequences will ensure equimolar amounts of opsin and luciferase are achieved, while (D) co-transfection allows us to introduce the two components at different ratios. Symbols are generic: "luciferase" can be GLuc, RLuc, etc.; "opsin" can be a channel or a pump, activating or silencing; and "fluorescent reporter" can be EYFP, dTomato, etc. The stippled lines indicate that the connected moiety will be released inside the cell after cleavage for the 2A sequence.

calcium-dependent luciferases, have to be expressed inside the cell. Rather than expressing a fusion protein, the intracellular individual moieties could be untethered by introducing 2A peptides (Figure 11.2C). In this configuration, opsin and luciferase will be expressed by the cell at equimolar concentrations. However, it is uncertain whether the light that is emitted will be close enough to the chromophore to activate the opsin. This could be compensated for by co-transfecting the opsin with an excess of luciferase (Figure 11.2D) in an attempt to maximize the chances of light emission near the opsin. A drawback of the two partners being expressed independently of each other is that it makes inter-experimental outcomes more variable.

In our current studies, we fused the luciferase and the opsin in a chimeric molecule. The use of brighter luciferase variants and more sensitive opsins will enable the introduction of larger distances between the light emitter and light sensor. This would considerably expand the applications of the bioluminescence-driven activation of light-sensing molecules (see next section).

11.3.4 Applications for Bioluminescence-activated Optogenetics

Activation of optogenetic sensors by physical light is a way to achieve temporal precision on a millisecond scale. This unique feature of optogenetic elements is fully preserved in LMOs, while the same elements are also accessible to systemic, non-invasive activation via the luciferase substrate. This allows for a more comprehensive analysis of neuronal circuits, as, for example, activation of all members of a genetically defined population can be complemented by the interrogation of spatially restricted subgroups of this population. In another application of this bimodal interrogation approach, neuronal circuits in the same experimental animals that have been stimulated *in vivo* daily for several months for behavioral testing can then be analyzed *ex vivo*, in brain slices, by detailed photostimulation mapping with laser point light (Wang *et al.*, 2007). Thus, LMOs expand the usage of optogenetics by making the same opsins accessible to non-invasive, systemic control, while preserving their ability to be activated by external light sources, thereby enabling the same probe to manipulate neuronal activity over a range of spatial and temporal scales.

Optogenetic and chemogenetic approaches make use of one genetically targeted component – a receptor, channel or pump – and one externally administered component – a small molecule or light. Unique to LMOs is that both the light-generating component (luciferase) and the light-sensing component (opsin) are genetically encoded. This opens up combinatorial possibilities that cannot be realized with other methods of genetically targeted neuronal manipulation. The simplest application is to take advantage of intersectional specificity by being able to express each component under the control of a different promoter (Figure 11.3). Pairing of luciferase and opsin to a functional LMO will only occur in neuronal subpopulations defined by the combinatorial expression of two genes, thereby increasing the specificity of neuronal interrogation. Reconstitution of functional LMOs might be enhanced by employing heterodimer elements (Figure 11.3).

A long-term unmet need in neuronal circuit analysis has been a method to functionally map long-range and local connectivity. While enormous progress has been made in anatomically mapping neuronal connections, from these data it is not obvious which of the possible connections are actually functionally used in a given context, such as a behavior or a disease state. A method that enables the interrogation of different genetically defined individual target populations receiving synaptic input from a genetically defined neuronal population for their role in behavioral outputs is not feasible as of now. When combining bioluminescence with optogenetics, both light production and light sensing are genetically encoded, so it will be possible to introduce the light-producing luciferase and the light-sensing opsin into pre- and post-synaptic cells, respectively, so as to enable the remote control and, thereby, interrogation of synaptic transmission between defined populations (Figure 11.4).

Yet another way to capitalize on a genetically encoded light source is to utilize calcium-sensing luciferases, such as the photoproteins aequorin or obelin. In this scenario, light emission in the presence of CTZ is dependent on calcium influx into the cell, and thus on neuronal activity (Figure 11.5). Combining calcium-sensing luciferases with light-sensing opsins such as channelrhodopsin (for

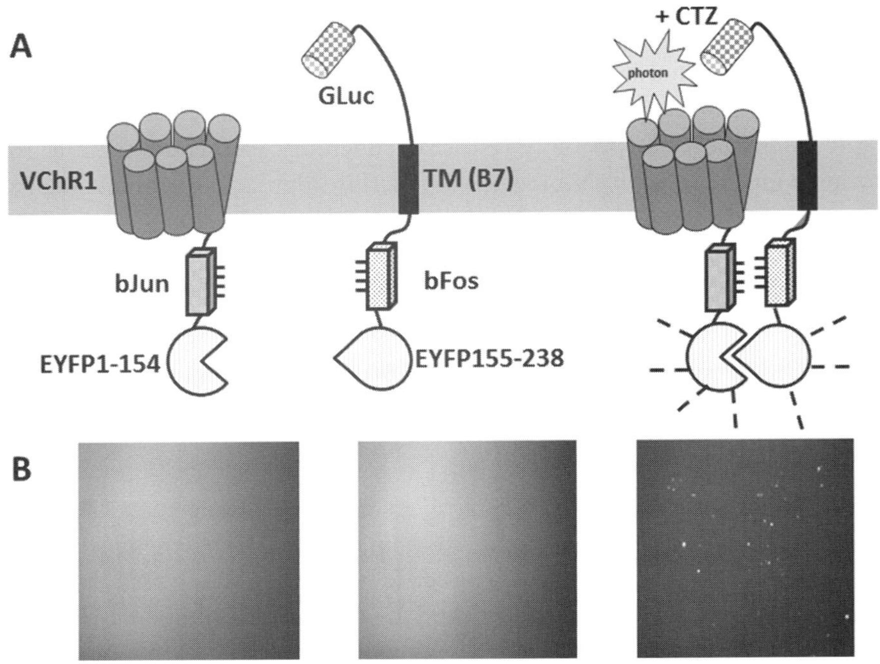

Figure 11.3 Expression of luciferase–opsin partners under the control of different promoters. (A) Increased intersectional specificity of expression can be achieved by using different promoters for the two partners whose activity intersects in selected subpopulation of cells. Close physical proximity of luciferase and opsin is achieved by introducing heterodimer sequences from bFos and bJun (Hu and Kerppola, 2003; Kerppola, 2006). Close proximity is monitored by fluorescence that is observed only when physical proximity is close enough to complement a split fluorescent reporter (N-terminal EYFP, aa 1–154; C-terminal EYFP, aa 155–238). The GLuc moiety is anchored to a B7 transmembrane sequence from the mouse CD80 antigen (Chou *et al.*, 1999). (B) A GLuc luciferase and VChR1 channelrhodopsin construct were transfected alone as well as co-transfected into HEK cells. Each construct by itself does not result in fluorescent cells, while co-transfection of both constructs shows fluorescence, indicating complementation of the split EYFP.

activation) or a proton pump (for silencing) would open the possibility of direct and non-invasive feedback control of neuronal activity.

A unique advantage of LMOs over any other genetic method for controlling neuronal activity is that light emission and the activation of the opsin can be simultaneously documented as luminescence. In mice, GLuc expressed in the brain produces detectable luminescence through the skull upon intravenous application of CTZ (Tannous *et al.*, 2005; Birkner *et al.*, 2014). As *in vivo* imaging of bioluminescence becomes increasingly powerful, it should be possible to correlate the intensity and location of bioluminescence with the modulation of neuronal activity and with behavioral phenotypes. This option for monitoring cells expressing LMOs would be especially useful in cell transplantation studies. Here, neural stem cells are transplanted at the site of neurodegeneration, ischemia or injury, with the ultimate goal of restoring neural function. There is increasing evidence that, in order for transplanted neurons to efficiently integrate into the brain, it might be highly beneficial to remotely stimulate activity in these neurons. LMOs will allow us to non-invasively modulate the neuro-restorative

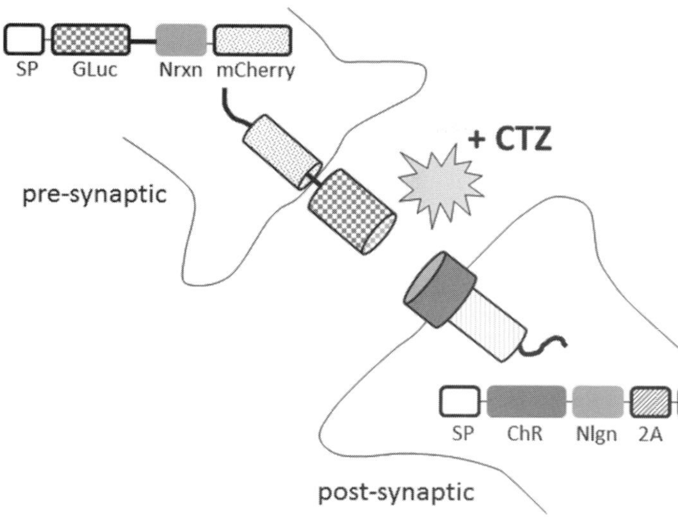

Figure 11.4 Expression of light-producing luciferase and light-sensing opsin across synaptic partners. Trans-synaptic information flow via photons can be achieved by expressing the light-emitting and light-transducing components at the pre- and post-synaptic sides. The principle design of our constructs closely follows those that have been developed for GFP reconstitution across synaptic partners (GRASP) (Feinberg *et al.*, 2008; Kim *et al.*, 2012; Yamagata and Sanes, 2012). The presynaptic component is tethered to neurexin (Nrxn) (Fairless *et al.*, 2008), and the postsynaptic component is tethered to neuroligin (Nlgn) (Ichtchenko *et al.*, 1995). Presynaptic component: signal peptide (SP), luciferase (GLuc), intracellular domain of the rat neurexin-1β (Nrxn), mCherry. Postsynaptic component: SP, channelrhodopsin (ChR), intracellular domain of mouse neuroligin-1 (Nlgn), self-cleaving 2A peptide (2A), EGFP. In the presence of the substrate CTZ, the luciferase emits photons, which activate the postsynaptic opsin.

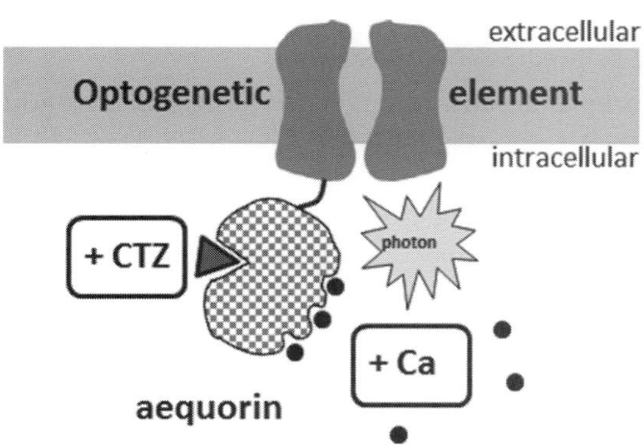

Figure 11.5 Self-regulation of neurons by making light emission calcium dependent. The photoprotein aequorin forms a complex with CTZ; binding to calcium ions is required to emit light. When combined with an optogenetic element, aequorin in the presence of CTZ emits light upon neuronal activation-induced calcium influx. Light emission will activate the opsin and, in case of a proton, chloride pump or channel, result in silencing of the neuron, or, when combined with a channelrhodopsin, cause activation of the neuron.

capacities of transplanted cells, while at the same time permitting the monitoring of their proliferation and integration into the host tissue.

Further possibilities for utilizing biological light sources lie in expanding bioluminescence-driven manipulation to other light-sensing tools. The optogenetic principle is applicable not only to light-sensing channels and pumps, but also to the rapidly expanding arsenal of light-sensing, broad-based cellular manipulation tools, such as plant phytochromes, cryptochromes and light oxygen voltage domain proteins for controlling protein localization, signaling pathways and DNA recombination and transcription. Bioluminescence-driven control of these tools would potentially enable the non-invasive control of basic cellular mechanisms.

11.4 Concluding Remarks

Using bioluminescence (i.e. luciferase-generated "biological" light) to activate optogenetic elements opens the entire optogenetic toolbox for complementation along a chemogenetic dimension, while preserving the unique features of opsins. LMOs enable manipulation over several timescales (milliseconds to tens of minutes) and over several spatial scales (localized to dispersed over the entire brain), thus providing tools for the analysis of both micro- and macro-circuits. Bioluminescent activation of optogenetic sensors has the potential to significantly improve the current capabilities of optogenetic tools by taking advantage of combining two genetically encoded modalities: light emission and light sensing. The luciferase and the optogenetic element can be expressed in the same cell under the control of different promoters for increased specificity, and in pre- and post-synaptic cells for functional mapping of synaptic partners. Employing calcium-sensing luciferases in combination with optogenetic elements could ultimately enable activity-dependent self-regulation of neurons.

This novel class of tools can be improved and extended in numerous ways and find applications beyond the photonic control of neurons in modifying many cell types and cellular processes.

REFERENCES

Atasoy, D., Aponte, Y., Su, H.H. and Sternson, S.M. (2008) A FLEX switch targets channelrhodopsin-2 to multiple cell types for imaging and long-range circuit mapping. *The Journal of Neuroscience*, **28**, 7025–30.

Baubet, V., Le Mouellic, H., Campbell, A.K., Lucas-Meunier, E., Fossier, P. and Brúlet, P. (2000) Chimeric green fluorescent protein-aequorin as bioluminescent Ca^{2+} reporters at the single-cell level. *Proceedings of the National Academy of Sciences of the United States of America*, **97**, 7260–5.

Berglund, K., Birkner, E., Augustine, G.J. and Hochgeschwender, U. (2013) Light-emitting channelrhodopsins for combined optogenetic and chemical-genetic control of neurons. *PLoS ONE*, **8**, e59759.

Berglund, K., Gutekunst, C.-A., Tung, J., Hochgeschwender, U. and Gross, R.E. (2015) Step-function luminopsin for prolonged activation of neurons by bioluminescence. *Society for Neuroscience Abstracts*.

Berglund, K., Tung, J.K., Higashikubo, B., Gross, R.E., Moore, C.I. and Hochgeschwender, U. (2016) Combined optogenetic and chemogenetic control of neurons. *Methods in Molecular Biology*, **1408**, 207–25.

Berndt, A., Lee, S.Y., Ramakrishnan, C. and Deisseroth, K. (2014) Structure-guided transformation of channelrhodopsin into a light-activated chloride channel. *Science (New York, N.Y.)*, **344**, 420–4.

Berndt, A., Schoenenberger, P., Mattis, J., Tye, K.M., Deisseroth, K., Hegemann, P., *et al.* (2011) High-efficiency channelrhodopsins for fast neuronal stimulation at low light levels. *Proceedings of the National Academy of Sciences of the United States of America*, **108**, 7595–600.

Berndt, A., Yizhar, O., Gunaydin, L.A., Hegemann, P. and Deisseroth, K. (2009) Bi-stable neural state switches. *Nature Neuroscience*, **12**, 229–34.

Birkner, E., Berglund, K., Klein, M.E., Augustine, G.J. and Hochgeschwender, U. (2014) Non-invasive activation of optogenetic actuators. *SPIE Proceedings*, **8928**, 89282F.

Boyden, E.S., Zhang, F., Bamberg, E., Nagel, G. and Deisseroth, K. (2005) Millisecond-timescale, genetically targeted optical control of neural activity. *Nature neuroscience*, **8**, 1263–8.

Campbell, A.K. (1974) Extraction, partial purification and properties of obelin, the calcium-activated luminescent protein from the hydroid *Obelia geniculata*. *The Biochemical Journal*, **143**, 411–8.

Chou, W.C., Liao, K.W., Lo, Y.C., Jiang, S.Y., Yeh, M.Y. and Roffler, S.R. (1999) Expression of chimeric monomer and dimer proteins on the plasma membrane of mammalian cells. *Biotechnology and Bioengineering*, **65**, 160–9.

Chow, B.Y., Han, X., Dobry, A.S., Qian, X., Chuong, A.S., Li, M., *et al.* (2010) High-performance genetically targetable optical neural silencing by light-driven proton pumps. *Nature*, **463**, 98–102.

Chuong, A.S., Miri, M.L., Busskamp, V., Matthews, G.A.C., Acker, L.C., Sørensen, A.T., *et al.* (2014) Noninvasive optical inhibition with a red-shifted microbial rhodopsin. *Nature Neuroscience*, **17**, 1123–9.

Clissold, K.A., Berglund, K., Klein, M.E., Prevosto, V., Koval, M., Abzug, Z.M., *et al.* (2014) *In vivo* bioluminescence-driven optogenetics for neuronal activation and inhibition. *Society for Neuroscience Abstracts*.

Deluca, M. (1976) Firefly luciferase. *Advances in Enzymology and Related Areas of Molecular Biology*, **44**, 37–68.

Ernst, O.P., Sánchez Murcia, P.A., Daldrop, P., Tsunoda, S.P., Kateriya, S. and Hegemann, P. (2008) Photoactivation of channelrhodopsin. *The Journal of Biological Chemistry*, **283**, 1637–43.

Fairless, R., Masius, H., Rohlmann, A., Heupel, K., Ahmad, M., Reissner, C., *et al.* (2008) Polarized targeting of neurexins to synapses is regulated by their C-terminal sequences. *The Journal of Neuroscience*, **28**, 12969–81.

Feinberg, E.H., Vanhoven, M.K., Bendesky, A., Wang, G., Fetter, R.D., Shen, K., *et al.* (2008) GFP reconstitution across synaptic partners (GRASP) defines cell contacts and synapses in living nervous systems. *Neuron*, **57**, 353–63.

Govorunova, E.G., Sineshchekov, O.A., Janz, R., Liu, X. and Spudich, J.L. (2015) Natural light-gated anion channels: a family of microbial rhodopsins for advanced optogenetics. *Science*, **349**, 647–50.

Haddock, S.H.D., Moline, M.A. and Case, J.F. (2010) Bioluminescence in the sea. *Annual Review of Marine Science*, **2**, 443–93.

Hall, M.P., Unch, J., Binkowski, B.F., Valley, M.P., Butler, B.L., Wood, M.G., *et al.* (2012) Engineered luciferase reporter from a deep sea shrimp utilizing a novel imidazopyrazinone substrate. *ACS Chemical Biology*, **7**, 1848–57.

Han, X. and Boyden, E.S. (2007) Multiple-color optical activation, silencing, and desynchronization of neural activity, with single-spike temporal resolution. *PLoS One*, **2**, e299.

Han, X., Chow, B.Y., Zhou, H., Klapoetke, N.C., Chuong, A., Rajimehr, R., *et al.* (2011) A high-light sensitivity optical neural silencer: development and application to optogenetic control of non-human primate cortex. *Frontiers in Systems Neuroscience*, **5**, 18.

Higashikubo, B., McDonnell, E., Hochgeschwender, U. and Moore, C.I. (2015) Multi-timescale In vivo regulation of the thalamic reticular nucleus using bioluminescent optogenetics (BL-OG). *Society for Neuroscience Abstracts*.

Hoshino, H., Nakajima, Y. and Ohmiya, Y. (2007) Luciferase-YFP fusion tag with enhanced emission for single-cell luminescence imaging. *Nature Methods*, **4**, 637–9.

Hu, C.-D. and Kerppola, T.K. (2003) Simultaneous visualization of multiple protein interactions in living cells using multicolor fluorescence complementation analysis. *Nature Biotechnology*, **21**, 539–45.

Ichtchenko, K., Hata, Y., Nguyen, T., Ullrich, B., Missler, M., Moomaw, C., *et al.* (1995) Neuroligin 1: a splice site-specific ligand for beta-neurexins. *Cell*, **81**, 435–43.

Kendall, J.M., Badminton, M.N., Dormer, R.L. and Campbell, A.K. (1994) Changes in free calcium in the endoplasmic reticulum of living cells detected using targeted aequorin. *Analytical Biochemistry*, **221**, 173–81.

Kerppola, T.K. (2006) Design and implementation of bimolecular fluorescence complementation (BiFC) assays for the visualization of protein interactions in living cells. *Nature Protocols*, **1**, 1278–1286.

Kim, S.B., Suzuki, H., Sato, M. and Tao, H. (2011) Superluminescent variants of marine luciferases for bioassays. *Analytical Chemistry*, **83**, 8732–40.

Kim, S.B., Torimura, M. and Tao, H. (2013) Creation of artificial luciferases for bioassays. *Bioconjugate Chemistry*, **24**, 2067–75.

Kim, J., Zhao, T., Petralia, R.S., Yu, Y., Peng, H., Myers, E., *et al.* (2012) mGRASP enables mapping mammalian synaptic connectivity with light microscopy. *Nature Methods*, **9**, 96–102.

Klapoetke, N.C., Murata, Y., Kim, S.S., Pulver, S.R., Birdsey-Benson, A., Cho, Y.K., *et al.* (2014) Independent optical excitation of distinct neural populations. *Nature Methods*, **11**, 338–46.

Kopparaju, R., Lin, S.H., Chen, Y.C., Hochgeschwender, U. and Chen, C.C. (2015) Intimate touch at a distance. *Society for Neuroscience Abstracts*.

Lanyi, J.K., Duschl, A., Hatfield, G.W., May, K. and Oesterhelt, D. (1990) The primary structure of a halorhodopsin from *Natronobacterium pharaonis*. Structural, functional and evolutionary implications for bacterial rhodopsins and halorhodopsins. *The Journal of Biological Chemistry*, **265**, 1253–60.

Li, X., Gutierrez, D. V, Hanson, M.G., Han, J., Mark, M.D., Chiel, H., *et al.* (2005) Fast noninvasive activation and inhibition of neural and network activity by vertebrate rhodopsin and green algae channelrhodopsin. *Proceedings of the National Academy of Sciences of the United States of America*, **102**, 17816–21.

Lin, J.Y., Knutsen, P.M., Muller, A., Kleinfeld, D. and Tsien, R.Y. (2013) ReaChR: a red-shifted variant of channelrhodopsin enables deep transcranial optogenetic excitation. *Nature Neuroscience*, **16**, 1499–508.

Martin, J.-R., Rogers, K.L., Chagneau, C. and Brûlet, P. (2007) *In vivo* bioluminescence imaging of Ca signalling in the brain of *Drosophila*. *PLoS One*, **2**, e275.

Le Masson, G., Przedborski, S. and Abbott, L.F. (2014) A computational model of motor neuron degeneration. *Neuron*, **83**, 975–88.

Morise, H., Shimomura, O., Johnson, F.H. and Winant, J. (1974) Intermolecular energy transfer in the bioluminescent system of aequorea. *Biochemistry*, **13**, 2656–62.

Nagel, G., Szellas, T., Huhn, W., Kateriya, S., Adeishvili, N., Berthold, P., *et al.* (2003) Channelrhodopsin-2, a directly light-gated cation-selective membrane channel. *Proceedings of the National Academy of Sciences of the United States of America*, **100**, 13940–5.

Naumann, E.A., Kampff, A.R., Prober, D.A., Schier, A.F. and Engert, F. (2010) Monitoring neural activity with bioluminescence during natural behavior. *Nature Neuroscience*, **13**, 513–20.

Rangaraju, V., Calloway, N. and Ryan, T.A. (2014) Activity-driven local ATP synthesis is required for synaptic function. *Cell*, **156**, 825–35.

Rogers, K.L., Picaud, S., Roncali, E., Boisgard, R., Colasante, C., Stinnakre, J., *et al.* (2007) Non-invasive *in vivo* imaging of calcium signaling in mice. *PLoS One*, **2**, e974.

Rogers, K.L., Stinnakre, J., Agulhon, C., Jublot, D., Shorte, S.L., Kremer, E.J., *et al.* (2005) Visualization of local Ca^{2+} dynamics with genetically encoded bioluminescent reporters. *The European Journal of Neuroscience*, **21**, 597–610.

Saito, K., Chang, Y.-F., Horikawa, K., Hatsugai, N., Higuchi, Y., Hashida, M., *et al.* (2012) Luminescent proteins for high-speed single-cell and whole-body imaging. *Nature Communications*, **3**, 1262.

Schnütgen, F., Doerflinger, N., Calléja, C., Wendling, O., Chambon, P. and Ghyselinck, N.B. (2003) A directional strategy for monitoring Cre-mediated recombination at the cellular level in the mouse. *Nature Biotechnology*, **21**, 562–5.

Shimomura, O. (1985) Bioluminescence in the sea: photoprotein systems. *Symposia of the Society for Experimental Biology*, **39**, 351–72.

Shimomura, O., Johnson, F.H. and Saiga, Y. (1962) Extraction, purification and properties of aequorin, a bioluminescent protein from the luminous hydromedusan, aequorea. *Journal of Cellular and Comparative Physiology*, **59**, 223–39.

Shimomura, O., Masugi, T., Johnson, F.H. and Haneda, Y. (1978) Properties and reaction mechanism of the bioluminescence system of the deep-sea shrimp *Oplophorus gracilorostris*. *Biochemistry*, **17**, 994–8.

Sternson, S.M. and Roth, B.L. (2014) Chemogenetic tools to interrogate brain functions. *Annual Review of Neuroscience*, **37**, 387–407.

Takai, A., Nakano, M., Saito, K., Haruno, R., Watanabe, T.M., Ohyanagi, T., *et al.* (2015) Expanded palette of nano-lanterns for real-time multicolor luminescence imaging. *Proceedings of the National Academy of Sciences of the United States of America*, **112**, 4352–6.

Tannous, B.A., Kim, D.-E., Fernandez, J.L., Weissleder, R. and Breakefield, X.O. (2005) Codon-optimized *Gaussia* luciferase cDNA for mammalian gene expression in culture and *in vivo*. *Molecular Therapy*, **11**, 435–43.

Teranishi, K. and Shimomura, O. (1997) Solubilizing coelenterazine in water with hydroxypropyl-.BETA.-cyclodextrin. *Bioscience, Biotechnology, and Biochemistry*, **61**, 1219–1220.

Tung, J.K., Gutekunst, C.-A. and Gross, R.E. (2015) Inhibitory luminopsins: genetically-encoded bioluminescent opsins for versatile, scalable, and hardware-independent optogenetic inhibition. *Scientific Reports*, **5**, 14366.

Verhaegen, M. and Christopoulos, T.K. (2002) Recombinant *Gaussia* luciferase. Overexpression, purification, and analytical application of a bioluminescent reporter for DNA hybridization. *Analytical Chemistry*, **74**, 4378–85.

Wang, H., Peca, J., Matsuzaki, M., Matsuzaki, K., Noguchi, J., Qiu, L., *et al.* (2007) High-speed mapping of synaptic connectivity using photostimulation in channelrhodopsin-2 transgenic mice. *Proceedings of the National Academy of Sciences of the United States of America*, **104**, 8143–8.

Ward, W.W. and Cormier, M.J. (1976) *In vitro* energy transfer in *Renilla* bioluminescence. *The Journal of Physical Chemsitry*, **80**, 2289–91.

Welsh, J.P., Patel, K.G., Manthiram, K. and Swartz, J.R. (2009) Multiply mutated *Gaussia* luciferases provide prolonged and intense bioluminescence. *Biochemical and Biophysical Research Communications*, **389**, 563–8.

Wen, L., Park, S.Y., Clissold, K.A., Berglund, K., Yin, H.H., Augustine, G.J., *et al.* (2015) Luminopsins allow neuronal activation over a range of spatial and temporal scales. *Society for Neuroscience Abstracts*.

Wietek, J., Beltramo, R., Scanziani, M., Hegemann, P., Oertner, T.G. and Simon Wiegert, J. (2015) An improved chloride-conducting channelrhodopsin for light-induced inhibition of neuronal activity *in vivo*. *Scientific Reports*, **5**, 14807.

Wietek, J., Wiegert, J.S., Adeishvili, N., Schneider, F., Watanabe, H., Tsunoda, S.P., *et al.* (2014) Conversion of channelrhodopsin into a light-gated chloride channel. *Science (New York, N.Y.)*, **344**, 409–12.

Yamagata, M. and Sanes, J.R. (2012) Transgenic strategy for identifying synaptic connections in mice by fluorescence complementation (GRASP). *Frontiers in Molecular Neuroscience*, **5**, 18.

Zhang, F., Prigge, M., Beyrière, F., Tsunoda, S.P., Mattis, J., Yizhar, O., *et al.* (2008) Red-shifted optogenetic excitation: a tool for fast neural control derived from *Volvox carteri*. *Nature Neuroscience*, **11**, 631–3.

Zhang, F., Wang, L.-P., Brauner, M., Liewald, J.F., Kay, K., Watzke, N., *et al.* (2007) Multimodal fast optical interrogation of neural circuitry. *Nature*, **446**, 633–9.

Zhao, H., Doyle, T.C., Wong, R.J., Cao, Y., Stevenson, D.K., Piwnica-Worms, D., *et al.* (2004) Characterization of coelenterazine analogs for measurements of *Renilla* luciferase activity in live cells and living animals. *Molecular Imaging*, **3**, 43–54.

Part III

Optogenetics in Neurobiology, Brain Circuits, and Plasticity

12 **Optogenetics for Neurological Disorders: What Is the Path to the Clinic?**

William F. Kaemmerer

12.1 Introduction

Since the development of optogenetic methods for the selective photostimulation of neurons (Zemelman *et al.*, 2002; Boyden *et al.*, 2005), the use of optogenetics as a research tool has grown tremendously, and has been very fruitful for elucidating the functions of neural circuits (Gradinaru *et al.*, 2009; Carr and Zachariou, 2014; Jin *et al.*, 2014; Lammel *et al.*, 2014). Given this success, it does not take a great leap of imagination to hope that optogenetics may also have therapeutic applications, helping address some of the many unmet medical needs for better treatments of neurological disorders. Optogenetic techniques are unlikely to provide a way to address the underlying pathology and progression of neurodegenerative diseases, or the diffuse biochemical or structural deficiencies underlying lysosomal storage disorders or schizophrenia. However, optogenetic control of neural circuitry may allow for better treatments of depression, obsessive–compulsive disorder, essential tremor, epilepsy, Tourette's syndrome or other disorders for which targeted modulation of neural circuitry may address the circuit dysfunctions underlying much of the person's suffering. Even in progressive diseases such as Parkinson's disease, more precise targeting of neuromodulation may have advantages over the present use of electrical stimulation to help manage a patient's symptoms.

12.2 Potential Advantages and Disadvantages of Optogenetics as a Therapeutic Modality

The potential advantages of optogenetics over other therapeutic modalities, such as drug delivery or electrical neuromodulation, are similar to those that have made optogenetics so valuable as a research tool. Light-responsive opsins allow for exquisite temporal specificity, with some capable of tracking light stimulation frequencies of as high as 200 Hz (Gunaydin *et al.*, 2010; Chiang *et al.*, 2014) and allowing virtually "instant on/instant off" control. Compared to electrical stimulation, optical stimulation offers the potential for superior spatial control over which portion of brain tissue is affected by the stimulation, not only by the placement of the light source, but also by virtue of the fact that only those cells transfected with an opsin are light responsive. This spatial control can be further enhanced by targeting opsin expression to specific cell types using cell type-specific promoters for driving opsin protein production, or by delivering the DNA encoding for the opsin to a region that is remote from the site of optical stimulation. The latter tactic involves relying on the intracellular transport of delivery vectors of opsins in order to result in light responsiveness in only those axons that are derived from the anatomical source of the DNA infusion. By this method, light stimulation of a fiber bundle can selectively modulate a specific circuit connection, while leaving other fibers of passage unaffected in a manner that is not possible with the steering of an electrical field. Furthermore, opsins that function as light-activated proton pumps or chloride ion pumps or channels can be used to suppress neuronal firing with a temporal and spatial precision that is not possible with pharmacological inhibitors or receptor antagonists. The use of excitatory and inhibitory opsins that are responsive to different wavelengths of light may provide bidirectional control of neural firing frequency in a manner that is unlikely ever to be achieved with electrical stimulation or drugs. Other opsins can be used to manipulate intracellular signaling pathways and transcription factors (see Krook-Magnuson and Soltesz [2015] for a review). Finally, as with electrical stimulation, optical stimulation can be applied without the pharmacological side effects and dosing considerations that invariably accompany drug interventions.

The potential disadvantages of optogenetics over other therapeutic modalities are largely unknown, but some can be anticipated. An optogenetic therapy for a disorder of the central nervous system will be invasive, involving at least the infusion of a vector to deliver the opsin and most likely a chronically implanted light delivery device. Furthermore, because the opsin is an ectopically expressed protein, there is the potential for the development of a limiting and possibly damaging immune response to the foreign protein. Even if the immune response is minimal, the safety of the chronic expression of the opsin protein vis-à-vis the continued viability and functioning of the host cell is an issue to be weighed against the potential benefits of the therapy.

Among the unknowns with respect to an optogenetic therapy are the phenomenological effects of the optogenetic activation or silencing of particular brain regions and circuits, which will vary depending upon the circuitry involved. What will it feel like to have areas of one's hippocampus silenced for a few seconds? Will it disrupt one's short-term memory or train of thought? Will it produce some other unusual experience or disruption of consciousness? These questions can be only imperfectly answered through animal studies; many unknowns will not be addressed until the first human clinical trials.

12.3 Practical Considerations for an Optogenetic Therapy

To get to first-in-human clinical trials, several practical considerations must be addressed. Unless a therapy proves effective when applied only on an intermittent and scheduled basis, the light source for the optogenetic stimulation of targets in the human brain will need to be contained within an implantable device. The light source itself could be in a device implanted outside of the cranium (e.g. in the pectoral region), and the light then delivered to the target tissue via optical fiber. Alternatively, miniature LEDs could be embedded in a lead implanted within the brain tissue itself. So far, engineering developments suggest that either approach may be feasible (see Ozden *et al.* [2013] and Montgomery *et al.* [2015], respectively). A greater consideration is the power requirement of the device. The power consumption of the therapy (with sufficient light power for efficacy) must be low enough for the device to have a longevity of several years if powered by a primary cell (battery), or at least several days if powered by a rechargeable cell (with recharging occurring transcutaneously). The overwhelming consideration for the longevity of the device will not be the complexity of the sensing and control circuitry, but the power drain by the light itself. For this reason, a therapy that need only be applied sporadically (i.e. with a low "duty cycle") will be more practical as a first clinical candidate than a therapy requiring either frequent or continuous application for efficacy.

By far the foremost practical consideration for an optogenetic therapy is the scale up of optogenetic techniques to a brain the size of a human's. The ability of optogenetic stimulation to modify neural functioning sufficiently to affect neurological performance and behavior has been demonstrated repeatedly in rodents. For example, optogenetic stimulation of the motor cortex in a mouse is sufficient to drive locomotion (Nature Video, 2010), and numerous studies have shown optogenetic stimulation to be capable of suppressing seizures in rodents (Krook-Magnuson *et al.*, 2013; Paz *et al.*, 2013; Chiang *et al.*, 2014; Krook-Magnuson *et al.*, 2014). However, optogenetic effects on behavior in primates have thus far been much more subtle. Optogenetic effects are readily observed in the primate brain at the level of single-unit recordings (Han, 2012), but demonstrations of effects on behavior have so far been limited to effects on saccades or other eye movements, requiring precise measurements to detect the effects (Cavanaugh *et al.*, 2012; Jazayeri *et al.*, 2012). The gap to be bridged between subtle effects in the primate and clinically useful effects in the human brain will require improvements in opsin efficiency, development of effective therapy delivery devices and procedures and judicious choice of therapeutic indications and anatomical targets.

These considerations suggest that the use of optogenetics to improve upon existing deep brain stimulation (DBS) therapies may not be the best first choice. Even with future closed-loop algorithms, the duty cycle of DBS for movement disorders might be 50% of the time or more. Too little is known about how stimulation can be most effectively applied for the treatment of other future indications, such as major depression, to know yet whether optogenetic stimulation could be effective with a low duty cycle.

Conversely, the use of electrical stimulation for the treatment of persons suffering from drug-resistant seizure disorders already utilizes non-continuous stimulation. To the extent that optical stimulation does not have disruptive side effects, the algorithm controlling therapy delivery could be tuned to allow a high rate of false-positive detections (to achieve a low rate of undetected seizures) and still have a low duty cycle, involving perhaps only several seconds of light delivery every several minutes or hours (<2%). Current neuromodulation therapies for epilepsy, though effective, leave much room for improvement. Therefore, epilepsy may be a good first choice for the application of optogenetics to a central nervous system disorder. Is there reason to believe this is a feasible choice?

12.4 Can Optogenetic Technology Provide a Clinically Relevant Treatment for Epilepsy?

Whether optogenetic technology will ever be able to provide a clinically relevant treatment for epilepsy is, at present, debatable. As noted before, optogenetic effects in the primate brain so far have been very subtle (e.g. producing saccadic reaction times that are 10% faster than those obtained in the absence of optical stimulation) (Gerits *et al.*, 2012). As of July 2013, reviewers noted that only three studies had reported any optogenetic-induced *behavioral* effects in non-human primates (Gerits and Vanduffel, 2013). There is also the danger that the effects of opsins and optical stimulation in pathological tissue may be unpredictably different from their effects in normal tissue. For example, a tempting choice for an opsin for the treatment of epilepsy may be a light-responsive chloride ion *channel*, such as SwiChR (Berndt *et al.*, 2014), which has silencing effects on neurons at light powers that are orders of magnitude lower than the power required for comparable silencing with opsins that are either chloride or proton *pumps*. However, even in normal tissue, use of the light-driven inward Cl^- pump eNpHR3.0, *Natronomonas pharaonis* halorhodopsin (Gradinaru *et al.*, 2010), can *increase* the probability of synaptically evoked spiking in a postsynaptic neuron harboring $GABA_A$ receptors in the seconds immediately following the end of light activation (Raimondo *et al.*, 2012), raising the possibility that attempts to control seizures using chloride pumps or ion channels might “backfire”.

Nevertheless, there are many reasons to believe an optogenetic therapy for epilepsy is feasible. Tonnesen *et al.* have demonstrated optogenetic control of the excessive hyperexcitability in organotypic hippocampal slice cultures that can be induced by electrical burst stimulation (Tonnesen *et al.*, 2009). Wykes *et al.* found that optogenetic activation of NpHR (a light-activated chloride ion pump) delivered to principle neurons in the rodent motor cortex could reduce the rate of occurrence of epileptiform events caused by tetanus toxin injection as a model of focal neocortical epilepsy (Wykes *et al.*, 2012). More importantly, Paz *et al.* found that reducing the firing rate of neurons in the ventrobasal thalamus of a rat with thalamocortical epilepsy caused by a cortical stroke was sufficient to interrupt and terminate an ongoing electrographic and *behavioral* seizure (Paz *et al.*, 2013). Taking a different strategy, Krook-Magnuson *et al.* used optogenetic techniques to activate a subpopulation of GABAergic interneurons in the intrahippocampal kainate mouse model of temporal lobe epilepsy and showed this approach is also capable of suppressing seizures (Krook-Magnuson *et al.*, 2013). Importantly, although this subpopulation of neurons represents less than 5% of hippocampal neurons, activation of these inhibitory interneurons was sufficient to rapidly stop seizures. More recently, Krook-Magnuson and her colleagues have found that either excitation or inhibition of parvalbumin-expressing neurons in the lateral or midline cerebellum results in a decrease in seizure duration, and optogenetic excitation of the midline cerebellum reduces spontaneous seizure frequency in that mouse model (Krook-Magnuson *et al.*, 2014).

Thus, substantial evidence that optogenetic stimulation can be a means for controlling seizures, at least in rodent-sized brains, exists. A reasonable next step toward a clinically relevant therapy for epilepsy is to determine whether optogenetic effects on the neural circuitry typically involved in epilepsy can be achieved in a brain that is much closer in size to that of a human.

12.5 Optogenetics and Epilepsy: Scaling Up Efforts to a Larger Brain

Even a rhesus monkey (*Macaca mulatta*) brain is only about a twelfth to a fifteenth the size of an adult human brain. Among mammals that can be readily obtained for research purposes, the domestic sheep (*Ovis aries*) provides a brain that is much closer to the volume of a human brain, being approximately a sixth the volume of the human brain. Stereotactic neurosurgery can be performed on sheep using the same surgical equipment and hardware as used in human neurosurgeries (see Figure 12.1). Preliminary work by the author and colleagues has explored the delivery of opsins to the sheep brain using adeno-associated viral vectors infused into the sheep hippocampus.

In a first animal, AAV serotype 9 was used to deliver the ArchT opsin (Han *et al.*, 2011). A total of 120 μl of AAV9-CAG-ArchT-eGFP (titer = 3.3×10^{12} vg mL^{-1}) were infused through a catheter targeted to the sheep's left hippocampus at a rate of 0.5 μl per minute. Three weeks later, stimulation and recording electrodes (DBS leads) were placed into the anterior nucleus of the thalamus and the hippocampus, respectively, and recordings of evoked potentials in the hippocampus in response to stimulation of the thalamus were confirmed. A fiber optic was inserted into the hippocampus to deliver an appropriate wavelength (595 nm) of light for the activation of ArchT (and the silencing of neurons expressing ArchT), or a control wavelength (460 nm), at varying power levels. No changes in the electrical recordings from the hippocampus in response to light delivered by the fiber optic were observed, even for 595-nm light delivered at the maximum power available (UHP-MIC-LED-595, Prizmatix, Givat-Shmuel, Israel). However, later examination of the sheep's brain sections (stained for eGFP to reveal where the viral vector had gone) showed that the lead and fiber optic that were positioned during the second surgery were not at all near to the infusion site of the viral vector (from the first surgery). Therefore, a study using a second sheep was planned in order to rectify this surgical issue.

We modified our approach in several ways. First, we altered our approach to include positioning and anchoring of the DBS leads at the time of the first surgery (the same surgery as for delivery of the viral vector). Second, we devised a way to

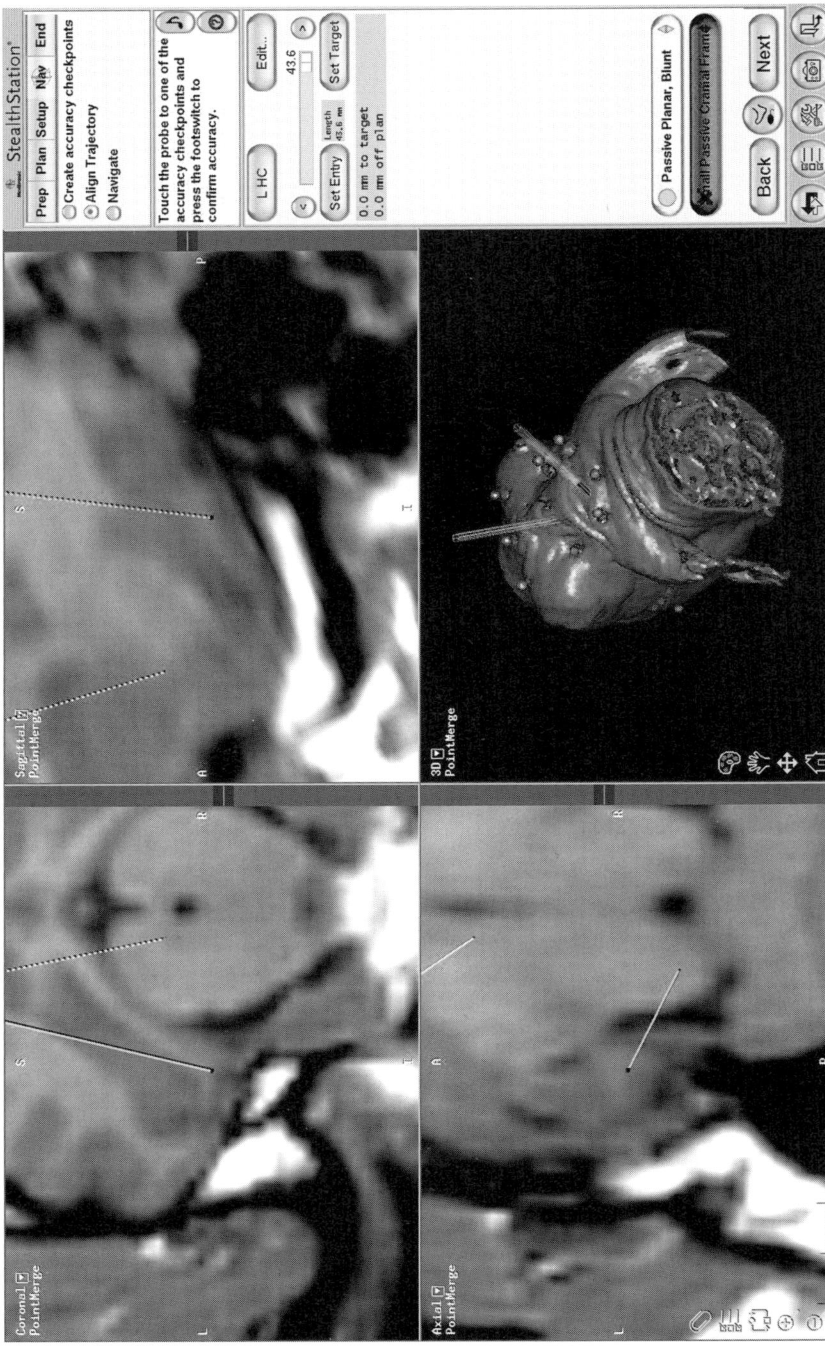

Figure 12.1 Screenshot of image-guided stereotactic neurosurgery in a sheep.

deliver the viral vector through the lumen of a modified (open-ended) DBS lead, ensuring that the lead and the viral vector would be in the same location in the hippocampus. Third, as part of the first surgery, we stimulated the anterior nucleus of the thalamus and recorded from the hippocampus to ensure that the

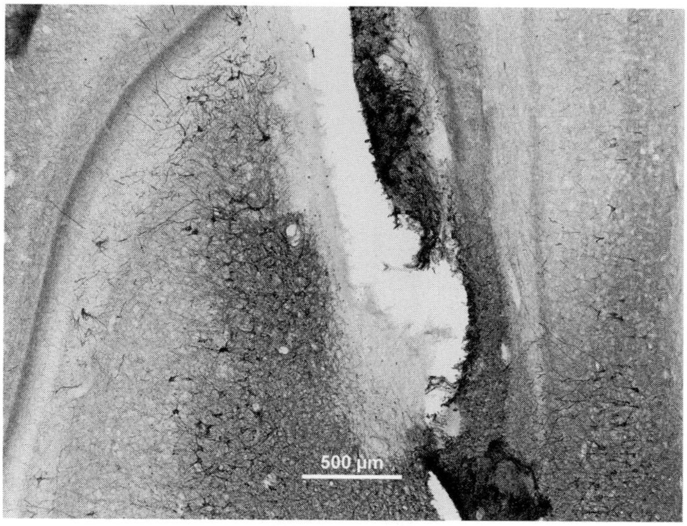

Figure 12.2 Co-located opsin-transduced tissue and DBS lead location in the sheep hippocampus.

viral vector was being delivered into hippocampal tissue. Fourth, we used an opsin ("JAWS") that is responsive to red light (630 nm); red light penetrates tissue more effectively than shorter wavelengths of light. Fifth, we used the "tried and true" serotype of AAV (AAV2) and a neuron-specific promoter (the synapsin promoter) to drive the expression of the opsin. Finally, we delivered the fiber optic (for delivery of the light to the transduced tissue) through the lumen of the DBS lead itself, again ensuring that the light, the opsin and the recording electrodes were all co-located.

Results with the second sheep were mixed. We again did not see any changes in the electrical signals from the hippocampus in response to light delivered by the fiber optic. However, histological examination of the sheep's brain sections showed that our new surgical approach of delivery of the viral vector and, later, the fiber optic through the DBS lead was highly successful in obtaining co-localization of the lead and the vector-transduced tissue (see Figure 12.2). Light delivered from the end of the lead or any of the translucent spaces between lead contacts up the lead body (all were tested) was co-located with the transduced cells.

Furthermore, we found that by delivering the AAV2 vector to the hippocampus at a site that was responsive to stimulation of the anterior nucleus of the thalamus, we effectively delivered the vector not only into hippocampal tissue, but also into the connections between the hippocampus and the rest of the circuit of Papez. Thus, not only hippocampal neurons, but also neurons in the fornix, mammillary bodies and the anterior nucleus of the thalamus itself were labeled by staining for the vector-delivered transgene (see Figure 12.3).

This "long-distance" staining of the circuit is significant not only for its potential for a future treatment of epileptic seizures involving the circuit of Papez, but also as a proof of principle for the use of optogenetics for circuit-selective neuromodulation (via "projection targeting") in the large brain, as well as in rodents.

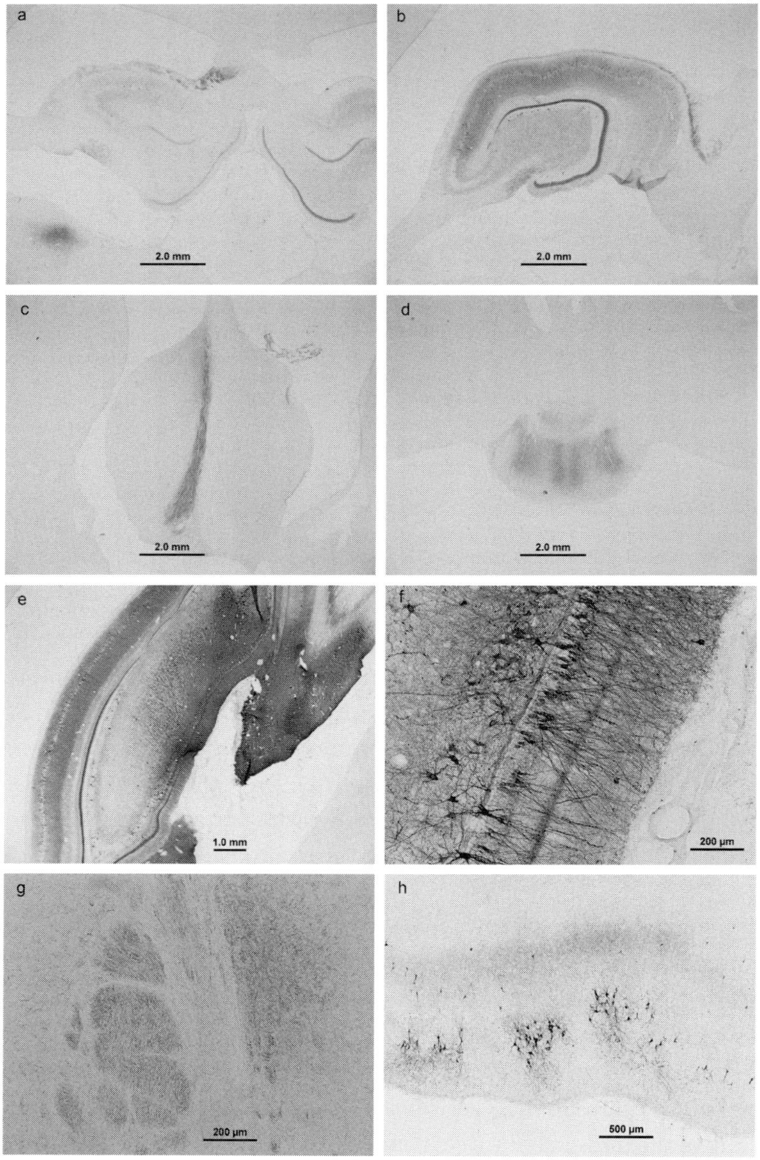

Figure 12.3 Transduction of much of the circuit of Papez achieved from a single infusion site in the sheep hippocampus. (A) Anterior nucleus of thalamus, (B) dorsal hippocampus, (C) ipsilateral fornix, (D) ventral fornix, (E) ventral hippocampus (infusion site), (F) higher magnification view of ventral hippocampus, (G) mammillary bodies and (H) entorhinal cortex. (A black-and-white version of this figure will appear in some formats. For the color version, please refer to the plate section.)

Projection targeting is accomplished by delivery of an opsin to one location, followed by delivering light to a second, connected location, in order to selectively activate or silence just those neurons that connect the two sites. Our "long-distance" transduction of the sheep's anterior nucleus centimeters away from a vector infusion into the hippocampus, serves as a proof of principle that such circuit-selective approaches may be feasible in the human brain. Future work should utilize the same surgical and recording techniques, but with more powerful and efficient light sources, recording methods (including single-unit recordings to verify the effects at the cellular level) and opsins.

12.6 Other Issues to Explore and Resolve

In addition to establishing the feasibility of modulating circuit functions in a large brain, there are several other issues to be considered with regards to the clinical application of optogenetics. The first concerns possible adverse effects of tissue heating occurring as a byproduct of the light delivery. However, the threshold for adverse thermogenic effects in mammalian brain tissue *in vivo* is about 43°C, which brain tissue can tolerate for 15–30 minutes. An advantage of epilepsy as a target for the clinical application of optogenetics is that even if the successful optogenetic control of seizures requires light that is powerful enough to raise the local tissue temperature to this extent, it is unlikely to require continuous illumination of the tissue for more than a few minutes.

A bigger issue is the safety of long-term opsin expression in the brain. Fortunately, most opsins are deliverable using adeno-associated viral vectors, which are known to result in long-term transgene expression in the primate brain. Thus, the safety of opsin expression over virtually any timeframe and anatomical target of the experimenter's choice can be readily tested by one-time AAV delivery into the primate brain via stereotactic surgery.

Finally, with regards to an optogenetic therapy for epilepsy, there is the issue of what tissue to target for the most efficacious seizure control. In DBS for epilepsy, the stimulation is applied to the anterior nucleus of the thalamus, regardless of the exact site from which seizures originate in the individual patient. In another electrical stimulation therapy for epilepsy (Bergey *et al.*, 2015), stimulation is applied via subdural or depth electrodes to one or two cortical sites identified as seizure foci. Perhaps an optogenetic therapy based on the silencing of neuronal firing would be most effective if applied directly to the seizure focus and/or immediately surrounding tissues. This strategy would also have the advantage that if unacceptable adverse effects of opsin expression were to emerge, the tissue could be removed as a last resort (assuming that tissue resection was an option for the individual in the first place), although some tissue transduced via axonal projections might remain. Alternatively, optogenetic modulation of the circuit of Papez or inhibition of the hippocampus via optogenetic stimulation of inhibitory interneurons in that structure may provide more "leverage" against the spread of seizure activity in the brain. Finally, Krook-Magnuson and colleagues have recently reported that optogenetic stimulation of the cerebellum provides seizure control in a rodent model of epilepsy (Krook-Magnuson *et al.*, 2014). This finding is further supported by a report that optogenetic stimulation of cerebellar nuclei abruptly stops generalized spike-and-wave discharges in mouse models of absence seizures (Kros *et al.*, 2015). An item for future research is to determine whether this target will be useful in larger species.

12.7 A Research Agenda: A Path to First-in-human Use

The following agenda may serve as a pathway to a first clinical trial of an optogenetic therapy. First and foremost, it is essential to establish that optogenetics can be used to modulate the firing rate of a brain region or circuit in a large mammalian brain, if not the overt behavior of the animal itself. Second, with regards to an application for epilepsy, it will be helpful to establish the limits of optogenetic efficacy in rodent models of seizure disorder. Can a rodent that is prone to having (or has been induced to have) seizures also be treated sufficiently by optogenetic methods to become effectively seizure-free (at least with respect to behavioral manifestations of seizure activity)? If not, is the extent of seizure reduction achievable in the rodent sufficient to suggest value for human patients relative to currently available treatments?

Once proof of scalability and a possibility of superior efficacy are established, development of an optogenetic therapy for epilepsy may follow an agenda similar to gene therapy development. The lead candidate for the opsin construct needs to be selected, including the choice not only of the opsin itself, but also the promoter driving opsin expression, and the incorporation of other DNA features for optimal transgene expression. Expression and appropriate intracellular trafficking of the opsin protein should be verified and characterized *in vitro* in neuronal cells. The serotype of the AAV vector to be used for transgene delivery should be chosen based on the cellular and tissue distribution desired.

Next, it will be necessary to conduct long-term toxicology studies in primates and rodents in order to obtain the preclinical evidence of safety necessary for regulatory approval for first-in-human trials. Ideally, the primate toxicology studies will utilize the same infusion techniques and implantable hardware as intended for the human, so that not only the safety of the viral vector and transgene expression, but also the biocompatibility of the chronically implanted sensing lead and light-delivery hardware to be used in patients can be established. Here is where guidance from studies in large animals, such as sheep, will be helpful for selecting the strategy for delivery of light to the brain tissue.

Because the focus of a Phase I trial is on safety rather than efficacy, the delivery of the light in the Phase I trial could be based on an open-loop schedule, as are present stimulation therapies. The first delivery of light for an optogenetic effect could be conducted in a controlled environment (e.g. an epilepsy monitoring unit) with light delivery triggered manually by an attending physician in order to determine its effects in the absence of a seizure, and perhaps its ability to alter the course of an observed seizure. A further possibility may be to allow the patient to self-administer the light delivery using a patient activator, which may provide some early data regarding therapy efficacy in those patients who experience auras warning them of impending seizures. Patients completing the Phase I trial without therapy-limiting safety issues would be offered implantation of the device connected to their intracranial lead, which up until this point could have remained externalized for the trial period.

12.8 Conclusion

Clearly, the path to the clinic will not be a short one. Nevertheless, the opportunity that optogenetic technology offers for a new therapy for epilepsy is too unique to be ignored. Also, regardless of whether or not an efficacious optogenetic therapy for epilepsy ever makes it across the finish line, what is learned along the way will be informative for other potential clinical applications of optogenetics, and may open the door to new therapeutic possibilities, providing hope for sufferers of many other neurological disorders.

ACKNOWLEDGMENTS

Thanks to Esther Krook-Magnuson for her review and comments for improvement of the manuscript; all opinions and any errors, however, remain my own.

REFERENCES

Bergey, G. K., M. J. Morrell, *et al.*, (2015). Long-term treatment with responsive brain stimulation in adults with refractory partial seizures. *Neurology* **84**(8): 810–817.

Berndt, A., S. Y. Lee, *et al.*, (2014). Structure-guided transformation of channelrhodopsin into a light-activated chloride channel. *Science* **344**(6182): 420–424.

Boyden, E. S., F. Zhang, *et al.*, (2005). Millisecond-timescale, genetically targeted optical control of neural activity. *Nat Neurosci* **8**(9): 1263–1268.

Carr, F. B. and V. Zachariou (2014). Nociception and pain: lessons from optogenetics. *Front Behav Neurosci* **8**: 69.

Cavanaugh, J., I. E. Monosov, *et al.*, (2012). Optogenetic inactivation modifies monkey visuomotor behavior. *Neuron* **76**(5): 901–907.

Chiang, C. C., T. P. Ladas, *et al.*, (2014). Seizure suppression by high frequency optogenetic stimulation using *in vitro* and *in vivo* animal models of epilepsy. *Brain Stimul* **7**(6): 890–899.

Gerits, A., R. Farivar, *et al.*, (2012). Optogenetically induced behavioral and functional network changes in primates. *Curr Biol* **22**(18): 1722–1726.

Gerits, A. and W. Vanduffel (2013). Optogenetics in primates: a shining future? *Trends Genet* **29**(7): 403–411.

Gradinaru, V., M. Mogri, *et al.*, (2009). Optical deconstruction of Parkinsonian neural circuitry. *Science* **324**(5925): 354–359.

Gradinaru, V., F. Zhang, *et al.*, (2010). Molecular and cellular approaches for diversifying and extending optogenetics. *Cell* **141**(1): 154–165.

Gunaydin, L. A., O. Yizhar, *et al.*, (2010). Ultrafast optogenetic control. *Nat Neurosci* **13**(3): 387–392.

Han, X. (2012). Optogenetics in the nonhuman primate. *Prog Brain Res* **196**: 215–233.

Han, X., B. Y. Chow, *et al.*, (2011). A high-light sensitivity optical neural silencer: development and application to optogenetic control of non-human primate cortex. *Front Syst Neurosci* **5**: 18.

Jazayeri, M., Z. Lindbloom-Brown, *et al.*, (2012). Saccadic eye movements evoked by optogenetic activation of primate V1. *Nat Neurosci* **15**(10): 1368–1370.

Jin, X., F. Tecuapetla, *et al.*, (2014). Basal ganglia subcircuits distinctively encode the parsing and concatenation of action sequences. *Nat Neurosci* **17**(3): 423–430.

Krook-Magnuson, E., C. Armstrong, *et al.*, (2013). On-demand optogenetic control of spontaneous seizures in temporal lobe epilepsy. *Nat Commun* **4**: 1376.

Krook-Magnuson, E. and I. Soltesz (2015). Beyond the hammer and the scalpel: selective circuit control for the epilepsies. *Nat Neurosci* **18**(3): 331–338.

Krook-Magnuson, E., G. G. Szabo, *et al.*, (2014). Cerebellar directed optogenetic intervention inhibits spontaneous hippocampal seizures in a mouse model of temporal lobe epilepsy. *eNeuro* **1**(1): e.2014.

Kros, L., O. H. Eelkman Rooda, *et al.*, (2015). Cerebellar output controls generalized spike-and-wave discharge occurrence. *Ann Neurol* **77**(6): 1027–1049.

Lammel, S., K. M. Tye, *et al.*, (2014). Progress in understanding mood d isorders: optogenetic dissection of neural circuits. *Genes Brain Behav* **13**(1): 38–51.

Montgomery, K. L., A. J. Yeh, *et al.*, (2015). Wirelessly powered, fully internal optogenetics for brain, spinal and peripheral circuits in mice. *Nat Methods* **12**(10): 969–974.

Nature Video. (2010). Method of the Year 2010: Optogenetics. from https://www.youtube.com/watch?v=I64X7vHSHOE.

Ozden, I., J. Wang, *et al.*, (2013). A coaxial optrode as multifunction write-read probe for optogenetic studies in non-human primates. *J Neurosci Methods* **219**(1): 142–154.

Paz, J. T., T. J. Davidson, *et al.*, (2013). Closed-loop optogenetic control of thalamus as a tool for interrupting seizures after cortical injury. *Nat Neurosci* **16**(1): 64–70.

Raimondo, J. V., L. Kay, *et al.*, (2012). Optogenetic silencing strategies differ in their effects on inhibitory synaptic transmission. *Nat Neurosci* **15**(8): 1102–1104.

Tonnesen, J., A. T. Sorensen, *et al.*, (2009). Optogenetic control of epileptiform activity. *Proc Natl Acad Sci U S A* **106**(29): 12162–12167.

Wykes, R. C., J. H. Heeroma, *et al.*, (2012). Optogenetic and potassium channel gene therapy in a rodent model of focal neocortical epilepsy. *Sci Transl Med* **4**(161): 161ra152.

Zemelman, B. V., G. A. Lee, *et al.*, (2002). Selective photostimulation of genetically chARGed neurons. *Neuron* **33**(1): 15–22.

13 **Optogenetic Control of Astroglia**

Anja G. Teschemacher and Sergey Kasparov

13.1 Introduction

Over the past couple of decades, it has become increasingly clear that astroglia play a fundamental role in neurotransmission, not only by fueling it and removing waste, but also by integrating, amplifying and modulating neuronal signals (Suzuki *et al.*, 2011; Verkhratsky *et al.*, 2015). Since astrocytes express a wide variety of receptors for neurotransmitters and can release gliotransmitters in response to stimulation, they are intimately linked with local neuronal activity. To begin to understand astrocytes' complex contributions to brain function, it has been crucial to be able to selectively manipulate them.

The initial driver behind the development of optogenetic actuators was the promise of non-invasive control over the electrical activity of selected populations of neurons and the ability to interrogate the function of specific neural circuits (Zemelman *et al.*, 2003; Nagel *et al.*, 2005a; Deisseroth *et al.*, 2006). More recently, it has become clear that optogenetic approaches are also proving invaluable for the study of the roles of astrocytes. Since changes in membrane potential are not thought to play a major role in astrocytic signal transduction, the optogenetic strategy is focused on modulation of the natural signaling modes of astrocytes (Figueiredo *et al.*, 2011; Li *et al.*, 2013; Figueiredo *et al.*, 2014). In particular, stimuli that modulate intracellular Ca^{2+}, cAMP or lactate levels appear to be highly relevant.

13.2 Expression Systems

The attraction of cell-selective manipulation rests on the ability to selectively express genes in specified cell types. As for neuronal applications, the selectivity as well as the strength of expression of actuator proteins are critical factors for achieving adequate optogenetic control. While both are often achieved in transgenic animals, expression via tetracycline-dependent promoters, due to their inherent leakiness, requires careful evaluation of the resulting expression patterns (Sasaki *et al.*, 2012; Chen *et al.*, 2013; Beppu *et al.*, 2014; Masamoto *et al.*, 2015). In this regard, it is probably preferable to use well-validated transgenic mouse lines where the expression of Cre recombinase is strictly limited to astrocytes. Other recent studies have relied on viral vectors to deliver optogenetic transgenes to astrocytes as this allows for greater flexibility and speed when investigating the intercommunication between astrocytes and various populations of neurons, each of which need to be targeted by different transgene constructs (Gourine *et al.*, 2010; Marina *et al.*, 2013; Tang *et al.*, 2014). Some studies utilize the cytomegalovirus (CMV) promoter and often achieve predominant or at least preferential expression in astrocytes, especially when using gliotropic viral vectors or in brain areas where neurons do not express CMV-driven transgenes well (Duale *et al.*, 2005; Li *et al.*, 2012; Chen *et al.*, 2013). However, glial fibrillary acidic protein (GFAP)-derived promoters are currently most commonly used for astrocyte-selective expression in transgenic animals, and so far they are used exclusively in context of viral vectors. Truncated versions of the natural GFAP promoter (e.g. GfaABC1D) have comparatively low transcriptional activity, and *per se* might be insufficiently strong for actuator expression levels that enable functional optogenetic stimulation. If light levels are increased in an attempt to compensate for low expression, there is a risk of laser heating artefacts (Gradinaru *et al.*, 2009; Christie *et al.*, 2013). Therefore, our group has developed a transcriptionally enhanced version of GfaABC1D that utilizes a chimeric transcriptional enhancer, GAL4BDp65, in combination with unique non-mammalian binding sites (GAL4BS) to boost transgene expression, while retaining high selectivity for astrocytes (Figure 13.1A) (Liu *et al.*, 2006; Liu *et al.*, 2008).

This promoter has been used (by us and others) in adenoviral and lentiviral backbones for a range of *in vitro* and *in vivo* studies (Gourine *et al.*, 2010; Figueiredo *et al.*, 2011; Marina *et al.*, 2013; Figueiredo *et al.*, 2014; Tang *et al.*, 2014). A caveat of the promoter is its poor expression when delivered by viral vectors to some brain areas (e.g. to the cortex of adult animals). Hence, there is scope for further testing and development of other astrocyte-selective promoters, such as Mlc1-, S100β-, GLT-1- or ALDH1L1- derived promoters, which may extend the range of astrocytic populations that can be studied using optogenetics.

13.3 Optogenetic Actuators

At present, most studies employing optogenetics for the stimulation of astrocytes use the already well-established channelrhodopsin-2 (ChR2) variant ChR2 (H134R). Upon absorption of blue light, ChR2-like proteins open non-selective cation channels, which are mainly permeable to Na^+, K^+, H^+ and, to some extent, also to Ca^{2+} (Nagel *et al.*, 2005b).

Since astrocytes, like neurons, have a strongly negative resting membrane potential and a low cytosolic Na^+ concentration, and, just as in neurons, stimulation of ChR2 with blue light produces time-locked depolarizations in astrocytes (Figure 13.1B). However, in contrast to neurons, where depolarizations trigger action potentials that have undisputable biological relevance, action potentials are not part of the astrocytic signaling repertoire (Figure 13.1B). While high-

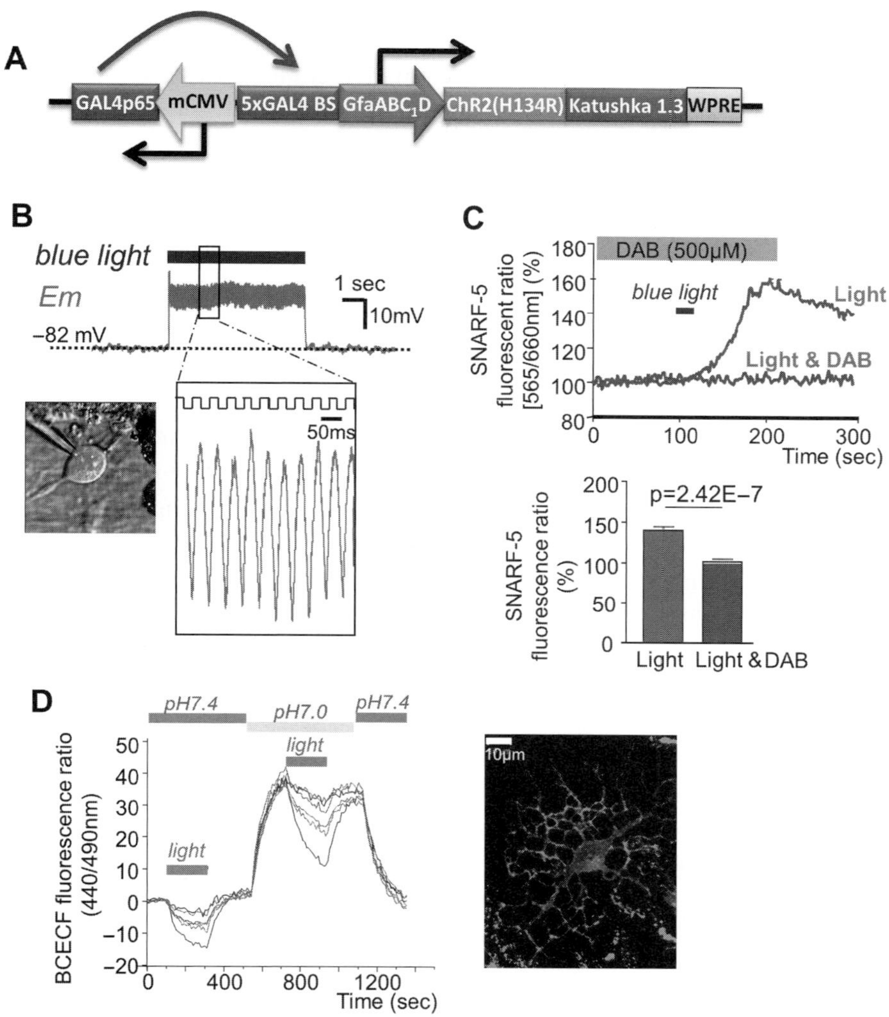

Figure 13.1 Application of optogenetic approaches to the stimulation of astroglia. (A) Layout of an optogenetic cassette for the stimulation of astrocytes via viral vector transgenesis. The chimeric transactivator GAL4p65, transcribed from a minimal cytomegalovirus promoter (mCMV), enhances transcription from GfaABC1D, a shortened version of the GFAP promoter, in an astrocyte-selective manner. ChR2(H134R) is fused to the red fluorescent protein Katushka1.3 to enable visualization in the absence of optogenetic stimulation. Woodchuck hepatitis virus response element (WPRE) is added to prolong the half-life of the transgene mRNA. (B) Astrocytic response to ChR2(H134R) activation. An astrocyte in organotypic brain slice culture expressing ChR2(H134R) was recorded in whole-cell patch clamp mode. Light flashes (470 nm; 20 ms, 25 Hz) caused time-locked depolarizations; note the absence of action potentials (adapted from Gourine *et al.* [2010]). (C) Optogenetic activation of astrocytes decreases intracellular pH. Light stimulation of optoβ2AR-expressing cultured astrocytes increased the emission ratio of the pH indicator SNARF-5. This intracellular acidification was sensitive to the blockade of glycogen metabolism by 1,4-dideoxy-1,4-imino-D-arabinitol (DAB; 500 μM); top – representative imaging trace; bottom – pooled data ($n = 32$) (adapted from Tang *et al.* [2014]). Stimulation using ChR2(H134R) yielded similar results (see Tang *et al.* [2014], supplement). (D) ArchT activation in astrocytes increases intracellular pH. Cortical astrocytes expressing the light-activated H^+ pump ArchT were loaded with the pH-sensitive dye BCECF and stimulated with green light and low pH. ArchT activation caused intracellular alkalinization and counteracted a pH drop induced by the acidic environment; right – BCECF-loaded astrocyte in culture (collaboration with Gourine A.V., unpublished data). (A black-and-white version of this figure will appear in some formats. For the color version, please refer to the plate section.)

frequency oscillations have been observed in cultured astrocytes, their physiological significance is currently unclear (Fleischer *et al.*, 2015). Expression of voltage-gated Ca^{2+} channel subunits has been reported in cultured astrocytes, but since physiologically plausible (10–20 mV) depolarization of astrocytes through an intracellular electrode fails to elicit appreciable time-locked $[Ca^{2+}]i$ rises, the functionality of such channels requires further investigation (Latour *et al.*, 2003; Steinhäuser *et al.*, 2013; Fleischer *et al.*, 2015).

On the other hand, optogenetically induced membrane potential changes are bound to affect reversal potentials for various ions and transporters across the astrocytic membrane. For example, depolarizations, if prolonged, would be expected to interfere with astrocytic control of extracellular ion homeostasis (Verkhratsky *et al.*, 2015) and may result in K^+ efflux. Because of ChR2's promiscuous cation permeability, K^+ will predictably exit the astrocytes, particularly when prolonged ChR2 openings are evoked. Increases in the extracellular K^+ concentration can potentially affect the properties of neighboring cells, such as blood vessel contractility (e.g. see Masamoto *et al.*, [2015]), and this should be taken into account when interpreting the physiological relevance of optogenetically stimulated effects. Step-function opsin variants of ChR2 with slow inactivation kinetics (e.g. ChR2[C128S]) require only brief illumination with blue light to open and stay in the open state for minutes unless they are subsequently closed with yellow (~590 nm) light (Berndt *et al.*, 2009). These variants offer the advantage of reducing potential phototoxicity, as they do not need sustained, intensive illumination to produce extended channel openings under conditions in which their expression is comparatively weak. However, when ChR2(C128S) is used to elicit prolonged cation conductance and to mimic protracted hyperactivation of astrocytes, K^+ efflux-related effects need to be considered (Beppu *et al.*, 2014).

The main reason why ChR2 is believed to successfully activate astrocytes is, however, the channel's (albeit limited) Ca^{2+} permeability. Due to the steep Ca^{2+} concentration gradient across the astrocytic membrane, the driving force for Ca^{2+} entry through open ChR2 is considerable. Therefore, light-induced channel opening permits Ca^{2+} influx, which mimics astrocytic secondary messenger signaling and initiates downstream events such as gliotransmitter release. Interestingly, several groups, including ours, have observed that light activation of CatCh, a ChR2 variant with increased Ca^{2+} permeability, is not more potent in increasing astrocytic Ca^{2+} levels when compared to ChR2, and that stimulation of either ChR2 variant results in Ca^{2+} transients that are largely due to Ca^{2+} release from intracellular stores (Kleinlogel *et al.*, 2011; Figueiredo *et al.*, 2014; Perea *et al.*, 2014; but see Li *et al.*, 2013). This suggests that the magnitude of the initial Ca^{2+} entry through ChR2 plays less of a role in astrocytic activation than the following signal amplification by intracellular Ca^{2+} mobilization. The mechanism of store activation is not fully clear at present, but could contain an element of Ca^{2+}-induced Ca^{2+} release. In addition, imaging of fusion constructs of ChR2 with a fluorescent protein tag suggests that a considerable proportion of the actuator protein is not successfully trafficked to the plasma membrane, but may be retained in transit in the endomembranes. In this case, light activation may open ChR2 located in the membranes of Ca^{2+}-accumulating organelles directly (Figueiredo *et al.*, 2014). One other, probably quite important aspect of ChR2-driven astrocytic Ca^{2+} responses is that they involve an element of the autocrine action of ATP, which astrocytes readily release when stimulated using ChR2 (Figure 13.2A) (Gourine *et al.*, 2010).

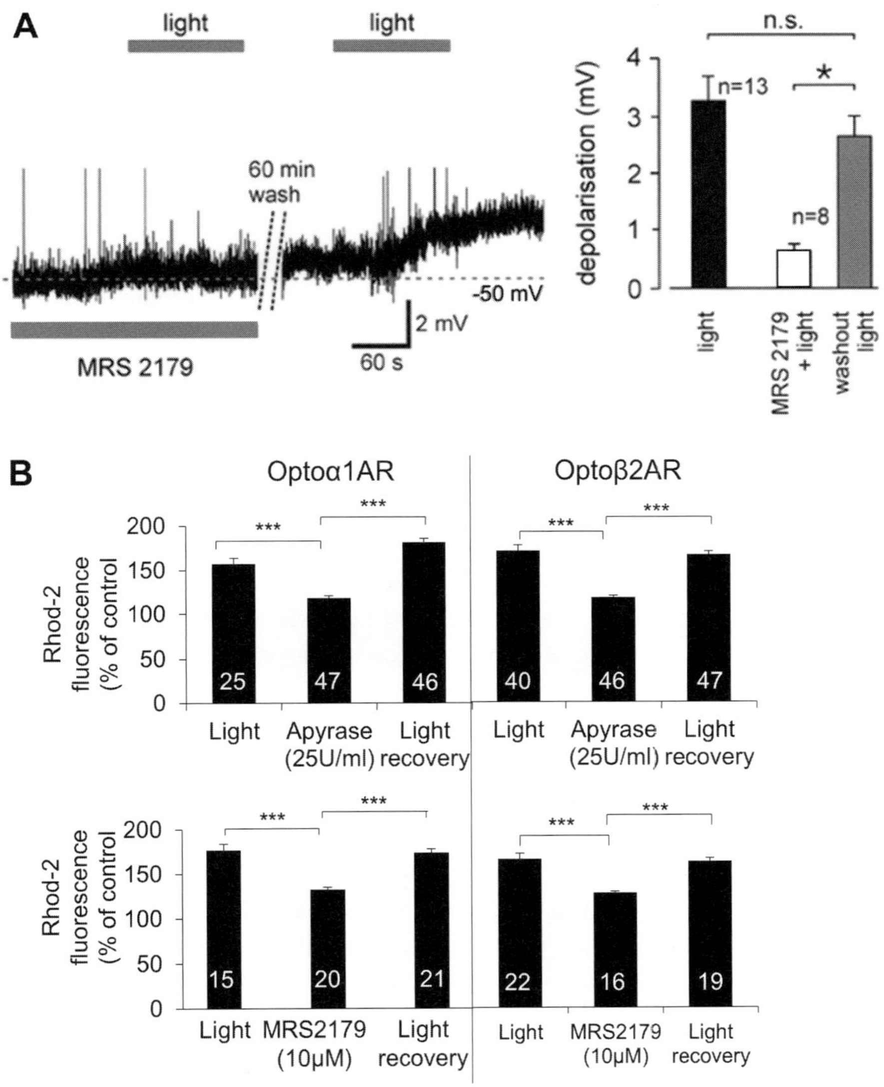

Figure 13.2 Astrocytic ATP release as a primary response to optogenetic stimulation. (A) Light stimulation of ChR2(H134R)-expressing astrocytes in the retrotrapezoid nucleus evoked depolarizations in neighboring neurons, which were dependent on purinergic signaling that is blocked in the presence of MRS 2179; left – representative whole-cell recording; right – pooled data ($^{*}p < 0.05$) (adapted from Gourine *et al.* [2010]). (B) Activation of astrocytes via opto-adrenoceptors α1 and β2 (optoα1AR and optoβ2AR) is amplified by the autocrine action of ATP. Intracellular Ca^{2+} elevations evoked by optoα1AR (left columns) and optoβ2AR (right columns) in cultured astrocytes were sensitive to extracellular pretreatment with the ATP-degrading enzyme apyrase, or to the ATP receptor inhibitor MRS 2179 ($^{***}p < 0.0001$) (adapted from Figueiredo *et al.* [2014]).

The non-selective cation permeability of channelrhodopsins also applies to protons (Nagel *et al.*, 2003). At around physiological pH levels (e.g. 7.3 intracellularly and 7.4 extracellularly), the proton concentrations are in the range of 50 and 40 nM (intra- and extra-cellular, respectively), and the equilibrium potential, according to the Nernst equation, is –6.15 mV. Therefore, the biophysical prediction is that, against the backdrop of extracellular Na^+ and intracellular K^+ concentrations (exceeding 100 mM), the proton gradient would make a negligible contribution to current flowing through the open ChR2 and would be directed outward. However, acidification through ChR2 activation can be successfully achieved under experimental conditions where the extracellular solution is strongly acidified (Li *et al.*, 2012; Beppu *et al.*, 2014). Nevertheless, intracellular acidification can be observed in astrocytes in response to strong activation, including neuronal hyperexcitation, oxidative stress or optogenetic stimulation (Beppu *et al.*, 2014; Tang *et al.*, 2014). According to our results, this acidification appears secondary to activity-dependent glycolysis and lactate production and is sensitive to inhibition of glycogenolysis in astrocytes (Figure 13.1C) (Tang *et al.*, 2014).

The range of available channelrhodopsin variants is constantly broadening and some may also be of interest for optogenetic studies on astrocytes (Lin, 2011; Li *et al.*, 2013). For example, improved features such as increased light sensitivity increase the chances of efficient stimulation while accommodating lower levels of specific transgene expression, avoiding overstimulation and phototoxicity (Berndt *et al.*, 2011; Pan *et al.*, 2014). For *in vivo* studies, it will be easier to work with actuators that can be stimulated with longer-wavelength light that penetrates further into the tissue. Here, recently developed variants with red-shifted activation spectra such as ReaChR are interesting options. Using ReaChR, it has been possible to activate neural circuits non-invasively through the skull in a mouse (Lin *et al.*, 2013). Because many *in vivo* imaging experiments employ multiphoton excitation, which is achieved using infrared pulsing lasers, efforts have been made to devise actuators that can be controlled with these specific wavelengths. Several studies have reported successful excitation of optogenetic actuators in two-photon excitation microscopy experiments (Mohanty *et al.*, 2008; Papagiakoumou *et al.*, 2010; Prakash *et al.*, 2012), but we are unaware of any studies in which multiphoton activation has been used for the optogenetic study of astrocytes.

Alternatively, a semisynthetic ultraviolet (UV) light-activated cation channel, LiGluR, is available, which is unrelated to channelrhodopsins, but instead derived from an ionotropic glutamate receptor (Volgraf *et al.*, 2006; Li *et al.*, 2012). Once a glutamate tether has been linked by UV illumination to the previously expressed receptor, LiGluR is activatable by UV light, inactivated by light above 480 nm and shows pronounced Ca^{2+} permeability. This channel has been used to evoke gliotransmitter release from cortical astrocytes *in vitro* (Li *et al.*, 2012), but has so far not been widely adopted, in spite of its attractive ionic profile (Li *et al.*, 2012).

Apart from ion channels, the available classes of light-sensitive proteins also include transporters, enzymes and metabotropic transducers. ArchT, for instance, is an example of an outward proton pump that alkalinizes the cytosol when switched on by yellow light (Han *et al.*, 2011). We observed efficient alkalinization of cortical astrocytes with light activation of ArchT (Figure 13.1D) (in collaboration with Gourine A.V., unpublished data). Beppu and coworkers expressed ArchT in Bergman glia, a cell type in the cerebellum that is related to astrocytes, and demonstrated that ArchT activation resulted in intracellular pH increases and was able to counteract the glial acidification in response to ischemia (Beppu *et al.*, 2014). Again, red-shifted variants such as Jaws may improve the applicability for *in vivo* experimentation (Chuong *et al.*, 2014).

Photosensitive G-protein-coupled "receptors" also have their place as astrocyte-relevant actuators since they allow modulation of the secondary messenger cascades through which astrocytes normally react to transmitter flux in their environment. Opto-adrenoceptors (opto-ARs) are chimeras engineered from the extracellular parts of rhodopsin that confer reactivity to blue light and the intracellular and transmembrane domains of Gq-coupled α1-adrenoreceptors or Gs-coupled β2-adrenoceptors (Airan *et al.*, 2009). We confirmed that opto-αAR activates Gq-protein signaling and increases intracellular Ca^{2+} in astrocytes (Figueiredo *et al.*, 2014). We also found that opto-βAR stimulation, as predicted for β-adrenoceptors, raises cAMP levels and, interestingly, additionally leads to Ca^{2+} signals in astrocytes, possibly via cross-talk between the Gq and Gs signaling pathways based on "exchange protein activated by cAMP" (Ferrero *et al.*, 2013; Figueiredo *et al.*, 2014). Using opto-ARs, we recently demonstrated that light stimulation of either G-protein-coupled pathway activates astrocytes, leading to gliotransmitter release, which modulates neuronal function (see below) (Tang *et al.*, 2014). Similar to ChR2-like constructs, these actuators trigger the release of ATP, which activates astrocytes in an autocrine manner. Hence, the end-point response is a combination of the intracellular effects of the actuator itself and events initiated via ATP receptors, which are abundantly expressed on astrocytes (Figure 13.2B) (Figueiredo *et al.*, 2014).

Since metabotropic receptors are central to astrocytic sensing of the brain microenvironment and homeostasis, expansion of the available range of optogenetically activatable constructs will be useful. For example, a recently introduced bleach-resistant Gs-coupled opsin, JellyOp, has not yet been tried in astrocytes, but holds promise for experiments requiring longer-term stimulation (Bailes *et al.*, 2012). Further photoactivatable transporters and enzymes that may potentially be of value for specifically probing astrocytic signaling and function include recent developments such as light-sensitive adenylate cyclases and nitric oxide synthase (Stierl *et al.*, 2011; Yarkoni *et al.*, 2012).

While many astrocyte-relevant intracellular signaling pathways are thus becoming accessible for direct optogenetic stimulation, the predominant problems with the currently available actuators are commonly the difficulties with controlling their expression levels and their dark activity levels. Therefore, developments that move toward fine-tuning these properties would improve the scope and applicability of optogenetic actuators for experimentation on astrocytes.

13.4 Astrocytic ATP-mediated Autocrine Effects and Ca^{2+} Waves

Irrespective of the optogenetic tool employed, one crucial feature of astrocytic activation is the intracellular Ca^{2+} response, which typically spreads through the astrocyte population, even where cells are not connected via gap junctions. Many studies, including our own, have shown that an early response to any stimulation of astrocytes is the release of gliotransmitters, particularly of ATP or glutamate, which may then lead to a self-perpetuating autocrine/paracrine Ca^{2+} wave spreading across the astrocytic network (Fields and Burnstock, 2006; Gourine *et al.*, 2010; Li *et al.*, 2012; Figueiredo *et al.*, 2014; Wells *et al.*, 2015). We therefore suggest that positive purinergic feedforward signaling may be an integral part of any experimentally or physiological astrocytic activation, and should be considered when interpreting the mechanisms of astrocyte activation (Figure 13.2).

13.5 Astrocytic Functions Elucidated by Optogenetic Stimulation

The utilization of optogenetic approaches has shed light on several vital roles through which astrocytic signaling in various different areas of the central nervous system is involved in brain function. Stimulation of astrocytes, in most cases using ChR2 variants, but also via LiGluR or opto-ARs, was shown to elicit the release of a range of substances with messenger functions, in particular ATP, glutamate and L-lactate (Gourine *et al.*, 2010; Li *et al.*, 2012; Sasaki *et al.*, 2012; Chen *et al.*, 2013; Marina *et al.*, 2013; Perea *et al.*, 2014; Beppu *et al.*, 2014; Tang *et al.*, 2014). As gliotransmitters, these not only cause spreading activity waves through astrocytic networks, but also serve to modulate specific neuronal circuits.

13.5.1 Astrocytes as Signal Transducers Regulating Autonomic Neuronal Circuits

13.5.1.1 Astrocytes and Regulation of Breathing

Astrocytes are part of the vascular–neuronal interface, and therefore are in an ideal position to serve as primary chemosensors of blood pH and partial pressure of oxygen and CO_2, relating the information to the relevant homeostatic circuits in the brain. The build-up of CO_2 (hypercapnia) in the blood is the main stimulus that triggers breathing. Apart from peripheral chemosensors, the ventral medulla of the medulla oblongata has long been acknowledged as a central CO_2-sensitive area, but the mechanisms of signal transduction were debated (Gourine *et al.*, 2010; Huckstepp *et al.*, 2010; Wang *et al.*, 2013). Hypercapnia acidifies the blood, and we found that the astrocyte population at the ventral brainstem surface is highly sensitive to small decreases in pH. It reacts with a wave of intracellular Ca^{2+} calcium signals, which propagates via autocrine purinergic excitation and carries through the local astrocytic network. Neurons belonging to the respiratory network that are more remote from the blood environment will, in turn, receive this amplified signal and stimulate breathing behavior.

In 2010, we were the first group to successfully activate astrocytes by light stimulation of ChR2(H134R), and we applied this approach to testing the hypothesis that ventral medullary surface astrocytes are primary chemosensors in the control of respiration (Gourine *et al.*, 2010). We were able to demonstrate that optogenetic activation of this "pre-inspiratory" astrocyte population mimics stimulation by hypercapnia, resulting in increased breathing frequency (Figure 13.3A) (Gourine *et al.*, 2010). Thus, ventral medullary surface astrocytes contribute to vital central chemosensitivity and should be considered in context of sleep apnea and the associated cardiovascular pathophysiology.

13.5.1.2 Astrocytes and Control of Sympathetic Discharge

Astrocytes in a neighboring area, the rostral ventro-lateral medulla (RVLM), surround the cell bodies of pre-sympathetic neurons, a mixed neuronal population of spinally projecting noradrenergic and glutamatergic phenotypes. The RVLM is the major driver for the pre-ganglionic neurons located in the spinal cord. Excessive sympatho-excitation is thought to be one of the pathogenic mechanisms of neurogenic hypertension, a major cause of mortality and morbidity in the developed world. In spite of many years of research, it is still unclear what triggers the gradual transition of the cardiovascular system toward higher-than-normal levels of blood pressure that, over time, leads to myocardial infarctions, strokes and other life-threatening events. Possibly, this results from hypoperfusion of the medullary centers that are responsible for blood pressure control. Sympathetic overactivity is also a signature of another highly prevalent cardiovascular condition – heart failure. It is thought that failing hearts provide insufficient perfusion to the tissues, including the brain, and that this may cause compensatory activation of the sympathetic drive to the heart. However, the high sympathetic drive proves to be maladaptive and is considered to be a major pathogenic factor in heart failure, and is clearly associated with a negative prognosis.

Given that astrocytes are highly sensitive to hypoxia, these cells could contribute to the activation of brainstem centers that control sympathetic outflow. Using optogenetic tools, we tested whether RVLM astrocytes provide an input for pre-sympathetic neurons and play a role in setting sympathetic tone. Similarly to the approach used for the study of "pre-inspiratory" astrocytes discussed above, ChR2(H134R) was expressed in RVLM astrocytes (Marina *et al.*, 2013). *In vitro*, light-activation resulted in ATP release and excitation of noradrenergic RVLM neurons, which was demonstrated by the sensitivity of evoked neuronal firing to apyrase, an ATPase. *In vivo*, light stimulation resulted in elevated sympathetic nerve activity, arterial blood pressure and heart rate. These experiments suggest that brainstem astrocytes may contribute to the increase in sympathetic outflow observed when oxygenation of the brainstem is compromised. This mechanism could underlie the development of essential hypertension and the maladaptive sympatho-excitation that is seen following myocardial infarction. Therefore, preventing the astrocytic purinergic drive within the RVLM may be a potential treatment strategy for cardiovascular disease (Figure 13.3B) (Marina *et al.*, 2013; Marina *et al.*, 2015).

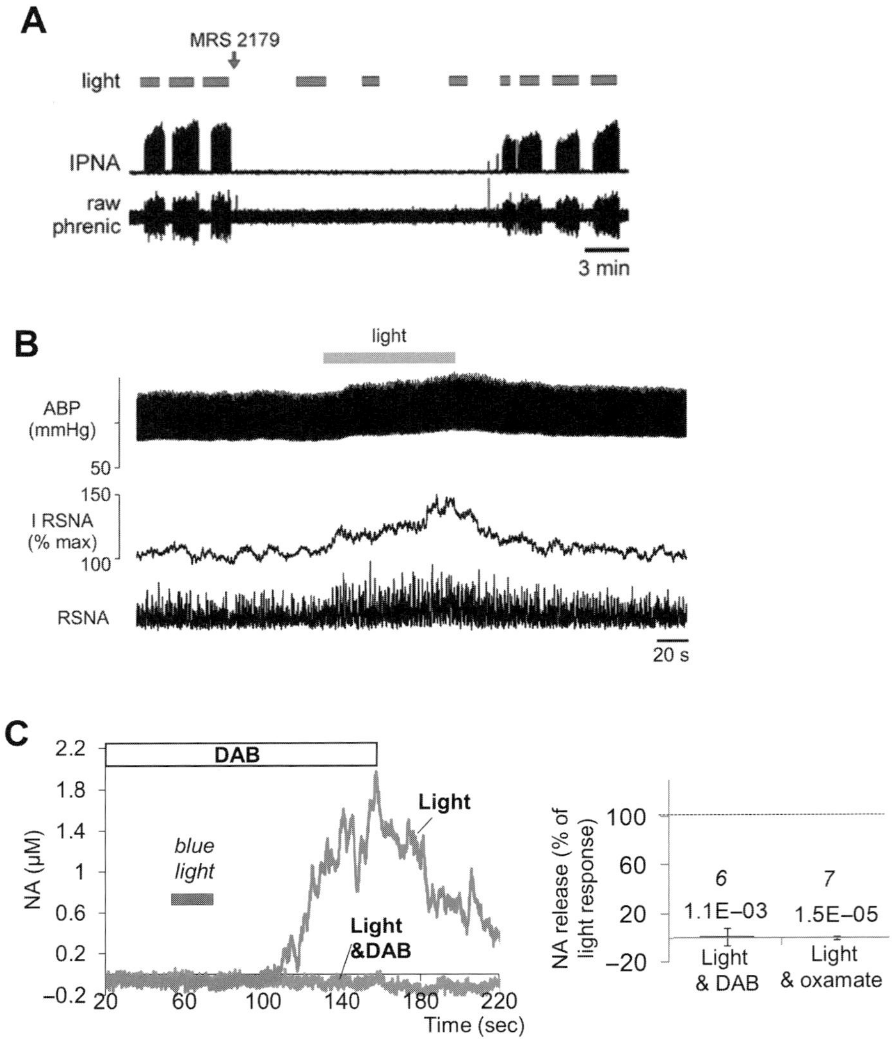

Figure 13.3 Astrocytic functions elucidated by optogenetic stimulation. (A) Astrocytes as primary central chemosensors regulating breathing. Light stimulation of ChR2(H134R)-expressing astrocytes at the ventral surface of the medulla oblongata triggered respiratory activation as represented by phrenic nerve activity (upper trace – integrated; lower trace – raw nerve activity). This effect was sensitive to the blockade of purinergic signaling by MRS 2179 (100 μM, 20 μl applied to the ventral surface) (adapted from Gourine *et al.* [2010]). (B) Astrocytes as central regulators of sympathetic outflow. Light stimulation of ChR2(H134R)-expressing astrocytes in the RVLM increased arterial blood pressure (ABP; top trace) and renal sympathetic nerve activity (RSNA; middle trace – integrated; lower trace – raw data) (adapted from Marina *et al.* [2013]). (C) Astrocytes as modulators of central noradrenergic tone. Light stimulation of optoβ2AR-expressing astrocytes in organotypic brain slice cultures including the locus coeruleus triggered the release of noradrenaline (NA) from locus coeruleus neurons, as demonstrated by voltammetry. This action was blocked by inhibition of glycogen breakdown by 1,4-dideoxy-1,4-imino-D-arabinitol (DAB; 500 μM) or the lactate dehydrogenase inhibitor oxamate (20 mM), indicating L-lactate release as an intermediate signaling step between astrocytes and neurons. Left – representative voltammetry traces; right – pooled data normalized to NA release in response to light stimulation (n and *p* values above columns). (adapted from Tang *et al.* [2014])

13.5.2 Astrocytes as Modulators of Synaptic Plasticity

Several studies using optogenetics in astrocytes have emphasized the importance of glutamate signaling in inter-astrocytic and astrocyte-to-neuron communication in the cerebral cortex. The ability of astrocytes to take up extracellular glutamate spill-over from synaptic transmission has long been recognized. In this role, astrocytes are essential for decreasing excitatory tone and, in addition, they help to recycle and restock neuronal glutamate stores (Hertz *et al.*, 1999). Recent *in vivo* work suggested that light activation of ChR2-expressing astrocytes from the visual cortex caused astrocytic glutamate release, enhancing local synaptic transmission via presynaptic metabotropic glutamate receptors (Perea *et al.*, 2014). Interestingly, synaptic inhibition and excitation were differentially modulated, allowing astrocytes to enhance the general level of cortical excitation and to alter the neuronal information processing of visual stimuli *in vivo*.

In the cerebellum, Bergman glia cells take the place of astrocytes. Optogenetic stimulation of Bergmann glia using the step-opsin ChR2(C128S) caused long-term depression of synaptic transmission *in vitro*, involving activation of metabotropic glutamate receptors (Sasaki *et al.*, 2012). On the other hand, *in vivo*, glial activation modulated optokinetic reflex behavior, which is cerebellum dependent. These experiments suggest that Bergmann glia cells release glutamate via anion channels in order to contribute to synaptic plasticity (Sasaki *et al.*, 2012). Under ischemic conditions, however, Bergman glia cells appear to develop intracellular acidosis, which leads to massive glutamate release and may cause excitotoxic neuronal damage, highlighting the importance of a balance in glial signaling for brain health (Beppu *et al.*, 2014). In contrast, in other brain areas, astrocytic release of ATP may play a prominent role in modulating synaptic transmission. For example, when ChR2-expressing astrocytes from the hippocampal CA1 layer were stimulated *in vitro*, long-term depression was evoked, presumably due to astrocytic ATP release (Chen *et al.*, 2013).

13.5.3 Astrocytes as Coordinators of Central Neurovascular Coupling and Noradrenergic Modulation

Astrocytes were proposed to act as primary recipients of central noradrenergic volume transmission. Since they tend to be the closest neighbors of noradrenergic varicosities, they express a complement of adrenoceptors and they respond to noradrenergic stimulation with Ca^{2+} excitation (Bekar *et al.*, 2008; Hertz *et al.*, 2010). Using optogenetic activation, we have recently discovered a novel signaling pathway by which astrocytes, in the reverse direction, communicate with noradrenergic neurons via the release of L-lactate (Tang *et al.*, 2014).

Light stimulation of ChR2(H134R)-expressing astrocytes in the locus coeruleus area evoked depolarizations in local noradrenergic neurons, which were recorded via patch clamp. The neuronal activation was blocked by inhibitors of glycogen breakdown and L-lactate production (Tang *et al.*, 2014). Since astrocytes are still generally thought to be the only glycogen-storing cells in the brain and are well known to produce and release L-lactate upon activation, our experiments suggested that astrocyte-derived L-lactate in turn stimulates noradrenergic neurons. Using opto-AR stimulation of astrocytes, we then went on to confirm that astrocytic L-lactate release results in neuronal noradrenaline release (Figure 13.3C). The detailed mechanism of this astrocyte-to-neuron communication is not yet understood, but appears to involve a receptor for L-lactate on noradrenergic neurons (Tang *et al.*, 2014). This not only implies that astroglia may play a role as master coordinators of noradrenergic modulation in the brain, but also provides a putative link between the body's metabolic state and central arousal, anti-nociception and affect.

13.5.4 Astrocytes as Mediators of Neurovascular Coupling and Blood Oxygen Level-dependent Contrast Imaging Signals

Activation of neurons leads to local changes in brain blood supply, which are directed toward activated areas. In fact, the influx of oxygenated blood is usually so great that there is a shift between deoxygenated and oxygenated hemoglobin in favor of the oxygenated form (super-compensation). This phenomenon is the basis of functional magnetic resonance imaging (fMRI). Astrocytes not only communicate within their networks and with neurons, but also essentially comprise the interface between the brain and the vasculature. Therefore, it is logical to expect astrocytes to be involved in neurovascular coupling (Attwell *et al.*, 2010). Recent work using ChR2 stimulation of cortical surface astrocytes has begun to investigate glia-to-vessel signaling. Focal laser stimulation of ChR2-expressing astrocytes *in vivo* resulted in an increase in local blood flow within seconds, which was insensitive to blockade of neuronal firing, but partially inhibited by barium ions (Masamoto *et al.*, 2015). This suggests the possibility that diffusible messengers, potentially K^+ ions, may be involved in the hyperemic response to increased neuronal activity. Interestingly, ChR2 has also recently been used to activate two other non-excitable cells in the brain: vascular smooth muscle cells and pericytes (Hill *et al.*, 2015). This study demonstrated that only smooth muscle cells are contractile.

At the same time, it is imperative to remember that fMRI is exquisitely sensitive to artefacts induced by light when it is used in optogenetic experiments. Indeed, when titered optical stimulation was delivered to the rat cortex, a "positive" signal could be detected in dead animals (Christie *et al.*, 2013). Therefore, great care is required when combining fMRI with optogenetic activation. Meanwhile, the role of purinergic (presumably astrocytic) signaling in the generation of fMRI responses has recently been demonstrated using local expression of an ATP-degrading enzyme in the rat cortex (Wells *et al.*, 2015). This procedure drastically reduced the amplitude of the signal in the areas where ATP-mediated signaling was disabled, pointing to an important role of astrocytes in this process. As yet, there is little understanding of the roles of astrocytes in regulating the vascular lumen in the brain, and there is an urgent need for further research in order to better understand the regulation of cerebral blood flow and the prevention of ischemia.

In conclusion, within the past few years alone, optogenetic approaches have catalyzed important breakthroughs in our understanding of the roles and signaling mechanisms of astrocytes. We have now begun to better appreciate how astrocytic populations in different brain areas appear to vary considerably in terms of their sensitivity to the microenvironment, their intracellular signaling pathways, their gliotransmitter complement and release mechanisms and their means of communication with neurons and other brain cells. The early success of applying optogenetic tools to the study of astrocytes will continue to drive these lines of research forward, and may potentially lead to therapeutic applications targeting astrocytes (Ji and Wang, 2015).

REFERENCES

Airan RD, Thompson KR, Fenno LE, Bernstein H, Deisseroth K (2009) Temporally precise *in vivo* control of intracellular signalling. *Nature* **458**:1025–1029.

Attwell D, Buchan AM, Charpak S, Lauritzen M, Macvicar BA, Newman EA (2010) Glial and neuronal control of brain blood flow. *Nature* **468**:232–243.

Bailes HJ, Zhuang LY, Lucas RJ (2012) Reproducible and sustained regulation of *Galphas* signalling using a metazoan opsin as an optogenetic tool. *PLoS One* **7**:e30774.

Bekar LK, He W, Nedergaard M (2008) Locus coeruleus alpha-adrenergic-mediated activation of cortical astrocytes *in vivo*. *Cereb Cortex* **18**:2789–2795.

Beppu K, Sasaki T, Tanaka KF, Yamanaka A, Fukazawa Y, Shigemoto R, Matsui K (2014) Optogenetic countering of glial acidosis suppresses glial glutamate release and ischemic brain damage. *Neuron* **81**:314–320.

Berndt A, Schoenenberger P, Mattis J, Tye KM, Deisseroth K, Hegemann P, Oertner TG (2011) High-efficiency channelrhodopsins for fast neuronal stimulation at low light levels. *Proc Natl Acad Sci U S A* **108**:7595–7600.

Berndt A, Yizhar O, Gunaydin LA, Hegemann P, Deisseroth K (2009) Bi-stable neural state switches. *Nat Neurosci* **12**:229–234.

Chen J, Tan Z, Zeng L, Zhang X, He Y, Gao W, Wu X, Li Y, Bu B, Wang W, Duan S (2013) Heterosynaptic long-term depression mediated by ATP released from astrocytes. *Glia* **61**:178–191.

Christie IN, Wells JA, Southern P, Marina N, Kasparov S, Gourine AV, Lythgoe MF (2013) fMRI response to blue light delivery in the naive brain: implications for combined optogenetic fMRI studies. *Neuroimage* **66**:634–641.

Chuong AS, *et al.* (2014) Noninvasive optical inhibition with a red-shifted microbial rhodopsin. *Nat Neurosci* **17**:1123–1129.

Deisseroth K, Feng G, Majewska AK, Miesenbock G, Ting A, Schnitzer MJ (2006) Next-generation optical technologies for illuminating genetically targeted brain circuits. *J Neurosci* **26**:10380–10386.

Duale H, Kasparov S, Paton JF, Teschemacher AG (2005) Differences in transductional tropism of adenoviral and lentiviral vectors in the rat brainstem. *Exp Physiol* **90**:71–78.

Ferrero JJ, Alvarez AM, Ramirez-Franco J, Godino MC, Bartolome-Martin D, Aguado C, Torres M, Lujan R, Ciruela F, Sanchez-Prieto J (2013) Beta-adrenergic receptors activate exchange protein directly activated by cAMP (Epac), translocate Munc13-1, and enhance the Rab3A–RIM1alpha interaction to potentiate glutamate release at cerebrocortical nerve terminals. *J Biol Chem* **288**:31370–31385.

Fields RD, Burnstock G (2006) Purinergic signalling in neuron-glia interactions. *Nat Rev Neurosci* **7**:423–436.

Figueiredo M, Lane S, Stout RF, Jr., Liu B, Parpura V, Teschemacher AG, Kasparov S (2014) Comparative analysis of optogenetic actuators in cultured astrocytes. *Cell Calcium* **56**:208–214.

Figueiredo M, Lane S, Tang F, Liu BH, Hewinson J, Marina N, Kasymov V, Souslova EA, Chudakov DM, Gourine AV, Teschemacher AG, Kasparov S (2011) Optogenetic experimentation on astrocytes. *Exp Physiol* **96**:40–50.

Fleischer W, Theiss S, Slotta J, Holland C, Schnitzler A (2015) High-frequency voltage oscillations in cultured astrocytes. *Physiol Rep* **3**:e12400.

Gourine AV, Kasymov V, Marina N, Tang F, Figueiredo MF, Lane S, Teschemacher AG, Spyer KM, Deisseroth K, Kasparov S (2010) Astrocytes control breathing through pH-dependent release of ATP. *Science* **329**:571–575.

Gradinaru V, Mogri M, Thompson KR, Henderson JM, Deisseroth K (2009) Optical deconstruction of parkinsonian neural circuitry. *Science* **324**:354–359.

Han X, Chow BY, Zhou H, Klapoetke NC, Chuong A, Rajimehr R, Yang A, Baratta MV, Winkle J, Desimone R, Boyden ES (2011) A high-light sensitivity optical neural silencer: development and application to optogenetic control of non-human primate cortex. *Front Syst Neurosci* **5**:18.

Hertz L, Dringen R, Schousboe A, Robinson SR (1999) Astrocytes: glutamate producers for neurons. *J Neurosci Res* **57**:417–428.

Hertz L, Lovatt D, Goldman SA, Nedergaard M (2010) Adrenoceptors in brain: cellular gene expression and effects on astrocytic metabolism and $[Ca^{2+}]$i. *Neurochem Int* **57**:411–420.

Hill RA, Tong L, Yuan P, Murikinati S, Gupta S, Grutzendler J (2015) Regional blood flow in the normal and ischemic brain is controlled by arteriolar smooth muscle cell contractility and not by capillary pericytes. Neuron **87**:95–110.

Huckstepp RT, id Bihi R, Eason R, Spyer KM, Dicke N, Willecke K, Marina N, Gourine AV, Dale N (2010) Connexin hemichannel-mediated CO_2-dependent release of ATP in the medulla oblongata contributes to central respiratory chemosensitivity. *J Physiol* **588**:3901–3920.

Ji ZG, Wang H (2015) Optogenetic control of astrocytes: is it possible to treat astrocyte-related epilepsy? *Brain Res Bull* **110**:20–25.

Kleinlogel S, Feldbauer K, Dempski RE, Fotis H, Wood PG, Bamann C, Bamberg E (2011) Ultra light-sensitive and fast neuronal activation with the Ca^{2+}-permeable channelrhodopsin CatCh. *Nat Neurosci* **14**:513–518.

Latour I, Hamid J, Beedle AM, Zamponi GW, Macvicar BA (2003) Expression of voltage-gated Ca^{2+} channel subtypes in cultured astrocytes. *Glia* **41**:347–353.

Li D, Agulhon C, Schmidt E, Oheim M, Ropert N (2013) New tools for investigating astrocyte-to-neuron communication. *Front Cell Neurosci* **7**:193.

Li D, Herault K, Isacoff EY, Oheim M, Ropert N (2012) Optogenetic activation of LiGluR-expressing astrocytes evokes anion channel-mediated glutamate release. *J Physiol* **590**:855–873.

Lin JY (2011) A user's guide to channelrhodopsin variants: features, limitations and future developments. *Exp Physiol* **96**:19–25.

Lin JY, Knutsen PM, Muller A, Kleinfeld D, Tsien RY (2013) ReaChR: a red-shifted variant of channelrhodopsin enables deep transcranial optogenetic excitation. *Nat Neurosci* **16**:1499–1508.

Liu B, Paton JF, Kasparov S (2008) Viral vectors based on bidirectional cell-specific mammalian promoters and transcriptional amplification strategy for use *in vitro* and *in vivo*. *BMC Biotechnol* **8**:49.

Liu BH, Yang Y, Paton JF, Li F, Boulaire J, Kasparov S, Wang S (2006) GAL4-NF-kappaB fusion protein augments transgene expression from neuronal promoters in the rat brain. *Mol Ther* **14**:872–882.

Marina N, Ang R, Machhada A, Kasymov V, Karagiannis A, Hosford PS, Mosienko V, Teschemacher AG, Vihko P, Paton JF, Kasparov S, Gourine AV (2015) Brainstem hypoxia contributes to the development of hypertension in the spontaneously hypertensive rat. *Hypertension* **65**:775–783.

Marina N, Tang F, Figueiredo M, Mastitskaya S, Kasimov V, Mohamed-Ali V, Roloff E, Teschemacher AG, Gourine AV, Kasparov S (2013) Purinergic signalling in the rostral ventro-lateral medulla controls sympathetic drive and contributes to the progression of heart failure following myocardial infarction in rats. *Basic Res Cardiol* **108**:317.

Masamoto K, Unekawa M, Watanabe T, Toriumi H, Takuwa H, Kawaguchi H, Kanno I, Matsui K, Tanaka KF, Tomita Y, Suzuki N (2015) Unveiling astrocytic control of cerebral blood flow with optogenetics. *Sci Rep* **5**:11455.

Mohanty SK, Reinscheid RK, Liu X, Okamura N, Krasieva TB, Berns MW (2008) In-depth activation of channelrhodopsin 2-sensitized excitable cells with high spatial resolution using two-photon excitation with a near-infrared laser microbeam. *Biophys J* **95**:3916–3926.

Nagel G, Brauner M, Liewald JF, Adeishvili N, Bamberg E, Gottschalk A (2005a) Light activation of channelrhodopsin-2 in excitable cells of *Caenorhabditis elegans* triggers rapid behavioral responses. *Curr Biol* **15**:2279–2284.

Nagel G, Szellas T, Huhn W, Kateriya S, Adeishvili N, Berthold P, Ollig D, Hegemann P, Bamberg E (2003) Channelrhodopsin-2, a directly light-gated cation-selective membrane channel. *Proc Natl Acad Sci U S A* **100**:13940–13945.

Nagel G, Szellas T, Kateriya S, Adeishvili N, Hegemann P, Bamberg E (2005b) Channelrhodopsins: directly light-gated cation channels. *Biochem Soc Trans* **33**:863–866.

Pan ZH, Ganjawala TH, Lu Q, Ivanova E, Zhang Z (2014) ChR2 mutants at L132 and T159 with improved operational light sensitivity for vision restoration. *PLoS One* **9**:e98924.

Papagiakoumou E, Anselmi F, Begue A, de Sars V, Gluckstad J, Isacoff EY, Emiliani V (2010) Scanless two-photon excitation of channelrhodopsin-2. *Nat Methods* **7**:848–854.

Perea G, Yang A, Boyden ES, Sur M (2014) Optogenetic astrocyte activation modulates response selectivity of visual cortex neurons *in vivo*. *Nat Commun* **5**:3262.

Prakash R, Yizhar O, Grewe B, Ramakrishnan C, Wang N, Goshen I, Packer AM, Peterka DS, Yuste R, Schnitzer MJ, Deisseroth K (2012) Two-photon optogenetic toolbox for fast inhibition, excitation and bistable modulation. *Nat Methods* **9**:1171–1179.

Sasaki T, Beppu K, Tanaka KF, Fukazawa Y, Shigemoto R, Matsui K (2012) Application of an optogenetic byway for perturbing neuronal activity via glial photostimulation. *Proc Natl Acad Sci U S A* **109**:20720–20725.

Steinhäuser C, Seifert G, Deitmer JW (2013) Physiology of astrocytes: Ion channels and ion transporters. In: *Neuroglia* (Kettenmann H, Ransom B, eds), pp. 185–196. Oxford: Oxford University Press.

Stierl M, Stumpf P, Udwari D, Gueta R, Hagedorn R, Losi A, Gartner W, Petereit L, Efetova M, Schwarzel M, Oertner TG, Nagel G, Hegemann P (2011) Light-modulation of cellular cAMP by a small bacterial photoactivated adenylyl cyclase, bPAC, of the soil bacterium *Beggiatoa*. *J Biol Chem* **286**:1181–1188.

Suzuki A, Stern SA, Bozdagi O, Huntley GW, Walker RH, Magistretti PJ, Alberini CM (2011) Astrocyte-neuron lactate transport is required for long-term memory formation. *Cell* **144**:810–823.

Tang F, Lane S, Korsak A, Paton JF, Gourine AV, Kasparov S, Teschemacher AG (2014) Lactate-mediated glia-neuronal signaling in the mammalian brain. *Nat Commun* **5**:3284.

Verkhratsky A, Nedergaard M, Hertz L (2015) Why are astrocytes important? *Neurochem Res* **40**:389–401.

Volgraf M, Gorostiza P, Numano R, Kramer RH, Isacoff EY, Trauner D (2006) Allosteric control of an ionotropic glutamate receptor with an optical switch. *Nat Chem Biol* **2**:47–52.

Wang S, Benamer N, Zanella S, Kumar NN, Shi Y, Bevengut M, Penton D, Guyenet PG, Lesage F, Gestreau C, Barhanin J, Bayliss DA (2013) TASK-2 channels contribute to pH sensitivity of retrotrapezoid nucleus chemoreceptor neurons. *J Neurosci* **33**:16033–16044.

Wells JA, Christie IN, Hosford PS, Huckstepp RT, Angelova PR, Vihko P, Cork SC, Abramov AY, Teschemacher AG, Kasparov S, Lythgoe MF, Gourine AV (2015) A critical role for purinergic signalling in the mechanisms underlying generation of BOLD fMRI responses. *J Neurosci* **35**:5284–5292.

Yarkoni O, Donlon L, Frankel D (2012) Creating a bio-hybrid signal transduction pathway: opening a new channel of communication between cells and machines. *Bioinspir Biomim* **7**:046017.

Zemelman BV, Nesnas N, Lee GA, Miesenbock G (2003) Photochemical gating of heterologous ion channels: remote control over genetically designated populations of neurons. *Proc Natl Acad Sci U S A* **100**:1352–1357.

14 Optogenetics for Neurohormones and Neuropeptides: Focus on Oxytocin

Yan Tang, Jérôme Wahis, Meggane Melchior, Valery Grinevich, and Alexandre Charlet

14.1 Introduction

To date, viral vectors represent the most popular means of delivering optogenetic tools to neural systems. Most advantageous is the rapidity of the experimental implementation and the capacity for temporally and spatially defined intervention. The recombinant adeno-associated viral (rAAVs) vectors are especially useful, as they display an extremely low immunogenicity that increases their potential to drive adequate expression for photocurrent stimulation due to multiplicity of infection. In addition, they exhibit high viral titers (usually ~10^{13} genomic copies per ml) (Grinevich *et al.*, 2015). Furthermore, rAAV is one of the smallest DNA viruses (20-nm diameter) (Grinevich *et al.*, 2015), which allows its efficient spread in extracellular spaces in order to cover a brain region of about 1 mm^3 (with a 300-nl injection volume). As a result, rAAV is able to infect a complete brain nucleus, such as the hypothalamic paraventricular nucleus (PVN) or supraoptic nucleus (SON), and render whole nuclei light sensitive (Knobloch *et al.*, 2012; Eliava *et al.*, 2016). Specificity of expression can be achieved in two ways: (1) directly by employing a selected gene promoter that drives viral expression of microbial opsins in a certain cell population of similar phenotype; or (2) indirectly by injecting a double-floxed, inverted open reading frame sequence in the hypothalamic nucleus of transgenic mice or rats expressing Cre recombinase in specific cell types, hence ensuring selective viral opsin expression in the neurons projecting to the nucleus of interest. Oxytocin (OT) neurons are among the unique cell types that can be directly targeted by viral vectors carrying short OT promoters (Atasoy *et al.*, 2012; Fields *et al.*, 2012; Knobloch *et al.*, 2012). In the following text, we will provide technical details of the optogenetics of OT neurons and discuss their limitations and advantages. In conclusion, we will highlight key questions regarding the physiology of OT and general peptidergic neurons and peptidergic signaling in the brain, which can be addressed by optogenetic means.

14.2 Specific Targeting of OT Neurons

According to the promoter analysis performed by the Harold Gainer group (Fields *et al.*, 2012), the introduction of a sequence of about 100 base pairs upstream of the start codon (ATG) to rAAV is sufficient to drive cell type-specific gene expression in OT neurons. In our work (Knobloch *et al.*, 2012; Eliava *et al.*, 2016), we used an evolutionarily conserved sequence of 1.9 kb long to achieve robust expression of genes of interests, such as ChR2, exclusively in OT neurons (Figure 14.1A).

In detail, for generating rAAVs with specific expression in OT cells, we used the software BLAT from UCSC (http://genome.ucsc.edu/cgi-bin/hgBlat) and selected a conserved 2.6-kb promoter directly upstream of the OT gene exon 1. This DNA was amplified from an EcoRI-linearized BAC clone RP24-388N9 (RPCI-24 Mouse, BACPAC Resources, CHORI, CA, USA) using a 5´-primer containing a NotI restriction site (5´-ATTAGCGGCCGCAGATGAGCTGGTGAGCATGTGAAGACATGC-3´) and a 3´-primer with a SalI restriction site (5´-ATTAGTCGACGGCGATGGTGCTCAGTCTGAGATCCGCTGT-3´), subcloned into pBlueScript SK and further cloned into the rAAV2 backbone, pAAV-αCaMKII-htTA, thereby substituting for the αCaMKII promoter. The resulting rAAV expression vector was used for the exchange of the htTA gene for ChR2-mCherry. As depicted in Figure 14.1A, injection of this rAAV into the hypothalamic PVN led to expression of ChR2-mCherry exclusively in OT neurons. In the following text, we will refer to this rAAV as "rAAV_pOT-ChR2mCherry".

It is important to note that, to our knowledge, the expression of ChR2-mCherry does not modify the physiological behavior of OT neurons in the conditions we reported. However, though crucial for a wide range of experiments involving behavior, such controls are rarely performed in the frame of optogenetic studies.

14.3 Stimulating and Recording OT Neuron Activity

The first advantage of optogenetics relates to gaining access to a specific neuronal population. Among them, neuropeptidergic neurons are of special interest thanks to the use of specific fragments of a given neuropeptide promoter. The use of targeted gene expression in a given neuron opens up a wide array of tools for studying different subtypes of cells in the brain. In the case of OT neurons, the ability to specifically express a fusion protein combining the now well-characterized channelrhodopsin-2 (ChR2) with a fluorescent protein, mCherry, eased the exploration and manipulation of neural circuits involving OT neurons. The following sections illustrate how combining *in vitro* patch clamp and *in vivo* extracellular techniques with optogenetics allows for efficient targeting and control of firing of OT neurons in the hypothalamus at the millisecond timescale.

14.3.1 Hypothalamus *In Vitro* Optical Stimulation and Electrophysiological Recordings

To perform *in vitro* optical stimulation of PVN OT neurons, rAAV_pOT-ChR2mCherry was bilaterally injected into the PVN (antero-posterior (AP) –1.8; medio-lateral (ML) 0.3; dorso-ventral (DV) 8.0). Two weeks after the virus was fully expressed, the animals were sacrificed and the brains swiftly removed and sliced, following a standard acute slicing protocol (Knobloch *et al.*, 2012). PVN was localized at four-times magnification under infrared light, and fluorescent OT cells were

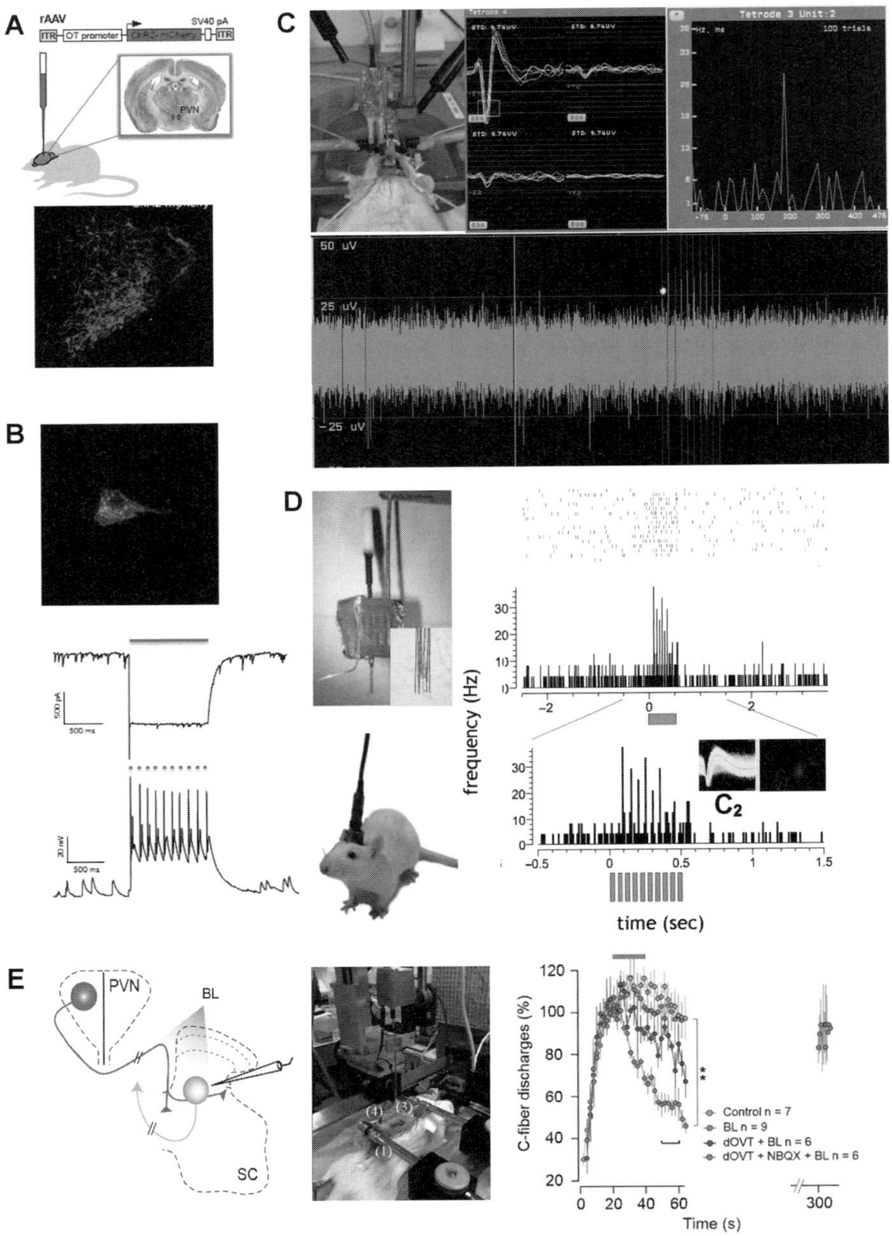

Figure 14.1 Optogenetics for oxytocin. (A) Specific expression of a transgene in oxytocinergic cells. (Top) Scheme showing the injection of the rAAV for the expression of ChR2-mCherry protein under control of the OT promoter. (Bottom) Validation of the approach. In red, direct visualization of mCherry 3 weeks after infection specifically in OT neurons (Knobloch *et al.*, 2012; Eliava *et al.*, 2016). (B) *In vitro* optogenetic manipulation and recording of an OT neuron. (Top) Two weeks after infection of the rAAV_pOT-ChR2mCherry, OT neurons were visualized in an acute slice of the PVN through a 40× objective. (Top trace) After successful whole-cell patch clamp of the OT neuron with a 4–9 MΩ borosilicate pipette filled with standard intracellular milieu, a 1s continuous blue light (475 ± 28 nm) pulse was applied through the 40× objective in the voltage clamp mode and the photocurrent was analyzed. An average of a 1.1-nA (± 0.1, n = 3) photocurrent was induced. (Bottom trace) The amplifier was then switched to the current clamp mode and the cell was maintained at a potential of around −60 mV, and a series of blue light pulses (10 ms width for 1 s at 10 Hz) was applied. (C) *In vivo*

targeted using a fluorescent microscope. Next, cells were patched in the whole-cell configuration and their responses to blue light (475 ± 28 nm) were studied (Figure 14.1B).

As demonstrated, this approach represents a unique tool allowing the precise control of the neuronal activity of OT neurons and, by extension, any neuropeptidergic neuron if the right sequence of the promoter is successfully designed. The *in vitro* approach offers ideal conditions, enabling the stimulation of distant axons without the risk of action potential (AP) back-propagation and subsequent cell body activation. In addition, it allows for optogenetics-driven circuit deconstruction coupled with pharmacological studies (Knobloch *et al.*, 2012).

14.3.2 Hypothalamus *In Vivo* Optical Stimulation and Electrophysiological Recordings in Anesthetized Animals

To perform *in vivo* optical stimulation of PVN OT neurons, rAAV_pOT-ChR2mCherry was unilaterally injected into the PVN. Two weeks after the virus was fully expressed, the animals were anesthetized in 1.5% isoflurane and placed in a stereotaxic frame. The same coordinates were used and the hole enlarged to Φ200 μm for opto-electrode implantation. The opto-electrodes were connected to a preamplifier for recording and an Opto-feurrl wire for laser stimulation. During implantation, the laser generated burst stimuli at 5-s intervals (20 Hz, 5 ms), while the electrodes were slowly moving deeper, towards the PVN. A successful response of OT neurons should occur within the laser spike delay in 10 ms. If a response

---

Caption for Figure 14.1 (cont.)

optogenetic manipulation and recording of OT neurons in anesthetized rats. (Top left) The experiment was carried out on rats fixed in a stereotaxic frame under isofluran anesthesia. The opto-electrodes were connected to a preamplifier for recording and an Opto-feurrl wire for laser stimulation. (Middle) Clear extracellular signals were detected on tetrodes by waveform threshold. (Top right) The per-stimuli time histogram (PSTH) online can quickly verify suspicious OT neurons; here is an example of an interneuron responding to OT neuron activation with a 180ms delay. (D) *In vivo* optogenetic manipulation and recording of OT neurons in freely moving rats. (Top left) The data acquisition system was built on OpenEphys Acquisition board (v2.2) and the head stage was connected to a 32-channel preamplifier and ultra-thin serial peripheral interface (SPI) cable (Intan Tech) with a sampling rate at 30 kHz, the low-pass filter set to 600 Hz and the high-pass filter set to 6000 Hz. An additional input/output board allowed synchronization of the digital signal with optogenetic stimulation. Data analysis: OpenEphys GUI 0.3.5 was used in online data analysis. MATLAB, Simpleclust 0.5 and Neuroexplorer 3.2 were used in offline data sorting and analyses. The 32-channel opto-electrodes were shielded by a copper net. (Bottom left) Implanted animals remained completely free to move without any trouble. (Top middle) The real OT neuron was activated by burst laser stimuli with a very short delay and stopped firing immediately after the stopping of the stimulation. (Top right) On the contrary, the network effect triggered local neuron to fire with a much longer onset and usually sustained firing after the end of the stimulation. (Bottom right) Cluster sorting illustrates two separated neurons. (E) *In vivo* long-range OT projection manipulation to decipher complex neuronal circuits. (Left) Scheme of the experiment illustrating an OT neuron (red) projecting to deep layers of the spinal cord. (Middle) Picture of the experimental setup showing an anesthetized rat laminectomized in the stereotaxic frame. (Right) Effect of the local release of OT triggered by blue light stimulation on the wind-up intensity of a wide dynamic range neuron. (A black-and-white version of this figure will appear in some formats. For the color version, please refer to the plate section.)

occurred after this delay, we considered that it belonged to the secondary network effect triggered by somatodendritic OT release (Figure 14.1C).

As demonstrated, this approach offers the unique advantages of manipulating and recording an OT neuron and neighboring network *in vivo*. It is an elegant approach to complement the *in vitro* one, allowing us to study the OT effect without perturbing the cerebral network or homeostasis (Eliava *et al.*, 2016). On the other hand, combined pharmacological studies, while feasible, are more difficult to perform than acute slicing. Finally, the approach in anesthetized animals does not allow the study of the effect of endogenous OT release on behavior.

14.3.3 Hypothalamus *In Vivo* Optical Stimulation and Electrophysiological Recordings in Freely Moving Animals

To bypass this last point, we performed *in vivo* recordings of OT neurons in freely moving animals. For this purpose, after rAAV_pOT-ChR2mCherry was unilaterally injected into the PVN, six screws (two with ground wires) were drilled into the skull around the injection site, together with dental cement to anchor the base for electrode implantation. Next, electrodes with assembled optic fibers were implanted 0.5 mm above the PVN. The eight nickel tetrodes were arranged in order, surrounded by the optic fiber (Φ200 um) and assembled in the 32-channel microdrive. Tetrode impedance was 0.3–0.5 MΩ before the surgery. Such a configuration allows simultaneous stimulation and recording of OT neuronal activity and, as previously stated, both the direct and indirect effects of blue light (BL) can be recorded (Figure 14.1D).

This approach is critical to understanding how an OT neuron reacts to a given stimulation and, in addition, to assessing how the surrounding neuronal network responds to such stimulation. Finally, the possibility of simultaneously manipulating specific OT neuronal activity will help us to understand the exact involvement of endogenous OT in a given behavioral situation.

14.4 Monitoring the Effect of Local OT Release on Distant Targets: Spinal Cord *In Vivo* Optical Stimulation and Electrophysiological Recordings in Anesthetized Animals

One of the most interesting features of optogenetics is that it allows the *in vivo* stimulation of long-distance axons from a specific neuronal population. Coupling long-range optical stimulation with extracellular electrophysiological recordings requires a combination of basic, but important, techniques, which are listed below. All of the different equipment is of crucial importance and should be carefully checked before starting a project.

Specifically, we used this approach to stimulate/record in the lumbar segment of the spinal cord, the most distant central structure from the hypothalamus. Gaining access to the spinal cord requires considerable surgery, making recordings of spinal neurons in freely moving animals extremely complicated. Thus, we limited our experiments on *in vivo* recordings to anesthetized animals (Figure 14.1E). As with the previously described approaches, rAAV_pOT-ChR2mCherry was injected into the PVN. In the case of spinal cord experiments, as the lumbar segment of the spinal cord is at a distance of about 10 cm from the hypothalamus in adult rats, it is recommended to wait a minimum of 4 weeks in order to allow significant levels of expression of the protein of interest (e.g. Chr2) at the extremity of the OT axons. To illustrate our goal, repetitive stimulation of the hind-paw was used to induce wind-up activity in the wide dynamic range (WDR) neuron (maximum wind-up = 100%) (Figure 14.1E). After 20 s of stable WDR discharge, a train pulse of blue light was applied to the spinal cord in order to induce OT release within the spinal cord, hence decreasing the intensity of wind-up to around 40%. This effect was blocked by oxytocin receptor (OTR) antagonist and NBQX, but not by vasopressin receptor 1a (V1aR) antagonist, which confirms that the OT effect on spinal pain processing is linked to OTR receptors and not to V1aR (Figure 14.1E).

This approach takes full advantage of optogenetics by manipulating the activity of the very distant axons of a very specific neuronal population. In our case, it offered the first evidence for a central release of OT supplemented by a systemic release of OT, both driven by the very same population of parvocellular neurons (Eliava *et al.*, 2016). It would not have been possible to obtain such conclusions without the appropriate combination of optogenetics.

In a general way, similar approaches, consisting of the study of a very local release of neuropeptides in an intact *in vivo* system, are mandatory in order to fully understand how the complex projections and networks formed by neuropeptidergic neurons can trigger and/or modulate elaborate processes such as anxiety, pain and sexual or maternal behaviors.

14.5 Discussion

14.5.1 Technical Considerations

Most optogenetic studies manipulate the activity of somas, not axons, in the regions of interest. Because neuropeptidergic cells carry bifurcating/multiple axons, the behavioral outcome of somatic stimulation involves multiple and largely unpredictable behavioral outcomes. On the other hand, we should consider the potential for back-propagation of axonal APs, mainly when using *in vivo* optogenetics (Tovote and Lüthi, 2012). Indeed, using capacitance measurement, de Kock *et al.* (2003) showed that a single antidromic AP is sufficient to elicit exocytosis of OT from large dense-core vesicles in dendrites. By contrast, Ludwig *et al.* (2002) showed that antidromic activation of SONs by electrical stimulation of the pituitary stalk did not elicit significant dendritic release of OT in virgin rats. Thus, although it is not known whether or not a back-propagated AP can effectively elicit neuropeptide release, one should always keep this possibility in mind. A possible solution relies on *in vitro* experiments involving: (1) dissecting somas from axons; and/or (2) operating with two opsins, enabling the inhibition of cell bodies and the simultaneous activation of distant axons. While ChR2 and HaloR do not work properly together in the same neuron due to the overlap of wavelengths, other light-activated pumps may do the job.

Intra-axonal stocks could easily be depleted. In addition, receptors of various neuropeptides are often subject to desensitization. While not of the same nature and very poorly documented, these two aspects must be considered when using optogenetics of neuropeptides by carefully characterizing the *in vitro* response with repeated stimulations. In addition, using minimal stimulation that still allows for the secretion of the peptide can be an elegant solution to the potentially rapid depletion of neuropeptide or desensitization of the receptor, although this is potentially very difficult to implement.

Hypothalamic neuropeptides are often subject to hormonal regulation (Mong and Pfaff, 2004). Thus, it is important to carefully address the question of sex in any such study (Scott *et al.*, 2015). In addition, this question can itself be of scientific interest for potential therapeutic application. In the same vein as the sex question, one should consider the importance of the hormonal status of a female, if females are chosen for the study. Indeed, they are subject to high hormonal regulation during different cycles, which can act on the expression of different neuropeptide receptors (Lee *et al.*, 2009). In addition, lactation can induce drastic changes in hypothalamic cellular organization (Oliet *et al.*, 2004; Ludwig *et al.*, 2005) and thus lead to potentially critical modifications in neuropeptidergic production and modulatory action.

Most neuropepetidergic systems are either influencing or influenced by the stress status of animals. As with the sex question, it is mandatory to standardize experiments and perform adequate controls in order to ensure that environmental conditions do not interfere with the planned experiment.

14.5.2 Open Questions

14.5.2.1 Which Neuropeptidergic Circuits Were Studied Optogenetically?

Despite a decade-long history of optogenetics, this technique was only applied to neuropeptidergic circuits relatively recently, with the first reports in 2012 (Atasoy *et al.*, 2012; Knobloch *et al.*, 2012). To date, five neuropeptidergic brain circuits have been explored by optogenetics – oxytocinergic (Knobloch *et al.*, 2012; Choe *et al.*, 2015; Eliava *et al.*, 2016; Marlin *et al.*, 2015), orexinergic (Sears *et al.*, 2013; Apergis-Schoute *et al.*, 2015;), galanin (Wu *et al.*, 2014; McCall *et al.*, 2015), vasoactive intestinal peptide (Park *et al.*, 2015) and neuropeptide Y (NPY)/agouti related neuropeptide (AgRP) and POMC circuits (Aponte *et al.*, 2011) – out of approximately 100 neuropeptides that have been identified. Despite the limited number of peptides studied, the optogenetic approach in these reports provided the functional architecture of the respective circuits and their contributions to fear, appetitive, feeding and parental behaviors and sleep–wakefulness.

14.5.2.2 What Is the Mechanism of Neuropeptide Release?

It has been known for many years that the release of neuropeptides from large dense-core vesicles is Ca^{2+} dependent (Dreifuss, 1975). Contrary to classical neurotransmitter release, which requires high but localized Ca^{2+} increases, neuropeptide release requires high, diffuse and sustained increases in Ca^{2+}, hence peptidergic neurons need stronger/longer stimulation in order to release neuropeptides. Blue light illumination induces massive Ca^{2+} entry into cell bodies and their axons, facilitating the release of large dense-core vesicles. This process seems to be much more intense than when it occurs in normal physiological situations. However, establishing the links between "normal-range" firing and Ca^{2+} entry into peptidergic axons requires novel imaging techniques employing pH-sensitive fluorophores (Miesenböck *et al.*, 1998), and this area remains poorly understood.

14.5.2.3 Where Are Neuropeptides Released in the Brain?

There is no evidence for synaptic neuropeptide release. Therefore, the term "postsynaptic action" of neuropeptides, which has been used in many cases, seems not to be fully correct. Due to reported delays (in a range of seconds) in the onset of neuropeptide-sensitive neuron responses, we recently proposed a model of "micro-volume" OT release in the addressed brain regions (Grinevich *et al.*, 2015). However, we are not excluding the possibility that neuropeptides may be released *en passant* of axons or perisynaptically. Nevertheless, the micro-volume release with slow diffusion, binding to G-coupled receptors, initiating intracellular cascades may underlie both delay with onset of effect, as well as prolonged duration of the effect on "postsynaptic" cells.

14.5.2.4 Do Neuropeptides Excite or Inhibit Sensitive Neurons?

Direct effects of neuropeptides depend on the composition of the inhibitory (G_i/G_o) or stimulatory (G_q) subunits of respective G-protein-coupled receptors. With respect to oxytocin, its effect in most forebrain regions is excitatory and addressed primarily to GABAergic interneurons (Knobloch *et al.*, 2012; Owen *et al.*, 2013; Marlin *et al.*, 2015), likely mediated by G_q subunits of OTR (Busnelli *et al.*, 2012). By contrast, our recent findings indicate that the same neuropeptide – OT – may exert an inhibitory action on sensory neurons of the spinal cord via the G_i subunit of OTR (Eliava *et al.*, 2016).

14.5.2.5 What Does the Coexistence of Neurotransmitters and Neuropeptides in the Same Neurons Mean?

Neuropeptidergic neurons co-express neurotransmitters, such as GABA and glutamate. Most of the time, vesicles in peptidergic neurons also contain many neuropeptides and/or neurotransmitters. In our case, oxytocin is co-expressed with encephalin and dynorphin in SONs (Martin *et al.*, 1983; Eriksson *et al.*, 1996), and probably glutamate in PVNs (Knobloch *et al.*, 2012; Eliava *et al.*, 2016). Vasopressin, which is often studied in parallel with oxytocin, is co-expressed with dynorphin, galanin and pituitary adenylate cyclase-activating peptide (PCAP) within hypothalamic nuclei (Watson *et al.*, 1982; Landry *et al.*, 2003). While single brief (of a few milliseconds in duration) light pulses induce GABA or glutamate release, this mode of stimulation is never sufficient to release neuropeptides (at least in terms of observing changes in electrical activity of presumably neuropeptide-sensitive cells). Therefore, it is tempting to speculate that neuropeptide release and neurotransmitter release may not be coupled processes and may occur in different types of neurons, which is especially true for OT neurons, which very rarely form classical axo-dendritic or axo-somatic synapses.

In conclusion, optogenetics provides a very powerful tool for dissecting peptidergic brain circuits and, at the same time, has raised fundamental cell biology questions relating to the mechanisms of peptide release and "postsynaptic" signaling, which could be further explored by progressively developing techniques that combine with optogenetics.

ACKNOWLEDGMENTS

This work was supported by the IASP Early Career Research grant 2012, FP7 Career Integration grant 334455, Initiative of Excellence (IDEX) Attractiveness grant W13RAT29, University of Strasbourg Institute for Advanced Study (USIAS) fellowship 2014–15, Foundation Fyssen research grant 2015 (to A.C.), Chica and Heinz Schaller Research Foundation (to V.G.), German Research Foundation (DFG) grant GR 3619/4-1, Human Frontier Science Program RGP0019/2015, DFG within the Collaborative Research Center (SFB) 1134 and 1158 (to V.G.) and PHC PROCOP program (DAAD and Campus France) (to V.G. and A.C.). The authors thank Anne Seller for proofreading the manuscript.

REFERENCES

Apergis-Schoute, J., Iordanidou, P., Faure, C., Jego, S., Schöne, C., Aitta-Aho, T., Adamantidis, A., Burdakov, D. (2015). Optogenetic evidence for inhibitory signaling from orexin to MCH neurons via local microcircuits. *Journal of Neuroscience*, **35**, 5435–5441.

Aponte, Y., Atasoy, D., Sternson, S.M. (2011). AGRP neurons are sufficient to orchestrate feeding behavior rapidly and without training. *Nature Neuroscience*, **14**, 351–355.

Atasoy, D., Betley, J.N., Su, H.H., Sternson, S.M. (2012). Deconstruction of a neural circuit for hunger. *Nature*, **488**, 172–177.

Busnelli, M., Saulière, A., Manning, M., Bouvier, M., Galés, C., Chini, B. (2012). Functional selective oxytocin-derived agonists discriminate between individual G protein family subtypes. *Journal of Biological Chemistry*, **287**, 3617–3629.

Choe, H.K., Reed, M.D., Benavidez, N., Montgomery, D., Soares, N., Yim, Y.S., Choi, G.B. (2015). Oxytocin mediates entrainment of sensory stimuli to social cues of opposing valence. *Neuron*, **87**, 152–163.

de Kock, C.P., Wierda, K.D., Bosman, L.W., Min, R., Koksma, J.J., Mansvelder, H.D., Verhage, M., Brussaard, A.B. (2003). Somatodendritic secretion in oxytocin neurons is upregulated during the female reproductive cycle. *Journal of Neuroscience*, **23**, 2726–2734.

Dreifuss, J.J. (1975). A review on neurosecretory granules: their contents and mechanisms of release. *Annals of New York Academy of Sciences*, **248**, 184–201.

Eliava, M., Melchior, M., Knobloch-Bollmann, H.S., Wahis, J., da Silva Gouveia M., Tang, Y., Ciobanu, A.C., Triana del Rio, R., Roth, L.C., Althammer, F., Chavant, V., Goumon, Y., Gruber, T., Busnelli, M., Chini, B., Tan, L., Mitre, M., Froemke, R.C., Chao, M.V., Giese, G., Sprengel, R., Kuner, R., Poisbeau, P., Seeburg, P.H., Stoop, R., Charlet, A., Grinevich, V. (2016). A new population of parvocellular oxytocin neurons controlling magnocellular neuron activity and inflammatory pain processing. *Neuron*, **89**, 1291–1304.

Eriksson, M., Ceccatelli, S., Uvnäs-Moberg, K., Iadarola, M., Hökfelt, T. (1996). Expression of fos-related antigens, oxytocin, dynorphin and galanin in the paraventricular and supraoptic nuclei of lactating rats. *Neuroendocrinology*, **63**, 356–367.

Fields, R.L, Ponzio, T.A., Kawasaki, M., Gainer, H. (2012). Cell-type specific oxytocin gene expression from AAV delivered promoter deletion constructs into the rat supraoptic nucleus *in vivo*. *PLoS One*, **7**, e32085.

Grinevich, V., Desarménien, M.G., Chini, B., Tauber, M., Muscatelli, F. (2015). Ontogenesis of oxytocin pathways in the mammalian brain: late maturation and psychosocial disorders. *Frontiers in Neuroanatomy*, **8**, 164.

Knobloch, H.S., Charlet, A., Hoffmann, L.C., Eliava, M., Khrulev, S., Cetin, A.H., Osten, P., Schwarz, M. K., Seeburg, P.H., Stoop, R., Grinevich, V. (2012). Evoked axonal oxytocin release in the central amygdala attenuates fear response. *Neuron*, **73**, 553–566.

Landry, M., Vila-Porcile, E., Hökfelt, T., Calas, A. (2003). Differential routing of coexisting neuropeptides in vasopressin neurons. *European Journal of Neuroscience*, **17**, 579–589.

Lee, H.J., Macbeth, A.H., Pagani, J.H., Young, W.S. 3rd. (2009). Oxytocin: the great facilitator of life. *Progress in Neurobiology*, **88**, 127–151.

Ludwig, M., Sabatier, N., Dayanithi, G., Russell, J.A., Leng, G. (2002). The active role of dendrites in the regulation of magnocellular neurosecretory cell behavior. *Progress in Brain Research*, **139**, 247–256.

Ludwig, M., Bull, P.M., Tobin, V.A., Sabatier, N., Landgraf, R., Dayanithi, G., Leng, G. (2005). Regulation of activity-dependent dendritic vasopressin release from rat supraoptic neurons. *Journal of Physiology*, **564**, 515–522.

Marlin, B.J., Mitre, M., D'amour, J.A., Chao, M.V., Froemke, R.C. (2015). Oxytocin enables maternal behaviour by balancing cortical inhibition. *Nature*, **520**, 499–504.

McCall, J.G., Al-Hasani, R., Siuda, E.R., Hong, D.Y., Norris, A.J., Ford, C.P., Bruchas, M.R. (2015). CRH engagement of the locus coeruleus noradrenergic system mediates stress-Induced anxiety. *Neuron*, **87**, 605–620.

Mendell, L.M., Wall, P.D. (1965). Responses of single dorsal cord cells to peripheral cutaneous unmyelinated fibers. *Nature*, **206**, 97–99.

Miesenböck, G., De Angelis, D.A., Rothman, J. E. (1998). Vizualizing secretion and synaptic transmission with pH-sensitive green fluorescent protein. *Nature*, **394**, 192–195.

Mong, J.A, Pfaff, D.W. (2004). Hormonal symphony: steroid orchestration of gene modules for sociosexual behaviors. *Molecular Psychiatry*, **9**, 550–556.

Oliet, S.H., Piet, R., Poulain, D.A, Theodosis, D.T. (2004). Glial modulation of synaptic transmission: insights from the supraoptic nucleus of the hypothalamus. *Glia*, **47**, 258–267.

Owen, S.F., Tuncdemir, S.N., Bader, P.L., Tirko, N.N., Fishell, G., Tsien, R.W. (2013). Oxytocin enhances hippocampal spike transmission by modulating fast-spiking interneurons. *Nature*, **500**, 458–462.

Park, S.J., Borghuis, B.G., Rahmani, P., Zeng, Q., Kim, I.J., Demb, J.B. (2015). Function and circuitry of VIP^+ interneurons in the mouse retina. *Journal of Neuroscience*, **35**, 10685–10700.

Tovote, P., Lüthi, A. (2012). Curbing fear by axonal oxytocin release in the amygdala. *Neuron*, **73**, 407–410.

Scott, N., Prigge, M., Yizhar, O., Kimchi, T. (2015). A sexually dimorphic hypothalamic circuit controls maternal care and oxytocin secretion. *Nature*, **525**, 519–522.

Sears, R.M., Fink, A.E., Wigestrand, M.B., Farb, C.R., de Lecea, L., Ledoux, J.E. (2013). Orexin/hypocretin system modulates amygdala-dependent threat learning through the locus coeruleus. *Proceeding of National Academy of Sciences of USA*, **110**, 20260–20265.

Watson, S.J., Akil, H., Fischli, W., Goldstein, A., Zimmerman, E., Nilaver, G., van wimersma Griedanus, T.B. (1982). Dynorphin and vasopressin: common localization in magnocellular neurons. *Science*, **216**, 85–87.

Wu, Z., Autry, A.E., Bergan, J.F., Watabe-Uchida, M., Dulac, C.G. (2014). Galanin neurons in the medial preoptic area govern parental behaviour. *Nature*, **509**, 325–330.

15 Optogenetic Approaches to Investigating Brain Circuits

Alexander M. Herman, Jay M. Patel, and Benjamin R. Arenkiel

15.1 Introduction

The brain, like much of the body, is a highly heterogeneous organ made up of a myriad of different cell types based on function, morphology, electrophysiological properties and cell type-specific markers. Yet within this seemingly highly heterogeneous tissue, it is obvious that the brain, and the cells that comprise it, is a highly structured, organized and meticulously patterned system capable of coordinating astoundingly complex physiological and behavioral roles. Thus, understanding the properties of individual cell types in the brain is critically important to forming a collective understanding of how the brain works and how various diseases and disorders arise when the components that constitute the brain are impaired or altered from their normal functions. To this end, it is critical to develop and employ methods for precisely manipulating specific cell types within the intact brain. Here, we will review the optogenetic toolkit and strategies that are available for targeting neurons with cell type specificity for activity manipulations with light, and highlight the benefits and limitations that they possess.

15.2 Bacterial Artificial Chromosome Transgenic, Conditional Genetic and Knock-in Mouse Models for Cell Type-specific Rhodopsin Expression

The mouse has firmly established itself as the prominent mammalian model organism. The relative ease in manipulating the mouse genome through classical approaches such as bacterial artificial chromosome (BAC) transgenesis and embryonic stem cell-based technologies, to more contemporary approaches that include the CRISPR/Cas9 system, zinc finger nucleases and transcription activator-like effector nucleases, has led to an explosion of engineered mouse lines for investigation of genes and the nervous system. Importantly, mouse lines have been generated to express genes in a heterologous manner for the purpose of conferring novel functions in order to manipulate specified cell types.

Channelrhodpsin-2 (ChR2), a natively expressed gene from the algae *Chlamydomonas reinhardtii*, encodes a light-gated ion channel that can be selectively expressed in neurons to drive neuronal activity when stimulated with blue light. In contrast, opsins such as halorhodopsin (Halo) and archaerhodopsin (Arch) can be used for targeted hyperpolarization and neuronal silencing. By targeting light-activated opsins to specified cell types in the brain, researchers can now begin to dissect the contributions of individual neurons, isolated circuits or elaborate networks in the contexts of disease, development and behavior.

To target rhodopsin expression in a cell type-specific manner, the most common approach is to drive expression under the control of a cell type-specific promoter, such that your reporter (e.g. ChR2, Halo or Arch) will only be expressed in neurons that normally utilize the promoter of interest. To date, a number of BAC transgenic, ChR2-expressing mouse lines are available that express ChR2 variants in a variety of cell types throughout the central nervous system (CNS) (Table 15.1). Cell type specificity in these mouse lines is promoter dependent, thereby restricting ChR2 expression to neuronal subtypes specified by their respective promoter activity. *Thy1-ChR2-EYFP* and *Vglut2-ChR2-EFYP* transgenic mice, for example, express ChR2 in subsets of excitatory cell types of the brain. Conversely, *Vgat-ChR2-EYFP* and *Pvalb-ChR2-EYFP* mice express ChR2 in inhibitory interneuron cell types. Whereas *Vgat* drives broad ChR2 expression in most interneurons of the brain, *Pvalb* drives expression in a more-restricted subpopulation of CNS interneurons. For ChR2 expression in either cholinergic or serotonergic neurons, *Chat-ChR2-EYFP* or *Tph2-ChR2-EYFP* mice are available, respectively. Other common ChR2 mouse lines include the cerebellar Purkinje cell-specific *Pcp2-ChR2-EYFP* BAC transgenics (commonly referred to as *L7-ChR2-EYFP*) and the olfactory sensory neuron-specific *OMP-ChR2-EYFP* transgenics, a targeted knock-in mouse line that broadly targets all olfactory sensory neurons. ChR2 BAC transgenic or targeted knock-in mouse lines, however, are still limited in availability with regards to the number of cell-specific mouse lines that exist. To overcome this problem and to facilitate opsin expression in other CNS cell types of interest, conditional genetic approaches have been of great utility.

Cre/Lox technology is widely used in mouse genetics to selectively drive gene expression in a promoter-dependent fashion, but unlike knock-in or BAC transgenic mice, gene expression using Cre/Lox methods is directly dependent upon the cell type-specific expression of Cre recombinase to turn on the expression of a gene of interest (GOI). A common approach to driving gene expression in this fashion involves placing a GOI downstream of a strong promoter that is followed by a pair of LoxP recombination sites that flank a transcriptional "stop" sequence (Figure 15.1). However, strong constitutive expression can also be achieved in targeted cell types by targeting a 5´-promoter–Lox–stop–Lox–GOI-3´ sequence to a ubiquitously expressed locus, such as ROSA26, or the more recently described TIGRE locus. By crossing this conditional "Lox–stop–Lox" allele into mice that express cell type-specific Cre recombinase, the offspring will selectively express a GOI only in cells that express Cre. Depending on the optogenetic response and

Table 15.1 Useful mice with light-activated channels.

Genomic modifying strategy	Technical name	Common name
Targeted locus ROSA26Sor		
	tm32(CAG-COP4*H134 R/EYFP)Hze	Rosa-LSL ChR2::EYFP
	tm35.1(CAG-aop3/GFP)Hze	Rosa-LSL Arch::EYFP
	tm39(CAG-hop/EYFP)Hze	Rosa-LSL HaloEYFP
	tm40.1(CAG-aop3/EGFP)Hze	Rosa-LSL ArchT::EGFP
	tm80.1(CAG-COP4*L132C/EYFP)Hze	Rosa-FSF-LSL CatCH::eyfp
	tm27.1(CAG-COP4*H134 R/tdTomato) Hze	Rosa-LSL ChR2::TdTom
	tm2.1Ksvo	Rosa-FSF-LSL ReaChR:: mCitrine
	tm1(CAG-COP4*E123T*H134 R,-tdTomato)Gfng	Rosa-LSL ChETA-p2a-tdtomato
Targeted Locus TIGRE (Igs7)		
	tm79.1(tetO-hop/EGFP)Hze	Igs7-TRE-LSL Jaws::EGFP
	tm90.1(tetO-COP4*/EGFP)Hze	Igs7-TRE-LSL Chronos::EGFP
Transgenic Thy1 promoter (TgThy1)		
	Tg(Thy1-hop/EYFP)4Gfng	eNpHR2.0::EYFP
	Tg(Thy1-COP4/EYFP)18Gfng	Chr2::EYFP
	Tg(Thy1-COP4*H134 R/EYFP)20Gfng	mhChR2::EYFP
	Tg(Thy1-COP4/EYFP)9Gfng	Chr2::EYFP
	Tg(Thy1-hop/EYFP)2Gfng	eNpHR2.0::EYFP
Bac transgenic promotor/enhancer		
mParvalbumin	Tg(Pvalb-COP4*H134 R/EYFP)15Gfng	Parvalbumin ChR2::EYFP
mSerotonin	Tg(Tph2-COP4*H134 R/EYFP)5Gfng	Serotoninergic ChR2::EYFP
Choline acetyltransferase	Tg(Chat-COP4*H134 R/EYFP,Slc18a3) 5Gfng	ChAT ChR2::EYFP
Purkinje cell protein 2	Tg(Pcp2-COP4*H134 R/EYFP)U126Isop	Pcp2 ChR2::EYFP
Targeted olfactory marker protein locus	Omptm1.1(COP4*/EYFP)Tboz	Omp ChR2::EYFP
Transgenic Tet response element	Tg(tetO-hop/EGFP,-COP4/mCherry) 6Kftnk	TRE Halo::EYFP and ChR2:: mCherry

color reporter desired, various fluorescently marked optogenetic reporter conditional mouse lines are available (Table 15.1). In this way, by crossing conditional optogenetic mice with cell type-specific Cre lines (Table 15.2), reporter expression can be selectively targeted to desired cells, lineages or brain regions. Because many more Cre-expressing mouse lines are available compared to transgenic optogenetic reporter mouse lines, opsins can be targeted to a multitude of CNS cell types using this conditional Cre/Lox strategy. However, it should be kept in mind that although many advertised Cre lines are available for cell type-restricted targeting of reporters, expected expression patterns should be verified in order to ensure that Cre expression drives the desired opsin reporters in the correct cell types. Not all Cre lines function as predicted, since transgenic approaches may not always recapitulate endogenous expression of utilized promoters or their expression elements. One major cause of this variability is that BAC transgenic Cre lines are generally not targeted to a specific locus, and so are randomly integrated in the

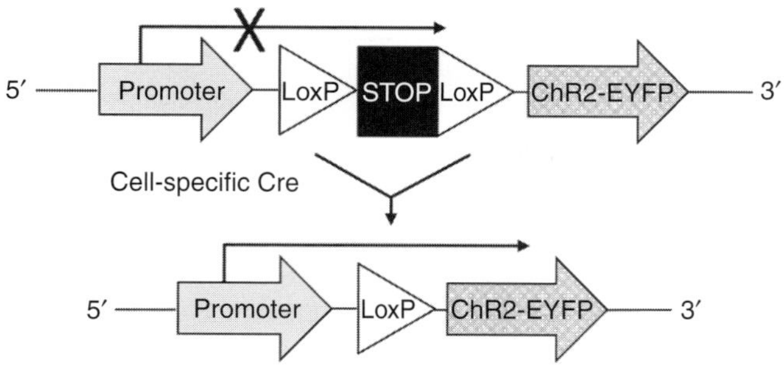

Figure 15.1 Conditional mouse genetics for driving optogenetic reporter expression with cell type specificity. A common approach for conditionally targeting transgene expression to a cell type of interest takes advantage of site-specific recombination systems such as the commonly used Cre/Lox system. A targeted transgene to a ubiquitous locus, such as the ROSA26 (R26) locus, includes a strong promoter and two homologous Lox sites with an intervening transcriptional "stop" cassette (Lox–stop–Lox), followed by a GOI (ChR2-EYFP shown). In this example, by crossing this R26-LSL-ChR2-EYFP mouse line with a Cre line of interest, Cre recombinase will mediate the excision of the "stop" cassette in order to allow for ChR2-EYFP expression only in cells expressing Cre recombinase.

genome. Thus, although Cre recombinase may be driven by a cell type-specific promoter in order to capture its specificity, not all promoter elements may be present, or off-target local genomic elements may affect true expression patterns. Even in lines in which Cre expression has been targeted to an endogenous locus, expression may not always recapitulate expected endogenous gene expression profiles. Furthermore, developmental programs often influence gene expression either dynamically and/or transiently in different cell types. As a conditional system, even very small amounts of Cre recombinase can lead to opsin expression in cell types that were not anticipated. Although these issues exist, Cre/Lox strategies for expressing optogenetic reporters in discrete CNS cell types or the inclusion of tamoxifen inducibility can be used to great effect.

In addition to Cre/Lox systems, other site-specific recombinase strategies are available for conferring cell type specificity, though their use and available resources may not be as ubiquitous. The Dre/Rox system and the FLP/FRT system act in similar fashions to Cre/Lox. Both systems utilize recombination of either Rox or FRT sites by their site-specific enzymes, Dre recombinase and Flippase (FLP), respectively, to drive a designed DNA recombination event. Overall, however, these systems also encounter similar potential limitations as Cre/Lox technology, since enzyme expression is still dependent on promoter activity and can lead to limitations in cell-specific expression. Similarly, precocious developmental expression may also lead to unanticipated recombination. As with Cre/Lox approaches, both Dre-expressing and FLP-expressing mouse lines should be validated to ensure expected cell type-specific expression profiles. It should be noted that an important benefit to the discovery and development of multiple site-specific recombinases is the ability to express multiple enzyme systems in a combinatorial approach to drive cell type-specific gene expression in more than one cell type, or in a subpopulation of cells that may express dual markers.

Table 15.2 Useful mouse lines for cell type-specific genetic manipulations.

Technical name	Common name	CNS cell subtype marker targeted by transgene	Expressed protein
Agrptm1(cre)Lowl	Agrp-Cre	Agouti-related protein	Cre recombinase
Avptm1.1(cre)Hze	Avp-Cre	Arginine vasopressin	Cre recombinase
Calb1tm1.1(folA/EGFP/cre)Hze	Calbindin1-Cre	Calbindin 1	Cre recombinase
Calb2tm1(cre)Zjh	Calbindin2-Cre	Calbindin 2	Cre recombinase
Tg(Pcp2-cre)2Mpin	Pcp2-Cre	Purkinje cell protein 2	Cre recombinase
Olig1tm1(cre)Rth	Olig1-Cre	Olig1	Cre recombinase
Ccktm1.1(cre)Zjh/J	CCK-Cre	Cholecystokinin	Cre recombinase
B6;129S6-Chattm1(cre)Lowl/J	Chat-Cre	Choline acetyltransferase	Cre recombinase
Corttm1(cre)Zjh/J	CST-Cre	Cortistatin	Cre recombinase
B6(Cg)-Crhtm1(cre)Zjh/J	Crh-Cre	Corticotropin-releasing hormone	Cre recombinase
Tg(dlx6a-cre)1Mekk/J	Dlx6a-Cre	Dlx5/6, forebrain GABAergic	Cre recombinase
B6.129S2-Emx1tm1(cre)Krj/J	Emx1-Cre	Emx1, most neocortex and hippocampus	Cre recombinase
Gad2tm2(cre)Zjh/J	Gad2-Cre	GABAergic	Cre recombinase
FVB-Tg(GFAP-cre)25Mes/J	GFAP-Cre	Glial fibrillary acidic protein	Cre recombinase
B6.Cg-Tg(Gfap-cre)73.12Mcs/J	GFAP-Cre	Glial fibrillary acidic protein	Cre recombinase
B6.Cg-Tg(Gfap-cre)77.6Mvs/J	GFAP-Cre	Glial fibrillary acidic protein	Cre recombinase
C57BL/6-Tg(Grik4-cre)G32-4Stl/J	Grik4-Cre	Glutamate receptor, ionotropic, kainate 4	Cre recombinase
B6.Cg-Tg(Ins2-cre)25Mgn/J	RIP Cre	Insulin	Cre recombinase
B6.129-Leprtm2(cre)Rck/J	LeptinR-Cre	Leptin receptor	Cre recombinase
B6.129S1-Mnx1tm4(cre)Tmj/J	HB9-Cre	Homeobox 9	Cre recombinase
C57BL/6 J-Tg(Nkx2-1-cre)2Sand/J	Nkx2 Cre	Nkx2.1 interneuron	Cre recombinase
FVB-Tg(Nr5a1-cre)2Lowl/J	Sf-1 Cre	SF-1 in hypothalamus	Cre recombinase
Tg(Pomc1-cre)16Lowl/J	Pomc-Cre	Pro-opiomelanocortin	Cre recombinase
B6;129P2-Pvalbtm1(cre)Arbr/J	Parvalbumin-Cre	Parvalbumin	Cre recombinase
B6;C3-Tg(Scnn1a-cre)3Aibs/J	Scnn1a Cre	Sodium channel, non-voltage-gated 1α	Cre recombinase
Tg(Sim1-cre)1Lowl/J	Sim1-Cre	Sim1	Cre recombinase
B6.SJL-Slc6a3tm1.1(cre)Bkmn/J	Dat-Cre	Dopamine transporter	Cre recombinase
Ssttm2.1(cre)Zjh/J	Somatostatin-Cre	Somatostatin	Cre recombinase
B6.Cg-Tg(Syn1-cre)671Jxm/J	Synapsin1-Cre	Synapsin 1	Cre recombinase
FVB/N-Tg(Thy1-cre)1Vln/J	Thy1-Cre	Thy1	Cre recombinase
Viptm1(cre)Zjh/J	VIP-Cre	Vasoactive intestinal peptide	Cre recombinase
B6.Cg-Tg(Wnt1-cre)11Rth Tg (Wnt1-GAL4)11Rth/J	Wnt1-Cre	Wingless mutant 1	Cre recombinase
Slc32a1tm2(cre)Lowl	Vgat-Cre	GABA vesicular transporter	Cre recombinase
Tg(Npy-cre)1Rck	Npy-Cre	Neuropeptide Y	Cre recombinase
Tg(Lhx6-icre)1Kess	Lhx6-Cre	Lhx6 interneuron	Cre recombinase

Table 15.2 (cont.)

Technical name	Common name	CNS cell subtype marker targeted by transgene	Expressed protein
Nos1tm1(cre)Mgmj	Nos1-Cre	Nitrous oxide synthase 1	Cre recombinase
Slc17a6tm2(cre)Lowl	Vglut-Cre	Glutamate vesicular transporter	Cre recombinase
B6(Cg)-Dlx5tm1(cre/ERT2)Zjh/J	DLx5-CreER	Dlx5/6, forebrain GABAergic	Estrogen receptor-sensitive Cre recombinase
B6.Cg-Tg(Plp1-cre/ERT)3Pop/J	Plp1-CreER	Proteolipid protein (myelin) 1	Estrogen receptor-sensitive Cre recombinase
B6;C3-Tg(Wfs1-cre/ERT2)3Aibs/J	Wfs1-CreER	Wolfram syndrome 1 homolog (human)	Estrogen receptor-sensitive Cre recombinase
Chattm1(cre/ERT)Nat	Chat-CreER	Choline acetyltransferase	Estrogen receptor-sensitive Cre recombinase
Pvalbtm1(cre/ERT2)Zjh	Parvalbumin-CreER	Parvalbumin	Estrogen receptor-sensitive Cre recombinase
Lhx6tm1(cre/ERT2)Zjh	Lhx6-CreER	Lhx6 interneuron	Estrogen receptor-sensitive Cre recombinase
Nos1tm1.1(cre/ERT2)Zjh	Nos1-CreER	Nitrous oxide synthase 1	Estrogen receptor-sensitive Cre recombinase
Calb2tm2.1(cre/ERT2)Zjh	Calbindin2-CreER	Calbindin 2	Estrogen receptor-sensitive Cre recombinase
Tg(Phox2b-Flpo)3276Grds	Phox2b-Flpo	Phox2b	Flp recombinase
Tg(mI56i-flpe)39Fsh	Dlx5/6-Flpe	Dlx5/6, forebrain GABAergic	Flpe recombinase
Gt(ROSA)26Sortm3(CAG-FLPo/ERT2)Alj	R26LSL-Flpo		Cre-dependent Flpo recombinase

Building from the mouse lines that are currently available, it is possible to generate models that express a desired recombinase in one cell type and a different recombinase in a completely different cell type. This would allow for the targeting of two independent cell types for optogenetic reporter expression, or even two different optogenetic reporters. Using this approach, appropriate mice first have to be bred in order to generate offspring that harbor both alleles for recombinase expression in specified cell types. In the most straightforward method, optogenetic reporters can then be selectively targeted to desired cell types and/or areas of the brain through the use of conditional recombinase-dependent viruses (described below). In fact, by utilizing intersectional genetics and multiple cell type-specific promoters, one can also drive the expression of one recombinase to allow the expression of another, such as driving the expression of FLP under the interneuron-specific parvalbumin (PV) promoter to recombine a DNA sequence encoding Cre recombinase that is driven by a somatostatin (SST) promoter. In this example, Cre would then recombine a Lox–stop–Lox sequence to express a GOI (such as an optogenetic reporter) only in subpopulations of interneurons that are both PV and SST positive. The obvious downside to this approach is the current limited availability of mouse lines, and the time-consuming mating schemes for producing mice that harbor the multiple desired alleles. However, it is worth noting that engineering mouse lines is becoming more high throughput, and obtaining new models that carry new or multiple alleles is more convenient than ever.

As mentioned above, non-specific expression of reporters through conditional activation in cell lineages during development remains a significant challenge. To circumvent non-specific developmental expression of recombination enzymes, and/or to enhance the temporal precision of gene expression, use of the CreER system affords great utility. In this system, Cre recombinase is fused to an estrogen receptor mutant that localizes Cre to the cytosol. Upon binding of the estrogen receptor domain by peripherally administered tamoxifen (the estrogen receptor mutant antagonist), the CreER fusion protein translocates to the nucleus in order to allow for site-specific recombination of Lox sites. Thus, the promoter that expresses Cre defines spatial control, and temporal control of gene expression is dependent on treatment with tamoxifen. There are currently many CreER mouse models available that target CreER expression to various neuronal cell types of the brain (Table 15.2), and the list is continually expanding. By crossing these mice with conditional optogenetic reporter mouse lines, temporal and cell type specificity of opsin expression can be achieved upon tamoxifen administration. It should be noted, however, that CreER mouse lines can at times also exhibit leaky Cre activity in the absence of tamoxifen, and should be tested beforehand to ensure appropriate levels of temporal and cell type selectivity. Temporal control of gene expression may also be generated by the use of tetracycline-responsive approaches. Several optogenetic reporter mouse lines are currently available that, when bred with appropriate driver lines, can be used for opsin expression under the control of the tetracycline promoter, which is only activated in the presence (Tet-On) or absence (Tet-Off) of doxycycline. Currently, however, the available mouse lines are limited in number for approaches that require tetracycline-based methods to drive opsin expression.

15.3 Viral Approaches for Cell Type-specific Targeting of Optogenetic Reporter Expression

Alongside traditional conditional genetic approaches, viral vectors are increasingly being used to drive opsin expression in specified cell types. Not only does this approach allow for cell type specificity, but also it affords temporal and site-specific control of expression. A challenge with many mouse lines is that even when employing conditional genetic crosses to restrict opsin expression to a specific cell type, often those same cell types are located throughout many parts of the brain. Thus, it may be difficult to discern the optogenetically induced outcomes of any particular brain region if other regions are also being inadvertently photoactivated. To further restrict opsin expression to a specific cell type in one part of the brain, a virus may be injected directly to the site of interest, so that only the targeted cell type within that specific region will express the optogenetic reporter. Cell type selectivity using a viral approach is typically generated in one of two ways. The first is through the use of promoters that are specific for one cell type or another. For example, the Ca^{2+}/calmodulin-dependent protein kinase II (CaMKII) promoter is the most widely used promoter for driving gene expression in excitatory neurons of the brain (Tighilet *et al.*, 1998). Short promoters, such as those for neuropeptide Y, PV, SST and vasoactive intestinal peptide can be used to drive relatively selective expression in their respective interneuron cell types (Nathanson *et al.*, 2009), though promiscuous gene expression in other cell types may occur if the promoter element does not contain all of the necessary components to drive specificity. Theoretically, other promoters can also be used in this fashion with a virus to drive cell-specific expression, so long as their nucleotide length is appropriate for the size limitation of the viral vector that is used. In fact, size limitation is one of the major concerns when using viruses to deliver genes. Recombinant adeno-associated viruses (AAVs) are the most widely used viruses for CNS-specific gene delivery, and are by far the most common viral tools for targeting opsin expression to specified cell types. Many AAV plasmids are cheaply and commercially available, can be used to drive prolonged gene expression and tend to be much less immunoreactive than other viruses that are used for gene delivery in the brain (Mastakov *et al.*, 2002). However, a major drawback of AAVs is their relatively small packaging limits (approximately <5 kb insert in between AAV inverted terminal repeat domains), which restricts the number and/or type of promoters, or other genomic elements one may want to express. Because of this limitation, many promoters cannot be used for efficient AAV vector production. To circumvent this problem, Cre-dependent AAVs have been engineered so that a GOI may be expressed in a Cre-dependent manner under the control of a strong promoter. Using this strategy, cell type specificity is conferred by Cre expression in a mouse line, rather than by promoter selectivity of the virus. Though vectors using a "Lox–stop–Lox" approach may be used to selectively drive the expression of a downstream gene, this approach also tends to be limited by size constraints, since intervening transcriptional "stop" cassettes are relatively large. In place of this strategy, Cre-dependent "FLEX" vectors have been developed (Atasoy *et al.*, 2008). This strategy implements an inverted GOI that is flanked by two non-homologous pairs of lox sites (Figure 15.2). Two serial recombination events are then required: the first to invert the GOI into its correct orientation; and the second to lock it into place. FLEX vectors are advantageous because they require much less of a total sequence length than "Lox–stop–Lox" vectors, allowing for more elements or GOIs to be sub-cloned into the viral plasmid. Furthermore, as previously mentioned, a plethora of Cre mouse lines are available for targeting virally expressed opsins throughout the CNS. To avoid potential confounding variability due to Cre expression, and to further ensure faithful gene expression in a specified cell type, it is also possible to use a viral vector whose gene products are both Cre dependent (i.e. FLEX) and driven by a cell type-specific promoter. With this approach, any promiscuity in Cre expression can be offset by the promoter's selectivity, and vice versa. One consideration, however, is that cell type-specific promoters generally do not drive high levels of gene expression compared to commonly used strong promoters. Therefore, for any given application, it is necessary to empirically determine beforehand if the viral

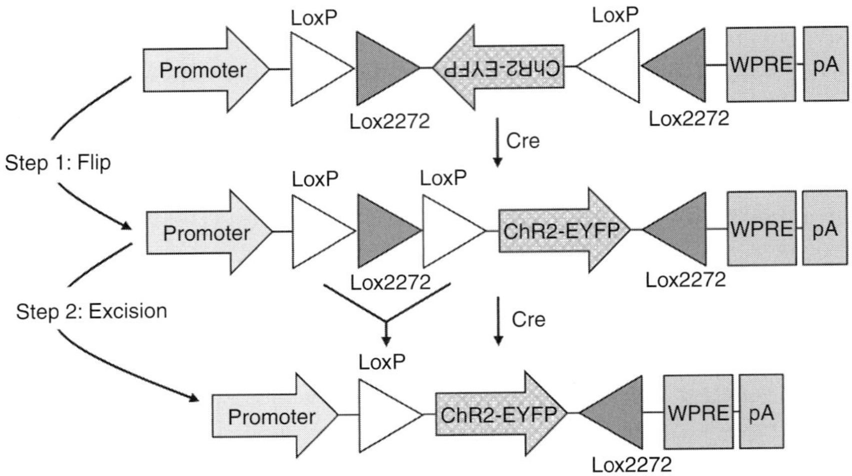

Figure 15.2 Cre-dependent "FLEX" vectors for conditionally targeting optogenetic reporter expression in a cell-specific manner. Flip-excision (FLEX) viral vectors allow for conditional gene expression only in cells expressing Cre recombinase, and only at a location that has been injected with the viral vector of choice, typically an AAV. The viral transgene includes a promoter – either a strong promoter or a cell-selective promoter – followed by two pairs of non-homologous Lox sites (LoxP and Lox2272 sites shown here). Intervening between the two pairs of Lox sites is a GOI (ChR2-EYFP shown here) in an inverted orientation. Elements such as the woodchuck post-transcriptional response element (WPRE) allow for stabilized long-term expression. A polyadenylation sequence is required as a transcriptional stop. It is important to note that homologous Lox sites facing in opposite directions facilitate DNA inversion, while homologous Lox sites facing in the same direction facilitate excision of intervening DNA. Following injection of the virus into a site of interest, Cre-mediated recombination first flips the GOI into the correct orientation. A second recombination event is then required to remove one Lox site (Lox2272 shown here), leaving only two non-homologous Lox sites on either side of the correctly oriented GOI. Because the two remaining Lox sites are non-homologous, no further Cre-mediated recombination can occur, allowing for stable promoter-driven gene expression.

vector confers appropriate expression levels in order to ensure high enough levels of expression in a targeted cell. In general, the expression from such promoters should be high enough for most applications, and may in fact be less toxic to a cell due to decreased protein burden, both from fluorescent reporters whose prolonged expression is known to be toxic (Shaner *et al.*, 2005; Shemiakina *et al.*, 2012) and from certain optogenetic reporters (Miyashita *et al.*, 2013).

Depending on experimental needs, other viruses may be more useful for gene delivery. Retroviruses, most notably lentiviruses, are good options for gene delivery when packaging limits cannot be met using a recombinant AAV. Lentiviral vectors have been successfully used to express optogenetic reporters in a variety of CNS cell types. Cell type specificity through lentiviral strategies is most often conferred using a cell-selective promoter (e.g. CaMKII) to drive opsin expression. While it may be possible to regulate expression using a Cre-dependent lentivirus, this approach is not common. One main reason for this may be the less predictable levels of conditional expression that might be obtained due to integration of the viral construct into the targeted cell genome, where recombination effects are more variable than what is afforded through AAV vectors. With respect to retroviruses, cell-specific biological factors may also be advantageous for selective targeting and expression. If, for example, the cell type that is targeted is at an immature stage of development, or is mitotically active, such as neural stem cells located in neurogenic niches of the brain, an AAV is generally a poor choice for gene delivery to these cell types. However, retroviruses such as lentiviruses are great candidates for delivering genes to mitotic or immature cell types, in addition to mature cell types. Because the retroviral genome will integrate into the host cell's genome, where it can be stably expressed, daughter cells born from transduced mother cells will inherit any integrated genomic viral elements, and this property should be considered when designing experiments. Moreover, retroviruses can also be utilized to infect only dividing cells. Under typical packaging protocols, however, retroviruses tend to have viral titers that are several orders of magnitude less than AAVs, which should also be considered when choosing the appropriate virus for gene delivery.

Genetically modified rabies virus (RV), which is growing in popularity for its use as a monosynaptic retrograde neuronal tracer, can also be used for expressing opsins, though to date only ChR2-expressing RV vectors have been described. Notably, such RV vectors have been engineered to be replication incompetent (Wickersham *et al.*, 2007), having substituted a necessary viral glycoprotein (G-protein) coding sequence with the sequence encoding a ChR2–mCherry fusion protein (Osakada *et al.*, 2011). These RVs are commonly referred to as SADΔG, denoting deletion of the G-protein, and numerous other variants exist that have replaced the endogenous G-protein coding sequence for various color and/or functional reporters (Osakada *et al.*, 2011). Without the viral G-protein, these viruses cannot spread to subsequent neurons after initial infection (Figure 15.3). Furthermore, these and similar enveloped viruses have been pseudo-typed with an avian-specific envelope protein (e.g. EnvA), such that they will only infect neurons that express the cognate receptor (e.g. TVA). To allow for initial infection and retrograde monosynaptic transfer, both the TVA receptor and the RV G-protein must be provided to the target cell in *trans*, which may be accomplished by different means, depending on necessity. For example, plasmid DNA encoding for both products can be introduced into neurons via electroporation, though this generally requires that the electroporation occurs either *in utero* or at early post-natal periods, since electroporation of DNA into adult nervous tissue is not easily achieved. Though this approach may provide adequate regional specificity of expression for many experiments, it may not afford adequate cell type specificity for applications that would benefit from precise neuronal subtype targeting. Recently, a mouse model has been engineered that allows for conditional Cre-dependent expression of both the G-protein and the TVA receptor from the ROSA26 locus (Takatoh *et al.*, 2013). By crossing this mouse line with a Cre driver line of interest, the G-protein and TVA receptor can be specified for expression exclusively in Cre-expressing neurons. Subsequent infection with the G-deleted, EnvA-pseudotyped RV (RV-SADΔG) will preferentially infect only Cre-expressing neurons that selectively express the G-protein and TVA receptor. Once a targeted neuron is infected with RV-SADΔG, competent virions will be generated due to the

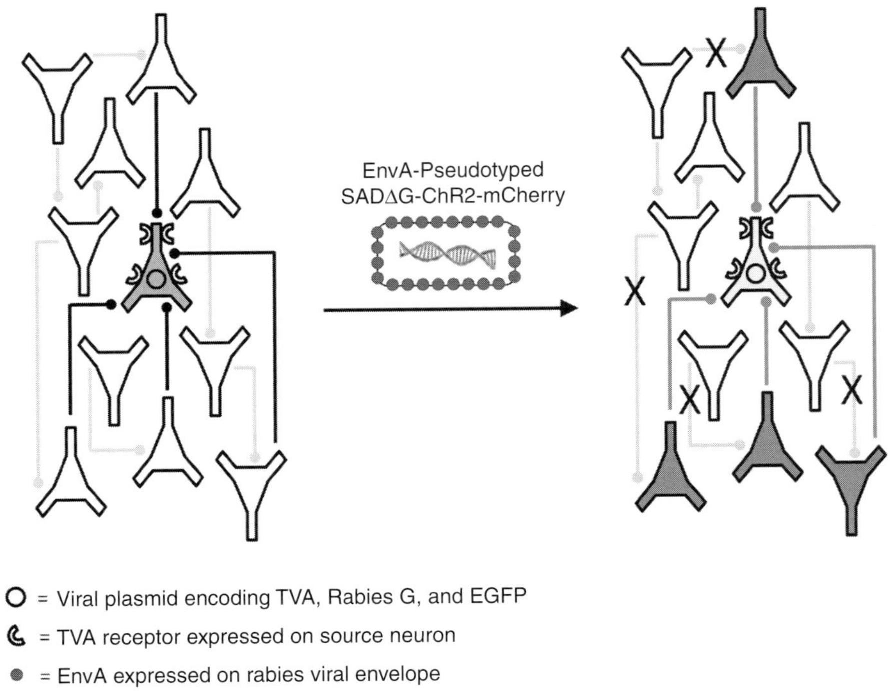

Figure 15.3 Monosynaptic RV tracing coupled with optogenetic reporter expression. RV is a neurotropic, retrograde, enveloped, single-stranded RNA virus. With several modifications, RV can be used as a monosynaptic neuronal tracer in order to identify presynaptic partners on a target source cell. Using any number of approaches, including conditional mouse genetics or viral delivery (shown here), Cre-expressing neurons of interest can be targeted for expression of the necessary RV tracing components. These include the Rabies G-protein, the avian-specific TVA receptor and an initial color reporter for easy identification of target source cells. To successfully target cells expressing the TVA receptor, the RV must be pseudotyped to express the avian-specific EnvA protein, which specifically binds to the TVA receptor. After infection of rabies G/TVA/EGFP-expressing source neurons with pseudotyped SADΔG RV, in which the rabies G coding sequence has been replaced with that of ChR2–mCherry, only cells expressing the TVA receptor will be targeted for infection. Because only infected neurons will express the rabies G-protein in *trans*, as is supplied initially by the viral plasmid, only in those neurons will more viral particles be able to successfully package and jump via their usual retrograde manner to a presynaptic neuron. Because presynaptic neurons do not express any of the RV tracing components, viral spread is halted after one jump. Because this RV variant expresses ChR2–mCherry, presynaptic neurons will be labeled red, distinguishing them from doubly labeled source neurons. ChR2 expression then allows for analysis of functional connectivity between a presynaptic neuron and its partner source cell.

presence of the G-protein. G-containing (infectious) viral particles can then spread in a retrograde fashion to an upstream presynaptic neuron. Once in the presynaptic neuron, viral spread is halted since only neurons that were initially targeted express the G-protein. For RV-SADΔG vectors that express ChR2–mCherry, functional connectivity can be tested between partner neurons after monosynaptic retrograde transfer has occurred. However, it should be cautioned that use of this mouse line requires proper controls in order to ensure fidelity of G-protein/TVA expression in specified cell types of interest so as to minimize the risk of promiscuous RV infection and non-specific tracing. Alternatively to using conditional alleles in mice, both G-protein and TVA receptor expression can be targeted to specific cell types using Cre-dependent AAVs. Using mice with restricted expression of Cre, conditional AAVs encoding the G-protein and/or TVA receptor can be injected into a region of interest, thereby targeting expression and thus rabies infection and subsequent trans-synaptic tracing to only neurons that express Cre recombinase. A downside of this approach is that it requires two independent injections of virus: one to express the G-protein and/or TVA receptor and another at a later time point to deliver RV-SADΔG to its target site. It is important to note that RV is cytotoxic after prolonged periods of infection. Therefore, it is generally recommended that brain tissue be analyzed within days of the RV injection, prior to neuronal death and/or circuit degradation. Though monosynaptic tracing using engineered RV is great for anatomical trans-synaptic tracing of presynaptic inputs, the addition of ChR2 allows for the simultaneous use as an anatomical tracer, as well as a light-triggered actuator, in order to validate patterns of functional connectivity.

Canine adenovirus type 2 (CAV-2), which is relatively new to the field, is a retrogradely transported virus that can be used to express a GOI by selective infection of a neuron's axon terminals (Soudais *et al.*, 2001). Upon virus uptake at the synaptic terminals, CAV-2 is transported in a retrograde fashion to a neuron's cell body. Previously, CAV-2 has been used to express Cre recombinase (CAV-Cre) (Schwarz *et al.*, 2015) in neuronal populations that can subsequently be targeted by a Cre-dependent AAV that expresses an optogenetic reporter. By first targeting neuron terminals at a projection site of a given neuronal population, Cre expression is restricted only to those neurons that project to that site. By subsequently targeting the cell bodies for Cre-dependent expression of an optogenetic reporter, functional connectivity can be tested. For example, assume that site "A" contains a mixed population of cell bodies that project to multiple areas of the brain, sites "B" and "C" (Figure 15.4). To test whether site "B" or site "C" is important for a given phenotype, CAV-Cre can be differentially injected into either site "B" or into site "C", and thus target Cre expression only to neurons projecting to those respective sites. Subsequent injection of a Cre-dependent AAV into site "A" will target optogenetic reporter expression only to those neurons projecting to either site "B" or to site "C". By subsequently stimulating the targeted neurons with light, behavioral outcomes associated with neurons projecting to one site or the other can be elucidated. Traditionally, using Cre-expressing mouse lines to target a given population of neurons for expression of a Cre-dependent optogenetic reporter and subsequently stimulating neuron terminals at a projection site has accomplished such mapping. Though this simpler strategy may also be effective, the use of retrogradely transported viruses such as CAV-2 is advantageous when a Cre-driver line is not available. Furthermore, when investigating the contribution of one cell type or another in a mixed population of neurons within a given region of the brain, the use of two or more Cre lines may not always be feasible. Additionally, the use of CAV-2 in this context allows one to study the role of specific projections without overtly stimulating all of the projections of a neuronal population that might produce non-specific results. However, it should be

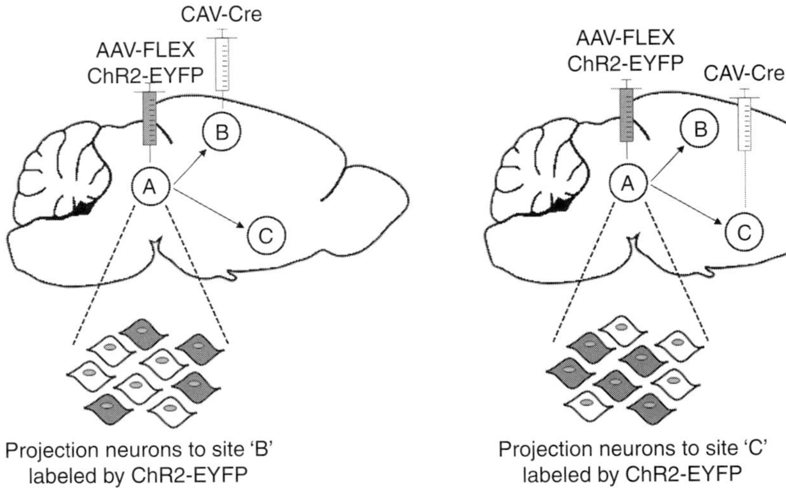

Figure 15.4 CAV2 for projection mapping and targeted gene expression. CAV-2 is a retrograde virus that is taken up at a synaptic terminal and transported to a neuron's soma. In this example, a mixed population of neurons from site "A" projects to two areas of the brain, site "B" and site "C". Neurons from site "A" projecting to site "B" can be targeted for Cre-dependent opsin expression by first injecting site "B" with CAV-Cre. Due to its retrograde transport, projection neurons in site "A" can be targeted for opsin expression using a FLEX AAV expressing an opsin of choice (ChR2-EYFP shown here). The same can be done with regards to neurons projecting to site "C". The contribution of either population of neurons to a particular phenotype can then be assessed either physiologically or behaviorally upon optogenetic stimulation.

kept in mind that, while useful, CAV-2 vectors are not currently as widely available as other commercially available viruses.

15.4 Other Considerations for Best Practices Using Optogenetics

Many light-inducible opsins are available for today's neuroscientist. However, it is important to remember that not all opsins exhibit identical properties, and it is therefore necessary to empirically determine the biophysical opsin characteristics per experimental need. References, tips and guidelines for optogenetic reporter variants can also be found online through sources such as the Optogenetics Resource Center at Stanford University and OpenOptogenetics.org. For excitation, many ChR2 variants are commercially available, but the biophysical properties of each variant should be referenced prior to application. These properties may include speed of ChR2 onset, time to channel closure, excitation wavelength, channel desensitization and light sensitivity, among others. The most commonly used ChR2 variant to date is the H134R variant, in which the histidine at position 134 of the ChR2 amino acid sequence has been mutated to an arginine. Compared to wild-type ChR2, this variant is more resistant to prolonged light desensitization, which was a major disadvantage of the originally described ChR2 in early studies (Nagel *et al.*, 2003; Lin, 2011). In addition, ChR2 (H134R) has a greater light sensitivity. However, this variant has slower off-kinetics compared to other ChR2 variants, meaning that the channel requires a longer time to close after light-induced activation. This latter property increases photo-induced currents into a cell, but at the expense of temporal precision. Nonetheless, ChR2 (H134R) is still the most widely used, and many existing ChR2 mouse lines and commercially available viral reagents utilize this ChR2 variant.

Since the generation of the H134R version, numerous ChR2 variants have been identified to improve on the kinetic properties of ChR2. Among these, the ChETA variant displays ultrafast response kinetics that allows for highly precise, sustained neuronal activation (Gunaydin *et al.*, 2010). For studies that require high-frequency stimulation parameters, such as those investigating fast-spiking interneurons, opsins such as ChETA are well suited. For applications that require prolonged neuronal activation, the use of step-function opsins (SFOs) (Berndt *et al.*, 2009) may be warranted. SFOs are ChR variants whose channels are stably opened after brief pulses of blue light; however, they may be reversibly closed by red-shifted light (e.g. yellow light). These channels remain open for lengthy periods of time (in the order of tens of minutes) by a single brief pulse of blue light. Generally, activation of SFOs leads to subthreshold depolarization of a neuron, meaning that SFOs expressed on a neuron do not directly elicit light-dependent action potentials, but rather increase the excitability of the neuron, allowing for synaptic inputs to drive depolarization the remainder of the way past the threshold so as to generate neuronal firing. In many cases, this property may be desirable for finer control of neuronal activity in a more physiological manner.

In addition to modifications aimed at changing the response kinetics of ChR2, other variants have been engineered that allow for activation of the channel at wavelengths that are different from typical blue light-responsive variants. Not only does this facilitate greater experimental flexibility, but also red-shifted variants are even more desirable due to the ability of longer wavelengths to penetrate tissues to greater depths. Early variants, namely C1V1 (Prakash *et al.*, 2012) and ReaChR (Lin *et al.*, 2013), are red-activatable excitatory ChRs. Though both are suitable for activation *in vivo*, both have their limitations. C1V1 is known to exhibit membrane-trafficking issues, particularly in neurons that have long-range synaptic terminals (Yizhar *et al.*, 2011; Rajasethupathy *et al.*, 2015). This is problematic if one wants to stimulate neuronal terminals that may not express high C1V1 levels due to poor trafficking. ReaChR improves upon C1V1's trafficking issues, but displays low light-induced photocurrents and diminished light-dependent spike fidelity (Lin *et al.*, 2013; Rajasethupathy *et al.*, 2015). Nonetheless, ReaChR currently enables experiments that require a red-activatable ChR. However, a further improved red-activatable ChR, called bReaChES (Rajasethupathy *et al.*, 2015), has been engineered to include mutations found in ChETA variants, among others. bReaChES displays no trafficking issues, has high spike fidelity and is capable of eliciting robust photocurrents. Going forward, bReaChES may best allow experiments that require a red-activatable ChR.

For studies that require independent activation of two distinct neuronal populations, the recently discovered ChR isolated from the algae *Chlamydomonas noctigama*, nicknamed Chrimson, has been shown to be an effective red-shifted opsin (Klapoetke *et al.*, 2014). Used in conjunction with the fast-responding, highly blue light-sensitive Chronos opsin, discovered at the same time from the algae *Stigeoclonium helveticum* (Klapoetke *et al.*, 2014), Chrimson and Chronos can be used in parallel to drive two distinct neuronal populations simultaneously with different wavelengths of light. Notably, Chrimson and Chronos exhibit minimal overlapping spectral properties, allowing independent activation of two distinct neuronal populations using two different wavelengths of light at the same time. Some of these variants are available as mouse models (Table 15.1), and all of these variants are available as viral vectors for targeting discrete cell types in the brain.

In addition to the numerous ChR variants that exist for driving neuronal excitation, various inhibitory opsins are available for neuronal silencing. Among these, halorhodopsin and archaerhodopsin are currently the most commonly used opsins for neuronal inhibition. Halorhodopsin is a light-gated chloride pump that allows for the influx of chloride ions into the cell in order to hyperpolarize the neurons in which it is expressed (Schobert and Lanyi, 1982; Zhang *et al.*, 2007). Older variants of halorhodopsin suffered from subcellular trafficking issues (Gradinaru *et al.*, 2008), but current variants eNpHR2.0 (Gradinaru *et al.*, 2008) and eNpHR3.0 (Gradinaru *et al.*, 2010) have largely resolved this issue in mammalian cell types. eNpHR3.0 exhibits a more robust inhibitory capacity than eNpHR2.0, and can be activated by longer wavelengths in red and far-red spectra, which may be more advantageous for tissue penetration. Archaerhodopsin is a proton pump that removes intracellular protons from the cell in order to drive hyperpolarization (Chow *et al.*, 2010). Generally, newer variants of archaerhodopsin, such as Arch3.0 (an enhanced proton pump from *Halobacterium sodomense*) (Mattis *et al.*, 2012) and ArchT3.0 (an enhanced proton pump from *Halorubrum* strain TP009) (Mattis *et al.*, 2012), tend to have greater inhibitory capacities compared to halorhodopsins, including eNpHR3.0 (Mattis *et al.*, 2012). The newly engineered "Jaws" cruxhalorhodopsin (Chuong *et al.*, 2014), from the archaea *Halobacterium salinarum*, is a red-shifted rhodopsin that exhibits robust inhibition in response to red light, allowing for deep tissue penetration.

Alternatively to using halorhodopsins or archaerhodopsins for neuronal silencing, ChRs have recently been engineered to allow for selective chloride influx when activated by blue light. The engineered iC1C2 opsin (a mutated chimeric opsin of ChR1 and ChR2) displays very quick, precise and robust optical inhibition compared to other inhibitory opsins that have been made available to date (Berndt *et al.*, 2014). Additional mutagenesis of iC1C2 led to the creation of SwiChR, a chloride-conducting ChR that exhibits step-function capability in order to drive prolonged inhibition of targeted neurons in response to brief light pulses, which can be reversed by red-shifted light (Berndt *et al.*, 2014). Given their quick and robust nature, these chloride-conducting inhibitory ChRs may be highly useful for most applications of neuronal inhibition; however, given their activation by blue light, they may not be as suitable for deep tissue penetration as other red light-activatable inhibitory opsins.

Given the number of opsin variants available and the various cell types in the brain that may be targeted for opsin expression, it is necessary to empirically test appropriate activation parameters in order to ensure faithful neuronal activity, particularly when using excitatory ChRs. Ideal stimulation parameters using opsins should elicit consistent and robust neuronal activation in a light-dependent manner. As a general rule, short pulses of light are sufficient in order for many cell types to elicit reproducible action potentials. Prolonged pulses, particularly in non-fast-spiking interneurons, may drive neurons into depolarization block (Herman *et al.*, 2014), thereby effectively inhibiting the cell of interest, rather than activating it. This appears to be less of a problem in fast-spiking interneurons and principal excitatory cell types (Herman *et al.*, 2014). Similarly, given the variation in cell properties between differing neuronal cell types, high-frequency stimulation should also be avoided in order to prevent depolarization block in cell types that are not known to be capable of high-frequency firing. For experiments that require prolonged periods of activation, short pulses of light can be delivered in prolonged trains. However, these trains should not be so long that they result in neurotransmitter depletion. An "off interval" between trains of light should be included in order to allow for transmitter replenishment and to prevent the possibility of transmitter or neuropeptide depletion.

15.5 Concluding Remarks

The ability to manipulate specified cell types in a targeted manner is essential for decoding the contributions of individual neuronal populations in various physiological states. Today, the toolkit to allow for cell type-specific manipulation is expansive and ever growing. Here, we reviewed the common strategies for marking selective cell types for optogenetic manipulation, namely through the available optogenetic reporter mouse lines, conditional mouse genetics and various viral approaches. Understanding the advantages and disadvantages of these approaches in the context of optogenetic manipulation will help with identifying the best strategies for experimental use. Furthermore, we discussed the differences between the most common optogenetic reporters available, as well as best practices for ensuring optimal optogenetic activation in given cell types. Overall, the fundamental concepts and ideas presented here should be minimal requisites for applying optogenetics in a cell type-specific manner, thus providing a general reference for applying the appropriate techniques for optimal experimental use to today's neuroscientist. The ever-expanding nature of technology will surely provide more tools and techniques both for targeting genes with high cell type specificity, and for further improving the optogenetic tools in order to allow for less invasive, highly precise light-induced modulation of the many important cell types that make up the nervous system.

REFERENCES

Atasoy, D., Aponte, Y., Su, H.H. *et al.* (2008). A FLEX switch targets channelrhodopsin-2 to multiple cell types for imaging and long-range circuit mapping. *The Journal of Neuroscience* **28**, 7025–7030.

Berndt, A., Lee, S.Y., Ramakrishnan, C. *et al.* (2014). Structure-guided transformation of channelrhodopsin into a light-activated chloride channel. *Science* **344**, 420–424.

Berndt, A., Yizhar, O., Gunaydin, L.A. *et al.* (2009). Bi-stable neural state switches. *Nature Neuroscience* **12**, 229–234.

Chow, B.Y., Han, X., Dobry, A.S. *et al.* (2010). High-performance genetically targetable optical neural silencing by light-driven proton pumps. *Nature* **463**, 98–102.

Chuong, A.S., Miri, M.L., Busskamp, V. *et al.* (2014). Noninvasive optical inhibition with a red-shifted microbial rhodopsin. *Nature Neuroscience* **17**, 1123–1129.

Gradinaru, V., Thompson, K.R., and Deisseroth, K. (2008). eNpHR: a *Natronomonas* halorhodopsin enhanced for optogenetic applications. *Brain Cell Biology* **36**, 129–139.

Gradinaru, V., Zhang, F., Ramakrishnan, C. *et al.* (2010). Molecular and cellular approaches for diversifying and extending optogenetics. *Cell* **141**, 154–165.

Gunaydin, L.A., Yizhar, O., Berndt, A. *et al.* (2010). Ultrafast optogenetic control. *Nature Neuroscience* **13**, 387–392.

Herman, A.M., Huang, L., Murphey, D.K. *et al.* (2014). Cell type-specific and time-dependent light exposure contribute to silencing in neurons expressing channelrhodopsin-2. *eLife* **3**, e01481.

Klapoetke, N.C., Murata, Y., Kim, S.S. *et al.* (2014). Independent optical excitation of distinct neural populations. *Nature Methods* **11**, 338–346.

Lin, J.Y. (2011). A user's guide to channelrhodopsin variants: features, limitations and future developments. *Experimental Physiology* **96**, 19–25.

Lin, J.Y., Knutsen, P.M., Muller, A. *et al.* (2013). ReaChR: a red-shifted variant of channelrhodopsin enables deep transcranial optogenetic excitation. *Nature Neuroscience* **16**, 1499–1508.

Mastakov, M.Y., Baer, K., Symes, C.W. *et al.* (2002). Immunological aspects of recombinant adeno-associated virus delivery to the mammalian brain. *Journal of Virology* **76**, 8446–8454.

Mattis, J., Tye, K.M., Ferenczi, E.A. *et al.* (2012). Principles for applying optogenetic tools derived from direct comparative analysis of microbial opsins. *Nature Methods* **9**, 159–172.

Miyashita, T., Shao, Y.R., Chung, J. *et al.* (2013). Long-term channelrhodopsin-2 (ChR2) expression can induce abnormal axonal morphology and targeting in cerebral cortex. *Frontiers in Neural Circuits* **7**, 8.

Nagel, G., Szellas, T., Huhn, W. *et al.* (2003). Channelrhodopsin-2, a directly light-gated cation-selective membrane channel. *PNAS* **100**, 13940–13945.

Nathanson, J.L., Jappelli, R., Scheeff, E.D. *et al.* (2009). Short promoters in viral vectors drive selective expression in mammalian inhibitory neurons, but do not restrict activity to specific inhibitory cell-types. *Frontiers in Neural Circuits* **3**, 19.

Osakada, F., Mori, T., Cetin, A.H. *et al.* (2011). New rabies virus variants for monitoring and manipulating activity and gene expression in defined neural circuits. *Neuron* **71**, 617–631.

Prakash, R., Yizhar, O., Grewe, B. *et al.* (2012). Two-photon optogenetic toolbox for fast inhibition, excitation and bistable modulation. *Nature Methods* **9**, 1171–1179.

Rajasethupathy, P., Sankaran, S., Marshel, J.H. *et al.* (2015). Projections from neocortex mediate top-down control of memory retrieval. *Nature* **526**, 653–659.

Schobert, B., and Lanyi, J.K. (1982). Halorhodopsin is a light-driven chloride pump. *The Journal of Biological Chemistry* **257**, 10306–10313.

Schwarz, L.A., Miyamichi, K., Gao, X.J. *et al.* (2015). Viral-genetic tracing of the input–output organization of a central noradrenaline circuit. *Nature* **524**, 88–92.

Shaner, N.C., Steinbach, P.A., and Tsien, R.Y. (2005). A guide to choosing fluorescent proteins. *Nature Methods* **2**, 905–909.

Shemiakina, II, Ermakova, G.V., Cranfill, P.J. *et al.* (2012). A monomeric red fluorescent protein with low cytotoxicity. *Nature Communications* **3**, 1204.

Soudais, C., Laplace-Builhe, C., Kissa, K. *et al.* (2001). Preferential transduction of neurons by canine adenovirus vectors and their efficient retrograde transport in vivo. *FASEB Journal* **15**, 2283–2285.

Takatoh, J., Nelson, A., Zhou, X. *et al.* (2013). New modules are added to vibrissal premotor circuitry with the emergence of exploratory whisking. *Neuron* **77**, 346–360.

Tighilet, B., Hashikawa, T., and Jones, E.G. (1998). Cell- and lamina-specific expression and activity-dependent regulation of type II calcium/calmodulin-dependent protein kinase isoforms in monkey visual cortex. *The Journal of Neuroscience* **18**, 2129–2146.

Wickersham, I.R., Lyon, D.C., Barnard, R.J. *et al.* (2007). Monosynaptic restriction of trans-synaptic tracing from single, genetically targeted neurons. *Neuron* **53**, 639–647.

Yizhar, O., Fenno, L.E., Prigge, M. *et al.* (2011). Neocortical excitation/inhibition balance in information processing and social dysfunction. *Nature* **477**, 171–178.

Zhang, F., Wang, L.P., Brauner, M. *et al.* (2007). Multimodal fast optical interrogation of neural circuitry. *Nature* **446**, 633–639.

16 Optogenetic Mapping of Neuronal Connections and their Plasticity

Michael M. Kohl and Dennis Kätzel

16.1 Optogenetic Mapping of Local Neural Circuits

16.1.1 The Birth of the "Opto" in "Optogenetics": Circuit Mapping via Glutamate Uncaging

A description of the structure of a neural network and the potential of plasticity at its nodes is prerequisite for a mechanistic conception of how neural signals flow within the circuit and ultimately realize the circuit's computational function (Martin, 2009). However, this endeavor is hampered by many problems. For example, descriptions of axonic and dendritic morphology (Binzegger *et al.*, 2004; Helmstaedter *et al.*, 2009a; Helmstaedter *et al.*, 2009b; Helmstaedter *et al.*, 2009c) do not capture the actual physiological connectivity (e.g. synaptic strength) between neurons. In contrast, approaches exclusively using electrophysiology (Thomson *et al.*, 1996; Gupta *et al.*, 2000; Thomson *et al.*, 2002; Brown and Hestrin, 2009) can only look at few neurons at a time (Scanziani and Häusser, 2009); they thus remain blind to the organization of the network as a whole.

Contrastingly, optical activation of neurons in combination with the electrophysiological recording of optically evoked inputs arriving at a postsynaptic cell has the potential to overcome these obstacles. It allows us to "scan" the connections in a circuit by moving an activating light beam from point to point and thereby probe the connections from hundreds of neurons (or small clusters of neurons) sequentially within a single experiment.

The key technology that enabled this optical mapping approach was the development of so-called caged neurotransmitters; that is, neurotransmitters covalently attached to chemical groups that inhibit their binding to endogenous synaptic receptors. When such biologically inert caged precursors are exposed to ultraviolet light, the "cage" group is photolytically removed and the transmitter freed in its active form. Although multiple caged agonists and antagonists of synaptic receptors have been developed, the most widely used compound deployed for mapping studies is caged glutamate. Its photo-cleavage leads to a sudden surge of the local glutamate concentration that activates all synapses and neurons in its immediate vicinity, irrespective of neuronal subtype. This method has been refined in order to optically activate local populations of neurons (Walker *et al.*, 1986; Wilcox *et al.*, 1990; Callaway and Katz, 1993; Wieboldt, *et al.*, 1994a; Wieboldt, *et al.*, 1994b) and map tens or hundreds of connections between neurons at high speed by rastering neural tissue with an uncaging light beam (Callaway and Katz, 1993; Katz and Dalva, 1994). In these experiments, scanning a stimulating beam across neural tissue will generate light-evoked excitatory postsynaptic currents (EPSCs) or inhibitory postsynaptic currents (IPSCs) in electrophysiologically recorded postsynaptic partners whenever the focal spot activates a presynaptically connected neuron. The resulting maps (Figure 16.1B) reflect the laminar and tangential distributions of sources of excitation or inhibition, depending on if EPSCs (at negative holding potentials) or IPSCs (at ~0 mV holding potentials) are recorded.

Using this technology, important principles relating to the excitatory connections to both inhibitory and excitatory neocortical cell types have been derived. Such findings include the differences between input connectivity patterns featured by distinct subtypes of interneurons (Dantzker and Callaway, 2000; Xu and Callaway, 2009; Yoshimura and Callaway, 2005), but also the more general question of whether neocortical connections are formed randomly or with high specificity (Schubert *et al.*, 2001; Shepherd *et al.*, 2005; Shepherd and Svoboda, 2005; Yoshimura *et al.*, 2005; Schubert *et al.*, 2007; Weiler *et al.*, 2008).

The use of non-linear optics to restrict photoactivation to small, diffraction-limited volumes ranging from the size of a single synapse to that of a neuron provided many more applications (Denk *et al.*, 1990; Denk, 1994). Such "two-photon" uncaging of glutamate (Matsuzaki *et al.*, 2001) provided enough spatial resolution to allow the activation of a single neuron within brain tissue, and – by extension – the simultaneous or sequential excitation of multiple predefined neurons by a complex spatiotemporal optical input pattern (Nikolenko *et al.*, 2007; Nikolenko *et al.*, 2008).

16.1.2 From Optical to Optogenetic Mapping of Neural Circuits

However, as photo-released glutamate activates all neurons, regardless of their type, the uncaging approach faced a fundamental limitation: since distinct cell types of neural circuits fulfill distinct, yet currently largely elusive functions, they need to be activated selectively in order to be studied properly, while leaving all other cell types unaltered. Thus, the advent of optogenetics, which means the ability to confer and restrict optical excitability genetically, represented a transformative step toward the functional mapping of neural circuits.

Surprisingly, however, neither the vague prediction of what later would become optogenetics by Francis Crick (Crick, 1999), nor its primary invention by the laboratory of Gero Miesenböck (Zemelman and Miesenböck, 2001; Zemelman *et al.*, 2002; Zemelman *et al.*, 2003) have emerged in the historical lineage of the older optical mapping approach. In fact, the original reason for encoding responsiveness to light genetically was to no longer depend on targeting the activation with a focused light beam, so that the light could be distributed through tissue like a broadcast signal, which only the genetically altered cell type would be able to pick up and transform into a neuronal message:

> Genetics and protein engineering are invading what not long ago was the exclusive domain of synthetic chemistry and pharmacology [...] A conspicuous void, however, exists in the genetic line-up at the position that photostimulants and caged neurotransmitters (that is, inactive precursors activated by light) occupy in the synthetic arsenal [...] Whereas conventional photostimulation must localize the stimulus in the form of the uncaging light beam, genetic schemes would localize the response (in the form of an encoded sensitivity to light), and illumination could then be broad. Patterns of distributed activity might be fed directly into a genetically circumscribed population of neurons, irrespective of the anatomical location of its members or their connection to sensory input (Zemelman and Miesenböck, 2001).

In line with this concept, the first optogenetic study in history (i.e. the first experiment in which an optogenetic actuator had been introduced into a genetically defined population of neurons to achieve their remote control) (Lima and Miesenböck, 2005) did not localize the light beam onto any particular circuit at all. Neither did subsequent pioneering optogenetic studies in worms (Nagel *et al.*, 2005), chicks (Li *et al.*, 2005) or mice (Bi *et al.*, 2006; Adamantidis *et al.*, 2007).

This development was not a coincidence: what was a given for mapping based on glutamate uncaging – namely the restriction of the locus of excitation to the somatodendritic compartment (since axons do not express glutamate receptors sufficiently) – was hard to achieve with optogenetics. The most widely used actuator, channelrhosopsin-2 (ChR2), accumulates strongly in axons and synapses. This made this technology principally unsuitable for the mapping of local circuits (i.e. circuits where pre- and post-synaptic neurons are intermingled). This is because with axonal expression of the actuator, a detected synaptic input could have been evoked optically not only at the somatodendritic compartment (revealing the localization of the presynaptic cell), but also all the way along the axon to the postsynaptic neuron – the decisive information on where the presynaptic partner of a postsynaptic neuron is situated is simply not obtainable.

From the beginning, this bug has been converted into a virtue by changing the application – in fact, the axonal and synaptic accumulation makes ChR2 an ideal tool for mapping long-range projections (i.e. detecting their existence, studying their input patterns and eventually analyzing their functional impact in the behaving animal) (Deisseroth, 2014). Thus, instead of mapping the location of nearby presynaptic neurons, the pioneering experiments of optogenetic or "ChR2-assisted" circuit mapping mapped the bare existence of a neuronal projection coming from outside the area of the postsynaptic target cell via optical stimulation of such projection fibers in order to elucidate innervation patterns at the target site. The first study using this technique delivered ChR2 via *in utero* electroporation to the supragranular layers (2/3) of the somatosensory cortex and analyzed the pattern of callosal (inter-hemispheric) as well as inter-laminar projections deriving from the principal cells of that layer (Petreanu *et al.*, 2007). A follow-up study from the same laboratory (of Karel Svoboda) was conducted under blockade of the ion channels that are necessary for action potential generation, so that the only measurable photoactivation was the direct triggering of synaptic release via activation of ChR2 molecules located in the presynapses of transfected neurons. In this way, anatomical principles regarding the relationship between the origin of the presynaptic cell (thalamic nuclei, motor cortex or local supragranular and granular layers) and the location of its presynaptic terminal along the dendritic tree of the postsynaptic cell were derived for principal neurons of layers 3 and 5 of the somatosensory cortex (Petreanu *et al.*, 2009). In this way, it could be revealed, for example, that projections onto layer 3 cells mediating sensory input (from layer 4 and the ventral posterior medial nucleus of the thalamus) arrived mainly close to the soma at proximal and basal dendrites, while projections delivering motor-related signals (from the motor cortex and the posterior medial nucleus of the thalamus) converged more at the distal dendrites.

It took the development of a novel transgenic mouse line in which ChR2 was expressed from a genomic locus in a conditional, *Cre*-dependent fashion (akin to *Cre* reporter lines) to overcome the problem of the strong axonal expression of ChR2 and to achieve an optogenetic equivalent to the circuit mapping based on glutamate uncaging (see Figure 16.1) (Kätzel *et al.*, 2011). This technology combined the genetic resolution of cell types with the perisomatic restriction of optical activation and a broad, uniform expression across the tissue. The latter is a second prerequisite for the mapping of local circuits, but is not achieved using focal transfection methods such as viral vectors or *in utero* electroporation, which necessarily suffer from concentration gradients of expression away from the site of transfection.

It is unclear why the expression from a genomic locus diminished axonal excitability, especially since a later, similar mouse line displayed some axonal excitability (Madisen *et al.*, 2012). A generally weaker expression of ChR2 compared to both the later mouse lines and viral or electroporation-based expression of opsins, which prevents ChR2-mediated depolarization from reaching the spiking threshold anywhere other than in the soma, is certainly part of the explanation. However, it might be that a difference in the N-terminal peptide being attached by accident before the coding sequence of ChR2 contributes to such differential sorting to the soma as well (Kätzel *et al.*, 2011).

By using this mouse line, the local connectivity of inhibitory neurons of the neocortex could be mapped for the first time (Figure 16.1). The optogenetic mapping using raster scanning of the laser across the brain slice is so efficient that the inhibitory connectivity of pyramidal cells in all principal layers could be derived in three functionally distinct areas, revealing various specializations (Kätzel *et al.*, 2011). Such connectivity patterns can be displayed in inputs maps and analyzed numerically (see Figure 16.2).

The optogenetic approach displayed a considerable improvement over previous attempts to map inhibitory connectivity using glutamate uncaging (Dantzker and Callaway, 2000; Brill and Huguenard, 2009; Xu and Callaway, 2009) in the following areas:

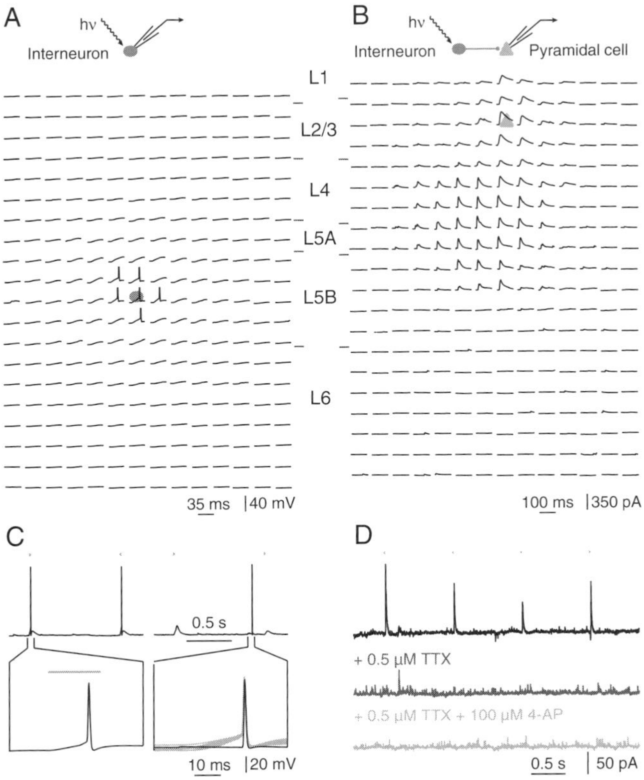

Figure 16.1 Optogenetic mapping of local connectivity requires a somatic locus of excitation in the channelrhodopsin-2 (ChR2)-expressing neuron (modified from Kätzel and Miesenböck [2014]). (A) In a mouse line with somatic expression of ChR2 in GABAergic interneurons, the membrane voltage of an interneuron in layer 5B of the neocortex is recorded using the patch-clamp technique. Simultaneously, the neocortical tissue is raster scanned at 14 × 20 locations with 60 μm grid spacing spread over all layers. Whole-cell current-clamp traces are plotted at the position in the brain slice to which the laser was targeted at the time of their measurement (i.e. from where they were evoked optically). The somatic position of the recorded interneuron is indicated by a gray ellipse. Note that only perisomatic illumination in relatively close proximity to the cell body evokes action potentials. (B) The same raster scanning optical stimulation of GABAergic interneurons and display as in (A), but here electrophysiological recordings from a voltage-clamped (0 mV) L2/3 pyramidal cell are shown, revealing the locations of presynaptically connected interneurons as inhibitory inputs (IPSCs) measured when the respective locations are photo-stimulated. Whole-cell voltage-clamp traces show light-evoked IPSCs as the focus of the stimulating beam is scanned across a grid of 14 × 20 locations. The somatic position of the recorded pyramidal cell is indicated in gray. (C) Somatic expression of ChR2 is also indicated by the absence of antidromic action potentials, which would indicate a light-induced axonal action potential and is commonly observed with a similar mouse line with a lack of peri-somatic restriction of ChR2 expression (see Figure 2 in Madisen *et al.* [2012]). Here, in contrast, no light-evoked antidromic spikes are detected: waveforms of light-evoked (left) and spontaneous (right) action potentials at different timescales (top and bottom) are shown. Note that a slow depolarizing ramp to threshold after the onset of a light pulse (20 ms, gray bars) distinguishes light-evoked (orthodromic) from spontaneous antidromic action potentials. The bottom right panel includes for comparison traces of all 15 light-evoked action potentials (gray traces). (D) Absence of synaptic expression of ChR2 is indicated by the sensitivity of optically evoked inputs to sodium channel block. Waveforms of light-evoked IPSCs in a pyramidal cell voltage-clamped at 0 mV are shown. IPSCs are blocked by bath application of 0.5 μM tetrodotoxin (TTX; dark gray) and fail to recover after additional application of 4-amino-pyridine (4-AP; light gray), which blocks repolarizing sodium channels and was shown to ensure optically evoked inputs under TTX when ChR2 is expressed in the pre-synapse (compare to Figure 1 in Petreanu *et al.* [2009]).

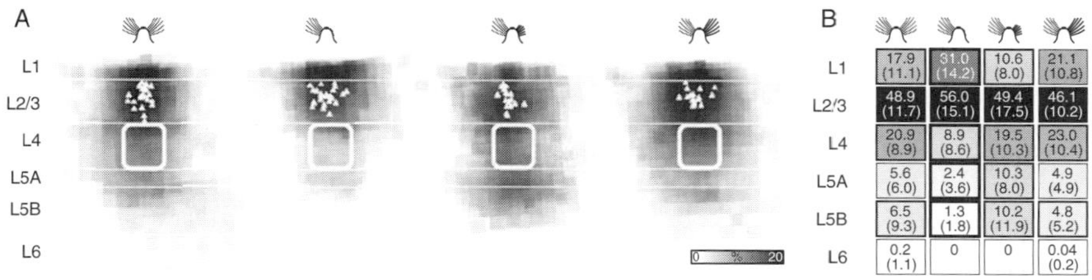

Figure 16.2 Optogenetic mapping of experience-dependent plastic changes of local neocortical inhibitory connectivity (modified from Kätzel and Miesenböck [2014]). (A) Maps of inhibitory inputs to excitatory neurons of layer 2/3 (L2/3) in barrels representing intact (left, n = 23), trimmed (center left, n = 23) or previously deprived whiskers after regrowth for 1 month (center right, n = 19) or 3 months (right, n = 17). The maps are scaled to the size of a standard barrel (thick white outline) and overlaid in order to depict the distribution of inhibitory input sources. The intensity of gray shading at each location indicates the cumulative inhibitory charge transfer. This normalized index measures the frequency with which IPSCs are elicited from corresponding locations in different slices, weighted by the average charge transfer per IPSC. Locations of postsynaptic excitatory neurons from which those input maps were recorded are shown as white triangles and layer borders as thin white lines. (B) Numerical analysis of the maps presented in (A): normalized inhibitory charge flow from the indicated source layers (rows) to L2/3 excitatory neurons in barrels representing intact (left), trimmed (center left) or previously deprived whiskers after regrowth for 1 month (center right) or 3 months (right). Values are represented numerically (± 1 SD) and in normalized grayscale. Thick black outlines mark significant differences associated with whisker trimming ($p < 0.05$; analysis of variance); thin black outlines indicate groups whose means differ from the whisker-trimmed state (Bonferroni-corrected t-test). No significant differences existed between control and regrowth conditions ($p > 0.05$; analysis of variance).

1) In contrast to inputs induced by glutamate uncaging, inputs evoked through ChR2-mediated stimulation are fast and highly stimulus locked (see Figure 16.1B), permitting their averaging and the measurement of their biophysical parameters, such as rise time, latency or peak amplitude.
2) It avoids confound due to the measurement of polysynaptic connections, which may arise with unspecific stimulation (e.g. when an inhibitory input derives from a low-threshold spiking interneuron) (Kozloski *et al.*, 2001; Wang *et al.*, 2004) that has been activated by a pyramidal cell, which in turn was activated by the stimulating light beam (potentially even at its apical dendritic tuft distant from its somatic location) (Dantzker and Callaway 2000)).
3) The optogenetic approach can be further refined regarding specificity for studying the connectivity of each of the many different subtypes of interneurons by capitalizing on the rich plethora of *Cre* driver lines that grant selective genetic access to them (Monyer and Markram, 2004; Kuhlman and Huang, 2008).

A principal disadvantage of the ChR2-based approach of local circuit mapping, however, is the lack of strong excitation of cells due to the low single-channel conductance of ChR2 of approximately 40 fs (Nagel *et al.*, 2003; Bamann *et al.*, 2008), in combination with the slowness of reaching high expression levels at the membrane (Boyden *et al.*, 2005). For example, this has prevented the use of ChR2-assisted circuit mapping in the developing neocortex (i.e. in the first 2 postnatal weeks). Therefore, another optogenetic approach has recently been developed that overcomes this problem by using the ion channel P2X2 (Anastasiades *et al.*, 2016), which is gated by uncaging of its agonist ATP as an optogenetic actuator, an approach that had been developed in cell culture and fruit flies previously (Zemelman *et al.*, 2003; Lima and Miesenböck, 2005; Clyne and Miesenböck, 2008). P2X2 displays 400-times the single-channel conductance of ChR2 (close to 20 pS), is expressed quickly (probably due to the fact that it is a mammalian protein), is largely confined to the perisomatic compartment and is not expressed endogenously in cortical circuits during the first 2–3 postnatal weeks (although it is expressed later). Using this technology, connectivity patterns of interneurons could be mapped from the first postnatal day onward, revealing previously unknown principles of dynamic (i.e. transient) integration of specific interneurons into neocortical circuits (Anastasiades *et al.*, 2016).

16.1.3 The Experimental Setup for Optogenetic Circuit Mapping

The setup for circuit mapping with optogenetic actuators or glutamate uncaging requires a normal patch-clamp setup enhanced by external laser optics connected to the optical pathway of the microscope (similar to an excitation light source used for epifluorescence) to ultimately align with the image plane of its transmission optics. The components of this setup are shown by example in Figure 16.3.

Custom-written control software (e.g. a LabView-based virtual instrument) is used for calibration and actual scanning. At the specimen, the activating beam usually has a diameter of a few micrometers and carries 0.5–2.5 mW of optical power. Optical stimulation spots are spaced at a lateral distance of 45–70 μm, reflecting the achievable optical resolution. Such spots are visited in a pseudo-

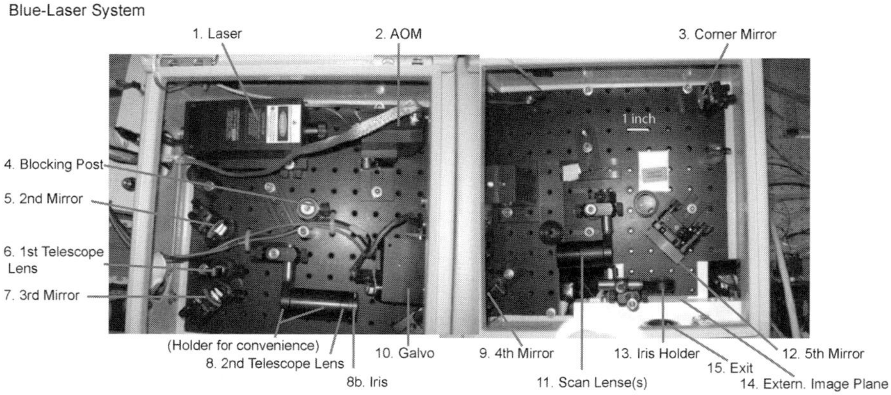

Figure 16.3 Beam path of a photostimulation setup for ChR2-based circuit mapping. (1) A laser – for channelrhodopsin-2 excitation, a digitally switched continuous-wave diode-pumped solid-state (DPSS) laser emitting 473 or 488 nm blue light, or (for better stability) diode lasers emitting 450 or 454 nm blue light at approximately 150–350 mW output is used. (2) An acousto-optic modulator to set the power of the beam. (3–14) An optical path consisting of various corner mirrors (3, 5, 7, 9 and 12), two telescope lenses (6 and 8) to adjust the diameter of the beam and a scan lens (11) to focus the beam onto the optical plane of the microscope optics. (10) A motor-controlled pair of galvanometric mirrors steers the beam to any point at the tissue level.

randomized order, which maximizes the time interval between the stimulation of two neighboring spots. Turn-key solutions including laser optics and control software are also available (e.g. from Rapp OptoElectronic).

16.1.4 Optogenetic Mapping with Single-cell Resolution (Two-photon Optogenetics)

The principal limitation of the optogenetic approaches detailed above is that they fail to determine how many presynaptic cells actually connect to the postsynaptic cells that the optically evoked inputs were recorded from. This also implies that the strength of connections can only be approximated and not really measured, since multiple presynaptic cells might contribute to one postsynaptic current (Kätzel *et al.*, 2011; Kätzel and Miesenböck, 2014). This problem scales with the density of optogenetically excitable cells: if a subtype is targeted that populates a circuit only sparsely, the likelihood that only a single cell is activated at every site increases.

To circumvent this problem, much focus has been invested in technological development for optogenetic activation with non-linear or multi-photon optics. The first result was an optogenetic approach that did not involve an optogenetic actuator at all, but was a clever extension of the mapping strategy using two-photon glutamate uncaging mentioned above (Nikolenko *et al.*, 2007). For this technology to work, the cells to be activated need to be localized (i.e. visualized) for the proper setup of the stimulation beam, anyway. Therefore, one can simply express a fluorescent protein like GFP under a cell type-specific promoter so that only cells of that type are visible in the tissue preparation and can be targeted for stimulation. In this way, the Yuste laboratory that pioneered this approach could determine the connectivity of two types of interneurons (parvalbumin and somatostatin) with single-cell resolution (Fino and Yuste, 2011; Packer and Yuste, 2011). However, it should be noted that even this approach relies on the relative sparseness of the cell population to be targeted – if cells are too densely packed (e.g. pyramidal cells in the neocortex), single-cell resolution cannot be guaranteed.

Two-photon-mediated excitation of actual optogenetic actuators (e.g. ChR2 or its derivatives) was less trivial, mainly due to a combination of two facts: the small single-channel conductance of such microbial opsins and the extremely small two-photon excitation spot, which is ideal for targeting individual synapses or dendritic branches, but is too small to cover a whole cell body. As a consequence, too few opsin molecules are usually activated by two-photon beams to produce sufficient inward current to drive a neuron past the spiking threshold. Interestingly, however, in zebrafish, expression levels of ChR2 were high enough – even when transgenically, conditionally and inducibly expressed using the Tet system – to allow reliable two-photon-mediated excitation *in vivo* (Zhu *et al.*, 2009). In mammalian tissue, however, this simple activation would not succeed, and in a first set of attempts, this problem was tackled by engineering the optical pathway, such as by placing multiple stimulation spots onto a cell body (e.g. spatiotemporal summation in spiral scans) or broadening the beam in the form of sculpted light (Mohanty *et al.*, 2008; Rickgauer and Tank, 2009; Andrasfalvy *et al.*, 2010; Papagiakoumou *et al.*, 2010). In a second wave (which corresponds to the current state of the art), researchers returned to biology for solutions and used opsins featuring a combination of higher single-channel conductance and longer activation and deactivation time constants, which allow some integration of activating photons over time (Packer *et al.*, 2012; Prakash *et al.*, 2012).

16.2 Optogenetic Investigation of the Plasticity of Neural Connections and Synapses

16.2.1 Optogenetic Mapping of Circuit Plasticity

The ability to detect neural connections to postsynaptic cells in a quantitative manner using the optogenetic stimulation of the presynaptic neurons or fiber afferents raises the potential to use the same approach for measuring changes in the strengths of such connections. This means using optogenetic stimulation and mapping to uncover the plasticity of individual neural connections and even of whole circuits.

The latter has been achieved using the same mouse line as was used for mapping the structure of neocortical inhibitory circuits in the study described above (Kätzel *et al.*, 2011). In that earlier study, large area-specific variations in the frequencies of certain translaminar inhibitory motifs had been found, suggesting that individual inhibitory connection motifs might adapt a neocortical area to its specific function. This notion was tested using optogenetic mapping: the fate of translaminar inhibitory connections in whisker-related somatosensory neocortex (barrel cortex) was mapped before, during and after transient deprivation from its sensory input. It was found that different inhibitory connections did indeed display experience-dependent plasticity, being specifically and reversibly introduced or removed with alterations in sensory input (Figure 16.2). This was the first demonstration that individual inhibitory circuit motifs are adjusted selectively in the adult neocortex to adapt a circuit to a changing function (Kätzel and Miesenböck, 2014).

16.2.2 Optogenetic Investigation of Synaptic Plasticity

Synaptic plasticity has long been of great interest: it is key in neural development and is considered to be a cellular substrate for learning and memory (for a review, see Martin *et al.* [2000]). Long-term plasticity was first demonstrated in the hippocampus by Bliss and Lømo (Bliss and Lømo, 1973), and has since been the focus of a vast number of studies, mainly in *in vitro* and *ex vivo* preparations, but increasingly also *in vivo* (Nabavi *et al.*, 2014). Although there are now more than 5000 publications on synaptic plasticity in the hippocampus alone, we still seem far from a solid understanding of the rules and mechanisms governing synaptic plasticity and its role in circuit function.

A key issue is that electrical stimulation of afferent fibers, which is commonly used to induce synaptic plasticity, is not selective for specific presynaptic inputs. For instance, electric stimulation in the stratum radiatum in the CA1 of the hippocampus will recruit not only Schaffer collateral input from ipsilateral CA3 to CA1 pyramidal cells, but also commissural fibers from the contralateral CA3,

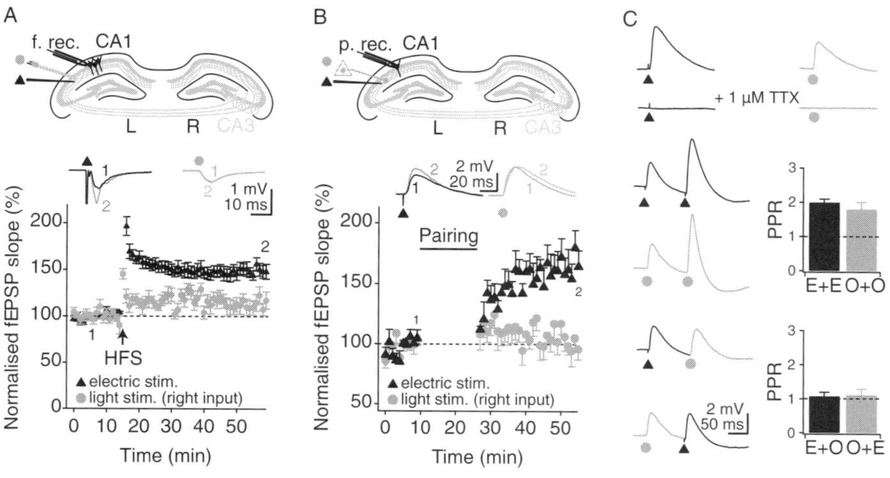

Figure 16.4 Optogenetic investigation of synaptic plasticity. (A) Fiber optics coupled to an LED can be inserted into acute brain slices (here, stratum radiatum in CA1 of the mouse hippocampus) just like conventional stimulation electrodes. Postsynaptic population responses are recorded with a field electrode (field recording [f. rec.]) placed into the pyramidal cell layer of CA1. Selective, optical stimulation (blue circles) of excitatory projections from CA3 to CA1 reveals that the right CA3 output is not plastic at the CA3–CA1 synapse. This lack of plasticity is overlooked under indiscriminate electric stimulation (black triangles). (B) Alternatively, a laser-based fixed or moving spot illumination system can be used to stimulate opsin-expressing fibers in acute brain slices. Here, patch-clamp recordings (p. rec.) from CA1 pyramidal cells demonstrated that optically recruited input from the right CA3 is unable to potentiate following a spike timing-dependent plasticity induction protocol (blue circles). In contrast, electrical stimulation led to robust synaptic potentiation (black triangles). (C) Electrically and optically stimulated postsynaptic responses are indistinguishable in terms of their kinetics and are blocked by 1 µM tetrodotoxin, demonstrating the electrical and optical recruitment of axonal compartments. Optical stimulation can be used to study short-term plasticity effects, such as paired pulse facilitation, and optical and electrical stimulation can be adjusted to be non-overlapping, if so desired. Note that optical stimulation of opsin expressing axons in thin, acute brain slices is possible (for more than an hour) without the cell body being present. PPR: paired pulse ratio; E: electric; O: optical stimulation. Broken lines represent the baseline. Error bars represent the standard error of the mean. Figures adapted, with permission, from Kohl *et al.* (2011) and Shipton *et al.* (2014). (A black-and-white version of this figure will appear in some formats. For the color version, please refer to the plate section.)

local inhibition and neuromodulatory input from subcortical areas. Using optogenetics to select the specific origin of the presynaptic input to be stimulated has started to reveal circuit-specific synaptic plasticity that was previously intractable by unspecific electrical stimulation (Figure 16.4).

For example, by injecting small amounts of an adeno-associated virus containing a floxed-opsin (ChR2) construct fused to a fluorescent reporter into either left or right dorsal CA3 of CamKII-Cre mice, it was possible to selectively target excitatory projections from either left or right CA3-to-CA1 pyramidal cell projections. This high level of selectivity revealed that only input from the left CA3, but not the right CA3, can undergo spike timing-dependent synaptic potentiation at CA1 synapses (Figure 16.4) (Kohl *et al.*, 2011). This hemispheric specialization had previously gone unnoticed, because electrical stimulation recruited the intermingled projections from left and right CA3 indiscriminately.

The setup required for the optogenetic stimulation of synaptic input *in vitro* or *ex vivo* is straightforward (see also Xiong and Jin [2012]): a 50–200 μm multi-mode fiber optic coupled to an LED light source is sufficient to activate monosynaptic input. Held on a manipulator, the fiber optic is placed directly into the tissue, akin to a stimulation electrode (Figure 16.4A). Finer control over the optical stimulation can be achieved with fixed spot or moving spot laser photostimulation systems as are used for circuit mapping. Because laser light is used and fed into the light-path either directly (see Figure 16.3) or via single-mode fiber optics (e.g. systems from Rapp OptoElectronic), light spots as small as 5 μm in diameter will have sufficient energy to reliably recruit presynaptic afferents expressing opsins (Figure 16.4B). Moving spot illumination systems can thus be used to rapidly switch the location of the light spot, enabling the recruitment of independent monosynaptic input. These systems permit the investigation of short-term plasticity (e.g. paired-pulse facilitation; see Figure 16.4C) or long-term plasticity where one input serves as an unpotentiated control pathway. The kinetics of the currently available optogenetic tools limit the frequency at which presynaptic input can be reliably recruited optically to about 20 Hz. While this is sufficient for most spike timing-dependent synaptic plasticity induction protocols, high-frequency induction protocols require modifications. For example, by using both a fiber optic and a stimulation electrode to recruit overlapping presynaptic inputs, the fiber optic can be used to selectively stimulate the area- and cell type-specific input. The electrical stimulation is then used to deploy high-frequency stimulation of various inputs, including the optogenetically labeled one. Optical stimulation is subsequently used to monitor synaptic plasticity in the area- and cell-type specific input only (Figure 16.4A).

Optogenetic control over specific synaptic inputs also enables the more rigorous combination of *ex vivo* and *in vivo* experiments in order to establish causal links between synaptic plasticity and behavior. In the hippocampus, selective optogenetic inhibition of CA3 output revealed a requirement for left CA3 activity, but not right CA3 activity, in long-term spatial memory tasks (Shipton *et al.*, 2014). Another study used optogenetics to induce long-term depression of synaptic connections from the infralimbic prefrontal cortex to D1 receptor-expressing medium-spiny neurons in the basal ganglia, which has relevance to addiction-induced behavioral adaptations (Pascoli *et al.*, 2012).

Most synaptic plasticity mechanisms are subject to neuromodulation (Seol *et al.*, 2007; Lesch and Waider, 2012; Teles-Grilo Ruivo and Mellor, 2013; Pignatelli and Bonci, 2015). Optogenetic tools permit precise spatial and temporal control of defined neuromodulatory inputs, thereby allowing one to dissociate tonic and phasic effects. For example, a recent study used optogenetics to selectively recruit dopaminergic input from the ventral tegmental area to the hippocampus, revealing a bidirectional, activity-dependent modulation of Schaffer collateral (CA3-to-CA1) synapses that has previously been undetected in experiments using exogenous dopamine application (Rosen *et al.*, 2015). Similarly, optogenetics enabled detailed studies of septal cholinergic input at the same Schaffer collateral synapse. Gu and Yakel found that when cholinergic input was active 10–100 ms before Schaffer collateral stimulation, nicotinic acetylcholine receptor-dependent long-term plasticity was observed (Gu and Yakel, 2011). In contrast, activating cholinergic input until 10 ms after the Schaffer collateral stimulation resulted in the induction of muscarinic acetylcholine receptor-dependent long-term synaptic potentiation (Gu and Yakel, 2011).

In summary, by making use of the rapidly expanding library of area- and cell type-specific *Cre*-transgenic mouse lines in order to target optogenetic activators to specific excitatory, inhibitory or modulatory presynaptic input streams, plasticity mechanisms can be dissected with a precision and resolution that was impossible to achieve with earlier technologies. In future, the combination of inducible (e.g. tetracycline transactivator system), activity-dependent (e.g. *cfos* promoter) (Reijmers *et al.*, 2007) and *Cre*-dependent expression of optogenetic tools will enable studies of synaptic plasticity and its causal link to behavior in spatially, genetically and temporally precisely defined neural circuit elements.

16.3 Conclusion

The mapping of neural circuits enables the high-throughput determination of the anatomical rules that govern the connections between the different nerve cells in a neural network. Although the invention of optical mapping of neural circuits predated the advent of optogenetics by a decade, the use of optogenetic tools for circuit mapping did not come easily to the field. The main reason for this was the difficulty of localizing optogenetic excitation to the somatodendritic compartment as opposed to axons and synapses. Here, we discussed strategies for how this feature has either been harnessed to map long-range connections or was overcome technologically to map local circuits. We further discussed how such mapping has been extended to the optogenetic dissection of the structural and functional plasticity of neural connections, including experience-dependent circuit alterations and long-term synaptic potentiation and depression.

REFERENCES

Adamantidis, A.R. *et al.*, 2007. Neural substrates of awakening probed with optogenetic control of hypocretin neurons. *Nature*, **450**(7168), pp.420–4.

Anastasiades, P.G. *et al.*, 2016. GABAergic interneurons form transient, layer-specific circuits in early postnatal neocortex. *Nat Commun*, **4**, p.7.

Andrasfalvy, B.K. *et al.*, 2010. Two-photon single-cell optogenetic control of neuronal activity by sculpted light. *Proc Natl Acad Sci U S A*, **107**(26), pp.11981–6.

Bamann, C. *et al.*, 2008. Spectral characteristics of the photocycle of channelrhodopsin-2 and its implication for channel function. *J Mol Biol*, **375**(3), pp.686–94.

Bi, A. *et al.*, 2006. Ectopic expression of a microbial-type rhodopsin restores visual responses in mice with photoreceptor degeneration. *Neuron*, **50**(1), pp.23–33.

Binzegger, T., Douglas, R.J. & Martin, K.A., 2004. A quantitative map of the circuit of cat primary visual cortex. *J Neurosci*, **24**(39), pp.8441–53.

Bliss, T.V.P. & Lømo, T., 1973. Long-lasting potentiation of synaptic transmission in the dentate area of the anaesthetized rabbit following stimulation of the perforant path. *The Journal of Physiology*, **232**(2), pp.331–56.

Boyden, E.S. *et al.*, 2005. Millisecond-timescale, genetically targeted optical control of neural activity. *Nat Neurosci*, **8**(9), pp.1263–8.

Brill, J. & Huguenard, J.R., 2009. Robust short-latency perisomatic inhibition onto neocortical pyramidal cells detected by laser-scanning photostimulation. *J Neurosci*, **29**(23), pp.7413–23.

Brown, S.P. & Hestrin, S., 2009. Intracortical circuits of pyramidal neurons reflect their long-range axonal targets. *Nature*, **457**(7233), pp.1133–6.

Callaway, E.M. & Katz, L.C., 1993. Photostimulation using caged glutamate reveals functional circuitry in living brain slices. *Proc Natl Acad Sci U S A*, **90**(16), pp.7661–5.

Clyne, J.D. & Miesenböck, G., 2008. Sex-specific control and tuning of the pattern generator for courtship song in *Drosophila*. *Cell*, **133**(2), pp.354–63.

Crick, F., 1999. The impact of molecular biology on neuroscience. *Philos Trans R Soc Lond B*, **354**, pp.2021–5.

Dantzker, J.L. & Callaway, E.M., 2000. Laminar sources of synaptic input to cortical inhibitory interneurons and pyramidal neurons. *Nat Neurosci*, **3**(7), pp.701–7.

Deisseroth, K., 2014. Circuit dynamics of adaptive and maladaptive behaviour. *Nature*, **505** (7483), pp.309–17.

Denk, W., 1994. Two-photon scanning photochemical microscopy: mapping ligand-gated ion channel distributions. *Proc Natl Acad Sci U S A*, **91**(14), pp.6629–33.

Denk, W., Strickler, J.H. & Webb, W.W., 1990. Two-photon laser scanning fluorescence microscopy. *Science*, **248**(4951), pp.73–6.

Fino, E. & Yuste, R., 2011. Dense inhibitory connectivity in neocortex. *Neuron*, **69**(6), pp. 1188–203.

Gu, Z. & Yakel, J.L., 2011. Timing-dependent septal cholinergic induction of dynamic hippocampal synaptic plasticity. *Neuron*, **71**(1), pp.155–65.

Gupta, A., Wang, Y. & Markram, H., 2000. Organizing principles for a diversity of GABAergic interneurons and synapses in the neocortex. *Science*, **287**(5451), pp.273–8.

Helmstaedter, M., Sakmann, B. & Feldmeyer, D., 2009a. L2/3 interneuron groups defined by multiparameter analysis of axonal projection, dendritic geometry, and electrical excitability. *Cereb Cortex*, **19**(4), pp.951–62.

Helmstaedter, M., Sakmann, B. & Feldmeyer, D., 2009b. Neuronal correlates of local, lateral, and translaminar inhibition with reference to cortical columns. *Cereb Cortex*, **19**(4), pp.926–37.

Helmstaedter, M., Sakmann, B. & Feldmeyer, D., 2009c. The relation between dendritic geometry, electrical excitability, and axonal projections of L2/3 interneurons in rat barrel cortex. *Cereb Cortex*, **19**(4), pp.938–50.

Katz, L.C. & Dalva, M.B., 1994. Scanning laser photostimulation: a new approach for analyzing brain circuits. *J Neurosci Methods*, **54**(2), pp.205–18.

Kätzel, D. *et al.*, 2011. The columnar and laminar organization of inhibitory connections to neocortical excitatory cells. *Nat Neurosci*, **14**(1), pp.100–7.

Kätzel, D. & Miesenböck, G., 2014. Experience-dependent rewiring of specific inhibitory connections in adult neocortex. *PLoS Biol*, **12**(2), p. e1001798.

Kohl, M.M. *et al.*, 2011. Hemisphere-specific optogenetic stimulation reveals left–right asymmetry of hippocampal plasticity. *Nat Neurosci*, **14**, pp.1413–5.

Kozloski, J., Hamzei-Sichani, F. & Yuste, R., 2001. Stereotyped position of local synaptic targets in neocortex. *Science*, **293**(5531), pp.868–72.

Kuhlman, S.J. & Huang, Z.J., 2008. High-resolution labeling and functional manipulation of specific neuron types in mouse brain by Cre-activated viral gene expression. *PLoS ONE*, **3**(4), p. e2005.

Lesch, K.-P. & Waider, J., 2012. Serotonin in the modulation of neural plasticity and networks: implications for neurodevelopmental disorders. *Neuron*, **76**(1), pp.175–91.

Li, X. *et al.*, 2005. Fast noninvasive activation and inhibition of neural and network activity by vertebrate rhodopsin and green algae channelrhodopsin. *Proc Natl Acad Sci U S A*, **102**(49), pp.17816–21.

Lima, S.Q. & Miesenböck, G., 2005. Remote control of behavior through genetically targeted photostimulation of neurons. *Cell*, **121**(1), pp.141–52.

Madisen, L. *et al.*, 2012. A toolbox of Cre-dependent optogenetic transgenic mice for light-induced activation and silencing. *Nat Neurosci*, **15**(5), pp.793–802.

Martin, K.A.C., 2009. The road ahead for brain-circuit reconstruction. *Nature*, **462**(7272), pp.411.

Martin, S.J., Grimwood, P.D., & Morris, R.G., 2000. Synaptic plasticity and memory: an evaluation of the hypothesis. *Annu Rev Neurosci*, **23**(1), pp.649–711.

Matsuzaki, M. *et al.*, 2001. Dendritic spine geometry is critical for AMPA receptor expression in hippocampal CA1 pyramidal neurons. *Nat Neurosci*, **4**(11), pp.1086–92.

Mohanty, S.K. *et al.*, 2008. In-depth activation of channelrhodopsin 2-sensitized excitable cells with high spatial resolution using two-photon excitation with a near-infrared laser microbeam. *Biophys J*, **95**(8), pp.3916–26.

Monyer, H. & Markram, H., 2004. Interneuron diversity series: molecular and genetic tools to study GABAergic interneuron diversity and function. *Trends Neurosci*, **27**(2), pp.90–7.

Nabavi, S. *et al.*, 2014. Engineering a memory with LTD and LTP. *Nature*, **511**(7509), pp.348–52.

Nagel, G. *et al.*, 2003. Channelrhodopsin-2, a directly light-gated cation-selective membrane channel. *Proc Natl Acad Sci U S A*, **100**(24), pp.13940–5.

Nagel, G. *et al.*, 2005. Light activation of channelrhodopsin-2 in excitable cells of Caenorhabditis elegans triggers rapid behavioral responses. *Curr Biol*, **15**(24), pp.2279–84.

Nikolenko, V. *et al.*, 2008. SLM microscopy: scanless two-photon imaging and photostimulation using spatial light modulators. *Front Neural Circuits*, **2**, p. 5.

Nikolenko, V., Poskanzer, K.E. & Yuste, R., 2007. Two-photon photostimulation and imaging of neural circuits. *Nat Methods*, **4**(11), pp.943–50.

Packer, A.M. *et al.*, 2012. Two-photon optogenetics of dendritic spines and neural circuits in 3D. *Nat Methods*, **9**(12), pp.1202–5.

Packer, A.M. & Yuste, R., 2011. Dense, unspecific connectivity of neocortical parvalbumin-positive interneurons: a canonical microcircuit for inhibition? *J Neurosci*, **31**(37), pp.13260–71.

Papagiakoumou, E. *et al.*, 2010. Scanless two-photon excitation of channelrhodopsin-2. *Nat Methods*, **7**(10), pp.848–54.

Pascoli, V., Turiault, M. & Luscher, C., 2012. Reversal of cocaine-evoked synaptic potentiation resets drug-induced adaptive behaviour. *Nature*, **481**(7379), pp.71–5.

Petreanu, L. *et al.*, 2007. Channelrhodopsin-2-assisted circuit mapping of long-range callosal projections. *Nat Neurosci*, **10**(5), pp.663–8.

Petreanu, L. *et al.*, 2009. The subcellular organization of neocortical excitatory connections. *Nature*, **457**(7233), pp.1142–5.

Pignatelli, M. & Bonci, A., 2015. Role of dopamine neurons in reward and aversion: a synaptic plasticity perspective. *Neuron*, **86**(5), pp.1145–57.

Prakash, R. *et al.*, 2012. Two-photon optogenetic toolbox for fast inhibition, excitation and bistable modulation. *Nat Methods*, **9**(12), pp.1171–9.

Reijmers, L.G. *et al.*, 2007. Localization of a stable neural correlate of associative memory. *Science*, **317**(5842), pp.1230–3.

Rickgauer, J.P. & Tank, D.W., 2009. Two-photon excitation of channelrhodopsin-2 at saturation. *Proc Natl Acad Sci U S A*, **106**(35), pp.15025–30.

Rosen, Z.B., Cheung, S. & Siegelbaum, S.A., 2015. Midbrain dopamine neurons bidirectionally regulate CA3–CA1 synaptic drive. *Nat Neurosci*, **18**(12), pp.1763–71.

Scanziani, M. & Häusser, M., 2009. Electrophysiology in the age of light. *Nature*, **461**(7266), pp.930–9.

Schubert, D. *et al.*, 2001. Layer-specific intracolumnar and transcolumnar functional connectivity of layer V pyramidal cells in rat barrel cortex. *J Neurosci*, **21**(10), pp.3580–92.

Schubert, D., Kötter, R. & Staiger, J., 2007. Mapping functional connectivity in barrel-related columns reveals layer- and cell type-specific microcircuits. *Brain Struct Funct*, **212**(2), pp.107–19.

Seol, G.H. *et al.*, 2007. Neuromodulators control the polarity of spike-timing-dependent synaptic plasticity. *Neuron*, **55**(6), pp.919–29.

Shepherd, G.M. *et al.*, 2005. Geometric and functional organization of cortical circuits. *Nat Neurosci*, **8**(6), pp.782–90.

Shepherd, G.M. & Svoboda, K., 2005. Laminar and columnar organization of ascending excitatory projections to layer 2/3 pyramidal neurons in rat barrel cortex. *J Neurosci*, **25**(24), pp.5670–9.

Shipton, O.A. *et al.*, 2014. Left–right dissociation of hippocampal memory processes in mice. *Proc Natl Acad Sci U S A*, **111**(42), pp.15238–43.

Teles-Grilo Ruivo, L. & Mellor, J., 2013. Cholinergic modulation of hippocampal network function. *Front Synaptic Neurosci*, **5**, p. 2.

Thomson, A.M. *et al.*, 1996. Single axon IPSPs elicited in pyramidal cells by three classes of interneurones in slices of rat neocortex. *J Physiol*, **496**(Pt 1), pp.81–102.

Thomson, A.M. *et al.*, 2002. Synaptic connections and small circuits involving excitatory and inhibitory neurons in layers 2–5 of adult rat and cat neocortex: triple intracellular recordings and biocytin labelling *in vitro*. *Cereb Cortex*, **12**(9), pp.936–53.

Thomson, A.M., Deuchars, J. & West, D.C., 1996. Neocortical local synaptic circuitry revealed with dual intracellular recordings and biocytin-filling. *J Physiol Paris*, **90**(3–4), pp.211–5.

Walker, J.W., McCray, J.A. & Hess, G.P., 1986. Photolabile protecting groups for an acetylcholine receptor ligand. Synthesis and photochemistry of a new class of o-nitrobenzyl derivatives and their effects on receptor function. *Biochemistry*, **25**(7), pp.1799–805.

Wang, Y. *et al.*, 2004. Anatomical, physiological and molecular properties of Martinotti cells in the somatosensory cortex of the juvenile rat. *J Physiol*, **561**(1), pp.65–90.

Weiler, N. *et al.*, 2008. Top-down laminar organization of the excitatory network in motor cortex. *Nat Neurosci*, **11**(3), pp.360–6.

Wieboldt, R. *et al.*, 1994a. Photolabile precursors of glutamate: synthesis, photochemical properties, and activation of glutamate receptors on a microsecond time scale. *Proc Natl Acad Sci U S A*, **91**(19), pp.8752–6.

Wieboldt, R. *et al.*, 1994b. Synthesis and photochemistry of photolabile derivatives of gamma-aminobutyric acid for chemical kinetic investigations of the gamma-aminobutyric acid receptor in the millisecond time region. *Biochemistry*, **33**(6), pp.1526–33.

Wilcox, M. *et al.*, 1990. Synthesis of photolabile precursors of amino acid neurotransmitters. *J Org Chem*, **55**(5), pp.1585–9.

Xiong, W. & Jin, X., 2012. Optogenetic field potential recording in cortical slices. *J Neurosci Methods*, **210**(2), pp.119–24.

Xu, X. & Callaway, E.M., 2009. Laminar specificity of functional input to distinct types of inhibitory cortical neurons. *J Neurosci*, **29**(1), pp.70–85.

Yoshimura, Y. & Callaway, E.M., 2005. Fine-scale specificity of cortical networks depends on inhibitory cell type and connectivity. *Nat Neurosci*, **8**(11), pp.1552–9.

Yoshimura, Y., Dantzker, J.L. & Callaway, E.M., 2005. Excitatory cortical neurons form fine-scale functional networks. *Nature*, **433**(7028), pp.868–73.

Zemelman, B.V. *et al.*, 2003. Photochemical gating of heterologous ion channels: remote control over genetically designated populations of neurons. *Proc Natl Acad Sci U S A*, **100**(3), pp.1352–7.

Zemelman, B.V. *et al.*, 2002. Selective photostimulation of genetically chARGed neurons. *Neuron*, **33**(1), pp.15–22.

Zemelman, B.V. & Miesenböck, G., 2001. Genetic schemes and schemata in neurophysiology. *Curr Opin Neurobiol*, **11**(4), pp.409–14.

Zhu, P. *et al.*, 2009. Optogenetic dissection of neuronal circuits in zebrafish using viral gene transfer and the Tet system. *Front Neural Circuits*, **3**, p. 21.

Part IV

Optogenetics in Learning, Neuropsychiatric Diseases, and Behavior

17 Optogenetics to Study Reward Learning and Addiction

Andrea L. Gutman and Ryan T. LaLumiere

17.1 Overview

The brain assigns positive or negative valence to environmental stimuli in order to maximize appetitive outcomes and avoid aversive outcomes. The mesocorticolimbic system, which includes ventral tegmental area (VTA) neurons and their projection targets such as the nucleus accumbens (NAc) and medial prefrontal cortex (mPFC), contributes to these neural computations and has, therefore, received much attention for its role in learning and reward-related behaviors. Although traditional techniques of lesions, pharmacology and electrophysiology have provided insight into the functioning of the mesocorticolimbic system, much remains unknown. Moreover, traditional techniques suffer from a number of drawbacks. For example, lesions can be imprecise and lack temporal control, whereas pharmacological manipulations have long time courses of action. Optogenetics has significantly advanced the study of reward-related learning by offering the ability to manipulate brain regions and classes of neurons with temporal and spatial precision that has heretofore been impossible. Moreover, optogenetics has made it possible to relate temporally specific patterns of activity in subpopulations of neurons to behavioral output.

This chapter is divided into two parts that address: (1) how optogenetic techniques have advanced our understanding of the role of the mesocorticolimbic system in learning; and (2) how optogenetics has been used to investigate this system in the context of drugs of abuse.

17.2 Optogenetics for Studying Reward and Aversion Learning

Beginning in the 1990s, Wolfram Schultz and colleagues published a series of electrophysiology papers suggesting that putative dopamine neurons in the VTA signal reward prediction errors (RPEs), or differences between anticipated reward and actual reward (Schultz *et al.*, 1993; Schultz *et al.*, 1997; Schultz, 1998). These neurons increased firing when an unexpected reward was presented. With multiple pairings of a reward-predictive cue and a reward, these neurons came to fire in response to the predictive cue rather than to the reward, as long as the reward arrived as expected. On the other hand, putative dopamine neurons decreased firing when an expected reward was omitted. Thus, these neurons signaled a RPE, or a teaching signal, indicating whether the actual reward was better or worse than the anticipated reward. Throughout the 1990s and early 2000s, the idea prevailed that dopamine neurons in the VTA contributed to learning by providing such a signal. However, many important questions remained open in the field. First, given that VTA neurons can release dopamine, GABA and glutamate (Fields *et al.*, 2007; Stuber *et al.*, 2010), the chemical identity of neurons signaling a RPE had not been resolved. Second, although the RPE appeared to reflect learning, whether this RPE was *necessary* for learning associations between reward-predictive cues and outcomes remained to be determined. Optogenetics made it possible to experimentally address these questions and to further elucidate the neural mechanisms of reward and aversion learning in the mesocorticolimbic system.

17.2.1 What Is the Identity of VTA Neurons that Encode a RPE?

Two optogenetic approaches made it possible to examine this question. First, the expression of light-sensitive opsins can be restricted to specific cellular populations. For example, the light-activated cation channel channelrhodopsin-2 (ChR2) can be expressed in dopamine neurons by injecting a Cre-inducible adeno-associated virus (AAV) vector carrying the gene encoding ChR2 into transgenic mice expressing Cre recombinase under the control of the promoter of the dopamine transporter (*DAT*) gene (Tsai *et al.*, 2009). When stimulated, these neurons elicit dopamine transients in the NAc comparable to those observed in response to natural reward (Roitman *et al.*, 2004; Tsai *et al.*, 2009). Similarly, expression of ChR2 can be restricted to GABA neurons in the VTA by expressing Cre recombinase under the control of the endogenous vesicular GABA transporter (*Vgat*) gene. Second, a "tagging" approach can be used to confirm the identity of neurons during electrophysiological recording (Lima *et al.*, 2009). Briefly, after restricted expression of ChR2 is achieved, these neurons can be activated by a blue laser, causing a short-latency action potential. Electrophysiological recording during behavioral sessions can then be used to examine the response properties of neurons that are optically responsive and, therefore, presumed to be specific types of neurons. This permits electrophysiological recording to be carried out in genetically identifiable neurons.

Cohen and colleagues used this approach to express ChR2 in either dopamine or GABA neurons in the VTA and then carry out *in vivo* extracellular recording while mice performed a task in which an odor cue predicted either a large or small water reward (Cohen *et al.*, 2012). Only dopamine neurons displayed firing patterns consistent with a RPE, increasing firing in response to reward-predictive cues or decreasing firing when an anticipated reward was omitted. Other dopamine neurons showed excitation in response to reward delivery, even after the task was well learned, or in response to both predictive cues and rewards. On the other hand, GABA neurons were modulated by reward expectation, typically showing an increase or decrease in firing between the presentation of the predictive cue and the delivery of the reward. These neurons, however, were not modulated by unexpected reward delivery or the omission of rewards. Thus, although not all dopaminergic neurons display canonical RPE signaling, only dopaminergic neurons appear to encode a RPE.

17.2.2 Does Dopamine Drive Reinforcement Learning?

In computational models of reinforcement learning (RL), the RPE drives reinforcement (Rescorla and Wagner, 1972; Sutton, 1988). Although electrophysiology studies suggested that putative dopamine neurons signal a RPE, these studies lacked causal evidence, and prior to the advent of optogenetics, it remained unknown whether the RPE was *necessary* for learning. Simultaneously, others suggested that reward learning does not require dopamine and that dopamine may serve other purposes, such as encoding incentive salience (Berridge and Robinson, 1998; Redgrave *et al.*, 1999; Horvitz, 2000; Wise, 2004; Berridge, 2007). By restricting opsin expression to dopamine neurons as described previously, studies could more closely examine the role of dopamine in learning. Phasic stimulation of dopamine neurons during a 30-minute exposure to a specific context induced a preference for the context, demonstrating that phasic dopaminergic activity is sufficient to elicit behavioral conditioning (Tsai *et al.*, 2009).

Although this finding supports a role for the dopaminergic system in Pavlovian conditioning, the RL theory of dopamine function suggests that transient activation of dopamine neurons provides the RPE and, consequently, must immediately follow a stimulus or action for RL to occur (Rescorla and Wagner, 1972; Sutton and Barto, 1998). To directly examine this issue, Kim and colleagues (2012) used 200-ms light pulses to activate VTA dopamine neurons expressing ChR2 immediately following self-initiated nose pokes and found that this temporally precise dopamine neuron activation was sufficient to reinforce the behavior (Kim *et al.*, 2012). Similarly, TH-Cre transgenic mice learned to lever press to receive phasic optogenetic stimulation of dopamine neurons and extinguished lever pressing when the operant response no longer produced optical stimulation (Ilango *et al.*, 2014a; Ilango *et al.*, 2014b). Consistent with these findings, optical activation of dopamine neurons coinciding with reward delivery enhanced learning about cue–reward associations (Adamantidis *et al.*, 2011; Steinberg and Janak, 2013). Conversely, optically stimulating dopamine neurons when a reward was omitted during extinction learning impaired such learning (Steinberg *et al.*, 2013). Such activation would have prevented the decrease in VTA dopamine neuron activity normally observed during the omission of an expected reward. Together, these findings provide direct evidence for the function of dopamine neuron activity, both in serving as a RPE signal and as a critical component in RL.

17.2.3 VTA and Dopamine in Aversive Learning

In contrast to the excitatory response elicited by dopamine neurons following the presentation of rewarding stimuli, a variety of responses has been reported in putative dopamine neurons following the presentation of aversive stimuli. Classic work suggested that the majority of putative dopamine neurons are inhibited or show no change in firing in response to aversive stimuli (Maeda and Mogenson, 1982; Schultz and Romo, 1987; Mantz *et al.*, 1989). Although electrophysiology studies provide correlative evidence suggesting that inhibition of putative dopamine neurons may contribute to learning about aversive stimuli, optogenetic approaches using inhibitory opsins that cause hyperpolarization of neurons in response to light were used to directly examine this issue. Using the light-activated chloride pump halorhodopsin (NpHR), Ilango and colleagues induced conditioned place aversion for a particular compartment in which mice had received optical inhibition of VTA dopamine neurons (Ilango *et al.*, 2014b). Consistent with these findings, other work indicates that optogenetic inactivation of VTA dopamine neurons using the hyperpolarizing Arch protein decreases dopamine release in the NAc and causes aversion for a previously preferred dark chamber (Danjo *et al.*, 2014). These studies, therefore, provide direct evidence that the inhibition of dopamine neurons contributes to aversive learning.

GABA neurons within the VTA contribute to RPE computations used in aversive learning by providing inhibitory input to dopamine neurons. VTA dopamine neurons are thought to be tonically inhibited by GABA neurons within the VTA and the rostromedial tegmental nucleus (RMTg) (Johnson and North, 1992; Jhou *et al.*, 2009). Thus, it would be expected that activation of these GABAergic neurons would decrease the firing of dopamine neurons, while inhibition of these neurons would increase firing. Studies using cell type-specific expression of ChR2 in GABA neurons in the VTA in combination with electrophysiology confirmed that activation of GABA neurons inhibits VTA dopamine neurons (Tan *et al.*, 2012). Moreover, activation of VTA GABAergic neurons elicits conditioned place aversion (Tan *et al.*, 2012) and disrupts reward consumption (van Zessen *et al.*, 2012), suggesting that GABA neurons in the VTA promote behaviors associated with aversion.

Excitations in putative dopamine neurons have also been reported in response to aversive stimuli (Brischoux *et al.*, 2009; Matsumoto and Hikosaka, 2009; Tan *et al.*, 2012), although it has been argued that these dopamine neurons were misidentified (Ungless and Grace, 2012). Others have suggested that phasic responses to aversive stimuli are due to the rewarding properties of terminating an aversive stimulus (Tanimoto *et al.*, 2004). Again, optogenetic tagging proved useful in settling this debate and showing that dopaminergic neurons in the VTA had some diversity in their responses to aversive stimuli (Cohen *et al.*, 2012). While the majority of dopamine neurons were inhibited by an aversive air puff, some were excited. Thus, optogenetics has significantly advanced the study of aversion learning by allowing cell type-specific manipulations of neurons within the mesocorticolimbic system.

17.2.4 Neural Circuitry Underlying Reward and Aversion

In addition to the classic RPE described above, a portion of putative dopamine neurons also shows phasic increases in firing to salient, novel and aversive stimuli (Horvitz, 2000; Matsumoto and Hikosaka, 2009; Ungless *et al.*, 2010). The variety

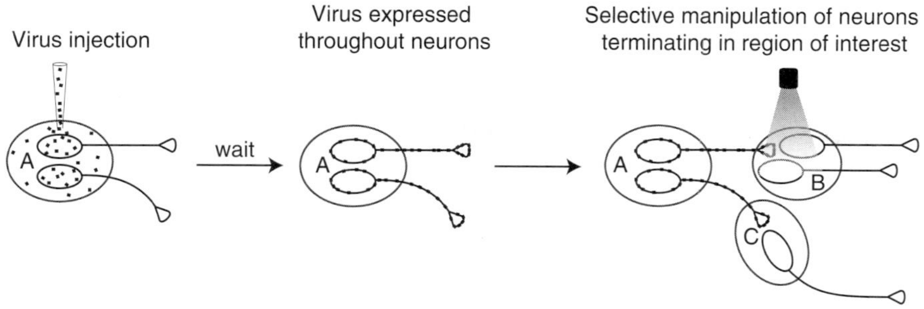

Figure 17.1 Optogenetic techniques are used to isolate specific neural pathways. Opsin expression is achieved in a specific class of neuron, represented here as originating from region A. Over time, the opsin is expressed throughout the neurons, including at the terminal ends. Fiber optics are implanted into a brain region, represented here as B, which receives input from region A. Optical stimulation in region B will affect the firing rate of neurons projecting from region A to region B, but not neurons projecting from region A to another brain region, represented here as region C.

of response properties of dopamine neurons suggests that these neurons may segregate into functional subsets, and evidence from optogenetic studies suggests that differences in anatomical pathways may underlie the heterogeneity in the activity patterns of dopamine neurons (Fernando *et al.*, 2012). Because midbrain dopamine neurons have a variety of afferents as well as widespread efferent projections, recent studies have focused on the role of dopamine afferents and efferents in learning. One of the powerful attributes of optogenetics is its ability to examine these neural pathways. Opsin expression can be targeted to a specific neuron type, and the opsin will be expressed throughout the neuron, including along axons that project to other brain regions. Specific pathways can then be isolated by implanting fiber optics in brain regions that receive input from neurons expressing the opsin (Figure 17.1). This approach has been used to examine the circuitry underlying reward- and aversion-related learning, and as described later in this chapter, has also been used to study the neural circuitry underlying drug-seeking behavior.

Laterodorsal tegmentum (LDT) neurons project to the VTA and promote burst firing of putative VTA dopamine neurons (Cornwall *et al.*, 1990; Lodge and Grace, 2006). Rabies virus-mediated expression of ChR2 in LDT projection neurons and a combination of *in vitro* and *in vivo* studies further established that LDT neurons send excitatory projections to dopamine neurons that subsequently project to the lateral shell of the NAc (Lammel *et al.*, 2012). Optogenetic stimulation of LDT terminals in the VTA produced conditioned place preference, and blockade of dopamine receptors in the NAc reversed this effect. Together, these studies suggest that the LDT-to-VTA dopamine-to-NAc shell pathway mediates reward-associated behaviors.

Just as optogenetics showed that the connectivity of dopamine neurons importantly mediates their functional roles in reward-seeking behavior, similar optogenetic techniques have been applied in order to examine the role of the mesolimbic dopamine system in aversion. The lateral habenula (LHb) projects to the VTA and has received attention for its role in mediating behavioral responses to aversive stimuli (Bromberg-Martin *et al.*, 2010). LHb neurons increase firing when expected rewards are omitted, providing a signal that is inverse to that in VTA dopamine neurons (Matsumoto and Hikosaka, 2007). Given the nature of LHb firing patterns and the onset of this firing relative to the firing of VTA dopamine neurons, scientists posited that this brain region contributes to learning by inhibiting dopamine neurons. Indeed, later experiments found that LHb neurons synapse onto inhibitory neurons in the RMTg (Kaufling *et al.*, 2009; Lammel *et al.*, 2012). Optical activation of LHb neurons in acute brain slices generated excitatory currents in GABAergic neurons of the RMTg, confirming the functionality of this pathway (Lammel *et al.*, 2012; Stamatakis and Stuber, 2012). Optogenetic stimulation of LHb neurons projecting to the RMTg was sufficient to induce behavioral avoidance and conditioned place aversion. These RMTg neurons synapse selectively onto dopamine neurons that project to the mPFC, and infusion of dopamine receptor antagonists into the mPFC blocked conditioned place aversion, supporting a role for this portion of the pathway in aversion learning (Lammel *et al.*, 2012). Together, these optogenetic studies have established a neural pathway mediating aversion learning that consists of LHb glutamate neurons projecting to RMTg GABA neurons projecting to VTA dopamine neurons projecting to mPFC glutamate neurons.

17.2.5 Non-dopaminergic Neural Circuitry Underlying Reward Learning

Although the mesolimbic dopamine system has received much attention for its role in reward learning, a variety of structures contribute to such learning. Just as optogenetics has brought about significant advances in our understanding of the role of the mesolimbic system, it has also contributed to our understanding of other neural circuitry underlying reward learning, although much remains to be investigated.

In one study, Stuber and colleagues injected the AAV encoding ChR2 or NpHR under the CaMKII promotor into the basolateral amygdala (BLA) (Stuber *et al.*, 2011). The preferential expression of CaMKII in glutamatergic neurons caused these opsins to be selectively expressed in the excitatory neurons of the BLA. An optical fiber was implanted into the NAc, which receives input from the BLA, permitting BLA terminals in the NAc to be stimulated. Mice were placed in an operant box and given the opportunity to nose poke for light stimulation. Activation of the BLA-to-NAc pathway increased the rate of nose poking (Stuber *et al.*, 2011). When mice were exposed to a light and tone that indicated sucrose availability, silencing of the BLA-to-NAc pathway following nose pokes decreased responding for the sucrose reward. These results indicate that activation of the BLA- to-NAc pathway reinforces behavior. Additional work found that optical stimulation of neurons projecting from the ventral hippocampus (vHipp), BLA or mPFC to the NAc was sufficient to reinforce nose-poking behavior (Britt *et al.*, 2012), although other work has not been in agreement on the mPFC inputs to the NAc (Stuber *et al.*, 2011). Moreover, optical stimulation of medium spiny neurons

(MSNs) alone supports self-stimulation (Britt *et al.*, 2012). These results suggest that activation of specific glutamatergic inputs to the NAc is sufficient to reinforce behavior.

17.3 Optogenetics for Studying Drug Addiction

Much evidence suggests that the mesolimbic system critically contributes to processing motivation, hedonia and decision making, and, therefore, it is unsurprising that these brain regions are involved in addiction, a pathological state in which such processes are disordered (Everitt and Robbins, 2005; Kalivas and Volkow, 2005). Optogenetics has significantly advanced the study of addiction by providing a technique that can isolate neural circuitry. As a result, optogenetic approaches can be combined with measures of plasticity and alterations in functional circuits to understand those circuits and then can be used again to interrogate those circuits (Figure 17.2). Studying addiction on both cellular and behavioral levels provides complementary lines of investigation and a more thorough understanding of the disease. The remainder of this section addresses how optogenetics has been used in addiction studies and illustrates the complementary roles that different techniques have provided in such studies.

17.3.1 The mPFC-to-NAc Pathway

As noted, a major advantage of optogenetics has been its ability to interrogate neural pathways, a strength that many addiction studies have used for great benefit. In particular, as prior work suggests that glutamatergic inputs to the NAc and alterations at those synapses are critically involved in relapse to drug use, studies have frequently focused on those glutamatergic afferents in the NAc. The NAc receives excitatory glutamatergic inputs from the mPFC, the BLA and the ventral subiculum of the hippocampus (vHipp). These projections provide distinct contextual, motivational and reward-related information. The NAc then integrates this information to drive appropriate behavioral output (Mogenson *et al.*, 1980; Bremner *et al.*, 1996).

Synaptic alteration in this circuitry as a consequence of long-term exposure to drugs of abuse likely contributes to prioritizing drug-seeking behaviors, resulting in relapse in humans or the reinstatement of drug-seeking behavior in rodent models of addiction (Russo *et al.*, 2010). However, until the development of optogenetics, it was difficult to isolate the different glutamatergic inputs in order to determine how each projection to the NAc differentially mediated addiction-related processes. Indeed, although prior studies pointed to the mPFC and its projections to the NAc in mediating the reinstatement of cocaine seeking (McFarland *et al.*, 2003), it had not been possible to directly control activity in this pathway. In order to examine the functional relationship between the mPFC and the NAc during reinstatement, Stefanik and colleagues optically inhibited the prelimbic portion of the mPFC, the NAc core or the prelimbic inputs to the NAc core during cocaine-seeking behavior (Stefanik *et al.*, 2013b; Stefanik *et al.*, 2016). Each of these manipulations effectively blocked reinstatement induced by a cocaine prime or a combination of

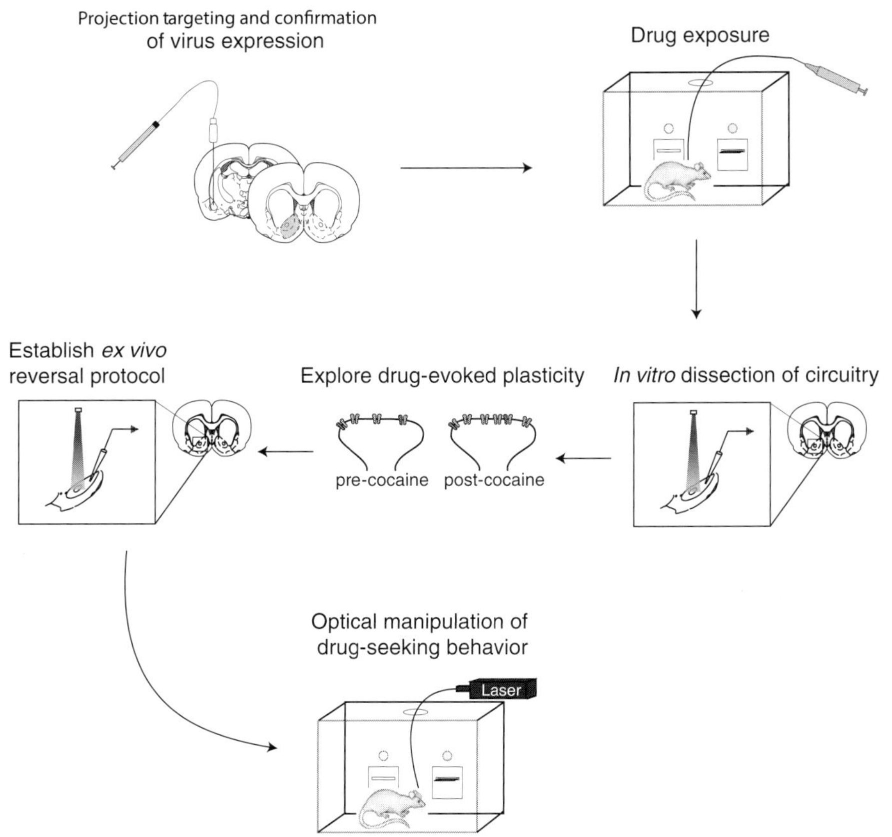

Figure 17.2 Projection targeting is used to confirm that virus injected in one brain region is expressed in the terminals of those neurons in another brain region. Animals are exposed to drugs of abuse and changes in neural circuitry are examined *in vitro*. A variety of electrophysiological techniques are used to characterize drug-evoked synaptic plasticity. Next, an *ex vivo* reversal protocol is established to reverse drug-related changes in synaptic plasticity. Finally, using information gained from *ex vivo* studies, optogenetics can be applied to awake, behaving animals in order to interrogate the role of the neural circuitry *in vivo* and, ultimately, to affect drug-seeking behaviors.

a cocaine prime and cues, suggesting that prelimbic projections to the NAc core critically mediate the reinstatement of drug seeking.

Other work has used optogenetics in both behavioral and physiological investigations of these pathways as part of drug addiction studies. Previous work was equivocal on whether chronic cocaine exposure caused presynaptic changes in glutamatergic inputs to NAc MSNs (Kourrich *et al.*, 2007). Moreover, if such changes occur, the identity of the glutamatergic afferents expressing such changes remained unknown. Therefore, Suska and colleagues used an optogenetic approach in order to target either the mPFC or BLA inputs to the NAc shell (Suska *et al.*, 2013). Their findings indicated that cocaine self-administration followed by withdrawal produced long-lasting enhanced release probabilities at PFC-to-NAc shell synapses, but not at BLA-to-NAc core synapses (Table 17.1) (Suska *et al.*, 2013). Alterations in release probabilities

Table 17.1 Pathway-specific synaptic changes following exposure to cocaine.

	Infralimbic-to-NAc	BLA-to-NAc	vHipp-to-NAc
Release probability	↑ Suska *et al.*, 2013	x Suska *et al.*, 2013	
CP-AMPAR	↑ Ma *et al.*, 2014; in D1+ neurons, Pascoli *et al.*, 2014	↑ Lee *et al.*, 2013, but see Pascoli *et al.*, 2014	x Britt *et al.*, 2012, Pascoli *et al.*, 2014
AMPA/NMDA ratio	↓ In D1+ neurons, Pascoli *et al.*, 2014	x Britt *et al.*, 2012	↑ Britt *et al.*, 2012; in D1+ neurons, Pascoli *et al.*, 2014
EPSC amplitude	↑ In D1+ neurons, Pascoli *et al.*, 2014	x Britt *et al.*, 2012	↑ Britt *et al.*, 2012

CP-AMPAR: calcium-permeable α-amino-3-hydroxy-5-methyl-4-isoxazoleproprionic acid receptors; AMPA/NMDA: α-amino-3-hydroxy-5-methyl-4-isoxazoleproprionic acid/*N*-methyl-D-aspartate; EPSC: excitatory postsynaptic current.

indicate changes in presynaptic functioning, thus providing direct evidence of such changes. Although the effects of cocaine on presynaptic changes had been previously examined, the use of optogenetics in this study made it possible to isolate this particular pathway and examine electrophysiological changes resulting from cocaine exposure.

Optogenetic tools have also been used to examine postsynaptic changes at mPFC-to-NAc synapses following cocaine exposure. Cocaine exposure generates silent glutamatergic synapses, which are immature glutamatergic synapses lacking α-amino-3-hydroxy-5-methyl-4-isoxazoleproprionic acid receptors (AMPARs) or containing highly labile AMPARs (Brown *et al.*, 2011; Koya *et al.*, 2012; Lee *et al.*, 2013). Because AMPAR-mediated signaling is absent, these synapses are silent under normal conditions. During extended periods of withdrawal, animals experience incubation of cocaine craving, or time-dependent increases in cue-evoked cocaine-seeking behavior during the first months following withdrawal from cocaine self-administration. Optogenetic techniques have been used to elucidate specific postsynaptic changes in NAc neurons receiving input from different regions of the mPFC that contribute to the incubation of cocaine craving. Specifically, during extended periods of withdrawal, silent synapses in the NAc mature, or become unsilenced, depending on the region from which the input originates. Infralimbic-to-NAc neurons become unsilenced by the accumulation of GluR2-lacking, calcium-permeable AMPARs (CP-AMPARs), whereas prelimbic-to-NAc neurons become unsilenced by the insertion of non-CP-AMPARs (Ma and Morilak, 2005; Pascoli *et al.*, 2014).

AMPAR transmission in mPFC neurons projecting to the NAc correlates with cocaine seeking after withdrawal, suggesting a functional significance for this plasticity (Pascoli *et al.*, 2014).

The maturation of silent synapses can be reversed by optically stimulating intralimbic neurons that terminate in the NAc to create long-term depression (LTD) in NAc shell neurons. This reversal process potentiates cocaine seeking after a withdrawal period, suggesting: (1) that the infralimbic to NAc shell projections exert inhibitory control of cocaine seeking; and (2) that maturation of silent synapses underlies this inhibitory control (Ma *et al.*, 2014). These findings extend pharmacological work showing that the infralimbic cortex and the NAc shell inhibit cocaine-seeking behavior (Peters *et al.*, 2008; LaLumiere *et al.*, 2012). On the other hand, optically stimulating prelimbic-to-NAc core neurons in order to cause LTD partially restores silent synapses and leads to an attenuation of cocaine seeking (Ma *et al.*, 2014). Together, these optogenetic studies have more clearly elucidated the mechanisms of two competing neural circuits: one projecting from the infralimbic cortex to the NAc shell, which inhibits cocaine-seeking behavior, and another projecting from the prelimbic cortex to the NAc core, which promotes cocaine-seeking behavior.

A second type of synaptic plasticity commonly seen following cocaine self-administration involves changes in AMPA glutamate receptor currents, which are estimated by the AMPA/*N*-methyl-D-aspartate (NMDA) receptor ratio. Again, cocaine self-administration has opposite effects on the infralimbic-to-NAc shell pathway compared to the prelimbic-to-NAc core pathway. Pascoli and colleagues showed that the infralimbic-to-NAc shell pathway demonstrated a reduction in the AMPA/NMDA ratio following cocaine self-administration, and normalizing cocaine-induced synaptic plasticity abolished cocaine-seeking behavior (Pascoli *et al.*, 2014). On the other hand, the infralimbic-to-NAc core pathway showed an enhanced AMPA/NMDA ratio, and optical inhibition of this pathway blocked cue-induced reinstatement of cocaine seeking, as well as decreased the AMPA/NMDA ratio (Stefanik *et al.*, 2016). Thus, changes in glutamatergic input at different inputs in the NAc are differentially affected by exposure to cocaine, and optogenetic normalization of this synaptic plasticity can affect ongoing cocaine-seeking behavior.

17.3.2 The BLA-to-NAc Pathway

Prior work has suggested that functional projections from the BLA to the NAc are necessary for cue-induced reward seeking (Setlow *et al.*, 2002; Di Ciano and Everitt, 2004; Stuber *et al.*, 2011). However, these studies have relied largely on disconnection techniques, which do not exclude the possibility of non-specific disruption of activity in other pathways. Indeed, although the BLA projects to the NAc, it also sends projections to the mPFC, which, as noted, is also involved in driving drug-seeking behavior. Thus, optogenetics has provided a tool for understanding the precise role of the BLA projections in regulating cocaine-seeking behavior. Indeed, recent work addressing this issue found that optically inhibiting BLA inputs to either the NAc or the prelimbic cortex prevents cue-induced reinstatement of cocaine seeking (Stefanik and Kalivas, 2013). These findings not only give insight into the roles of different BLA pathways during drug seeking, but also indicate the critical importance of pathway targeting in resolving controversies about neural circuits governing behavior.

As with the mPFC, recent work has used optogenetic control of BLA projections to examine plasticity following drug exposure. As noted above, by using optogenetic and electrophysiological methods to isolate this pathway, Suska and colleagues found no presynaptic changes following long-term withdrawal from cocaine exposure (Suska *et al.*, 2013). However, the BLA-to-NAc pathway showed postsynaptic changes. Optical stimulation of the BLA-to-NAc glutamatergic pathway caused down-regulation of the CP-AMPARs that are thought to mediate the incubation of craving, leading to decreased cue-evoked cocaine-seeking behavior (Lee *et al.*, 2013; but see also Pascoli *et al.*, 2014). Thus, while presynaptic changes in the BLA-to-NAc pathway are not present, postsynaptic changes may contribute to cocaine-seeking behavior.

17.3.3 The vHipp-to-NAc Pathway

Immunolabeling has shown that the medial NAc shell receives more dense projections from the vHipp than from the BLA or the mPFC (Britt *et al.*, 2012). These glutamatergic projections contribute to cocaine-seeking behavior (Sun and Rebec, 2003; Atkins *et al.*, 2008), and optogenetic experiments have uncovered unique plasticity in this pathway relative to the mPFC-to-NAc pathway. Specifically, optical stimulation of vHipp fibers terminating in the NAc shell elicits considerably larger excitatory postsynaptic currents (EPSCs) than optical stimulation of the BLA-to-NAc or mPFC-to-NAc pathways (Britt *et al.*, 2012). Chronic (non-contingent) cocaine exposure or withdrawal from cocaine self-administration enhanced AMPA/NMDA ratios in the vHipp-to-NAc shell pathway (Britt *et al.*, 2012; Pascoli *et al.*, 2014). This plasticity has functional significance for cocaine-associated behaviors, as optical activation or inhibition of the vHipp-to-NAc shell pathway enhances or inhibits, respectively, cocaine-induced locomotion (Britt *et al.*, 2012). Similarly, reversal of the enhanced AMPA/NMDA ratio decreases cocaine-seeking behavior after withdrawal from cocaine self-administration. Although relatively little research has focused on the vHipp-to-NAc shell pathway, optogenetic studies provide additional evidence to suggest that synaptic changes in this pathway occur during drug exposure, and that such changes contribute to cocaine-seeking behavior.

17.3.4 The NAc and Downstream Structures in Cocaine-seeking Behavior

The NAc is comprised of >95% MSNs, which can be divided into two subtypes based on their expression of D1 or D2 receptors (D1+ or D2+, respectively). Cocaine exposure results in long-term cellular and molecular changes in the striatum, many of which are specific to either D1+ or D2+ neurons (Lobo and Nestler, 2011). The roles of these classes of MSNs in cocaine-associated learning and memory have been examined more thoroughly in recent years by using genetically engineered mice that express Cre recombinase in a cell type-specific manner. In one study, ChR2 was expressed in either D1+ or D2+ MSNs in the NAc using D1-Cre and D2-Cre BAC transgenic mice, respectively, combined with Cre-dependent viral expression (Lobo *et al.*, 2010). The authors found that optical stimulation of D1+ NAc neurons during exposure to a cocaine-associated context enhanced the preference for that context, whereas stimulation of D2+ NAc neurons produced the opposite effect.

Other work has used optogenetic approaches for studying sensitization, whereby exposure to psychostimulants enhances locomotor activity in response to subsequent drug exposures. Repetitive optical stimulation of D2+ MSNs in the NAc during withdrawal from cocaine attenuates subsequent expression of cocaine-induced behavioral sensitization (Song *et al.*, 2014). These findings are not unprecedented: the direct striatonigral pathway, which contains D1+ neurons, has an established role in reward learning and drug sensitization, while the indirect striatonigral pathway, which contains D2+ neurons, negatively regulates drug sensitization (Hikida *et al.*, 2010; Ferguson *et al.*, 2011; Kravitz *et al.*, 2012). These optogenetic studies contribute to a body of work suggesting that D1+ NAc neurons promote drug-related learning, while activity in D2+ NAc neurons inhibits such learning.

17.3.5 NAc Projections in Cocaine-seeking Behavior

Much research is built upon a framework of striatal function consisting of "direct" pathway striatonigral neurons and "indirect" pathway striatopallidal neurons (Albin *et al.*, 1989; Gerfen, 1992). In neurons originating from the dorsal striatum, the expression of D1 and D2 receptors segregates along striatonigral and striatopallidal projections in neurons. However, this segregation is not present in the ventral striatum, where striatopallidal neurons express both D1 and D2 receptors (Lu *et al.*, 1998). Thus, targeting the NAc projections themselves is a critical advance in understanding the NAc-based neural circuits in addiction processes. Recent work examined the roles of the NAc core projections to the ventral pallidum and substantia nigra in cocaine seeking. Specifically, the authors found that inhibition of the indirect pathway projections from the NAc core to the dorsolateral ventral pallidum prevented the reinstatement of cocaine-seeking behavior, whereas inhibition of direct pathway projections from the NAc core to the medial substantia nigra had no effect (Stefanik *et al.*, 2013a). Thus, these findings indicate that the direct and indirect pathways differentially mediate cocaine-seeking behavior.

17.4 Summary

Traditional methods have provided a strong groundwork for understanding the neural correlates of drug addiction. However, the development of optogenetic approaches has significantly advanced the study of addiction by affording us the ability to more closely interrogate neural circuitry and pathway-specific synaptic adaptations with far greater temporal and spatial precision than was possible in the past. The contributions of optogenetic studies to the field's understanding of the neural correlates of addiction will undoubtedly inform the identification and development of treatments for this disease.

REFERENCES

Adamantidis, A. R., Tsai, H. C., Boutrel, B., Zhang, F., Stuber, G. D., Budygin, E. A., Tourino, C., Bonci, A., Deisseroth, K. & De Lecea, L. 2011. Optogenetic interrogation of dopaminergic modulation of the multiple phases of reward-seeking behavior. *J Neurosci*, **31**, 10829–35.

Albin, R. L., Young, A. B. & Penney, J. B. 1989. The functional anatomy of basal ganglia disorders. *Trends Neurosci*, **12**, 366–75.

Atkins, A. L., Mashhoon, Y. & Kantak, K. M. 2008. Hippocampal regulation of contextual cue-induced reinstatement of cocaine-seeking behavior. *Pharmacol Biochem Behav*, **90**, 481–91.

Berridge, K. C. 2007. The debate over dopamine's role in reward: the case for incentive salience. *Psychopharmacology (Berl)*, **191**, 391–431.

Berridge, K. C. & Robinson, T. E. 1998. What is the role of dopamine in reward: hedonic impact, reward learning, or incentive salience? *Brain Res Brain Res Rev*, **28**, 309–69.

Bremner, J. D., Krystal, J. H., Southwick, S. M. & Charney, D. S. 1996. Noradrenergic mechanisms in stress and anxiety: II. Clinical studies. *Synapse*, **23**, 39–51.

Brischoux, F., Chakraborty, S., Brierley, D. I. & Ungless, M. A. 2009. Phasic excitation of dopamine neurons in ventral VTA by noxious stimuli. *Proc Natl Acad Sci U S A*, **106**, 4894–9.

Britt, J. P., Benaliouad, F., Mcdevitt, R. A., Stuber, G. D., Wise, R. A. & Bonci, A. 2012. Synaptic and behavioral profile of multiple glutamatergic inputs to the nucleus accumbens. *Neuron*, **76**, 790–803.

Bromberg-Martin, E. S., Matsumoto, M. & Hikosaka, O. 2010. Distinct tonic and phasic anticipatory activity in lateral habenula and dopamine neurons. *Neuron*, **67**, 144–55.

Brown, T. E., Lee, B. R., Mu, P., Ferguson, D., Dietz, D., Ohnishi, Y. N., Lin, Y., Suska, A., Ishikawa, M., Huang, Y. H., Shen, H., Kalivas, P. W., Sorg, B. A., Zukin, R. S., Nestler, E. J., Dong, Y. & Schluter, O. M. 2011. A silent synapse-based mechanism for cocaine-induced locomotor sensitization. *J Neurosci*, **31**, 8163–74.

Cohen, J. Y., Haesler, S., Vong, L., Lowell, B. B. & Uchida, N. 2012. Neuron-type-specific signals for reward and punishment in the ventral tegmental area. *Nature*, **482**, 85–8.

Cornwall, J., Cooper, J. D. & Phillipson, O. T. 1990. Projections to the rostral reticular thalamic nucleus in the rat. *Exp Brain Res*, **80**, 157–71.

Danjo, T., Yoshimi, K., Funabiki, K., Yawata, S. & Nakanishi, S. 2014. Aversive behavior induced by optogenetic inactivation of ventral tegmental area dopamine neurons is mediated by dopamine D2 receptors in the nucleus accumbens. *Proc Natl Acad Sci U S A*, **111**, 6455–60.

Di Ciano, P. & Everitt, B. J. 2004. Direct interactions between the basolateral amygdala and nucleus accumbens core underlie cocaine-seeking behavior by rats. *J Neurosci*, **24**, 7167–73.

Everitt, B. J. & Robbins, T. W. 2005. Neural systems of reinforcement for drug addiction: from actions to habits to compulsion. *Nat Neurosci*, **8**, 1481–9.

Ferguson, S. M., Eskenazi, D., Ishikawa, M., Wanat, M. J., Phillips, P. E., Dong, Y., Roth, B. L. & Neumaier, J. F. 2011. Transient neuronal inhibition reveals opposing roles of indirect and direct pathways in sensitization. *Nat Neurosci*, **14**, 22–4.

Fernando, A. B., Economidou, D., Theobald, D. E., Zou, M. F., Newman, A. H., Spoelder, M., Caprioli, D., Moreno, M., Hipolito, L., Aspinall, A. T., Robbins, T. W. & Dalley, J. W. 2012. Modulation of high impulsivity and attentional performance in rats by selective direct and indirect dopaminergic and noradrenergic receptor agonists. *Psychopharmacology (Berl)*, **219**, 341–52.

Fields, H. L., Hjelmstad, G. O., Margolis, E. B. & Nicola, S. M. 2007. Ventral tegmental area neurons in learned appetitive behavior and positive reinforcement. *Annu Rev Neurosci*, **30**, 289–316.

Gerfen, C. R. 1992. The neostriatal mosaic: multiple levels of compartmental organization in the basal ganglia. *Annu Rev Neurosci*, **15**, 285–320.

Hikida, T., Kimura, K., Wada, N., Funabiki, K. & Nakanishi, S. 2010. Distinct roles of synaptic transmission in direct and indirect striatal pathways to reward and aversive behavior. *Neuron*, **66**, 896–907.

Horvitz, J. C. 2000. Mesolimbocortical and nigrostriatal dopamine responses to salient non-reward events. *Neuroscience*, **96**, 651–6.

Ilango, A., Kesner, A. J., Broker, C. J., Wang, D. V. & Ikemoto, S. 2014a. Phasic excitation of ventral tegmental dopamine neurons potentiates the initiation of conditioned approach behavior: parametric and reinforcement-schedule analyses. *Front Behav Neurosci*, **8**, 155.

Ilango, A., Kesner, A. J., Keller, K. L., Stuber, G. D., Bonci, A. & Ikemoto, S. 2014b. Similar roles of substantia nigra and ventral tegmental dopamine neurons in reward and aversion. *J Neurosci*, **34**, 817–22.

Jhou, T. C., Fields, H. L., Baxter, M. G., Saper, C. B. & Holland, P. C. 2009. The rostromedial tegmental nucleus (RMTg), a GABAergic afferent to midbrain dopamine neurons, encodes aversive stimuli and inhibits motor responses. *Neuron*, **61**, 786–800.

Johnson, S. W. & North, R. A. 1992. Two types of neurone in the rat ventral tegmental area and their synaptic inputs. *J Physiol*, **450**, 455–68.

Kalivas, P. W. & Volkow, N. D. 2005. The neural basis of addiction: a pathology of motivation and choice. *Am J Psychiatry*, **162**, 1403–13.

Kaufling, J., Veinante, P., Pawlowski, S. A., FREUND-MERCIER, M. J. & Barrot, M. 2009. Afferents to the GABAergic tail of the ventral tegmental area in the rat. *J Comp Neurol*, **513**, 597–621.

Kim, K. M., Baratta, M. V., Yang, A., Lee, D., Boyden, E. S. & Fiorillo, C. D. 2012. Optogenetic mimicry of the transient activation of dopamine neurons by natural reward is sufficient for operant reinforcement. *PLoS One*, **7**, e33612.

Kourrich, S., Rothwell, P. E., Klug, J. R. & Thomas, M. J. 2007. Cocaine experience controls bidirectional synaptic plasticity in the nucleus accumbens. *J Neurosci*, **27**, 7921–8.

Koya, E., Cruz, F. C., Ator, R., Golden, S. A., Hoffman, A. F., Lupica, C. R. & Hope, B. T. 2012. Silent synapses in selectively activated nucleus accumbens neurons following cocaine sensitization. *Nat Neurosci*, **15**, 1556–62.

Kravitz, A. V., Tye, L. D. & Kreitzer, A. C. 2012. Distinct roles for direct and indirect pathway striatal neurons in reinforcement. *Nat Neurosci*, **15**, 816–8.

Lalumiere, R. T., Smith, K. C. & Kalivas, P. W. 2012. Neural circuit competition in cocaine-seeking: roles of the infralimbic cortex and nucleus accumbens shell. *Eur J Neurosci*, **35**, 614–22.

Lammel, S., Lim, B. K., Ran, C., Huang, K. W., Betley, M. J., Tye, K. M., Deisseroth, K. & Malenka, R. C. 2012. Input-specific control of reward and aversion in the ventral tegmental area. *Nature*, **491**, 212–7.

Lee, B. R., Ma, Y. Y., Huang, Y. H., Wang, X., Otaka, M., Ishikawa, M., Neumann, P. A., Graziane, N. M., Brown, T. E., Suska, A., Guo, C., Lobo, M. K., Sesack, S. R., Wolf, M. E., Nestler, E. J., Shaham, Y., Schluter, O. M. & Dong, Y. 2013. Maturation of silent synapses in amygdala-accumbens projection contributes to incubation of cocaine craving. *Nat Neurosci*, **16**, 1644–51.

Lima, S. Q., Hromadka, T., Znamenskiy, P. & Zador, A. M. 2009. PINP: a new method of tagging neuronal populations for identification during *in vivo* electrophysiological recording. *PLoS One*, **4**, e6099.

Lobo, M. K., Covington, H. E., 3RD, Chaudhury, D., Friedman, A. K., Sun, H., DAMEZ-WERNO, D., Dietz, D. M., Zaman, S., Koo, J. W., Kennedy, P. J., Mouzon, E., Mogri, M., Neve, R. L., Deisseroth, K., Han, M. H. & Nestler, E. J. 2010. Cell type-specific loss of BDNF signaling mimics optogenetic control of cocaine reward. *Science*, **330**, 385–90.

Lobo, M. K. & Nestler, E. J. 2011. The striatal balancing act in drug addiction: distinct roles of direct and indirect pathway medium spiny neurons. *Front Neuroanat*, **5**, 41.

Lodge, D. J. & Grace, A. A. 2006. The laterodorsal tegmentum is essential for burst firing of ventral tegmental area dopamine neurons. *Proc Natl Acad Sci U S A*, **103**, 5167–72.

Lu, X. Y., Ghasemzadeh, M. B. & Kalivas, P. W. 1998. Expression of D1 receptor, D2 receptor, substance P and enkephalin messenger RNAs in the neurons projecting from the nucleus accumbens. *Neuroscience*, **82**, 767–80.

Ma, S. & Morilak, D. A. 2005. Norepinephrine release in medial amygdala facilitates activation of the hypothalamic-pituitary-adrenal axis in response to acute immobilisation stress. *J Neuroendocrinol*, **17**, 22–8.

Ma, Y. Y., Lee, B. R., Wang, X., Guo, C., Liu, L., Cui, R., Lan, Y., BALCITA-PEDICINO, J. J., Wolf, M. E., Sesack, S. R., Shaham, Y., Schluter, O. M., Huang, Y. H. & Dong, Y. 2014. Bidirectional modulation of incubation of cocaine craving by silent synapse-based remodeling of prefrontal cortex to accumbens projections. *Neuron*, **83**, 1453–67.

Maeda, H. & Mogenson, G. J. 1982. Effects of peripheral stimulation on the activity of neurons in the ventral tegmental area, substantia nigra and midbrain reticular formation of rats. *Brain Res Bull*, **8**, 7–14.

Mantz, J., Thierry, A. M. & Glowinski, J. 1989. Effect of noxious tail pinch on the discharge rate of mesocortical and mesolimbic dopamine neurons: selective activation of the mesocortical system. *Brain Res*, **476**, 377–81.

Matsumoto, M. & Hikosaka, O. 2007. Lateral habenula as a source of negative reward signals in dopamine neurons. *Nature*, **447**, 1111–5.

Matsumoto, M. & Hikosaka, O. 2009. Two types of dopamine neuron distinctly convey positive and negative motivational signals. *Nature*, **459**, 837–41.

Mcfarland, K., Lapish, C. C. & Kalivas, P. W. 2003. Prefrontal glutamate release into the core of the nucleus accumbens mediates cocaine-induced reinstatement of drug-seeking behavior. *J Neurosci*, **23**, 3531–7.

Mogenson, G. J., Jones, D. L. & Yim, C. Y. 1980. From motivation to action: functional interface between the limbic system and the motor system. *Prog Neurobiol*, **14**, 69–97.

Pascoli, V., Terrier, J., Espallergues, J., Valjent, E., O'CONNOR, E. C. & Luscher, C. 2014. Contrasting forms of cocaine-evoked plasticity control components of relapse. *Nature*, **509**, 459–64.

Peters, J., Vallone, J., Laurendi, K. & Kalivas, P. W. 2008. Opposing roles for the ventral prefrontal cortex and the basolateral amygdala on the spontaneous recovery of cocaine-seeking in rats. *Psychopharmacology (Berl)*, **197**, 319–26.

Redgrave, P., Prescott, T. J. & Gurney, K. 1999. Is the short-latency dopamine response too short to signal reward error? *Trends Neurosci*, **22**, 146–51.

Rescorla, R. A. & Wagner, A.R. 1972. *A Theory of Pavlovian Condition: Variations in the Effectiveness of Reinforcement and Nonreinforcement*. New York, Appleton-Century-Crofts.

Roitman, M. F., Stuber, G. D., Phillips, P. E., Wightman, R. M. & Carelli, R. M. 2004. Dopamine operates as a subsecond modulator of food seeking. *J Neurosci*, **24**, 1265–71.

Russo, S. J., Dietz, D. M., Dumitriu, D., Morrison, J. H., Malenka, R. C. & Nestler, E. J. 2010. The addicted synapse: mechanisms of synaptic and structural plasticity in nucleus accumbens. *Trends Neurosci*, **33**, 267–76.

Schultz, W. 1998. Predictive reward signal of dopamine neurons. *J Neurophysiol*, **80**, 1–27.

Schultz, W., Apicella, P. & Ljungberg, T. 1993. Responses of monkey dopamine neurons to reward and conditioned stimuli during successive steps of learning a delayed response task. *J Neurosci*, **13**, 900–13.

Schultz, W., Dayan, P. & Montague, P. R. 1997. A neural substrate of prediction and reward. *Science*, **275**, 1593–9.

Schultz, W. & Romo, R. 1987. Responses of nigrostriatal dopamine neurons to high-intensity somatosensory stimulation in the anesthetized monkey. *J Neurophysiol*, **57**, 201–17.

Setlow, B., Gallagher, M. & Holland, P. C. 2002. The basolateral complex of the amygdala is necessary for acquisition but not expression of CS motivational value in appetitive Pavlovian second-order conditioning. *Eur J Neurosci*, **15**, 1841–53.

Song, S. S., Kang, B. J., Wen, L., Lee, H. J., Sim, H. R., Kim, T. H., Yoon, S., Yoon, B. J., Augustine, G. J. & Baik, J. H. 2014. Optogenetics reveals a role for accumbal medium spiny neurons expressing dopamine D2 receptors in cocaine-induced behavioral sensitization. *Front Behav Neurosci*, **8**, 336.

Stamatakis, A. M. & Stuber, G. D. 2012. Activation of lateral habenula inputs to the ventral midbrain promotes behavioral avoidance. *Nat Neurosci*, **15**, 1105–7.

Stefanik, M. T. & Kalivas, P. W. 2013. Optogenetic dissection of basolateral amygdala projections during cue-induced reinstatement of cocaine seeking. *Front Behav Neurosci*, **7**, 213.

Stefanik, M. T., Kupchik, Y. M., Brown, R. M. & Kalivas, P. W. 2013a. Optogenetic evidence that pallidal projections, not nigral projections, from the nucleus accumbens core are necessary for reinstating cocaine seeking. *J Neurosci*, **33**, 13654–62.

Stefanik, M. T., Kupchik, Y. M. & Kalivas, P. W. 2016. Optogenetic inhibition of cortical afferents in the nucleus accumbens simultaneously prevents cue-induced transient synaptic potentiation and cocaine-seeking behavior. *Brain Struct Funct*, **221**, 1681–9.

Stefanik, M. T., Moussawi, K., Kupchik, Y. M., Smith, K. C., Miller, R. L., Huff, M. L., Deisseroth, K., Kalivas, P. W. & Lalumiere, R. T. 2013b. Optogenetic inhibition of cocaine seeking in rats. *Addict Biol*, **18**, 50–3.

Steinberg, E. E. & Janak, P. H. 2013. Establishing causality for dopamine in neural function and behavior with optogenetics. *Brain Res*, **1511**, 46–64.

Steinberg, E. E., Keiflin, R., Boivin, J. R., Witten, I. B., Deisseroth, K. & Janak, P. H. 2013. A causal link between prediction errors, dopamine neurons and learning. *Nat Neurosci*, **16**, 966–73.

Stuber, G. D., Hnasko, T. S., Britt, J. P., Edwards, R. H. & Bonci, A. 2010. Dopaminergic terminals in the nucleus accumbens but not the dorsal striatum corelease glutamate. *J Neurosci*, **30**, 8229–33.

Stuber, G. D., Sparta, D. R., Stamatakis, A. M., Van Leeuwen, W. A., Hardjoprajitno, J. E., Cho, S., Tye, K. M., Kempadoo, K. A., Zhang, F., Deisseroth, K. & Bonci, A. 2011. Excitatory transmission from the amygdala to nucleus accumbens facilitates reward seeking. *Nature*, **475**, 377–80.

Sun, W. & Rebec, G. V. 2003. Lidocaine inactivation of ventral subiculum attenuates cocaine-seeking behavior in rats. *J Neurosci*, **23**, 10258–64.

Suska, A., Lee, B. R., Huang, Y. H., Dong, Y. & Schluter, O. M. 2013. Selective presynaptic enhancement of the prefrontal cortex to nucleus accumbens pathway by cocaine. *Proc Natl Acad Sci U S A*, **110**, 713–8.

Sutton, R. S. 1988. Learning to predict by the methods of temporal differences. *Machine Learning*, **3**, 9–44.

Sutton, R. S. & Barto, A. G. 1998. *Reinforcement Learning: An Introduction*. Cambridge, MA: MIT Press.

Tan, K. R., Yvon, C., Turiault, M., Mirzabekov, J. J., Doehner, J., Labouebe, G., Deisseroth, K., Tye, K. M. & Luscher, C. 2012. GABA neurons of the VTA drive conditioned place aversion. *Neuron*, **73**, 1173–83.

Tanimoto, H., Heisenberg, M. & Gerber, B. 2004. Experimental psychology: event timing turns punishment to reward. *Nature*, **430**, 983.

Tsai, H. C., Zhang, F., Adamantidis, A., Stuber, G. D., Bonci, A., De Lecea, L. & Deisseroth, K. 2009. Phasic firing in dopaminergic neurons is sufficient for behavioral conditioning. *Science*, **324**, 1080–4.

Ungless, M. A., Argilli, E. & Bonci, A. 2010. Effects of stress and aversion on dopamine neurons: implications for addiction. *Neurosci Biobehav Rev*, **35**, 151–6.

Ungless, M. A. & Grace, A. A. 2012. Are you or aren't you? Challenges associated with physiologically identifying dopamine neurons. *Trends Neurosci*, **35**, 422–30.

Van Zessen, R., Phillips, J. L., Budygin, E. A. & Stuber, G. D. 2012. Activation of VTA GABA neurons disrupts reward consumption. *Neuron*, **73**, 1184–94.

Wise, R. A. 2004. Dopamine, learning and motivation. *Nat Rev Neurosci*, **5**, 483–94.

18 Optogenetics and the Dissection of Neural Circuits Underlying Depression and Substance-use Disorders

Barbara Juarez, Allyson K. Friedman, and Ming-Hu Han

18.1 Introduction to Reward-related Mood Disorders

Neuropsychiatric disorders encompass a range of diseases that create a huge burden on society. In the last 60 years, research into the biological basis of these disorders has progressed tremendously because of technological precision in studying cellular neurobiology, advances in the scale of genome-wide studies, refined animal models and the tuned dissection of neural circuits. Of particular interest is the study of the causes, mechanisms and treatments of major depressive disorder (MDD) and pathological substance-use disorders. The serendipitous discovery in the 1950s of tricyclic antidepressants has created a foundation for the generation of receptor- and transporter-based pharmacotherapeutics for MDD and substance-use disorders.

Of particular interest in the study of MDD and substance-use disorders are the neurons involved in the emotion- and reward-related midbrain dopaminergic system. This circuit consists of ventral tegmental area (VTA) dopamine neurons projecting to the nucleus accumbens (NAc), the medial prefrontal cortex (mPFC) and the amygdala, regions involved with encoding reward, modulating executive control and associating context and cues (Figure 18.1) (Hyman, 2007; Russo and Nestler, 2013). Neurons in these downstream regions are dopaminoceptive and express either dopamine receptor-1 (D1) neurons or dopamine receptor-2 (D2) neurons. Many pharmacotherapeutics that are used to treat MDD and substance-use disorders, such as monoamine oxidase inhibitors (MAOIs), selective serotonin and norepinephrine reuptake inhibitors and antagonists of opioid receptors, have been found to have robust actions on neural substrates in the dopaminergic reward system (Berton and Nestler, 2006). However, with the success of deep brain stimulation in the treatment of MDD and substance-use disorders (a technique that uses chronically implanted stimulating electrodes to specifically target brain regions in the reward system for electrical stimulation), there is an ever-evolving hypothesis that neuropsychiatric disorders are neural circuit disorders (Lobo *et al.*, 2012; Volkow and Koob, 2015). Thus, the importance of identifying how neural

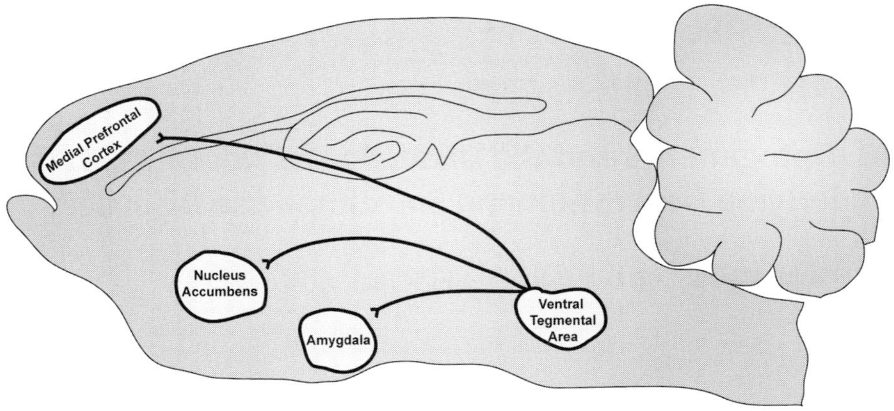

Figure 18.1 Neural circuit schematic of the rodent midbrain dopaminergic reward system. The emotion- and reward-related midbrain dopaminergic system is composed of VTA dopamine neurons projecting to the mPFC, the NAc and the amygdala.

substrates of this circuit are pathologically altered in MDD and substance-use disorders is critical.

Optogenetics has granted researchers the unique ability to selectively target genetically distinct cell populations or neural circuits for temporally precise stimulation or inhibition in awake and behaving animal models. This has shed light on the neural circuit mechanisms involved in depression and substance abuse. Optogenetics, in conjunction with validated animal models for neuropsychiatric disorders, has dramatically altered how researchers approach and interpret the study of depression and substance-use disorders.

18.2 Identifying Pathological Adaptations of the Dopaminergic Reward System to Reveal Causal Mechanisms of Neuropsychiatric Disorders

18.2.1 Animal Models of Neuropsychiatric Disorders

During the development of animal models for human neuropsychiatric disorders, researchers try to use a framework of construct validity, face validity and predictive validity to judge the effectiveness of the model (Nestler and Hyman, 2010). The judgment of construct validity attempts to recreate the scenario leading to the emergence of a disorder in order to model it in animals. Face validity of animal models for disorders refers to the model exhibiting symptoms that are similar to the human state. Predictive validity refers to the ability of the animal model to respond to known treatments employed in humans. The criteria for which these three validations are met vary across a number of animal models of neuropsychiatric disorders, particularly because of the complexity of causes, symptoms and responsiveness to treatments seen across depression and substance-use disorders. Thus, when interpreting results from such studies, one must be aware of the implications, particularly when set against the model that the investigators utilized.

Table 18.1 Criteria for MDD and substance use disorder (DSM-5)[a].

MDD		Substance-use disorder	
Five or more of the symptoms below for at least 2 weeks		The presence of at least two of these symptoms indicates a substance-use disorder	
One of the symptoms must be depressed mood or anhedonia		The severity is defined as: mild (2–3 symptoms), moderate (4–5 symptoms) or severe (≥6 symptoms)	
Depressed mood	Loss of interest or pleasure (anhedonia)	Longer than intended drug taking	Inability to reduce or stop drug taking
Change in weight (5%) or daily appetite	Insomnia or hypersomnia	Drug taking causes reckless behavior	Constant thinking about or craving for drug
Loss of energy or fatigue	Psychomotor retardation or agitation	Drug taking interferes with responsibilities	Abandonment of activities in order to take drug
Feelings of worthlessness or guilt	Impaired concentration or indecisiveness	Degradation of interpersonal relationships because of drug taking	Degradation of interpersonal relationships because of drug taking
Thoughts of death or suicidal tendencies or attempts		Drug taking despite health issues	Drug taking uses up a lot of time
		Tolerance to drug	Withdrawal from drug

[a] *Criteria for both disorders derived from the Diagnostic and Statistical Manual of mental disorders-5 (DSM-5).*

18.2.2 Studying MDD in a Rodent Model

MDD afflicts 2–5% of the population and is the second leading cause of disability worldwide according to the Worldwide Health Organization (Nestler *et al.*, 2002; Ferrari *et al.*, 2013). MDD is a multifaceted disorder with a multitude of symptoms and levels of severity that varies across individuals (Table 18.1).

In the 1950s, two separate research groups serendipitously discovered two classes of centrally acting antidepressants, the tricyclics and the MAOIs (Schildkraut, 1965; Coppen, 1967). These discoveries have since helped researchers hone in on the specific neural substrates that are affected by these drugs, such as the regions involved in the midbrain dopaminergic reward system. Additionally, the development of modern antidepressants, such as inhibitors of neurotransmitters, have more targeted actions on these systems.

A challenge in identifying and modeling the causes of depression is the absence of consistent genetic mutations across patients (Nestler and Hyman, 2010). Thus, the main focus on creating animal models of depression had relied on the development of models with high predictive validity. Two classic paradigms used in the study of antidepressants in mice are the forced swim test (FST) and the tail suspension test (TST) (Figure 18.2A) (Nestler and Hyman, 2010). These tests are used in the screening of drugs for antidepressant-like effects and rely on the mouse transitioning from a passive to an active state. It was found that when investigators administered classic antidepressants to a mouse before placing them in a beaker of water (FST) or suspending them by their tail (TST), the mouse swam or struggled for a longer amount of time when compared to control, respectively.

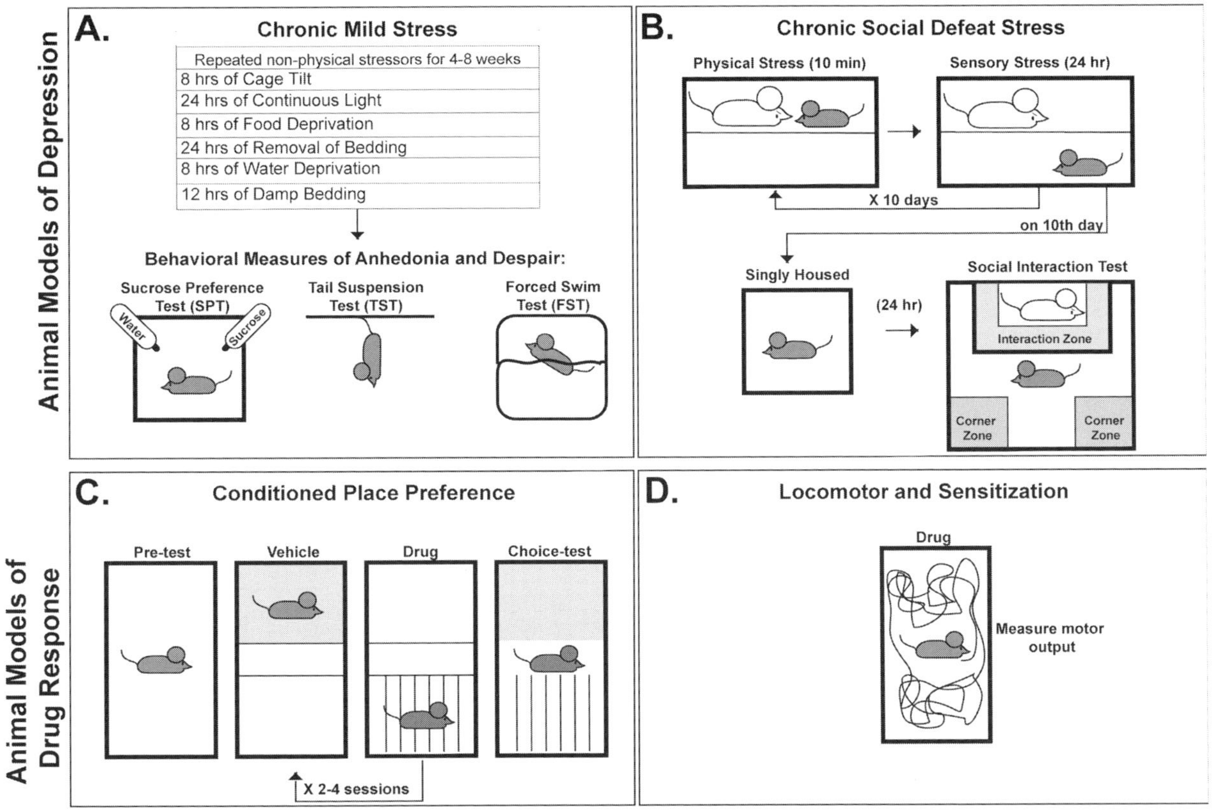

Figure 18.2 Animal models for studying MDD and substance-use disorders. (A) Chronic mild stress involves submitting a rodent to a series of daily mild stressors for 4–8 weeks. Following this paradigm, mice exhibit low sucrose preference (anhedonia) and increased immobility in the TST and the FST. (B) Chronic social defeat stress involves the daily physical and sensory subordination of a small, C57BL/6J test mouse to a larger CD1 aggressor mouse for 10 days. On the last day, the test mouse is singly housed for 24 hours and then undergoes a social interaction test with a novel CD1 aggressor in an arena with designated zones. Stress-nave and resilient mice show increased time spent in the interaction zone, but susceptible mice show decreased time spent in the interaction zone. (C) Conditioned place preference involves multiple rounds of conditioning a rodent to vehicle or drug in chambers with specific cues. On the last day, and with no drug on board, the mouse performs a choice test. Drugs of abuse often cause increased time to be spent in the drug-paired chamber. (D) Locomotor tests measure the motor output and sensitization that repeated drug administrations can induce.

This decrease in immobility has been interpreted as a decrease in "behavioral despair" phenotypes. However, this paradigm is not considered a representative model of depression. Currently, researchers have developed animal models that have high construct validity with representative depressive-like symptoms measured across a number of behavioral assays. Here, we will focus on two very different animal models for depression. Later on, we will discuss how using optogenetics not only identified the neural substrates that are causally involved in the expression of depression-like behaviors, but also how optogenetics elucidated the diverse nature of the expression of depression-like behaviors even further.

The chronic mild stress (CMS) model employs an array of mild physically stressful events to induce depressive-like phenotypes in mice, and this approach has strong construct, face and predictive validity (Nestler and Hyman, 2010). CMS involves a series of daily unpredictable stressors lasting for between 4 and 8 weeks (Figure 18.2A) (Hill *et al.*, 2013; Tye *et al.*, 2013). After CMS, mice exhibit anhedonia, as measured by decreased sucrose preference. Additionally, the mice also display increased immobility in the FST and the TST. Interestingly, the depressive-like symptoms of CMS can be reversed with chronic antidepressant treatments, showing great predictive validity (Nestler and Hyman, 2010).

Another animal model of MDD that is often employed in research is the chronic social defeat stress (CSDS) paradigm (Berton *et al.*, 2006; Krishnan *et al.*, 2007; Krishnan and Nestler, 2008; Nestler and Hyman, 2010; Golden *et al.*, 2011). This is thought to be a more severe form of stress, exposing the smaller black "test" mouse to repeated bouts of subordination. CSDS involves exposing a C57BL/6J subordinate mouse to 10-minute bouts of physical stressor from a larger, retired breeder aggressor CD1 white mouse for 10 days (Figure 18.2B). Immediately after the 10-minute bout of physical stress, the subordinate mouse is separated from the aggressor by a perforated partition to allow for 24 hours of psychological stress (Figure 18.2B). The process is then repeated on the next day with a different aggressor CD1 mouse. After 10 days of repeated physical and sensory defeat, the subordinate mouse is singly housed for 24 hours (Figure 18.2B). One day later, the subordinate test mouse undergoes a social interaction test in which a novel CD1 aggressor white mouse is placed in an interaction zone behind a mesh barrier (Figure 18.2B). The time that the subordinate test mouse spends in the interaction zone with the novel CD1 aggressor mouse and the time that it spends in the corner zones are measured. CSDS often causes social isolation (decreased time in the interaction zone and increased time in the corned zones) in the subordinate mouse, a phenotype that is described as "susceptible". However, a unique quality about CSDS is the emergence of a behavioral phenotype in some mice that maintain high social interaction (increased time in the interaction zone and decreased time in the corner zones), despite having undergone the same stressors as the susceptible mouse. These mice are deemed "resilient". Along with social isolation, susceptible mice also display anhedonia and a variety of metabolic alterations (Krishnan *et al.*, 2007).

18.2.3 Studying the Effects of Drugs of Abuse in a Rodent Model

Substance-use disorder refers to the pathological consumption of, and eventual dependence on, a drug that can lead to major mental and physical impairments (Hyman *et al.*, 2006). Substance-use disorder can range from mild to severe based on the presence of a number of symptoms (Table 18.1). Drugs of abuse have psychoactive effects that are highly addictive. The progression into an addicted state is marked by repeated transitions through a three-stage cycle that recruits specific neural circuits (Volkow and Morales, 2015). The first is the binge or intoxication phase. Here, the reinforcing effects of drugs of abuse cause an increase of dopamine release from the VTA onto the NAc, a key regulator of reward processing, as well as other brain regions in the dopamine reward system. The next stage is known as the withdrawal phase. This stage is marked by the expression of a negative emotional state through the recruitment of the extended amygdala neural circuit. The third stage is known as the preoccupation stage. This stage involves the activation of cortical structures and the temporal lobe, which are involved in executive control, and contextual information that cues the subject to the rewarding aspects of drug use and leads to craving. This stage often leads the subject back into the binge phase, and the cycle repeats. Because of the many studies analyzing human brains involved in different stages of this cycle, the idea of addiction as a neural circuit disorder is widely recognized.

There are a number of animal models for the study of substance-use disorders that uniquely capture certain stages of the drug abuse cycle. To study the initial rewarding effects of a drug, researchers often employ the conditioned place preference (CPP) paradigm in rodents. This paradigm utilizes the nature of Pavlovian learning and conditioning to measure a rodent's motivation to seek an environment where a drug was presented before (Cunningham *et al.*, 2006). Briefly, in a two-chambered arena that has a middle zone, the mouse is administered the drug of abuse in one chamber with distinguishing contextual cues (stripped walls and gridded floors). During the next session, the mouse is administered a vehicle in the second chamber, which has different contextual cues (gray walls and mesh floors). This conditioning is repeated a number of times. On the test day, the mouse, without any drug administered, is allowed to freely choose which chamber to spend its time in during a choice test (Figure 18.2C). A preference for spending time in the drug-paired chamber is interpreted as increased motivation for the drug due to increased reward signaling. While this paradigm offers a chance to investigate the rewarding mechanisms of a drug of abuse, a disadvantage of this paradigm is the fact that the mouse receives the drug passively, instead of self-administering.

In addition to observing the mouse's inclination to seek an environment in which it received a drug of abuse, researchers can also measure a rodent's response to a drug, particularly stimulants such as cocaine, through a locomotor paradigm (Lobo *et al.*, 2010; Walker *et al.*, 2015). Here, the mouse is given a drug of abuse and its locomotion is measured in a box with specialized sensors (Figure 18.2D). Higher locomotor activity signifies higher response. With stimulants, this paradigm can also elucidate sensitization to the drug after repeated administrations, which often occurs through specific neuroadaptations that contribute to multiple stages of the drug abuse cycle (Robinson and Berridge, 2003; Robinson and Berridge, 2008).

18.3 Optogenetics Can Target Cell Type-specific Populations in the Reward System

The dopaminergic reward system consists of VTA dopamine neurons projecting to the NAc, mPFC and the amygdala. Many classical studies have shown that intracranial electrical self-stimulation of these regions modulates reward (Carlezon and Chartoff, 2007). However, these neural substrates of the reward system consist of heterogeneous neuronal populations that, with electrical stimulation, cannot be dissociated. The VTA includes projecting dopamine cells, inhibitory GABA interneurons and projecting cells and excitatory glutamate projecting cells (Walsh and Han, 2014). The NAc consists of projecting dopaminoceptive medium spiny neurons (MSNs) and an array of GABAergic and acetylcholine interneurons (Shirayama and Chaki, 2006; Russo and Nestler, 2013). By employing a combination of transgenic mouse lines and viral-mediated gene transfer, researchers have been able to target specific cell populations for manipulation with optogenetics.

To target the expression of a viral light-activated opsin to a specific neuronal population, many laboratories have used transgenic mouse lines that express cre recombinase (Cre) in a promoter that is genetically unique to a neuronal subtype (Han and Friedman, 2011). For instance, dopamine cells express tyrosine hydroxylase (TH), a rate-limiting enzyme in the production of dopamine. By having Cre expressed under the TH promoter (TH-Cre), researchers can target Cre-driven viral vectors to TH-expressing cells uniquely (Gong *et al.*, 2007).

Light-activated opsins can be packaged in different types of viruses including adeno-associated viral (AAV) vectors, which allow for long-term expression, and herpes simplex viral vectors, which allow for faster and short-term expression (Neve and Lim, 2013). To allow for specific cell expression in a promoter-driven Cre mouse line, researchers flank the gene with Cre-inducible lox sites (DIO or LS1L) that allow for the transcription of the gene of interest only in the presence of Cre (Han and Friedman, 2011).

18.4 Optogenetics Can Find Causal Links in Pathological Firing Adaptations in Neuropsychiatric Disorders

As researchers view MDD and substance-use disorder as neural circuit diseases, many investigations have focused on determining how brain regions are physiologically altered. Neuronal firing can be measured using *in vivo* and *in vitro* electrophysiological techniques. Using *in vivo* techniques can allow the researchers to measure neural activity with the advantage of having intact synaptic and circuit inputs. Using *in vitro* techniques while dissociating these inputs allows researchers to measure the neural activity of spontaneously active cells and to measure changes in ion channel function and synaptic strength. When combined with animal models for neuropsychiatric disorders, electrophysiology can help researchers pinpoint pathological firing adaptations that are correlated with behavior. Because of the role that the dopaminergic reward system plays in MDD and substance-use disorders, determining alterations of activity in neural substrates in animal models has been of particular interest.

Changes in neuronal firing in VTA dopamine neurons have been linked to regulation of reward prediction errors, encoding valence of stimuli and creating incentive salience due to downstream changes in dopamine release (Robinson and Berridge, 2003; Grace *et al.*, 2007; Schultz, 2015). VTA dopamine neurons display and transition between two types of firing patterns *in vivo* – a slow, single-spike tonic pattern and a fast, multi-spike phasic firing pattern (Grace and Bunney, 1984a; Grace and Bunney, 1984b). It has been shown that phasic firing increases dopamine release in downstream targets (Di Chiara and Imperato, 1988; Wise and Rompre, 1989). This region has been of particular interest when trying to determine pathological firing states in neuropsychiatric disorders.

18.4.1 Major Depressive Disorder

With regards to MDD, there have been a number of electrophysiological investigations of VTA dopamine neurons performed in mice that have undergone the CSDS model (Krishnan *et al.*, 2007; Cao *et al.*, 2010; Chaudhury *et al.*, 2013; Friedman *et al.*, 2014). Because of the CSDS's ability to parse out two populations of mice (those susceptible to defeat and those resilient to defeat), this has made it an ideal model for comparing neurophysiological alterations of the reward system. Researchers first identified differences in neuronal firing between susceptible and resilient mice using an *in vitro* approach. It was demonstrated that VTA dopamine neurons of susceptible mice had higher firing activities when compared to stress-naïve mice in a preparation of VTA brain slices. Interestingly, resilient mice seemed to have undergone no changes in firing, displaying firing activities that were similar to stress-naïve mice (Krishnan *et al.*, 2007). However, VTA dopamine neurons display firing patterns *in vivo* that cannot be measured in an *in vitro*, acute brain slice preparation. When researchers recorded the *in vivo* activity of VTA dopamine neurons from anesthetized mice, they discovered that mice that were susceptible to defeat displayed increased firing rates and bursting activities when compared to resilient mice and stress-naïve mice. These activity patterns were inversely correlated with social interaction behavior (Cao *et al.*, 2010). These findings together indicated a possible role for VTA dopamine neurons in the regulation of social behavior after stress.

In order to causally link the above findings to a depressive state, researchers employed *in vivo* optogenetics in order to mimic the firing adaptations observed in a physiologically relevant manner. To specifically target the dopamine cells of the VTA, researchers used a TH-Cre mouse line and injected a Cre-inducible AAV-DIO-ChR2-eYFP into the VTA (Chaudhury *et al.*, 2013). It was demonstrated that increasing VTA dopaminergic activity during a social interaction test by using stimulation patterns that mirrored what was observed *in vivo* rapidly induced antisocial behaviors and anhedonia in mice that had either received a subthreshold defeat or mice that had been originally deemed resilient after the 10-day CSDS. Additionally, decreasing the pathological increase in VTA dopaminergic firing by using a Cre-inducible halorhodopsin (AAV-DIO-NpHR3.0-eYFP) during a social interaction test effectively rescued social interaction behaviors. These studies were the first to show a causal link for physiologically relevant VTA dopamine neuron activity in the regulation of depressive-like phenotypes following CSDS.

18.4.2 Opiate Reward

Opiates have indirect mechanisms that increase the neural activity of VTA dopamine neurons and cause increased release of dopamine in the NAc (Di Chiara and Imperato, 1988). They are known to increase dopaminergic release in the NAc by inhibiting local GABA neurons in the VTA, thus disinhibiting VTA dopamine neurons (Gysling and Wang, 1983). Mice will spend an increased amount of time in a morphine-paired chamber during CPP (Olmstead and Franklin, 1997). Interestingly, chronic morphine, administered either through subcutaneous pellets or through intraperitoneal injections, downregulates brain-derived neurotrophic factor (BDNF) in the VTA (Koo *et al.*, 2012). Infusing BDNF into the VTA during morphine conditioning significantly reduced the preference for the morphine-paired chamber. Chronic morphine was observed to increase *in vivo* VTA dopaminergic firing activity and bursting activity. When mice are administered chronic morphine with BDNF infusions into the VTA, their dopaminergic activity returns to baseline. This also decreases preference for the drug-paired chamber in CPP. These findings suggested a role for BDNF as a negative modulator of VTA dopaminergic activity in mediating the rewarding effects of morphine (Koo *et al.*, 2012).

To determine the interplay between dopaminergic firing and BDNF signaling in mediating morphine reward, the same investigators stimulated VTA dopaminergic ChR2-containing terminals in the NAc in a physiologically relevant manner (as determined by the *in vivo* work) during CPP conditioning with sub-rewarding doses of morphine (Koo *et al.*, 2012). This increase in dopamine activity in the NAc caused a significant induction of morphine CPP. This stimulation was also observed to ablate the behavioral effects of BDNF infusions during CPP, increasing the rewarding aspects of morphine in those treated animals. This optical induction of CPP via stimulation of dopaminergic terminals was determined to be dependent on dopamine signaling through D1 receptors. By first identifying what the firing adaptations were *in vivo* after chronic morphine and with BDNF, and by applying them in a behavioral model, these investigators parsed out an intricate role of BDNF in the VTA as a negative modulator of morphine reward.

18.5 Optogenetics Has Revealed Diverse Actions of Dopamine Neurons and Dopaminoceptive Neurons in Neuropsychiatric Disorders

The ability to target specific populations of cells using optogenetics to mimic pathological firing adaptations *in vivo* has provided insight into the mechanisms of neuropsychiatric disorders. Additionally, optogenetics has shed light on the functional diversity of neuronal populations that display similar genetic makeup. For instance, dopamine neurons in the VTA express the TH promoter. However, there have been reports of heterogeneous roles for VTA dopamine neurons that project to certain areas of the reward circuit (Lammel *et al.*, 2008; Margolis *et al.*, 2008). With the advances in optogenetic tracing techniques, researchers have been able to further investigate the functions of projection-specific VTA dopaminergic neuronal populations. These tools have also been used to dissect the functional diversity of different dopaminoceptive neurons in the NAc during drug reward.

18.5.1 Heterogeneous Dopamine Neuronal Populations Have Projection-specific Neuroadaptations Following CSDS

There have been a number of investigations noting differences in the ion channel compositions and physiological characteristics of the dopamine neurons of the VTA (Ford *et al.*, 2006; Lammel *et al.*, 2008; Margolis *et al.*, 2008). Thus, when researchers first found that mice susceptible to CSDS had an increase in the *in vivo* firing properties of VTA dopamine neurons, they turned to an *ex vivo*, retrograde fluorescent bead investigation in order to determine whether this characteristic was attributable to all VTA dopamine projections, or just a subpopulation (Chaudhury *et al.*, 2013). Indeed, when they compared the firing properties of VTA neurons that projected to the NAc (VTA–NAc) with those neurons that project to the mPFC (VTA–mPFC), they discovered that there were drastic differences in the functional properties following CSDS. Susceptible mice displayed increased firing rates of the VTA–NAc neurons when compared to stress-naïve and resilient mice. However, in the VTA–mPFC subpopulation, susceptible mice had decreased firing activity when compared to stress-naïve and resilient mice.

Having identified these projection-specific neuronal alterations of VTA dopamine neurons, the investigators then turned to circuit-dissecting optogenetic techniques in order to specifically manipulate these dopaminergic subpopulations. To specifically isolate a pathway, these investigators used a combination of retrograding viruses that had a Cre recombinase vector injected into the target of interest (NAc or mPFC) and Cre-inducible AAV-DIO-ChR2-eYFP or AAV-DIO-NpHR-eYFP injected into the VTA (Chaudhury *et al.*, 2013). Thus, only those neuronal terminals from the injected target of interest that had retrograded the viral Cre vector could express the light-activated opsin. Mimicking the hyperactivity of the VTA–NAc pathway in mice that had undergone a subthreshold defeat induced social avoidance behaviors and decreased sucrose preference. Additionally, decreasing the activity of the VTA–NAc pathway in susceptible mice induced resilience. Interestingly, mimicking the hypoactivity in the VTA–mPFC pathway only decreased social avoidance, and had no effects on sucrose preference. These projection-specific manipulations thus highlighted unique neuronal substrates involved in different aspects of depressive-like behaviors.

18.5.2 Dopaminoceptive Neuronal Populations Encode the Rewarding Properties of Drugs of Abuse Differently

In the field of drug abuse, many investigators have sought to understand the roles that the two types of dopaminoceptive neurons of the NAc have in reward processing. The projecting MSNs in the NAc express either D1 or D2 receptors almost exclusively. These subpopulations of dopaminoceptive MSNs send distinct projections throughout the basal ganglia circuit. While it was generally known that dopaminoceptive neurons of the NAc were involved in the rewarding aspects of a drug, little was known of the contributions the distinct subpopulations regulated. Using transgenic mouse lines that express Cre recombinase using either the D1 promoter or the D2 promoter has allowed researchers to record or manipulate these subpopulations in the NAc in the face of drugs of abuse.

Knocking down BDNF's receptor TrkB in D1 or D2 MSNs of the NAc differentially modulated the rewarding properties of cocaine by causing increased or decreased CPP performance, respectively (Lobo *et al.*, 2010). This knockdown of TrkB in D1 MSNs of the NAc also increased locomotor activity and sensitization to cocaine, while the knockdown in D2 MSNs decreased locomotor activity and produced no sensitization. These investigators discovered that knocking down TrkB in D2 MSNs of the NAc caused increased neurophysiological excitability of these neurons, while there was only a trend toward increased excitability in D1 MSNs of the NAc with the TrkB knockdown. Now, with an understanding of the neurophysiological basis of BDNF-TrkB regulation of altered reward in these neuronal subtypes, the investigators then selectively expressed ChR2 in either D1 or D2 MSNs of the NAc. Selectively stimulating D1 MSNs during cocaine (sub-rewarding dose) conditioning in the CPP paradigm caused significant preferences for the cocaine-paired chamber during the choice test. In contrast, stimulating D2 MSNs selectively during cocaine (rewarding dose) conditioning in the CPP paradigm caused significant decreases in preference for the cocaine-paired chamber during the choice test. Both of these neuronal manipulations caused molecular changes in pathways that are known to be involved in reward. Therefore, this study revealed cell-specific roles for dopaminoceptive neurons in the rewarding properties of cocaine.

18.5.3 How Circuit-specific Optogenetics Elucidated Diverse Forms of Depression

Optogenetics has also elucidated possible neurophysiological mechanisms underlying the diversity of depressive-like symptoms. Simultaneous to the investigations discussed in Sections 18.4.1 and 18.5.1 that revealed that hyperactivity of VTA–NAc neurons leads to susceptibility to CSDS, another group discovered that hypoactivity of this pathway leads to a different set of depressive-like phenotypes during behavioral measures of despair and anhedonia (Tye *et al.*, 2013). Using Cre-inducible NpHR to express specifically in VTA dopamine neurons of TH-Cre mice, these investigators inhibited dopaminergic activity in VTA–NAc terminals. Inhibition of VTA–NAc terminals during TSTs increased time spent immobile (more despair). Additionally, inhibiting this pathway also decreased sucrose preference (anhedonia). These researchers then investigated how increasing VTA–NAc activity after CMS would affect performance in these measures of despair and anhedonia. Mice that had undergone CMS displayed depressive-like phenotypes in TSTs, FSTs and sucrose preference tests.

However, optogenetically increasing the VTA–NAc pathway in CMS-treated mice relieved these behaviors. These two contrasting optogenetic studies of MDD revealed complex interactions in which different forms of stress and neurophysiological alterations encode depressive-like behaviors.

18.6 Optogenetics and the Importance of Context Specificity

As mentioned throughout this chapter, optogenetic manipulation of neuronal firing is a powerful tool for linking neural circuit activity to disease states. In addition to firing activity, the brain contains a tremendous amount of information regarding the state of the neural system. This information is encoded by neuropeptides, hormones and growth factors, providing a context regarding previous experience and behavioral state. This complex overlay of information is crucial when utilizing optogenetics in order to probe a neural system so as to determine causal links between neural activity and behavioral outcomes.

The importance of neural state, induced by experience or substance context, is clearly demonstrated by the diverse actions of the dopamine system. Within the mesolimbic dopaminergic reward circuit, a multi-layer neuromodulatory system enables dopamine neurons to respond to both rewarding and aversive stimuli, and can mediate divergent behavioral outputs based upon the specificity of the context (Koob *et al.*, 2004; Grace *et al.*, 2007; Krishnan *et al.*, 2007; Cao *et al.*, 2010; Lammel *et al.*, 2011; Tye *et al.*, 2013; Wanat *et al.*, 2013).

One context that is often comorbid with mood and substance-use disorders is stress. Stress has been shown both to reduce and to upregulate BDNF. It has been shown to be reduced in the hippocampus, while simultaneously upregulated by CSDS in the mesolimbic dopaminergic reward pathway. Specifically, it is increased in VTA neurons projecting to the NAc (Grace *et al.*, 2007; Lobo *et al.*, 2010; Schultz, 2010; Lammel *et al.*, 2011). BDNF deletion in this pathway alters the same genes and signaling cascades that are regulated by antidepressants, suggesting a role for BDNF in mood disorders (Berton *et al.*, 2006). Building on this work, BDNF was shown to play a key role in determining susceptibility versus resilience to social defeat-induced depression (Krishnan *et al.*, 2007). To more deeply understand the importance of the regulation of BDNF in this pathway, future studies should utilize targeted optogenetics.

As mentioned previously, increased phasic VTA dopaminergic activity was found to be causal of susceptibility to CSDS. Optogenetically mimicking this phasic firing in the VTA–NAc pathway of mice that had undergone a subthreshold defeat induced an increase in BDNF in the NAc and depressive-like behaviors, including social avoidance (Chaudhury *et al.*, 2013). This was surprising because it was thought that increased VTA–NAc activity was associated with reward. The question that arose is how phasic firing patterns and BDNF regulation could encode context specificity and induce a depressive state instead of encoding a rewarding signal. Utilizing targeted projection and cell type optogenetic manipulations, Walsh *et al.* (2014) tackled this question by performing a series of optogenetic stimulations in parallel with blocking established stress signals in this pathway.

To examine whether the phasic stimulation-induced increases in BDNF levels in NAc were contextually dependent, Walsh *et al.* (2014) performed phasic optical stimulation of the VTA–NAc pathway in stress-naïve mice. They found no change in behavioral phenotype, as well as no up-regulation of BDNF signals. This demonstrated that phasic optical stimulation-induced increases in BDNF in the NAc were contextually dependent on having been stressed, and that a context signal combined with phasic firing of mesolimbic DA neurons was critical in order for behaviorally relevant adaptations to occur. This work demonstrates the importance of context when optically manipulating neuronal firing, as well as the importance of understanding the molecular signals that establish different neural states.

This work raised the question of what the stress signal was within this VTA–NAc pathway. Stress is known to elevate a series of stress signals in the brain, including neurotransmitters, peptides and steroidal hormones. Corticotrophin-releasing factor (CRF), a neuropeptide that is released in response to acute stressors and arousing stimuli, has been shown to be active in the NAc and to increase during motivation for cued rewards (Lemos *et al.*, 2012). In addition, its actions have been shown to regulate dopamine release in the VTA.

To determine whether CRF was the necessary signal for encoding the stress context, Lemos *et al.* (2012) used the optical phasic stimulation of the VTA–NAc pathway in combination with pharmacological infusions of a CRF antagonist in order to block the stress signal. By using a subthreshold stress paradigm, these researchers infused a CRF receptor antagonist in the NAc prior to the phasic stimulation. They found that with the antagonist on board, the phasic firing of the VTA–NAc pathway no longer induced a depressive-like phenotype. Furthermore, these animals did not demonstrate an increase in BDNF in the NAc. This work demonstrated the necessity of a context signal for upregulating relevant molecular signals in response to optical stimulation. Phasic firing of VTA dopamine neurons alone is not sufficient to induce BDNF regulation in this circuit, and further signals within the mesolimbic circuit play a role in context detection.

Phasic firing of VTA dopamine neurons also plays an essential role in mediating reward behaviors (Tsai *et al.*, 2009; Koo *et al.*, 2012). The release of BDNF from neuronal processes depends on neural activity (Matsuda *et al.*, 2009; Park and Poo,

2013) and is a key regulator in promoting the action of drugs of abuse such as cocaine. By utilizing optogenetics, Koo *et al.* (2012) determined that morphine-induced BDNF release and action on VTA dopamine neurons are important for BDNF's influence on behavioral responses to morphine. This study directly links the importance of the morphine drug context, VTA BDNF expression and VTA dopamine neuron excitability in morphine reward. This further demonstrates the context-dependent flexibility of the dopamine system for encoding either stress or reward depending on the molecular signals.

18.7 Intrinsic and Extrinsic Plasticity Mechanisms Underlying the Manifestation of Depression

Optogenetics enables less invasive optical control of specific neuronal types and projections in order to examine firing activity-induced plasticity in freely moving animals. Utilizing this precise and targeted tool allows for the further understanding of both the intrinsic and extrinsic circuit plasticity involved in depressive-susceptible and -resilient phenotypes.

18.7.1 The Role of Homeostatic Plasticity in Mediating Resilience

To explore the neurophysiological mechanisms of resilience following CSDS, Friedman *et al.* (2014) utilized optogenetics in order to induce a natural form of plasticity in the VTA–NAc pathway that they observed in animals that were resilient to CSDS. Following CSDS, increases in firing rates and bursting events were found in the VTA dopamine neurons in susceptible mice, but not in the resilient subgroup (Krishnan *et al.*, 2007; Cao *et al.*, 2010). Using phasic optical stimulation, it was shown that this hyperactivity was causally linked to the susceptible phenotype (Chaudhury *et al.*, 2013; Walsh *et al.*, 2014). Underlying the VTA dopamine neuron hyperactivity of susceptible mice was an up-regulation of a hyperpolarization-activated current (I_h), which is a key current in the up-regulation of firing activity (Neuhoff *et al.*, 2002; Arencibia-Albite *et al.*, 2007). Surprisingly, despite resilient mice exhibiting stable, normal firing of these neurons, resilient mice had an even larger I_h. This increase in excitatory current was observed in parallel with increased inhibitory potassium channel currents. In order to test the hypothesis that the inhibitory potassium channel up-regulation in the VTA dopamine neurons occurred in response to stress-induced hyperactivity, chronic phasic optical stimulation of the VTA–NAc pathway in susceptible mice was performed. It was found that optogenetically increasing the hyperactivity of VTA dopamine neurons in susceptible mice completely reversed depression-related behaviors, an antidepressant effect that was achieved through resilience-like homeostatic plasticity. This demonstrated that resilient animals homeostatically maintain healthy dopamine neuron activity through compensatory up-regulation of potassium channels in response to excess hyperactivity. This work utilized targeted optogenetic manipulation of the VTA–NAc pathway and revealed the intrinsic plasticity-induced up-regulation in potassium currents in VTA dopamine neurons. Importantly, these authors found that this induced plasticity is sufficient to reduce the stress-induced hyperactivity and pathogenic firing, introducing a novel therapeutic strategy for depression (Friedman *et al.*, 2014).

18.7.2 Extrinsic Adaptations of Diverse Glutamatergic Synapses that Mediate Susceptibility or Resilience to CSDS

Chronic stress-induced depression produces a variety of pathway-specific alterations that can be explored in order to understand depressive-like behaviors. Since glutamatergic transmission in medial NAc is increased in mice exhibiting depression-like behaviors after CSDS, investigations into these pathways have provided information regarding the extrinsic plasticity of these pathways (Vialou *et al.*, 2010; Christoffel *et al.*, 2011). The NAc receives glutamatergic inputs from the mPFC, ventral hippocampus (vHIP) and basolateral amygdala, substrates that are involved in a depressive phenotype. Bagot *et al.* investigated whether there was differential regulation of the glutamatergic projections from the vHIP, mPFC and amygdala following CSDS (Bagot *et al.*, 2014). They examined immediate early gene expression as an indicator of neuronal activity and electrophysiological correlates of presynaptic neurotransmitter release and found afferent-specific CSDS-induced adaptations in vHIP and mPFC that were specific to susceptibility or resilience, respectively. Using optogenetic manipulations in order to bidirectionally control afferent-specific synaptic function, they demonstrated a unique role for vHIP–NAc in driving depressive-like behavioral phenotypes. They found that that vHIP–NAc synaptic transmission is selectively increased in mice that are susceptible to CSDS. To determine whether decreasing glutamatergic synaptic transmission at vHIP–NAc synapses could promote resilience in previously defeated mice, they utilized a optogenetic low-frequency stimulation (LFS; 473 nm, 10 minutes, 1 Hz, 4ms pulse width) of ChR2-infected terminals in NAc in order to induce long-term depression (LTD) of MSN excitatory postsynaptic currents. By utilizing projection-specific *in vivo* optogenetic manipulations, they demonstrated that reducing vHIP–NAc synaptic transmission with LFS 45 minutes before social interaction testing could reverse the susceptible phenotype. In contrast, they found that mPFC–NAc synaptic transmission to be decreased in susceptible animals, and they showed that increasing either mPFC or amygdala synaptic transmission by acute stimulation *in vivo* promotes resilience. These findings highlight the diversity of pathways to be explored when examining the complex etiologies of psychiatric disorders and the advantage of being able to induce plasticity, such as LTD, in a freely moving animal.

In further examinations into the postsynaptic remodeling of glutamatergic synapses, Christoffel *et al.* found that following chronic social defeat, susceptible mice demonstrated increased synaptic strength at intralaminar thalamus (ILT), but not prefrontal cortex (PFC) inputs to MSNs of the NAc (Christoffel *et al.*, 2015). By utilizing targeted optogenetics, these researchers demonstrated that modulation of ILT–ventral striatum (vSTR) versus PFC–vSTR neuronal activity differentially regulated dendritic spine plasticity and social avoidance. They showed that stimulation of ILT–vSTR somas reduced social interaction, whereas stimulation of

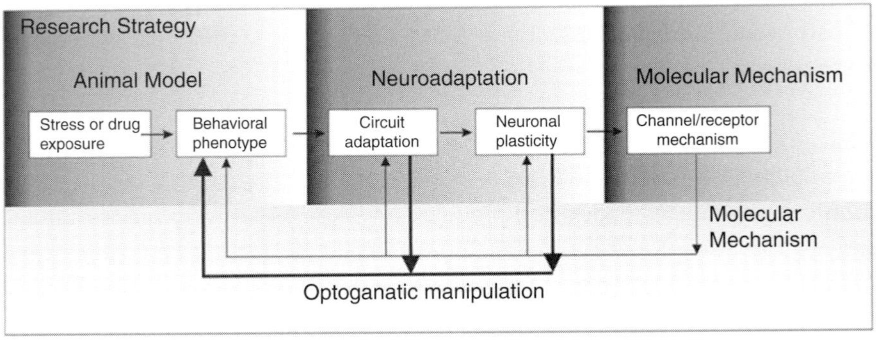

Figure 18.3 Systematic optogenetic research strategy for exploring the mechanisms of neuropsychiatric diseases. Strong or persistent environmental stimulation such as stress and drugs of abuse can induce behavioral dysfunctions. This dysfunction can be mediated by neuroadaptations at the levels of neural circuitry and neuronal plasticity. The systematic research strategy approach involves first investigating neural circuit changes, firing activity alterations and ionic mechanisms that are altered in diseased states. After they are identified by utilizing a variety of methods, including optogenetics, the adaptations are determined to be necessary or sufficient to induce the pathogenic behaviors observed. This method is essential to advancing our understanding of neuropsychiatric diseases and substance abuse disorders.

PFC–vSTR had no effect. This work suggests that two distinct glutamatergic inputs to the vSTR have divergent roles in the behavioral response to stress.

18.8 Concluding Remarks

Optogenetics is an outstanding tool for mapping out disease-related circuits. Its usage requires extreme care in both experimental design and interpretation of results. By utilizing a systematic research strategy, it is possible to link behavioral phenomena with cellular/circuit mechanisms and to return back to the behavioral level and identify the molecular and cellular adaptations that are both sufficient and necessary to underlie the behavioral phenotypes of interest (Figure 18.3). This approach has revealed a large amount of targeted information regarding susceptibility and resilience to chronic social defeat and substance abuse models.

REFERENCES

Arencibia-Albite, F., Paladini, C., Williams, J. T. *et al.*, (2007) Noradrenergic modulation of the hyperpolarization-activated cation current (Ih) in dopamine neurons of the ventral tegmental area. *Neuroscience*, **149**, 303–314.

Bagot, R. C., Parise, E. M., Pena, C. J. *et al.*, (2014) Ventral hippocampal afferents to the nucleus accumbens regulate susceptibility to depression. *Nat Commun*, **6**:7062.

Berton, O., Mcclung, C. A., Dileone, R. J. *et al.*, (2006) Essential role of BDNF in the mesolimbic dopamine pathway in social defeat stress. *Science*, **311**, 864–868.

Berton, O. and Nestler, E. J. (2006) New approaches to antidepressant drug discovery: beyond monoamines. *Nat Rev Neurosci*, **7**, 137–151.

Cao, J.-L., Covington, H. E., Friedman, A. K. *et al.*, (2010) Mesolimbic dopamine neurons in the brain reward circuit mediate susceptibility to social defeat and antidepressant action. *J Neurosci*, **30**, 16453–16458.

Carlezon, W. A. and Chartoff, E. H. (2007) Intracranial self-stimulation (ICSS) in rodents to study the neurobiology of motivation. *Nat Protoc*, **2**, 2987–2995.

Chaudhury, D., Walsh, J. J., Friedman, A. K. *et al.*, (2013) Rapid regulation of depression-related behaviours by control of midbrain dopamine neurons. *Nature*, **493**, 532–536.

Christoffel, D. J., Golden, S. A., Dumitriu, D. *et al.*, (2011) IkB kinase regulates social defeat stress-induced synaptic and behavioral plasticity. *J Neurosci*, **31**, 314–321.

Christoffel, D. J., Golden, S. A., Walsh, J. J. *et al.*, (2015) Excitatory transmission at thalamo-striatal synapses mediates susceptibility to social stress. *Nat Neurosci*, **18**, 962–964.

Coppen, A. (1967) The biochemistry of affective disorders. *Br J Psychiatry*, **113**, 1237–1264.

Cunningham, C. L., Gremel, C. M. and Groblewski, P. A. (2006) Drug-induced conditioned place preference and aversion in mice. *Nat Protoc*, **1**, 1662–1670.

Di Chiara, G. and Imperato, A. (1988) Opposite effects of mu and kappa opiate agonists on dopamine release in the nucleus accumbens and in the dorsal caudate of freely moving rats. *J Pharmacol Exp Ther*, **244**, 1067–1080.

Ferrari, A. J., Charlson, F. J., Norman, R. E. *et al.*, (2013) Burden of depressive disorders by country, sex, age, and year: findings from the Global Burden of Disease Study 2010. *PLoS Med*, **10**, e1001547.

Ford, C. P., Mark, G. P. and Williams, J. T. (2006) Properties and opioid inhibition of mesolimbic dopamine neurons vary according to target location. *J Neurosci*, **26**, 2788–2797.

Friedman, A. K., Walsh, J. J., Juarez, B. *et al.*, (2014) Enhancing depression mechanisms in midbrain dopamine neurons achieves homeostatic resilience. *Science*, **344**, 313–319.

Golden, S. A., Covington, H. E., Berton, O. *et al.*, (2011) A standardized protocol for repeated social defeat stress in mice. *Nat. Protoc*, **6**, 1183–1191.

Gong, S., Doughty, M., Harbaugh, C. R. *et al.*, (2007) Targeting Cre recombinase to specific neuron populations with bacterial artificial chromosome constructs. *J Neurosci*, **27**, 9817–9823.

Grace, A. A. and Bunney, B. S. (1984a) The control of firing pattern in nigral dopamine neurons: burst firing. *J Neurosci*, **4**, 2877–2890.

Grace, A. A. and Bunney, B. S. (1984b) The control of firing pattern in nigral dopamine neurons: single spike firing. *J Neurosci*, **4**, 2866–2876.

Grace, A. A., Floresco, S. B., Goto, Y. *et al.*, (2007) Regulation of firing of dopaminergic neurons and control of goal-directed behaviors. *Trends Neurosci*, **30**, 220–227.

Gysling, K. and Wang, R. Y. (1983) Morphine-induced activation of A10 dopamine neurons in the rat. *Brain Res*, **277**, 119–127.

Han, M. H. and Friedman, A. K. (2011) Virogenetic and optogenetic mechanisms to define potential therapeutic targets in psychiatric disorders. *Neuropharmacology*, **62**, 89–100.

Hill, M. N., Hellemans, K. G., Verma, P. *et al.*, (2013) Neurobiology of chronic mild stress: parallels to major depression. *Neurosci Biobehav Rev*, **36**, 2085–2117.

Hyman, S. E. (2007) How mice cope with stressful social situations. *Cell*, **131**, 232–234.

Hyman, S. E., Malenka, R. C. and Nestler, E. J. (2006) Neural mechanisms of addiction: the role of reward-related learning and memory. *Annu Rev Neurosci*, **29**, 565–598.

Koo, J. W., Mazei-Robison, M. S., Chaudhury, D. *et al.*, (2012) BDNF is a negative modulator of morphine action. *Science*, **338**, 124–128.

Koob, G. F., Ahmed, S. H., Boutrel, B. *et al.*, (2004) Neurobiological mechanisms in the transition from drug use to drug dependence. *Neurosci Biobehav Rev*, **27**, 739–749.

Krishnan, V., Han, M. H., Graham, D. L. *et al.*, (2007) Molecular adaptations underlying susceptibility and resistance to social defeat in brain reward regions. *Cell*, **131**, 391–404.

Krishnan, V. and Nestler, E. J. (2008) The molecular neurobiology of depression. *Nature*, **455**, 894–902.

Lammel, S., Hetzel, A., Hackel, O. *et al.*, (2008) Unique properties of mesoprefrontal neurons within a dual mesocorticolimbic dopamine system. *Neuron*, **57**, 760–773.

Lammel, S., Ion, D. I., Roeper, J. *et al.*, (2011) Projection-specific modulation of dopamine neuron synapses by aversive and rewarding stimuli. *Neuron*, **70**, 855–862.

Lemos, J. C., Wanat, M. J., Smith, J. S. *et al.*, (2012) Severe stress switches CRF action in the nucleus accumbens from appetitive to aversive. *Nature*, **490**, 402–406.

Lobo, M. K., Covington, H. E., 3rd, Chaudhury, D. *et al.*, (2010) Cell type-specific loss of BDNF signaling mimics optogenetic control of cocaine reward. *Science*, **330**, 385–390.

Lobo, M. K., Nestler, E. J. and Covington, H. E. (2012) Potential utility of optogenetics in the study of depression. *Biol Psychiatry*, **71**, 1068–1074.

Margolis, E. B., Mitchell, J. M., Ishikawa, J. *et al.*, (2008) Midbrain dopamine neurons: projection target determines action potential duration and dopamine D(2) receptor inhibition. *J Neurosci*, **28**, 8908–8913.

Matsuda, N., Lu, H., Fukata, Y. *et al.*, (2009) Differential activity-dependent secretion of brain-derived neurotrophic factor from axon and dendrite. *J Neurosci*, **29**, 14185–14198.

Nestler, E. J., Gould, E. and Manji, H. (2002) Preclinical models: status of basic research in depression. *Biol Psychiatry*, **52**, 503–528.

Nestler, E. J. and Hyman, S. E. (2010) Animal models of neuropsychiatric disorders. *Nat Neurosci*, **13**, 1161–1169.

Neuhoff, H., Neu, A., Liss, B. *et al.*, (2002) I(H) channels contribute to the different functional properties of identified dopaminergic subpopulations in the midbrain. *J Neurosci*, **22**, 1290–1302.

Neve, R. L. and Lim, F. (2013) Generation of high-titer defective HSV-1 vectors. *Curr Protoc Neurosci*, Chapter 4, Unit 4.13.

Olmstead, M. C. and Franklin, K. B. (1997) The development of a conditioned place preference to morphine: effects of lesions of various CNS sites. *Behav Neurosci*, **111**, 1313–1323.

Park, H. and Poo, M. M. (2013) Neurotrophin regulation of neural circuit development and function. *Nat Rev Neurosci*, **14**, 7–23.

Robinson, M. J. and Berridge, K. C. (2008) Instant transformation of learned repulsion into motivational "wanting". *Curr Biol*, **23**, 282–289.

Robinson, T. E. and Berridge, K. C. (2003) Addiction. *Annu Rev Psychol*, **54**, 25–53.

Russo, S. J. and Nestler, E. J. (2013) The brain reward circuitry in mood disorders. *Nat Rev Neurosci*, **14**, 609–625.

Schildkraut, J. J. (1965) The catecholamine hypothesis of affective disorders: a review of supporting evidence. *Am J Psychiatry*, **122**, 509–522.

Schultz, W. (2010) Dopamine signals for reward value and risk: basic and recent data. *Behav Brain Funct*, **6**, 24.

Schultz, W. (2015) Neuronal reward and decision signals: from theories to data. *Physiol Rev*, **95**, 853–951.

Shirayama, Y. and Chaki, S. (2006) Neurochemistry of the nucleus accumbens and its relevance to depression and antidepressant action in rodents. *Curr Neuropharmacol*, **4**, 277–291.

Tsai, H. C., Zhang, F., Adamantidis, A. *et al.*, (2009) Phasic firing in dopaminergic neurons is sufficient for behavioral conditioning. *Science*, **324**, 1080–1084.

Tye, K. M., Mirzabekov, J. J., Warden, M. R. *et al.*, (2013) Dopamine neurons modulate neural encoding and expression of depression-related behaviour. *Nature*, **493**, 537–541.

Vialou, V., Robison, A. J., Laplant, Q. C. *et al.*, (2010) [Delta]FosB in brain reward circuits mediates resilience to stress and antidepressant responses. *Nat Neurosci*, **13**, 745–752.

Volkow, N. D. and Koob, G. (2015) Brain disease model of addiction: why is it so controversial? *Lancet Psychiatry*, **2**, 677–679.

Volkow, N. D. and Morales, M. (2015) The brain on drugs: from reward to addiction. *Cell*, **162**, 712–725.

Walker, D. M., Cates, H. M., Heller, E. A. *et al.*, (2015) Regulation of chromatin states by drugs of abuse. *Curr Opin Neurobiol*, **30**, 112–121.

Walsh, J. J., Friedman, A. K., Sun, H. *et al.*, (2014) Stress and CRF gate neural activation of BDNF in the mesolimbic reward pathway. *Nat Neurosci*, **17**, 27–29.

Walsh, J. J. and Han, M. H. (2014) The heterogeneity of ventral tegmental area neurons: projection functions in a mood-related context. *Neuroscience*, **282c**, 101–108.

Wanat, M. J., Bonci, A. and Phillips, P. E. (2013) CRF acts in the midbrain to attenuate accumbens dopamine release to rewards but not their predictors. *Nat Neurosci*, **16**, 383–385.

Wise, R. A. and Rompre, P. P. (1989) Brain dopamine and reward. *Annu Rev Psychol*, **40**, 191–225.

19 Optogenetics Research in Behavioral Neuroscience: Insights into the Brain Basis of Reward Learning and Goal-directed Behavior

Adam C. G. Crego, Stephen E. Chang, William N. Butler, and Kyle S. Smith

19.1 Introduction

In recent years, there has been a rapid expansion of insights into brain function related to reward learning and goal-directed behavior through the use of optogenetics technologies. This, along with related research tools, has arrived at a critical time: we have a sense of the brain circuits and neural dynamics related to reward and goal-directed behavior in learning and motivational processes, but little understanding of how specific patterns of neural activity or specific circuits causally control these behaviors. This has been a major gap in our scientific knowledge. However, optogenetics has begun to help research fill in this gap by employing highly precise interventions in neural activity as applied to specific questions.

Optogenetic methods have been extensively reviewed, and we refer readers to several articles for details on the methodology (Bernstein and Boyden 2011; Fenno *et al.*, 2011; Mattis *et al.*, 2012). Through viral vectors and genetic engineering, genes for light-sensitive opsins can be inserted into neurons, where they are expressed through endogenous transcription and translation machinery. Once expressed, light can then be applied directly in order to activate these exogenous receptors and change the neuronal membrane voltage at the sub-second timescale. The optogenetic toolbox is deep: technologies exist for inhibition, excitation and subthreshold membrane voltage changes; molecular genetic targeting approaches can lead to optogenetic expression in diverse neuronal types and pathways; and variants exist for inhibition or excitation at precise light wavelengths. There have been many successful applications combining optogenetics with other neuroscience approaches, including imaging, electrophysiology, immunocytochemistry, chemogenetics, etc. In short, the myriad of options that optogenetics carries with it opens the door for research into neuronal information processing related to learning and complex behavior in the patterned dynamics of activity and circuit communication. Here, we outline several advantages and considerations for when using optogenetics in behavioral tasks, and then review our progress of using optogenetics approaches in order to understand the neural basis of goal-directed learning and reward-related behavior.

19.2 Spatial and Temporal Advantages of an Optogenetics Approach to Reward Learning and Goal-directed Behavior Research

Perhaps the two greatest advantages to this technology are: (1) its ability to manipulate specific cell types or pathways within a greater brain circuit; and (2) its ability to manipulate sub-second patterns of neural activity related to a particular behavioral event. Relating the former to the latter, one of the most exciting advances in behavioral neuroscience is based on the possibility of targeting specific types of cells or brain pathways as animals perform tasks. At the forefront of this is the use of Cre/Lox recombination technology. Cre recombinase is an enzyme that is used to genetically catalyze a specific gene for recombination between two LoxP sites. By engineering genetically defined types of brain cells to express Cre and introducing Cre-dependent optogenetic constructs, these cells can then be placed under the causal control of light, allowing neuroscientists to increase or decrease the activity of specific neurons without directly affecting other cell types in a defined area.

For instance, Cre driven by a tyrosine hydroxylase (TH-Cre) promoter can be genetically engineered in rodents, and Cre-dependent optogenetic constructs can be introduced in order to selectively target dopamine cells with either excitation or inhibition procedures. This strategy has been used in several studies to test the contribution and role of dopamine in motivation and behavior. Figure 19.1A gives an example of the expression of an optogenetic construct in dopamine neurons in the substantia nigra pars compacta (SNc) using this approach. In one study assessing the influence of nigrostriatal dopamine activity on instrumental learning, TH-Cre mice were injected with a channelrhodopsin-2 (ChR2)-containing virus into the SNc, where, due to the TH-Cre technology, ChR2 was expressed exclusively in dopamine-producing neurons (Rossi *et al.*, 2013). Interestingly, these mice had to learn to press a lever to receive direct stimulation of their SNc (an action–outcome contingency) and showed no such decline in pressing of the active (stimulating) lever, even after self-stimulation over two consecutive sessions. The press rate for the active lever remained the same in both sessions. In a contingency reversal (switching inactive and active levers) paradigm for SNc laser simulation, mice readily adjusted to pressing the new active lever, suggesting the role of the SNc in highly sensitive action–outcome contingencies. These findings suggests that SNc dopamine neurons are sufficient to stand in as a reward for an instrumental action (Rossi *et al.*, 2013; Ilango *et al.*, 2014), bolstering the notion that reinforcing properties are not restricted to mesolimbic dopamine neurons contained within the ventral tegmental area (VTA). As another example, a recent study using Cre recombinase technology evaluated the role of the dorsal medial habenula (dMHb) in its regulation of voluntary motor control, reinforcement and hedonic reward states (Hsu *et al.*, 2014). Using Cre-mediated expression of ChR2, dMHb stimulation was found to support self-stimulation. In contrast, eNpHR (halorhodopsin [Halo])-mediated inhibition of the dMHb resulted in aversive effects, as shown in the acute place preference/aversion paradigm, in which mice spent significantly less time in the

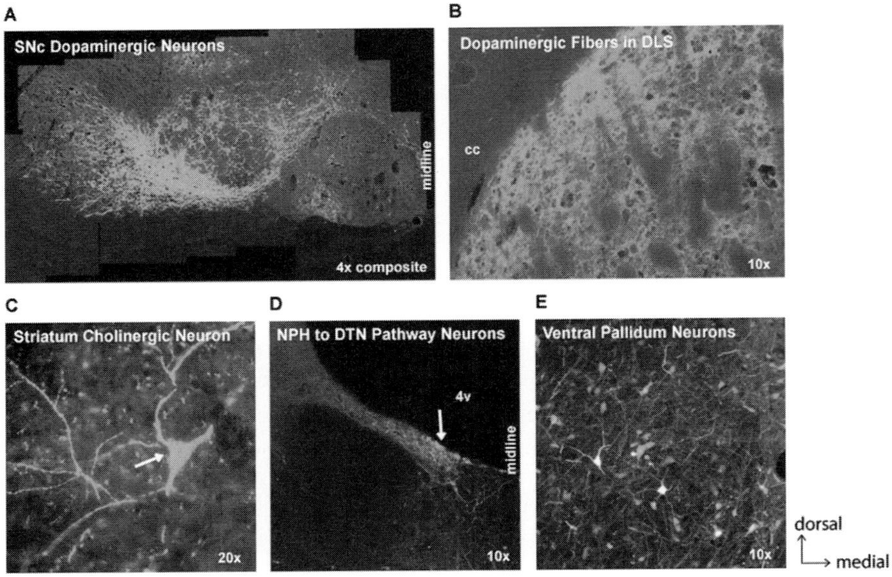

Figure 19.1 Examples of ChR2 and eNpHR expression in the rat brain. (A) Expression of ChR2-EYFP in dopaminergic neurons of the SNc in a transgenic TH-Cre rat. (B) Dopaminergic fibers in the DLS expressing ChR2-EYFP from the same rat. (C) ChR2-EYFP expression in cholinergic neurons of the striatum in a ChAT-Cre rat. (D) Expression of eNpHR -EYFP in DTN neurons after an injection of CAV-Cre in the DTN and Cre-dependent eNpHR in the NPH. (E) Generic expression of eNpHR-EYFP in the ventral pallidum using a synapsin promoter. cc: corpus callosum; 4v: fourth ventricle; DTN: dorsal tegmental nucleus; DLS: dorsolateral striatum; NPH: nucleus prepositus hypoglossi; SNc: substantia nigra pars compacta. (A black-and-white version of this figure will appear in some formats. For the color version, please refer to the plate section.)

eNpHR laser-paired side as compared to the non-laser-paired side, demonstrating bidirectional control over self-stimulation behavior by this structure (Hsu *et al.*, 2014). As these examples show, the ability to virally or transgenically introduce Cre into different cell types – and there are many rodent Cre driver lines (see also Figure 19.1C) and viral approaches available for doing so – raises profound new research possibilities.

Further concerning this spatial advantage, optogenetic constructs can also be used to manipulate specific brain pathways. One approach exploits the fact that opsins are not limited to the cell body, but can be expressed throughout axons even at the terminals. Figure 19.1B shows an example of this, involving expression of ChR2 in dopaminergic fibers arriving in the dorsal striatum from the SNc. Thus, light can be delivered to the terminals in order to manipulate a projection pathway of interest. Many reward studies have illuminated axon terminals for pathway specificity, with highly interesting results (Stuber *et al.*, 2012; Tye and Deisseroth, 2012). There are, however, several considerations to take into account when using this approach. First, targeting a specific axon terminal in this way could also inadvertently affect fibers of passage arising from the same set of opsin-expressing neurons in the same area. Second, this technique requires careful control in order to verify whether any effects on behavior are due to backward propagation of inhibition/excitation to the cell body, which could conceivably affect axon collaterals of that cell in other brain areas (Stuber *et al.*, 2011; Tye *et al.*, 2011). Finally, another consideration to take into account is that light delivery at axon terminals may not have as robust of an effect on neuronal activity and behavior as cell body illumination, though this potential issue is less resolved (Yizhar *et al.*, 2011).

An alternative method of pathway-defined optogenetic targeting involves using viruses that lead to retrograde gene delivery (i.e. transportation of optogenetic constructs from axon terminals back to cell bodies). Some of the more common retrograde viruses used to accomplish this form of projection targeting are strains of pseudorabies virus (PRV) (Wickersham *et al.*, 2007; Luo *et al.*, 2008; Wall *et al.*, 2013). For example, one approach involves injecting PRV conjugated with Cre recombinase in order to generate Cre-dependent expression in upstream neurons, both for visualization and for optogenetic manipulation of particular neuronal pathways. Similarly, herpes simplex virus (HSV) (Neve and Carlezon Jr., 2002) has been used for a projection-specific optogenetic manipulation of a neuronal subpopulation upstream from the injection site (Lima *et al.*, 2009). However, both of these retrograde viruses have potential drawbacks. HSV and PRV might in some cases affect glial cells, so their effects could be non-neuron specific (Ugolini, 2010). Additionally, PRV has been shown to induce neuronal damage and degeneration, causing cell death by approximately 2 weeks post-transduction (Wickersham *et al.*, 2007). Thus, HSVs tend to produce rapid but time-limited expression of opsins, at least in cell bodies (Han, 2012). These issues may limit the use of these viruses in some applications, such as tasks requiring manipulations over long periods (e.g., learning tasks). However, HSV-mediated delivery of Cre recombinase followed by a Cre-dependent optogenetic construct can circumvent this in some preparations, leading to longer-lasting opsin expression in the targeted cells.

One promising viral vector in this regard is canine adenovirus 2 (CAV-2). *In vivo*, the level of retrograde transport of CAV-2 is comparable to HSV vectors and maintains 60–70% of baseline expression after 6 months (Soudais *et al.*, 2004), and it is not thought to be deleterious to neuronal health. The utility of a CAV-2 approach was shown in a recent work by Senn *et al.* (2014), who used this vector to selectively and retrogradely target the projections from subpopulations of basal amygdala neurons to both the infralimbic (IL) or prelimbic (PL) divisions of the medial prefrontal cortex using ChR2 and eNpHR in separate mouse groups. Optogenetic manipulations of these pathway-specific circuits demonstrated a causal balance between these two pathways in fear conditioning and extinction. In a similar manner, Namburi *et al.* (2015) used CAV-2-Cre in conjunction with a Cre-dependent eNpHR construct to target neurons projecting from the basolateral amygdala (BLA) to either the nucleus accumbens (NAc) or the central nucleus of the amygdala; both of these BLA projections were separately and optogenetically manipulated in order to test their differential roles in fear and reward conditioning. We, and our colleagues, are also having early success with this technique in the field of spatial cognition in order to select for neurons projecting to the head direction circuit, and to test whether they play a role in guiding navigational behavior (Figure 19.1D) (Butler *et al.*, 2015).

The second major advantage to optogenetics research concerns its temporal properties, an advantage that behavioral neuroscience is only beginning to leverage. Diverse phasic patterns of activity in the neurons of a given brain area can be used to represent different aspects of reward learning and behavior. In principle, these patterns can now be targeted causally in order to complement correlational readouts of the activity in relation to behavior (e.g. through electrophysiological recordings). Different frequencies can be played into neurons of interest in order to establish optimal gain-of-function effects for reward-related behaviors (Tsai *et al.*, 2009; Rothwell *et al.*, 2015), though generally it remains unclear for most brain systems related to reward as to how the second-to-second dynamics of activity that reflect reward processes causally control them. For example, in the ventral pallidum, a brain area that is critical for reward (Smith *et al.*, 2009; Root *et al.*, 2015), phasic bursts of activity have been shown to at least correlate with the motivational value of reward-predictive cues, as reflected in behavioral reactions (Tindell *et al.*, 2006; Tindell *et al.*, 2009; Smith *et al.*, 2011). Questions that are now addressable by expressing optogenetic molecules in this area (see Figure 19.1E) include whether this activity triggers attraction to reward cues, the ability of the cues to invigorate reward seeking or other potential aspects of cue processing, such as associative representations of reward outcome.

This kind of logic leads us to suggest an ideal hypothetical experimental setup in which there would be a known pattern of activity that occurs in a brain area or subset of neurons related to reward learning that would be the target of perturbation. There are several examples of such a physiology-inspired approach (Tsai *et al.*, 2009; Smith and Graybiel, 2013a; Steinberg *et al.*, 2013; Pascoli *et al.*, 2015). However, generally, the normal brain dynamics related to a particular aspect of reward, learning or behavior requires much further characterization. This is particularly true of areas such as the ventral striatum that often exhibit a robust heterogeneity of responses during reward tasks (Taha and Fields, 2005; Day and Carelli, 2007; Nicola, 2007). One consideration in applying excitatory manipulations to neurons when their task-related activity patterns are unclear is the extent to which the manipulation is increasing or perturbing their functionality. Not unlike many other common methods for neuronal perturbation, optogenetic stimulations introduce a potentially artificial type of neuronal activity (e.g., entrainment of firing of all affected cells at a selected frequency). Moreover, a depolarization block in neurons can occur during continuous illumination of, for example, ChR2. In this case, solutions could involve low-power continuous illumination (Kravitz *et al.*, 2013). Though gain-of-function effects of optical stimulation are often reasonably clear in the context of the extant literature, a loss-of-function complement can be added in order to corroborate such effects, such as through eNpHR or archaerhodopsin (Arch) in order to suppress neural activity. Nevertheless, with either stimulation or inhibition, it may be difficult to recapitulate the natural dynamics of neuronal firing and pauses in firing (Deng *et al.*, 2014). In this context, it is actually quite remarkable that clear enhancements of function can occur following stimulation patterns that may be rather foreign to an ensemble of neurons (Aravanis *et al.*, 2007).

Finally, we note that, depending on the given research question when using optogenetics to modulate a brain area or circuit, there are numerous options to consider that can impact the specificity and efficacy of the manipulation. These include, for example, opsin choice (e.g. ChR2, eNpHR or Arch), viral vector choice (e.g. virus, serotype, promoter or fluorophore), animal subject (e.g. wild-type or transgenic), light delivery hardware (e.g. lasers or LEDs) and light delivery protocol (e.g. pulsed or sustained, or task event-related or non-event-related). At present, it is not quite a "one-size-fits-all" scenario for reward learning research. For example, while numerous transgenic mouse lines exist for targeting optogenetic manipulations to specific cell types, there are very few transgenic rat options available at present (Witten *et al.*, 2011). This raises a barrier to studies requiring rats for more complex learning tasks, or for carrying hardware implants of greater weight (e.g. for multi-site, chronic neural recordings alongside optogenetic manipulations). Similarly, while LEDs offer an attractive alternative to laser sources, being inexpensive and smaller, their low power output in most models may be a limitation for some *in vivo* applications requiring greater light coverage in the brain (factors governing coverage include fiber implant diameter and numerical aperture, light delivery power, light wavelength and tissue absorption) (Aravanis *et al.*, 2007). On this point, it is often unknown how much coverage of a brain area is required for perturbing its function; but again, despite this, stunning behavioral effects can be seen after manipulating even only a small proportion of neurons. Still, it is plausible that a restricted area of manipulation in a brain region may not be sufficient for revealing some behavioral functions. Major efforts are underway in order to engineer methods that will enable greater coverage in the brain for optogenetic manipulations, such as through the use of red-shifted opsins. For instance, Tsien and colleagues were able to demonstrate transracial excitation with optogenetics (Lin *et al.*, 2013). By utilizing a variant of ChR2, Tsien's group was able to apply orange to red wavelengths (560–630 nm) directly onto the skull surface of a mouse and produce vibrissa motor cortex excitation. Moreover, Boyden and colleagues also created a non-invasive optical method for inhibition, in which a red-shifted halorhodopsin called "Jaws" was used for transcranial inhibition of neurons in the awake mouse (Chuong *et al.*, 2014).

19.3 Reward Learning Applications

Several studies using Pavlovian conditioning paradigms have utilized optogenetics not only based on the advantages described above, but also in order to uncover the roles of various reward-related brain regions. Much of this work has focused on the NAc and its dopamine input from the VTA. Functionally, phasic dopamine release from the VTA to the NAc is observed when unexpected rewards are presented, and this phasic activation of dopaminergic neuronal activity is shifted from the reward to a cue that predicts reward delivery once the reward is fully predicted by the cue (Schultz 2006; Day *et al.*, 2007; Flagel *et al.*, 2010). This difference between what is received and what was expected is often termed a "reward prediction error", which is reflected by the degree of dopaminergic neuronal firing. To determine whether such dopamine signaling has a causal link to learning about reward-paired cues, Steinberg *et al.* (2013) investigated the effects of stimulating VTA dopamine neurons utilizing a modified blocking procedure, a task in which prior training of one cue paired with a food reward can block learning about a second cue if presented in compound with the first cue and paired with the same reward (e.g. Kamin, 1968; Kamin, 1969). First, rats were trained to enter a reward port during an auditory cue in order to earn deliveries of a liquid sucrose reward. Second, rats were presented with a compound cue consisting of the auditory cue and a novel light cue that was paired with the same sucrose reward. Finally, rats were given presentations of the light cue alone in a probe test session in the absence of sucrose delivery in order to assess the degree by which rats had learned to associate the light cue with sucrose. Because the sucrose reward was fully predicted by prior training with the auditory cue, there was no prediction error when the auditory/light compound was paired with the sucrose reward. Thus, the association between the light cue and sucrose reward should be minimal. As expected, control rats showed minimal levels of responding to the light cue during the test probe session. However, rats that received optogenetic stimulation of VTA dopamine neurons paired with sucrose reward delivery during auditory/light compound presentations showed elevated levels of time spent in the reward port compared to control rats during the test probe session, suggesting that the laser VTA stimulation simulated an "artificial-like" prediction error signal that resulted in learning about the light cue. Importantly, this increase in learning was not simply due to laser stimulation increasing the value of the sucrose reward. In a separate task, laser stimulation did not alter the preferences of two distinctly flavored and equally preferred sucrose solutions in which one was paired with laser stimulation and the other was not, suggesting that laser stimulation had no effect on the intrinsic value of the reward. The authors thus concluded that phasic dopamine activity was causally relevant to prediction error-based reward learning. Alternative interpretations are also possible, such as a modulation of the motivational value of the reward-predictive cues (Berridge and Robinson, 1998). Although such results may be interpreted in multiple ways, they nevertheless demonstrate how timed manipulations of specific cell types in the brain can profoundly influence reward-seeking behaviors.

In addition to investigating the role of VTA to NAc dopamine projections in reward learning, other investigations applying optogenetics have focused on NAc projections from the BLA. For instance, Stuber *et al.* (2011) investigated the effects of inhibiting BLA to NAc projections in mice using a Pavlovian conditioning task. Mice were given presentations of a tone/house-light conditioned stimulus (CS) that was paired with the delivery of a liquid sucrose unconditioned stimulus (US). Control mice acquired robust licking responses upon CS presentations, as well as to US delivery. In contrast, mice that received laser stimulation in which BLA to

NAc projections were inhibited during CS presentations showed reduced licking responses not only during the CS, but also when the US was delivered. BLA to NAc projections have also been found to be important in the reinstatement of reward-seeking following extinction. Stefanik and Kalivas (2013) first trained rats to press a lever that was reinforced with intravenous infusions of cocaine and a light/tone compound stimulus. Following several extinction training sessions in which lever presses were unrewarded, rats were given reinstatement sessions in which lever press responding resulted in presentation of the light/tone compound stimulus only. Inhibition of the BLA to NAc pathway either directly or indirectly through the PL prefrontal cortex throughout the session reduced cue-induced reinstatement of cocaine seeking in terms of lever press responding. Taken together, optogenetics has demonstrated the importance of the BLA to NAc circuit in mediating the ability of reward-paired cues to motivate reward-seeking behaviors.

A related study used a delay-discounting task in which rats learned to associate lever press responding during one cue for a small, immediate reward and another cue for a large, delayed reward (Saddoris *et al.*, 2015). This work found that the level of dopamine release in the NAc during cue presentations reflected the relative values of the cues, which positively correlated with choice performance when rats were presented with both cues and given the choice to earn either the large, delayed or small, immediate reward. In a separate experiment, rats learned to associate lever press responding during one cue for a small, immediate reward and another cue for the same small reward, but with a delay for delivery. When presented with both cues, control rats pressed the lever associated with the small, immediate reward more than the small, delayed reward, as was expected. However, rats that received stimulation of dopamine terminals in the NAc during cue presentation for the small, delayed reward during training responded more to the lever associated with the small, delayed reward compared to control group rats when both cues were presented. These results demonstrate that stimulation of dopamine terminals in the NAc can enhance the value of reward-paired cues, as reflected by the changes observed in choice behavior.

Such examples highlight how optogenetics can be used to understand the causal mechanisms of Pavlovian reward learning and motivation. Concerning the latter process, the importance of cues in driving motivated behavior is patently clear in the Pavlovian conditioning paradigm known as autoshaping or sign-tracking, in which, for example, the insertion of a lever CS signals the delivery of a food US upon retraction. After repeated CS–US presentations, rats often acquire a conditioned response (CR) in which they will approach, press and bite the lever CS, even though no response is required for US delivery. Many researchers have argued that the occurrence of sign-tracking is the result of the CS acquiring incentive salience, a process by which the incentive motivational value of the US is transferred to the CS (Berridge, 2004; Flagel *et al.*, 2010; Robinson *et al.*, 2014). There has thus been a great focus on the neural circuitry involved in the acquisition and expression of sign-tracking (Mahler and Berridge, 2009; Flagel *et al.*, 2010; Chang *et al.*, 2012; Saunders and Robinson, 2012; Chang *et al.*, 2015). Sign-tracking offers a particularly compelling example of a research area that could benefit from an optogenetics approach. Most of these studies utilized techniques such as intracranial drug injections or neurotoxic lesions, which target specific brain regions or transmitters but lack the ability to affect only single events during training (e.g. CS presentation or US delivery). Additionally, neurotoxic lesions cause permanent damage that may lead to compensatory adaptations, and intracranial injections, if done repeatedly, can cause physical damage. Utilizing optogenetics circumvents these obstacles and allows us to consider whether a brain region is involved in processing the CS or US. For example, inhibiting a brain region during CS presentation and observing a reduction in sign-tracking may be evidence that the region is involved in the attribution of incentive value to the CS. In addition, observing a deficit in sign-tracking when inhibiting that same brain region during US delivery may be evidence that the region is important for encoding the reward value of the US. Alternatively, inhibition during presentation of the CS or US only may lead to dissociable effects, with inhibition during one event resulting in a reduction in sign-tracking, while inhibition during the other event may have no effect. Importantly, observing such a double dissociation during training would not be possible using more standardized methodologies such as intracranial injections or neurotoxic lesions. In addition, sign-tracking deficits may only be observed when a brain region is inhibited during the entire CS–US interval, which would suggest that the region is important for acquiring/expressing the CS–US association, rather than attributing incentive value to the CS or encoding US value.

19.4 Instrumental Learning: Goal-directed Actions and Habits

A related area of research that has benefited from optogenetics concerns the brain mechanisms underlying flexible versus fixed action patterns that arise from instrumental conditioning. Over a century ago, Edward Thorndike developed his "Law of Effect" hypothesis, which set the stage for understanding the structure of instrumental behaviors: some actions lead to positive effects, strengthen behavior and tend to be repeated, while other actions lead to negative effects, weaken behavior and tend not to be repeated (Thorndike, 1898). Reward in this domain comes to be understood based on its reinforcement properties of strengthening or weakening antecedent actions. While there are important distinctions to be made between reward (hedonic impact) and this sort of reinforcement signaling in the brain (Berridge, 2001), the Thorndikian framework for understanding goal-directed behavior learning has been used to make tremendous strides in understanding the brain mechanisms that are responsible for forming action plans aimed at achieving rewards. In particular, experimental procedures that manipulate the value of the expected reward for an action, or its contingency with the action, have led to a two-process model by which behavior can be controlled either through underlying action–outcome (A–O) associations or stimulus–response (S–R) associations (Adams, 1982; Dickinson, 1985; Balleine and Dickinson, 1998). These processes are explanations for why some actions can, in some circumstances, be directly sensitive to changes in expected reward value, while others are not. The former are regarded as A–O-based, goal-directed behaviors, and the latter as S–R-based habits. This framework provides a strong example of how operational definitions for types of learning, established through behavioral studies, can serve as references for understanding behavior as action or habit, and what brain areas might be involved in either.

There are currently very few studies using optogenetics in order to address questions about the neural processes underlying the transition from A–O behaviors to S–R behaviors, although the potential in this area is high (Smith and Graybiel, 2013b). For instance, Gremel and Costa (2013) highlight this well in a mouse instrumental lever pressing paradigm, where optogenetic manipulations were used to identify the role of the orbitofrontal cortex (OFC) within the basal ganglia circuit of decision making. They found that neural activity in the OFC positively correlates with the goal-directed action of lever pressing for a reward, and that optogenetically stimulating the OFC resulted in increased goal-oriented pressing, as though suppressing habitual aspects of the behavior. This finding complements growing evidence that the OFC is critical for representing expected outcomes, whether in Pavlovian or instrumental settings, and that a loss of this representation renders behavior insensitive to outcome changes (which is a feature of habits) (Schoenbaum and Roesch, 2005).

Another recent study by Smith *et al.* (2012) took advantage of this technique in order to perform real-time manipulations of the brain, finding an area called the IL cortex within the medial prefrontal cortex that exerts a remarkably high level of online control over habits. Prior studies on the IL cortex had used lesions and pharmacologic manipulations to show that it is necessary in expressing habits once they are acquired using an outcome-sensitivity measure (Coutureau and Killcross, 2003; Killcross and Coutureau, 2003; Hitchcott *et al.*, 2007; Barker *et al.*, 2014). Using this information, Smith, Graybiel and colleagues conducted a series of studies in order to establish whether the IL cortex was important for acquiring and controlling habits in real time (Smith *et al.*, 2012; Smith and Graybiel, 2013a). By combining modified electrophysiology headstages with optogenetic fibers, Smith and colleagues injected halorhodopsin within the IL of rats and compared these rats with controls lacking the opsin (Smith *et al.*, 2012). Rats were then trained to perform a behavior of turning in a specified direction on a T-maze for a reward (e.g. chocolate milk for left turn, sugar reward for right turn) based on an auditory cue (e.g. high tone for chocolate milk, low tone for sugar) at the start of the task. Rats were over-trained until turning became a habit by the operational definition. Devaluation was administered for one (chocolate milk) of the two rewards by pairing it with LiCl. The probe trial revealed that control rats were no longer sensitive to the outcome reward and ran by habit to the devalued arm. When halorhodopsin was used to inhibit the IL, rats began to avoid the devalued goal, demonstrating a reinstatement of a goal-directed behavior. After this, rats underwent a post-probe training period, in which they developed a new strategy of traversing the non-devalued arm when cued to go to the devalued arm, despite not being rewarded for such wrong-way runs. Additional perturbations of the IL blocked this new behavior as well, and actually reinstated the old routine of running to both goals when so instructed. This habit reinstatement effect, found by leveraging the repeatability of optogenetics manipulations, suggests that the IL cortex somehow maintains newly established habits at the expense of prior habit strategies. In another study taking advantage of the ability to manipulate the IL during training, and then testing for learning without further manipulations, Smith and Graybiel (2013a) found that IL activity during task performance was necessary for acquiring a habit in the first place. Complementary studies using optogenetics in this domain could address the more fine timescale aspects of habitual behavior with respect to the neural dynamics seen in both IL and basal ganglia circuits as habits are formed and maintained (Graybiel, 2008; Gremel and Costa, 2013; Smith and Graybiel, 2014). In ongoing research, we are using optogenetics to explore how the timing of phasic activity in these areas dictates behavioral rigidity and habits (Crego *et al.*, 2015). Early evidence suggests not only that behaviors can be made to be more or less habitual using subsecond stimulations or inhibitions of neural activity, but also that when these phasic activity patterns occur makes all of the difference.

Similar work using optogenetics in order to study the balance of behavioral flexibility/inflexibility in reward seeking has been done in other brain reward areas, such as the NAc. A recent study by Aquili *et al.*, (2014) suppressed the NAc shell using eNpHR during an operant chamber lever pressing win–stay/lose–shift task. In this paradigm, optogenetic inhibition was administered at specific time points during the switching of four different stimulus–reward contingencies, progressively becoming more difficult with each stage: (1) pressing one of two levers for an immediate reward; (2) between-session reversal; (3) single reversal within a session; and (4) multiple reversals within a session. They found that errors were reduced after a stimulus–reward contingency switch when the NAc shell was inhibited at the time between action selection and outcome. This suggests that the NAc shell integrates either past error or reward feedback at specific times during an instrumental behavior.

Along these same lines, an optogenetics approach to understanding the mechanisms of behavioral flexibility and fixity has also seen progress in research on the acquisition of skills (Ferster and Skinner, 1957). For example, Jin *et al.* (2014) investigated the distinction between striatal D1- and D2-expressing medium spiny neurons (MSNs) by injecting ChR2 into the striatum of Cre-dependent mice in order to investigate the roles of striatonigral and striatopallidal MSN projection activities in a rapid sequence task combining an electrophysiology array with optogenetic fibers. First, their results demonstrated that rapid whole action sequences could be encoded as single actions in the striatum. This was confirmed through electrophysiology recordings revealing MSN activity at the initiation and termination of a sequence. Second, during action sequence initiation, direct and indirect pathways were concomitantly active, as indicated by similar percentages of D1- and D2-expressing MSN signaling sequence starts versus stops in their rapid sequence task. Lastly, these two distinct pathways operated in another fashion during action performance. First, direct pathway

MSNs preferred to fire at the onset of an action and often remained active throughout the action sequence; in contrast, indirect pathway MSNs were active at the start of the action, but the activity was depressed during the sequence. These results suggest that both basal ganglia pathways behave differently during the execution of motor action sequences, but show concomitant activity at initiation. These general conclusions were supported by findings from a recent study by Rothwell *et al.* (2015) using serial order of actions. In this study, optogenetic stimulation of a projection pathway was used in an anterograde fashion in a serial order task of pressing two sequential levers (press "A" then "B"). ChR2 was bilaterally injected into the secondary motor (M2) cortex of mice, in conjunction with the implantation of optic fibers in the dorsolateral striatum (DLS) just above ChR2-expressing axon terminals. Optogenetic stimulation anterogradely of M2 projections to the DLS was conducted after 30 days of training on this serial order task, with stimulation occurring either at the start of the sequence or between the first and second presses. Their results revealed that ChR2-mediated stimulation of direct connections from M2 to the DLS increases (promotes) the likelihood of initiating of a learned motor action sequence. This can either impair or improve sequence performance depending on when the stimulation is applied.

19.5 Looking Forward

These examples provide a compelling demonstration of the utility of an optogenetic approach to reward learning research. Many interesting questions can be addressed by taking advantage of its high spatiotemporal precision, such as the autoshaping phenomenon we highlighted above. In conclusion, it goes without saying that the level of complexity within both instrumental learning (goal-directed actions and habits) and Pavlovian reward learning is vast, though research continues to make headway. Certain questions, such as those related to the function of temporal dynamics of activity in a brain area and the differential roles of brain pathways related to reward-directed behavior, may particularly benefit from an optogenetics approach, as the above examples illustrate.

ACKNOWLEDGMENTS

This work was supported by NIH grant MH106178 (SEC) and a Whitehall Foundation research grant (to KSS). We thank Fabian Stocek and Kenneth Amaya for work leading to Figure 19.1.

REFERENCES

Adams CD. 1982. Variations in the sensitivity of instrumental responding to reinforcer devaluation. *Quart J Exp Psychol B* **34**: 77–98.

Aquili L, Liu AW, Shindou M, Shindou T, Wickens JR. 2014. Behavioral flexibility is increased by optogenetic inhibition of neurons in the nucleus accumbens shell during specific time segments. *Learn Mem* **21**: 223–231.

Aravanis AM, Wang LP, Zhang F, Meltzer LA, Mogri MZ, Schneider MB, Deisseroth K. 2007. An optical neural interface: *in vivo* control of rodent motor cortex with integrated fiberoptic and optogenetic technology. *J Neural Eng* **4**: S143–156.

Balleine BW, Dickinson A. 1998. Goal-directed instrumental action: contingency and incentive learning and their cortical substrates. *Neuropharmacology* **37**: 407–419.

Barker JM, Taylor JR, Chandler LJ. 2014. A unifying model of the role of the infralimbic cortex in extinction and habits. *Learn Mem* **21**: 441–448.

Bernstein JG, Boyden ES. 2011. Optogenetic tools for analyzing the neural circuits of behavior. *Trends Cogn Sci* **15**: 592–600.

Berridge KC. 2001. Reward learning: Reinforcement, incentives, and expectations. In *The Psychology of Learning and Motivation* (ed. DL Medin), pp. 223–278. Academic Press, New York.

Berridge KC. 2004. Motivation concepts in behavioral neuroscience. *Physiol Behav* **81**: 179–209.

Berridge KC, Robinson TE. 1998. What is the role of dopamine in reward: hedonic impact, reward learning, or incentive salience? *Brain Res Brain Res Rev* **28**: 309–369.

Butler WN, Smith KS, Taube JS. 2015. Shifting the neural compass: reversible optical disruption of the head direction signal *in vivo*. *Society for Neuroscience Abstracts* 444.08.

Chang SE, Todd TP, Bucci DJ, Smith KS. 2015. Chemogenetic manipulation of ventral pallidal neurons impairs acquisition of sign-tracking in rats. *Eur J Neurosci* **42**: 3105–3116.

Chang SE, Wheeler DS, Holland PC. 2012. Roles of nucleus accumbens and basolateral amygdala in autoshaped lever pressing. *Neurobiol Learn Mem* **97**: 441–451.

Chuong AS, Miri ML, Busskamp V, Matthews GA, Acker LC, Sorensen AT, Young A, Klapoetke NC, Henninger MA, Kodandaramaiah SB *et al.*, 2014. Noninvasive optical inhibition with a red-shifted microbial rhodopsin. *Nat Neurosci* **17**: 1123–1129.

Coutureau E, Killcross S. 2003. Inactivation of the infralimbic prefrontal cortex reinstates goal-directed responding in overtrained rats. *Behav Brain Res* **146**: 167–174.

Crego AC, Marchuk AG, Smith KS. 2015. Investigating the role of striatum in habits with optogenetics in a plus maze paradigm. *Society for Neuroscience Abstracts* DP09.05/DP05.

Day JJ, Carelli RM. 2007. The nucleus accumbens and Pavlovian reward learning. *Neuroscientist* **13**: 148–159.

Day JJ, Roitman MF, Wightman RM, Carelli RM. 2007. Associative learning mediates dynamic shifts in dopamine signaling in the nucleus accumbens. *Nat Neurosci* **10**: 1020–1028.

Deng W, Goldys EM, Farnham MM, Pilowsky PM. 2014. Optogenetics, the intersection between physics and neuroscience: light stimulation of neurons in physiological conditions. *Am J Physiol Regul Integr Comp Physiol* **307**: R1292–R1302.

Dickinson A. 1985. Actions and habits: the development of behavioral autonomy. *Philos Trans R Soc Lond B Biol Sci* **308**: 67–78.

Fenno L, Yizhar O, Deisseroth K. 2011. The development and application of optogenetics. *Annu Rev Neurosci* **34**: 389–412.

Ferster CB, Skinner BF. 1957. *Schedules of Reinforcement*. Appleton-Century-Crofts, New York.

Flagel SB, Clark JJ, Robinson TE, Mayo L, Czuj A, Willuhn I, Akers CA, Clinton SM, Phillips PE, Akil H. 2010. A selective role for dopamine in stimulus–reward learning. *Nature* **469**: 53–57.

Graybiel AM. 2008. Habits, rituals, and the evaluative brain. *Annu Rev Neurosci* **31**: 359–387.

Gremel CM, Costa RM. 2013. Orbitofrontal and striatal circuits dynamically encode the shift between goal-directed and habitual actions. *Nat Commun* **4**: 2264.

Han X. 2012. *In vivo* application of optogenetics for neural circuit analysis. *ACS Chem Neurosci* **3**: 577–584.

Hitchcott PK, Quinn JJ, Taylor JR. 2007. Bidirectional modulation of goal-directed actions by prefrontal cortical dopamine. *Cereb Cortex* **17**: 2820–2827.

Hsu YW, Wang SD, Wang S, Morton G, Zariwala HA, de la Iglesia HO, Turner EE. 2014. Role of the dorsal medial habenula in the regulation of voluntary activity, motor function, hedonic state, and primary reinforcement. *J Neurosci* **34**: 11366–11384.

Ilango A, Kesner AJ, Keller KL, Stuber GD, Bonci A, Ikemoto S. 2014. Similar roles of substantia nigra and ventral tegmental dopamine neurons in reward and aversion. *J Neurosci* **34**: 817–822.

Jin X, Tecuapetla F, Costa RM. 2014. Basal ganglia subcircuits distinctively encode the parsing and concatenation of action sequences. *Nat Neurosci* **17**: 423–430.

Kamin LJ. 1968. "Attention-like" processes in classical conditioning. In *Miami Symposium on the Prediction of Behavior: Aversive Stimulation* (ed. MR Jones), pp. 9–31. University of Miami Press Coral Gables, Florida.

Kamin LJ. 1969. Predictability, surprise, attention, and conditioning. In *Punishment and Aversive Behavior* (ed. BA Campbell, RM Church), pp. 279–296. Appleton-Century-Crofts, New York.

Killcross S, Coutureau E. 2003. Coordination of actions and habits in the medial prefrontal cortex of rats. *Cereb Cortex* **13**: 400–408.

Kravitz AV, Owen SF, Kreitzer AC. 2013. Optogenetic identification of striatal projection neuron subtypes during *in vivo* recordings. *Brain Res* **1511**: 21–32.

Lima SQ, Hromadka T, Znamenskiy P, Zador AM. 2009. PINP: a new method of tagging neuronal populations for identification during *in vivo* electrophysiological recording. *PLoS One* **4**: e6099.

Lin JY, Knutsen PM, Muller A, Kleinfeld D, Tsien RY. 2013. ReaChR: a red-shifted variant of channelrhodopsin enables deep transcranial optogenetic excitation. *Nat Neurosci* **16**: 1499–1508.

Luo L, Callaway EM, Svoboda K. 2008. Genetic dissection of neural circuits. *Neuron* **57**: 634–660.

Mahler SV, Berridge KC. 2009. Which cue to "want?" Central amygdala opioid activation enhances and focuses incentive salience on a prepotent reward cue. *J Neurosci* **29**: 6500–6513.

Mattis J, Tye KM, Ferenczi EA, Ramakrishnan C, O'Shea DJ, Prakash R, Gunaydin LA, Hyun M, Fenno LE, Gradinaru V *et al.*, 2012. Principles for applying optogenetic tools derived from direct comparative analysis of microbial opsins. *Nat Methods* **9**: 159–172.

Namburi P, Beyeler A, Yorozu S, Calhoon GG, Halbert SA, Wichmann R, Holden SS, Mertens KL, Anahtar M, Felix-Ortiz AC *et al.*, 2015. A circuit mechanism for differentiating positive and negative associations. *Nature* **520**: 675–678.

Neve RL, Carlezon Jr WA. 2002. Gene delivery into the brain using viral vectors. In *Neuropsychopharmacology: The Fifth Generation of Progress* (ed. KL Davis, D Charney, JT Coyle, C Nemeroff). Lippincott, Williams, & Wilkins, Pennsylvania.

Nicola SM. 2007. The nucleus accumbens as part of a basal ganglia action selection circuit. *Psychopharmacology (Berl)* **191**: 521–550.

Pascoli V, Terrier J, Hiver A, Luscher C. 2015. Sufficiency of mesolimbic dopamine neuron stimulation for the progression to addiction. *Neuron* **88**: 1054–1066.

Robinson TE, Yager LM, Cogan ES, Saunders BT. 2014. On the motivational properties of reward cues: Individual differences. *Neuropharmacology* **76**: 450–459.

Root DH, Melendez RI, Zaborszky L, Napier TC. 2015. The ventral pallidum: subregion-specific functional anatomy and roles in motivated behaviors. *Prog Neurobiol* **130**: 29–70.

Rossi MA, Sukharnikova T, Hayrapetyan VY, Yang L, Yin HH. 2013. Operant self-stimulation of dopamine neurons in the substantia nigra. *PLoS One* **8**: e65799.

Rothwell PE, Hayton SJ, Sun GL, Fuccillo MV, Lim BK, Malenka RC. 2015. Input- and output-specific regulation of serial order performance by corticostriatal circuits. *Neuron* **88**: 345–356.

Saddoris MP, Sugam JA, Stuber GD, Witten IB, Deisseroth K, Carelli RM. 2015. Mesolimbic dopamine dynamically tracks, and is causally linked to, discrete aspects of value-based decision making. *Biol Psychiatry* **77**: 903–911.

Saunders BT, Robinson TE. 2012. The role of dopamine in the accumbens core in the expression of Pavlovian-conditioned responses. *Eur J Neurosci* **36**: 2521–2532.

Schoenbaum G, Roesch M. 2005. Orbitofrontal cortex, associative learning, and expectancies. *Neuron* **47**: 633–636.

Schultz W. 2006. Behavioral theories and the neurophysiology of reward. *Annu Rev Psychol* **57**: 87–115.

Senn V, Wolff SB, Herry C, Grenier F, Ehrlich I, Grundemann J, Fadok JP, Muller C, Letzkus JJ, Luthi A. 2014. Long-range connectivity defines behavioral specificity of amygdala neurons. *Neuron* **81**: 428–437.

Smith KS, Berridge KC, Aldridge JW. 2011. Disentangling pleasure from incentive salience and learning signals in brain reward circuitry. *Proc Natl Acad Sci U S A* **108**: E255–E264.

Smith KS, Graybiel AM. 2013a. A dual operator view of habitual behavior reflecting cortical and striatal dynamics. *Neuron* **79**: 361–374.

Smith KS, Graybiel AM. 2013b. Using optogenetics to study habits. *Brain Res* **1511**: 102–114.

Smith KS, Graybiel AM. 2014. Investigating habits: strategies, technologies and models. *Front Behav Neurosci* **8**: 39.

Smith KS, Tindell AJ, Aldridge JW, Berridge KC. 2009. Ventral pallidum roles in reward and motivation. *Behav Brain Res* **196**: 155–167.

Smith KS, Virkud A, Deisseroth K, Graybiel AM. 2012. Reversible online control of habitual behavior by optogenetic perturbation of medial prefrontal cortex. *Proc Natl Acad Sci U S A* **109**: 18932–18937.

Soudais C, Skander N, Kremer EJ. 2004. Long-term *in vivo* transduction of neurons throughout the rat CNS using novel helper-dependent CAV-2 vectors. *FASEB J* **18**: 391–393.

Stefanik MT, Kalivas PW. 2013. Optogenetic dissection of basolateral amygdala projections during cue-induced reinstatement of cocaine seeking. *Front Behav Neurosci* **7**: 213.

Steinberg EE, Keiflin R, Boivin JR, Witten IB, Deisseroth K, Janak PH. 2013. A causal link between prediction errors, dopamine neurons and learning. *Nat Neurosci* **16**: 966–973.

Stuber GD, Britt JP, Bonci A. 2012. Optogenetic modulation of neural circuits that underlie reward seeking. *Biol Psychiatry* **71**: 1061–1067.

Stuber GD, Sparta DR, Stamatakis AM, van Leeuwen WA, Hardjoprajitno JE, Cho S, Tye KM, Kempadoo KA, Zhang F, Deisseroth K *et al.*, 2011. Excitatory transmission from the amygdala to nucleus accumbens facilitates reward seeking. *Nature* **475**: 377–380.

Taha SA, Fields HL. 2005. Encoding of palatability and appetitive behaviors by distinct neuronal populations in the nucleus accumbens. *J Neurosci* **25**: 1193–1202.

Thorndike EL. 1898. *Animal Intelligence: An Experimental Study of the Associative Processes in Animals*. Macmillan, New York.

Tindell AJ, Smith KS, Berridge KC, Aldridge JW. 2009. Dynamic computation of incentive salience: "wanting" what was never "liked". *J Neurosci* **29**: 12220–12228.

Tindell AJ, Smith KS, Peciña S, Berridge KC, Aldridge JW. 2006. Ventral pallidum firing codes hedonic reward: when a bad taste turns good. *J Neurophysiol* **96**: 2399–2409.

Tsai HC, Zhang F, Adamantidis A, Stuber GD, Bonci A, de Lecea L, Deisseroth K. 2009. Phasic firing in dopaminergic neurons is sufficient for behavioral conditioning. *Science* **324**: 1080–1084.

Tye KM, Deisseroth K. 2012. Optogenetic investigation of neural circuits underlying brain disease in animal models. *Nat Rev Neurosci* **13**: 251–266.

Tye KM, Prakash R, Kim SY, Fenno LE, Grosenick L, Zarabi H, Thompson KR, Gradinaru V, Ramakrishnan C, Deisseroth K. 2011. Amygdala circuitry mediating reversible and bidirectional control of anxiety. *Nature* **471**: 358–362.

Ugolini G. 2010. Advances in viral transneuronal tracing. *J Neurosci Methods* **194**: 2–20.

Wall NR, De La Parra M, Callaway EM, Kreitzer AC. 2013. Differential innervation of direct- and indirect-pathway striatal projection neurons. *Neuron* **79**: 347–360.

Wickersham IR, Finke S, Conzelmann KK, Callaway EM. 2007. Retrograde neuronal tracing with a deletion-mutant rabies virus. *Nat Methods* **4**: 47–49.

Witten IB, Steinberg EE, Lee SY, Davidson TJ, Zalocusky KA, Brodsky M, Yizhar O, Cho SL, Gong S, Ramakrishnan C *et al.*, 2011. Recombinase-driver rat lines: tools, techniques, and optogenetic application to dopamine-mediated reinforcement. *Neuron* **72**: 721–733.

Yizhar O, Fenno LE, Davidson TJ, Mogri M, Deisseroth K. 2011. Optogenetics in neural systems. *Neuron* **71**: 9–34.

20 Toward an Optogenetic Therapy for Epilepsy

Jan Tønnesen, Marco Ledri, and Merab Kokaia

20.1 Introduction

Since its breakthrough a decade ago, optogenetics has offered the opportunity to instantly change the excitability of targeted neural populations in a bidirectional and dose-dependent manner (Boyden *et al.*, 2005; Zhang *et al.*, 2007). As such, it has already revolutionized research in experimental neuroscience, as it allows neural cell types to be controlled optically after genetic targeting through promoter-regulated expression of photosensitive membrane proteins. This unprecedented opportunity to tightly control neuronal activity optically immediately caught the attention of researchers exploring novel therapeutic approaches in the treatment of neurological disorders (reviewed in [Tye and Deisseroth *et al.*, 2012; Tønnesen, 2013]). Epilepsy was among the first diseases to be explored in this context, as it appears feasible to intercept the neural activity underlying seizures through an optogenetic approach. While the term "epilepsy" covers an array of various pathophysiological neurologic conditions, the main focus of the optogenetic approach was on focal epilepsies, such as temporal lobe epilepsy (TLE), where aberrant excitatory activity spreads from a spatially defined area within the temporal lobes and may become generalized (Berg, 2008). TLE is frequently medically, as well as surgically, intractable and has devastating consequences for quality of life in patients, thereby urgently calling for innovative new treatment strategies to be developed (Schuele and Luders, 2008).

In this chapter, we first introduce epilepsy and the difficulties in treating it by conventional means. We then outline putative optogenetic strategies that could be applied as therapeutic approaches and present experimental work from the literature to support these. Finally, we identify the challenges that we need to address before initiating clinical trials.

20.2 Epilepsy and Current Treatment Options

20.2.1 Epilepsy: A Heterogeneous Group of Diseases

Epilepsy is one of the most common neurological disorders, affecting more than 60 million people worldwide (Ngugi *et al.*, 2011). According to the International

League Against Epilepsy, the practical (clinical) definition of epilepsy is "(i) At least two unprovoked (or reflex) seizures occurring >24 h apart; or (ii) one unprovoked (or reflex) seizure and a probability of further seizures similar to the general recurrence risk (at least 60%) after two unprovoked seizures, occurring over the next 10 years; or (iii) diagnosis of an epilepsy syndrome" (Fisher *et al.*, 2014). Epilepsy can occur at all ages, but is most common in children and the elderly, and, importantly, epilepsy is not a single disease, but has to be considered a family of disorders with different etiologies (Shorvon, 2011). Independently of the cause, epilepsy manifests with the occurrence of epileptic seizures, which can be partial (focal) or generalized (Fisher *et al.*, 2005; Seino, 2006). Partial seizures usually involve only one confined brain area where the abnormal activity occurs; generalized seizures, on the other hand, can start in one brain area, but quickly spread to the entire brain.

One of the most common forms of epilepsy in adults is TLE, characterized by partial complex seizures usually with secondary generalization. In TLE, the brain area where seizures begin is confined within limbic structures, such as the amygdala, temporal neocortex or the hippocampus (Engel, 2001). About two-thirds of patients with TLE develop hippocampal sclerosis, which consists of cell loss in several sub-regions within the hippocampal formation (Wieser, 2004).

20.2.2 Pharmacological Treatments and their Shortcomings

The current treatment against epilepsy relies on symptomatic pharmacological suppression of seizures by antiepileptic drugs (AEDs). Most people with epilepsy eventually become seizure free on AEDs, but 20–30% of patients, and up to 40% of those with TLE, do not respond to the treatment and develop pharmacoresistance (Annegers *et al.*, 1979). It is obvious that the possibility of suffering uncontrolled and unprovoked seizures has major consequences for the quality of life of people with epilepsy. Moreover, epilepsy carries an increased risk of morbidity and premature mortality, estimated as two- or three-times higher than the general population (Cockerell *et al.*, 1994). The increased mortality is partially due to the underlying cause of epilepsy, but is also a direct consequence of seizures, such as increased accidents and sudden unexpected death in epilepsy (Tomson *et al.*, 2005).

Because of their nature, AEDs only control symptoms and are not disease modifying, while at the same time they inevitably give rise to side effects because of their systemic administration. Some new classes of AEDs are being developed, but their efficiency has not yet been demonstrated to be superior to the older drugs, despite having generally better pharmacokinetics and tolerance (Prunetti and Perucca, 2011). For these reasons, the development of novel therapeutic strategies is highly needed. In an ideal scenario, the treatment should directly target the brain area where seizures develop or propagate and be specific to involved neural cell types, while leaving other cell populations unaffected. Optogenetics-based alternatives are therefore highly attractive as possible strategies against seizures, and, in particular, for inoperable TLE and other focal epilepsies, where the seizure focus is usually defined and identifiable.

20.2.3 Non-conventional Clinical Treatment Options

Aside from pharmacological treatments, electrical stimulation has been explored for the treatment of pharmacoresistant epilepsy, including deep brain stimulation (DBS) (reviewed by Fisher *et al.*, 2014) and vagus nerve stimulation (reviewed by Ben-Menachem, 2002).

In those pharmacoresistant cases where the focus is identifiable and confined, the last therapeutic option is surgical resection of the epileptic focus. Therefore, resection can be performed only in a well-selected cohort of candidates, which represent a minority of patients (Engel *et al.*, 2003). In these patients, surgery is still the most successful treatment for refractory TLE, with randomized clinical trials reporting up to 70% of patients becoming seizure free (Wiebe *et al.*, 2001; Tanriverdi *et al.*, 2008). Despite its efficacy, some negative side effects of surgery occur in around 10% of patients, with defects in a small area of their visual field being the most prominent, but also including mood disturbances and memory deficits (Spencer and Huh, 2008; Georgiadis *et al.*, 2013). Nevertheless, and most importantly, in the majority of cases, the benefits of seizure freedom outweigh the relatively modest risk of these unwanted side effects (Tanriverdi *et al.*, 2008).

20.3 Optogenetic Strategies for Treating Epilepsy

As outlined above, epilepsy can be simplistically viewed as an imbalance between excitatory and inhibitory drive that is leaning towards excitation and thereby gives rise to excessive neural firing. This imbalance can conceivably be counteracted by the strengthening or weakening of inhibition and excitation, respectively. Optogenetics is compatible with either scenario, as both depolarizing and hyperpolarizing tools are available. Thus, there are two obvious ways in which optogenetics can intercept seizure activity: (1) by shutting down excitatory neurons; or (2) by boosting the intrinsic inhibitory interneuron drive on these cells. As these scenarios involve different cell populations, namely excitatory and inhibitory neurons, they can be applied simultaneously for an even stronger seizure-suppressant effect. A schematic illustration of the optogenetic strategies currently being explored experimentally in models of epilepsy is provided in Figure 20.1.

Optogenetic treatment of epilepsy can conceivably be achieved by locating the seizure-triggering epileptic focus or, alternatively, the nuclei that modulate seizure activity, and then targeting neurons here with genetic vectors carrying optogenetic probe constructs. After expression of the vectors in the cells of interest, the targeted area would need illumination with defined wavelengths, which seems possible only via implanted optical fibers or nano-/micro-LEDs. In addition, a seizure detection system is necessary in order to automatically detect aberrant neural activity and switch on optogenetic stimulation. While this may appear far-fetched, it is not inconceivable, since the optical fiber is not more invasive than the chronic DBS electrodes that are currently implanted in patients. Such optogenetic DBS would, however, be more spatially precise and cell population tailored compared to electric DBS.

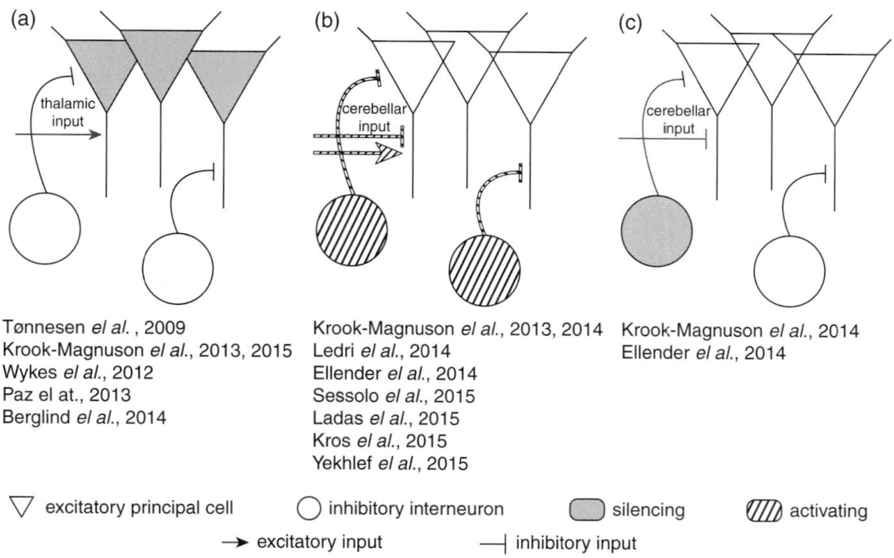

Figure 20.1 Optogenetic strategies currently being explored in experimental models of seizures and epilepsy. (A) Direct silencing of excitatory principal neurons or of incoming excitatory input. (B) Activation of local inhibitory interneurons targeting somatic or dendritic regions and activation of inhibitory or excitatory cerebellar fibers onto hippocampal pyramidal cells. (C) Silencing of perisomatic-targeting interneurons and inhibitory cerebellar inputs onto principal cells of the hippocampus.

20.3.1 Targeting Excitatory Neurons

TLE, like most epilepsies, is characterized by bouts of excessive excitatory firing, manifested as seizures. Probably the most obvious way to intercept seizures optogenetically is to shut down (hyperpolarize) the excitatory neurons of the epileptic focus by optogenetic tools (e.g. NpHR). In the hippocampus, where the epileptic focus is often located, excitatory neurons account for 95% of the total number of neurons, further qualifying them as a key therapeutic target for optogenetic interventions (Woodson *et al.*, 1989). Fortunately, excitatory neurons are easily discriminated from inhibitory neurons based on selective promoter activity, with CaMKIIα probably being the most widely utilized for expressing optogenetic constructs in these neurons by means of viral vectors (Zhang *et al.*, 2007).

20.3.2 Targeting Inhibitory Interneurons

Inhibitory interneurons represent a widely heterogeneous population of cells in the brain that inhibit pyramidal neurons by releasing GABA as the main neurotransmitter. There are almost 20 different types of interneurons in cortical areas, including the hippocampus, and their classification is based on their different anatomical and neurochemical properties (Freund and Buzsaki, 1996; Somogyi and Klausberger, 2005). Functionally, interneurons targeting different domains of pyramidal cells contribute in fundamentally different ways to the control of network function and pyramidal cell excitability. In this context, interneurons are broadly divided into those that target pyramidal cell dendritic or perisomatic regions (Freund and Katona, 2007; Klausberger, 2009; Mullner *et al.*, 2015).

In tissue from animal models of epilepsy and in epileptic tissue from human patients, the inhibitory system provided by GABAergic interneurons undergoes dramatic changes, as some cell types die and others alter their connectivity. Dendrite-targeting interneurons expressing neuropeptide Y and somatostatin (SST) seem to be the most vulnerable (Sperk *et al.*, 1992; Schwarzer *et al.*, 1995), while perisomatic inhibition, and in particular that provided by parvalbumin (PV)-expressing interneurons, is to some extent more preserved in TLE (Wyeth *et al.*, 2010). As the perisomatic inhibition provided by some PV-expressing cells is very efficient in controlling the generation of action potentials by pyramidal cells, one possibility is that the preservation of PV terminals compensates increased tissue excitability and limits the generation or propagation of seizures. However, the ability of perisomatic interneurons to generate and control oscillatory activity might also indicate that the preservation of PV-expressing axons could instead contribute to the generation of abnormal synchrony and therefore promote seizures (Wittner *et al.*, 2005; Marchionni and Maccaferroi, 2009). Optogenetic techniques are therefore highly suited for functionally dissecting and better understanding the roles of different types of inhibitory interneurons in ictogenesis, as they provide the unique possibility of selectively stimulating or silencing specific interneuron populations and studying the impact on epileptiform activity and seizures. Therefore, optogenetic activation of interneurons (e.g. via ChR2 expression) in order to release GABA may result in mixed and opposite functional outcomes, depending on the specificity of the transgene targeting.

20.4 Experimental Studies Providing Proof of Principle

A number of experimental studies have applied optogenetics in animal models of epilepsy, ranging from brain slice preparations to intact freely behaving animals. As we outline here, most of these studies can be classified based on whether they target excitatory or inhibitory neurons, though some apply both, and others explore modulating oscillatory signaling rather than simply switching neurons on or off.

20.4.1 Silencing Excitatory Neurons in the Epileptiform Focus

In 2009, we provided the first experimental proof of principle for the optogenetic control of seizures by showing that electrically induced epileptiform burst-firing in hippocampal slice cultures could be inhibited by the optogenetic silencing of excitatory neurons at the time of onset of the seizure-inducing stimulus (Tønnesen *et al.*, 2009). We used the hyperpolarizing orange light-driven inward chloride pump NpHR expressed lentivirally in excitatory neurons, and illuminated the preparation via a microscope objective through a simple wide-field approach. The NpHR probe has since been developed into much more photoefficient versions that provide stronger hyperpolarization for a given illumination intensity (Gradinaru *et al.*, 2008; Gradinaru *et al.*, 2010). Moving into intact animals and illuminating genetically transduced cells via surgically implanted optical fibers, Wykes *et al.* recorded electroencephalographic activity in a rat tetanus toxin injection model of focal epilepsy, and compared spontaneous high-frequency discharge events during periods with focal NpHR activation in the seizure focus to periods without NpHR activation (Wykes *et al.*, 2012). In such experimental conditions, NpHR-mediated hyperpolarization of neurons attenuated high-frequency electroencephalographic activity, suggesting modulation of epileptiform activity in live animals. Following this study, a closed-loop monitoring/stimulation system was developed that enabled electrophysiological detection of stroke-induced spontaneous cortical seizures and their acute optogenetic interception by hyperpolarization of excitatory thalamic neurons in rats (Paz *et al.*, 2013). Impressively, this approach reliably and immediately stopped both the electrographic and behavioral seizures. Almost at the same time, very similar results were published based on a mouse model of TLE, where kainic acid injection into the hippocampus forms an epileptic seizure focus (Krook-Magnuson *et al.*, 2013). In addition to reducing epileptic activity through silencing of excitatory neurons, this study successfully demonstrated that activation of inhibitory neurons using the depolarizing ChR2 transgene is an equally effective strategy, which we describe further below.

Following these initial studies, several publications have described attenuation of epileptiform activity in intact animals based on various models of epilepsy, including the pilocarpine model (Sukhotinsky *et al.*, 2013) and the picrotoxin model (Berglind *et al.*, 2014), all of them utilizing variants of the NpHR probe to silence excitatory neurons through illumination via implanted optical fibers.

The general strategy for intercepting seizure activity by targeting and hyperpolarizing excitatory neurons through sustained illumination and NpHR activation appears sound, as it has been applied in various epilepsy models with convincing results (Figure 20.2).

A somewhat different approach is *activating* excitatory neurons to intercept seizures, which has proven feasible in a model of absence epilepsy. Absence epilepsy is most prevalent among children and is very different from TLE in being manifested as temporarily impaired consciousness and behavioral arrest. The underlying electroencephalographic activity is a generalized spike-and-wave discharge pattern throughout the cortex and thalamus (Snead, 1995). Early on, stimulation of the cerebellar cortex was used to alleviate generalized seizures clinically, though with inconsistent results (Cooper *et al.*, 1976). Kros *et al.* showed that the generalized spike-and-wave discharge pattern observed in mouse models of absence epilepsy was associated with phase-locked firing in subsets of cerebellar nuclei neurons, which receive input from the cerebellar cortex (Kros *et al.*, 2015). A brief <300 ms optogenetic activation of excitatory neurons of the cerebellar nuclei via ChR2 was able to reset the aberrant discharge pattern and immediately intercept seizures on a behavioral level. The authors conclude that stimulating the cerebellar nuclei is able to interrupt absence seizures and should be explored for interrupting other types of generalized seizures. This study is noteworthy because it controls seizures in genetic epilepsy models without a well-defined seizure focus. The approach is conceptually very interesting clinically, especially when a seizure focus is ill-defined and/or cannot be targeted for surgical removal or implantation of optical fibers.

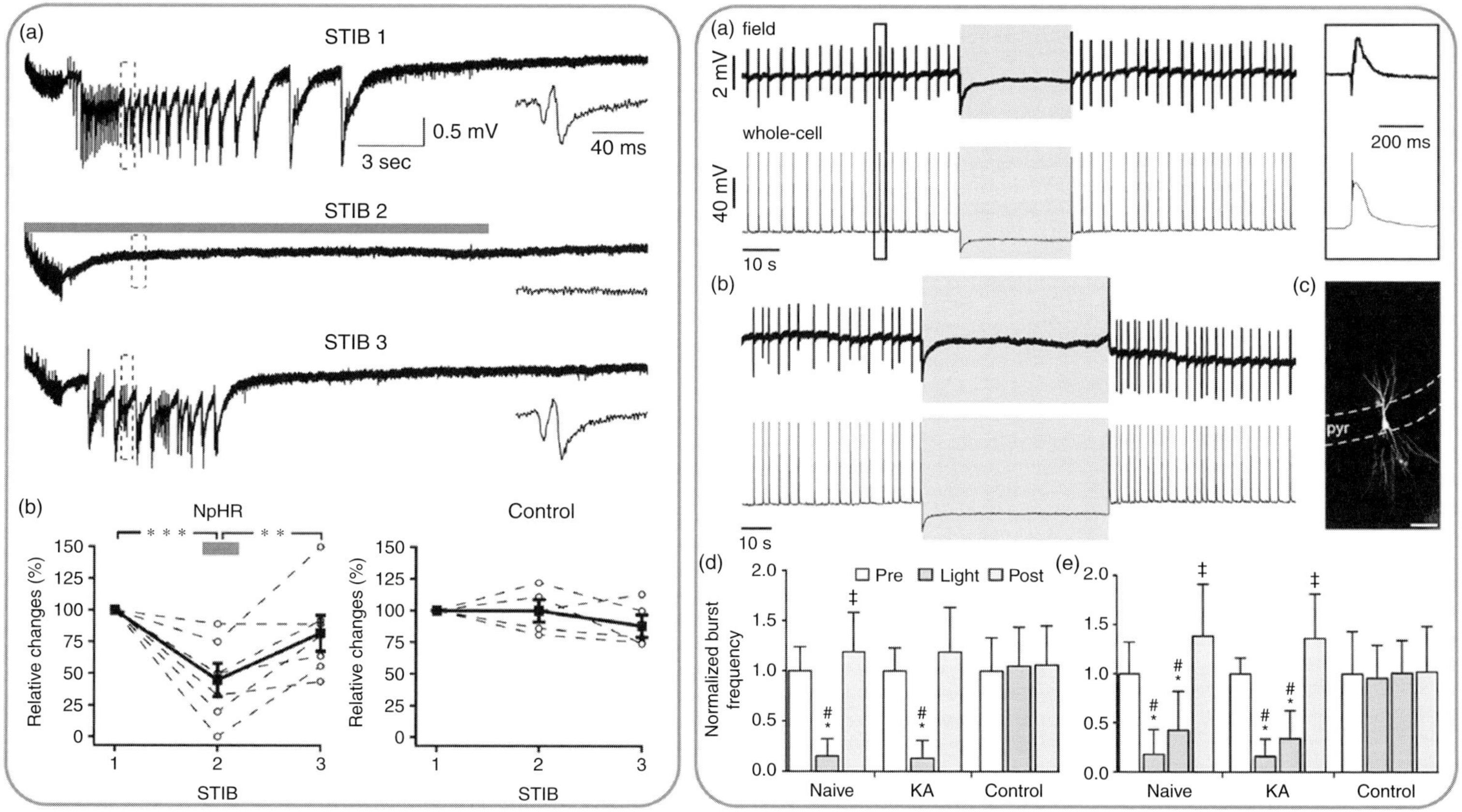

Figure 20.2 Optogenetic silencing of excitatory principal neurons to decrease epileptiform activity and seizures. Illumination of NpHR-expressing pyramidal neurons (gray bars) is able to decrease epileptiform activity in the form of stimulation-induced bursting in organotypic hippocampal slices (top left) or in the presence of low magnesium and 4-aminopyridine in acute hippocampal preparations (top right). *In vivo*, direct silencing of NpHR-expressing principal neurons by closed-loop systems can interrupt kainite-induced seizures in the hippocampus (bottom left). Cortical seizures developing after stroke can be intercepted by the silencing of thalamocortical projections (bottom right). Figures reproduced with permissions from Tønnesen *et al.* (2009), Krook-Magnuson *et al.* (2013), Paz *et al.* (2013) and Berglind *et al.* (2014).

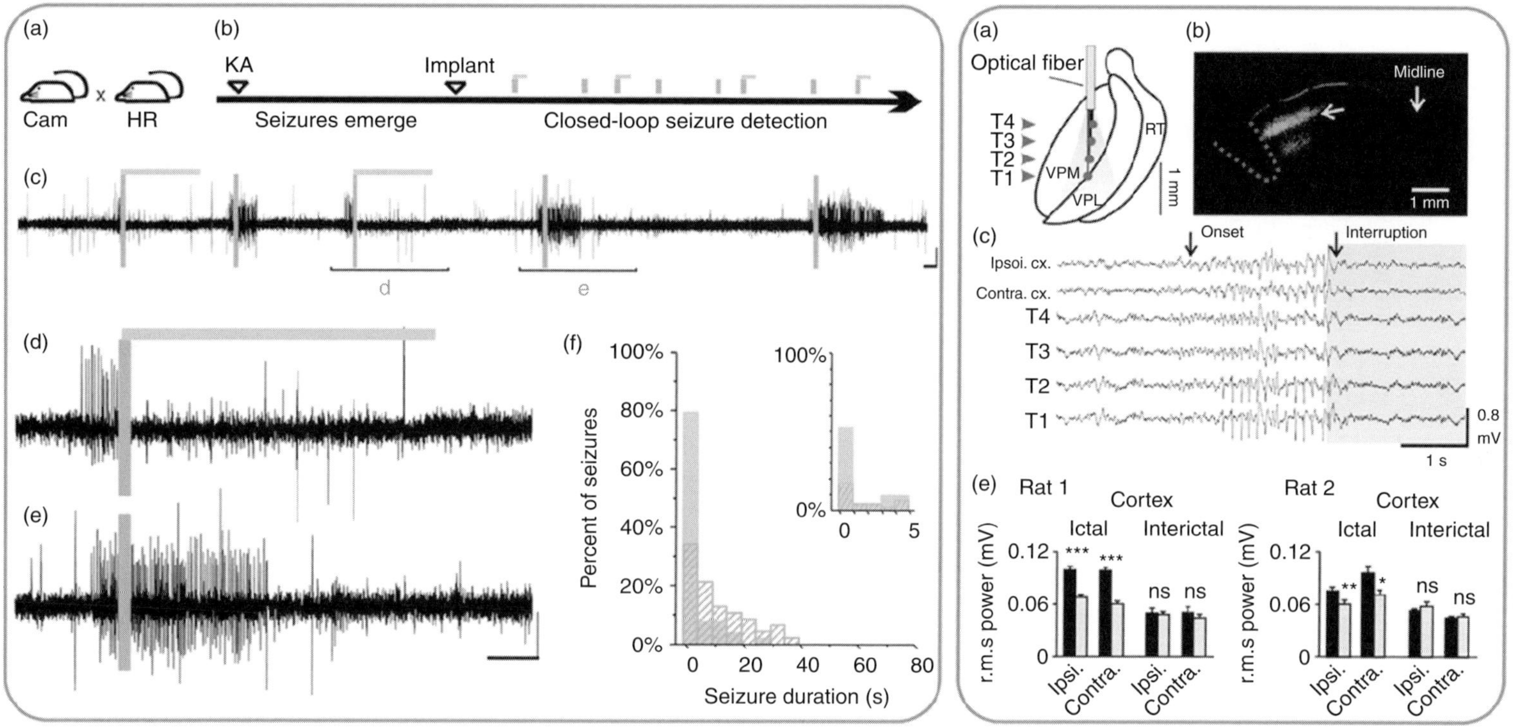

Figure 20.2 (cont.)

Although the above findings are encouraging, there is room for improvement in the efficacy of the optogenetic approach, and the results need to be confirmed in larger species, which we address later in this chapter.

20.4.2 Activating Inhibitory Neurons to Suppress Excessive Network Excitability

Our group was among the first to show that in acute hippocampal slices, optogenetic activation of interneurons at 20- and 50-Hz frequencies can reduce the epileptiform activity induced by bath application of the potent convulsant 4-aminopyridine (4-AP) (Ledri *et al.*, 2014). Moreover, we demonstrated that global activation of all interneurons by expressing ChR2 in glutamic acid decarboxylase cells was more effective than activation of PV- or SST-expressing cells alone (Figure 20.3). In a similar model, another study reported that 1-Hz

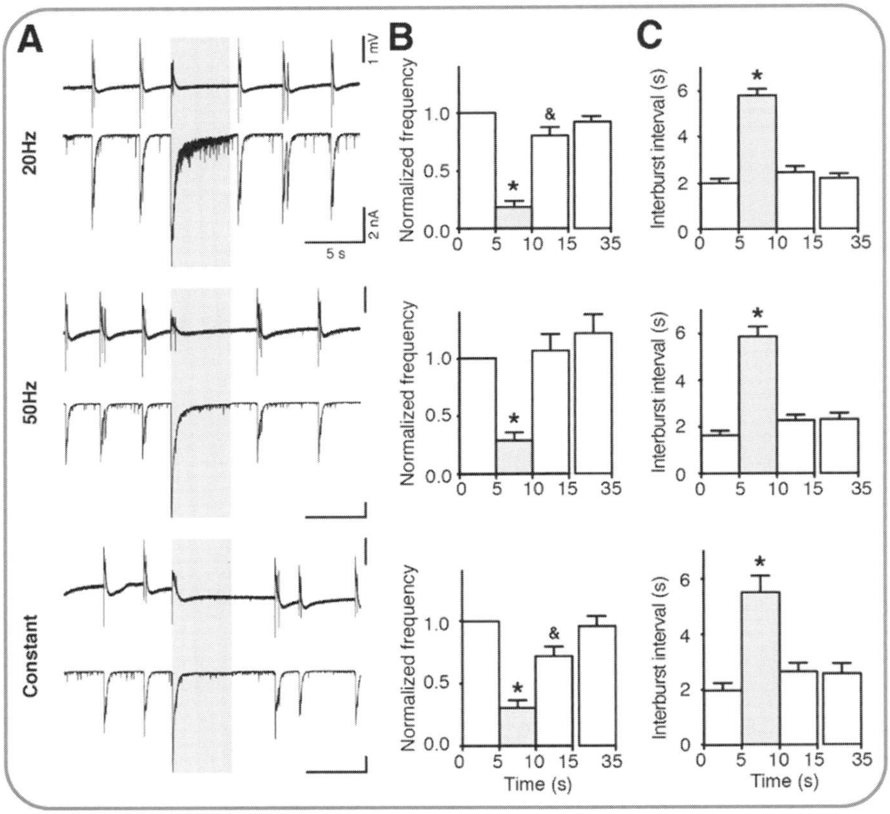

Figure 20.3 Optogenetic activation of inhibitory interneurons is able to reduce the frequency of epileptiform activity and seizures. Activation of ChR2-expressing GABAergic interneurons by illumination (gray bars) can reduce the frequency of epileptiform discharges evoked by 4-AP in acute hippocampal slices (top). Moreover, activation of PV-expressing interneurons *in vivo* by closed-loop optogenetics has proven effective at blocking seizures induced by intrahippocampal kainic acid injection (bottom). Figures reproduced with permission from Krook-Magnuson *et al.* (2013) and Ledri *et al.* (2014).

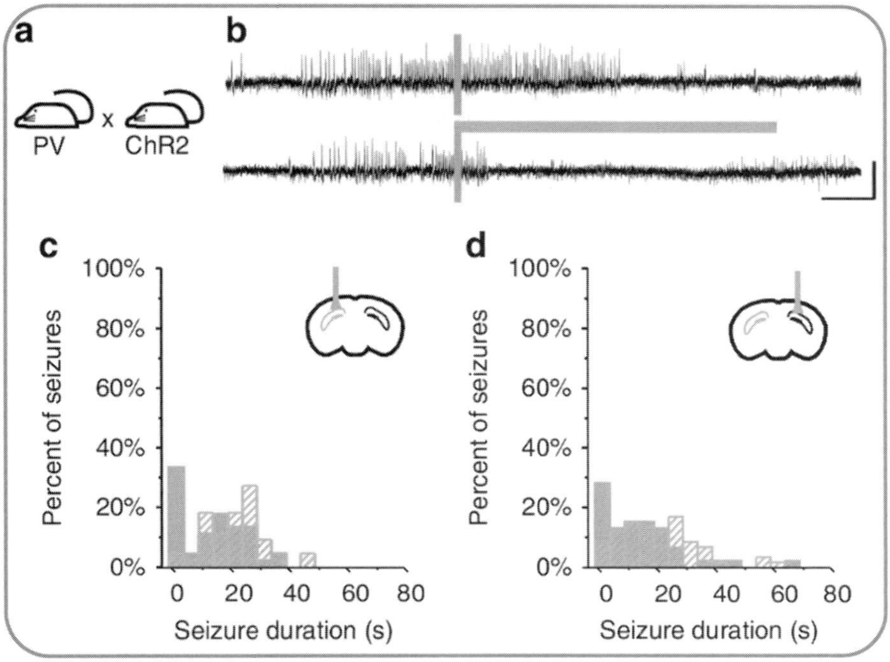

Figure 20.3 (cont.)

optogenetic activation of interneurons expressing the vesicular GABA transporter was effective at reducing the occurrence of interictal-like activity in hippocampal slices (Ladas *et al.*, 2015). However, in both cases, optogenetic stimulation of interneurons and the concomitant synchronous release of large amounts of GABA caused different degrees of pyramidal cell firing, posing questions as to the mechanisms that are underlying the effectiveness of controlling epileptiform activity.

In line with these observations, an *in vitro* study using focal N-methyl-D-aspartate (NMDA) delivery and 4-AP application in cortical slices as an *in vitro* model of ictal events found that optogenetic activation of PV-expressing cells before or after NMDA application was unable to stop the generation of ictal-like activity. On the contrary, it was responsible for increasing the duration of ictal events (Sessolo *et al.*, 2015). The authors were then able to demonstrate that activation of PV interneurons promotes synchronous firing of pyramidal cells by inducing post-inhibitory rebound spiking, arguing that this approach would not be suited for controlling epileptiform activity. However, when activation of PV interneurons was performed in regions distant from the area of focal NMDA application, light was able to block the propagation and shorten the duration of ictal events.

Other studies support the idea that PV interneurons contribute to the generation of seizure-like events *in vitro* and are ineffective at reducing epileptiform discharges (Ellender *et al.*, 2014; Yekhlef *et al.*, 2015). However, it seems that applying this approach *in vivo* could be fundamentally different. Using a closed-loop system and intrahippocampal kainic acid injection as an epilepsy model, Krook-Magnuson and colleagues showed that selective stimulation of PV interneurons both ipsi- and contra-lateral to the epileptic focus was very effective at reducing the duration of electrographic seizures (Figure 20.3) (Krook-Magnuson *et al.*, 2013). Using the same model, they went on to demonstrate that optogenetic activation of cerebellar Purkinje cells was effective at decreasing electrographic seizures in the hippocampus, arguing that increased inhibitory drive from PV interneurons, independently of their location, is highly effective at reducing seizures *in vivo* (Krook-Magnuson *et al.*, 2014).

The discrepancy between the results obtained *in vitro* and *in vivo* underlines the importance of recognizing the different model systems that are used when interpreting experimental data. One could perhaps argue that, in this specific case, *in vitro* models of epileptiform activity based on 4-AP or low magnesium create conditions that are somewhat different from what happens in the epileptic tissue of the animal *in vivo*. Nevertheless, the importance of *in vitro* studies should not be diminished, as only in such closely controlled conditions is it possible to unveil specific mechanisms that may be fundamental to seizure generation, propagation or control, and should be carefully considered when choosing the most appropriate strategies for further *in vivo* verification.

Most of the initial studies utilizing this approach focused primarily on interneurons that synaptically target the perisomatic compartment of the principal neurons, but the potential for increasing the inhibitory drive of dendrite-targeting inhibitory cells remains poorly explored. Despite being located far away from the cell soma and axon initial segment, dendritic inhibitory synapses might be more effective at controlling the action potential output of the principal neurons by shunting excitatory currents and effectively reducing the summation of excitatory inputs (Lovett-Barron *et al.*, 2012). This strategy should therefore be considered for the optogenetic control of hyperexcitability in epilepsy.

20.4.3 Alternative Optogenetic Approaches in Epilepsy Therapy

Apart from the direct targeting of neural populations by sustained NpHR or ChR2 transgene activation, as described above, alternative seizure-suppressing strategies are being currently explored. For example, it is plausible that neuronal excitability can be modulated indirectly by optogenetically controlling non-neuronal cells (e.g. glia). Astrocytes are known to influence the functional state of neurons and are targetable with optogenetic probes such as ChR2 (Figueiredo *et al.*, 2011). It may be feasible to modulate astrocyte function and thereby effectively control seizures, although this approach may operate on a slower timescale compared to neuronal modulation and therefore not be feasible as an acute closed-loop interception strategy. Nevertheless, it could be worth exploring whether modulating the ionic composition of the extracellular space via optogenetic probes in the membranes of astrocytes may lead to lowered excitability of neurons.

20.5 Developing Optogenetic Approaches toward Clinical Use

It is clear that animal models of epilepsy have provided proof of concept for the potential use of optogenetics in epilepsy therapy. As already outlined, the studies have further proven the feasibility of self-contained devices offering integrated seizure detection and optogenetic stimulation, and such closed-loop systems could be developed for clinical use. Nevertheless, there are major obstacles that need be overcome before experimental findings can be translated into clinical trials.

20.5.1 The Scaling Problem: From Mouse to Human

The scaling problem refers to the difference in scale between the brain size of humans as compared to the animals used for disease modeling, and the difficulties this poses for translating experimental findings to the human situation. For optogenetic approaches, the scaling problem is particularly evident, as the efficacy of the treatment relates directly to the tissue volume that is controllable optogenetically and the volume of the epileptiform focus. Optogenetic therapy is most effective where these overlap. This puts certain demands on the delivery vector efficacy in covering an area to transduce sufficient cells, and on the ability of light to penetrate the brain tissue and address these (Figure 20.4).

Coverage of larger brain structures by transgenes can be resolved by multiple viral vector injections. As to the limitations related to illumination, photons in the visible spectrum of light that are normally used to activate opsins penetrate poorly in the brain tissue, where they are typically scattered and absorbed well within 1 mm of the light source (Al-Juboori *et al.*, 2013; Azimipour *et al.*, 2014). Blood vessels may further complicate this scenario and limit the propagation of light in the brain (Azimipour *et al.*, 2015). While a single cubic millimeter is a notable part of a mouse hippocampus, it is a several-fold smaller volume of a human hippocampus, and it may not be sufficient for optically controlling TLE (Figure 20.5).

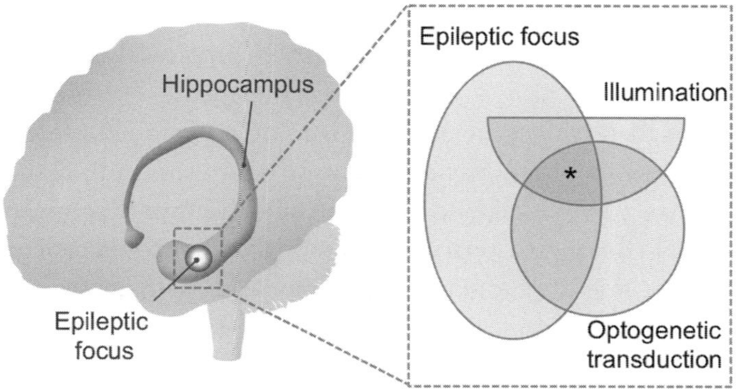

Figure 20.4 The effective optogenetic treatment volume in tissue is defined by the overlap of the epileptic focus, the volume of optogenetically transduced cells and the illumination volume. Only where all three overlap (marked by an asterisk) is optogenetic intervention of seizure activity efficient.

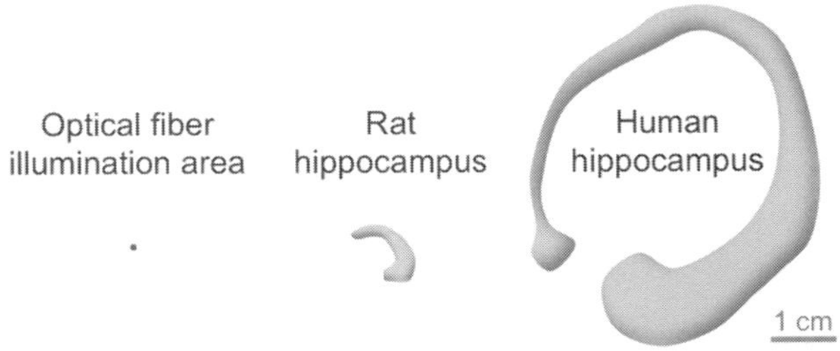

Figure 20.5 The scaling problem is evident from comparing the relative sizes of the area that an optical fiber can typically illuminate in brain tissue to the respective areas of a rodent hippocampus and a human hippocampus. Whereas the illumination area may target a notable part of the rodent hippocampus, the corresponding area represents a much smaller fraction of the human hippocampus, which will therefore be more difficult to control optogenetically. As illustrated in Figure 20.4, the efficacy of optogenetic therapy will depend not only on the illumination area, but also on the size of the epileptic focus and accurate optogenetic transfection of cells in this focus.

A conceivable strategy for overcoming this problem is to use multiple optical fibers in order to address the relevant tissue volume. Another strategy is to develop opsins that are activated by photons that are further in the red light spectrum, which penetrate much longer distances in the tissue. The latter approach has been explored, enabling optogenetic control at several millimeters of tissue depth by illumination at 635-nm wavelength (Chuong *et al.*, 2014). The authors of this study even reported non-invasive optical cell control by transcranial illumination through the intact skull in mice, though again this finding will likely be difficult to translate to humans, who have much thicker skulls.

While the several rodent experiments are very encouraging and useful, it will be interesting to attempt to replicate the findings of these in larger animal models, such as minipigs or non-human primates.

20.6 Concluding Remarks

Finally, it is important to consider what would be a successful outcome of optogenetic therapy for epilepsy. This treatment does not seem to provide a cure or disease modification, but rather offers a prosthetic approach, suppressing seizures when they are about to occur or have started. A given therapeutic outcome of optogenetics must be weighed against the potential risks from viral vector injections, foreign protein expression and implantations of optical fibers into the brain. Moreover, the efficacy of optogenetic treatment should prove superior to currently available AEDs. In this regard, pharmacoresistant focal epilepsies may be the primary targets of the use of the optogenetic approach. With respect to therapeutic effects, it is worth considering how given experimental findings translate into clinical impact. Here, it is clear that a reduction in seizure frequency seems to be far more beneficial to the patient than a reduction in seizure duration. Any outcome that is partial in terms of decreased seizure frequency and/or seizure duration may not be sufficient from a patient's perspective, and only a seizure-free condition may justify the risk/benefit ratio.

Whether optogenetics, after resolving all potential translational hurdles, can provide such a therapeutic outcome remains to be proven, though experimental research is continuously providing encouraging data and motivating further efforts in this direction.

REFERENCES

Al-Juboori, S. I., Dondzillo, A., Stubblefield, E. A., *et al.* (2013). Light scattering properties vary across different regions of the adult mouse brain. *PLoS One*, **8**, e67626.

Annegers, J. F., Hauser, W. A. and Elveback, L. R. (1979). Remission of seizures and relapse in patients with epilepsy. *Epilepsia*, **20**, 729–737.

Azimipour, M., Atry, F. and Pashaie, R. (2015). Effect of blood vessels on light distribution in optogenetic stimulation of cortex. *Opt Lett*, **40**, 2173–2176.

Azimipour, M., Baumgartner, R., Liu, Y., *et al.* (2014). Extraction of optical properties and prediction of light distribution in rat brain tissue. *J Biomed Opt*, **19**, 75001.

Ben-Menachem, E. (2002). Vagus-nerve stimulation for the treatment of epilepsy. *Lancet Neurol*, **1**, 477–482.

Berg, A. T. (2008). The natural history of mesial temporal lobe epilepsy. *Curr Opin Neurol*, **21**, 173–178.

Berglind, F., Ledri, M., Sorensen, A. T., *et al.* (2014). Optogenetic inhibition of chemically induced hypersynchronized bursting in mice. *Neurobiol Dis*, **65**, 133–141.

Boyden, E. S., Zhang, F., Bamberg, E., *et al.* (2005). Millisecond-timescale, genetically targeted optical control of neural activity. *Nat Neurosci*, **8**, 1263–1268.

Chuong, A. S., Miri, M. L., Busskamp, V., *et al.* (2014). Noninvasive optical inhibition with a red-shifted microbial rhodopsin. *Nat Neurosci*, **17**, 1123–1129.

Cockerell, O. C., Johnson, A. L., Sander, J. W., *et al.* (1994). Mortality from epilepsy: results from a prospective population-based study. *Lancet*, **344**, 918–921.

Cooper, I. S., Amin, I., Riklan, M., *et al.* (1976). Chronic cerebellar stimulation in epilepsy. Clinical and anatomical studies. *Arch Neurol*, **33**, 559–570.

Ellender, T. J., Raimondo, J. V., Irkle, A., *et al.* (2014). Excitatory effects of parvalbumin-expressing interneurons maintain hippocampal epileptiform activity via synchronous after discharges. *J Neurosci*, **34**, 15208–15222.

Engel, J., Jr. (2001). A proposed diagnostic scheme for people with epileptic seizures and with epilepsy: report of the ILAE Task Force on Classification and Terminology. *Epilepsia*, **42**, 796–803.

Engel, J., Jr., Wiebe, S., French, J., *et al.* (2003). Practice parameter: temporal lobe and localized neocortical resections for epilepsy. *Epilepsia*, **44**, 741–751.

Figueiredo, M., Lane, S., Tang, F., *et al.* (2011). Optogenetic experimentation on astrocytes. *Exp Physiol*, **96**, 40–50.

Fisher, R. S., Acevedo, C., Arzimanoglou, A., *et al.* (2014). ILAE official report: a practical clinical definition of epilepsy. *Epilepsia*, **55**, 475–482.

Fisher, R. S., van Emde Boas, W., Blume, W., *et al.* (2005). Epileptic seizures and epilepsy: definitions proposed by the International League Against Epilepsy (ILAE) and the International Bureau for Epilepsy (IBE). *Epilepsia*, **46**, 470–472.

Fisher, R. S. and Velasco, A. L. (2014). Electrical brain stimulation for epilepsy. *Nat Rev Neurol*, **10**, 261–270.

Freund, T. F. and Buzsaki, G. (1996). Interneurons of the hippocampus. *Hippocampus*, **6**, 347–470.

Freund, T. F. and Katona, I. (2007). Perisomatic inhibition. *Neuron*, **56**, 33–42.

Georgiadis, I., Kapsalaki, E. Z. and Fountas, K. N. (2013). Temporal lobe resective surgery for medically intractable epilepsy: a review of complications and side effects. *Epilepsy Res Treat*, 2013, 752195.

Gradinaru, V., Thompson, K. R. and Deisseroth, K. (2008). eNpHR: a *Natronomonas* halorhodopsin enhanced for optogenetic applications. *Brain Cell Biol*, **36**, 129–139.

Gradinaru, V., Zhang, F., Ramakrishnan, C., *et al.* (2010). Molecular and cellular approaches for diversifying and extending optogenetics. *Cell*, **141**, 154–165.

Klausberger, T. (2009). GABAergic interneurons targeting dendrites of pyramidal cells in the CA1 area of the hippocampus. *Eur J Neurosci*, **30**, 947–957.

Krook-Magnuson, E., Armstrong, C., Oijala, M., *et al.* (2013). On-demand optogenetic control of spontaneous seizures in temporal lobe epilepsy. *Nat Commun*, **4**, 1376.

Krook-Magnuson, E., Szabo, G. G., Armstrong, C., *et al.* (2014). Cerebellar directed optogenetic intervention inhibits spontaneous hippocampal seizures in a mouse model of temporal lobe epilepsy. *eNeuro*, **1**, e.2014.

Kros, L., Eelkman Rooda, O. H., Spanke, J. K., *et al.* (2015). Cerebellar output controls generalized spike-and-wave discharge occurrence. *Ann Neurol*, **77**, 1027–1049.

Ladas, T. P., Chiang, C. C., Gonzalez-Reyes, L. E., *et al.* (2015). Seizure reduction through interneuron-mediated entrainment using low frequency optical stimulation. *Exp Neurol*, **269**, 120–132.

Ledri, M., Madsen, M. G., Nikitidou, L., *et al.* (2014). Global optogenetic activation of inhibitory interneurons during epileptiform activity. *J Neurosci*, **34**, 3364–3377.

Lovett-Barron, M., Turi, G. F., Kaifosh, P., *et al.* (2012). Regulation of neuronal input transformations by tunable dendritic inhibition. *Nat Neurosci*, **15**, 423–430, S421–423.

Marchionni, I. and Maccaferri, G. (2009). Quantitative dynamics and spatial profile of perisomatic GABAergic input during epileptiform synchronization in the CA1 hippocampus. *J Physiol*, **587**, 5691–5708.

Mullner, F. E., Wierenga, C. J. and Bonhoeffer, T. (2015). Precision of inhibition: dendritic inhibition by individual GABAergic synapses on hippocampal pyramidal cells is confined in space and time. *Neuron*, **87**, 576–589.

Ngugi, A. K., Kariuki, S. M., Bottomley, C., *et al.* (2011). Incidence of epilepsy: a systematic review and meta-analysis. *Neurology*, **77**, 1005–1012.

Paz, J. T., Davidson, T. J., Frechette, E. S., *et al.* (2013). Closed-loop optogenetic control of thalamus as a tool for interrupting seizures after cortical injury. *Nat Neurosci*, **16**, 64–70.

Prunetti, P. and Perucca, E. (2011). New and forthcoming anti-epileptic drugs. *Curr Opin Neurol*, **24**, 159–164.

Schuele, S. U. and Luders, H. O. (2008). Intractable epilepsy: management and therapeutic alternatives. *Lancet Neurol*, **7**, 514–524.

Schwarzer, C., Williamson, J. M., Lothman, E. W., *et al.* (1995). Somatostatin, neuropeptide Y, neurokinin B and cholecystokinin immunoreactivity in two chronic models of temporal lobe epilepsy. *Neuroscience*, **69**, 831–845.

Seino, M. (2006). Classification criteria of epileptic seizures and syndromes. *Epilepsy Res*, **70** (Suppl. 1)., S27–33.

Sessolo, M., Marcon, I., Bovetti, S., *et al.* (2015). Parvalbumin-positive inhibitory interneurons oppose propagation but favor generation of focal epileptiform activity. *J Neurosci*, **35**, 9544–9557.

Shorvon, S. D. (2011). The etiologic classification of epilepsy. *Epilepsia*, **52**, 1052–1057.

Snead, O. C., 3rd (1995). Basic mechanisms of generalized absence seizures. *Ann Neurol*, **37**, 146–157.

Somogyi, P. and Klausberger, T. (2005). Defined types of cortical interneurone structure space and spike timing in the hippocampus. *J Physiol*, **562**, 9–26.

Spencer, S. and Huh, L. (2008). Outcomes of epilepsy surgery in adults and children. *Lancet Neurol*, **7**, 525–537.

Sperk, G., Marksteiner, J., Gruber, B., *et al.* (1992). Functional changes in neuropeptide Y- and somatostatin-containing neurons induced by limbic seizures in the rat. *Neuroscience*, **50**, 831–846.

Sukhotinsky, I., Chan, A. M., Ahmed, O. J., *et al.* (2013). Optogenetic delay of status epilepticus onset in an *in vivo* rodent epilepsy model. *PLoS One*, **8**, e62013.

Tanriverdi, T., Poulin, N. and Olivier, A. (2008). Life 12 years after temporal lobe epilepsy surgery: a long-term, prospective clinical study. *Seizure*, **17**, 339–349.

Tomson, T., Walczak, T., Sillanpaa, M., *et al.* (2005). Sudden unexpected death in epilepsy: a review of incidence and risk factors. *Epilepsia*, **46**(Suppl. 11), 54–61.

Tønnesen, J. (2013). Optogenetic cell control in experimental models of neurological disorders. *Behav Brain Res*, **255**, 35–43.

Tønnesen, J., Sorensen, A. T., Deisseroth, K., *et al.* (2009). Optogenetic control of epileptiform activity. *PNAS*, **106**, 12162–12167.

Tye, K. M. and Deisseroth, K. (2012). Optogenetic investigation of neural circuits underlying brain disease in animal models. *Nat Rev Neurosci*, **13**, 251–266.

Wiebe, S., Blume, W. T., Girvin, J. P., *et al.* (2001). A randomized, controlled trial of surgery for temporal-lobe epilepsy. *N Engl J Med*, **345**, 311–318.

Wieser, H. G. (2004). ILAE Commission Report. Mesial temporal lobe epilepsy with hippocampal sclerosis. *Epilepsia*, **45**, 695–714.

Wittner, L., Eross, L., Czirjak, S., *et al.* (2005). Surviving CA1 pyramidal cells receive intact perisomatic inhibitory input in the human epileptic hippocampus. *Brain*, **128**, 138–152.

Woodson, W., Nitecka, L. and Ben-Ari, Y. (1989). Organization of the GABAergic system in the rat hippocampal formation: a quantitative immunocytochemical study. *J Comp Neurol*, **280**, 254–271.

Wyeth, M. S., Zhang, N., Mody, I., *et al.* (2010). Selective reduction of cholecystokinin-positive basket cell innervation in a model of temporal lobe epilepsy. *J Neurosci*, **30**, 8993–9006.

Wykes, R. C., Heeroma, J. H., Mantoan, L., *et al.* (2012). Optogenetic and potassium channel gene therapy in a rodent model of focal neocortical epilepsy. *Sci Transl Med*, **4**, 161ra152.

Yekhlef, L., Breschi, G. L., Lagostena, L., *et al.* (2015). Selective activation of parvalbumin- or somatostatin-expressing interneurons triggers epileptic seizurelike activity in mouse medial entorhinal cortex. *J Neurophysiol*, **113**, 1616–1630.

Zhang, F., Wang, L. P., Brauner, M., *et al.* (2007). Multimodal fast optical interrogation of neural circuitry. *Nature*, **446**, 633–639.

21 Using Optogenetics and Stem Cell-derived Neural Engraftment Techniques to Restore Lost Motor Function

J. Barney Bryson and Linda Greensmith

21.1 Introduction

The major function of the nervous system in all higher organisms is to receive and process sensory information from the environment and, in turn, to produce an appropriate response in order to adapt to changes in the environment. In most cases, the elicited response takes the form of movement or "motor function". Almost all human behavioral output is underpinned by motor functions, ranging from locomotion and articulated hand movements, to speech and expression of emotions. The neural circuits that govern motor function are extremely complex, and damage to almost any part of the motor system, either as a result of trauma or disease, can result in an inability to control muscles, termed paralysis. In this chapter, we will focus on two such insults that can critically affect the motor system: traumatic spinal cord injury (SCI) and the progressive neurodegenerative disease amyotrophic lateral sclerosis (ALS). Given the importance of the motor system, even minor impairment of its function can significantly affect the quality of life of affected individuals, whilst severe loss of motor function can be immediately life-threatening. Currently, there are no conventional therapeutic approaches that are capable of inducing spontaneous regeneration of trauma-induced lesions affecting motor circuits within the central nervous system (CNS) and there are no effective therapies that can delay or reverse the fatal progressive loss of motor neurons that occurs in ALS. Consequently, alternative strategies are now being sought to repair the neural circuits that mediate motor control and to artificially restore function to specific muscle groups in order to enable essential motor tasks. This chapter will discuss recent advances toward the translational application of stem cell-based neural replacement techniques and artificial control of motor function using optogenetics as therapeutic strategies to restore lost motor function.

21.1.1 Overview of the Motor System and Paralytic Disorders

In order to fully appreciate the challenges that have so far hindered attempts to repair the damaged motor system, an understanding of the basic anatomy and function of the voluntary motor system is required. Normally, neuronal commands to produce a voluntary muscle movement are initiated in the brain in the primary motor cortex. These signals are relayed by specialized projection neurons (cortical motor neurons or Betz cells) to the appropriate level of the brainstem or spinal cord and are either transmitted directly to α-motor neurons within the ventral horn of the spinal cord in the case of corticomotor neuron synaptic transmission (Lemon, 2008) or indirectly following additional processing by local neural circuits (see Figure 21.1). The spinal motor neurons then convey the motor command out of the spinal cord and along an axon within a peripheral nerve to the appropriate effector muscle that executes the intended movement. In addition, feedback of sensory information about the elicited movement is continuously relayed back to the CNS by proprioceptive neurons that can dynamically modulate muscle contraction in order to match the intended motor function. The neural circuits that are required to accomplish this process are extremely complex; for example, there are known to be at least 21 first-order inputs onto spinal motor neurons that control and modulate their function (Brownstone and Bui, 2010), and the formation of these complex motor circuits is governed by an intricate interplay of molecular and genetic programs during development (Ladle *et al.*, 2007).

Damage to any part of this neural circuit arising from traumatic injury or a wide range of other insults, including stroke, tumors and multiple sclerosis, can result in permanent paralysis of muscles that lie distal to the lesion, as often occurs following traumatic SCI. The extent of SCI can vary greatly, ranging from a discrete transection injury (e.g. in the case of a knife injury to the spinal cord) or, more commonly, to contusion injuries following vertebral facture that can span many spinal levels and often result in the irreversible death of neurons within the site of the injury (Nicaise *et al.*, 2012). Typically, the higher the site of the injury, the greater the area of the body that is paralyzed. Thus, in the case of cervical SCI, there can be a loss of function, or even destruction, of phrenic motor neurons that control the diaphragm muscle (Nicaise *et al.*, 2012), thereby necessitating artificial ventilation in order to maintain respiration and indeed survival. The neurodegenerative condition ALS (Peters *et al.*, 2015) also results in permanent muscle paralysis; however, in this case, *all* components of the motor system are affected in a progressive manner (Figure 21.2), eventually leading to the loss of all motor function. The site of disease onset in ALS can vary considerably, with some patients first exhibiting limb weakness and others experiencing upper motor and bulbar motor neuron deficits, such as difficulty in swallowing and speech early in the disease. Ultimately, ALS patients face the prospect of an inability to breathe and even to communicate – a condition termed "locked-in syndrome" when experienced in totality – and the disease is invariably fatal. With the lack of any effective drug treatment for ALS, there is therefore an urgent need to develop novel strategies to overcome muscle paralysis.

21.1.2 Centrally Targeted Stem Cell-based Strategies to Restore Motor Function

In the absence of effective, conventional pharmacological or surgical interventions to restore lost motor function in SCI and ALS, research into stem cell-based

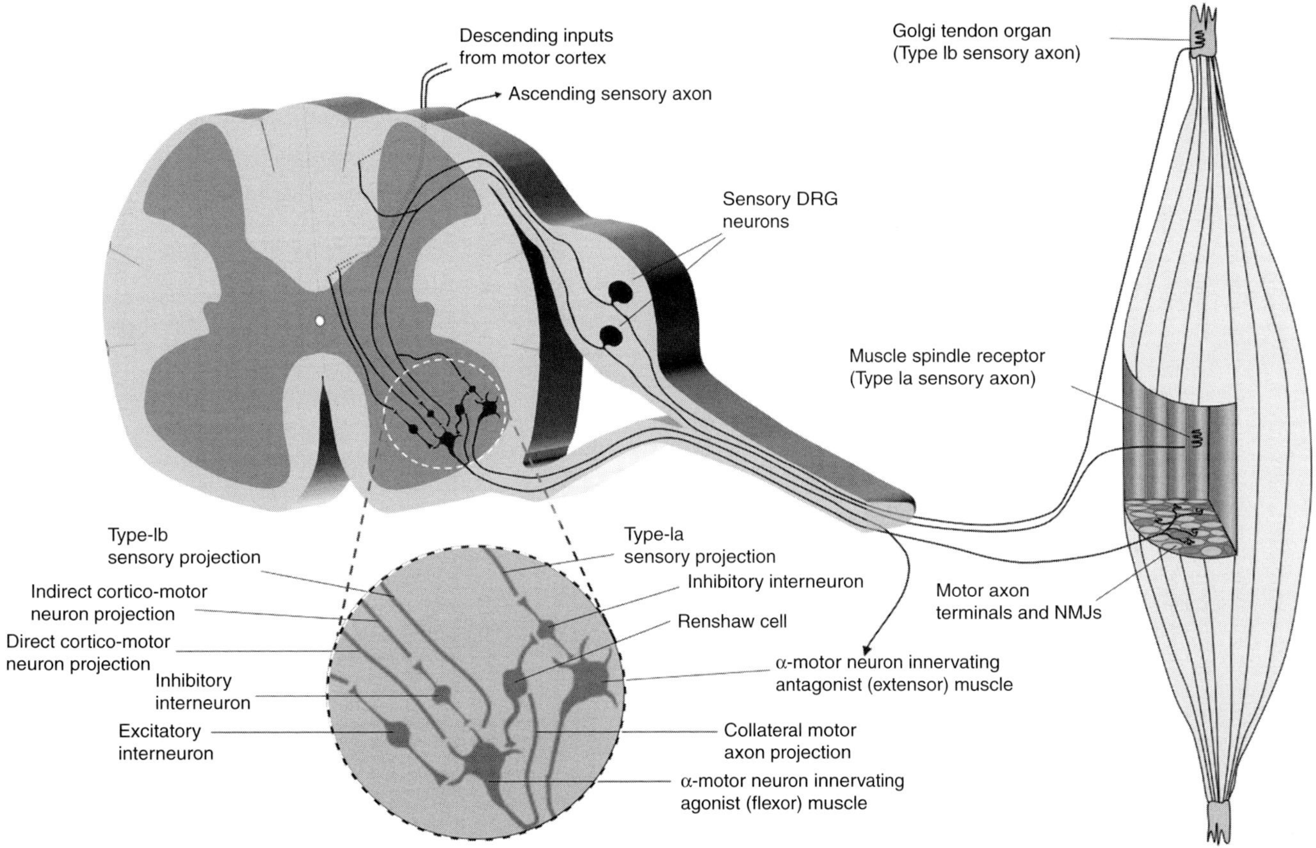

Figure 21.1 Simplified schematic of spinal motor circuits and muscle innervation. Briefly, motor signals from the primary cortex are transmitted along the corticospinal tract (shown for rodents rather than humans). These motor signals then either directly (in the case of corticomotor neuron connections) or indirectly via interneurons (in the case of corticospinal connections) activate the appropriate activity pattern in the intended αmotor neuron. The α-motor neuron activity is then relayed along a peripherally projecting axon within a nerve to the target muscle, where excitatory postsynaptic potentials at specialized synapses with the muscles fibers – neuromuscular junctions – initiate muscle fiber contraction. Sensory feedback information from the muscle is then relayed back to the spinal cord in order to regulate the activity of the initial α-motor neuron, as well as the activity of motor neurons controlling antagonist muscles. The inset represents an enlargement of the ventral horn area, showing some of the first-order connections (i.e. direct synaptic inputs) that are essential for controlling the activity of motor neurons (for a full description, see Brownstone and Bui, 2010).

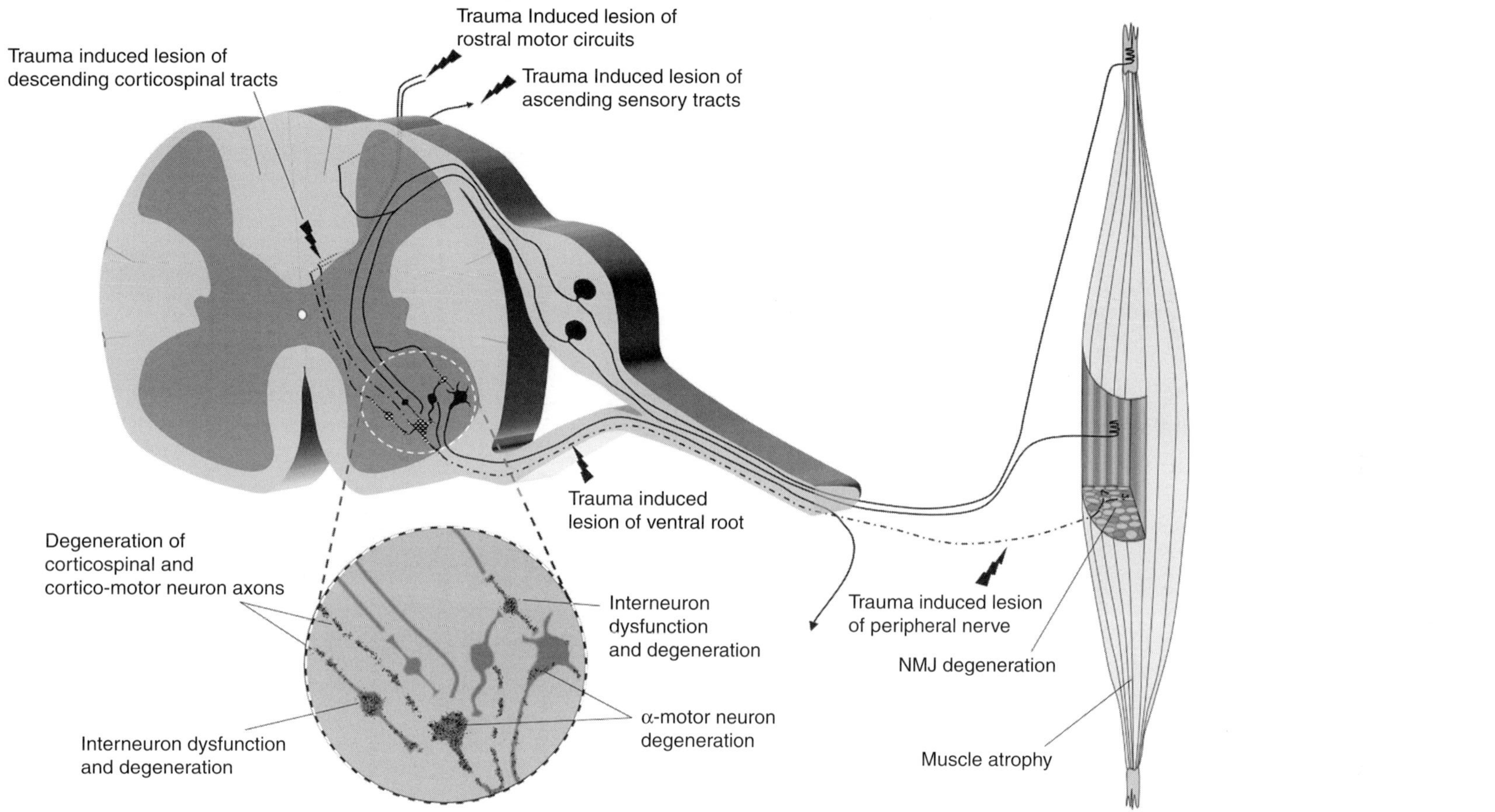

Figure 21.2 Schematic illustration of pathologies that block the transmission of motor signals, resulting in muscle paralysis. Trauma-induced lesions to descending motor tracts results in the blockade of synaptic transmission to distal spinal motor circuits, whilst injury to ventral roots (e.g. avulsion injury) or peripheral nerves can directly interrupt the output of α-motor neurons, causing muscle denervation and paralysis. The inset shows degenerative changes that occur in ALS, which results in not only the degenerative loss of α-motor neurons, but also the degeneration of descending inputs (corticospinal axons), as well as interneurons, thus profoundly affecting the whole motor circuit. Degenerative changes in ALS often begin focally, initially affecting a discrete region of the spinal cord, but pathology progressively spreads throughout the spinal cord to affect all spinal levels.

strategies has been conducted in an effort to provide possible alternative therapies for paralytic disorders (Thomsen *et al.*, 2014). Early evidence of lifespan extension in transgenic rodent models of ALS following intraspinal transplantation of human neuronal precursor cells (hNPCs) (Xu *et al.*, 2006) has recently progressed to Phase 1 clinical trials in ALS patients, where it has been shown to be safe and well tolerated (Feldman *et al.*, 2014). However, it is now widely accepted that transplanted hNPCs cannot restore the anatomical connectivity of spinal motor circuits or replace lost motor neurons, but, rather, they provide trophic support that delays the loss of endogenous motor neurons. This beneficial effect will clearly be restricted to the motor neuron cell body within the CNS, whilst motor axon integrity and muscle innervation are not preserved (Suzuki *et al.*, 2007).

Similarly, efforts to repair damaged spinal motor circuits following SCI have largely focused on the intraspinal transplantation of non-neuronal cell types, with the aim of promoting the spontaneous regeneration of surviving neuronal projections across the lesion site in order to re-innervate and restore the control of distal motor functions. Whilst experimental results from rodent models using these approaches have been promising, evidence from numerous clinical trials suggests that they provide limited benefit in SCI patients. Moreover, most SCIs are caused by contusion injuries, which may span multiple vertebral levels and can result in the death of spinal motor neurons. Indeed, in the case of cervical-level SCIs, contusion injuries to phrenic motor neurons can lead to the irreversible loss of these neurons and their peripheral axon projections, resulting in permanent denervation of the diaphragm muscle and severe impairment and even complete loss of normal respiration. Therefore, in cases of paralytic disorders where motor function is lost as a result of the death of post-mitotic neurons in the spinal cord, more targeted strategies are required to either replace the lost motor neurons themselves or to artificially compensate for their lost function.

The ability to differentiate murine embryonic stem cells (ESCs) into spinal motor neurons was first described by Wichterle and co-workers (Wichterle *et al.*, 2002), and since then, a variety of refined protocols have been developed that enable specific subtypes of motor neurons to be generated from different types of stem cells, including human pluripotent stem cells (Chambers *et al.*, 2009; Amoroso *et al.*, 2013; Maury *et al.*, 2015). Although this ability does raise the prospect of more targeted neuronal replacement strategies for repairing damaged spinal motor circuits using CNS stem cell-derived neural grafts (Thompson and Bjorklund, 2015), this approach is severely impeded by major difficulties, which are discussed below.

21.1.3 Barriers to Repairing Damaged Motor Circuits within the CNS

Firstly, the partial or complete absence of developmentally restricted molecular and genetic programs responsible for the formation of extremely complex spinal motor circuits (Ladle *et al.*, 2007; Brownstone and Bui, 2010) in adults makes it unlikely that grafted motor neurons would be able to functionally integrate into existing, damaged spinal motor circuits. Furthermore, even if engrafted motor neurons were able to integrate into local spinal motor circuits, it is likely that they would be cut off from the supraspinal signals from the brain that are necessary to produce coordinated motor function. Additionally, in ALS, the engraftment of replacement motor neurons into the spinal cord would expose them to the same toxic environment that plays a role in the degeneration of the endogenous motor neurons in the first place. Moreover, engrafted motor neurons would have to extend a nascent axon out from the ventral horn parenchyma, across the white matter tract of the ventral root exit zone and then across the CNS–peripheral nervous system boundary (Harper *et al.*, 2004), and these motor neurons do not normally support axon outgrowth. Indeed, the release of inhibitory molecules from CNS-derived myelin debris following axonal damage or degeneration is a major impediment to the regeneration of endogenous axons, whilst the reactive gliosis and scarring associated with CNS pathology also potently inhibit axon regeneration (Filbin, 2003).

Assuming that replacement motor neurons engrafted within the CNS could survive and extend axons out into the appropriate peripheral nerve, they would then have to grow over great distances and past many decision or branch points to reach the appropriate target muscle – again in the absence of the developmental guidance factors that normally regulate the directional growth of axons (Harel and Strittmatter, 2006). Although it is well accepted that peripheral nerves, unlike the CNS, can robustly support axonal regeneration following denervation injuries, this ability is limited to a finite period, and after chronic denervation, the Schwann cells within the nerve stump begin to lose their ability to support axonal regeneration (Arthur-Farraj *et al.*, 2012). Since the longest human nerves can be over 1 m long and the growth rate of axons within adult nerves is estimated to be between 1 and 3 mm per day (Kang and Lichtman, 2013), it would take approximately 1 year for the most distal muscles to be re-innervated, by which time the nerves would be very unlikely to still support axon growth. Importantly, in ALS, this timeframe may be longer than the lifespan of many patients. Finally, during the intervening period between loss of functional innervation and growth of the grafted motor axons, target muscles will atrophy, possibly irreversibly, due to chronic denervation or inactivity and, consequently, cease producing re-innervation cues (Gordon *et al.*, 2011). Therefore, strategies that depend on the transplantation of stem cell-derived motor neurons directly into the CNS are unlikely to be successful as therapies for restoring lost motor function.

21.2 Peripherally Targeted Stem Cell-based Strategies to Restore Motor Function

Although the soma of spinal motor neurons normally resides within the ventral horn of the spinal cord, it has been previously shown that it is possible to engraft stem cell-derived motor neurons into the peripheral nerve environment and that these ectopically located motor neurons are capable of re-innervating distal target muscles (Yohn *et al.*, 2008). We have recently used this approach to restore specific motor functions in mice (see Section 21.3.1 for detailed description)

(Bryson *et al.*, 2014). Importantly, this approach circumvents many of the challenges described above.

Engraftment of motor neurons into peripheral nerves not only avoids the requirement for the grafted motor neurons to integrate into complex spinal motor circuits, but the neurons are also located outside the toxic/inhibitory environment of the spinal cord, which is particularly important in ALS. Moreover, the transplanted cells can be placed close to the target muscle, greatly accelerating the time taken to re-innervate the muscle, thereby avoiding the problems caused by diminished Schwann cell support of axon growth and muscle atrophy. Furthermore, by engrafting motor neurons near the motor nerve entry point of individual muscles, multiple muscles can be specifically targeted without the requirement for complex guidance by developmentally restricted molecular cues. However, since peripherally engrafted motor neurons are ectopically located outside the CNS and therefore lack supraspinal input, their activity, and consequently the activity of the muscle with which they communicate, must be controlled in an artificial manner. There are now two options by which such peripherally engrafted motor neurons can be activated.

21.2.1 Artificial Control of Motor Function: Electrical Stimulation

Since Luigi Galvani's experiments in the eighteenth century, it has been known that electrical stimulation of peripheral nerves can elicit motor responses from muscles. In recent times, technological advances in the application of this technique in order to control muscle contraction via functional electrical stimulation (FES) of peripheral motor nerves in patients suffering from paralysis have led to the ability to accurately stimulate muscle contraction. Indeed, the development of implantable electrodes enables the differential control of multiple muscles that can drive coordinated, complex motor programs, such as hand grasping (Ethier and Miller, 2015), and efforts are underway to develop more sophisticated stimulation devices that can execute locomotor function in paralyzed humans (King *et al.*, 2015).

However, FES does have fundamental limitations, particularly in the context of ALS. For example, FES is dependent on the integrity of the motor neuron axons within the peripheral nerve in order to induce muscle contraction. In ALS, as well as in cases of contusion-induced loss of motor neurons in SCI, motor axons degenerate, resulting in loss of muscle innervation, rendering FES of peripheral nerves ineffective. Although direct electrical stimulation of muscle can be used to induce contraction, even following muscle denervation, the stimulus amplitude required is prohibitively high for sustained use, not least because direct stimulation results in rapid muscle fatigue and can be painful. Indeed, even when muscles remain innervated, electrical stimulation can result in the indiscriminate activation of both the efferent motor axons and sensory afferents that are present in peripheral nerves; depending on the intensity of the electrical stimulus, this may not only induce muscle contraction, but can also simultaneously activate axons of nociceptive sensory neurons, which is perceived as pain. In the case of ALS patients, the sensory system remains largely intact, and in cases of SCI, although the transmission of pain signals may be completely blocked, local activation of pain circuits by high-frequency or high-amplitude electrical stimulation may have unforeseen consequences.

The most important limitation of FES as a means of controlling sustained muscle function over extended periods is the known reversal or random recruitment of motor units that results from electrical stimulation of peripheral nerves (Hamada *et al.*, 2004). A motor unit consists of a single motor neuron axon and all of the muscles fibers that it innervates, which typically number between 10 and 100. Normally, motor units are recruited according to Henneman's size principle, whereby the smaller-diameter axons of slow-firing motor neurons that innervate relatively few slow-twitch, fatigue-resistant muscle fibers are activated by the smallest-amplitude stimulus from the brain. In contrast, larger-diameter motor neurons and their axons, which innervate a large number of fast-twitch, rapidly fatigable muscle fibers, are only activated by supraspinal inputs of increasingly greater magnitude in proportion to the strength of the intended motor output (Henneman and Olson, 1965). However, in the case of electrical stimulation, the largest motor axons and the most fatigable muscles that they innervate are recruited at the lowest stimulus intensity, whilst smaller motor units are activated by higher-intensity stimuli. This non-physiological reversal in the graded recruitment of motor units by FES has very significant consequences, most critically that muscles rapidly fatigue following sustained FES (Hamada *et al.*, 2004). Thus, for long-term stimulation of muscles, in particular of critical muscles such as the diaphragm, the use of electrical stimulation may be inappropriate, as it is unlikely to support the long-term rhythmic contractions that are essential to maintaining respiration.

Pertinently, the early termination of a recent Phase 1 clinical trial that was designed to assess the safety and tolerability of electrical pacing of the diaphragm muscle in ALS patients was based on evidence of markedly decreased survival of patients fitted with the device (Di and Di, 2015). Although the underlying cause of the negative effect on patient survival due to electrical pacing of the diaphragm is unclear from this study, it did not appear to be associated with the surgical implantation of the device. It is possible that electrical stimulation of the diaphragm may have resulted in enhanced muscle fatigue, which in turn may have exacerbated the degeneration of the declining population of phrenic motor neuron axons and accelerated diaphragm muscle denervation. This example highlights the need to develop alternative strategies for the artificial control and restoration of muscle function that are more physiological, particularly in cases where sustained contraction of muscles that are essential for survival, such as the diaphragm, is required.

The now-established technique of optogenetics provides just such an alternative approach that can overcome the significant drawbacks of FES discussed above through the ability to enable specific and physiological control of motor function.

21.2.2 Artificial Control of Motor Function: Optogenetic Stimulation

The first report of optogenetic control of motor activity was in transgenic mice that express channelrhodopsin-2 (ChR2) under the neuronal *Thy1* promoter (*Thy1::ChR2* mice), in which optical stimulation was applied to the primary motor cortex via a tethered optical fiber in order to induce motor activity (Aravanis *et al.*, 2007). More recently, also using *Thy1::ChR2* transgenic mice that express ChR2 in spinal motor neurons, it was shown that motor axons in the peripheral nerve can be optically stimulated by a nerve cuff coupled to a laser light source, resulting in highly controlled muscle contractions (Llewellyn *et al.*, 2010). Importantly, this study demonstrated that motor units activated by optical stimulation are recruited in the normal physiological order, with smaller, fatigue-resistant motor units being recruited at lower optical stimulus intensities and larger, fatigable motor units only being activated at higher intensities (Llewellyn *et al.*, 2010). The same physiological recruitment of ChR2-expressing motor units and the prevention of muscle fatigue were also verified in our recent study (Bryson *et al.*, 2014), which is discussed in detail below. Theoretical modeling of the orderly recruitment of motor units in peripheral optogenetic neural stimulation (PONS) suggests that the reduced internodal distance and resulting increase in the density of ChR2 exposure at the more frequent nodes in small-diameter myelinated motor axons are essential parameters underlying this phenomenon (Arlow *et al.*, 2013). In addition to the major advantage of physiological, graded recruitment of motor units, optical stimulation also has the significant advantage that it only induces activity in neurons that express the light-responsive opsin. This makes it possible to specifically activate motor axons using PONS, avoiding the indiscriminate activation of non-targeted motor axons as well as nociceptive afferent axons (Iyer *et al.*, 2014). However, a translational method to target opsin expression to the desired population of motor neuron axons in order to mediate fine optical control of motor function is a remaining hurdle that needs to overcome and is particularly challenging in the face of the progressive degeneration of motor axons that occurs in ALS.

21.3 Translational Strategies for Optical Control and Restoration of Lost Motor Function

Optogenetics has now been an available experimental technique for over a decade, during which time it has yielded invaluable insights into the physiological function of neural circuits, including those underlying motor function. However, the clinical translation of optogenetics, although highly anticipated for the treatment of a wide range of disorders, has yet to be realized in terms of using the optical control of opsin-expressing tissue in patients in order to control any physiological function or ameliorate any disease symptoms.

To date, the investigation of the optogenetic control of motor function *in vivo* has primarily relied upon either transgenic mouse models or viral transduction of neurons in rodents and non-human primates in order to express opsins. Undoubtedly, important experimental information *has* been obtained from the use of optogenetics in transgenic mice in particular, but this is not a viable translational strategy for humans. On the other hand, adeno-associated viruses, as well as other viruses, have shown promise as potential gene therapy delivery vectors and are capable of targeting opsin expression to specific neuronal populations (Ji *et al.*, 2013); for example, to sensory versus motor axons in peripheral nerves in rodents (Towne *et al.*, 2013). However, whilst viral transduction is a suitable means of expressing opsins in experimental models, it is less attractive as a mechanism to enable opsin expression in a therapeutic environment, as it has several disadvantages and risks, not least the biosafety concerns that arise with the use of viral vectors in humans. In addition, it has recently been shown that high-level expression of ChR2 in rats, following *in utero* electroporation and, to a lesser extent, viral expression, can result in axonal pathology (Miyashita *et al.*, 2013), indicating that the level of opsin expression in neurons must be carefully controlled. Moreover, in the case of ALS, even if viral transduction could safely and effectively deliver appropriate levels of opsins to surviving endogenous motor axons as a therapeutic strategy for restoring motor function, the ongoing degeneration of these axons and resulting muscle denervation would rapidly render the approach ineffective.

Perhaps the most translationally viable option for enabling the optogenetic control of paralyzed muscles would be to take advantage of recent advances in gene targeting technology, such as the highly specific CRISPr/Cas9 method (Cong *et al.*, 2013; Mali *et al.*, 2013), along with advances in induced pluripotent stem cell technology (Thomsen *et al.*, 2014), in order to generate optogenetically modified neural grafts that can be targeted to peripheral nerves (see Figure 21.3 and Section 21.2). For example, human ESC-derived neurons that stably express ChR2 have recently been shown to survive for >6 months following transplant into severe combined immunodeficiency mice (Weick *et al.*, 2010), which supports our observations of long-term (>1year) survival of murine ESC-derived motor neuron that express ChR2 following engraftment into the peripheral nerves of wild-type mice (unpublished data).

21.3.1 Using Optogenetics in Combination with Neural Engraftment to Restore Muscle Function

Indeed, in a recent study, we developed a combinatorial strategy that utilizes the advantages of both stem cell-derived neural engraftment into denervated peripheral motor nerves and optogenetic control of motor neuron function as a translationally relevant approach to restoring lost muscle function in mice following peripheral nerve injury (Bryson *et al.*, 2014). To accomplish this, we first generated murine ESCs that were genetically targeted to express ChR2, as well as the neurotrophic factor glial-derived neurotrophic factor (*Gdnf*), both under control of the high-expression cytomegalovirus chicken β-actin (GAGs) promoter. Thus, motor neurons derived from these ESCs are optically excitable and also secrete their own trophic support, promoting their long-term survival. These ESC-derived motor neurons were then engrafted into denervated peripheral nerves of adult mice, which showed that they not only survive in this peripheral

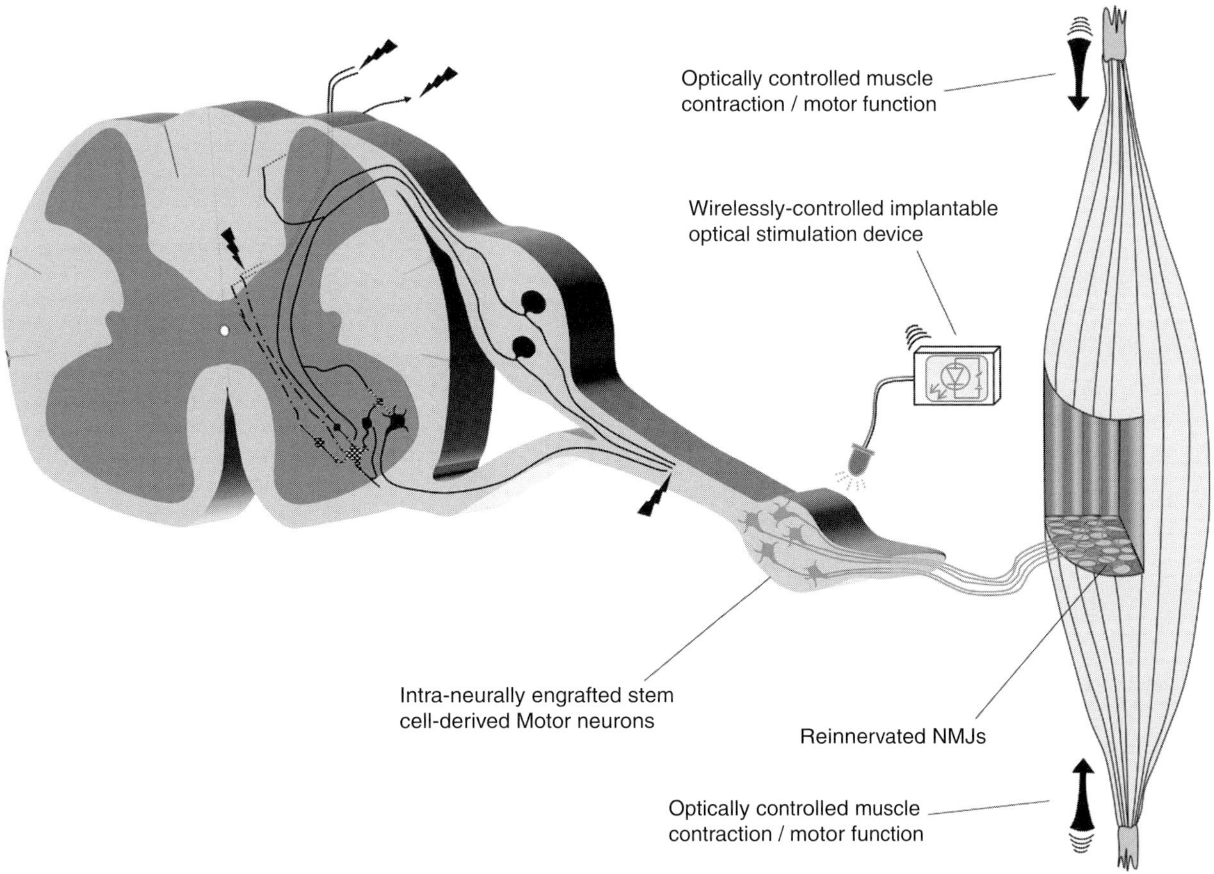

Figure 21.3 Schematic representation of an approach to restoring controllable function to paralyzed muscles. Irrespective of the cause of muscle paralysis, stem cell-derived motor neurons that are modified to express light-sensitive opsins can be engrafted into peripheral nerves, close to the motor nerve entry point of the muscle, resulting in robust innervation of the muscle. Activity of the engrafted motor neurons and thereby function of the muscle(s) that they innervate can then be controlled by an implanted optical stimulator device. (A black-and-white version of this figure will appear in some formats. For the color version, please refer to the plate section.)

environment, but also extend axons that robustly re-innervate target muscles in the hind-limb. Specifically, motor neurons engrafted into the common peroneal nerve branch innervated the extensor muscles (tibialis anterior and extensor digitorum longus), whilst motor neurons engrafted into the tibial nerve branch innervated the large flexor muscle group of the hind-limb (the triceps surae).

Moreover, we demonstrated that optical stimulation of these ChR2-expressing grafted motor neurons with blue (470-nm) light resulted in controlled contraction of the target muscles. Importantly, this optically controlled muscle function avoided the rapid muscle fatigue associated with electrical neuromuscular stimulation (Bryson *et al.*, 2014), since optical stimulation of the grafted neurons resulted in the normal, physiological recruitment of motor unit, thus confirming the findings of other groups (Llewellyn *et al.*, 2010). This approach may be ideally suited as a translational strategy for enabling optical control of the diaphragm muscle in ALS patients by using an optical pace-maker to maintain respiratory function. This would avoid the need for mechanical ventilation and the problems associated with electrical pacing of the diaphragm muscle (Di and Di, 2015). Indeed, the ability to experimentally control diaphragm function in rodents by using optogenetics has already been demonstrated (Alilain *et al.*, 2008).

The combinatorial approach described above demonstrates the advantages of optogenetic control of motor function, along with the ability to functionally re-innervate muscles that have been paralyzed as a result of denervation. However, the next major hurdle to overcome is to develop suitable optical stimulation devices in order to enable chronic control of muscle activity, which is essential if this approach is to have any therapeutic value for the restoration of motor function. Furthermore, the ability to chronically activate ChR2-expressing motor neurons is essential for the maintenance of the structure and function of neuromuscular junctions (NMJs), which are known to be highly dependent on synaptic activity (Darabid *et al.*, 2014), without which the quiescent motor nerve terminal begins to detach from the NMJ and the muscle fiber begins to undergo atrophy. Thus, optical stimulation devices that can deliver chronic, appropriately patterned activity are required in order to maximize the re-innervation and muscle strength that are achievable through using this approach.

21.3.2 Development of Implantable Optical Stimulators to Control Motor Function

The significant benefits of optical versus electrical stimulation warrant the rapid development of more sophisticated technological and bioengineering solutions in order to expedite the clinical application of the approach described above. Indeed, major advances have been made toward the development of optical stimulation methods in rodent models in the past few years. Initial optical stimulation experiments employed tethered optical fibers connected to a laser light source (Aravanis *et al.*, 2007). However, whilst this approach has been elegantly used to control muscle function in awake, freely moving rats following viral transduction in order to express ChR2 in peripheral nerves (Towne *et al.*, 2013), it remains technically challenging and impractical for large stimulation experiments requiring long-term stimulation; this approach also prohibits normal social behavior (Iwai *et al.*, 2011; Wentz *et al.*, 2011).

However, a recent, elegant study has described the development of fully implantable, wirelessly powered mini-LED devices that are small enough for use in mice and for the stimulation of peripheral nerves, weighing as little as 20 mg (Montgomery *et al.*, 2015). These devices are likely to greatly facilitate the development of optogenetic techniques to control motor function in translationally relevant model systems. This is particularly important for the investigation of the motor neuron activation of muscle function, since the formation and maintenance of NMJs is dependent on chronic, long-term stimulation of the transplanted neurons and subsequent muscle activity, which is not provided by the intermittent activation paradigms afforded by tethered systems. A remaining engineering requirement for these devices, in terms of enabling normal control of muscle function, is to enable gradual ramping of light intensity in order to recruit motor units in a normal, graded, physiological order and thereby avoid muscle fatigue. Nonetheless, this technology represents a significant advance toward the ability to reliably control motor function using optogenetics. Indeed, we believe that the ability to optically control more complex motor functions is largely restricted at present by the current sophistication of optical stimulation devices, since either neural replacement or viral transduction strategies can readily confer optogenetic control of spatially discrete, opposable muscle groups, at least experimentally (Towne *et al.*, 2013).

21.4 Further Advances toward Optogenetic Control of Motor Function

An additional requirement for the translational application of optogenetics in order to control motor function is the development and refinement of optimized opsins. As noted above, too high an expression level of ChR2 has been shown to induce axonopathy; therefore, opsins that enable greater (more selective) cation flow and respond to weaker optical stimuli are beneficial, such as the channelrhodopsin Cheriff (Hochbaum *et al.*, 2014) and the red-shifted channelrhodopsins ReaChR (Lin *et al.*, 2013) and Chrimson (Klapoetke *et al.*, 2014). These red-shifted opsins have the advantage of requiring less energetic activation wavelengths in the orange–red spectrum that have greater tissue penetration, thereby enabling more flexibility in terms of optical stimulator development and avoiding potential cellular damage from comparatively high-energy blue light (Hockberger *et al.*, 1999). Moreover, further characterization of existing optogenetic actuators remains to be undertaken. For example, it was recently shown that the well-established neuronal activation by ChR2 is actually reversed at lower temperatures and causes inhibition of motor activity in mice under such conditions (Liske *et al.*, 2013). Indeed, the use of halorhodopsins in order to block motor activity represents an additional means of therapeutically inhibiting motor function, which may be of use, for example, in cases of spasticity.

Finally, a recent development in the optogenetic control of muscle function is the use of the direct optogenetic control of muscle contraction, using both transgenic ChR2-expressing mice and viral transduction of muscle *in vivo* to optically control the muscles of the larynx (Bruegmann *et al.*, 2015). This approach could be a complementary therapy for the prevention of muscle wasting whilst waiting for regenerating axons to grow and so re-innervate the muscle (Magown *et al.*, 2015).

21.5 Conclusions

Recent advances in a range of different fields have resulted in the development of novel methods to restore motor function that circumvent the need to repair damage to the CNS, which, despite intensive research, has so far proven to be an elusive goal. By combining the potential of stem cell differentiation and the ability to optogenetically control motor neurons, together with the development of more sophisticated optical stimulation devices, it may eventually be possible to couple this approach with advanced methods that can interpret intended motor output. Indeed, a synthesis of brain–machine interface (BMI) technology with FES in order to control muscle function (Ethier and Miller, 2015) has recently been shown to be a feasible approach to enabling overground walking in SCI (King *et al.*, 2015). Thus, in the long term, a combination of intraneural stem cell-derived motor neuron engraftment and optogenetics together with BMIs may enable paralyzed patients to exert control over their own musculature. Recent developments in BMIs from the neurotechnology company BrainGate have produced a device that can wirelessly transmit intended motor commands collected from a brain implant in order to steer a wheelchair or robotic arm (Homer *et al.*, 2013). Moreover, paralyzed patients, some with ALS, are currently taking part in trials of similar technologies (Jarosiewicz *et al.*, 2015).

Great progress has therefore been made in the development of the biological and technological components that would enable a BMI to be constructed that would enable optical control of less complex, but essential motor functions, such as respiration, swallowing and bowel/bladder function. The continued development of more sophisticated optical control devices, which has rapidly evolved during the short history of optogenetics, could see the goal of restoring more complex motor functions come to fruition for patients in the foreseeable future.

REFERENCES

Alilain, W. J., Li, X., Horn, K. P., Dhingra, R., Dick, T. E., Herlitze, S. & Silver, J. 2008. Light-induced rescue of breathing after spinal cord injury. *J Neurosci*, **28**, 11862–11870.

Amoroso, M. W., Croft, G. F., Williams, D. J., O'keeffe, S., Carrasco, M. A., Davis, A. R., Roybon, L., Oakley, D. H., Maniatis, T., Henderson, C. E. & Wichterle, H. 2013. Accelerated high-yield generation of limb-innervating motor neurons from human stem cells. *J Neurosci*, **33**, 574–586.

Aravanis, A. M., Wang, L. P., Zhang, F., Meltzer, L. A., Mogri, M. Z., Schneider, M. B. & Deisseroth, K. 2007. An optical neural interface: *in vivo* control of rodent motor cortex with integrated fiberoptic and optogenetic technology. *J Neural Eng*, **4**, S143–S156.

Arlow, R. L., Foutz, T. J. & Mcintyre, C. C. 2013. Theoretical principles underlying optical stimulation of myelinated axons expressing channelrhodopsin-2. *Neuroscience*, **248**, 541–551.

Arthur-Farraj, P. J., Latouche, M., Wilton, D. K., Quintes, S., Chabrol, E., Banerjee, A., Woodhoo, A., Jenkins, B., Rahman, M., Turmaine, M., Wicher, G. K., Mitter, R., Greensmith, L., Behrens, A., Raivich, G., Mirsky, R. & Jessen, K. R. 2012. c-Jun reprograms Schwann cells of injured nerves to generate a repair cell essential for regeneration. *Neuron*, **75**, 633–647.

Brownstone, R. M. & Bui, T. V. 2010. Spinal interneurons providing input to the final common path during locomotion. *Prog Brain Res*, **187**, 81–95.

Bruegmann, T., Van Bremen, T., Vogt, C. C., Send, T., Fleischmann, B. K. & Sasse, P. 2015. Optogenetic control of contractile function in skeletal muscle. *Nat Commun*, **6**, 7153.

Bryson, J. B., Machado, C. B., Crossley, M., Stevenson, D., Bros-Facer, V., Burrone, J., Greensmith, L. & Lieberam, I. 2014. Optical control of muscle function by transplantation of stem cell-derived motor neurons in mice. *Science*, **344**, 94–97.

Chambers, S. M., Fasano, C. A., Papapetrou, E. P., Tomishima, M., Sadelain, M. & Studer, L. 2009. Highly efficient neural conversion of human ES and iPS cells by dual inhibition of SMAD signaling. *Nat Biotechnol*, **27**, 275–280.

Cong, L., Ran, F. A., Cox, D., Lin, S., Barretto, R., Habib, N., Hsu, P. D., Wu, X., Jiang, W., Marraffini, L. A. & Zhang, F. 2013. Multiplex genome engineering using CRISPR/Cas systems. *Science*, **339**, 819–823.

Darabid, H., Perez-Gonzalez, A. P. & Robitaille, R. 2014. Neuromuscular synaptogenesis: coordinating partners with multiple functions. *Nat Rev Neurosci*, **15**, 703–718.

Di, P. W. C. & Di, P. S. G. C. 2015. Safety and efficacy of diaphragm pacing in patients with respiratory insufficiency due to amyotrophic lateral sclerosis (DiPALS): a multicentre, open-label, randomised controlled trial. *Lancet Neurol*, **14**, 883–892.

Ethier, C. & Miller, L. E. 2015. Brain-controlled muscle stimulation for the restoration of motor function. *Neurobiol Dis*, **83**, 180–190.

Feldman, E. L., Boulis, N. M., Hur, J., Johe, K., Rutkove, S. B., Federici, T., Polak, M., Bordeau, J., Sakowski, S. A. & Glass, J. D. 2014. Intraspinal neural stem cell transplantation in amyotrophic lateral sclerosis: Phase 1 trial outcomes. *Ann Neurol*, **75**, 363–373.

Filbin, M. T. 2003. Myelin-associated inhibitors of axonal regeneration in the adult mammalian CNS. *Nat Rev Neurosci*, **4**, 703–713.

Gordon, T., Tyreman, N. & Raji, M. A. 2011. The basis for diminished functional recovery after delayed peripheral nerve repair. *J Neurosci*, **31**, 5325–534.

Hamada, T., Kimura, T. & Moritani, T. 2004. Selective fatigue of fast motor units after electrically elicited muscle contractions. *J Electromyogr Kinesiol*, **14**, 531–538.

Harel, N. Y. & Strittmatter, S. M. 2006. Can regenerating axons recapitulate developmental guidance during recovery from spinal cord injury? *Nat Rev Neurosci*, **7**, 603–616.

Harper, J. M., Krishnan, C., Darman, J. S., Deshpande, D. M., Peck, S., Shats, I., Backovic, S., Rothstein, J. D. & Kerr, D. A. 2004. Axonal growth of embryonic stem cell-derived motoneurons in vitro and in motoneuron-injured adult rats. *Proc Natl Acad Sci U S A*, **101**, 7123–7128.

Henneman, E. & Olson, C. B. 1965. Relations between structure and function in the design of skeletal muscles. *J Neurophysiol*, **28**, 581–598.

Hochbaum, D. R., Zhao, Y., Farhi, S. L., Klapoetke, N., Werley, C. A., Kapoor, V., Zou, P., Kralj, J. M., Maclaurin, D., Smedemark-Margulies, N., Saulnier, J. L., Boulting, G. L., Straub, C., Cho, Y. K., Melkonian, M., Wong, G. K., Harrison, D. J., Murthy, V. N., Sabatini, B. L., Boyden, E. S., Campbell, R. E. & Cohen, A. E. 2014. All-optical electrophysiology in mammalian neurons using engineered microbial rhodopsins. *Nat Methods*, **11**, 825–833.

Hockberger, P. E., Skimina, T. A., Centonze, V. E., Lavin, C., Chu, S., Dadras, S., Reddy, J. K. & White, J. G. 1999. Activation of flavin-containing oxidases underlies light-induced production of H_2O_2 in mammalian cells. *Proc Natl Acad Sci U S A*, **96**, 6255–6260.

Homer, M. L., Nurmikko, A. V., Donoghue, J. P. & Hochberg, L. R. 2013. Sensors and decoding for intracortical brain computer interfaces. *Annu Rev Biomed Eng*, **15**, 383–405.

Iwai, Y., Honda, S., Ozeki, H., Hashimoto, M. & Hirase, H. 2011. A simple head-mountable LED device for chronic stimulation of optogenetic molecules in freely moving mice. *Neurosci Res*, **70**, 124–127.

Iyer, S. M., Montgomery, K. L., Towne, C., Lee, S. Y., Ramakrishnan, C., Deisseroth, K. & Delp, S. L. 2014. Virally mediated optogenetic excitation and inhibition of pain in freely moving nontransgenic mice. *Nat Biotechnol*, **32**, 274–278.

Jarosiewicz, B., Sarma, A. A., Bacher, D., Masse, N. Y., Simeral, J. D., Sorice, B., Oakley, E. M., Blabe, C., Pandarinath, C., Gilja, V., Cash, S. S., Eskandar, E. N., Friehs, G., Henderson, J. M., Shenoy, K. V., Donoghue, J. P. & Hochberg, L. R. 2015. Virtual typing by people with tetraplegia using a self-calibrating intracortical brain–computer interface. *Sci Transl Med*, **7**, 313ra179.

Ji, Z. G., Ishizuka, T. & Yawo, H. 2013. Channelrhodopsins – their potential in gene therapy for neurological disorders. *Neurosci Res*, **75**, 6–12.

Kang, H. & Lichtman, J. W. 2013. Motor axon regeneration and muscle reinnervation in young adult and aged animals. *J Neurosci*, **33**, 19480–19491.

King, C. E., Wang, P. T., Mccrimmon, C. M., Chou, C. C., Do, A. H. & Nenadic, Z. 2015. The feasibility of a brain–computer interface functional electrical stimulation system for the restoration of overground walking after paraplegia. *J Neuroeng Rehabil*, **12**, 80.

Klapoetke, N. C., Murata, Y., Kim, S. S., Pulver, S. R., Birdsey-Benson, A., Cho, Y. K., Morimoto, T. K., Chuong, A. S., Carpenter, E. J., Tian, Z., Wang, J., Xie, Y., Yan, Z., Zhang, Y., Chow, B. Y., Surek, B., Melkonian, M., Jayaraman, V., Constantine-Paton, M., Wong, G. K. & Boyden, E. S. 2014. Independent optical excitation of distinct neural populations. *Nat Methods*, **11**, 338–346.

Ladle, D. R., Pecho-Vrieseling, E. & Arber, S. 2007. Assembly of motor circuits in the spinal cord: driven to function by genetic and experience-dependent mechanisms. *Neuron*, **56**, 270–283.

Lemon, R. N. 2008. Descending pathways in motor control. *Annu Rev Neurosci*, **31**, 195–218.

Lin, J. Y., Knutsen, P. M., Muller, A., Kleinfeld, D. & Tsien, R. Y. 2013. ReaChR: a red-shifted variant of channelrhodopsin enables deep transcranial optogenetic excitation. *Nat Neurosci*, **16**, 1499–1508.

Liske, H., Qian, X., Anikeeva, P., Deisseroth, K. & Delp, S. 2013. Optical control of neuronal excitation and inhibition using a single opsin protein, ChR2. *Sci Rep*, **3**, 3110.

Llewellyn, M. E., Thompson, K. R., Deisseroth, K. & Delp, S. L. 2010. Orderly recruitment of motor units under optical control in vivo. *Nat Med*, **16**, 1161–1165.

Magown, P., Shettar, B., Zhang, Y. & Rafuse, V. F. 2015. Direct optical activation of skeletal muscle fibres efficiently controls muscle contraction and attenuates denervation atrophy. *Nat Commun*, **6**, 8506.

Mali, P., Yang, L., Esvelt, K. M., Aach, J., Guell, M., Dicarlo, J. E., Norville, J. E. & Church, G. M. 2013. RNA-guided human genome engineering via Cas9. *Science*, **339**, 823–826.

Maury, Y., Come, J., Piskorowski, R. A., Salah-Mohellibi, N., Chevaleyre, V., Peschanski, M., Martinat, C. & Nedelec, S. 2015. Combinatorial analysis of developmental cues efficiently converts human pluripotent stem cells into multiple neuronal subtypes. *Nat Biotechnol*, **33**, 89–96.

Miyashita, T., Shao, Y. R., Chung, J., Pourzia, O. & Feldman, D. E. 2013. Long-term channelrhodopsin-2 (ChR2) expression can induce abnormal axonal morphology and targeting in cerebral cortex. *Front Neural Circuits*, **7**, 8.

Montgomery, K. L., Yeh, A. J., Ho, J. S., Tsao, V., Mohan Iyer, S., Grosenick, L., Ferenczi, E. A., Tanabe, Y., Deisseroth, K., Delp, S. L. & Poon, A. S. 2015. Wirelessly powered, fully internal optogenetics for brain, spinal and peripheral circuits in mice. *Nat Methods*, **12**, 969–974.

Nicaise, C., Putatunda, R., Hala, T. J., Regan, K. A., Frank, D. M., Brion, J. P., Leroy, K., Pochet, R., Wright, M. C. & Lepore, A. C. 2012. Degeneration of phrenic motor neurons induces long-term diaphragm deficits following mid-cervical spinal contusion in mice. *J Neurotrauma*, **29**, 2748–2760.

Peters, O. M., Ghasemi, M. & Brown, R. H., Jr. 2015. Emerging mechanisms of molecular pathology in ALS. *J Clin Invest*, **125**, 1767–1779.

Suzuki, M., Mchugh, J., Tork, C., Shelley, B., Klein, S. M., Aebischer, P. & Svendsen, C. N. 2007. GDNF secreting human neural progenitor cells protect dying motor neurons, but not their projection to muscle, in a rat model of familial ALS. *PLoS One*, **2**, e689.

Thompson, L. H. & Bjorklund, A. 2015. Reconstruction of brain circuitry by neural transplants generated from pluripotent stem cells. *Neurobiol Dis*, **79**, 28–40.

Thomsen, G. M., Gowing, G., Svendsen, S. & Svendsen, C. N. 2014. The past, present and future of stem cell clinical trials for ALS. *Exp Neurol*, **262**(Pt B), 127–137.

Towne, C., Montgomery, K. L., Iyer, S. M., Deisseroth, K. & Delp, S. L. 2013. Optogenetic control of targeted peripheral axons in freely moving animals. *PLoS One*, **8**, e72691.

Weick, J. P., Johnson, M. A., Skroch, S. P., Williams, J. C., Deisseroth, K. & Zhang, S. C. 2010. Functional control of transplantable human ESC-derived neurons via optogenetic targeting. *Stem Cells*, **28**, 2008–2016.

Wentz, C. T., Bernstein, J. G., Monahan, P., Guerra, A., Rodriguez, A. & Boyden, E. S. 2011. A wirelessly powered and controlled device for optical neural control of freely-behaving animals. *J Neural Eng*, **8**, 046021.

Wichterle, H., Lieberam, I., Porter, J. A. & Jessell, T. M. 2002. Directed differentiation of embryonic stem cells into motor neurons. *Cell*, **110**, 385–397.

Xu, L., Yan, J., Chen, D., Welsh, A. M., Hazel, T., Johe, K., Hatfield, G. & Koliatsos, V. E. 2006. Human neural stem cell grafts ameliorate motor neuron disease in SOD-1 transgenic rats. *Transplantation*, **82**, 865–875.

Yohn, D. C., Miles, G. B., Rafuse, V. F. & Brownstone, R. M. 2008. Transplanted mouse embryonic stem-cell-derived motoneurons form functional motor units and reduce muscle atrophy. *J Neurosci*, **28**, 12409–12418.

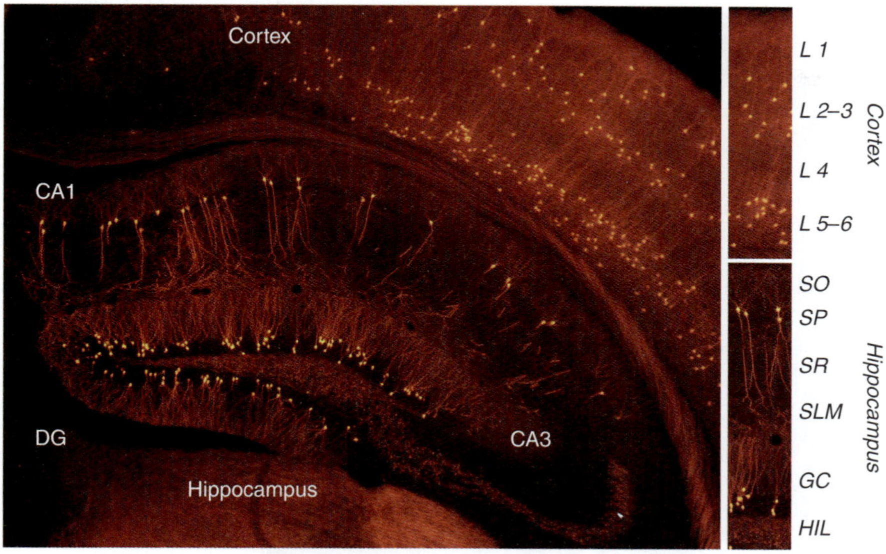

Figure 2.1 Activity-dependent cortical and dorsal hippocampal neuron labeling in an Arc-CreER mouse. Representative coronal brain section from the progeny of Arc-CreER and CAG-(tdTomato)^{Cre} reporter mice (Ai14) (Madisen *et al.*, 2009) following exposure to an enriched environment. Hydroxytamoxifen was administered 2 hours prior to the start of the experiment. Subsequently, animals were returned to home cages and sacrificed 2 weeks later. Note the absence of label in the CA3 region and cortical layer IV. In addition, inhibitory interneurons that are normally abundant in the SO and HIL are not labeled. *DG:* dentate gyrus; *SO:* stratum oriens; *SP:* stratum pyramidale; *SR:* stratum radiatum, *SLM:* stratum lacunosum-moleculare; *GC:* granule cell layer of the dentate gyrus; *HIL:* hilus.

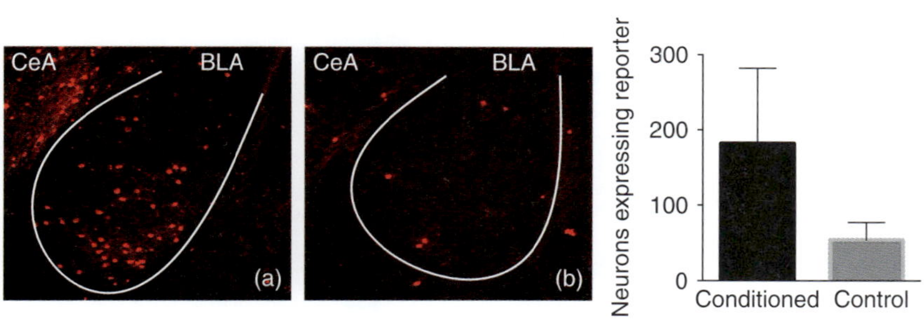

Figure 2.2 Activity-dependent amygdala labeling in Arc-CreER mice. (A) Animals were subjected to appetitive conditioning, with labeling initiated using hydroxytamoxifen at asymptotic task performance (n = 4). (B) Control animals were labeled in home cages (n = 3). Representative amygdala sections and aggregate cell counts are shown. The experimental preparation was similar to that described in Figure 2.1. *CeA:* central amygdala; *BLA:* basolateral amygdala (based on Maya *et al.*, 2016, in preparation).

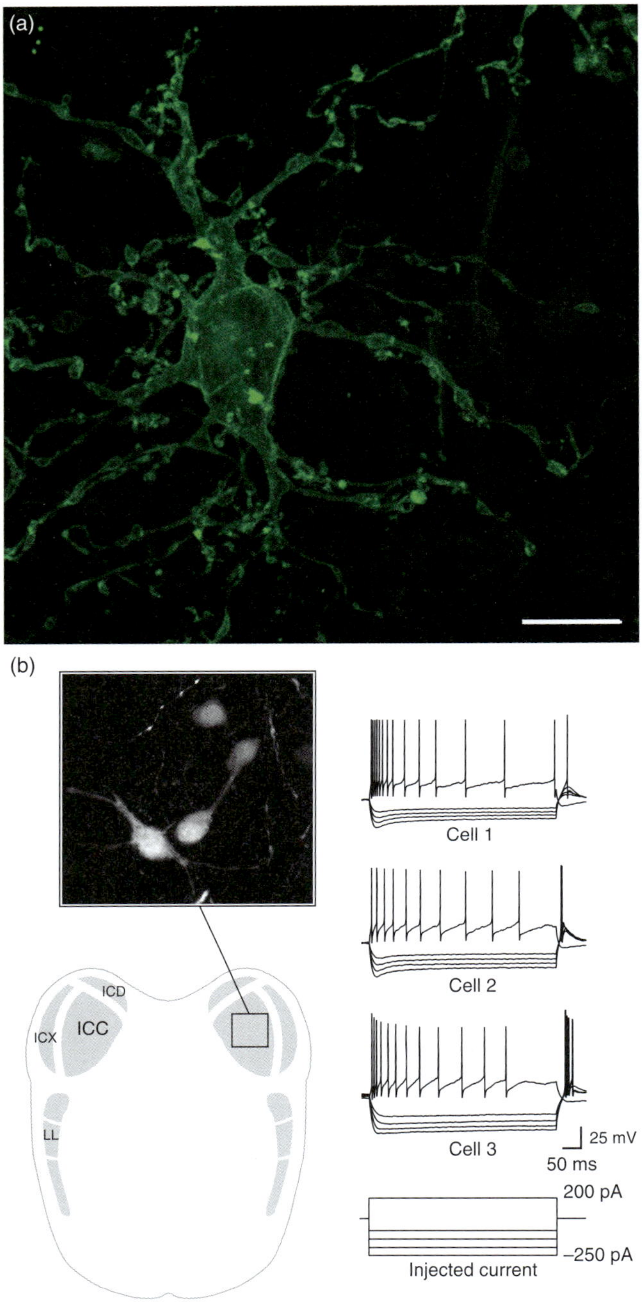

Figure 2.4 Cell type-specific promoters drive the expression of virus-encoded transgenes in rodent neurons. (A) A promoter composed of conserved non-coding regions of the *MGluR1* gene exclusively labels retinal amacrine AII cells (based on Borghuis *et al.*, 2011). Scale bar: 10 µm. (B) Interdependent viruses comprising truncated calbindin and calcium/calmodulin-dependent protein kinase II (CaMKII) non-coding regions restrict fluorophore expression to a morphologically and electrically homogeneous population of neurons within the gerbil inferior colliculus. Injected current steps are indicated (based on Kreeger *et al.*, 2016, in preparation).

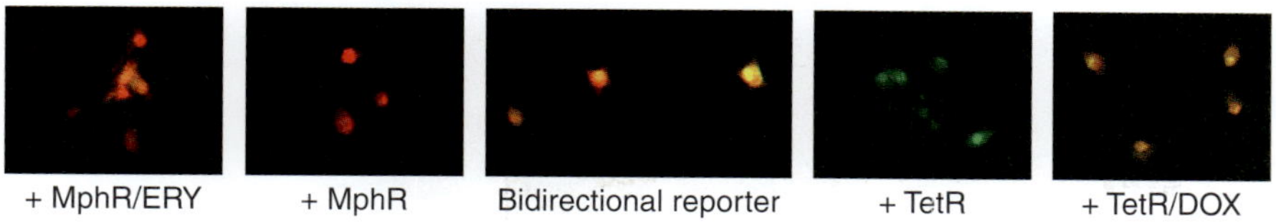

Figure 2.3 A bidirectional virus-based neuronal activity reporter. *Middle panel:* Cultured hippocampal neurons stimulated with forskolin express red and green fluorophores, producing yellow–orange labeling. *+MphR/ +TetR panels:* The indicated bacterial repressor blocks expression of green or red fluorophores. *ERY/DOX panels:* The addition of respective ligands restores two-color expression in activated neurons.

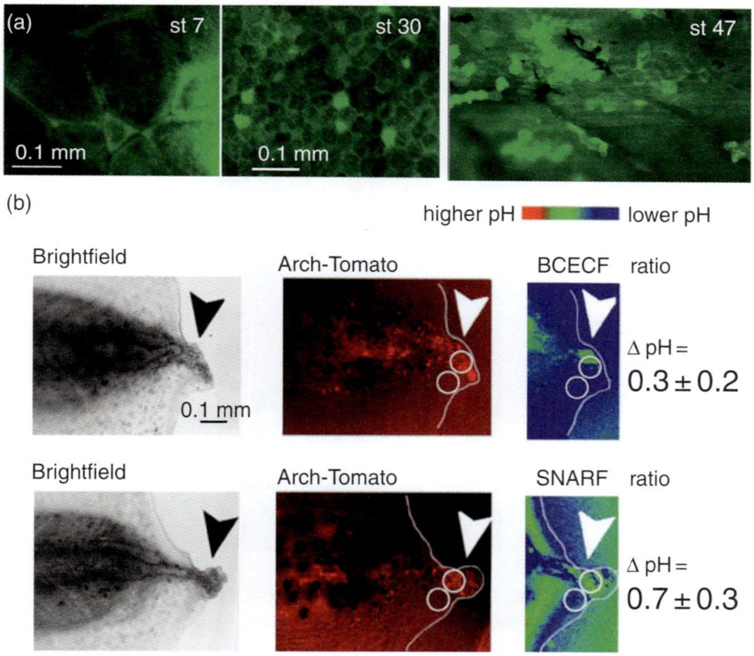

Figure 5.2 GFP-Arch mRNA was expressed and Arch is active in *Xenopus* cells. (A) By stage 7, expression is visible in some cells. Expression in cell membranes is clear by stage 30, and still present at stage 47, the middle of the refractory stage. (B) The change of pH in Arch-expressing cells is consistent with the presence of Arch activity. Two dyes were used to measure pH: one that is only weakly activated by the wavelengths that activate arch, BCECF, and another that is strongly activated by the light that activates Arch. Thus, BCECF reveals the pH of dark-maintained Arch and SNARF-5F reveals the pH of light-activated Arch. As predicted, pH is higher when Arch is activated by light. All figures were adapted from images originally published in *Biology Open* (Adams *et al.*, 2013) and are freely available under a Creative Commons License (http://creativecommons.org/licenses/by/3.0 /legalcode); copyright belongs to the authors. Images have been adjusted for color and print size only.

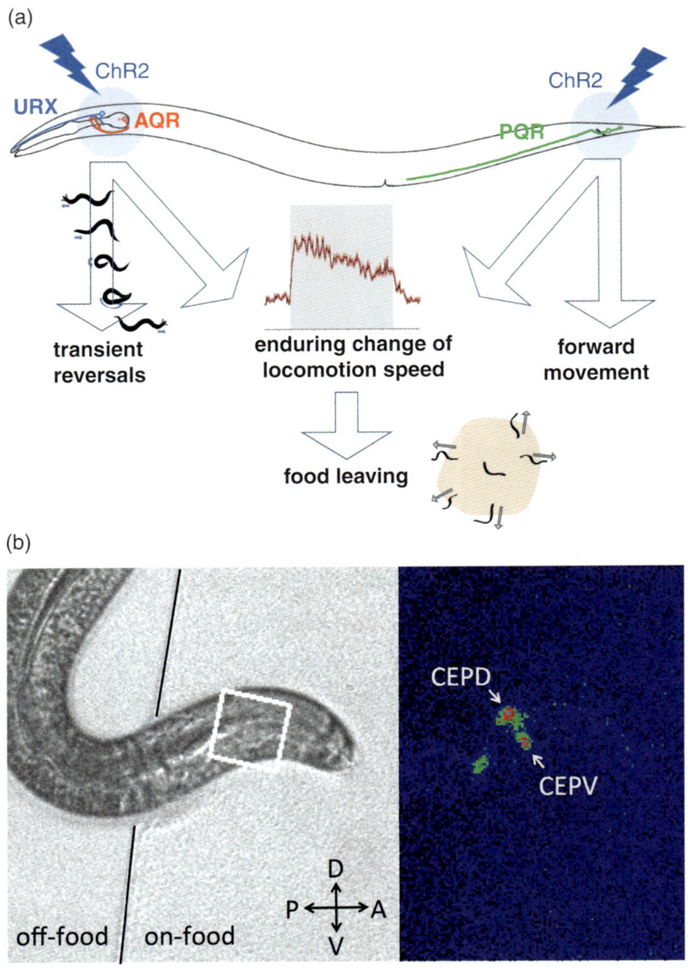

Figure 3.2 Neuronal targets for optogenetics in *C. elegans*. (A) Selective optogenetic activation of the URX, AQR (head) and PQR (tail) O_2-sensing neurons showed that the neurons contribute differently to multiple behavioral outputs. (B) Asymmetric activation of CEP neuron pairs. A bright-field image of *C. elegans* upon entering a lawn of bacteria (left) and fluorescent calcium signals from CEPD and CEPV (right) are shown.

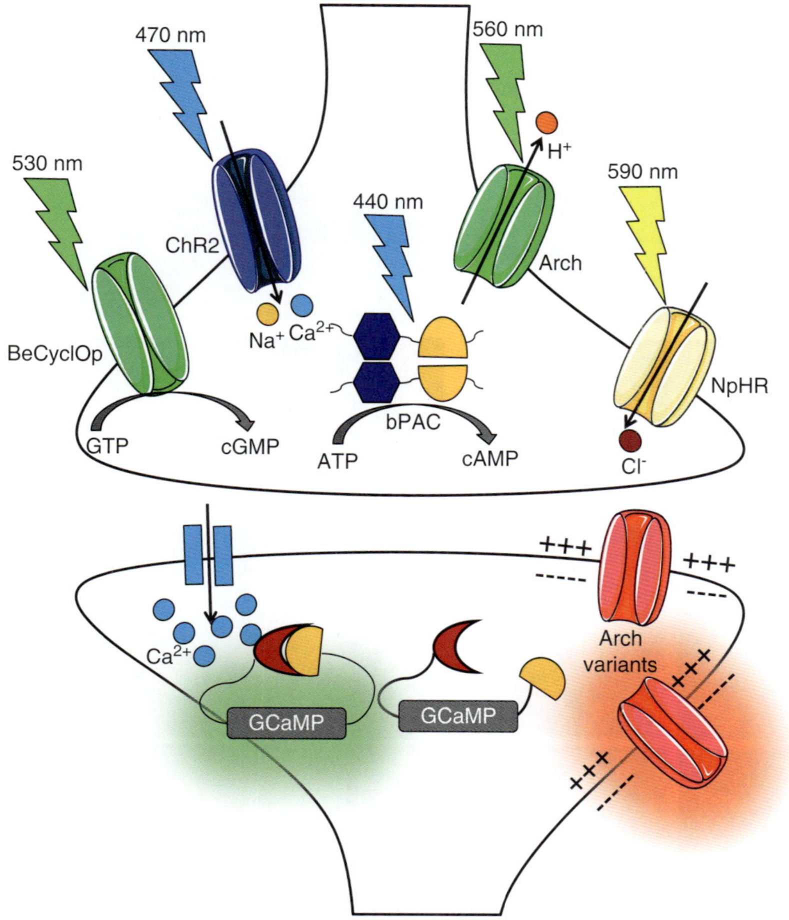

Figure 4.1 An illustration of the synapse featuring the prominent optogenetic actuators and sensors discussed in the review. The blue light-activated cation channel, channelrhodopsin-2 (ChR2), is the primary optogenetic depolarizer activating the excitable cell, while yellow light-activated chloride pump, halorhodopsin (NpHR), and green–yellow light-activated proton pump, archaerhodopsin (Arch), have been used as optogenetic hyperpolarizers, thereby inhibiting the cell. *Beggiatoa*-derived photoactivated adenylyl cyclase (bPAC) is a blue light sensing using flavin (BLUF) domain-containing enzyme catalyzing the synthesis of cAMP upon blue light illumination. Blue–green light control of cGMP synthesis has been achieved with a guanylyl cyclase opsin from *Blastocladiella emersonii* (BeCyclOp). The genetically encoded calcium indicator GCaMP and the voltage sensor Arch, which are the optogenetic sensors reporting on neural activation, have been found in the postsynaptic compartment.

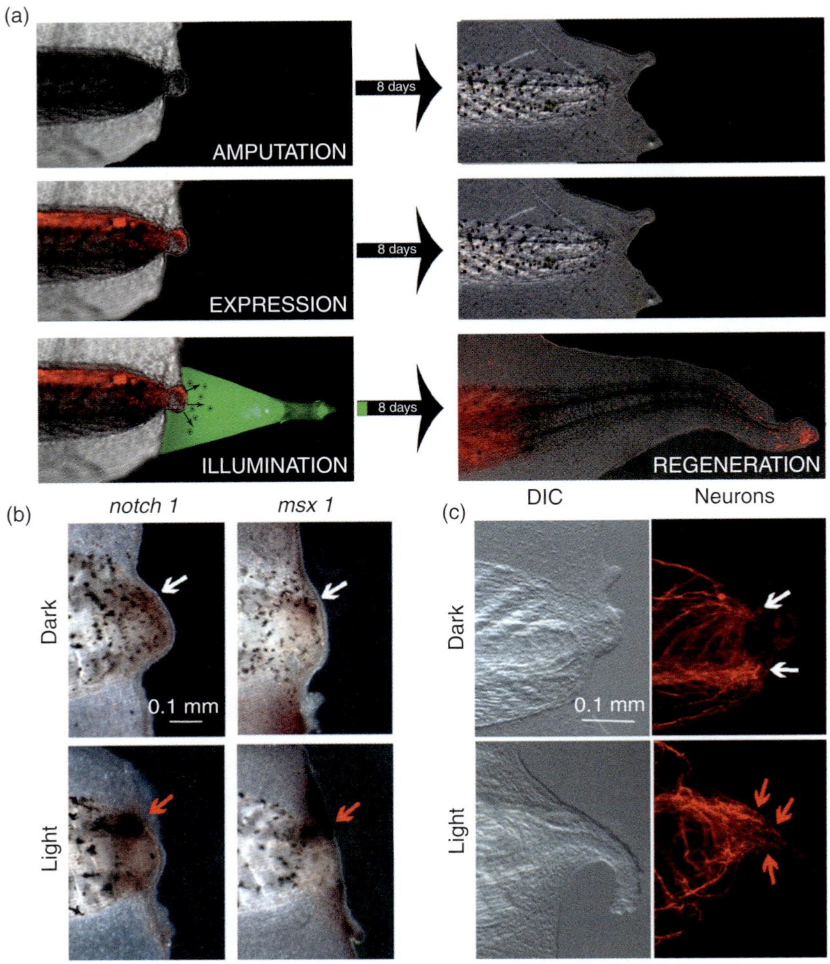

Figure 5.4 Light-activated arch triggers regeneration along the normal pathways. (A) A summary of the main results. Tails amputated at stage 47 do not grow back. Expression of Arch alone does not trigger regeneration, indicating that the presence of the exogenous protein does not, on its own, affect regeneration. However, when Arch-expressing tadpoles are exposed to light for 48 hours, many refractory tails regenerate in the normal 8day time period. (B) Two markers of standard regeneration, notch1 and msx1, are both expressed in the right place – the tip of the tail – only when Arch is light activated. (C) Arch-expressing embryos maintained in the dark have neurons that terminate at the amputation plane, like refractory tails. The neurons in light-activated, Arch-expressing tails, however, grow into the tip of the tail, just as they do in normally regenerating tails. All figures were adapted from images originally published in *Biology Open* (Adams *et al.*, 2013) and are freely available under a Creative Commons License (http://creativecommons.org/licenses/by/3.0/legalcode); copyright belongs to the authors. Images have been adjusted for color and print size only.

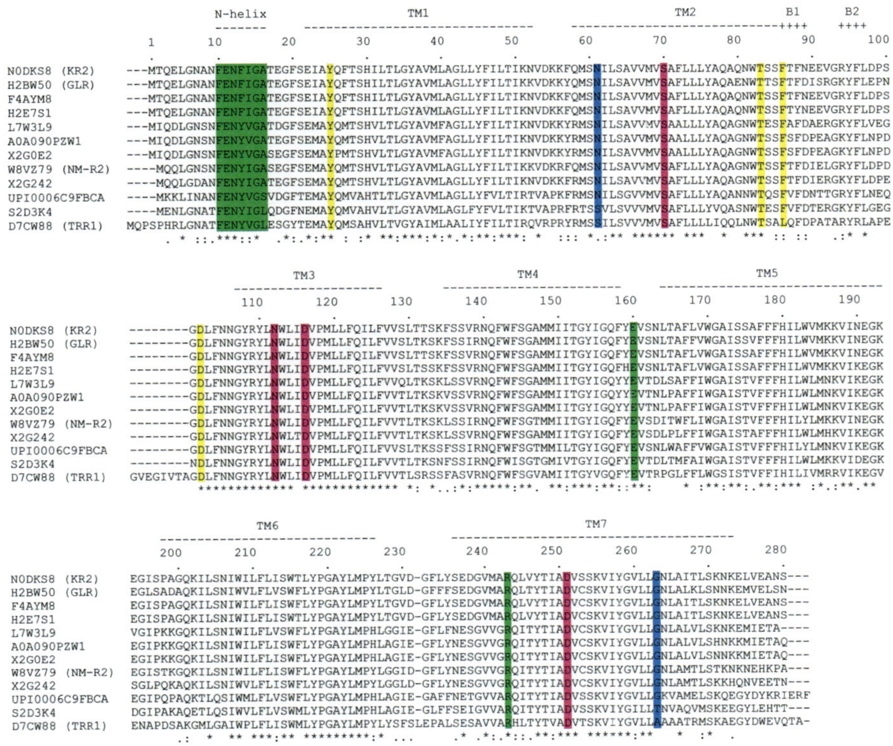

Figure 6.2 Structure-based sequence alignment of putative sodium-pumping rhodopsins. In order to precisely specify the proteins, we indicate the UniProtIDs. However, some of the proteins have "established" common names in the literature, so we cite them as well. TM indicates transmembrane regions, B shows the location on beta-sheets. The ion uptake cavity (cyan) is represented by an N61–G263 pair, the two latter members have an S61–T/A263 pair. The cavity in proximity to retinal (magenta) is completely conserved in the class (i.e. S70, N112, D116 and D251). The ion release cavity contains a cluster of three conserved ionizable residues: E11, E160 and R243 (green). All members possess an N-terminal α-helix that caps the putative ion release cavity. The oligomerization interface is also well conserved (yellow) among the members of sodium-pumping rhodopsins (Y25, T83, F/L86 and D102).

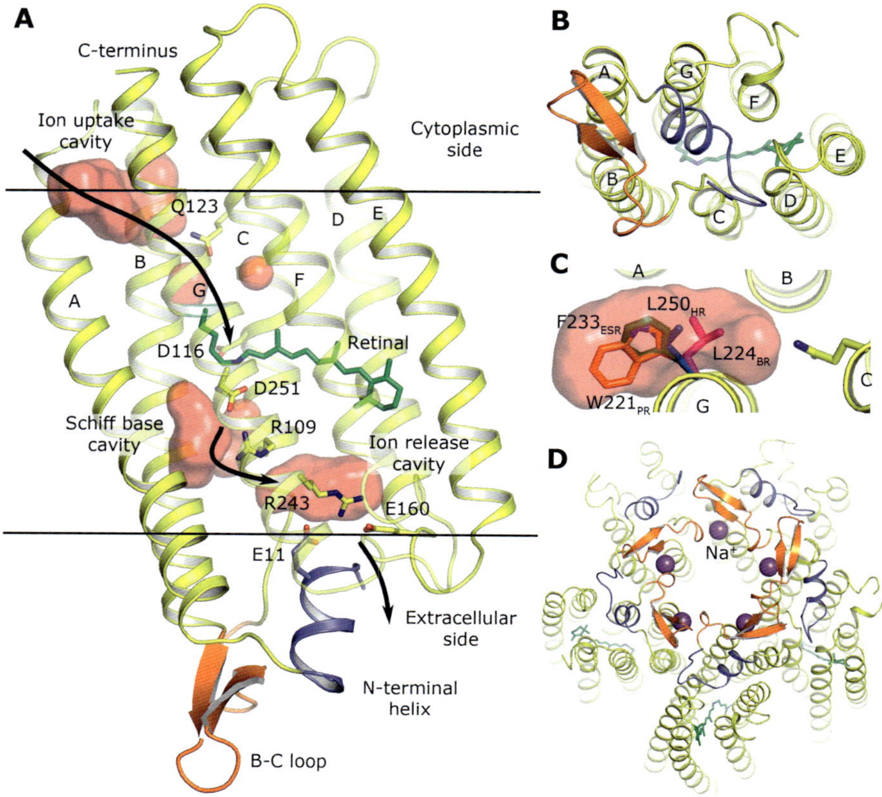

Figure 6.3 Structure of KR2. The N-terminal helix is shown in blue and the B–C loop is shown in orange. Water-accessible cavities are shown as red surfaces. (A) Side view. The ion path is shown with arrows. (B) View from the extracellular side. (C) KR2 ion uptake cavity. Hypothetical effects of the G263L, G263F and G263W mutations are shown using the conformations of the homologous residues L224 of bacteriorhodopsin (BR), L250 of [20], F233 of *Exiguobacterium sibiricum* rhodopsin (ESR) and W221 of blue proteorhodopsin (PR). (D) Pentameric assembly of KR2. Sodium ions are shown in violet.

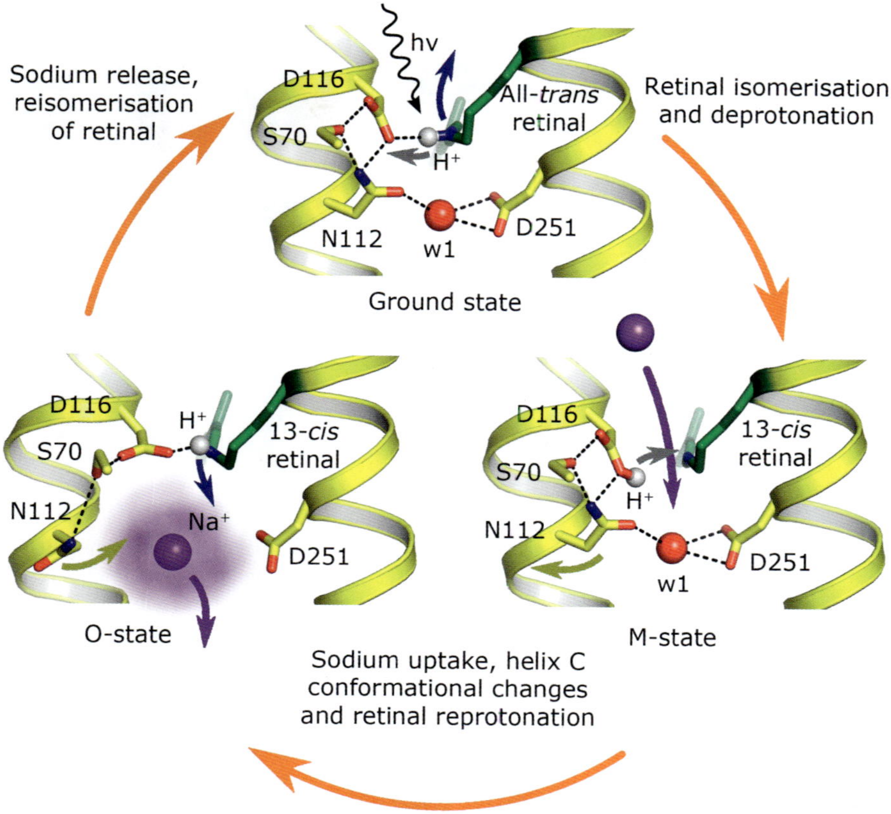

Figure 6.5 Proposed model of the structural changes during sodium translocation.

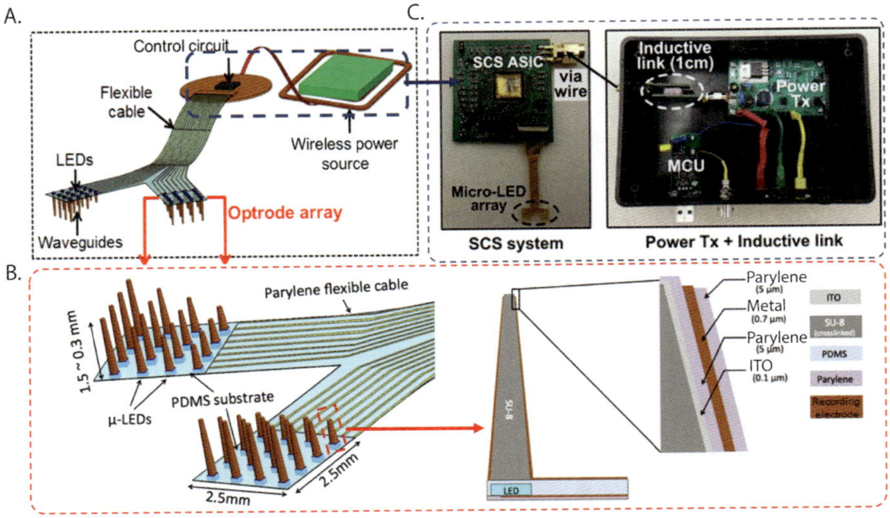

Figure 7.4 Wireless powered optrode array. (A) Conceptual illustration of the wireless optrode neural interface. (B) Overall design of the LED-coupled optrode array with 32 μLEDs, microwaveguides, and microelectrodes integrated on a Parylene-C substrate. (C) The inductively-powered telemetry and switched-capacitor based stimulating (SCS) system used to wirelessly drive the μLED array. Adapted from Kwon *et al.*, 2015 with permission.

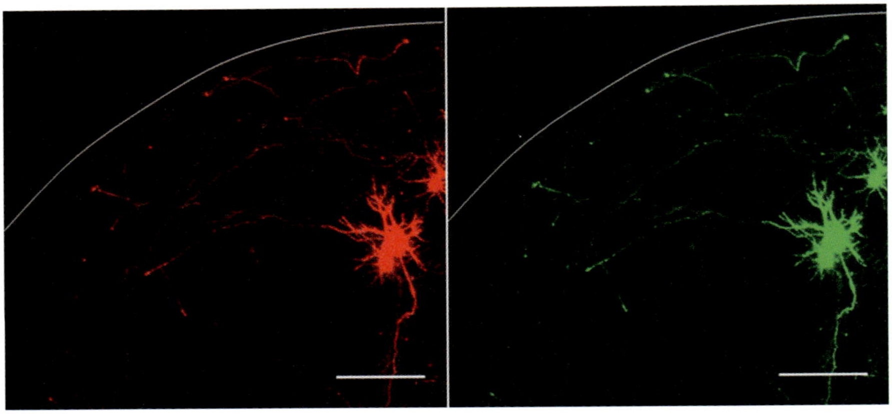

Figure 8.3 Cellular morphology of cortical upper layer neurons and horizontal axons in slice culture. A small number of upper layer neurons were transfected with DsRed and ChR2 plasmids. Expression of fluorescent proteins shows a pyramidal-shaped cell with a primary axon extending into the deeper layers and horizontally elongating collaterals extending into the upper layers. Scale bar: 100 μm. ChR2 is expressed in soma and axons (right)

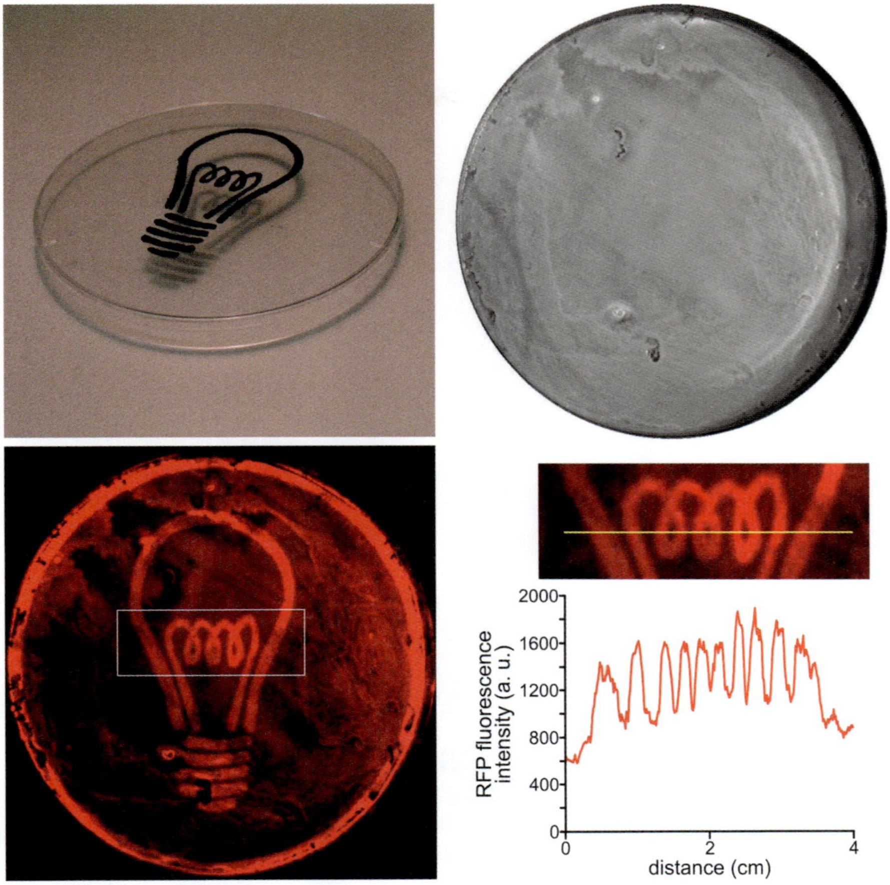

Figure 9.2 Construction of a light sensor using the psd module. Yeast cells expressing P_{TDH3}-*RFP-psd* (plasmid based) were spread on solid medium and incubated at 30°C for 24 hours. Blue light (465 nm, 10 μmol m^{-2} s^{-1}) was applied on parts of the plate using a mask (upper left panel). Please note that the rim of the plate was not illuminated due to the usage of a hood to block lateral light. RFP fluorescence (lower left image) and yeast cell growth (upper right image) were imaged with a fluorescence image analyzer and a digital camera, respectively. The graph (lower right panel) shows the line plot of fluorescence (measured along the yellow line) after background subtraction. The magnified area is indicated in the fluorescence image. Background fluorescence was obtained from cell-free areas.

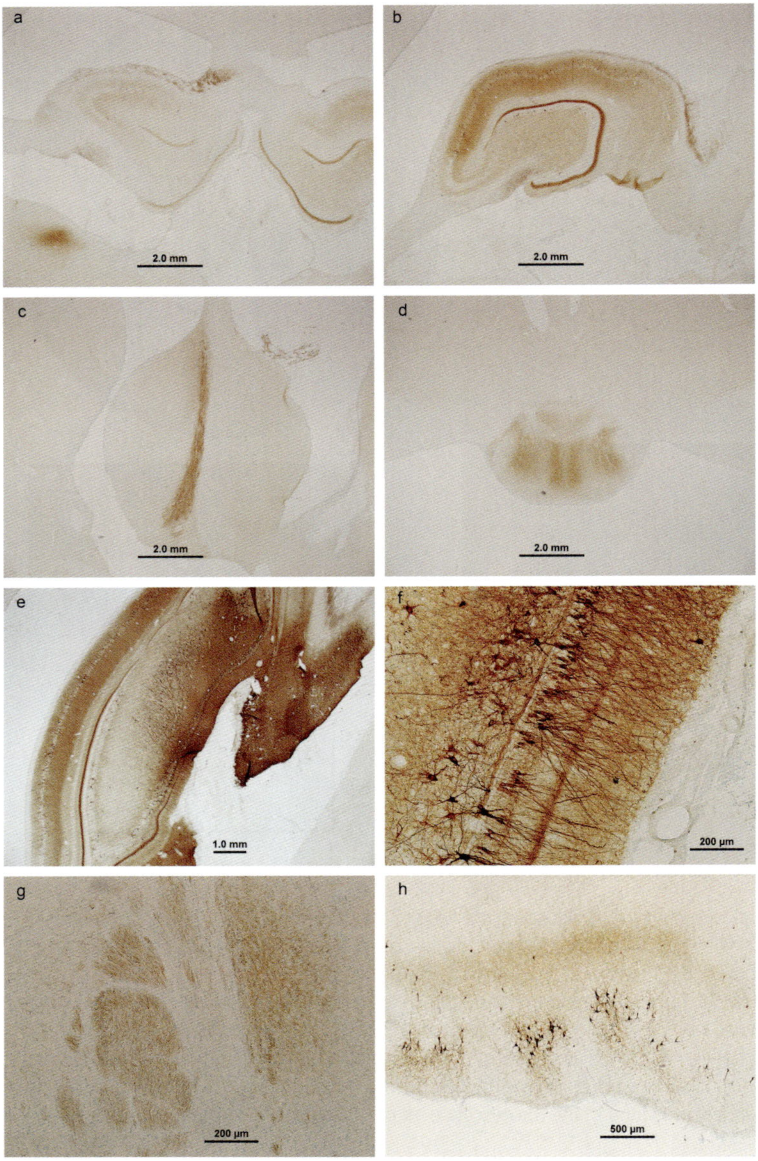

Figure 12.3 Transduction of much of the circuit of Papez achieved from a single infusion site in the sheep hippocampus. (A) Anterior nucleus of thalamus, (B) dorsal hippocampus, (C) ipsilateral fornix, (D) ventral fornix, (E) ventral hippocampus (infusion site), (F) higher magnification view of ventral hippocampus, (G) mammillary bodies and (H) entorhinal cortex.

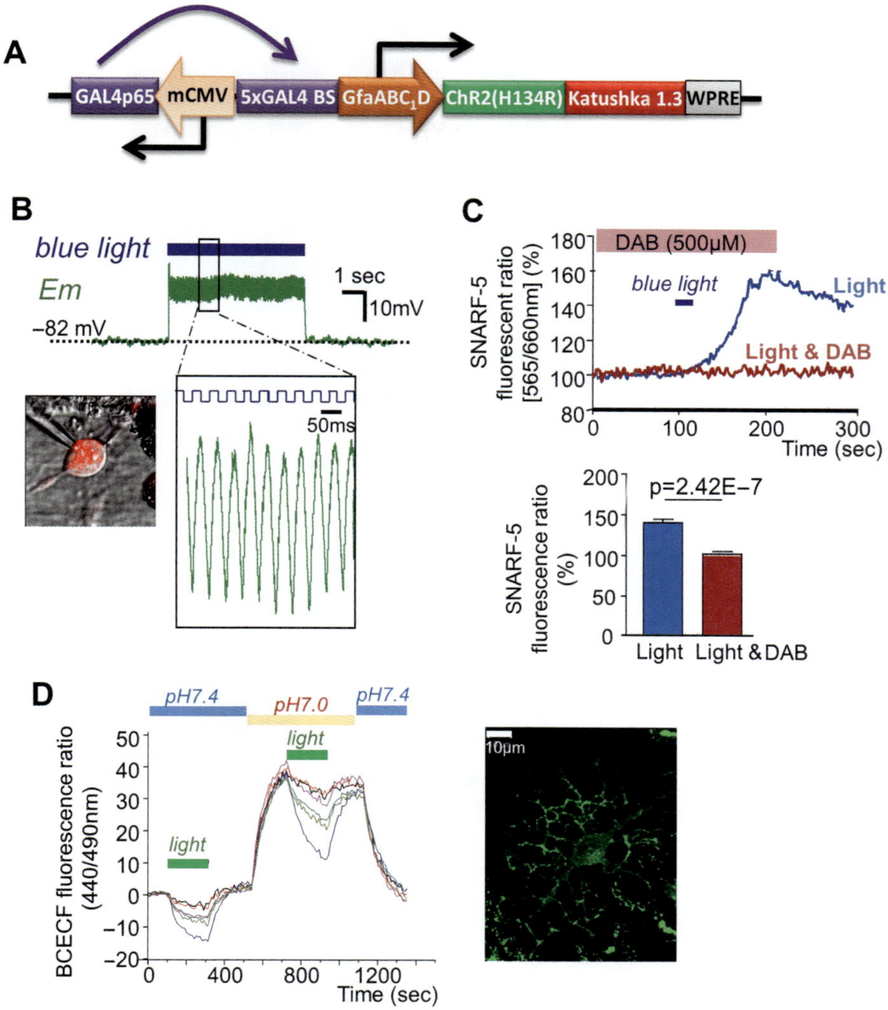

Figure 13.1 Application of optogenetic approaches to the stimulation of astroglia. (A) Layout of an optogenetic cassette for the stimulation of astrocytes via viral vector transgenesis. The chimeric transactivator GAL4p65, transcribed from a minimal cytomegalovirus promoter (mCMV), enhances transcription from GfaABC1D, a shortened version of the GFAP promoter, in an astrocyte-selective manner. ChR2(H134R) is fused to the red fluorescent protein Katushka1.3 to enable visualization in the absence of optogenetic stimulation. Woodchuck hepatitis virus response element (WPRE) is added to prolong the half-life of the transgene mRNA. (B) Astrocytic response to ChR2(H134R) activation. An astrocyte in organotypic brain slice culture expressing ChR2(H134R) was recorded in whole-cell patch clamp mode. Light flashes (470 nm; 20 ms, 25 Hz) caused time-locked depolarizations; note the absence of action potentials (adapted from Gourine *et al.* [2010]). (C) Optogenetic activation of astrocytes decreases intracellular pH. Light stimulation of optoβ2AR-expressing cultured astrocytes increased the emission ratio of the pH indicator SNARF-5. This intracellular acidification was sensitive to the blockade of glycogen metabolism by 1,4-dideoxy-1,4-imino-D-arabinitol (DAB; 500 μM); top– representative imaging trace; bottom – pooled data ($n = 32$) (adapted from Tang *et al.* [2014]). Stimulation using ChR2(H134R) yielded similar results (see Tang *et al.* [2014], supplement). (D) ArchT activation in astrocytes increases intracellular pH. Cortical astrocytes expressing the light-activated H^+ pump ArchT were loaded with the pH-sensitive dye BCECF and stimulated with green light and low pH. ArchT activation caused intracellular alkalinization and counteracted a pH drop induced by the acidic environment; right – BCECF-loaded astrocyte in culture (collaboration with Gourine A.V., unpublished data).

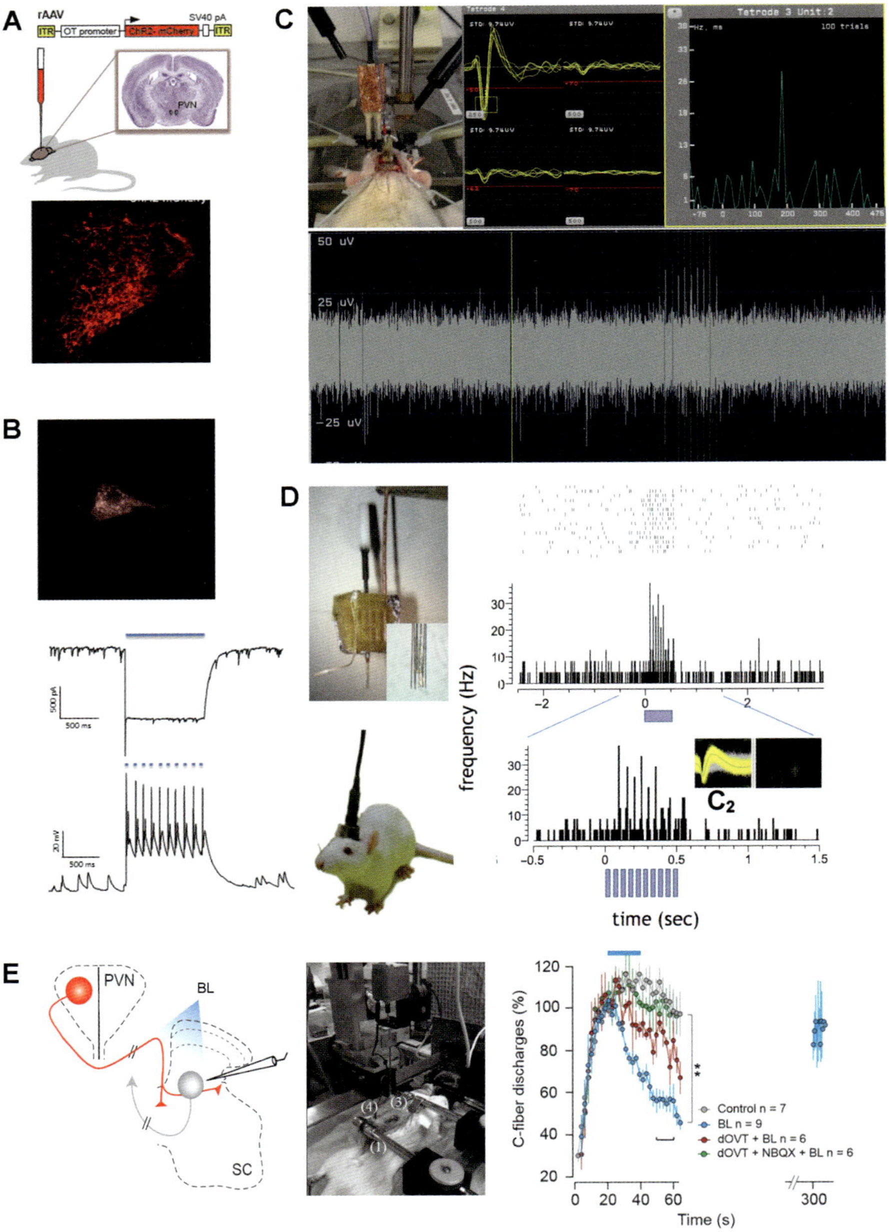

Figure 14.1 Optogenetics for oxytocin. (A) Specific expression of a transgene in oxytocinergic cells. (Top) Scheme showing the injection of the rAAV for the expression of ChR2-mCherry protein under control of the OT promoter. (Bottom) Validation of the approach. In red, direct visualization of mCherry 3 weeks after infection specifically in OT neurons (Knobloch *et al.*, 2012; Eliava *et al.*, 2016). (B) *In vitro* optogenetic manipulation and recording of an OT neuron. (Top) Two weeks after infection of the rAAV_pOT-ChR2mCherry OT neuron was visualized in an acute slice of the PVN through a 40× objective. (Top trace) After successful whole-cell patch clamp of the OT neuron with a 4–9 MΩ borosilicate pipette filled with standard intracellular milieu, a 1s continuous blue light (475 ± 28 nm) pulse was applied through the 40× objective in the voltage clamp mode and the photocurrent was analyzed. An average of a 1.1-nA (± 0.1, n = 3) photocurrent was induced. (Bottom trace) The amplifier was then switched to the current clamp mode and the cell was maintained at a potential of around −60 mV, and a series of blue light pulses (10 ms width for 1 s at 10 Hz) was applied. (C) *In vivo* optogenetic manipulation and recording of OT neurons in anesthetized rats. (Top left) The experiment was carried out on rats fixed in a stereotaxic frame under isofluran anesthesia. The opto-electrodes were connected to a preamplifier for recording and an Opto-feurrl wire for laser stimulation. (Middle) Clear extracellular signals were detected on tetrodes by waveform threshold. (Top right) The per-stimuli time histogram (PSTH) online can quickly verify

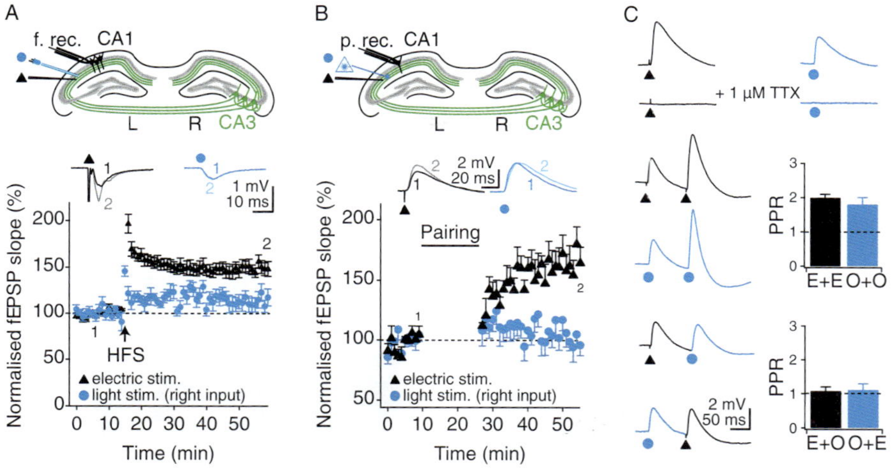

Figure 16.4 Optogenetic investigation of synaptic plasticity. (A) Fiber optics coupled to an LED can be inserted into acute brain slices (here, stratum radiatum in CA1 of the mouse hippocampus) just like conventional stimulation electrodes. Postsynaptic population responses are recorded with a field electrode (field recording [f. rec.]) placed into the pyramidal cell layer of CA1. Selective, optical stimulation (blue circles) of excitatory projections from CA3 to CA1 reveals that the right CA3 output is not plastic at the CA3–CA1 synapse. This lack of plasticity is overlooked under indiscriminate electric stimulation (black triangles). (B) Alternatively, a laser-based fixed or moving spot illumination system can be used to stimulate opsin-expressing fibers in acute brain slices. Here, patch-clamp recordings (p. rec.) from CA1 pyramidal cells demonstrated that optically recruited input from the right CA3 is unable to potentiate following a spike timing-dependent plasticity induction protocol (blue circles). In contrast, electrical stimulation led to robust synaptic potentiation (black triangles). (C) Electrically and optically stimulated postsynaptic responses are indistinguishable in terms of their kinetics and are blocked by 1 μM tetrodotoxin, demonstrating the electrical and optical recruitment of axonal compartments. Optical stimulation can be used to study short-term plasticity effects, such as paired pulse facilitation, and optical and electrical stimulation can be adjusted to be non-overlapping, if so desired. Note that optical stimulation of opsin expressing axons in thin, acute brain slices is possible (for more than an hour) without the cell body being present. PPR: paired pulse ratio; E: electric; O: optical stimulation. Broken lines represent the baseline. Error bars represent the standard error of the mean. Figures adapted, with permission, from Kohl *et al.* (2011) and Shipton *et al.* (2014).

Caption for Figure 14.1 (cont.)

suspicious OT neurons; here is an example of an interneuron responding to OT neuron activation with a 180ms delay. (D) *In vivo* optogenetic manipulation and recording of OT neurons in freely moving rats. (Top left) The data acquisition system was built on OpenEphys Acquisition board (v2.2) and the head stage was connected to a 32-channel preamplifier and ultra-thin SPI cable (Intan Tech) with a sampling rate at 30 kHz, the low-pass filter set to 600 Hz and the high-pass filter set to 6000 Hz. An additional I/O board allowed synchronized digital signal of optogenetic stimulation in data acquire board. Data analysis: OpenEphys GUI 0.3.5 was used in online data analysis. MATLAB, Simpleclust 0.5 and Neuroexplorer 3.2 were used in offline data sorting and analyses. The 32-channel opto-electrodes were shielded by a copper net. (Bottom left) Implanted animals remained completely free to move without any trouble. (Top middle) The real OT neuron was activated by burst laser stimuli with a very short delay and stopped firing immediately after the stopping of the stimulation. (Top right) On the contrary, the network effect triggered local neuron to fire with a much longer onset and usually sustained firing after the end of the stimulation. (Bottom right) Cluster sorting illustrates two separated neurons. (E) *In vivo* long-range OT projection manipulation to decipher complex neuronal circuits. (Left) Scheme of the experiment illustrating an OT neuron (red) projecting to deep layers of the spinal cord. (Middle) Picture of the experimental setup showing an anesthetized rat laminectomized in the stereotaxic frame. (Right) Effect of the local release of OT triggered by blue light stimulation on the wind-up intensity of a wide dynamic range neuron.

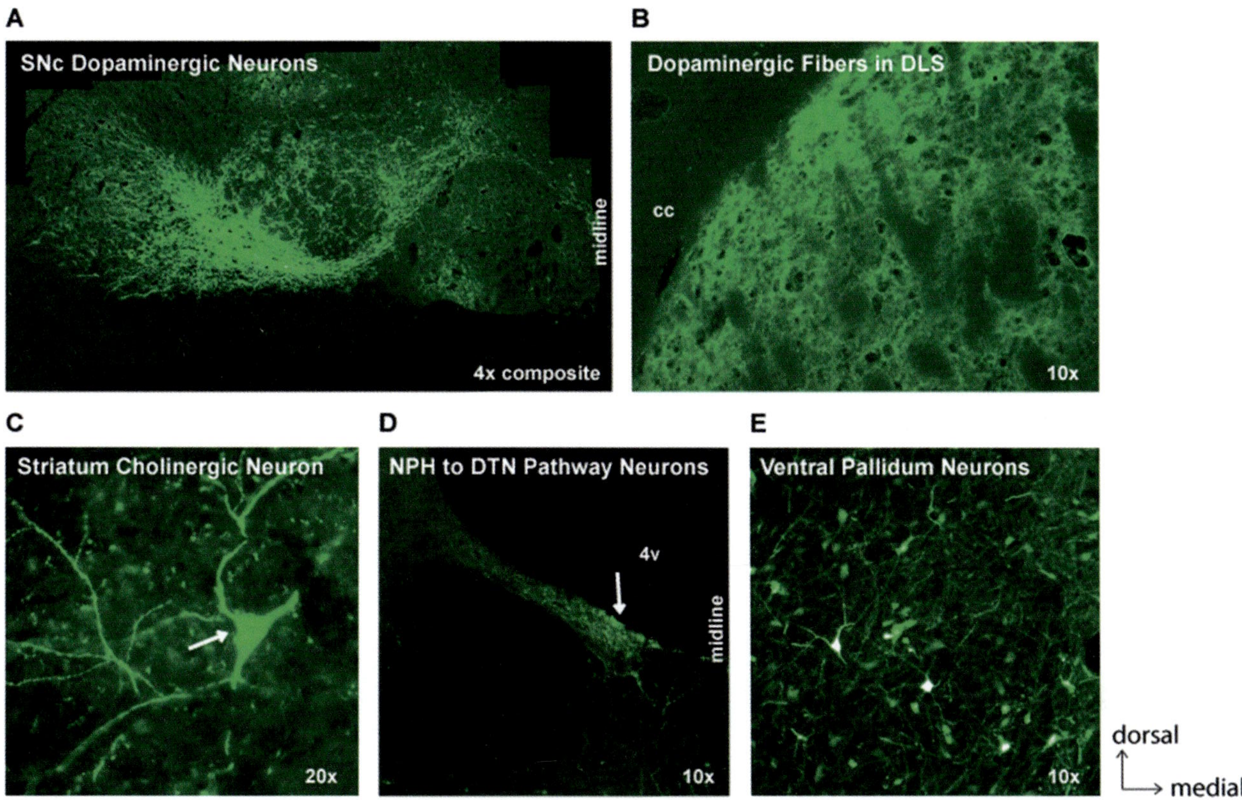

Figure 19.1 Examples of ChR2 and eNpHR expression in the rat brain. (A) Expression of ChR2-EYFP in dopaminergic neurons of the SNc in a transgenic TH-Cre rat. (B) Dopaminergic fibers in the DLS expressing ChR2-EYFP from the same rat. (C) ChR2-EYFP expression in cholinergic neurons of the striatum in a ChAT-Cre rat. (D) Expression of eNpHR -EYFP in DTN neurons after an injection of CAV-Cre in the DTN and Cre-dependent eNpHR in the NPH. (E) Generic expression of eNpHR-EYFP in the ventral pallidum using a synapsin promoter. cc: corpus callosum; 4v: fourth ventricle; DTN: dorsal tegmental nucleus; DLS: dorsolateral striatum; NPH: nucleus prepositus hypoglossi; SNc: substantia nigra pars compacta.

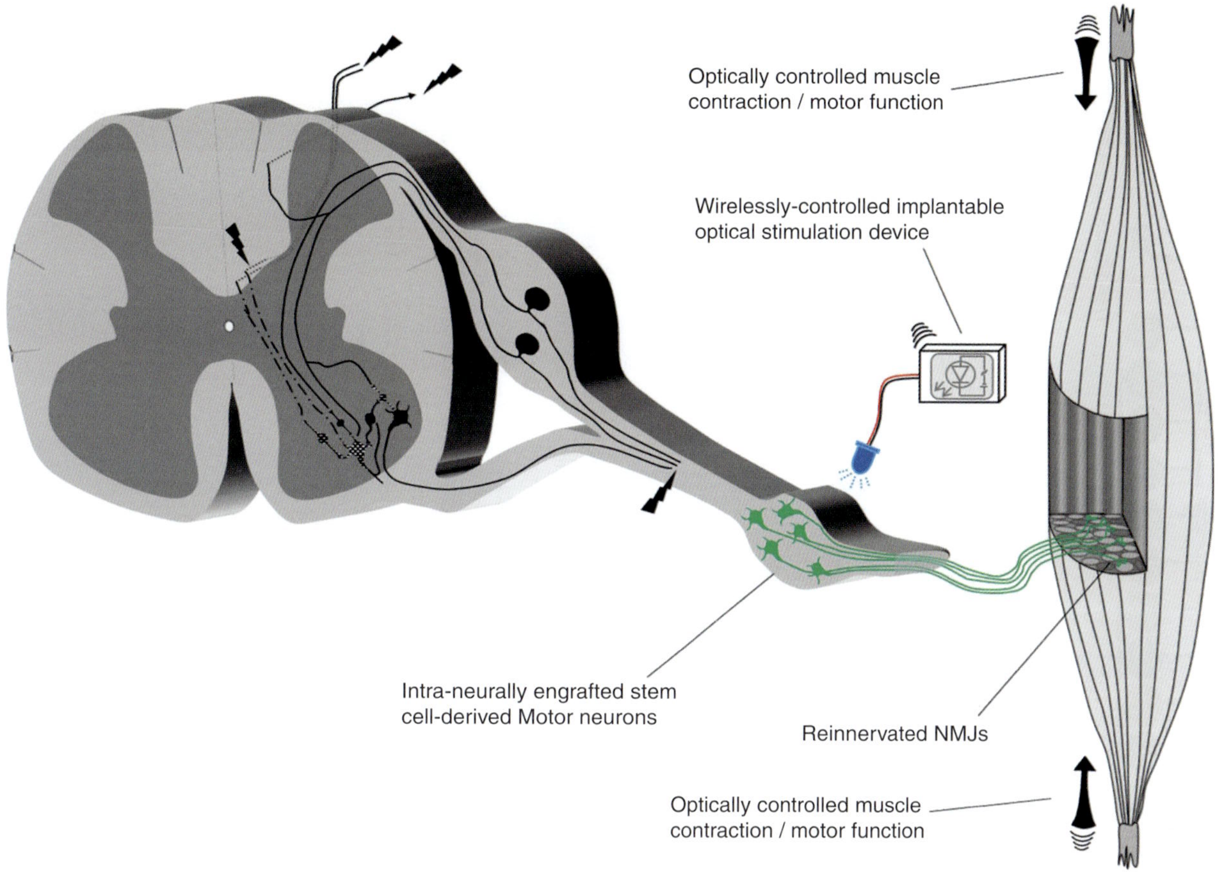

Figure 21.3 Schematic representation of an approach to restoring controllable function to paralyzed muscles. Irrespective of the cause of muscle paralysis, stem cell-derived motor neurons that are modified to express light-sensitive opsins can be engrafted into peripheral nerves, close to the motor nerve entry point of the muscle, resulting in robust innervation of the muscle. Activity of the engrafted motor neurons and thereby function of the muscle(s) that they innervate can then be controlled by an implanted optical stimulator device.

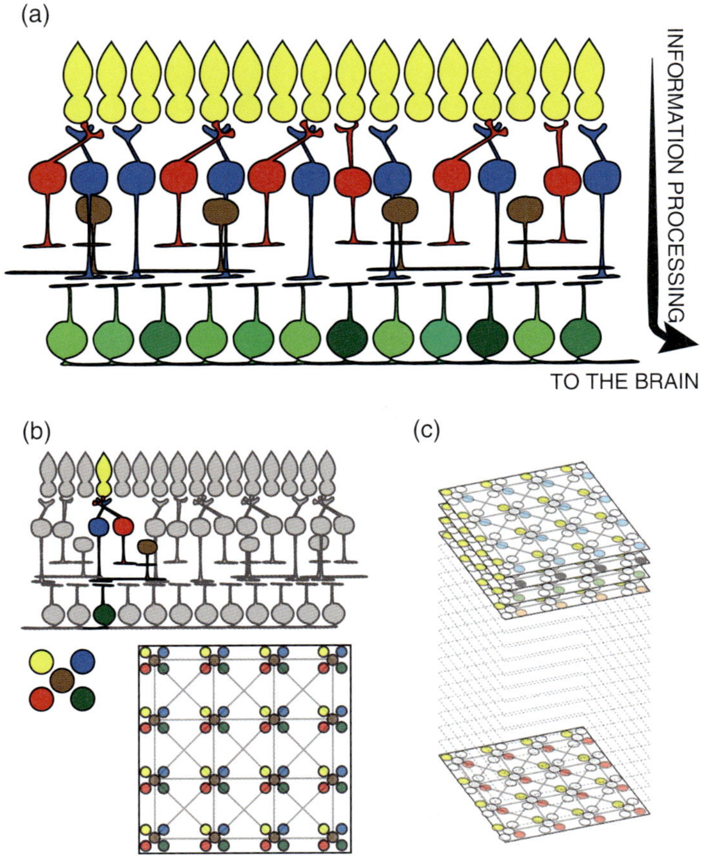

Figure 22.1 (A) General structure of the vertebrate retina. After light detection by photoreceptors (yellow), visual information is passed along to different types of bipolar cells (red and blue). This information is modulated by inhibitory amacrine cells (brown) and is finally transmitted to different classes of ganglion cells (green). Ganglion cells deliver processed visual information to the brain. (B) Each class of ganglion cells transmits a different version of the visual scene to the brain, since it receives information extracted by a specific processing microcircuit that is formed by a combination of bipolar and amacrine cells. Microcircuits performing the same type of processing tile the visual space evenly. (C) Taken together, the ensemble of the approximately 20 different human ganglion cell classes projects to the brain 20 different version of the visual scene.

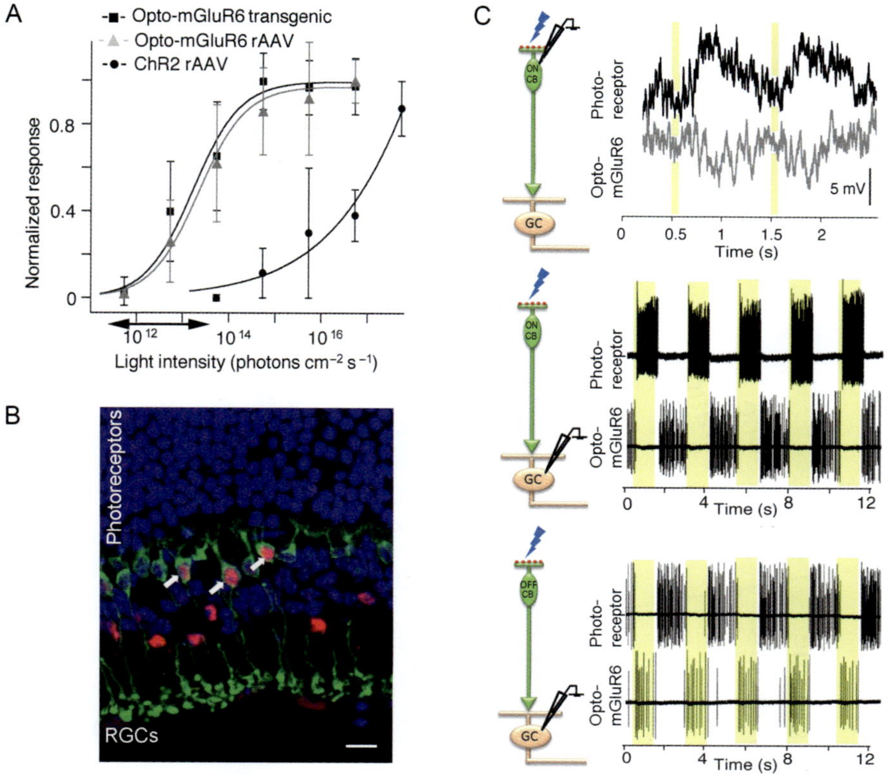

Figure 23.4 Performance of Opto-mGluR6 in the blind mouse retina. (A) Average light intensity response curves of seven RGCs from transgenic Opto-mGluR6 retinas and six RGCs from AAV-transduced Opto-mGluR6 retinas compared to published values for ChR2 (Wässle, 2004; Lagali *et al.*, 2008; Doroudchi *et al.*, 2011). The Opto-mGluR6 sensitivity curve lies in the range of cone photoreceptor sensitivity (indicated by the double-sided arrow); that is, Opto-mGluR6 mediates vision in daylight. In comparison, ChR2's activation threshold is equivalent to Opto-mGluR6 saturation intensities. The graph was deduced from the light-induced changes of ON-RGC spike rates that were normalized to a maximum of one for each RGC. (B) Confocal image of a longitudinal section through a healthy, photoreceptor-containing retina transduced with AAV particles containing the Opto-mGluR6 transgene. Opto-mGluR6 (red) is specifically expressed in ON-BPCs (arrows), co-labeled in green. Scale bar: 20 μm. (C) Electrophysiological patch clamp experiments on bipolar and ganglion cells of Opto-mGluR6 (red dots) – treated healthy retinas before (top traces) and after (bottom traces) the photoreceptor responses have been pharmacologically blocked (i.e. top traces are photoreceptor-mediated and bottom traces are Opto-mGluR6-mediated light responses of the same cell). Representative responses to a blue light stimulus (yellow highlights) demonstrate: (1) that the fast, native response is retained when Opto-mGluR6 mediates light responsiveness; (2) that Opto-mGluR6 recovers ON and OFF responses; and (3) that Opto-mGluR6 induces a response of inverse polarity compared to the photoreceptor-mediated responses, as light hyperpolarizes photoreceptors, leading to a decrease in glutamate release and, consequently, less mGluR6 activity. By contrast, Opto-mGluR6 is directly activated by light.

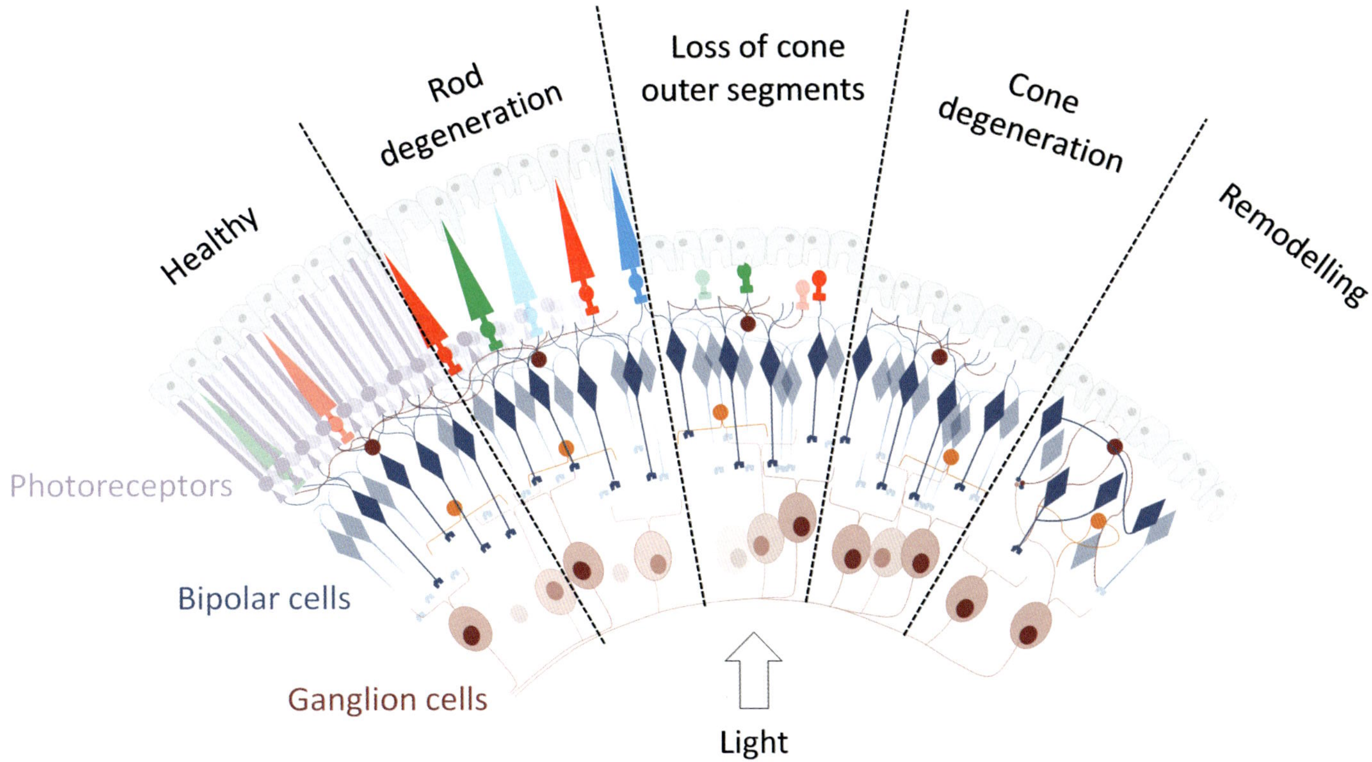

Figure 24.1 Schematic diagram of a healthy compared to a degenerated retina. Rod photoreceptors are the first to degenerate. Their loss induces the consequent loss of the cone outer segments and eventually cell bodies. Important circuitry remodeling occurs after deafferentation. The three main populations of cells are labeled, but others are also present: retinal pigment epithelium, horizontal cells and amacrine cells in gray, dark red and orange, respectively. Modification of connectivity after photoreceptor degeneration is not pictured.

Figure 25.1 Holographic optical neural interfacing concept depicted in an illustration of the future optical-based prosthesis: the camera video stream is processed and fed to a holographic projection system. A spatial light modulator (placed in the corners of the glasses) phase modulates a laser beam wave-front and excitation patterns are projected onto the photosensitized retina.

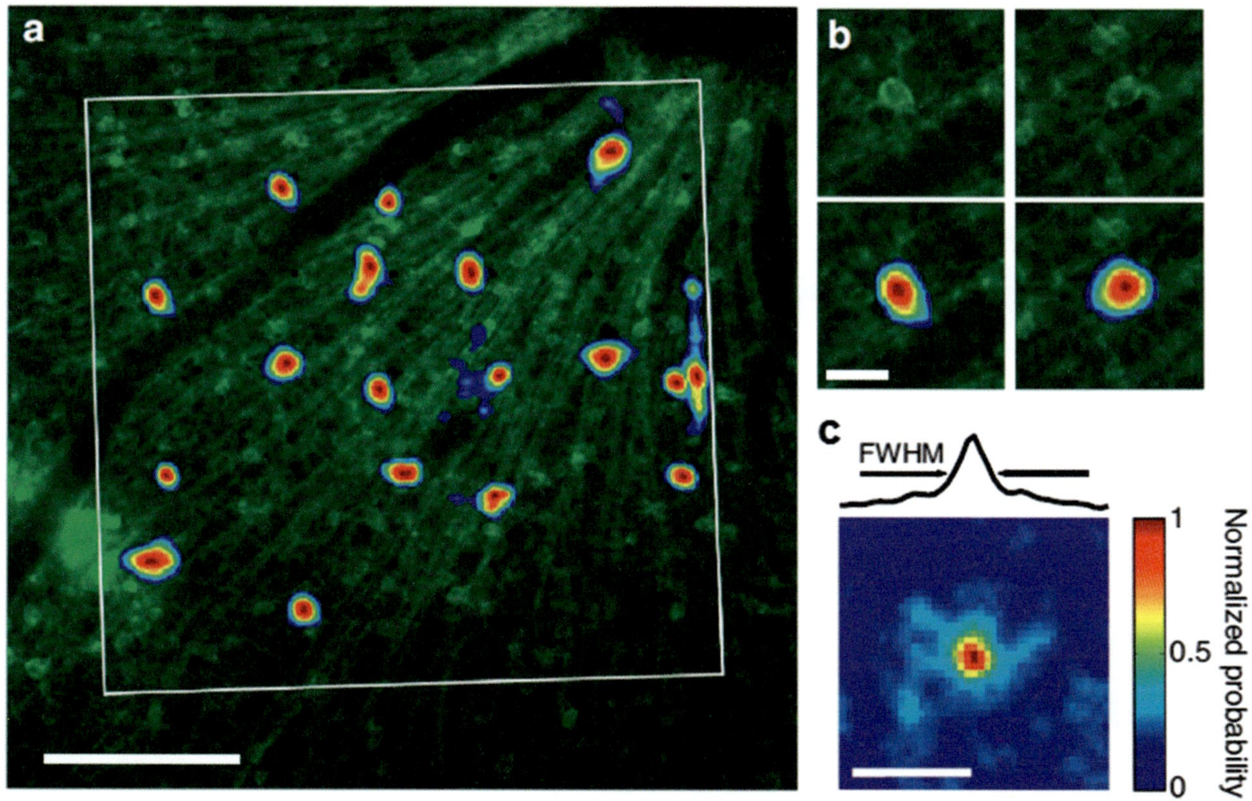

Figure 25.2 Single-cell resolution control of ChR2-expressing RGCs. (A) Superimposed images of a retina expressing ChR2-eYFP in the RGCs flattened on an MEA (black dots) and representative stimulation fields. Scale bar = 200 μm. (B) The stimulation fields match the underlying visualized RGCs. Scale bar = 20 μm. (C) Mean spatial distribution of the stimulation fields calculated for 202 units from 11 retinas. Scale bar = 50 μm. Figure adapted from Reutsky-Gefen *et al.* (2013).

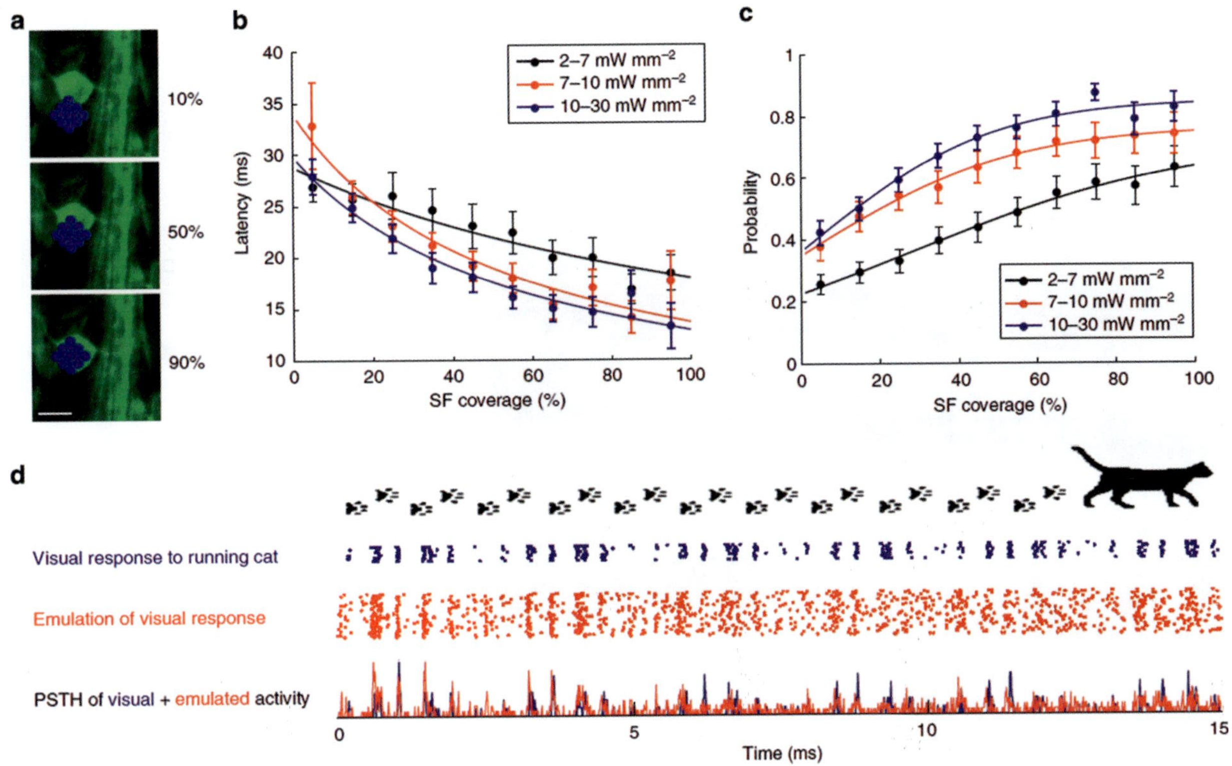

Figure 25.3 HONI performance and emulation. (A) Illustration of different cell coverages using a projection of spot patches (after smoothing in order to obtain a speckle-free spot). Scale bar = 10 μm. (B and C) Spike latency and firing probability as a function of the cell coverage and intensity of each patch. (D) Raster graph of optogenetically mediated RGC activity in blind ChR2-expressing retinas (red) following the projection of holographic patterns for reproducing photoreceptor-mediated visual responses in a sighted retina (blue) to a running cat. Bottom: Overlaid PSTH graphs illustrating the resemblance of the responses. Figure adapted from Reutsky-Gefen *et al.* (2013).

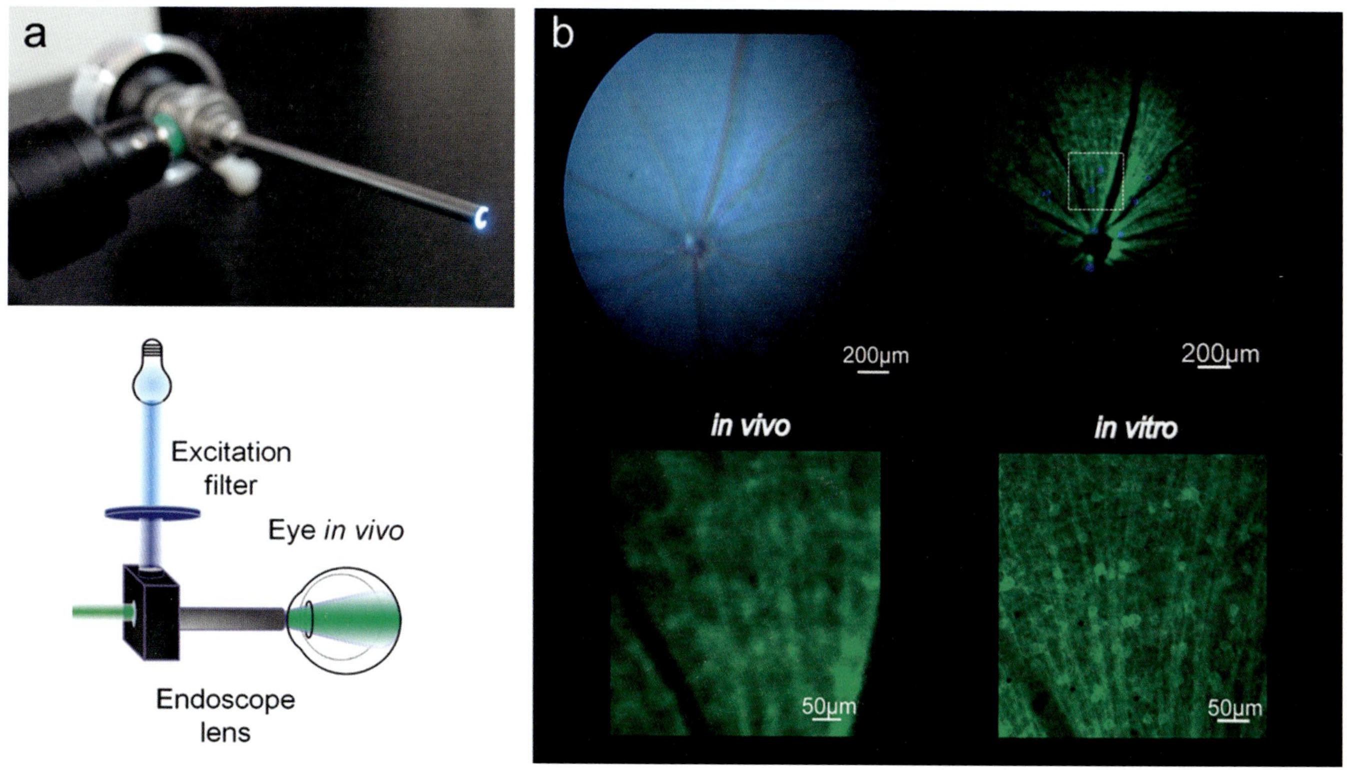

Figure 25.4 *In vivo* fundoscope system. (A) Photograph of the endoscope and custom excitation filter add-on (top). Scheme of excitation light illuminating the retina in a crescent shape and emitted fluorescence being collected by the endoscope (bottom). (B) Bright field (top left) and fluorescent fundus image with an overlay of holographic pattern projection (blue spots) on the retina (top right) of ChR2-expressing RGCs in a B6.Cg-Tg(Thy1-COP4/eYFP) mouse. (Bottom panels) Comparison between the magnification of an *in vivo* image and an image of a similar retina acquired *in vitro*. Figure adapted from Schejter *et al.* (2012).

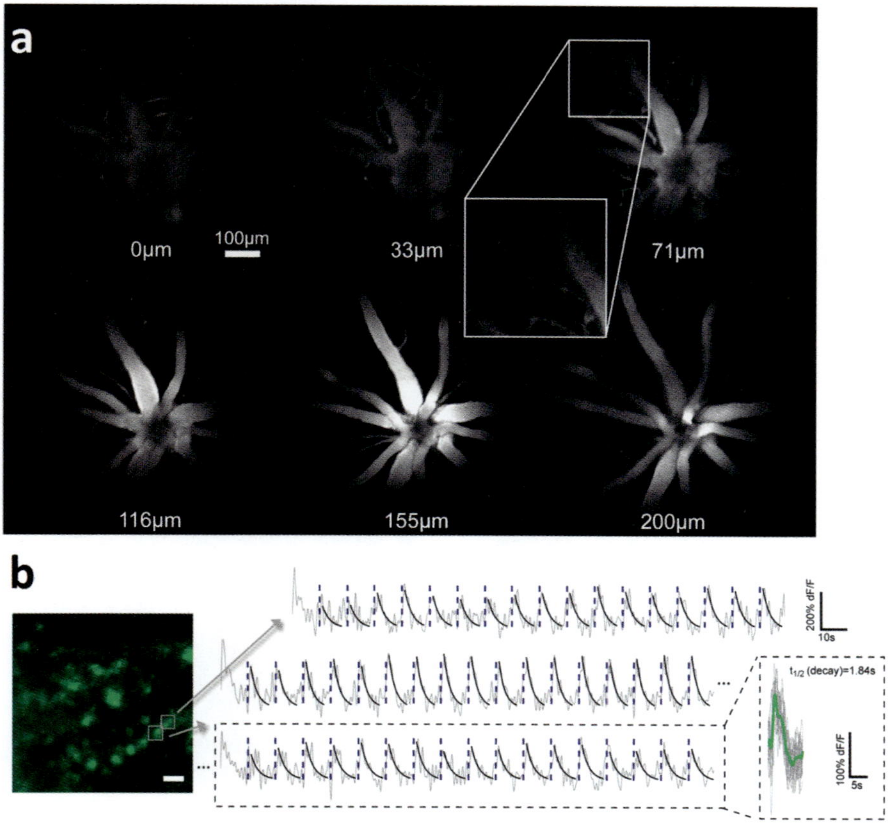

Figure 25.5 Two-photon *in vivo* mouse retinal imaging. (A) Optically sectioned, fluorescein-labeled retinal vasculature acquired at different depths using varying ETL focal lengths. The marked axial shifts were calculated from the paraxial model. (B) Calcium traces (gray) from two representative cells (squares in the average image on the left) in response to flashes of blue light (blue dashed lines) at 10-s intervals acquired at 6.7 fps. The middle and bottom traces were acquired from the same cell and 5 minutes apart. The black lines represent the exponential decay functions that were fitted to the average response (marked by a green line on the right) and fitted to the magnitude of each single response. Scale bar = 20 μm. Adapted from Schejter Bar-Noam *et al.* (2016).

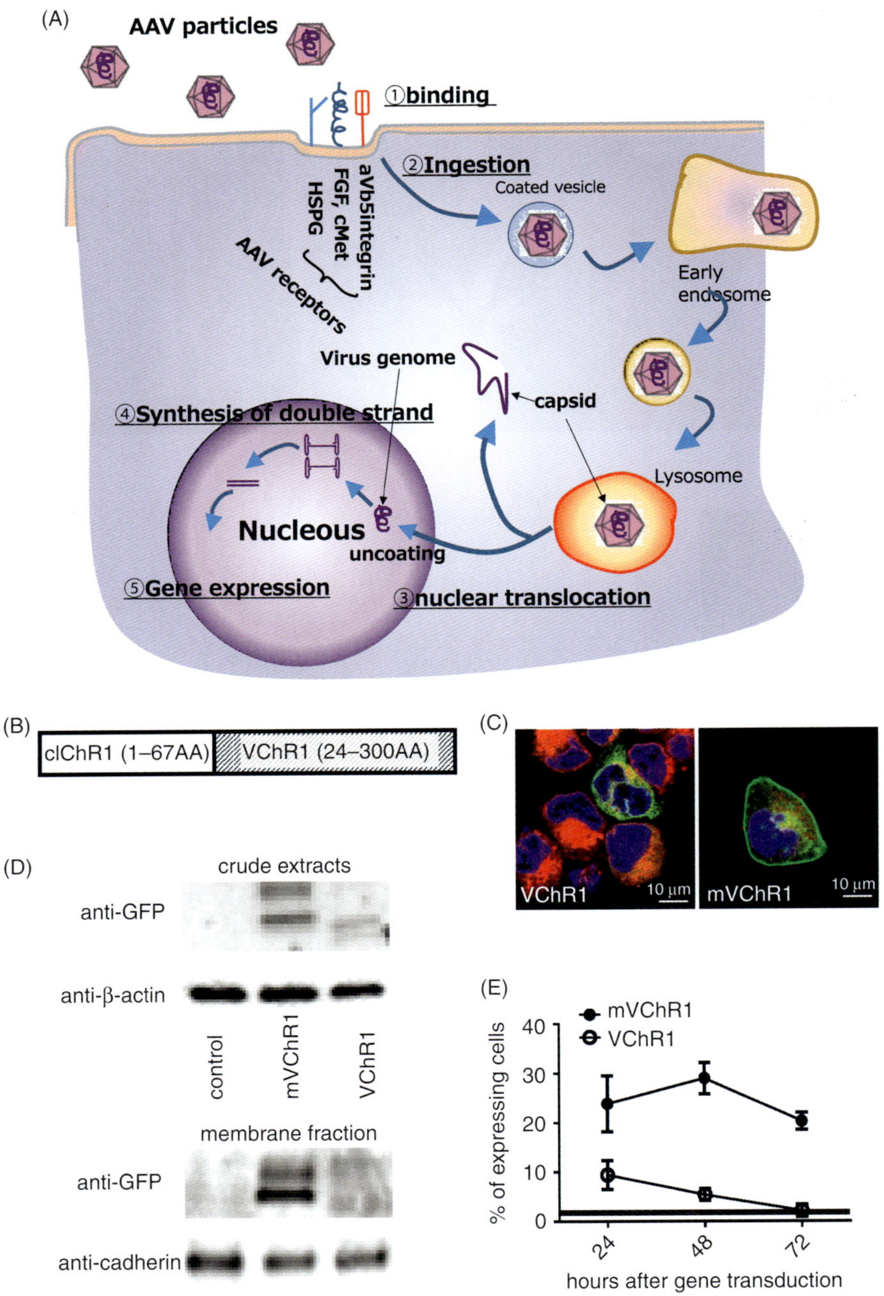

Figure 26.1 (A) Schematic depiction of AAV infection. The AAV-2-type virus binds to the heparan sulfate receptors in the host cell and is absorbed into the cell by endocytosis. The receptors involved also include αVβ5-integrin receptors and fibroblast growth factor receptors as co-receptors. The fusion of the endosomal membrane and viral membrane causes the absorbed virus to shed its capsid protein, transferring only its DNA into the nucleus. Subsequently, double-stranded DNA is formed using the polymerase in the nucleus, leading to gene expression. (B) mVChR1 structure. (C)

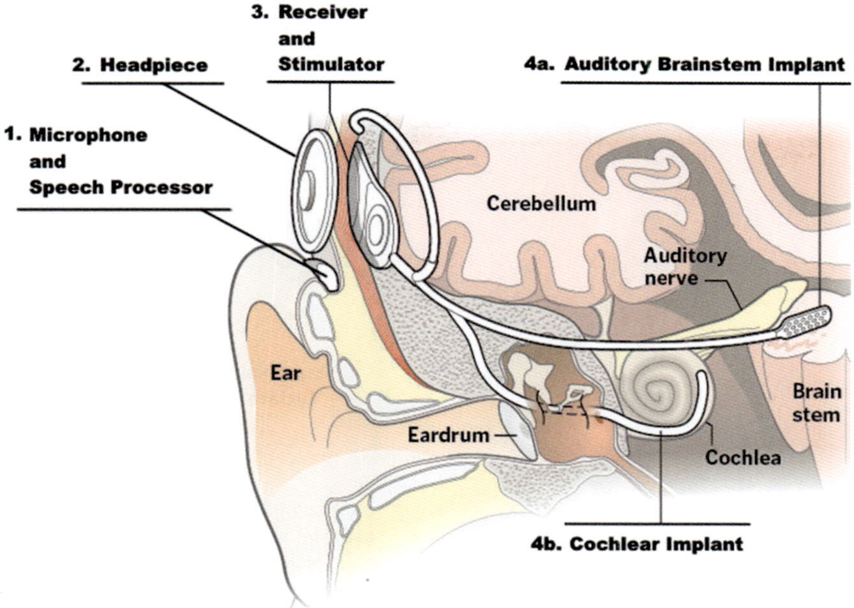

Figure 29.1 Contemporary auditory implants stimulate peripheral (CI) or central (ABI) neurons using multichannel electrode arrays. Sound from the external environment is collected by (1) the microphone and transformed into an electrical code in the speech processor that also contains the battery. This information is sent from (2) an external transmitter, called the headpiece, to (3) a surgically implanted receiver/simulator that is powered through induction from the speech processor. The external and internal components are brought into close proximity across the skin using magnets of opposite polarity in the headpiece and the receiver/stimulator. The receiver/stimulator is connected by wires leading to an electrode array. The CI array, receiver/stimulator and wires are placed during a 2hour out-patient procedure called a mastoidectomy. The electrode array is placed into the scala tympani of the cochlea (4a). In contrast, the ABI is placed during a more extensive surgical approach called a craniotomy, which exposes the region of the auditory nerve termination in the brainstem center called the cochlear nucleus. The ABI uses the same speech processor as the CI and activates (4b) an array of electrodes that stimulate cochlear nucleus neurons. Image adapted from (Neff, 2014).

Caption for Figure 26.1 (cont.)

Fluorescence microscope image of the HEK293 cells expressing VChR1 (left) and mVChR1 (right). (D) Western blot of the cell extract and cell membrane fractions. Given that VChR1 and mVChR1 bind to the fluorescent protein Venus, green fluorescent protein (GFP) antibodies that cross-react with Venus were used for detection. (E) After transduction of each gene, a Tali Image-based Cytometer (Life Technologies, Tokyo, Japan) was used to quantify the ratio of cells expressing fluorescent proteins to the total cell count. HSPG: heparan sulfate proteoglycan. Panels (C) and (D) were modified from Tomita *et al.* (2014).

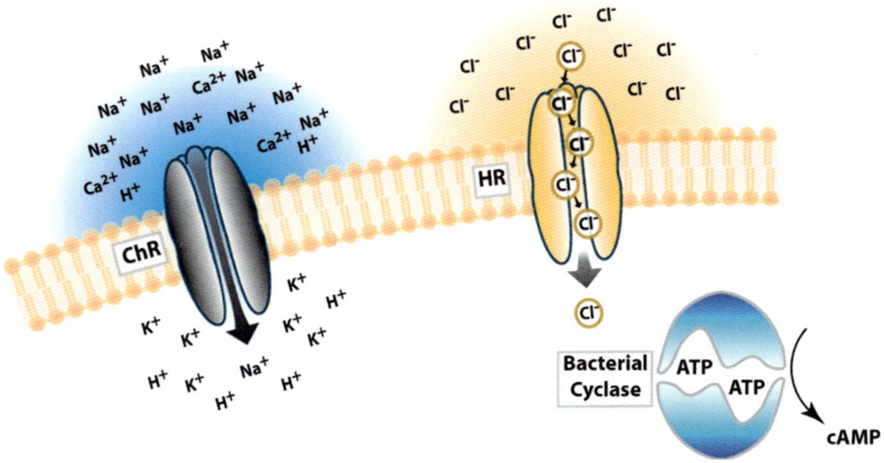

Figure 29.2 Fundamentals of optogenetics. Radiant exposure of opsins to specific wavelengths of light opens these channels to allow the influx of cations into the interior of the neuron. In the case of ChR2 (left), the influx of cations depolarizes the neuron, whereas in the case of halorhodopsin (HR; right), the influx of anions hyperpolarizes the neuron. Expression of opsins on peripheral or central auditory neurons requires transgenic, viral or stem cell-mediated approaches prior to stimulation with light. Figure adapted from Yizhar *et al.* (2011).

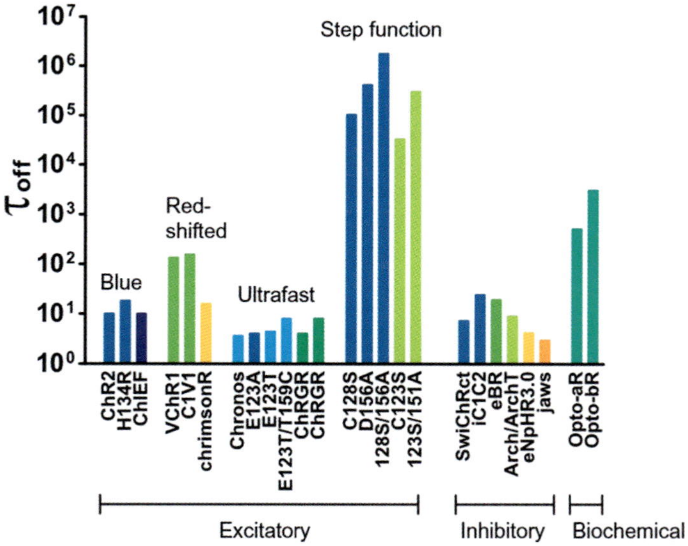

Figure 29.3 Classes of opsins. Chart showing the off-time constants (in milliseconds) of a variety of opsins. Since the discovery of ChR2, a range of opsins has been discovered or bioengineered from existing opsins, including red-shifted, ultrafast, on/off step function and inhibitory opsins. The ideal opsins for the auditory system are low threshold with ultrafast on/off kinetics and a brief refractory period for encoding the high pulse trains that are required for speech perception. Figure adapted from Guru *et al.* (2015).

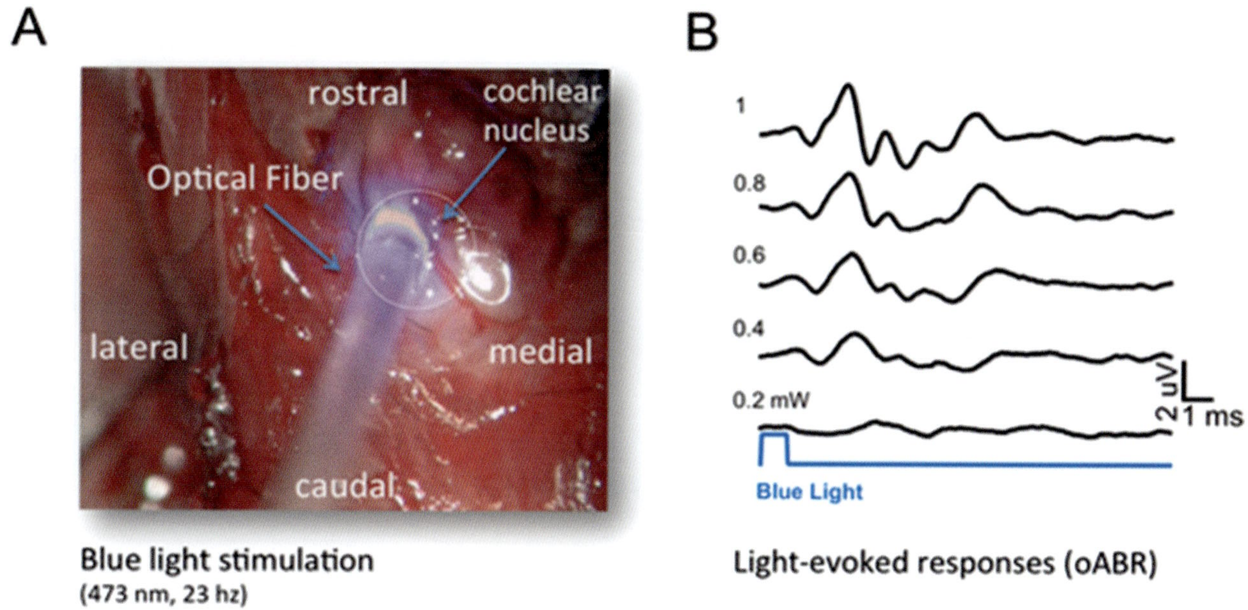

Figure 29.4 Optogenetic model of the ABI. (A) Experiment to stimulate the mouse cochlear nucleus with light conveyed by an optical fiber (400 μm diameter). Transduction of auditory brainstem neurons was achieved by direct inoculation of the cochlear nucleus via posterior craniotomy with an AAV vector with ChR2 or Chronos under the control of a ubiquitous promoter (CAG). Following a 1 month incubation period, the cochlear nucleus was re-exposed and stimulated with pulsed blue light (laser wavelength 475 nm for ChR2 or 500 nm for Chronos; 14–448 pulses s^{-1}, train durations 300–500 ms, light intensity 0–13 mW). (B) Resulting far-field responses are optically evoked auditory brainstem responses. These multi-peaked waveforms resembled sound-evoked auditory brainstem responses. No responses were observed from postmortem subjects or from control mice without cochlear nucleus opsin expression. Figure adapted from Hight *et al.* (2015b).

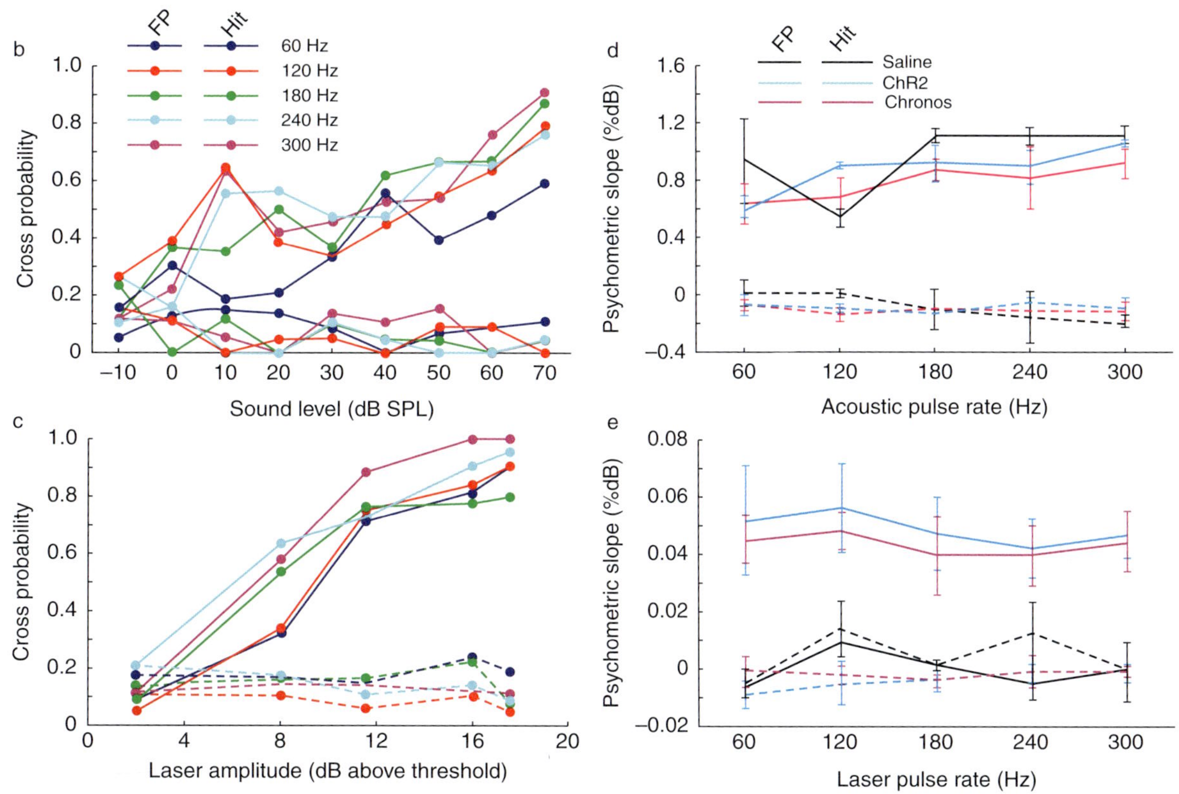

Figure 29.6 Behavioral detection of photostimulation mediated by ChR2 is similar to sound stimulation, but was not superior in Chronos-expressing mice. Animals with Chronos, ChR2 or saline injected into the IC (midbrain) were trained to perform a behavioral detection task in response to pulsed noise. Animals cross the shuttle box to avoid a foot shock. After successful training, animals were implanted with chronic optical fibers at the site of prior injections and the detection task was repeated with photostimulation. Correct and false-positive crossing probabilities are calculated as a function of (B) sound level and (C) laser amplitude. The slope of the linear fit of psychometric curves is used as a proxy for pulse detection. The psychometric slopes for ChR2- and Chronos-expressing mice are not significantly different across changing rates of (D) acoustic or (E) laser stimulation. Figure adapted from Guo *et al.* (2015).

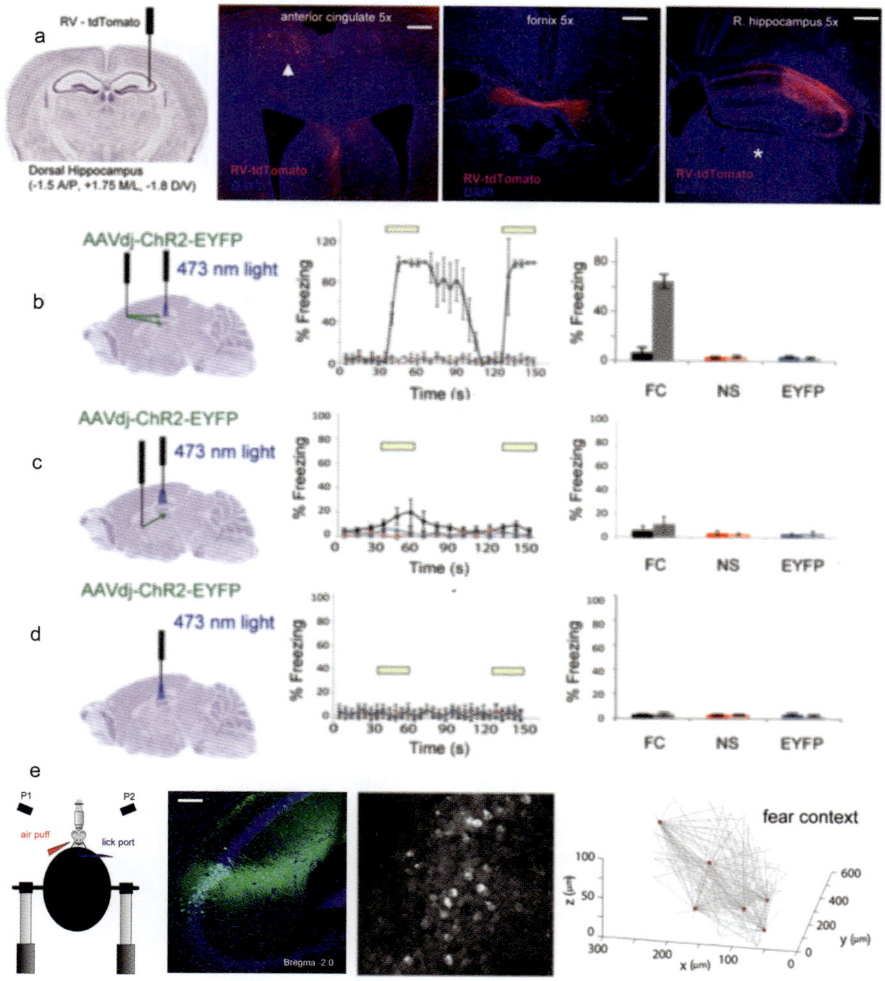

Figure 27.1 (A) Retrograde rabies tracing from the hippocampus reveals afferent fibers originating in the AC (arrowhead). There was no leak of rabies virus into the medial dorsal thalamus (asterisk), a known projection target of PFC. Scale bar: 300 μm. (B) Stimulating AC–CA projections resulted in significant light-induced memory retrieval (freezing) that was not observed in mice without opsin (EYFP) or mice that had not been fear conditioned (no shock). (C) Stimulating medial septal projections to the hippocampus demonstrated a strikingly smaller trend. (D) Directly stimulating the hippocampus demonstrated no light-dependent freezing. (E) Visualizing population dynamics in CA2/CA3 during contextual fear conditioning in a virtual environment: the emergence of hub neurons during learning.

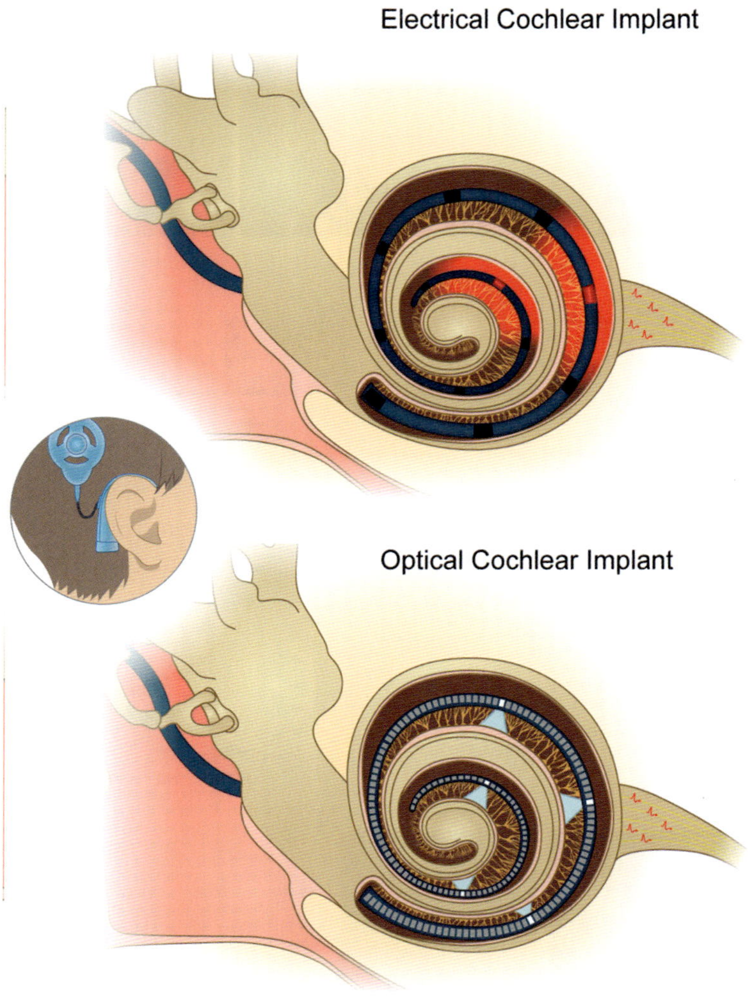

Figure 30.1 Electrical versus optical spiral ganglion (SGN) neuron stimulation. Top: electrical cochlear implants (blue) use between 8 and 24 electrode contacts (black and red) to stimulate SGNs (yellow). The very low number of electrodes and the current spread around an electrode contact (red halo) dramatically reduce the stimulation along the cochlear tonotopic axis, in this way limiting the frequency resolution of electrical coding. Bottom: focused optical stimulation promises spatially confined activation of SGNs, allowing for a higher number of independent stimulation channels and improving the frequency and intensity resolution of sound coding.

Part V

Optogenetics in Vision Restoration and Memory

22 Optogenetics in Treating Retinal Disease

Dasha Nelidova and Federico Esposti

22.1 Introduction

The retina is our most complex sensory organ. Blindness often results from dysfunction of this complex neural network. At present, degenerative diseases of the retina remain incurable. Patients with these conditions suffer progressive visual decline resulting from the ongoing dysfunction or loss of retinal cells, with photoreceptors being particularly prone to degeneration. Despite photoreceptor loss, downstream neural circuitry is often sufficiently intact to make prospective sight-restoring therapies feasible, if the appropriate cells can be made to respond to light. One of the main goals of translational retinal research is to develop strategies to repair the retina. Given the great complexity of this tissue, any attempt to restore retinal function requires quantitative knowledge of retinal neural circuits, as well as methods to target these circuits specifically.

No conventional medical or surgical treatment has the ability to restore high-acuity vision, but low-resolution visual perception has been produced in clinical trials of implantable retinal prosthetic devices in blind human subjects. For full visual perception, however, and in light of the fact that most of the patients seeking visual restoration do have some degree of remaining visual function, a major challenge remains the reinstatement of high-quality vision that is able to discriminate subtle details on a par with native vision. In this respect, viral delivery of optogenetic tools to retinal cell types is perhaps best able to approximate natural vision. It is noteworthy that delivery of Adeno-associated virus (AAV) to the retinas of human patients has been shown to be safe in multiple clinical trials (Bainbridge *et al.*, 2008; Cideciyan *et al.*, 2008; Hauswirth *et al.*, 2008; Maguire *et al.*, 2008).

22.2 Retinal Circuits

The mammalian retina is composed of more than 60 cells types. The retina can be viewed as a parallel image processor that acquires images via a mosaic of photoreceptor cells (Figure 22.1). Photoreceptors are the natural photon detectors. Rod

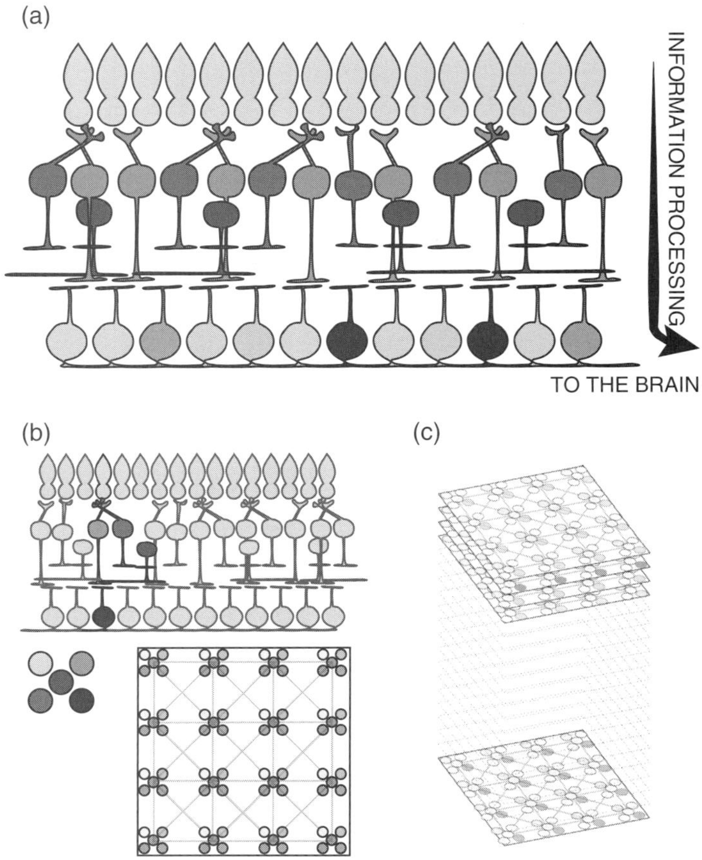

Figure 22.1 (A) General structure of the vertebrate retina. After light detection by photoreceptors (yellow), visual information is passed along to different types of bipolar cells (red and blue). This information is modulated by inhibitory amacrine cells (brown) and is finally transmitted to different classes of ganglion cells (green). Ganglion cells deliver processed visual information to the brain. (B) Each class of ganglion cells transmits a different version of the visual scene to the brain, since it receives information extracted by a specific processing microcircuit that is formed by a combination of bipolar and amacrine cells. Microcircuits performing the same type of processing tile the visual space evenly. (C) Taken together, the ensemble of the approximately 20 different human ganglion cell classes projects to the brain 20 different version of the visual scene. (A black-and-white version of this figure will appear in some formats. For the color version, please refer to the plate section.)

photoreceptors greatly outnumber cone photoreceptors, but are saturated at high daytime light intensities. Cones respond during daylight, preferentially to light of either long (red), medium (green) or short (blue) wavelengths, giving rise to our ability to discriminate color. Both rods are cones respond to light by graded hyperpolarization.

Downstream of photoreceptors, retinal cells are assembled into local circuit modules, which tile the retina as highly ordered mosaics with characteristic spacing between circuits of the same type (Figure 22.1). Retinal circuits carry out a phenomenal amount of image processing, computing more than a dozen neural representations of the image in parallel, which are then sent to higher brain regions. Each local circuit extracts a particular feature from the visual scene, such as information about the direction of image motion or perception of image edges (Azeredo da Silveira and Roska, 2011).

Parallel processing begins at the first visual synapse. Each cone photoreceptor in the retina is connected to around ten types of cone bipolar cells. Bipolar cells can be grouped into two categories: ON bipolar cells depolarize in response to light increments, while OFF bipolar cells depolarize to light decrements. Each of these bipolar cells connects to several types of retinal ganglion cell (RGC). There are ON, OFF and ON–OFF ganglion cells depending on the input they receive from bipolar cells. Inhibitory interneurons modulate synaptic transmission between photoreceptors and bipolar cells (horizontal cells) and between bipolar cells and ganglion cells (amacrine cells). Each ganglion cell type incorporates a few bipolar and amacrine cell types in its local retinal circuit (Sahel and Roska, 2013).

In humans, the spatial layout of the retinal circuit organization described above is modified in a specialized retinal area – the fovea. The fovea is the place of high-acuity vision and is therefore critical for human visual activities requiring high resolution, such as reading. Cone to cone spacing (density) is the physical determinant of our ability to resolve fine detail, and cone density is highest at the fovea. In order to sense light without dispersion or noise, all other cell types in the fovea are pushed to the side, forming a concentric ring of cell bodies, while leaving cone outer segments unobstructed (Hendrickson, 1992). The convergence of foveal cones to ganglion cells is much lower in the fovea than in the periphery, with some ganglion cell types connected to only a single cone, leading to small receptive fields that are well suited to discriminating fine spatial detail. Foveal cones are the only neurons in direct contract with the vitreous body, a clear gel that fills the space between the inner retina and the lens, and hence cones are the only neurons that can be infected after intravitreal injection of a viral vector. Inner retinal neurons to the side of the fovea are covered by the inner limiting membrane, a basement membrane that is not easily penetrated by AAVs.

Since the functional unit of the retina is a mosaic of cell types assembled into local circuits, cell type-specific interventions are infinitely preferable if vision restoration is to be high resolution. Therapy that does not take into account retinal cell type-based computation will only provide a limited ability to reconstruct a natural scene. If possible, foveal cones should be stimulated, but ultimately, the choice of which cell type to stimulate is dictated by the state of the retina in a given patient at a given time. The closer to the photoreceptors we stimulate cells in the neuronal chain, the more normal retinal information processing will be preserved, which greatly improves image quality. If the photoreceptor cells are degenerated, downstream retinal neurons can be considered.

22.3 Emerging Therapeutic Options for Retinal Disease

Optogenetics is one of several emerging therapies for the treatment of retinal disease. Some of these emerging therapies take into account the mutated gene or even the type of mutation leading to the disease. Others are mutation independent, concentrating more on the structural and functional state of the remaining retinal circuit.

A key barrier preventing the successful translation of therapy that takes into account the mutated gene lies in the large number of mutations present. While being an entirely reasonable approach to correcting a genetic disease of the eye, one must note that thousands of disease-causing mutations have been reported. Each would require synthesis of custom-made DNA and viral vectors and its own clinical trial. Further complicating matters are dominant mutations. These require not only the delivery of the correct copy of a gene, but also a second construct to eradicate the incorrect version of the gene that is antagonizing the wild-type allele. Finally, only a few disease-associated genes fit the modest packaging capabilities of gene delivery vectors that are suitable for use in humans, which makes the correction of many gene defects challenging. Hence, it is mainly the single-gene recessive hereditary diseases caused by mutations in small genes that are suitable candidates for simple gene supplementation therapy (Sahel and Roska, 2013).

An alternative to gene therapy comes in the form of stem cell or retinal prosthetic device implantation. Stem cell approaches (Singh and MacLaren, 2011; Ong and da Cruz, 2012; Tibbetts *et al.*, 2012) attempt to derive new photoreceptor cells after implantation of neural progenitor cells. Here, correct differentiation of progenitor cells followed by establishing precise synaptic connectivity within existing retinal circuits is required, but, to date, has been difficult to achieve. Retinal prosthetic devices operate by injecting a charge into excitable tissues (Zrenner *et al.*, 2011; Humayun *et al.*, 2012; Hadjinicolaou *et al.*, 2015) through an array of electrodes. The main difficulty faced by RGC-based prostheses is their inability to stimulate the different parallel retinal circuits. Sub-retinal prostheses, which attempt to stimulate surviving photoreceptors or bipolar cells after implantation of the device in the sub-retinal compartment above the photoreceptors, have been shown to accelerate the rate of retinal degeneration (Lorach *et al.*, 2015). This is undoubtedly because of the permanent retinal detachment that ensues when the photoreceptor layer is physically separated from the overlying supporting tissues. This detachment prevents the outer retina from receiving the nourishment that is normally provided by the overlying layers.

22.4 Optogenetic Considerations

22.4.1 Targeting the Right Circuit Element

A key advantage of optogenetics over other methods is that it allows the "right" neurons, as opposed to a broad set of neurons, to be activated or silenced (Packer *et al.*, 2013). The state of the retina of a particular patient at a given time cannot be predicted, but can be well characterized – down to single-cell resolution – during the course of clinical investigation. Selection of a specific cell type for treatment depends on the state of the retina in a given patient at a given time and on the type of therapy being considered. If one is attempting restoration of color vision using optogenetics, therapy could involve stimulation of a single cell type (e.g. blue cone photoreceptors). If one is attempting a general improvement in mobility and orientation, a broader cell class (e.g. all ON cone bipolar cells) could be considered. In certain cases, manipulation of functionally defined neurons could be appropriate, such as ganglion cells responding to horizontal image motion. Finally, treatment could require focusing on subcellular localization (e.g. the dendrites of starburst amacrine cells, which normally provide asymmetric inhibition to direction-selective ganglion cells). Targeting the right neuron still poses practical difficulties, particularly when moving from mouse to primate to human studies, where levels of expression drop off significantly with each step.

So far, optogenetic approaches have been used in rodent models of the human hereditary rod/cone degenerative disease retinitis pigmentosa (Busskamp and Roska, 2011; Busskamp *et al.*, 2012). The basic idea of optogenetics in retinal disease is to provide the ability to signal light increments or decrements to photoreceptors that have lost the ability to do so because of disease, or else to equip downstream neurons with the ability to respond to light, making them into artificial photoreceptors. The most used optogenetic tools are mutant forms of channelrhodopsin-2 (ChR2) and halorhodopsin (NpHR). Optogenetic tools can operate with high spatial and temporal resolution at multiple wavelengths. They continue to undergo significant improvement and refinement of features such as light sensitivity, channel conductance and spectral specification. Additionally, these light-gated proteins are single-component compounds, requiring no externally supplied cofactor, which greatly simplifies their application. It is this relative ease of application that is rapidly pushing optogenetics into clinical trials in human patients.

22.4.2 Viral Delivery of Optogenetic Tools

For optogenetics to work well, it is of course necessary to optimize delivery. The most common delivery approach is to use an AAV to deliver optogenetic tools to specific cell types within the eye by intraocular injection. Delivery of AAV to the retina of human patients has been shown to be safe in multiple clinical trials. Viruses are small, they can be injected at any time into any ocular compartment and they can lead to high-level, long-term expression of optogenetic tools. Viruses, like optogenetic tools, can be engineered to improve their properties. For example, AAVs have coat proteins that exist in hundreds of different variants in nature, while thousands more can be made synthetically by directed evolution and rational mutagenesis approaches (Kienle *et al.*, 2012). Coat protein engineering can improve particular properties (e.g. making entry into inner retinal layers from intravitreal injections more efficient) (Dalkara *et al.*, 2013). Virus components can be exchanged between viruses of the same family or even between viruses of different families in order to gain additional desirable properties.

In general, permissive AAV serotypes are preferred (Sahel and Roska, 2013). Specificity of expression is then achieved via a cell type-specific promoter. The challenge for further developing AAV-based optogenetic therapy lies in the fact that, in the primate, the inner limiting membrane, which shields RGCs from the vitreous body, acts as a serious barrier to AAV diffusion. In these cases, the inner limiting membrane needs to be removed, or non-viral transfer may be attempted. Non-viral vectors such as nanoparticles or liposomes have also been developed for drug, gene (Han *et al.*, 2012) and trophic factor delivery (Jiang *et al.*, 2007). Unfortunately, despite significant engineering efforts, these non-viral vectors still lag behind viral vectors in terms of transfection efficiencies and ability to generate long-term expression. Another way to get around the problem of the inner limiting membrane is to perform sub-retinal injections of virus. Sub-retinal injection is an excellent way to deposit a large number of AAV particles right next to the photoreceptor layer, and when done by an experienced clinician, this approach is associated with a much lower risk of retinal detachment than retinal prosthetics. If only foveal cones need to be transduced, a virus may be injected intravitreally, although a significant dilution will occur due to the large size of the vitreous body.

The major drawback in using viruses with promoters as delivery tools comes from the variation in the number of AAV particles present in different cells, and even within cells of the same type, together with the variation in the mean and variance of optogenetic tool expression, even when the same promoter is used (Sahel and Roska, 2013). This is a major limitation since it precludes quantitative interpretation of results and makes comparisons between constructs difficult.

22.4.3 Selecting an Optogenetic Tool

The next step is to consider which opsin to use for optogenetic manipulation. Here, one needs to consider the polarity of the expected response (depolarizing, hyperpolarizing or producing bidirectional shifts in membrane potential) and the wavelength preference of the tool (Packer *et al.*, 2013). For visual restoration, red-shifted tools are preferred, since blue light causes constriction of the pupil and will decrease the total number of photons reaching the retina. Red light also penetrates deeper into tissue than other wavelengths. Both red-shifted channelrhodopsins and a red-shifted halorhodopsin are now available (Chuong *et al.*, 2014). The optogenetic tools that have been engineered so far have fast response kinetics, which is important for temporal resolution. Recent developments have also resulted in new tools with significantly increased sensitivities (Kleinlogel *et al.*, 2011), but even with these improvements, the light requirement for stimulation would saturate any residual opsin response of the cone, making them incompatible with residual vision. This incompatibility is probably the single biggest obstacle facing the clinical application of optogenetics for visual restoration purposes, since it limits optogenetic applications to treating only those patients who have close to zero visual perception.

22.4.4 Light Adaptation, Gain Control and Color

Retinal processing is influenced strongly by ambient light levels. This adaptation is essential since the range of light intensities the retina has to deal with changes over approximately 12 orders of magnitude and the eye can only sense a contrast ratio of about 1% (Pelli and Bex, 2013). Adaptation can occur either in the photoreceptor phototransduction cascade in order to control the light sensitivity of photoreceptors or downstream of photoreceptors in order to optimize the gain and the signal-to-noise ratio in particular light conditions. For example, if light levels are very low, high gain is critical in order to detect the few photons that are available. This is expected to increase noise, so other systems are activated to reduce the noise. At high light levels, the gain can be reduced and thus noise levels are lower.

If photoreceptors are degenerated, the pattern of stimulation light falling on optogenetically transduced bipolar cells in the absence of photoreceptors elicits different responses to those that would be elicited in healthy retinas, since the gain control and adaptation mechanisms of the outer retina are missing. Additionally, no optogenetic sensor that has been made to date is able to work over even half the range of light intensities that is encountered in a natural scene. Retinal cells that have been made light sensitive by optogenetic tools are active in a window spanning not more than four log units of light intensities. For this reason, optogenetic therapy, whether photoreceptor, bipolar or ganglion cell based, needs to be combined with an external goggle device worn by the patient over the eye, which can: (1) adapt the mean brightness of the projected image; and (2) project the image with a dynamic intensity range matching the sensitivity range of the optogenetically transduced cell. When ganglion cells act as light sensors, this external device should not only perform adaptation and stimulation, but also inner retinal processing operations such as gain control and edge enhancement,

A wide-angle, high-resolution camera mounted on a goggle that mimics the photoreceptor layer can tune the brightness and wavelengths of a natural scene in order to meet the requirements of the optogenetic tool. However, transducing all cone photoreceptor types with the same optogenetic tool will eliminate color vision. In this case, different optogenetic tools, each with their own wavelength preference targeted to specific cone types, may re-establish at least basic color perception. This would require promoters that are able to target red, green and blue cones specifically.

22.4.5 ON and OFF Channels

Optogenetic tools have been expressed in cone photoreceptor cell bodies (Busskamp *et al.*, 2010), a subset of ON bipolar cells (Lagali *et al.*, 2008) and in RGCs in mouse models of the disease retinitis pigmentosa (Bi *et al.*, 2006). Since two pathways discriminating between increases in illumination (ON bipolar and ON ganglion cells) and decreases in illumination (OFF bipolar and OFF ganglion cells) work to produce center–surround receptive fields, which is a key feature of retinal processing responsible for spatial contrast, future non-photoreceptor-based optogenetics will need to demonstrate the ability to independently stimulate both ON and OFF cells in the same retinal region. Furthermore, because ON and OFF channels inhibit each other in a push–pull manner in the retina and in the visual cortex, simultaneously stimulating them with the same spatio-temporal pattern may be confusing for visual signal processing. Creating artificial photoreceptors from cells that are downstream of photoreceptors then requires promoters that are specific not only to cell class, such as bipolar or ganglion cell, but also to the ON versus OFF response property of the cell. To this end, separate center and surround stimulation of RGCs has been attempted using protein-targeting motifs in order to achieve differential targeting of depolarizing ChR2 and hyperpolarizing NpHR to the center versus peripheral regions of the RGC dendritic tree (Wu *et al.*, 2013) or to the somatic versus dendritic regions of RGCs (Greenberg *et al.*, 2011).

22.4.6 Image Distortion

Unless optogenetic tools are targeted to dendrites only, optical stimulation of optogenetically transduced cells will excite the neurons proximally at their dendrites and cell bodies, as well as distally at their axons. This will undeniably cause image distortions, particularly when axons are long, spreading over large areas of the retina (Sahel and Roska, 2013).

Additional distortion can be anticipated if perifoveal bipolar and ganglion cells are targeted. Around the fovea, these cells are not arranged in mosaics; instead, they are pushed to the side and lie heaped over each other. If optogenetic stimulation of these cells is attempted, this piling will complicate the development of spatial maps, which are normally delivered by mosaics, and will likely produce a somewhat distorted image. One could perhaps consider targeting bipolar cells and RGCs in the periphery, but here acuity will be low.

For peripheral retinal optogenetics to really work, it will be crucial to take eye movements into account, since normal eye movements would work to catch and hold images over the fovea, rather than the newly light-sensitive optogenetically transduced region. Here, goggles worn over the eye would need to track the patient's eye movements and redirect the goggle-projected image over the part of the retina that received the optogenetic therapy. As a result, a person who has some remaining foveal function would now theoretically have the use of a second, optogenetically transduced retinal locus for aid in performing visual tasks. Indeed, this concept of a "pseudo-fovea" already exists. Some patients with macular (foveal) degeneration have been reported to use a healthy retinal locus in order to perform fovea-based activities, such as reading and tracking objects (Schuchard, 2005).

22.5 Conclusion

Ideally, high-resolution vision restoration would determine the physiological spatiotemporal patterns of activity in retinal circuits and would aim to reproduce the same spike patterns in each of the 20 ganglion cell mosaics as would normally be produced during natural vision, but, at present, this is perhaps unrealistic, if only because retinal remodeling during the course of degeneration is expected to change not just the cell type composition of the retina, but also many other dynamic processes, which will make it difficult to perfectly reproduce the retinal neural code. It is perhaps more reasonable to consider which circuit elements are most important when it comes to crucial daily tasks such as object recognition, and to focus optogenetic therapy on these more relevant circuit components. All of the preclinical reports thus far have demonstrated that optogenetic therapy is safe and is capable of inducing light-driven activity in mouse models of retina degeneration, irrespective of the cell type chosen for therapy. The next step will be to see how the different optogenetic approaches will fare in clinical trials, which are currently being arranged by several groups and companies.

ACKNOWLEDGMENT

We thank Botond Roska for many valuable discussions and advice on the material in this chapter.

REFERENCES

Azeredo da Silveira R, Roska B. 2011. Cell types, circuits, computation. *Curr Opin Neurobiol* **21**: 664–671.

Bainbridge JW, Smith AJ, Barker SS, Robbie S, Henderson R, *et al.* 2008. Effect of gene therapy on visual function in Leber's congenital amaurosis. *N Engl J Med* **358**: 2231–2239.

Bi A, Cui J, Ma YP, Olshevskaya E, Pu M, *et al.* 2006. Ectopic expression of a microbial-type rhodopsin restores visual responses in mice with photoreceptor degeneration. *Neuron* **50**: 23–33.

Busskamp V, Duebel J, Balya D, Fradot M, Viney TJ, *et al.* 2010. Genetic reactivation of cone photoreceptors restores visual responses in retinitis pigmentosa. *Science* **329**: 413–417.

Busskamp V, Picaud S, Sahel JA, Roska B. 2012. Optogenetic therapy for retinitis pigmentosa. *Gene Ther* **19**: 169–175.

Busskamp V, Roska B. 2011. Optogenetic approaches to restoring visual function in retinitis pigmentosa. *Curr Opin Neurobiol* **21**: 942–946.

Chuong AS, Miri ML, Busskamp V, Matthews GA, Acker LC, *et al.* 2014. Noninvasive optical inhibition with a red-shifted microbial rhodopsin. *Nat Neurosci* **17**: 1123–1129.

Cideciyan AV, Aleman TS, Boye SL, Schwartz SB, Kaushal S, *et al.* 2008. Human gene therapy for RPE65 isomerase deficiency activates the retinoid cycle of vision but with slow rod kinetics. *Proc Natl Acad Sci U S A* **105**: 15112–15117.

Dalkara D, Byrne LC, Klimczak RR, Visel M, Yin L, *et al.* 2013. *In vivo*-directed evolution of a new adeno-associated virus for therapeutic outer retinal gene delivery from the vitreous. *Sci Transl Med* **5**: 189ra76.

Greenberg KP, Pham A, Werblin FS. 2011. Differential targeting of optical neuromodulators to ganglion cell soma and dendrites allows dynamic control of center-surround antagonism. *Neuron* **69**: 713–720.

Hadjinicolaou AE, Meffin H, Maturana MI, Cloherty SL, Ibbotson MR. 2015. Prosthetic vision: devices, patient outcomes and retinal research. *Clin Exp Optom* **98**: 395–410.

Han Z, Conley SM, Makkia RS, Cooper MJ, Naash MI. 2012. DNA nanoparticle-mediated ABCA4 delivery rescues Stargardt dystrophy in mice. *J Clin Invest* **122**: 3221–3226.

Hauswirth WW, Aleman TS, Kaushal S, Cideciyan AV, Schwartz SB, *et al.* 2008. Treatment of Leber congenital amaurosis due to RPE65 mutations by ocular subretinal injection of adeno-associated virus gene vector: short-term results of a Phase I trial. *Hum Gene Ther* **19**: 979–990.

Hendrickson A. 1992. A morphological comparison of foveal development in man and monkey. *Eye (Lond)* **6**(Pt 2): 136–144.

Humayun MS, Dorn JD, da Cruz L, Dagnelie G, Sahel JA, *et al.* 2012. Interim results from the international trial of Second Sight's visual prosthesis. *Ophthalmology* **119**: 779–788.

Jiang C, Moore MJ, Zhang X, Klassen H, Langer R, Young M. 2007. Intravitreal injections of GDNF-loaded biodegradable microspheres are neuroprotective in a rat model of glaucoma. *Mol Vis* **13**: 1783–1792.

Kienle E, Senis E, Borner K, Niopek D, Wiedtke E, *et al.* 2012. Engineering and evolution of synthetic adeno-associated virus (AAV) gene therapy vectors via DNA family shuffling. *J Vis Exp* **62**: 3819.

Kleinlogel S, Feldbauer K, Dempski RE, Fotis H, Wood PG, *et al.* 2011. Ultra light-sensitive and fast neuronal activation with the Ca^{2+}-permeable channelrhodopsin CatCh. *Nat Neurosci* **14**: 513–518.

Lagali PS, Balya D, Awatramani GB, Munch TA, Kim DS, *et al.* 2008. Light-activated channels targeted to ON bipolar cells restore visual function in retinal degeneration. *Nat Neurosci* **11**: 667–675.

Lorach H, Goetz G, Mandel Y, Lei X, Kamins TI, *et al.* 2015. Performance of photovoltaic arrays in-vivo and characteristics of prosthetic vision in animals with retinal degeneration. *Vision Res* **111**: 142–148.

Maguire AM, Simonelli F, Pierce EA, Pugh EN, Jr., Mingozzi F, *et al.* 2008. Safety and efficacy of gene transfer for Leber's congenital amaurosis. *N Engl J Med* **358**: 2240–2248.

Ong JM, da Cruz L. 2012. A review and update on the current status of stem cell therapy and the retina. *Br Med Bull* **102**: 133–146.

Packer AM, Roska B, Hausser M. 2013. Targeting neurons and photons for optogenetics. *Nat Neurosci* **16**: 805–815.

Pelli DG, Bex P. 2013. Measuring contrast sensitivity. *Vision Res* **90**: 10–14.

Sahel JA, Roska B. 2013. Gene therapy for blindness. *Annu Rev Neurosci* **36**: 467–488.

Schuchard RA. 2005. Preferred retinal loci and macular scotoma characteristics in patients with age-related macular degeneration. *Can J Ophthalmol* **40**: 303–312.

Singh MS, MacLaren RE. 2011. Stem cells as a therapeutic tool for the blind: biology and future prospects. *Proc Biol Sci* **278**: 3009–3016.

Tibbetts MD, Samuel MA, Chang TS, Ho AC. 2012. Stem cell therapy for retinal disease. *Curr Opin Ophthalmol* **23**: 226–234.

Wu C, Ivanova E, Zhang Y, Pan ZH. 2013. rAAV-mediated subcellular targeting of optogenetic tools in retinal ganglion cells in vivo. *PLoS One* **8**: e66332.

Zrenner E, Bartz-Schmidt KU, Benav H, Besch D, Bruckmann A, *et al.* 2011. Subretinal electronic chips allow blind patients to read letters and combine them to words. *Proc Biol Sci* **278**: 1489–1497.

23 Optogenetics for Vision Recovery: From Traditional to Designer Optogenetic Tools

Sonja Kleinlogel

23.1 Introduction

Most forms of inherited blindness result from photoreceptor cell death caused by mutations in genes specifically expressed in photoreceptor cells. Despite the loss of photoreceptors, inner retinal neurons were shown to survive for an extended time period and retain their ability to send visual information to the brain, both in human patients and in retinitis pigmentosa mouse models (Strettoi and Pignatelli, 2000; Chang *et al.*, 2002). To date, the only therapy available for late-stage photoreceptor degeneration is based on electronic chips that are implanted in close proximity to the retina. These chips eclectically stimulate the surviving retinal neurons according to the input they receive from a camera viewing the surrounding. Although retinal implants have been shown to restore some degree of vision, this is, in the best case, restricted to crude, high-contrast objects (e.g. a single white letter on a black computer monitor) (Zrenner *et al.*, 2011). The chips are also costly and, due to their small size, recover vision only locally with crude and non-physiological electrical activation.

For this reason, multiple laboratories are currently exploring optogenetic therapies that collectively aim to render surviving inner retinal cells directly sensitive to light; in other words, to turn the cells of the inner retina into "replacement photoreceptors". This optogenetic approach consists of the introduction of a gene encoding a light-sensitive membrane protein into the preserved inner retinal neurons. The most commonly used optogenetic proteins are members of the channelrhodopsin family, which are blue light-gated (λ_{max} = 472 nm), non-selective cation channels that cause a depolarization of the cell (Nagel, 2003). In an alternative approach, referred to as "photopharmacology", a light-sensitive small molecule, termed a "photoswitch", is tethered to either a native or an engineered ion channel, with the latter allowing genetic targeting to specific cells (Polosukhina *et al.*, 2012; Tochitsky *et al.*, 2014). Absorption of light by the photoswitch mediates opening of the ion channel. Photoswitching kainite channels in the retina recovered light sensitivity in mice (Caporale *et al.*, 2011; Gaub *et al.*, 2014). Both the optogenetic and photopharmacological approaches for vision restoration hold considerable therapeutic potential as they impart direct light sensitivity to surviving cells, making these much more physiological approaches than, for example, retinal implants. Other potential advantages of optogenetic vision recovery are that treatment is ambulant, long lived, low cost and has the theoretical potential to recover high-resolution vision across the entire visual field.

In this chapter, I will focus on the optogenetic approach using opsin-based light sensors. The majority of optogenetic studies use the *rd1* (Keeler, 1966) retinitis pigmentosa mouse model with a null mutation in the gene encoding the β-subunit of the cGMP phosphodiesterase (PDE), which phenotypically resembles some forms of retinitis pigmentosa in humans. Pioneering studies have shown that channelrhodopsin-2 (ChR2) expression can reinstate light responsiveness in inner retinal neurons (Bi *et al.*, 2006; Lagali *et al.*, 2008; Busskamp *et al.*, 2010; Doroudchi *et al.*, 2011). However, there are obstacles to using ChR2 as a therapeutic in humans. The primary drawback is the extremely low light sensitivity of ChR2. To exemplify this, ChR2 only starts being activated at light intensities equivalent to a bright, sunny day on a snowfield at high altitude. It is therefore easily conceivable that chronic activation of ChR2 will be toxic to any remaining photoreceptors and potentially other retinal tissue (Remé *et al.*, 1998; The Lasker/IRRF Initiative for Innovation in Vision Science, 2014). Such a toxic effect is of particular concern, as retinitis pigmentosa patients often possess remnant cones (Busskamp *et al.*, 2010) that may be preserved by pharmacological and miRNA treatment (Busskamp *et al.*, 2014). Recently, optogenetic vision recovery entered a new era in which light-sensing proteins are being engineered according to the needs of a specific application. This chapter will discuss the directions taken to design optimized optogenetic tools for potential vision restoration in blind patients suffering from photoreceptor degeneration.

23.2 Considerations for Engineering an Optogenetic Tool for the Retina

23.2.1 Which Cells to Target?

The direct retinal pathway runs from the photoreceptors to the bipolar cells (BPCs) on to the retinal ganglion cells (RGCs), which project via the optic nerve to the brain (Figure 23.1A). At each level of this pathway, there exists a highly organized mosaic arrangement of cells. Because of this vertical arrangement, elements of the visual scene "map" in straightforward fashion to locations on the retina. In the dark, the photoreceptors release glutamate, which inhibits (hyperpolarizes) ON-BPCs and excites (depolarizes) OFF-BPCs, segregating the retinal signal into ON and OFF pathways (Figure 23.1A). In light, the ON-BPC loses its inhibition and becomes active (depolarized), while the OFF-BPC loses its excitation and becomes silent (hyperpolarized). BPCs receive synaptic input from either rods or cones, but not both, and are accordingly designated rod BPCs or cone BPCs, respectively. Cone BPCs exist as ON- and OFF-BPCs, whereas rod BPCs are all ON-BPCs. ON-BPCs receive glutamatergic input from the photoreceptors

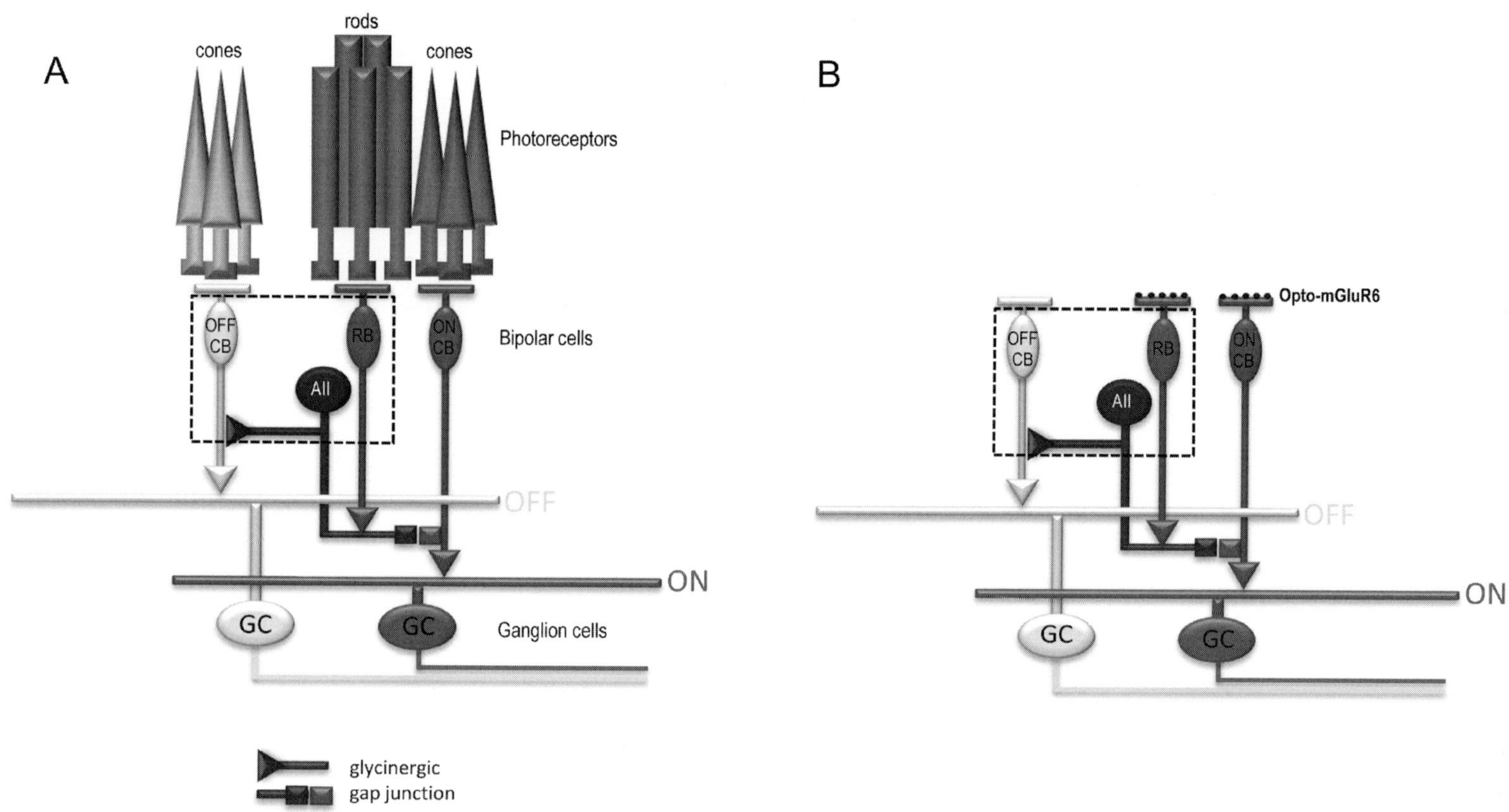

Figure 23.1 Retinal circuitry in health and disease. The retinal cell types are clearly structured into a highly organized mosaic arrangement. There are two types of photoreceptors – rods and cones – and rods outnumber cones. Photoreceptors transmit the light signal to bipolar cells, of which there are again two classes – ON-bipolar cells and OFF-bipolar cells – which provide parallel information channels responding to increases versus decreases in light intensity and project to ON and OFF retinal ganglion cells (GC), respectively. Whilst cone bipolar cells (receiving input from cones) are divided into ON and OFF types, rod bipolar cells (RB), receiving input from rods, are all of the ON type. However, rod bipolar cells do not directly project to the retinal ganglion cells, instead feeding the signal into the cone bipolar pathway. (A) The intact rod pathway (caged): upon illumination, the rod bipolar

via a metabotropic glutamate receptor (mGluR6) and OFF-BPCs receive glutamatergic input via ionoptropic glutamate receptors (AMPA/kainate). Whilst cone BPCs transmit their signals directly to RGCs, rod BPCs do not synapse directly onto RGCs, but instead onto AII amacrine cells, mediating a horizontal network between rod and cone BPCs (Figure 23.1) (Wässle, 2004). Depolarization of rod BPCs excites AII cells, which depolarize cone ON-BPCs via gap junctions and inhibit cone OFF-BPCs via a sign-inverting, glycinergic synapse. This connectivity feeds the rod signal into the cone pathway. Depending on their cone BPC input, RGCs are also separated into ON- and OFF-RGCs, responding to light with an increase or decrease in action potential firing, respectively. Mirror-symmetric light responses in multiple ON- and OFF-RGC subtypes imply antagonistic comparison of these signals at higher visual centers (Thyagarajan *et al.*, 2010).

When optogenetically targeting RGCs, the following problems arise: (1) the axonal processes of RGCs run across the retina toward the optic disc, which may result in non-localized responsiveness over a large area of the retina outside the native RGC receptive field, compromising visual resolution. (2) Important retinal processing takes place at each synapse, particularly at the BPC-to-RGC synapse, which is bypassed when RGCs are targeted directly with optogenetic tools. (3) As no ON- or OFF-RGC-specific promoters yet exist, ON- and OFF-RGCs cannot be separately targeted with an excitatory and inhibitory optogenetic tool, respectively. Consequently, expression of ChR2 in all RGCs ubiquitously turned all RGCs into the ON type, and ON and OFF pathway segregation was lost (Bi *et al.*, 2006; Lin *et al.*, 2008). In an attempt to re-establish center–surround concentric RGC receptive fields, which is a key feature of edge detection and contrast enhancement, optogenetic actuators and inhibitors have been differentially expressed in the soma and dendrites (Greenberg *et al.*, 2011) or center and peripheral dendritic fields (Wu *et al.*, 2013) of RGCs. This has been achieved by identifying amino acid segments ("motifs") of native proteins that mediate targeting to distinct sites within the cell, and then incorporating these motifs, as an appendage, into the gene encoding the optogenetic tool. This method re-established center–surround receptive fields; however, as mentioned before, the optogenetic actuators and inhibitors need to be differentially targeted to ON- and OFF-RGCs in order to fully mimic ON and OFF pathways. (4) An additional concern for clinical application is the projections of RGCs to the brain, which bear the risk of a potential immunogenic spread.

Caption for Figure 23.1 (cont.)

cell (RB) is depolarized, which subsequently depolarizes the AII amacrine cell (AII). In turn, the cone ON-bipolar cell (ON CB) is also depolarized through a sign-conserving electrical synapse, while the cone OFF-bipolar cell (OFF CB) is hyperpolarized through a sign-inverting glycinergic synapse. Finally, ON ganglion cells (GCs, gray) are depolarized and OFF ganglion cells (GCs, white) are hyperpolarized by light. (B) The degenerated pathway recovered through Opto-mGluR6, which is expressed in all ON-bipolar cells. The inner retinal circuitries remain functional after the photoreceptors have degenerated. As a consequence, Opto-mGluR6 not only activates the ON pathway, but also indirectly the OFF pathway through horizontal signal transfer via the AII amacrine cell.

To avoid the above concerns altogether, BPCs have become the preferred target for optogenetic vision recovery. As specific promoters for ON-BPCs exist (Kim *et al.*, 2008) and rod ON-BPC activation (if strong enough) is transferred via the AII amacrine cell to both ON and OFF pathways (Wässle, 2004), separation of ON and OFF pathways is retained. Further, since BPCs are located upstream in the retinal circuitry, many aspects of retinal processing can be preserved. Additional advantages are a minimal risk of immunogenic spread, as BPCs terminate within the retina and there is a potentially high photon catch as a consequence of the tight tiling of BPCs throughout the retina. Alternatively, remnant cones (Busskamp *et al.*, 2010) or AII amacrine cells could also be candidates, but early intervention is required for cone-directed regeneration and no specific promoter is yet available for AII amacrine cells.

Pioneering studies expressed ChR2 in ON-BPCs and demonstrated the feasibility of restoring light sensitivity to a degenerated retina of an *rd1* mouse (Lagali *et al.*, 2008; Doroudchi *et al.*, 2011). However, in these early studies, only responses to light increments were restored, indicating that the drive of rod BPCs was not sufficient to activate the OFF pathway. This problem has been recently overcome by using engineered, more light-sensitive ChR2 variants, ChR2(H134 R) and Catch (Kleinlogel *et al.*, 2011) in combination with engineered viral adeno-associated virus (AAV) capsids that enhanced ON-BPC transduction efficiency from below 10% to approximately 80% (Cronin *et al.*, 2014; Macé *et al.*, 2015).

23.2.2 What Qualities Are Needed from the Ideal Optogenetic Tool?

Although cell-specific targeting of ON-BPCs and separate activation of ON and OFF pathways with fast kinetics has been achieved (Cronin *et al.*, 2014; Macé *et al.*, 2015), some caveats regarding the clinical applicability of ChR2 still remain.

First, the optogenetic therapeutic should not elicit an immune response in the patient. As ChR2s are algal and not human proteins, they remain potentially antigenic (The Lasker/IRRF Initiative for Innovation in Vision Science, 2014; van Wyk *et al.*, 2015). The extremely high ChR2 expression levels that are needed to sufficiently drive the retinal circuitry further increase the likelihood of causing an immune response and cellular toxicity (The Lasker/IRRF Initiative for Innovation in Vision Science, 2014). Why do we need such high ChR2 expression? (1) The single-channel conductance (rate of ion travel through a single ChR2 channel upon activation: ~60 fS) (Feldbauer *et al.*, 2009) is 10–1000-times smaller than that of a typical ion channel and cannot be markedly increased (Kleinlogel *et al.*, 2011). (2) The photopigment usually resides in the tightly packed discs of the outer segments of the photoreceptors. Such high expression densities are impossible to achieve on the flat membrane surface of a RGC or BPC. (3) ChR2 is a directly light-gated ion channel and not, as with vertebrate visual pigments, coupled to intracellular G-protein cascades that amplify the incoming signal (Figure 23.2).

Second, the therapeutic should activate the target cell physiologically. Introduction of the non-selective cation channel ChR2 into the membrane of an ON-BPC, for example, will mediate a light-triggered depolarization, but the

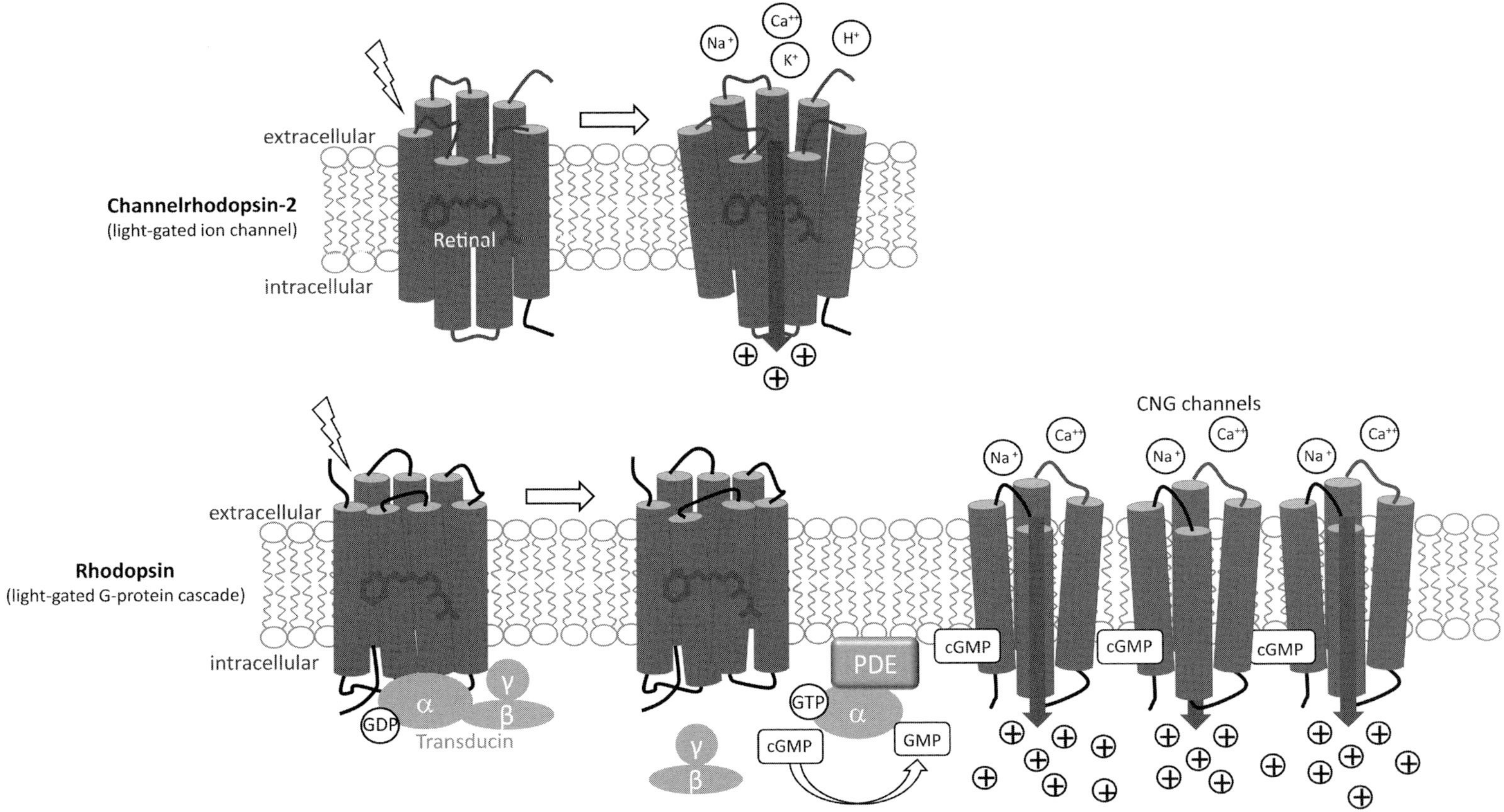

Figure 23.2 The principle of signal amplification through G-proteins. Despite the structural similarity between microbial (e.g. ChR2) and vertebrate visual pigments (e.g. rhodopsin), they quite markedly differ mechanistically. ChR2 is a light-triggered, non-selective cation channel that transmits a few ions upon light activation. In contrast, rhodopsin and other vertebrate visual pigments are G-protein-coupled receptors (GPCRs), which activate a heterotrimeric G-protein consisting of α-, β- and γ-subunits on the intracellular side of the membrane; in the case of rhodopsin, this is transducin. The activated G-protein promotes the exchange of bound GDP for GTP on the G-protein α-subunit. GTP binding changes the conformation of the α-subunit, which allows the bound Gprotein to be released from the receptor and to dissociate into a catalytic α-subunit (GTP-bound) and

nature of this light response is not native to the BPC and can consequently lead to substantial alterations in the way that the BPC processes the light signal. As a result, the spatial and temporal properties of RGC responses to a given visual scene are most likely different from the responses produced by the same scene in a healthy retina.

Third, and importantly, the optogenetic therapeutic has to be sufficiently light sensitive. All ChR2-based optogenetic tools provide inadequate light sensitivity to function under environmental light conditions and, in fact, the light intensities required for activation are well above the safety limit for retinal illumination (European Parliament, Council of the European Commission, 2006). Chronic ChR2 activation is consequently predetermined to damage surviving retinal cells, particularly remnant photoreceptors (The Lasker/IRRF Initiative for Innovation in Vision Science, 2014), which would be an unacceptable side effect of any clinical therapy. The compromised light sensitivity of ChR2 again results from the small single-channel conductance and the direct channel gating rather than light-induced activation of a signal-amplifying G-protein cascade (Figure 23.2).

23.3 Approaches that Have Been Taken to Engineer Next-generation Optogenetic Tools

The obvious goal of recent development was to increase light sensitivity while avoiding complications with associated antigenicity. As extensive molecular engineering has shown that the light sensitivity of ChR2 is limited and can be maximally increased by 1.5 log units (Kleinlogel *et al.*, 2011; Pan *et al.*, 2014), exploration has turned away from microbial to mammalian opsins.

23.3.1 Mammalian Opsins

ChR2 and mammalian opsins are structurally similar, despite the fact that the ChR2 forms a directly light-gated ion channel and mammalian opsins couple to G-proteins, activating signal amplifying intracellular pathways (Figure 23.2). In mammalian opsins, a single photon activates multiple G-proteins, each triggering large changes in secondary messengers acting on distinct downstream effectors, including ion channels. This amplification cascade renders mammalian opsins 2–3 log units more light sensitive than ChR2s (van Wyk *et al.*, 2015). In other

Caption for Figure 23.2 (cont.)

a βγ-dimer. The α-subunit and the βγ-dimer go on to activate distinct downstream effectors, such as adenylyl cyclase, phosphodiesterase (PDE), phospholipase C and ion channels. Transducin binds to and activates a PDE, which converts cGMP to GMP. The concentration of cGMP decreases below what is required to open cGMP-gated sodium/calcium (CNG) channels. The bulk charge that is translocated through CNG channels upon rhodopsin activation is, due to intracellular signal amplification, 100–1000-fold larger than the charge transmitted through a single ChR2 channel. G-protein signaling is terminated by the hydrolysis of the α-subunit-bound GTP to GDP, resulting in the re-association of the α-subunit and βγ-subunits and their binding to the GPCR. GTP binding to the α-subunit is further regulated by regulator of G-protein signaling (RGS) proteins.

words, mammalian opsins are activated by indoor light intensities as opposed to the laser light intensities that are required for activating ChR2.

The nature of the intracellular signaling of a G-protein-coupled receptor (GPCR) is determined by the class of G-protein to which it couples. G-proteins are heterotrimeric complexes consisting of three subunits (α, β and γ), of which the α-subunit specifies the classification of the GPCR and exhibits the catalytic activity (Figure 23.2). There exist four main families of GPCRs: the G_i, G_q, G_s and $G_{12,13}$ types. Mammalian opsins activate $G\alpha$ subunits of the G_i class, leading to a reduction in cyclic nucleotide concentrations (e.g. rhodopsin) (Figure 23.2), the G_s class, increasing cyclic nucleotide concentrations, and the G_q class, stimulating a phospholipase C-dependent cascade leading to a range of physiological events, including changes in intracellular calcium (Koyanagi and Terakita, 2014). The visual pigments of mammalian rod and cone photoreceptors are rhodopsin and cone opsin, respectively, which both couple to the G_i-type G-protein transducin, also referred to as G_t (Figure 23.2). Light activation of the visual pigment activates G_t by causing it to release GDP and bind GTP. GTP-bound transducin binds to and activates a PDE, which converts cGMP to GMP. As a consequence, the concentration of cGMP decreases below what is required to open cGMP-gated sodium and calcium channels, so that the photoreceptor hyperpolarizes (Yau and Hardie, 2009). A third class of mammalian retinal opsin, blue light-sensitive melanopsin (mouse λ = 467 nm) (Sekharan *et al.*, 2012), which couples to G_q, was recently discovered in a small group of intrinsically photosensitive RGCs. Intrinsically photosensitive RGCs in mammals are specialized for tasks such as measuring ambient illumination, controlling the pupillary light reflex, synchronizing the circadian clock, modulating sleep and suppressing pineal melatonin, to name but a few (Hattar *et al.*, 2002). Additionally, they have recently been shown to contribute to visual discrimination. There are currently no known opsins that couple to the fourth Gprotein family, $G_{12,13}$.

Rhodopsin and melanopsin have been used for vision recovery in *rd1* mice (Cehajic-Kapetanovic *et al.*, 2015; Gaub *et al.*, 2015). The two photopigments differ from each other in that melanopsin is bistable and rhodopsin is not (Sexton *et al.*, 2012). In other words, whilst rhodopsin bleaches and becomes non-responsive under repeated illumination, the amplitude of the melanopsin response remains unchanged. The reason for this is the intrinsic retinal re-isomerization capability of melanopsin, which isomerizes retinal back from the inactive to the active form. By contrast, rhodopsin releases retinal to be re-isomerized externally in the retinal pigment epithelium. This property makes melanopsin, as opposed to rhodopsin, highly resistant to strong light and allows successive light activation without a response rundown.

Gaub *et al.* (2015) and Cehajic-Kapetanovic *et al.* (2015) introduced rat and human rhodopsin, respectively, into the ON-BPCs of *rd1* mice. Their rationale was based on the ability of rhodopsin to activate other G-proteins of the G_i family in the absence of G_t (Cao *et al.*, 2012). As the native mGluR6 receptor of ON-BPCs couples to G_o, which is also a G-protein of the G_i family, the groups assumed that rhodopsin would be able to "hijack" the pre-existing mGluR6 messaging system in ON-BPCs. Both studies proved the above hypothesis right and showed that rhodopsin restores light responsiveness in the blind *rd1* mouse retina and brain (Cehajic-Kapetanovic *et al.*, 2015; Gaub *et al.*, 2015). As opposed to the studies using ChR2, ON and OFF responses were recovered and treated mice showed improved visual capabilities, such as the ability to distinguish moving from static stimuli (Gaub *et al.*, 2015), to recognize coarse spatial patterns and to respond to elements of a natural movie (Cehajic-Kapetanovic *et al.*, 2015). The treated mouse retina could resolve flicker frequencies of 4 Hz (flicker fusion for a human eye is 24 Hz) and, most importantly, rhodopsin-driven responses could be elicited by moderate daylight illumination, allowing visual discrimination under natural viewing conditions.

However, a few obstacles still remain to using rhodopsin as an optogenetic tool: (1) rhodopsin-mediated OFF responses were slower (seconds) than those of a healthy retina (milliseconds), which sets a limit on the temporal resolution of recovered vision (Gaub *et al.*, 2015). (2) Rhodopsin bleached upon repeated light exposure, resulting in a complete loss of light responsiveness in retinal explants and similar, but less dramatic, effects *in vivo*.

To avoid the problem of bleaching, Lin *et al.* (2008) used bistable melanopsin instead. The team targeted RGCs, as melanopsin is known to couple to the G_q-type of G-protein that is present in intrinsically photosensitive RGCs natively expressing melanopsin. Lin and colleagues found that melanopsin was expressed and functional not only in intrinsically photosensitive RGCs, but also in the great majority of RGC subtypes. These results show that the signaling system that couples melanopsin to membrane depolarization is ubiquitous or at least very widespread in RGCs, and clearly not restricted to the native melanopsin-expressing cells. Visual responses could be elicited at light intensities that were comparable to those in studies using rhodopsin. However, vision recovered by melanopsin was far less sophisticated: melanopsin only re-established the pupillary light reflex and light-avoidance behavior to near normal, but the mice were not able to carry out more demanding visual discriminations. One reason for this may be the simultaneous targeting of ON- and OFF-RGCs, while another may be the slow response kinetics of melanopsin, with response onsets of several hundreds of milliseconds to several seconds and persistence for tens of seconds after termination of the stimulus (Hattar *et al.*, 2002). The latter problem clearly excludes melanopsin from practical use for vision of moving objects and visually guided motility.

Whilst the kinetics of melanopsin-mediated responses are known to be slow, the reason for the slowed kinetics of rhodopsin-mediated light responses when expressed ectopically in ON-BPCs is a good example of the importance of cell-specific regulatory proteins, such as regulators of G-protein activity proteins (regulator of G-protein signaling [RGS] proteins) and cell-specific GPCR kinases (GRKs) and arrestins, which modify GPCR activity. In other words, rhodopsin kinase and arrestin are not found within ON-BPCs, significantly slowing down the return from the light-activated to the dark-adapted state of rhodopsin.

23.3.2 Cell-tailored Chimeric Optogenetic GPCRs: Opto-mGluR6

To engineer a kinetically fast, bistable melanopsin that activates the ON-BPC native G_O pathway, we used the following rationale: (1) ON-BPCs receive glutamatergic input from the photoreceptors via the mGluR6 receptor, the downstream signaling cascade of which we would like to reactivate when photoreceptor input is missing. (2) Our novel optogenetic protein needs to contain the intracellular domains of mGluR6, which bind to G_O, and the transmembrane domains of melanopsin, which bind to the photoswitch retinal (Figure 23.2). (3) The clear functional compartmentalization and the structural homology between melanopsin and mGluR6 allow us to engineer a chimeric protein based on both of them.

Opto-mGluR6 is identical to melanopsin, but has intracellular loops 2 and 3 and the C-terminus (CT), which mediate G-protein selectivity (Kobilka *et al.*, 1988; Verrall *et al.*, 1997; Audet and Bouvier, 2012), replaced by those of mGluR6 (Figure 23.3A). When expressed in ON-BPCs of blind *rd1* mice, Opto-mGluR6 restored bleach-resistant light responses with native mGluR6 kinetics at indoor light intensities (Figure 23.4) (van Wyk *et al.*, 2015). The high light sensitivity conforms with Opto-mGluR6's ability to activate the G_O-coupled pathway, ultimately closing hundreds of TRPM1 cation channels (Morgans *et al.*, 2009). The fast response kinetics conform with Opto-mGluR6's ability to activate ON-BPC-specific, pre-existing G-protein complexes, including RGS, GRKs and arrestins (Dhingra *et al.*, 2004; Masuho *et al.*, 2010). Bleach resistance is mediated through the melanopsin light switch.

23.3.2.1 Structural Considerations Underlying the Opto-mGluR6 Design

Based on structural and functional similarity, mammalian GPCRs are divided into five families or classes, with the light-sensitive opsins belonging to class A and the mGluRs to class C (Schiöth and Frederiksson, 2005). Although GPCRs of all families share a highly conserved tertiary structure and, ultimately, a similar mode of receptor activation, virtually no sequence homology exists between classes (Choe *et al.*, 2011; Doré *et al.*, 2014). All GPCRs comprise seven transmembrane helices (TM1–7) connected via intracellular loops (ILs) and extracellular loops (ELs) of varied lengths (Figure 23.3A). Structural alignment of the three major GPCR classes A, B and C reveals a particularly strong consensus across the intracellular halves of the TM domains (Rasmussen *et al.*, 2011; Hollenstein *et al.*, 2013), exhibiting a set of conserved micro-domains that are important in signal transduction (Figure 23.3A) (Schwartz *et al.*, 2006). These include the highly conserved E(D)RY motif at the junction of TM3 with IL2 and the NP*xx*Y (*x*PK*x*Y in mGluRs) motif at the junction of TM7 and the C-terminus (CT) required for the structural integrity of the receptor (Pin *et al.*, 2003; Schwartz *et al.*, 2006; Hollenstein *et al.*, 2013). The crystal structures of the active states of class A GPCRs, including rhodopsin (Choe *et al.*, 2011), provided important insights into the conformational changes that occur upon activation. All crystallized GPCRs appear to share the same activation scheme: an "ionic lock" formed between the arginine residue of the E(D)RY motif (e.g. R_{167} in melanopsin and R_{668} in mGluR6) and a glutamate residue at the border of IL3 and TM6 (e.g. E_{288} in

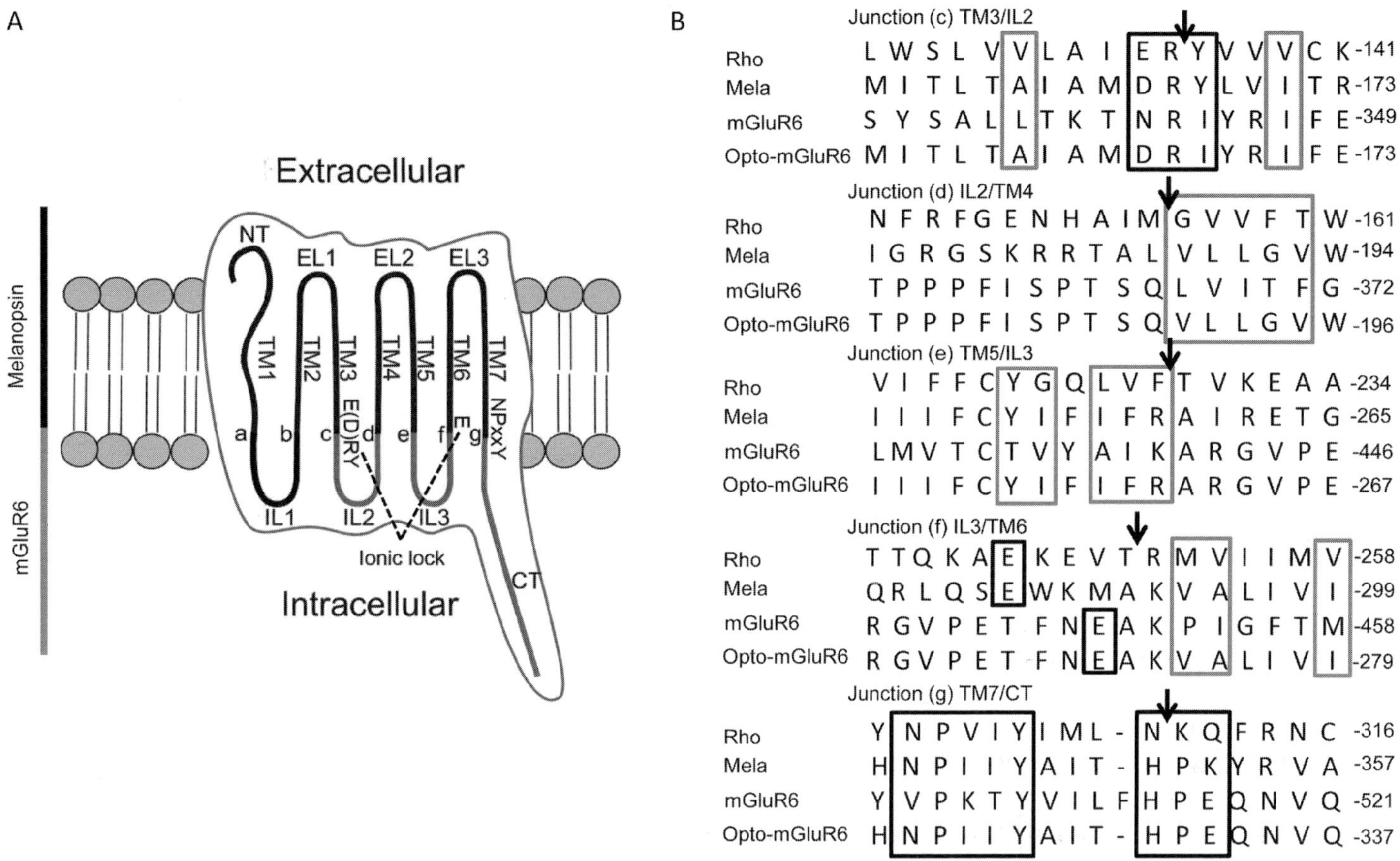

Figure 23.3 Design of Opto-mGluR6. (A) A sketch of Opto-mGluR6 composed of the N-terminus (NT), transmembrane domains (TM1–TM7) and extracellular loops 1–3 (EL1–EL3) of melanopsin and the intracellular loops (ILs) 2 and 3 and the C-terminus (CT) of mGluR6. Locations of the splice sites (a–g, connection sites between the melanopsin and mGluR6 peptides) and conserved micro-domains that are important in signal transduction ("ionic lock" and NP*xx*Y sequence) are indicated. Bound retinal between TM3 and TM7 is not

melanopsin and E_{775} in mGluR6) stabilizes the inactive state (Figure 23.3). In mGluRs, the inactive state is further stabilized by hydrogen bonds between residues in IL1, TM3, TM6 and TM7. Reorganization of the hydrogen-bonding networks and disruption of the "ionic lock" upon receptor activation lead to a large outward movement of TM6, exposing the Gprotein binding sites on IL2, IL3 and the CT (Choe *et al.*, 2011; Rasmussen *et al.*, 2011).

The most striking differences between class A and class C GPCRs are the positions of TM5 (Doré *et al.*, 2014) and the identity of the longest IL, which determines Gprotein selectivity and is IL3 in class A GPCRs and IL2 in class C GPCRs (Cotecchia *et al.*, 1990; Yamashita *et al.*, 2001). For opsins, the integrity of the retinal binding pocket is crucial for proper function. The binding pocket comprises a conserved lysine residue (K_{337} in melanopsin) in TM7 that is covalently linked to the chromophore retinal via a Schiff base and is stabilized by a negative counterion in TM3 (Y_{145} for melanopsin). Absorption of a photon isomerizes *cis*-retinal to the all-*trans* form, inducing helix opening and exposure of the cytoplasmic G-protein binding sites. The negatively charged residue in the E(D)RY motif (e.g. D_{166} in melanopsin) additionally stabilizes the inactive opsin molecule (Figure 23.3B).

Generally, G-protein selectivity is conferred by the IL2 and IL3 (Kobilka *et al.*, 1988; Verrall *et al.*, 1997; Audet and Bouvier, 2012), whilst important regulatory functions such as GPCR trafficking and activity regulation are mediated by the CT (Swift *et al.*, 2006; Hampson *et al.*, 2008;Audet and Bouvier, 2012). We therefore substituted IL2, IL3 and the CT of melanopsin with those of mGluR6 in order to generate Opto-mGluR6. Apart from these modifications, the ELs and TM domains, including the chromophore binding pocket of melanopsin (Sekharan *et al.*, 2012), were left intact in order to keep Opto-mGluR6 "invisible" to the immune system and to conserve the light-activated photocycle of melanopsin. Suitable cutting and ligation sites between mGluR6 and melanopsin were primarily based on computer modeling of secondary and tertiary protein structures in order to identify the borders of ILs and ELs, as well as the primary sequence alignment at the N- and C-terminal ends of any particular domain (Figure 23.3B).

23.3.2.2 Quality of Recovered Vision

We were able to show that Opto-mGluR6 is approximately 5000-times more light sensitive than ChR2 and, when targeted to retinal ON-BPCs of *rd1* mice, recovers both ON and OFF responses with fast kinetics similar to light responses recorded

---

Caption for Figure 23.3 (cont.)

depicted for reasons of clarity; please refer to Figure 23.2. (B) Amino acid sequence alignment of mouse rhodopsin, melanopsin, mGluR6 and Opto-mGluR6; splice sites (c–g) for exchange of IL2, IL3 and the CT are indicated by arrows. Amino acid numbering is indicated to the right. Cages indicate residues forming the "ionic lock" and the conserved NP*xx*Y and NR(K)Q sequences (black), as well as conserved residues on the basis of polarity and acidity (gray). Adapted from van Wyk *et al.* (2015).

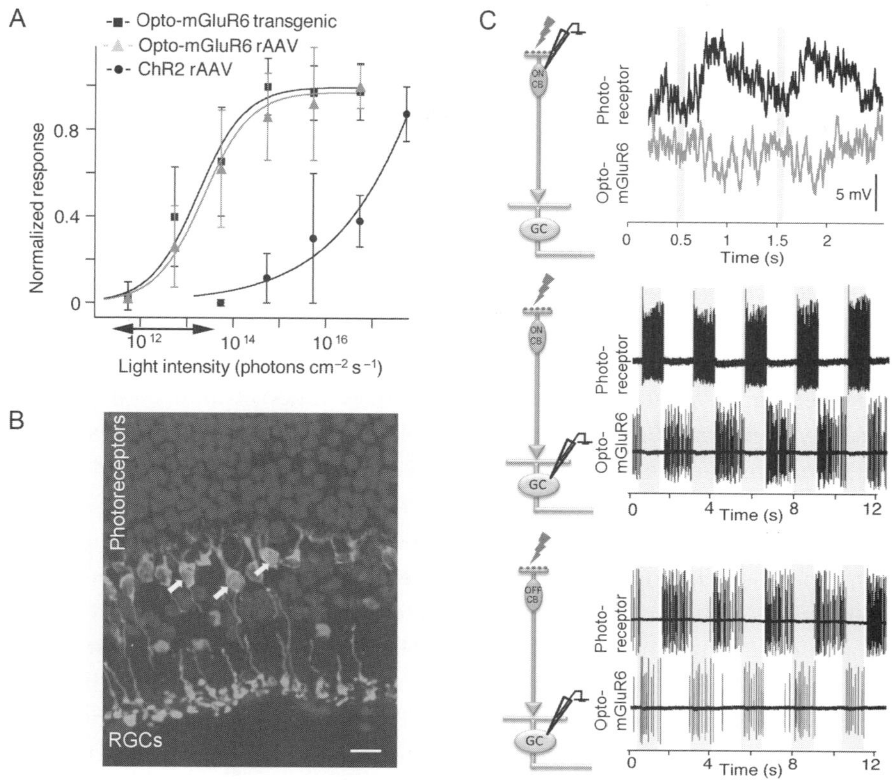

Figure 23.4 Performance of Opto-mGluR6 in the blind mouse retina. (A) Average light intensity response curves of seven RGCs from transgenic Opto-mGluR6 retinas and six RGCs from AAV-transduced Opto-mGluR6 retinas compared to published values for ChR2 (Wässle, 2004; Lagali *et al.*, 2008; Doroudchi *et al.*, 2011). The Opto-mGluR6 sensitivity curve lies in the range of cone photoreceptor sensitivity (indicated by the double-sided arrow); that is, Opto-mGluR6 mediates vision in daylight. In comparison, ChR2's activation threshold is equivalent to Opto-mGluR6 saturation intensities. The graph was deduced from the light-induced changes of ON-RGC spike rates that were normalized to a maximum of one for each RGC. (B) Confocal image of a longitudinal section through a healthy, photoreceptor-containing retina transduced with AAV particles containing the Opto-mGluR6 transgene. Opto-mGluR6 (red) is specifically expressed in ON-BPCs (arrows), co-labeled in green. Scale bar: 20 μm. (C) Electrophysiological patch clamp experiments on bipolar and ganglion cells of Opto-mGluR6 (red dots) – treated healthy retinas before (top traces) and after (bottom traces) the photoreceptor responses have been pharmacologically blocked (i.e. top traces are photoreceptor-mediated and bottom traces are Opto-mGluR6-mediated light responses of the same cell). Representative responses to a blue light stimulus (yellow highlights) demonstrate: (1) that the fast, native response is retained when Opto-mGluR6 mediates light responsiveness; (2) that Opto-mGluR6 recovers ON and OFF responses; and (3) that Opto-mGluR6 induces a response of inverse polarity compared to the photoreceptor-mediated responses, as light hyperpolarizes photoreceptors, leading to a decrease in glutamate release and, consequently, less mGluR6 activity. By contrast, Opto-mGluR6 is directly activated by light. (A black-and-white version of this figure will appear in some formats. For the color version, please refer to the plate section.)

from healthy wild-type retinas (Figure 23.4). This suggests that Opto-mGluR6 is trafficked to the same cellular locations as native mGluR6 and is able to assemble with pre-existing G-protein/RGS/TRPM1 complexes (Dhingra *et al.*, 2004; Masuho *et al.*, 2010). Opto-mGluR6 also re-established diverse RGC light responses comprising ON, OFF, ON–OFF, sustained and transient responses. This suggests that

Opto-mGluR6 is able to sufficiently drive ON-BPCs and thereby postsynaptic AII amacrine cells, as well as inhibitory amacrine feedback loops, dissecting the ON-BPC signal into ON, OFF, transient and sustained components (Nirenberg and Meister, 1997).

We further proved that re-established retinal light responses were conveyed to the visual cortex by presenting a periodic moving bar to an anesthetized mouse and concomitantly imaging V1 responses optically (Figure 23.5A) (Kalatsky and Stryker, 2003). Behaviorally, we were able to show that treated mice recovered an optomotor reflex (Prusky *et al.*, 2004). The optomotor tracking reflex enables mice to follow a stripe pattern and is indicative of their ability to respond to a spatially organized stimulus. Opto-mGluR6 mice responded to spatial frequencies of 0.17 cycles per degree, which corresponds to a spatial resolution of about 50% of that of a healthy, seeing mouse. To obtain a measure of perceptual vision, we designed a visually guided swim task in which mice were trained to swim toward a light stimulus (blue LED) indicating a submerged escape platform in a large circular water tank filled with opaque water (Figure 23.5B). The platform was moved in each experiment and swim path lengths were taken as a measure of performance. After 5 consecutive days, Opto-mGluR6-treated and seeing mice directly targeted the platform, whilst blind littermates displayed entirely random swim paths (Figure 23.5B and C). Behavioral performance was not compromised by non-saturating training or test light intensities (Figure 23.5D), as long as intensities remained within the sensitivity range of Opto-mGluR6 (see Figure 23.4A).

23.4 Concluding Remarks

An ideal therapy for patients suffering from photoreceptor degeneration will not only restore the light sensitivity of the retina, but also: (1) will function at moderate daylight intensities; (2) will be physiologically compatible with the surviving inner retina; (3) will conserve a natural range of RGC trigger features; and (4) will be devoid of toxic and immunogenic side effects. All of this should be accomplished with a minimally invasive and safe clinical technology. Opto-mGluR6, which overcomes most of the shortfalls of existing optogenetic tools, meets most of these criteria and enhances the clinical feasibility of optogenetic vision recovery.

23.4.1 What Vision Can an Optogenetic Therapy so far Recover?

Recent studies with rhodopsin and Opto-mGluR6 demonstrated that optogenetics can recover some useful vision: treated *rd1* mice regained the ability to associate light with a reward (Gaub *et al.*, 2015; van Wyk *et al.*, 2015) and to distinguish dynamic light stimuli from intensity-matched static stimuli (Gaub *et al.*, 2015). In particular, treated mice recovered the ability to recognize spatial patterns of modest contrast that are typical of natural scenes (contrast ratio of approximately 1:7.5) and reached a temporal resolution of recovered vision of up to 10 Hz (compared to 35 Hz in a healthy mouse eye) (Cehajic-Kapetanovic *et al.*, 2015; van Wyk *et al.*, 2015). What we currently do not know is whether treated

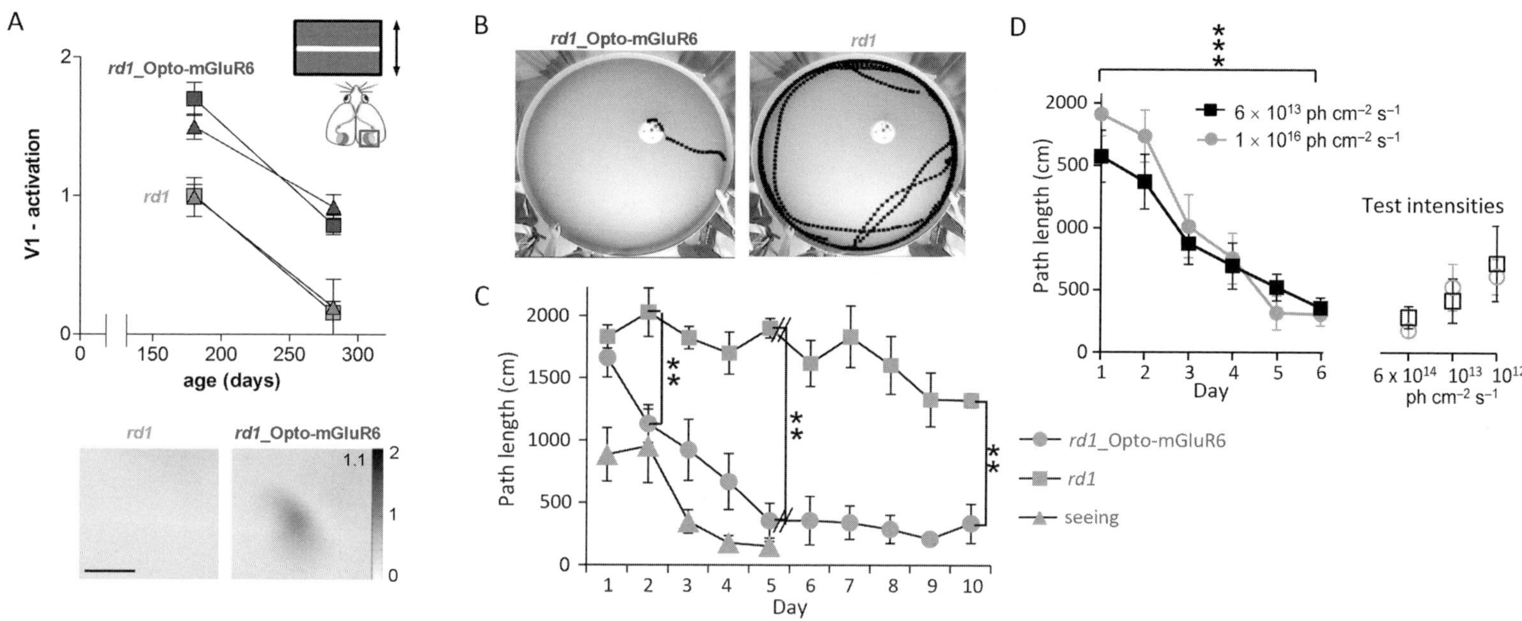

Figure 23.5 Typical experiment to test for vision recovery *in vivo*. (A) Optical imaging of cortical responses while the mouse is presented with a high-contrast moving bar on a computer screen. (Top) Blind *rd1* mice that express Opto-mGluR6 in their ON-BPCs show significantly higher activation in the visual cortex (V1) than their blind littermates, both at 180 days of age (n = 5) and at 280 days of age (n = 40), when all photoreceptors have degenerated and *rd1* mice are cortically blind (mean ± standard error of the mean, analysis of variance, $^{***}p < 0.001$). (Bottom) Typical examples of binocular V1 activity maps at 280 days of age. The grayscale-coded response magnitude corresponds to a fractional change in reflection of $\times 10^{-4}$; that is, the darker the area, the more active the response. Scale bar: 1 mm. (B–D) Visual performance determined in a behavioral experiment in which mice have to learn to associate blue light emitted from a LED tower positioned approximately 5 cm above the water surface with the location of a submerged platform in a circular pool (1.8-m diameter) filled with opaque water. (B) Online analyzed images from an overhead high-resolution camera giving examples of swim paths at training day 10. Clearly, only the *rd1*_Opto-mGluR6 mouse can perform the task. (C) Diagram showing total swim distance as a function of training days. From training day 2, *rd1*_Opto-mGluR6 mice (n = 6) learned to find the platform significantly faster than their blind *rd1* littermates (n = 5), but not significantly slower than seeing mice (n = 3). (D) Sensitivity of behavioral vision. *rd1*_Opto-mGluR6 mice were trained for 6 days at saturating and threshold intensities (left) and subsequently tested for their visual performance at three intensities spanning the Opto-mGluR6 sensitivity range (right) (c.f. Figure 23.4A). Non-saturating training light intensities did not compromise learning, and the performance at the three test intensities did not differ significantly. Error bars indicate standard error of the mean in (C) and (D).

mice can actually discriminate two different shapes or distinguish two temporal pulse patterns (flash sequences). Such advanced visual qualities can be tested by presenting short movies of natural scenes repeatedly to a retina or a mouse *in vivo* and extracting the temporal and spatial trigger features that elicit a response. Comparison with data from a healthy retina will give an idea of the quality of vision that can be optogenetically recovered in a mouse, and maybe even in a human patient.

23.4.2 How Could Optogenetic Tools for Vision Recovery Be Further Improved?

The present focus clearly lies on the improvement of delivery and expression, such as designing more efficient AAV vectors and finding OFF-BPC-specific short enhancer–promoter sequences to independently activate the OFF pathway and to guarantee stable long-term expression of the optogenetic transgene. However, in the future, recovery of color and not just monochromatic vision may be attempted. Presently, all ON-BPCs are targeted by the same optogenetic tool. However, there exist 12 types of cone BPCs in the mammalian retina with different morphologies, physiologies and gene expression patterns, some of them carrying color-selective information (Mariani, 1984; Neitz and Neitz, 2011; Puller and Haverkamp, 2011; Euler *et al.*, 2014). If one could target two cone BPC types selectively with spectrally orthogonal optogenetic tools, color vision might be (at least partially) re-installed.

REFERENCES

Audet, M., & Bouvier, M. (2012). Restructuring G-protein-coupled receptor activation. *Cell*, **151**, 14–23.

Bi, A., Cui, J., Ma, Y., *et al.* (2006). Ectopic expression of a microbial-type rhodopsin restores visual responses in mice with photoreceptor degeneration. *Neuron*, **50**(1), 23–33.

Busskamp, V., Duebel, J., Balya, D., *et al.* (2010). Genetic reactivation of cone photoreceptors restores visual responses in retinitis pigmentosa. *Science*, **329**(5990), 413–417.

Busskamp, V., Krol, J., Nelidova, D., *et al.* (2014). miRNAs 182 and 183 are necessary to maintain adult cone photoreceptor outer segments and visual function. *Neuron*, **83**, 586–600.

Cao, P., Sun, W., Kramp, K., *et al.* (2012). Light-sensitive coupling of rhodopsin and melanopsin to G(i/o) and G(q) signal transduction in *Caenorhabditis elegans*. *FASEB J*, **26**, 480–491

Caporale, N., Kolstad, K., Lee, T., *et al.* (2011). LiGluR restores visual responses in rodent models of inherited blindness. *Mol Ther*, **19**, 1212–1219.

Cehajic-Kapetanovic, J., Eleftheriou, C., Allen, A., *et al.* (2015). Restoration of vision with ectopic expression of human rod opsin. *Curr Biol*, **25**, 2111–2122.

Chang, B., Hawes, N., Hurd, R., *et al.* (2002). Retinal degeneration mutants in the mouse. *Vision Res*, **42**, 517–525.

Choe, H., Kim, Y., Park, J., *et al.* (2011). Crystal structure of metarhodopsin II. *Nature*, **471**, 651–655.

Cotecchia, S., Exum, S., Caron, M., *et al.* (1990). Regions of the alpha 1-adrenergic receptor involved in coupling to phosphatidylinositol hydrolysis and enhanced sensitivity of biological function. *Proc Natl Acad Sci U S A*, **87**, 2896–2900.

European Parliament, Council of the European Commission (2006). Directive 2006/25/EC of the European Parliament and of the Council on the minimum health and safety requirements regarding the exposure of workers to the risks arising from physical agents (artificial optical radiation) 19th individual directive within the meaning of Article 16(1) of Directive 89/391/EEC). *Off. J. Eur. Union* 114, 38–59.

Cronin, T., Vandenberghe, L., Hantz, P., *et al.* (2014). Efficient transduction and optogenetic stimulation of retinal bipolar cells by a synthetic adeno-associated virus capsid and promoter. *EMBO Mol Med*, **6**, 1175–1190.

Dhingra, A., Faurobert, E., Dascal, N., *et al.* (2004). A retinal-specific regulator of G-protein signaling interacts with Gαo and accelerates an expressed metabotropic glutamate receptor 6 cascade. *J Neurosci*, **24**, 5684–5693.

Doré, A., Okrasa, K., Patel, J., *et al.* (2014). Structure of class C GPCR metabotropic glutamate receptor 5 transmembrane domain. *Nature*, **511**, 557–562.

Doroudchi, M., Greenberg, K., & Liu, J. (2011). Virally delivered channelrhodopsin-2 safely and effectively restores visual function in multiple mouse models of blindness. *Mol Ther*, **19**, 1220–1229.

Euler, T., Havekamp, S., Schubert, T., *et al.* (2014). Retinal bipolar cells: elementary building blocks of vision. *Nature Rev*, **15** (507–519)

Feldbauer, K., Zimmermann, D., Pintschovius, V., *et al.* (2009). Channelrhodopsin-2 is a leaky proton pump. *Proc Natl Acad Sci USA*, **106**, 12317–12322.

Gaub, B., Berry, M., Holt, A., *et al.* (2015). Optogenetic vision restoration using rhodopsin for enhanced sensitivity. *Mol Ther*, **23**(10), 1562–1571.

Gaub, B., Berry, M., Holt, A., *et al.* (2014). Restoration of visual function by expression of a light-gated mammalian ion channel in retinal ganglion cells or ON-bipolar cells. *Proc Natl Acad Sci U S A*, **111**, E5574–5583.

Greenberg, K., Pham, A., & Werblin, F. (2011). Differential targeting of optical neuromodulators to ganglion cell soma and dendrites allows dynamic control of centre–surround antagonism. *Neuron*, **69**, 713–720.

Hampson, D., Rose, E., & Antflick, J. (2008). *The Structures of Metabotropic Glutamate Receptors*. (pp. 363–386). Totowa, NJ: Human Press.

Hattar, S., Liao, H., Takao, M., *et al.* (2002). Melanopsin-containing retinal ganglion cells: architecture, projections, and intrinsic photosensitivity. *Science*, **295**, 1065–1070.

Hollenstein, K., Kean, J., Bortolato, A., *et al.* (2013). Structure of class B GPCR corticotropin-releasing factor receptor 1. *Nature*, **499**, 438–443.

Kalatsky, V., & Stryker, M. (2003). New paradigm for optical imaging: temporally encoded maps of intrinsic signal. *Neruon*, **38**, 529–545.

Keeler, C. (1966). Retinal degeneration in the mouse is rodless retina. *J Hered*, **57**(2), 47–50.

Kim, D.S., Matsuda, T., & Cepko, C.L. (2008). A core paired-type and POU homeodomain-containing transcription factor program drives retinal bipolar cell gene expression. *J Neurosci*, **28**(31), 7748–7764.

Kleinlogel, S., Feldbauer, K., Dempski, R., *et al.* (2011). Ultra light-sensitive and fast neuronal activation with the Ca^{2+}-permeable channelrhodopsin CatCh. *Nat Neurosci*, **14**(4), 513–518.

Kobilka, B., Kobilka, T., Daniel, K., *et al.* (1988). Chimeric alpha 2-beta 2-adrenergic receptors: delineation of domains involved in effector coupling and ligand binding specificity. *Science*, **240**, 1310–1316.

Koyanagi, M., & Terakita, A. (2014). Diversity of animal opsin-based pigments and their optogenetic potential. *Biochim Biophys Acta*, **1837**, 710–716.

Lagali, P., Balya, D., Awatramani, G., *et al.* (2008). Light-activated channels targeted to ON bipolar cells restore visual function in retinal degeneration *Nat Neurosci*, **11**(6), 667–675.

Lin, B., Koizumi, A., Tanaka, N., *et al.* (2008). Restoration of visual function in retinal degeneration mice by ectopic expression of melanopsin. *Proc Natl Acad Sci USA*, **105**, 16009–16014.

Macé, E., Caplette, R., Marre, O., *et al.* (2015). Targeting channelrhodopsin-2 to ON-bipolar cells with vitreally administered AAV Restores ON and OFF visual responses in blind mice. *Mol Ther*, **23**, 7–16.

Mariani, A. (1984). Bipolar cells in monkey retina selective for the cones likely to be blue-sensitive. *Nature*, **308**, 184–186.

Masuho, I., Celver, J., Kovoor, A., *et al.* (2010). Membrane anchor R9AP potentiates GTPase-accelerating protein activity of RGS11·Gβ$_5$ complex and accelerates Inactivation of the mGluR6-G$_o$ signaling. *J Biol Chem*, **285**, 4781–4787.

Morgans, C., Zhang, J., Jeffrey, B., *et al.* (2009). TRPM1 is required for the depolarizing light response in retinal ON-bipolar cells. *Proc Natl Acad Sci USA* **106**, 19174–19178.

Nagel, G. (2003). Channelrhodopsin-2, a directly light-gated cation-selective membrane channel. *Proc Natl Acad Sci USA*, **100**, 13940–13945.

Neitz, J., & Neitz, M. (2011). The genetics of normal and defective color vision. *Vision Res*, **51**, 633–651.

Nirenberg, S., & Meister, M. (1997). The light response of retinal ganglion cells is truncated by a displaced amacrine circuit. *Neuron*, **18**, 637–650.

Pan, Z., Ganjawala, T., Lu, Q., *et al.* (2014). ChR2 mutants at L132 and T159 with improved operational light sensitivity for vision restoration. *PLoS ONE*, **9**, e98924.

Pin, J., Galvez, T., & Prézeau, L. (2003). Evolution, structure and activation mechanism of family 3/C G-protein-coupled receptors. *Pharmacol Ther*, **98**, 325–354.

Polosukhina, A., Litt, J., Tochitsky, I., *et al.* (2012). Photochemical restoration of visual responses in blind mice. *Neuron*, **75**(2), 271–282.

Prusky, G., Alam, N., Beekman, S., *et al.* (2004). Rapid quantification of adult and developing mouse spatial vision using a virtual optomotor system. *Invest Opthalmol Vis Sci*, **45**, 4611–4616.

Puller, C., & Haverkamp, S. (2011). Bipolar cell pathways for color vision in non-primate dichromats. *Vis Neurosci*, **28**, 51–60.

Rasmussen, S., DeVree, B., Zou, Y., *et al.* (2011). Crystal structure of the β2 adrenergic receptor–Gs protein complex. *Nature*, **477**, 549–555.

Remé, C., Grimm, C., Hafezi, F., *et al.* (1998). Apoptotic cell death in retinal degenerations. *Prog Retin Eye Res*, **17**, 443–464.

Schiöth, H., & Frederiksson, R. (2005). The GRAFS classification system of G-protein coupled receptors in comparative perspective. *Gen Comp Endocrinol*, **142**, 94–101.

Schwartz, T., Frimurer, T., Holst, B., *et al.* (2006). Molecular mechanism of 7TM receptor activation – a global toggle switch model. *Annu Rev Pharmacol Toxicol*, **46**, 481–519.

Sekharan, S., Wei, J., & Batista, V. (2012). The active site of melanopsin: the biological clock photoreceptor. *J Am Chem Soc*, **134**, 19536–19539.

Sexton, T., Buhr, E., & Van Gelder, R. (2012). Melanopsin and mechanisms of non-visual ocular photoreception. *J Biol Chem*, **287**, 1649–1656.

Strettoi, E., & Pignatelli, V. (2000). Modifications of retinal neurons in a mouse model of retinitis pigmentosa. *Proc Natl Acad Sci USA*, **97**, 11020–11025.

Swift, S., Leger, A., Talavera, J., *et al.* (2006). Role of the PAR1 receptor 8th helix in signaling: the 7-8-1 receptor activation mechanism. *J Biol Chem*, **281**, 4109–4116.

The Lasker/IRRF Initiative for Innovation in Vision Science (2014). Restoring Vision to the Blind: The Lasker/IRRF Initiative for Innovation in Vision Science. *Transl Vis Sci Technol*, **3**(7), 1.

Thyagarajan, S., van Wyk, M., Lehmann, K., *et al.* (2010). Visual function in mice with photoreceptor degeneration and transgenic expression of channelrhodopsin 2 in ganglion cells. *J Neurosci*, **30**(26), 8745–8758.

Tochitsky, I., Polosukhina, A., Degtyar, V., *et al.* (2014). Restoring visual function to blind mice with a photoswitch that exploits electrophysiological remodeling of retinal ganglion cells. *Neuron*, **81**, 800–813.

van Wyk, M., Pielecka-Fortuna, J., Löwel, S., *et al.* (2015). Restoring the ON-switch in blind retinas: Opto-mGluR6, a next-generation, cell-tailored optogenetic tool. *PLoS Biol*, **13** (5), e1002143.

Verrall, S., Ishii, M., Chen, M., *et al.* (1997). The thrombin receptor second cytoplasmic loop confers coupling to Gq-like G proteins in chimeric receptors. Additional evidence for a common transmembrane signaling and G protein coupling mechanism in G protein-coupled receptors. *J Biol Chem*, **272**, 6898–6902.

Wässle, H. (2004). Parallel processing in the mammalian retina. *Nat Rev Neurosci*, **5**, 747–757.

Wu, C., Ivanova, E., Zhang, Y., *et al.* (2013). rAAV-mediated subcellular targeting of optogenetic tools in retinal ganglion cells *in vivo*. *PLoS ONE*, **8**, e66332.

Yamashita, T., Terakita, A., & Shichida, Y. (2001). The second cytoplasmic loop of metabotropic glutamate receptor functions at the third loop position of rhodopsin. *J Biochem*, **130**, 149–155.

Yau, K., & Hardie, R. (2009). Phototransduction motifs and variations. *Cell*, **139**, 246–264.

Zrenner, E., Bartz-Schmidt, K., Benav, H., *et al.* (2011). Subretinal electronic chips allow blind patients to read letters and combine them to words. *Proc Biol Sci*, **278**, 1489–1497.

24 A Promise of Vision Restoration

Grégory Gauvain, Antoine Chaffiol, Jens Duebel, and Serge A. Picaud

24.1 Introduction: Optogenetics for the Restoration of Vision

Optogenetics has revolutionized biology by giving scientists the ability to control neuronal activity with high spatiotemporal precision using only light. Its applications in fundamental science are boundless, potentially enabling the study of the role of any cell type in a specific neuronal circuit. However, therapeutic approaches based on its fundamental applications are not yet available.

Multiple pathological conditions could benefit from the development of treatments based on optogenetic strategies, but none show greater promise than those for retinal diseases. Optical stimulation of retinal neurons can be easily implemented as the eye provides a natural window to the retina. This approach for visual restoration is currently under development as it may provide higher cellular resolution than other approaches, such as electrical stimulation by retinal prostheses.

Degeneration of photoreceptors – the cells that are responsible for the canonical image-forming photo-transduction – is the cause of the two most common blinding diseases, retinitis pigmentosa (RP) and age-related macular degeneration (AMD), which affect millions of people worldwide. It would thus be highly exciting to reintroduce light sensitivity, using optogenetic therapy in surviving cells, in order to restore vision.

Before looking in detail at the various possible optogenetic strategies, we will explore the pathological context and other possible therapeutic solutions for these diseases.

24.1.1 Retinal Degenerative Diseases

24.1.1.1 Retinitis Pigmentosa

RP regroups numerous inherited dystrophies presenting similar clinical characteristics: the progressive degeneration of retinal cells and the loss of visual function (Figure 24.1). The first cells to be impaired are rod photoreceptors, resulting in nyctalopia (night blindness). As the disease progresses, an increasing

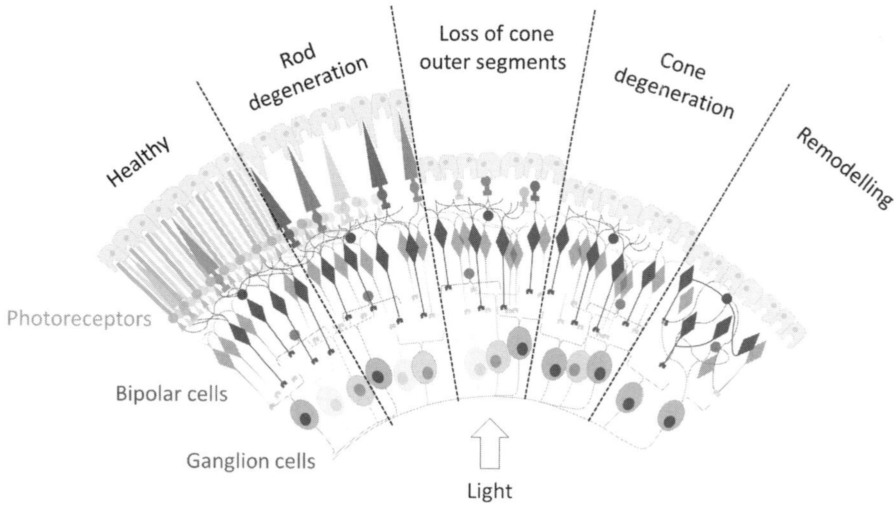

Figure 24.1 Schematic diagram of a healthy compared to a degenerated retina. Rod photoreceptors are the first to degenerate. Their loss induces the consequent loss of the cone outer segments and eventually cell bodies. Important circuitry remodeling occurs after deafferentation. The three main populations of cells are labeled, but others are also present: retinal pigment epithelium, horizontal cells and amacrine cells in gray, dark red and orange, respectively. Modification of connectivity after photoreceptor degeneration is not pictured. (A black-and-white version of this figure will appear in some formats. For the color version, please refer to the plate section.)

number of rods are affected and die. The loss of rod photoreceptors is followed by cone photoreceptor degeneration, even though cone photoreceptors do not express the mutated gene that causes the disease. This cone loss is attributed to the loss of the rod-derived cone viability factor (Léveillard *et al.*, 2004; Aït-Ali *et al.*, 2015). The progressive loss of cone photoreceptors, which are responsible for diurnal and color vision, results in a constriction of the visual field, also called tunnel vision. The loss of cones finally affects the central area, where they are much more numerous, leading to a complete loss of color and central vision.

Having more than 3000 identified mutations (Daiger *et al.*, 2013), RP displays considerable genetic heterogeneity that affects both its inheritance pattern and phenotypic severity. Furthermore, the genetic heterogeneity of RP is also manifested in the variety of biological functions served by the mutated proteins. These patients represent the first population target in clinical trials for visual restoration because they have no remaining visual function at late stages.

24.1.1.2 Age-related Macular Degeneration

AMD is a degenerative condition that affects the central part of the retina (macula). It is the leading cause of irreversible blindness in industrialized countries (Lim *et al.*, 2012) and predominantly affects individuals aged 65 years and above. The development of the pathology from the early to intermediate stages accounts for 80% of the identified cases of AMD. It is characterized by progressive photoreceptor loss associated with retinal pigment epithelium dysfunction and

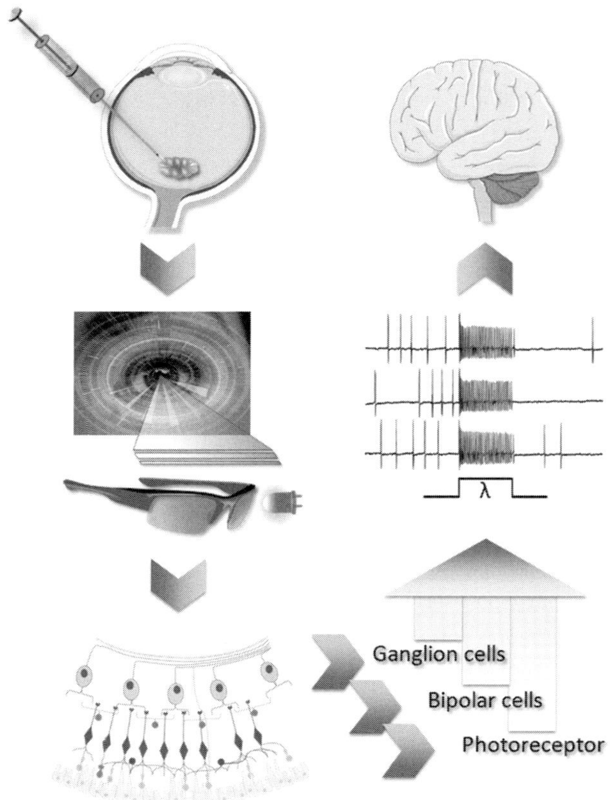

Figure 24.2 Overview of the optogenetic vision restoration strategy. The first step is an intravitreal injection of the engineered viral vector into the patient's eye. Following adequate time passing for expression of the opsin, an LED light source is used to stimulate the transfected cells. The patients can use enhancement goggles, if necessary. The integration of the incoming information begins at the target cells and leads to the generation of a spike train by retinal ganglion cells. An example of an extracellular recording of a ganglion cell expressing ChR2[a] is shown. The firing rate increases as the light intensity increases. The pattern of activity is then transmitted to the brain through the optic nerve.
[a] *rd1* mice retinal ganglion cells were infected using AAV2-CAG-ChR2-GFP. There were three repetitions of similar blue light flashes (λ).

degeneration. Later stages of the disease include two different phenotypes: neovascularization, or so called "wet" AMD; and geographical atrophy, or "dry" AMD. Although both late-stage forms of AMD only account for 20% of AMD cases, they account for most of the visual loss (~90%). Current treatments for neovascular AMD, which demonstrate variable efficacy (from slower progression to improved vision) (Holz *et al.*, 2014), all necessitate regular drug administration through intravitreal or trans-scleral injections or insertions of drug-diffusing implants. There is no approved treatment for "dry" AMD except for dietary supplements. The presence of peripheral vision in "dry" AMD patients excludes them from clinical trials for visual restoration. However, when an approach will restore central vision above the threshold for defining legal vision, blind AMD patients will greatly benefit from this strategy.

24.2 Alternative Vision Restoration Strategies

To put optogenetic vision restoration in perspective, it is important to take into account the various strategies that all have the same goal: restoring sight. This is especially true for clinical trials of retinal prostheses that have resulted in some restoration of useful vision, demonstrating the ability of the visual system of blind patients to interpret new artificial visual information despite retinal remodeling. Providing useful vision to blind patients will therefore depend on the resolution of retinal-stimulating strategies. The cellular resolution of optogenetic therapy could therefore offer great potential for visual restoration.

The most elegant strategy for vision restoration would likely be to replace the missing photoreceptors with new functional cells. A growing number of studies have used engineered stem cells as a renewable source of photoreceptors (Lamba *et al.*, 2009). Indeed, induced pluripotent stem cells can be differentiated into photoreceptor precursors and inserted into the retinal network of healthy rodents. However, important challenges remain: this strategy works on healthy tissue, but must be adapted for use on tissues in advanced degenerative states. Furthermore, the number of transplanted cells that integrate into the retinal network remains very limited, even in normal animals (MacLaren *et al.*, 2006; Bartsch *et al.*, 2008), Finally, stem cell treatment would require custom-made cell production for patients, or at least for the most common human leukocyte antigen types.

The most advanced strategy for vision restoration is the use of retinal prosthetic devices. This approach relies on the implantation of a small array of electrodes in the eye of the patient. These electrodes are used to directly or indirectly stimulate ganglion cells. There are various implant technologies and implantation strategies (for reviews, see Bertschinger *et al.*, 2008; Chuang *et al.*, 2014). Multiple clinical trials have already demonstrated the efficacy of these devices for restoring sight to a certain extent (Zrenner *et al.*, 2011; Humayun *et al.*, 2012; da Cruz *et al.*, 2013). The company Second Sight has obtained a CE mark and FDA approval for its Argus II device, and Retina Implant AG has obtained a CE mark. This approach benefits from the outstanding success of cochlear implants (Yawn *et al.*, 2015) and attempts to transfer that success to another sensory modality, vision. However, contrary to the relatively simple spatial coding of frequency in the cochlea, artificial retinas require at least 600 independent pixels in order to enable facial recognition, text reading and independent locomotion (Cha *et al.*, 1992a; Cha *et al.*, 1992b; Sommerhalder *et al.*, 2004). Even though devices currently in clinical trials require complex surgeries for implantation and wires entering the eye, new photovoltaic implants will be wireless with distant optical activation (Lorach *et al.*, 2015).

Another recently developed approach uses a photoswitchable K^+ channel blocker to restore light sensitivity (Polosukhina *et al.*, 2012; Tochitsky *et al.*, 2014). Retinal ganglion cells (RGCs) become light sensitive in blind mice (*rd1*) after ocular injection of the compound. It is not known yet how the encouraging results obtained in mice will transfer to primates or humans, but this approach would require regular intravitreal injections or the insertion of slow-diffusing implants as the compound washes away after a couple of days.

A wide range of gene defects can cause RP or AMD. Within this spectrum, it is possible to identify individual genes for which the mutations are responsible for the most prevalent forms of these diseases. These specific mutations are excellent candidates for the development of gene therapy, and multiple clinical trials are already underway. This is notably the case for four retinal dystrophies that share many similarities with RP: Leber's congenital amaurosis (Cideciyan *et al.*, 2009), Stargardt disease (Sanofi, 2015), choroideremia (MacLaren *et al.*, 2014) and Usher syndrome (Sanofi, 2016). One clinical trial has also started for a form of wet AMD (GenVec, 2011).

Results from the first phases of these clinical trials are encouraging, but it is important to keep in mind that the different dystrophies targeted here represent only a small fraction of these syndromes, which are considered to be rare diseases, making it difficult to envision parallel approaches for all of the genetic conditions behind this particular spectrum of dystrophies. Furthermore, gene therapy can only be used as a protective measure; it will prevent visual deterioration, but not restore sight once degeneration has taken place.

24.3 The Optogenetic Approach

The optogenetic strategy relies on the use of light-sensitive proteins (e.g. microbial opsins) to transform neurons, especially non-naturally photosensitive retinal neurons, into photoreceptors. Microbial opsins are ion channels or pumps that are activated by light. When expressed on the surface of a neuron, they trigger an ionic current through the plasma membrane upon light stimulation. Any neuron can thus be transformed into an artificial photoreceptor. This system is not as efficient as the phototransduction cascade present in the healthy retina that is composed of two amplification levels, because it is based on a single protein. Therefore, high light levels are required for light activation.

The employment of opto-genes to cure blindness is a potentially useful treatment. It has a one-size-fits-all aspect, as it would work for any type of photoreceptor degeneration at any stage of the disease. There are diverse strategies for optogenetic vision restoration in this emerging field (Tables 24.1 and 24.2), but there are generally two choices that must be made: the specific opsin to be expressed and the cell to be targeted. These two choices will appreciably affect the outcome of the light sensitization and may even change the performance of the proposed strategy.

The healthy retina possesses an incredible dynamic range, from bright daylight to starlight, encompassing nearly ten orders of magnitude. This is achieved by the use of numerous mechanisms for adaptation and light amplification.

Optogenetic strategies can be improved by increasing the light sensitivity of microbial opsins by using molecular engineering. They can also be improved by restoring light sensitivity in the retinal network as early as possible, thus conserving at least some convergence-induced light amplification. However, adaptation would be limited, even in these improved scenarios, and the dynamic range would be small. Biomimetic stimulation devices have been designed to deliver appropriate light levels to the retina in order to expand this range, independent of environmental lighting conditions.

Biomimetic stimulation devices, in the form of glasses, will be composed of: (1) a visual sensor to acquire visual information; (2) a pocket PC to compute optical stimulation; and (3) an optical stimulator to deliver the proper light levels to the retina.

We will first discuss the consequences of selecting microbial or vertebrate opsins and then consider different possible cellular targets and their repercussions on reactivation.

24.3.1 Selection of the Photoactivatable Protein

Many vision restoration studies have used channelrhodopsin-2 (ChR2) and its mutants (Bi *et al.*, 2006; Tomita *et al.*, 2007; Lagali *et al.*, 2008; Ivanova and Pan, 2009; Zhang *et al.*, 2009; Ivanova *et al.*, 2010; Tomita *et al.*, 2010; Doroudchi *et al.*, 2011; Macé *et al.*, 2015). ChR2 is also the opsin that was selected for the first human trials (RetroSense Therapeutics, 2016). Unfortunately, the ChR2 absorption peak is in the blue light range (~460 nm), even though blue light can become photo-toxic (Tosini *et al.*, 2016). The guidelines on limits of exposure to incoherent radiation (International Commission on Non-Ionizing Radiation Protection, 2013) clearly emphasize the photo-toxic properties of blue light, leading to an order of magnitude difference in the recommended amount of blue light to be used for stimulation versus that for orange/red light. It is thus necessary to select alternatives to ChR2 in order to maximize cellular responses while minimizing health risks.

Two trends are emerging in the vision restoration field: selection of the best-suited microbial opsins and engineering of vertebrate opsins.

24.3.1.1 Microbial Opsins

Microbial opsins present important advantages over vertebrate opsins. They are simpler systems because they are directly photoactivatable ion channels or pumps and do not require any cellular machinery, unlike Gprotein-coupled vertebrate opsins. They autonomously recycle their visual pigment, all-*trans* retinal (Haupts *et al.*, 1997), freeing them from the transcellular recycling pathway of vertebrate opsins involving retinal pigmental epithelium and Muller cells. Finally, they have very rapid kinetics, often faster than intrinsic responses (Busskamp *et al.*, 2010; Macé *et al.*, 2015).

There are now numerous optogenetic tools of microbial origin that may be better suited than ChR2 for the particular challenges of vision restoration (Zhang *et al.*, 2011; Lin *et al.*, 2013; Klapoetke *et al.*, 2014). The search continues for new microbial opsins by screening the microbial diversity (Klapoetke *et al.*, 2014) in order to complement current efforts to improve the amplitude of the ChR2 current, as in CatCh (Kleinlogel *et al.*, 2011), or the spectral range of ChR2, as in ReaChR (Lin *et al.*, 2013).

The ideal opsin for vision restoration depends on the selected cellular strategy: it may be a depolarizing opsin for bipolar cell or RGC targeting, but should be hyperpolarizing for dormant cone reactivation (Busskamp *et al.*, 2010). The biophysical properties of the opsin should also be tailored to the strategy that is selected. For example, fast onset and offset kinetics are critical for the generation of highly temporally precise spike trains for the RGC strategy. Furthermore, shifting the spectral sensitivity of the opsin to the red range should reduce light toxicity at equivalent intensities. Red-shifted opsins have already been identified (Lin *et al.*, 2013; Klapoetke *et al.*, 2014), and work is underway to assess their functionality for vision restoration.

24.3.1.2 Vertebrate Opsins and Photoactivatable Proteins

The primary concern for the use of microbial opsins is their possible immunogenicity. Considerable strides have thus been made in the vison restoration field to switch from microbial opsin variants to vertebrate opsins (Lin *et al.*, 2008; Cehajic-Kapetanovic *et al.*, 2015; Gaub *et al.*, 2015). There are various advantages and drawbacks concerning this approach. The first advantage is that vertebrate opsins are more light sensitive than microbial opsins due to amplification of the light responses through G-protein-coupled cascades. Treated patients should thus be able to detect light levels that are comparable to ambient daylight or the indoor environment without the need for stimulation devices. The second advantage of using vertebrate opsins is their physiological compatibility. The use of vertebrate protein sequences should reduce the risk of potential adverse immune responses. This is particularly true for human protein sequences, such as the use of the human rhodopsin reported by Cehajic-Kapetanovic *et al.* (2015). Results show that reactivation using rhodopsin in bipolar cells (Cehajic-Kapetanovic *et al.*, 2015; Gaub *et al.*, 2015) or melanopsin in RGCs (Lin *et al.*, 2008) restores light sensitivity and some visual function in blind mouse models, as demonstrated at the retinal, cortical and behavioral levels under moderate-intensity illumination.

However, contrary to microbial opsins (Ernst *et al.*, 2014), the photo-pigment of mammalian opsins must be renewed after each event of photo-isomerization. These strategies must thus be validated *in vivo* and in a non-human primate model.

The major drawback of this approach is that restored light responses are slow relative to the rapid response kinetics that have been obtained with recently engineered microbial opsins (Klapoetke *et al.*, 2014), which would dramatically affect the temporal resolution of a reactivated eye. This is particularly true for melanopsin and its very slow kinetics.

Another approach has been to engineer vertebrate proteins to bring together the advantages of microbial and vertebrate opsins. One group has developed and used a genetically and chemically engineered light-gated ionotropic glutamate receptor (LiGluR) (Volgraf *et al.*, 2006; Caporale *et al.*, 2011; Gaub *et al.*, 2015). Some concern remains regarding its human application, despite the promising results obtained: regular injection of a photoswitchable ligand would be required, and light sensitivity is in the blue or UV range (depending on the ligand injected). Another group chose to confer the light sensitivity of melanopsin to the specific glutamate receptor of ON-bipolar cells, mGluR6 (van Wyk *et al.*, 2015). The resulting protein, called Opto-mGluR6, was introduced into ON-bipolar cells and restored ON/OFF responses from RGCs and had a behavioral impact. This method shows great promise and should be applied to a non-human primate model in the near future.

24.3.2 Layer-specific Strategies

Beyond the selection of the photoactivatable protein, there is another choice to make that will greatly affect the efficiency of the overall strategy: which target cell layer in the retina to choose.

24.3.2.1 Conferring Light Sensitivity to RGCs

The first study to report the reactivation of degenerated retina using a microbial opsin (ChR2) was published in 2006 by the Zhuo-Hua Pan group (Bi *et al.*, 2006). This study demonstrated the restoration of visual responses in RGCs and the cortex of blind *rd1* mice. They used intravitreal injection of AAV vectors with a non-specific promoter to induce preferential expression in ganglion cells, but expression was identified in other cell types as well (bipolar, horizontal and amacrine cells). In this study, all cells were found to have the same ON responses, regardless of their cell type. This paper triggered the emergence of optogenetic vision restoration. It demonstrated the potential of opto-genes to confer light sensitivity to RGCs and showed that the recovered visual information is sent to the visual cortex. These results were confirmed and extended using other animal models of blindness (see Table 24.1) (Bi *et al.*, 2006; Tomita *et al.*, 2007; Zhang *et al.*, 2009).

This strategy is already being considered for clinical trials (RetroSense Therapeutics, 2016). The greatest advantage of this strategy relates to the relative ease of targeting RGCs by viral particles using intravitreal injection. Intravitreal injection is a common procedure in modern ophthalmology that encompasses little risk relative to sub-retinal delivery. The strategy was validated on the marmoset, a non-human primate model, in order to test clinical translatability (Ivanova *et al.*, 2010). However, marmosets do not exhibit AAV gene delivery restricted to the peripheral foveal ring of RGCs as do non-human primates that are closer to humans, such as macaques (Yin *et al.*, 2011). Furthermore, immune responses in marmosets are less similar to those of humans than those of macaques. Validation of the strategy in a non-human primate that is closer to humans is likely to be necessary prior to any clinical trial in order to demonstrate efficacy and safety of the strategy. We have obtained positive data in our laboratory with this strategy when using different opsins in macaques. However, only clinical trials will determine whether the loss of retinotopy at the perifoveal ring can be compensated for by cortical information processing.

Direct RGs stimulation at the retinal output bypasses all information processing by the outer retinal circuitry. This natural computation from photoreceptors to RGCs could be recapitulated using a biomimetic stimulation device, or so-called enhancement goggles (Lorach *et al.*, 2012). There is, however, no convergence, and the light that is needed to activate the photoactivatable protein could be greater than for other strategies.

24.3.2.2 Conferring Light Sensitivity to Bipolar Cells

Bipolar cells, which are relatively well preserved in advanced degenerative states (Santos *et al.*, 1997), have also been studied. Bipolar cells are found in the inner nuclear layer of the retina in two different forms: ON- and OFF-type bipolar. All of the studies on bipolar cells have focused on ON-bipolar cells (see Table 24.2) (Lagali *et al.*, 2008; Doroudchi *et al.*, 2011; Cronin *et al.*, 2014; Gaub *et al.*, 2014; Cehajic-Kapetanovic *et al.*, 2015; Gaub *et al.*, 2015; Macé *et al.*, 2015; van Wyk *et al.*, 2015). This was possible because of a promoter sequence from the *grm6* gene (encoding the ON-bipolar cell-specific metabotropic glutamate receptor, mGluR6) (Masu *et al.*, 1995). The specific targeting of ON-bipolar cells has the advantage of not sending conflicting information to the RGCs and has been sufficient to generate ON as well as OFF responses in RGCs in a number of these studies (Cronin *et al.*, 2014; Macé *et al.*, 2015; van Wyk *et al.*, 2015). Furthermore, one study successfully recorded ON and OFF responses in multi-unit recordings at the cortex (Macé *et al.*, 2015). It is most likely that these OFF signals emerge from rod ON-bipolar cells, which are connected through AII amacrine cells to both cone ON- and cone OFF-bipolar cells (Macé *et al.*, 2015).

Restoring the OFF component of the visual response alongside the ON component is undoubtedly a major advantage of this strategy relative to RGCs targeting strategies that generate ON responses in all cell types. There are no published behavioral studies on the effects of selective blockade of the OFF pathway, but selectively blocking the ON pathway in rhesus monkeys leads to impaired perception, particularly concerning contrast sensitivity (Schiller *et al.*, 1986). It is highly possible that restoring only the ON component of the visual response may lead to similar impairment.

Unfortunately, the retina of human RP patients undergoes considerable reorganization, particularly in the inner nuclear layer where the bipolar cells reside (Jones *et al.*, 2016). This may jeopardize the efficacy of this otherwise promising strategy.

24.3.2.3 Reactivation of Surviving Cones

Although retinal dystrophies promote photoreceptor loss of function and/or apoptosis, a fraction of cones appear to survive even at advanced stages of disease. These surviving photoreceptors are called "dormant" as they have lost their light sensitivity. Dormant cones possess important morphological alterations as they have lost the membrane protrusions that are characteristic of photoreceptors: their outer segment and, to a lesser extent, their inner segment. The outer segment is a specialized cellular compartment that photoreceptors constantly renew and that encompass the entire phototransduction pathway; without it, the cell is no longer light sensitive.

In 2010, Busskamp and collaborators proposed to reactivate these dormant cones using an optogenetic approach (Busskamp *et al.*, 2010). Dormant cones were transduced with viral vectors resulting in the restoration of all visual functions at the RGC level. The RGC population exhibited classic responses: ON, OFF and ON–OFF receptor field properties and a center–surround opposition with normally expressed lateral inhibition. Even direction selectivity was preserved,

Table 24.1 Overview of the literature on the RGC strategy.

Publication	Light-sensitive protein	Expressing cells	RGC response	Cortical response	Behavior	Models	Interest
Bi *et al.* 2006	ChR2	RGCs (mostly)	ON only	VEP	None	Mice (*rd1*[a])	***
Tomita *et al.*, 2007	ChR2	RGCs (mostly)	None	VEP	None	Rats (*rdy/rdy*[b])	***
Lin et at., 2008	Melanopsin	RGCs (mostly)	ON (long-lasting)	None	Pupillary reflex, light avoidance	Mice (*rd1*)	*
Zhang *et al.*, 2009	ChR2 and NpHR	RGCs	ON–OFF	None	None	Mice (*rd1*)	**
Ivanova *et al.*, 2009, 2010	ChR2	RGCs (mostly)	ON only	None	None	Mice (*rd1*) and marmosets	**
Tomita *et al.*, 2010	ChR2	RGCs (mostly)	None	VEP	Optomotor response	Rats (*rdy/rdy*)	*
Greenberg *et al.*, 2011	ChR2 and NpHR	Soma and dendrites of RGCs	Restore center–surround	None	None	Rabbit	**
Caporale *et al.*, 2011	LiGluR (+ MAG)[c]	RGCs	ON only	VEP	Pupillary reflex, light avoidance	Mice (*rd1*)	**
Wu *et al.*, 2013	ChR2 and NpHR	RGCs (mostly)	ON or OFF	None	None	Mice (*rd1*)	*
Tomita *et al.*, 2014	mVChR1	RGCs (mostly)	None	VEP	Optomotor response	Rats (*rdy/rdy*)	*

Interest: * small; ** keen; *** great.

[a] *rd1*: mutant mouse for gene *Pde6b*.

[b] *rdy/rdy*: retinal dystrophy homologous locus in Royal College of Surgeons rats (Vollrath *et al.*, 2001).

[c] LiGluR (+ MAG): the engineered light-gated ionotropic glutamate receptor and its required ligand, maleimide-azobenzene-glutamate. 7-.

ChR2: channelrhodopsin-2; NpHR: *Natronomonas pharaonis* halorhodopsin; mVChR1: modified *Volvox*-derived channelrhodopsin-1; RGC: retinal ganglion cell; VEP: visually evoked potential.

Table 24.2 Overview of the literature on the cone and bipolar cells strategy.

Publication	Light-sensitive protein	Expressing cells	RGC response	Cortical response	Behavior	Models	Interest
			Surviving cones				
Busskamp *et al.* 2010	eNpHR1	Surviving cones	Restore ON/ON–OFF/OFF receptor field, DS2	None	Light avoidance, optomotor response	Mice (*rd1*[a])	***
			Bipolar Cells				
Lagali *et al.*	ChR2	ON-bipolar	ON only (+ lateral inhibition)	VEP	Light/dark box test, optomotor response	Mice (*rd1*)	***
Doroudchi *et al.*, 2011	ChR2	ON-bipolar	ON only	None	Visually guided behavior	Mice (*rd1*)	**
Cronin *et al.*, 2014	CatCh	ON-bipolar	ON–OFF responses	None	None	Mice (*rd1*)	*
Gaub *et al.*, 2014	LiGluR (+ MAG)[b]	RGCs or ON-bipolar	ON only	None	Light avoidance, associative learning	Mice (*rd1*) and dogs (*rod 1*[c])	**
Gaub *et al.*, 2015	Rhodopsin	ON-bipolar	ON only	VEP	Light avoidance, associative learning	Mice (*rd1*)	*
Mace *et al.*, 2015	ChR2	ON-bipolar	ON–OFF responses	ON–OFF responses	Light avoidance	Mice (*rd1*)	**
van Wyk *et al.*, 2015	Opto-mGluR6[d]	ON-bipolar	ON–OFF responses	Intrinsic imaging	OKR, light-induced locomotor modulation	Mice (*rd1*)	**
Cehajic-Kapetanovic *et al.*, 2015	Rhodopsin (human)	ON-bipolar and non-specific	ON (sluggish)	LGN multi-electrode recording	Light avoidance, acuity test	Mice (*rd1*)	**

Interest: * small; ** keen; *** great.

[a] *rd1*: mutant mouse for gene *Pde6b*.

[b] LiGluR (+ MAG): the engineered light-gated ionotropic glutamate receptor and its required ligand maleimide-azobenzene-glutamate.

[c] *rcd1*: dog mutant for gene *Pde6b*.

[d] Opto-mGluR6: fusion protein of melanopsin and the ON-bipolar-specific metabotropic glutamate receptor.

CatCh: calcium translocating channel (Kleinlogel *et al.*, 2011); ChR2: channelrhodopsin-2; DS: direction selectivity; eNpHR: enhanced *Natronomonas pharaonis* halorhodopsin; LGN: lateral geniculate nucleus; OKR: opto-kinetic reflex; RGC: retinal ganglion cell; VEP: visually evoked potential albeit at a reduced level. This strategy takes advantage of the tremendous natural retinal filtering system in order to increase its effective range. The expression and functions of microbial opsins were also demonstrated in human photoreceptors on cultured retinal explants from human postmortem retinas.

Dormant cones in blind patients have also been shown to be present in the macular area (Busskamp *et al.*, 2010). Delivery of AAV vectors will most likely require sub-retinal injections, which is a more delicate procedure than intravitreal injection. Additionally, this delivery method will lead to the formation of a bolus of vector that will not diffuse over the entire retina, limiting the extent of the treated area. It is possible to counteract this issue by targeting the injection to the foveal region, which is the zone of higher cone density and therefore of higher visual spatial resolution.

24.4 Concluding Remarks

The use of genetically encoded, light-activated proteins has revolutionized all fields of neuroscience by enabling the control of the excitation state of neurons with high spatial and temporal precision. These new, powerful methods have had a major impact on every research field and the optogenetic toolbox is still evolving as it becomes more highly optimized and diversified.

There are still major challenges to overcome before it will be possible to use optogenetics in the treatment of retinal cell diseases and vision restoration. First and foremost, safety concerns need to be reduced by thorough studies in non-human primates. These preclinical tests will need to carefully assess the immunogenicity of the gene candidates, the correct trafficking of the proteins and the stability of expression, as well as the biocompatibility of the light stimulation if a device is required.

The first human trial will become a reality once a satisfying strategy emerges, but not before. This trial will represent an important milestone, not only for optogenetics, but also for medicine in general. Contrary to other gene therapy strategies, this will not be a "gene surgery" aiming to repair a mutation or a gene defect, but the introduction of an engineered protein conferring new properties to a cell population.

The use of optogenetic tools in order to restore the sight of blind patients has an extremely promising future. The growing number of scientific groups worldwide working in this specific field attest to the feasibility of the approach.

REFERENCES

Aït-Ali, N., Fridlich, R., Millet-Puel, G. *et al.* (2015). Rod-derived cone viability factor promotes cone survival by stimulating aerobic glycolysis. *Cell*, **161**, 817–832.

Bartsch, U., Oriyakhel, W., Kenna, P.F. *et al.* (2008). Retinal cells integrate into the outer nuclear layer and differentiate into mature photoreceptors after subretinal transplantation into adult mice. *Exp. Eye Res.*, **86**, 691–700.

Bertschinger, D.R., Beknazar, E., Simonutti, M. *et al.* (2008). A review of *in vivo* animal studies in retinal prosthesis research. *Graefes Arch. Clin. Exp. Ophthalmol.*, **246**, 1505–1517.

Bi, A., Cui, J., Ma, Y.-P. *et al.* (2006). Ectopic expression of a microbial-type rhodopsin restores visual responses in mice with photoreceptor degeneration. *Neuron*, **50**, 23–33.

Busskamp, V., Duebel, J., Balya, D. *et al.* (2010). Genetic reactivation of cone photoreceptors restores visual responses in retinitis pigmentosa. *Science*, **329**, 413–417.

Caporale, N., Kolstad, K.D., Lee, T. *et al.* (2011). LiGluR restores visual responses in rodent models of inherited blindness. *Mol. Ther. J. Am. Soc. Gene Ther.*, **19**, 1212–1219.

Cehajic-Kapetanovic, J., Eleftheriou, C., Allen, A.E. *et al.* (2015). Restoration of vision with ectopic expression of human rod opsin. *Curr. Biol.*, **25**, 2111–2122.

Cha, K., Horch, K.W., Normann, R.A., (1992a). Mobility performance with a pixelized vision system. *Vision Res.*, **32**, 1367–1372.

Cha, K., Horch, K.W., Normann, R.A. *et al.* (1992b). Reading speed with a pixelized vision system. *J. Opt. Soc. Am. A.*, **9**, 673–677.

Chuang, A.T., Margo, C.E., Greenberg, P.B. (2014). Retinal implants: a systematic review. *Br. J. Ophthalmol.*, **98**, 852–856.

Cideciyan, A.V., Hauswirth, W.W., Aleman, T.S. *et al.* (2009). Human *RPE65* gene therapy for Leber congenital amaurosis: persistence of early visual improvements and safety at 1 year. *Hum. Gene Ther.*, **20**, 999–1004.

Cronin, T., Vandenberghe, L.H., Hantz, P. *et al.* (2014). Efficient transduction and optogenetic stimulation of retinal bipolar cells by a synthetic adeno-associated virus capsid and promoter. *EMBO Mol. Med.* 6, 1175–1190.

da Cruz, L., Coley, B.F., Dorn, J. *et al.* (2013). The Argus II epiretinal prosthesis system allows letter and word reading and long-term function in patients with profound vision loss. *Br. J. Ophthalmol.*, **97**, 632–636.

Daiger, S.P., Sullivan, L.S., Bowne, S.J., (2013). Genes and mutations causing retinitis pigmentosa. *Clin. Genet.* **84**, 132–141.

Doroudchi, M.M., Greenberg, K.P., Liu, J. *et al.* (2011). Virally delivered channelrhodopsin-2 safely and effectively restores visual function in multiple mouse models of blindness. *Mol. Ther.*, **19**, 1220–1229.

Ernst, O.P., Lodowski, D.T., Elstner, M. *et al.* (2014). Microbial and animal rhodopsins: structures, functions, and molecular mechanisms. *Chem. Rev.*, **114**, 126–163.

Gaub, B.M., Berry, M.H., Holt, A.E. *et al.* (2015). Optogenetic vision restoration using rhodopsin for enhanced sensitivity. *Mol. Ther.*, **23**, 1562–1571.

Gaub, B.M., Berry, M.H., Holt, A.E. *et al.* (2014). Restoration of visual function by expression of a light-gated mammalian ion channel in retinal ganglion cells or ON-bipolar cells. *Proc. Natl. Acad. Sci. U. S. A.*, **111**, E5574–E5583.

GenVec, (2011). Study of AdGVPEDF.11D in neovascular age-related macular degeneration (AMD). URL https://clinicaltrials.gov/ct2/show/NCT00109499

Greenberg, K.P., Pham, A., Weblin, F.S. (2011). Differential targeting of optical neuromodulators to ganglion cell soma and dendrites allows dynamic control of center-surround antagonism. *Neuron*, 69, 713–720.

Haupts, U., Tittor, J., Bamberg, E. *et al.* (1997). General concept for ion translocation by halobacterial retinal proteins: the isomerization/switch/transfer (IST) model. *Biochemistry*, **36**, 2–7.

Holz, F.G., Schmitz-Valckenberg, S., Fleckenstein, M., (2014). Recent developments in the treatment of age-related macular degeneration. *J. Clin. Invest.*, **124**, 1430–1438.

Humayun, M.S., Dorn, J.D., da Cruz, L. *et al.* (2012). Interim results from the international trial of Second Sight's visual prosthesis. *Ophthalmology*, **119**, 779–788.

International Commission on Non-Ionizing Radiation Protection, (2013). ICNIRP guidelines on limits of exposure to incoherent visible and infrared radiation. *Health Phys.* 105, 74–96.

Ivanova, E., Hwang, G.-S., Pan, Z.-H. *et al.* (2010). Evaluation of AAV-mediated expression of Chop2-GFP in the marmoset retina. *Invest. Ophthalmol. Vis. Sci.*, **51**, 5288–5296.

Ivanova, E., Pan, Z.-H., (2009). Evaluation of the adeno-associated virus mediated long-term expression of channelrhodopsin-2 in the mouse retina. *Mol. Vis.*, **15**, 1680–1689.

Jones, B., Pfeiffer, R., Ferrell, W. *et al.* (2016). Retinal remodeling in human retinitis pigmentosa. *Exp. Eye Res.*, 150, 149–165.

Klapoetke, N.C., Murata, Y., Kim, S.S. *et al.* (2014). Independent optical excitation of distinct neural populations. *Nat. Methods*, **11**, 338–346.

Kleinlogel, S., Feldbauer, K., Dempski, R.E. *et al.* (2011). Ultra light-sensitive and fast neuronal activation with the Ca^{2+}-permeable channelrhodopsin CatCh. *Nat. Neurosci.*, **14**, 513–518.

Lagali, P.S., Balya, D., Awatramani, G.B. *et al.* (2008). Light-activated channels targeted to ON bipolar cells restore visual function in retinal degeneration. *Nat. Neurosci.*, **11**, 667–675.

Lamba, D., Gust, J., Reh, T., (2009). Transplantation of human embryonic stem cells derived photoreceptors restores some visual function in Crx deficient mice. *Cell Stem Cell*, **4**, 73–79.

Léveillard, T., Mohand-Saïd, S., Lorentz, O. *et al.* (2004). Identification and characterization of rod-derived cone viability factor. *Nat. Genet.*, **36**, 755–759.

Lim, L.S., Mitchell, P., Seddon, J.M. *et al.* (2012). Age-related macular degeneration. *Lancet Lond. Engl.*, **379**, 1728–1738.

Lin, B., Koizumi, A., Tanaka, N. *et al.* (2008). Restoration of visual function in retinal degeneration mice by ectopic expression of melanopsin. *Proc. Natl. Acad. Sci. U. S. A.*, **105**, 16009–16014.

Lin, J.Y., Knutsen, P.M., Muller, A. *et al.* (2013). ReaChR: a red-shifted variant of channelrhodopsin enables deep transcranial optogenetic excitation. *Nat. Neurosci.*, **16**, 1499–1508.

Lorach, H., Benosman, R., Marre, O. *et al.* (2012). Artificial retina: the multichannel processing of the mammalian retina achieved with a neuromorphic asynchronous light acquisition device. *J. Neural Eng.*, 9, 066004.

Lorach, H., Goetz, G., Smith, R. *et al.* (2015). Photovoltaic restoration of sight with high visual acuity. *Nat. Med.*, **21**, 476–482.

Macé, E., Caplette, R., Marre, O. *et al.* (2015). Targeting channelrhodopsin-2 to ON-bipolar cells with vitreally administered AAV restores ON and OFF visual responses in blind mice. *Mol. Ther.*, **23**, 7–16.

MacLaren, R.E., Groppe, M., Barnard, A.R. *et al.* (2014). Retinal gene therapy in patients with choroideremia: initial findings from a Phase 1/2 clinical trial. *Lancet*, **383**, 1129–1137.

MacLaren, R.E., Pearson, R.A., MacNeil, A. *et al.* (2006). Retinal repair by transplantation of photoreceptor precursors. *Nature*, **444**, 203–207.

Masu, M., Iwakabe, H., Tagawa, Y. *et al.* (1995). Specific deficit of the ON response in visual transmission by targeted disruption of the mGIuR6 gene. *Cell*, **80**, 757–765.

Polosukhina, A., Litt, J., Tochitsky, I. *et al.* (2012). Photochemical restoration of visual responses in blind mice. *Neuron*, **75**, 271–282.

RetroSense Therapeutics, (2016). RST-001 Phase I/II trial for retinitis pigmentosa. URL https://clinicaltrials.gov/ct2/show/NCT02556736?term=optogenetic&rank=1

Sanofi, (2015). Phase I/IIa study of SAR422459 in patients with Stargardt's macular degeneration. URL https://clinicaltrials.gov/ct2/show/NCT01367444

Sanofi, (2016). Study of UshStat in patients with retinitis pigmentosa associated with usher syndrome type 1B. URL https://clinicaltrials.gov/ct2/show/NCT01505062

Santos, A., Humayun, M.S., de Juan, E. *et al.* (1997). Preservation of the inner retina in retinitis pigmentosa. A morphometric analysis. *Arch. Ophthalmol.*, **115**, 511–515.

Schiller, P., Sandell, J., Maunsell, J. (1986). Functions of the ON and OFF channels of the visual system. *Nature*, 322, 824–825.

Sommerhalder, J., Rappaz, B., de Haller, R. *et al.* (2004). Simulation of artificial vision: II. Eccentric reading of full-page text and the learning of this task. *Vision Res.*, **44**, 1693–1706.

Tochitsky, I., Polosukhina, A., Degtyar, V.E. *et al.* (2014). Restoring visual function to blind mice with a photoswitch that exploits electrophysiological remodeling of retinal ganglion cells. *Neuron*, **81**, 800–813.

Tomita, H., Sugano, E., Isago, H. *et al.* (2010). Channelrhodopsin-2 gene transduced into retinal ganglion cells restores functional vision in genetically blind rats. *Exp. Eye Res.*, **90**, 429–436.

Tomita, H., Sugano, E., Murayama, N. *et al.* (2014). Restoration of the majority of the visual spectrum by using modified *Volvox* channelrhodopsin-1. *Mol. Ther.*, 22, 1434–1440.

Tomita, H., Sugano, E., Yawo, H. *et al.* (2007). Restoration of visual response in aged dystrophic RCS rats using AAV-mediated channelopsin-2 gene transfer. *Invest. Ophthalmol. Vis. Sci.*, **48**, 3821–3826.

Tosini, G., Ferguson, I., Tsubota, K., (2016). Effects of blue light on the circadian system and eye physiology. *Mol. Vis.*, **22**, 61–72.

van Wyk, M., Pielecka-Fortuna, J., Löwel, S. *et al.* (2015). Restoring the ON switch in blind retinas: opto-mGluR6, a next-generation, cell-tailored optogenetic tool. *PLoS Biol.*, **13**, 31002143.

Volgraf, M., Gorostiza, P., Numano, R. *et al.* (2006). Allosteric control of an ionotropic glutamate receptor with an optical switch. *Nat. Chem. Biol.*, **2**, 47–52.

Vollrath, D., Feng, W., Duncan, J.L. *et al.* (2001). Correction of the retinal dystrophy phenotype of the RCS rat by viral gene transfer of Mertk. *Proc. Natl. Acad. Sci. U. S. A.*, **98**, 12584–12589.

Wu, C., Ivanova, E., Zhang, Y. *et al.* (2013). rAAV-mediated subcellular targeting of optogenetic tools in retinal ganglion cells *in vivo*. *PLoS One*, 8, 66332.

Yawn, R., Hunter, J.B., Sweeney, A.D. *et al.* (2015). Cochlear implantation: a biomechanical prosthesis for hearing loss. *F1000Prime Rep.*, **7**, 45.

Yin, L., Greenberg, K., Hunter, J.J. *et al.* (2011). Intravitreal injection of AAV2 transduces macaque inner retina. *Invest. Ophthalmol. Vis. Sci.*, **52**, 2775–2783.

Zhang, F., Vierock, J., Yizhar, O. *et al.* (2011). The microbial opsin family of optogenetic tools. *Cell*, **147**, 1446–1457.

Zhang, Y., Ivanova, E., Bi, A. *et al.* (2009). Ectopic expression of multiple microbial rhodopsins restores ON and OFF light responses in retinas with photoreceptor degeneration. *J. Neurosci.*, **29**, 9186–9196.

Zrenner, E., Bartz-Schmidt, K.U., Benav, H. *et al.* (2011). Subretinal electronic chips allow blind patients to read letters and combine them to words. *Proc. Biol. Sci.*, **278**, 1489–1497.

25 Holographic Optical Neural Interfacing with Retinal Neurons

Adi Schejter Bar-Noam, Inna Gefen, and Shy Shoham

25.1 Introduction

By affecting the most prominent source of sensory information in humans, blindness inflicts a disability with profound personal and societal implications. Some of the most common causes of blindness are degenerative diseases of the outer retina, such as age-related macular degeneration (AMD) and retinitis pigmentosa (RP). Each year, about 700 000 new patients are diagnosed with AMD, and there are a total of 1.5 million individuals affected by RP worldwide (Margalit *et al.*, 2003). Diseases of the outer retina result in photoreceptor loss, while the inner retinal neurons, and, in particular, the retinal ganglion cells (RGCs) and their optic nerve projections, largely maintain their functionality. Artificial stimulation of these relatively well-preserved nerve cells is one of the main approaches being pursued for vision restoration.

A retinal neuroprosthesis is a device that aims to provide an artificial sense of vision by translating visual scenes into appropriate spatiotemporal patterns of neuronal activity. Current retinal prostheses largely rely on microelectrode array implants (Weiland *et al.*, 2014); however, although early devices in this category are already being used to aid blind human subjects (recently gaining FDA and CE regulatory approval), it appears that their ultimate resolution may be severely limited by their current spread. For example, recent clinical studies report best-case acuities of approximately 20/1200 for the 60-electrode epi-retinal system (Humayun *et al.*, 2012) and approximately 20/550 for the 1500-electrode sub-retinal system (Zrenner *et al.*, 2011). Other concerns with this general approach include long-term interface stability between electrode arrays and the fragile retina tissue, risks associated with extended surgery and the challenges of scaling up to interfaces with thousands of channels in an attempt to approach highly functional vision.

Alternative, non-contact approaches based on direct light activation for artificially controlling neural activity in a vision prosthesis are not expected to have this limitation and will ideally allow cellular-resolution, rapid, massively parallel, light-efficient stimulation of retinal cells across macroscopic (millimeter-scale) coverage

Figure 25.1 Holographic optical neural interfacing concept depicted in an illustration of the future optical-based prosthesis: the camera video stream is processed and fed to a holographic projection system. A spatial light modulator (placed in the corners of the glasses) phase modulates a laser beam wave-front and excitation patterns are projected onto the photosensitized retina. (A black-and-white version of this figure will appear in some formats. For the color version, please refer to the plate section.)

areas. A number of studies by several research groups have explored the fundamental feasibility of using ChR2 and other optogenetic probes in an optical retinal prosthesis, clearly suggesting that this technology may provide a viable path to vision restoration (see Roska *et al.* [2014] for a recent review). These studies primarily focused on stably transducing RGCs (Bi *et al.*, 2006; Nirenberg *et al.*, 2012), bipolar cells (Lagali *et al.*, 2008; Doroudchi *et al.*, 2011; Gaub *et al.*, 2015; Mace *et al.*, 2015; van Wyk *et al.*, 2015) or surviving photoreceptors (Busskamp *et al.*, 2010) in blind rodent retinas using a recombinant adeno-associated virus (rAAV) vector and measured responses to non-specific light flashes using electrophysiological measures such as visual evoked potentials (Bi *et al.*, 2006; Lagali *et al.*, 2008) or by looking at light-seeking behaviors (Lagali *et al.*, 2008; Doroudchi *et al.*, 2011; Gaub *et al.*, 2015; Mace *et al.*, 2015; van Wyk *et al.*, 2015). Combining optogenetics with advanced light projection technologies can potentially allow a relatively direct route toward non-contact restoration of high-acuity vision (Figure 25.1).

In this chapter, we describe how spatial light modulation technology can be harnessed as a powerful tool for large-scale, cellular-resolution, massively parallel artificial control of neuron population activity in isolated blinded retinas through holographic optical neural interfaces (HONIs). In addition, we discuss the challenges and recent progress gained in testing this approach in the intact retina.

25.2 HONIs: Background

Controlled projection of light patterns for scan-less, parallel optical neuronal stimulation can be obtained using optical devices known as spatial light modulators (SLMs). SLMs generally rely on one of two modulation approaches: intensity (amplitude) or phase modulations of the beam wave-front. Amplitude modulation SLMs (e.g. digital mirror devices [DMDs] [Farah *et al.*, 2007; Wang *et al.*, 2007] or micro-LEDs [Grossman *et al.*, 2010]) shape the intensity of light by either blocking the light for the "off" pixels or directing light onto the sample for the "on" pixels. DMD-type devices provide high temporal and lateral resolution, but are very inefficient for optogenetic applications since the "off" regions block the major portion of the incoming light. In contrast, phase SLMs modulate the phase of the incoming beam, thereby diffractively creating predetermined light patterns, termed computer generated holograms, where the light is diffractively divided between the "on" spots in the hologram, resulting in minimal light loss. Holographic illumination has many advantages over other light patterning techniques in addition to its high efficiency: it enables flexible generation of arbitrary illumination patterns and can provide a relatively large excitation field of view (FOV) and a dynamic correction of the projected pattern. A prominent advantage is its unique ability to produce three-dimensional light patterns while other methods are restricted to two dimensions (Golan *et al.*, 2009; Anselmi *et al.*, 2011; Yang *et al.*, 2011). Moreover, due to their high efficiency and intensity, holographically generated light patterns can excite neurons using a number of different mechanisms, including single- or multi-photon optogenetic or photothermal (Farah *et al.*, 2013) excitation modes. Holographic light patterns can be coupled to the target neurons using either free space projection or through fiber bundles (Farah *et al.*, 2015) or other waveguides when the target tissue is not directly accessible.

25.3 Methods and Results

25.3.1 HONI Projection Systems

In our work, we developed and applied systems for rapid holographic photostimulation for *in vitro* and *in vivo* studies in ChR2-expressing retinas (Golan *et al.*, 2009; Reutsky-Gefen *et al.*, 2013). These systems project high-intensity patterns with high spatial and temporal resolution (2000 frames s^{-1}) using a rapid ferroelectric liquid-crystal SLM (SXGA-R3, ForthDD, UK) illuminated by a diode-pumped solid-state laser (473 nm; VA-I-N-473, 50 mW, Viasho, China). In addition to enabling rapid pattern projection in the order of milliseconds, the high refresh rate of the projection system can be used for averaging out holographic noise from a single hologram, for multiplexing large numbers of different holograms (Golan *et al.*, 2009) and/or to project multiple different colors for the excitation of multiple optogenetic probes (Golan *et al.*, 2009).

In these systems, polarized laser light from a blue laser is expanded using a beam expander and then phase-modulated by the SLM. The phase hologram, which is projected on the SLM and controlled by a computer, is imaged onto the back aperture of an imaging lens following demagnification in order to optimally fill the lens. This results in the projection of light patterns onto retinas, whose RGCs express ChR2, in the lens focal plane.

In computer-generated holography, since the total power delivered to the sample is distributed solely between the projected points, it is possible to simultaneously illuminate hundreds to several thousands of points at effective ChR2 stimulation intensities with as low as approximately 1 mW of total projected power.

25.3.2 HONI of Isolated Retinas from Optogenetically Transduced Blind Mice

Due to its high efficiency advantages, HONIs have the potential to control spatiotemporal responses across many RGCs with single-cell resolution. In order to investigate the characteristics of optogenetically mediated responses, we projected holographic patterns onto isolated retinas from two different transgenic mouse strains whose RGCs express ChR2. The first strain was a sighted transgenic mouse line expressing ChR2-eYFP in their RGCs under a Thy1.2 promoter; because it had natural vision, it was used for comparing photoreceptor-mediated responses with optogenetic-only responses following a "blinding" procedure using pharmacological agents. The second strain was made up of crossbred blind mice expressing both ChR2-eYFP and a mutant gene *Pde6brd1* that leads to congenital outer retinal degeneration. These mice provide a direct animal model of optogenetic vision restoration and were used for studying the spatiotemporal characteristics of the RGC responses following direct optical stimulation. In order to record the responses from the RGCs, the isolated retinas were laid flat on a transparent multi-electrode array (MEA) (100/10-ITO, Multi-Channel Systems MCS GmbH, Germany) with the RGCs facing downwards so that they were in close contact with the recording electrodes. The MEA, with 60 10μm diameter electrodes spaced by 100 μm, recorded evoked activity from the RGCs at a 10KHz sampling rate.

By projecting patterns with subcellular dimensions onto the ChR2-expressing RGCs using ×4 and ×10 objective lenses (Nikon, Japan), it was possible to map the effective "stimulation fields" (i.e. the coverage area of the RGC where light elicits a response). The resulting stimulation fields have both the size and shape of single identified ChR2-expressing RGC somas (Figure 25.2), demonstrating the ability to achieve single-cell control of RGCs.

In the next step, we partially or fully covered the cell somas with holographic multi-pixel patches of light (Figure 25.3A) and explored how coverage affects stimulation efficiency. The results of these experiments effectively showed that it is possible to efficiently control millimeter-scale (large) populations of RGCs in blind retinas with single-cell resolution and millisecond temporal precision (Figure 25.3B and C).

Furthermore, we demonstrated the ability to emulate natural vision using optogenetically mediated stimulation of RGCs (Figure 25.3D): first, we projected low-intensity movies on a sighted ChR2-expressing retina and recorded the response to natural stimulation mediated by the photoreceptors. Next, we artificially blinded the retina using pharmacological blockers and performed stimulation field mapping in order to determine the spot dimensions that are required for eliciting an optogenetically mediated response. Finally, we replicated the photoreceptor-mediated responses from the initial movie through

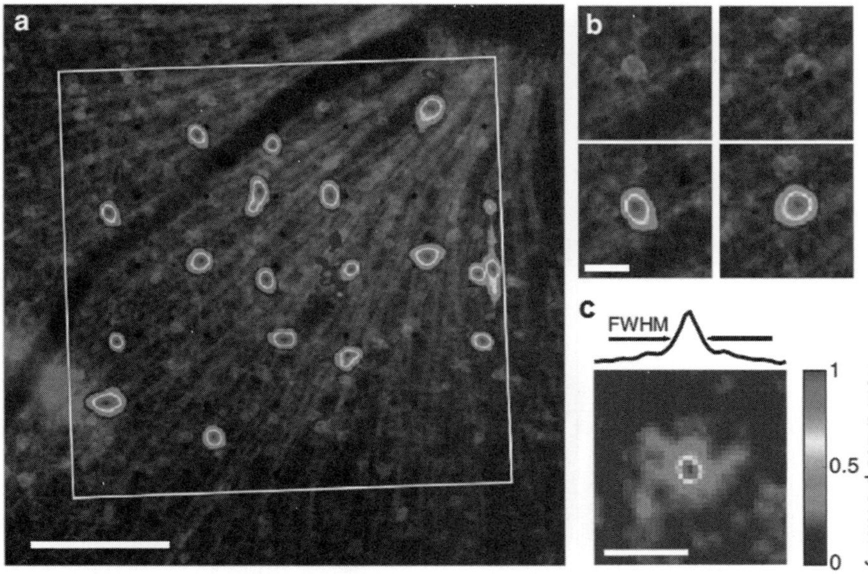

Figure 25.2 Single-cell resolution control of ChR2-expressing RGCs. (A) Superimposed images of a retina expressing ChR2-eYFP in the RGCs flattened on an MEA (black dots) and representative stimulation fields. Scale bar = 200 μm. (B) The stimulation fields match the underlying visualized RGCs. Scale bar = 20 μm. (C) Mean spatial distribution of the stimulation fields calculated for 202 units from 11 retinas. Scale bar = 50 μm. Figure adapted from Reutsky-Gefen *et al.* (2013). (A black-and-white version of this figure will appear in some formats. For the color version, please refer to the plate section.)

optogenetic stimulation of the RGCs, validating a first step toward cellular-scale vision restoration using optogenetic stimulation.

25.3.3 Toward *In Vivo* HONIs

Our *in vitro* results show that the HONI strategy has the potential to restore vision to nearly natural levels through a highly controlled and cell-specific concerted activation of RGCs. Nevertheless, progress toward translation requires extensive *in vivo* tests and validation, as was previously accomplished for earlier strategies of artificial retinal stimulation (Bertschinger *et al.*, 2008). In these studies, visual responses were obtained on a large- or intermediate-scale using electrical recordings and intrinsic optical imaging (Wilms *et al.*, 2003; Walter *et al.*, 2005; Mandel *et al.*, 2013; van Wyk *et al.*, 2015); however, these methods do not enable us to truly investigate the spatiotemporal characteristics of elicited responses. Techniques for obtaining cellular-resolution functional information of large neuronal populations can rely on functional calcium imaging methods for monitoring neuronal activity using genetically encoded calcium indicators, such as GCaMP6 probes (Chen *et al.*, 2013), as demonstrated in experiments carried out in the visual cortex (Chen *et al.*, 2013; Dana *et al.*, 2014) and isolated retinas (Borghuis *et al.*, 2011). Functional calcium imaging relies on the fluorescent imaging of transient changes in Ca^{2+} concentrations resulting from neural activity.

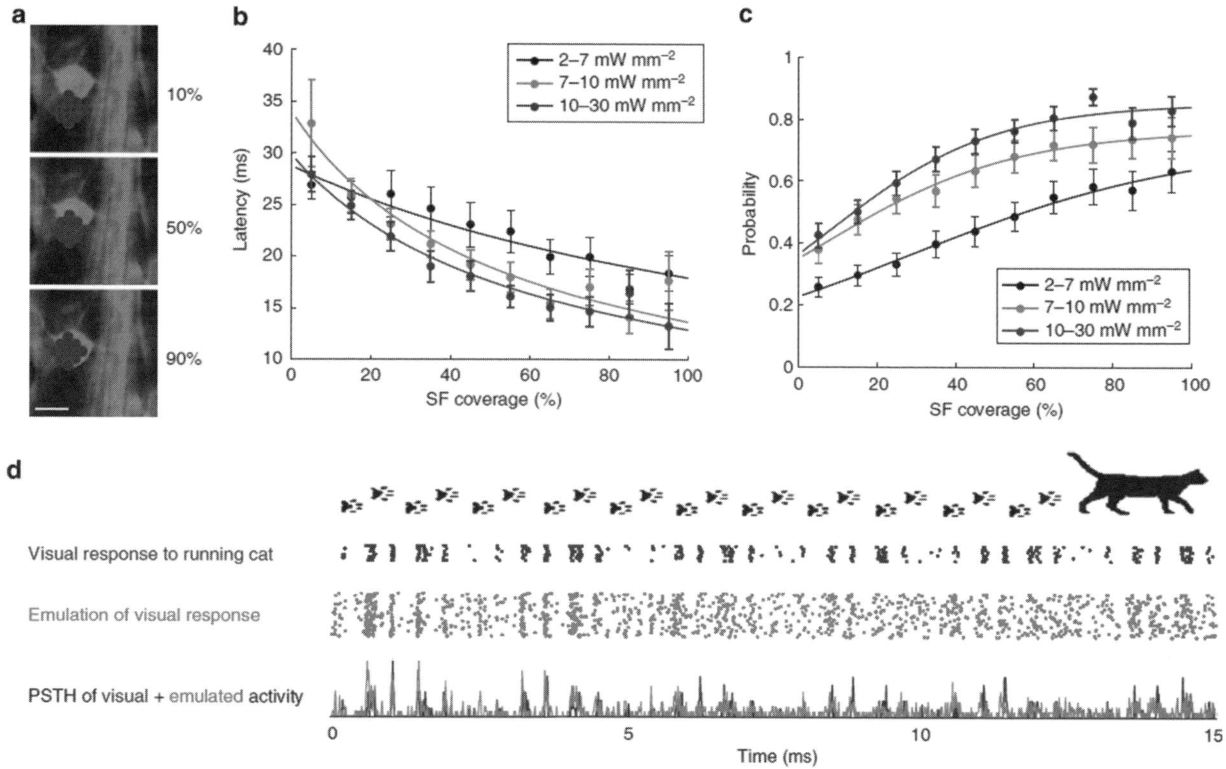

Figure 25.3 HONI performance and emulation. (A) Illustration of different cell coverages using a projection of spot patches (after smoothing in order to obtain a speckle-free spot). Scale bar = 10 μm. (B and C) Spike latency and firing probability as a function of the cell coverage and intensity of each patch. (D) Raster graph of optogenetically mediated RGC activity in blind ChR2-expressing retinas (red) following the projection of holographic patterns for reproducing photoreceptor-mediated visual responses in a sighted retina (blue) to a running cat. Bottom: Overlaid PSTH graphs illustrating the resemblance of the responses. Figure adapted from Reutsky-Gefen *et al.* (2013). (A black-and-white version of this figure will appear in some formats. For the color version, please refer to the plate section.)

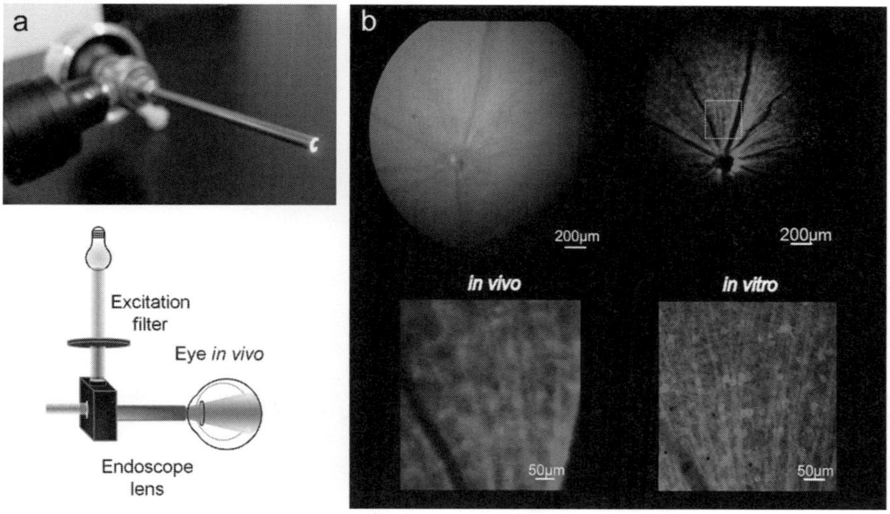

Figure 25.4 *In vivo* fundoscope system. (A) Photograph of the endoscope and custom excitation filter add-on (top). Scheme of excitation light illuminating the retina in a crescent shape and emitted fluorescence being collected by the endoscope (bottom). (B) Bright field (top left) and fluorescent fundus image with an overlay of holographic pattern projection (blue spots) on the retina (top right) of ChR2-expressing RGCs in a B6.Cg-Tg(Thy1-COP4/eYFP) mouse. (Bottom panels) Comparison between the magnification of an *in vivo* image and an image of a similar retina acquired *in vitro*. Figure adapted from Schejter *et al.* (2012). (A black-and-white version of this figure will appear in some formats. For the color version, please refer to the plate section.)

In order to record RGC responses *in vivo*, it is first necessary to be able to visualize these cells at a cellular resolution. Our initial attempt to address this challenge was based on adapting a fundoscope system that was first described by Paques *et al.* (2007), demonstrating that it enables high-resolution *in vivo* fundus imaging of rodent retinas expressing optogenetic probes (Schejter *et al.*, 2012) for both the monitoring and stimulation of neuronal activity (Figure 25.4). This relatively simple setup consists of a rigid endoscope connected to a custom-made adapter for the introduction of light and spectral filtering (Figure 25.4A) and a camera. High-quality fundus images and movies (with a resolution of <20 μm) can be acquired even with a consumer-grade SLR camera (Figure 25.4B).

In order to artificially stimulate ChR2-expressing RGCs, we projected high-intensity light patterns onto RGCs by integrating a precise spatiotemporal holographic photostimulation system into the imaging setup, employing the design that was used for the *in vitro* experiments. Using this combined system, it is possible to project cellular-resolution patterns, estimated at <40 μm FWHM (in a non-optimized system) (Figure 25.4B, top right) (Reutsky-Gefen *et al.*, 2013) and record the responses of neurons in the retina (using stimulus-triggered imaging and specific dichroic filters) (Schejter *et al.*, 2012). This constitutes a first proof of concept of our ability to translate the results obtained in isolated retinas to the *in vivo* setup.

In addition to the fundoscope-based system, which naturally provides a very high depth of field leading to a large FOV, we continue to actively explore

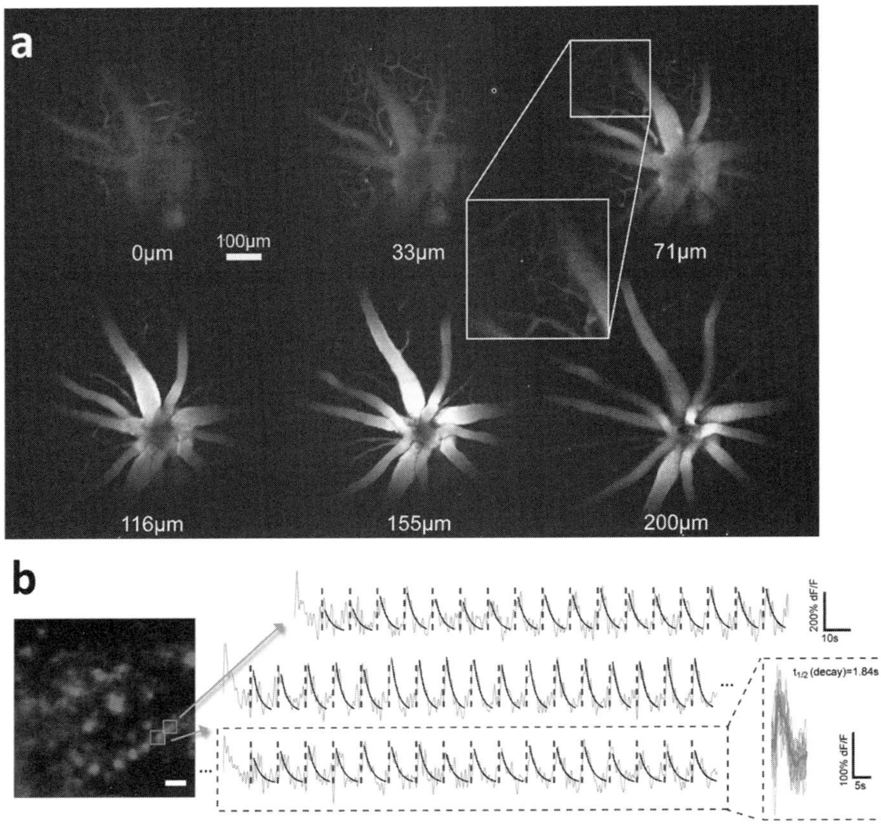

Figure 25.5 Two-photon *in vivo* mouse retinal imaging. (A) Optically sectioned, fluorescein-labeled retinal vasculature acquired at different depths using varying ETL focal lengths. The marked axial shifts were calculated from the paraxial model. (B) Calcium traces (gray) from two representative cells (squares in the average image on the left) in response to flashes of blue light (blue dashed lines) at 10-s intervals acquired at 6.7 fps. The middle and bottom traces were acquired from the same cell and 5 minutes apart. The black lines represent the exponential decay functions that were fitted to the average response (marked by a green line on the right) and fitted to the magnitude of each single response.
Scale bar = 20 μm. Adapted from Schejter Bar-Noam *et al.* (2016). (A black-and-white version of this figure will appear in some formats. For the color version, please refer to the plate section.)

additional optical elements and strategies for optimally approaching this research question. The use of multiphoton excitation for imaging the retinal physiology has a major advantage in our experiments – since it uses near-infrared light, it can enable imaging of the responses in sighted retinas and has the potential to combine functional imaging with single-photon optogenetic stimulation without an inherent spectral overlap (bidirectional optogenetic stimulation and imaging remains a major challenge) (Emiliani *et al.*, 2015). Moreover, two-photon excitation provides intrinsic optical sectioning, enabling us to image specific retinal layers without the use of additional spatial filtering (Gualda *et al.*, 2010).

With these advantages in mind, we developed in recent work a new solution for obtaining two-photon mouse retinal imaging. Our solution relies on the preselection of specific objective lenses and a relatively simple insertion of an offset lens and an electrically tunable lens into the imaging path of the microscope (Schejter Bar-Noam *et al.*, 2016). This breakthrough – the first such demonstration not requiring the use of a highly complex adaptive optics system (Sharma *et al.*, 2013; Palczewska *et al.*, 2014) – relied largely on the development of a detailed approximate model of the compound optical system of the microscope and mouse eye, which neatly predicts its experimental behavior. Our analysis suggests that the key hurdle regarding two-photon retinal imaging without adaptive optics is likely to be the paired selection of a combined offset lens with a large focal distance objective that can synergistically leverage the strong refraction power of the mouse crystalline lens in order to achieve a reliable focus on the retina. An additional model based on the Zemax ray tracing software was used to validate the theoretical results and provide an estimate of the attainable imaging resolution (>1 μm). Using this system, we were able to acquire high-quality two-photon fluorescence images of the retinal vasculature and optogenetic probes in the mouse retina *in vivo*, which were remotely depth-scanned using an electrically tunable lens in order to obtain a volumetric stack (Figure 25.5A). To our knowledge, the new system also provided the first demonstration of two-photon *in vivo* imaging of calcium responses in mice. Figure 25.5B shows recorded calcium dynamics from multiple RGCs expressing GCaMP6 *in vivo* following short (10–20-ms), repetitive, wide-field illumination of sighted mouse retinas.

25.4 Concluding Remarks

In summary, establishing HONIs as robust tools for controlling RGCs with a cellular resolution presents new opportunities for large-scale neuronal interfacing with the retina. This projection technique is uniquely suited for optogenetic excitation since it enables parallel illumination of many spots at high intensities with millisecond-scale temporal precision. This solution provides multiple advantages for the generation of sparse spatiotemporal RGC activity patterns over other commonly used projection systems based on mini-digital light processor displays (e.g. a micromirror-based digital light processor) (Nirenberg *et al.*, 2012). The advantages of this approach include a high power efficiency when projecting sparse patterns and the inherent capability of three-dimensional projection, which is required for the simultaneous stimulation of multi-layered RGCs around the foveal pit in human retinas (two to five layers). In addition, it is possible to integrate multiple lasers with various wavelengths into these holographic projection systems for stimulating optogenetic probes with different sensitivity spectra. This can be used, for example, in order to excite favorable red-shifted alternatives to ChR2, whose excitation peak unfortunately coincides with wavelengths that are quite effectively absorbed in humans by pigments of the macula lutea, or alternatively to enable multicolor excitation of multiple probes expressed in parallel.

Our *in vitro* results indicate that illuminating RGCs with patches of light that match the soma dimensions (coverage >70%) produces effective excitation of ChR2-expressing cells (90% spiking probability with ~10ms latencies). Therefore, in order to pursue the possibility of efficiently restoring high-acuity vision *in vivo*, we have begun exploring cellular-resolved *in vivo* retinal optogenetic targeting; combining this stimulation with functional retinal imaging provides important validation and the ability to optimize excitation patterns. Future research will be based on the integration of the retinal holographic stimulation and imaging capabilities, together with functional imaging of the primary visual cortex, in order to investigate the spatiotemporal characteristics of the visual response in downstream areas. Furthermore, utilizing HONI capabilities for researching neuronal network activity and connectivity offers a wide range of applications for shedding new light on fundamental problems in neuroscience research, as well as in medical applications.

REFERENCES

Anselmi, F., Ventalon, C., Begue, A., *et al.* (2011). "Three-dimensional imaging and photostimulation by remote-focusing and holographic light patterning." *Proc Natl Acad Sci U S A* **108**: 19504–19509.

Bertschinger, D. R., Beknazar, E., Simonutti, M., *et al.* (2008). "A review of *in vivo* animal studies in retinal prosthesis research." *Graefes Arch Clin Exp Ophthalmol* **246**: 1505–1517.

Bi, A., Cui, J., Ma, Y. P., *et al.* (2006). "Ectopic expression of a microbial-type rhodopsin restores visual responses in mice with photoreceptor degeneration." *Neuron* **50**: 23–33.

Borghuis, B. G., Tian, L., Xu, Y., *et al.* (2011). "Imaging light responses of targeted neuron populations in the rodent retina." *J Neurosci* **31**: 2855–2867.

Busskamp, V., Duebel, J., Balya, D., *et al.* (2010). "Genetic reactivation of cone photoreceptors restores visual responses in retinitis pigmentosa." *Science* **329**: 413–417.

Chen, T. W., Wardill, T. J., Sun, Y., *et al.* (2013). "Ultrasensitive fluorescent proteins for imaging neuronal activity." *Nature* **499**: 295–300.

Dana, H., Chen, T. W., Hu, A., *et al.* (2014). "Thy1-GCaMP6 transgenic mice for neuronal population imaging *in vivo*." *PLoS One* **9**: e108697.

Doroudchi, M. M., Greenberg, K. P., Liu, J., *et al.* (2011). "Virally delivered channelrhodopsin-2 safely and effectively restores visual function in multiple mouse models of blindness." *Mol Ther* **19**: 1220–1229.

Emiliani, V., Cohen, A. E., Deisseroth, K., *et al.* (2015). "All-optical interrogation of neural circuits." *J Neurosci* **35**: 13917–13926.

Farah, N., Levinsky, A., Brosh, I., *et al.* (2015). "Holographic fiber bundle system for patterned optogenetic activation of large-scale neuronal networks." *Neurophotonics* **2**: 045002.

Farah, N., Reutsky, I. and Shoham, S. (2007). "Patterned optical activation of retinal ganglion cells." *Conf Proc IEEE Eng Med Biol Soc* **2007**: 6368–6370.

Farah, N., Zoubi, A., Matar, S., *et al.* (2013). "Holographically patterned activation using photo-absorber induced neural-thermal stimulation." *J Neural Eng* **10**: 056004.

Gaub, B. M., Berry, M. H., Holt, A. E., *et al.* (2015). "Optogenetic vision restoration using rhodopsin for enhanced sensitivity." *Mol Ther* **23**: 1562–1571.

Golan, L., Reutsky, I., Farah, N., *et al.* (2009). "Design and characteristics of holographic neural photo-stimulation systems." *J Neural Eng* **6**: 066004.

Golan, L. and Shoham, S. (2009). "Speckle elimination using shift-averaging in high-rate holographic projection." *Opt Express* **17**: 1330–1339.

Grossman, N., Poher, V., Grubb, M. S., *et al.* (2010). "Multi-site optical excitation using ChR2 and micro-LED array." *J Neural Eng* **7**: 16004.

Gualda, E. J., Bueno, J. M. and Artal, P. (2010). "Wavefront optimized nonlinear microscopy of *ex vivo* human retinas." *J. Biomed. Opt.* **15**: 026007.

Humayun, M. S., Dorn, J. D., da Cruz, L., *et al.* (2012). "Interim results from the international trial of Second Sight's visual prosthesis." *Ophthalmology* **119**: 779–788.

Lagali, P. S., Balya, D., Awatramani, G. B., *et al.* (2008). "Light-activated channels targeted to ON bipolar cells restore visual function in retinal degeneration." *Nat Neurosci* **11**: 667–675.

Mace, E., Caplette, R., Marre, O., *et al.* (2015). "Targeting channelrhodopsin-2 to ON-bipolar cells with vitreally administered AAV Restores ON and OFF visual responses in blind mice." *Mol Ther* **23**: 7–16.

Mandel, Y., Goetz, G., Lavinsky, D., *et al.* (2013). "Cortical responses elicited by photovoltaic subretinal prostheses exhibit similarities to visually evoked potentials." *Nat Commun* **4**: 1980.

Margalit, E., Weiland, J. D., De Juan, E., *et al.* (2003). *Chapter 7.5 in Neuroprosthetics: Theory and Practice. World Scientific Publishers, New Jersey.*

Nirenberg, S. and Pandarinath, C. (2012). "Retinal prosthetic strategy with the capacity to restore normal vision." *Proc Natl Acad Sci U S A* **109**: 15012–15017.

Palczewska, G., Dong, Z., Golczak, M., *et al.* (2014). "Noninvasive two-photon microscopy imaging of mouse retina and retinal pigment epithelium through the pupil of the eye." *Nat Med* **20**: 785–789.

Paques, M., Guyomard, J. L., Simonutti, M., *et al.* (2007). "Panretinal, high-resolution color photography of the mouse fundus." *Invest Ophthalmol Vis Sci* **48**: 2769–2774.

Reutsky-Gefen, I., Golan, L., Farah, N., *et al.* (2013). "Holographic optogenetic stimulation of patterned neuronal activity for vision restoration." *Nat Commun* **4**: 1509.

Roska, B. and Pepperberg, D. (2014). "Restoring vision to the blind: optogenetics." *Transl Vis Sci Technol* **3**: 4.

Schejter, A., Tsur, L., Farah, N., *et al.* (2012). "Cellular resolution panretinal imaging of optogenetic probes using a simple funduscope." *Trans Vis Sci Tech* **1**: 4.

Schejter Bar-Noam, A., Farah, N. and Shoham, S. (2016). "Correction-free remotely scanned two-photon *in vivo* mouse retinal imaging." *Light: Science & Applications* **5**: e16007.

Sharma, R., Yin, L., Geng, Y., *et al.* (2013). "*In vivo* two-photon imaging of the mouse retina." *Biomed Opt Express* **4**: 1285–1293.

van Wyk, M., Pielecka-Fortuna, J., Lowel, S., *et al.* (2015). "Restoring the ON switch in blind retinas: opto-mGluR6, a next-generation, cell-tailored optogenetic tool." *PLoS Biol* **13**: e1002143.

Walter, P., Kisvarday, Z. F., Gortz, M., *et al.* (2005). "Cortical activation via an implanted wireless retinal prosthesis." *Invest Ophthalmol Vis Sci* **46**: 1780–1785.

Wang, S., Szobota, S., Wang, Y., *et al.* (2007). "All optical interface for parallel, remote, and spatiotemporal control of neuronal activity." *Nano Lett* **7**: 3859–3863.

Weiland, J. D. and Humayun, M. (2014). "Retinal prosthesis." *IEEE Trans Biomed Eng* **61**: 1412–1424.

Wilms, M., Eger, M., Schanze, T., *et al.* (2003). "Visual resolution with epi-retinal electrical stimulation estimated from activation profiles in cat visual cortex." *Vis Neurosci* **20**: 543–555.

Yang, S., Papagiakoumou, E., Guillon, M., *et al.* (2011). "Three-dimensional holographic photostimulation of the dendritic arbor." *J Neural Eng* **8**: 046002.

Zrenner, E., Bartz-Schmidt, K. U., Benav, H., *et al.* (2011). "Subretinal electronic chips allow blind patients to read letters and combine them to words." *Proc Biol Sci* **278**: 1489–1497.

26 Strategies for Restoring Vision by Transducing a Channelrhodopsin Gene into Retinal Ganglion Cells

Hiroshi Tomita and Eriko Sugano

26.1 Background

Globally, cataracts are the principal cause of vision loss. Currently, the diffusion of intraocular lenses and the ability to treat cataracts through surgery have placed age-related macular degeneration, glaucoma and diabetic retinopathy as the leading causes of blindness. Research has been conducted on various pharmaceutical preparations, and although a complete cure for these conditions is yet to be developed, drugs that can delay their progression have been developed and implemented. Meanwhile, another disease among the leading causes of vision loss is retinitis pigmentosa, a hereditary condition for which an effective treatment method has not yet been developed, despite the identification of several causative genes.

Retinitis pigmentosa occurs due to mutations in genes that encode proteins related to phototransduction pathways in retinal pigment epithelial cells or photoreceptor cells. These gene mutations cause degeneration of photoreceptor cells, resulting in a loss of photoreceptive ability and leading to blindness. Even after a patient has become blind, the retinal neurons other than the photoreceptor cells are relatively well preserved (Santos *et al.*, 1997), and techniques to use these surviving neurons in order to restore visual function are being investigated. The typical method is retinal prosthesis, a system in which images of the external environment are captured with a camera and electrical stimulation corresponding to the image data is applied to the surviving retinal cells by using an electrode implanted in the eye so as to provide a simulated optic sense. Clinical trials of retinal prostheses are already underway (Rizzo *et al.*, 2014), and while high-resolution imagery is not expected, results indicate that the technique may be able to act as a substitute for visual function. Further, as these clinical trials of retinal prostheses have clarified that the surviving retinal cells continue to function even after the degeneration of photoreceptor cells, it is believed that restoration of visual function might be possible with the right technique transmitting optical information to the remaining retinal cells.

In addition to its involvement in photoreception, channelrhodopsin-2 (ChR2) derived from the green algae *Chlamydomonas* functions as a light-gated, cation-selective channel that permits the intracellular passage of cations (Nagel *et al.*, 2003). Using these distinctive functions, neurons can be endowed with photoreceptive ability through the transduction and expression of the *ChR2* gene in the cells. In this chapter, we present a gene therapy for vision regeneration by using ChR2 and a modified *Volvox*-derived channelrhodopsin-1 (mVChR1) that was developed in our laboratory.

26.2 Phototransduction in the Retina

A variety of different neurons with differing functions exists in the retina; these work in concert to produce advanced visual function. Images (light) from the external environment pass through the cornea and the lens to focus on the retina, and then the entering light passes through the inner retina to arrive at the photoreceptor cells in the posterior region of the retina. The photoreceptor cells contain rhodopsin, a photoreceptor-specific, light-sensitive receptor protein that absorbs photons via the chromophore retinal. Rhodopsin, which is activated by retinal isomerization, activates G-protein, which causes activation of the cGMP-hydrolyzing enzyme phosphodiesterase (PDE). In darkness, cGMP-gated sodium channels are open, keeping photoreceptor cells depolarized. Activation of PDE causes a reduction in cGMP concentration, leading cGMP-gated sodium channels to close and photoreceptor cells to hyperpolarize. Thus, photoreception by photoreceptor cells happens due to a chain reaction involving photoreceptor-specific proteins. While there are reports documenting the successful provision of photoreceptive ability by expressing these photoreceptor-specific proteins in cultured cells (Zemelman *et al.*, 2002), it is difficult to translate multiple proteins in a single cell, and these results have not yet progressed to the development of a treatment strategy for the blind. Because ChR2 is a single protein that functions as a light-gated, cation-selective channel, transducing the *ChR2* gene into surviving neurons of a blind retina has the potential to restore vision. While other neurons survive even after the onset of blindness caused by photoreceptor cell degeneration, the synapse structure of bipolar cells (secondary neurons) and ganglion cells (tertiary neurons) is reconstructed after the onset of blindness into a configuration that is reported to be different from the normal synapse structure (Strettoi *et al.*, 2003). Therefore, we used the retinal ganglion cells comprising the optic nerve as target cells in order to investigate the visual function obtained by gene transduction into ganglion cells.

26.3 Adeno-associated Virus Vector-mediated Gene Therapy

26.3.1 Preparation of the Adeno-associated Virus Vector

Adeno-associated virus (AAV) was selected as the vector for gene transduction into retinal ganglion cells. The reasons for this selection derive from the avirulent nature of AAV, and include efficient gene transduction into non-dividing neurons and long-term gene expression (measured in years).

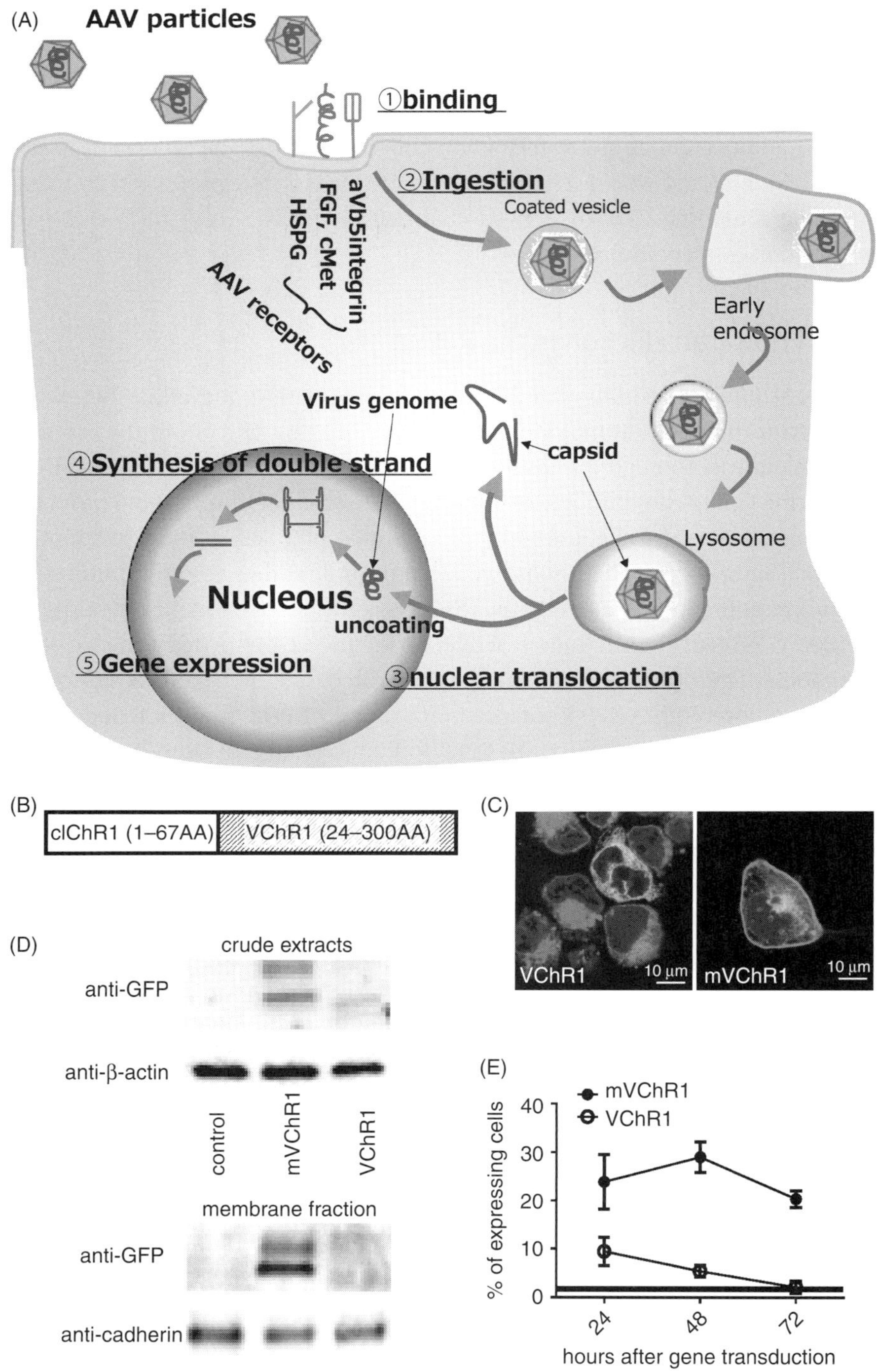

Figure 26.1 (A) Schematic depiction of AAV infection. The AAV-2-type virus binds to the heparan sulfate receptors in the host cell and is absorbed into the cell by endocytosis. The receptors involved also include αVβ5-integrin receptors and fibroblast growth factor receptors as co-receptors. The fusion of the endosomal membrane and viral membrane causes the absorbed virus to shed its capsid protein, transferring only its DNA into the nucleus. Subsequently, double-stranded DNA is formed using the polymerase in the nucleus, leading to gene expression. (B) mVChR1 structure. (C) Fluorescence microscope image of the HEK293 cells expressing VChR1 (left) and mVChR1 (right). (D) Western blot of

AAV belongs to the family Parvoviridae, which comprises the smallest linear, single-stranded DNA viruses among animal viruses. Of the ten different existing AAV serotypes (1–10), serotype 2 was used in the experiments. Serotype 2 has a proven usage history and is considered very safe owing to its previous use in gene therapy for the treatment of conditions such as Parkinson's disease, hemophilia B and cystic fibrosis (Kantor *et al.*, 2014) and its use in the eye in clinical trials for Leber's congenital blindness caused by the *RPE65* gene mutation (Bainbridge *et al.*, 2015). Additionally, it reportedly has a strong affinity for retinal ganglion cells following intravitreal injection of AAV.

The AAV used in these experiments had the genes related to self-propagation removed, and the gene sequences related to the production of the viral proteins (Rep and Cap) that are necessary for replication or formation of viral particles were eliminated. This ensured that using AAV for gene transduction *in vivo* did not allow AAV self-propagation in the body. Additionally, as the AAV vector does not contain genes encoding virus-specific proteins, it also has the benefit of being unlikely to elicit an immune response against cells containing transduced genes (Figure 26.1A) (Schnepp *et al.*, 2003).

When performing gene therapy, in addition to vector choice, the selection of a promoter to control the expression of the target gene is also important. In this experiment, the highly active CAG promoter was used to induce target gene expression. The structure of the CAG promoter consisted of a cytomegalovirus enhancer linked to a chicken β-actin promoter (Niwa *et al.*, 1991), and was capable of strongly inducing the expression of the target gene in a cell-non-specific manner. Further, a woodchuck hepatitis virus post-transcriptional regulatory element sequence was added in order to stabilize the expression of the target gene. Analysis and identification of this sequence were originally conducted because infection of a woodchuck with the hepatitis virus resulted in the development of cancer with 100% certainty, and it is used as a strong inducer of gene expression (Higashimoto *et al.*, 2007). ChR2 or mVChR1 was incorporated into a plasmid vector with this basic structure in order to create an AAV vector for the purpose of gene transduction.

26.3.2 Construction of mVChR1

In our experiments, we selected *Chlamydomonas*-derived ChR2 (GenBank accession number: AF461397) and mVChR1 cDNA as the genes encoding proteins that

Caption for Figure 26.1 (cont.)

the cell extract and cell membrane fractions. Given that VChR1 and mVChR1 bind to the fluorescent protein Venus, green fluorescent protein (GFP) antibodies that cross-react with Venus were used for detection. (E) After transduction of each gene, a Tali Image-based Cytometer (Life Technologies, Tokyo, Japan) was used to quantify the ratio of cells expressing fluorescent proteins to the total cell count. HSPG: heparan sulfate proteoglycan. Panels (C) and (D) were modified from Tomita *et al.* (2014). (A black-and-white version of this figure will appear in some formats. For the color version, please refer to the plate section.)

function as light-gated, cation-selective channels. The mVChR1 is a chimeric protein (Figure 26.1B) formed from amino acid region 1–67 of *Chlamydomonas*-derived channelrhodopsin-1 (GenBank accession number: AF385748) and amino acid region 24–300 of VChR1 (GenBank accession number: EU285658).

To visualize the expression of both the *ChR2* and *mVChR1* genes, their termination codons were removed. The cDNA encoding the fluorescent protein Venus was added to the C-terminal domain, and AAV-ChR2V and AAV-mVChR1V were created and expressed as fusion proteins. In contrast to the VChR1 protein, which was mainly localized in the cytoplasm (Figure 26.1C, left), the genetic modification of the mVChR1 protein resulted in its localization around the cell membrane (Figure 26.1C, right). The protein must be expressed in the cell membrane to function as a light-gated ion channel; hence, this localization is important. To confirm the localization site, membrane fractions were prepared from cells into which VChR1 or mVChR1 had been transduced, and the expression of each protein was investigated by western blot. This resulted in the detection of clear bands in the membrane fractions obtained from cells expressing mVChR1, and mVChR1 was confirmed to be localized to the cell membrane (Figure 26.1D). Meanwhile, compared to mVChR1, VChR1 exhibited lower expression in the cell extract and the membrane fractions (Figure 26.1D), and it became clear that the cells expressing VChR1 in culture were gradually diminishing (Figure 26.1E). Given this result, we considered that VChR1 expression in the cytoplasm might impose some kind of cellular disorder.

26.4 Ion Channel Properties of ChR2 and mVChR1

26.4.1 Whole-cell Recording

Cell strains to constitutively express each gene were established (ChR2-HEK293, VChR1-HEK293 and mVChR1-HEK293), and the light-induced ionic current was measured using the patch clamp method. While a response was recorded in the VChR1-HEK293 cells (Figure 26.2A), for all photic stimulation from 450 to 600 nm, the light-induced outward ionic current was approximately 1/30th of that observed in mVChR1-HEK293 cells (Figure 26.2B) (Tomita *et al.*, 2014). Meanwhile, ChR2-HEK293 cells responded to 450-nm and 500-nm photic stimulation, and a light intensity-dependent increase in membrane current was observed. While a minimal response was observed to 400-nm and 550-nm wavelength light, no response was recorded for photic stimulation at 600 nm (Figure 26.2C). On comparing the sensitive wavelength ranges of mVChR1 and ChR2, we found that mVChR1 had a smaller light-induced ionic current than did ChR2; however, an almost uniform ionic current was observed from 450 to 600 nm, indicating the ability of mVChR1 to respond to a broad range of light wavelengths (Figure 26.2D). The above results indicated that the modification of VChR1 increased its transferability to the cell membrane and enabled its effective functioning as an ion channel.

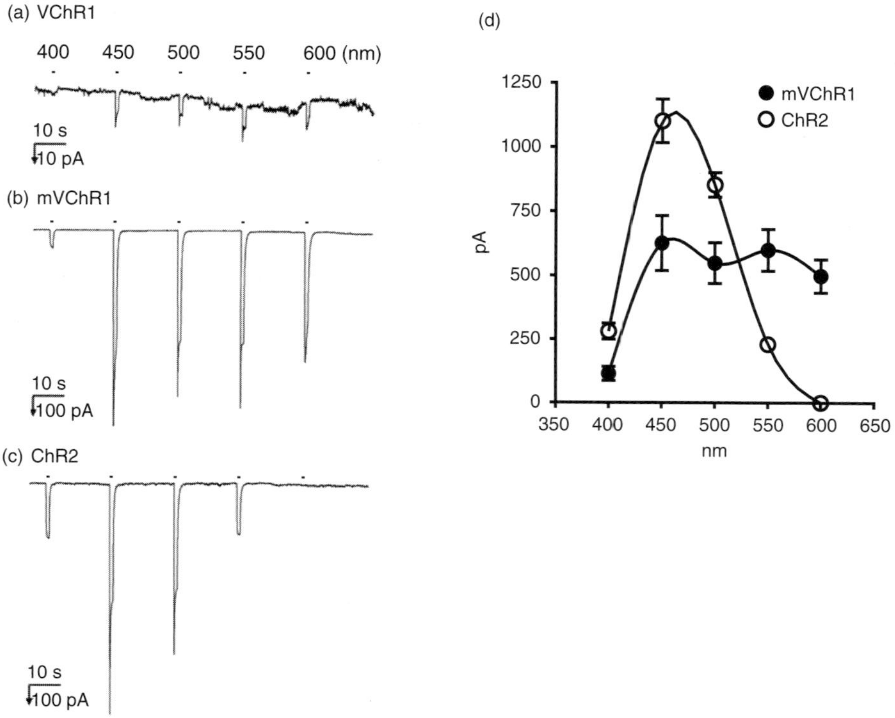

Figure 26.2 Measurement of light-induced ionic current in gene-expressing cells. (A) VChR1-HEK 293 cells. (B) mVChR1-HEK293 cells. (C) ChR2-HEK293 cells. (D) Comparison of sensitive wavelength ranges of mVChR1 and ChR2. This figure was modified from Tomita *et al.* (2014).

26.4.2 Visual Properties of Blind Rats in which Retinal Cells Were Transduced with the *ChR2* Gene

Royal College of Surgeons (RCS) rats are widely used in experiments as animal models of hereditary photoreceptor cell degeneration, as in retinitis pigmentosa. While retina formation in RCS rats initially proceeds normally after birth, degeneration of photoreceptor cells begins after approximately 20 days, and by 3 months of age, the cells regress and the rats become blind. We injected 5 μl AAV-ChR2V solution into the vitreous body of RCS rats at either 6 or 10 months of age and transduced the *ChR2* gene into the surviving retinal cells. As a result of the gene transduction using the AAV vector, *ChR2* expression was observed extensively in the retina, with a cell count of approximately 30% of the number of retinal ganglion cells. When a 480 ± 20-nm light source was used to measure visually evoked potentials (VEPs) at 1 week after gene transduction, a slight increase in VEP amplitude was observed. The amplitude subsequently increased steadily, reaching its maximum after 8 weeks. The recovered VEP was then maintained for 18 months without attenuation (Figure 26.3A and B) (Sugano *et al.*, 2010). Meanwhile, although the recorded VEP amplitude was small, reactivity to stimulation frequency was the same as that observed for rats with normal visual function, responding to photic stimulation from 1 to 20 Hz (Figure 26.3C) and

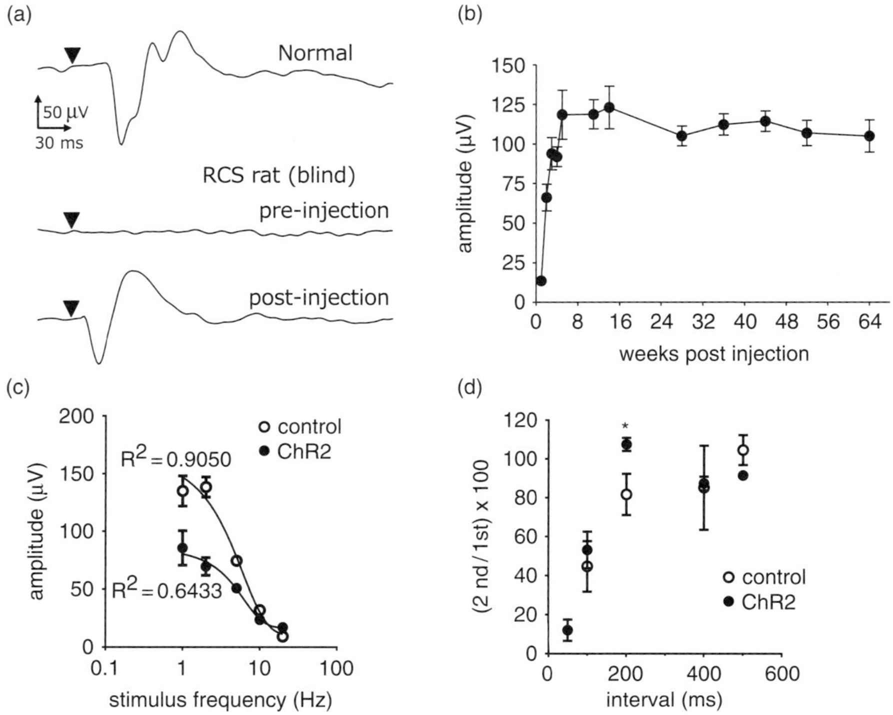

Figure 26.3 VEPs in genetically blind rats in which *ChR2* was transduced into the retinal cells. (A) Typical VEP waveforms for a normal rat (top), genetically blind rat prior to gene transduction (middle) and genetically blind rat 8 weeks after *ChR2* gene induction (bottom). (B) Changes in VEP after gene transduction. (C) Changes in VEP with different frequencies of photic stimulation. (D) Drop in amplitude due to differences in stimulation intervals. Panel (B) was modified from Sugano *et al.* (2010); panels (C) and (D) were modified from Tomita *et al.* (2010).

demonstrating the ability to avoid drops in amplitude due to continuous stimulation with photic stimulation intervals of 200 ms (Figure 26.3D) (Tomita *et al.*, 2010).

Visual function analysis using VEP as an indicator in this way demonstrated the same responsiveness as normal vision. However, the amplitude of VEP restored through gene transduction was smaller than that of normal visual function, and considering that the efficiency of gene transduction is approximately 30%, these results suggested that the recovered amplitude was dependent upon gene transduction efficiency.

26.4.3 Visual Properties of Blind Rats in which Retinal Cells Were Transduced with the *mVChR1* Gene

The *mVChR1* gene, which had demonstrated a broad light wavelength sensitivity using the patch clamp method, was transduced into the retinal cells of RCS rats using an AAV vector in a similar process to that undertaken with *ChR2*. Whereas almost no response was observed to photic stimulation over 587 nm in rats with the transduced *ChR2* gene, those with transduced *mVChR1* responded to the broad light

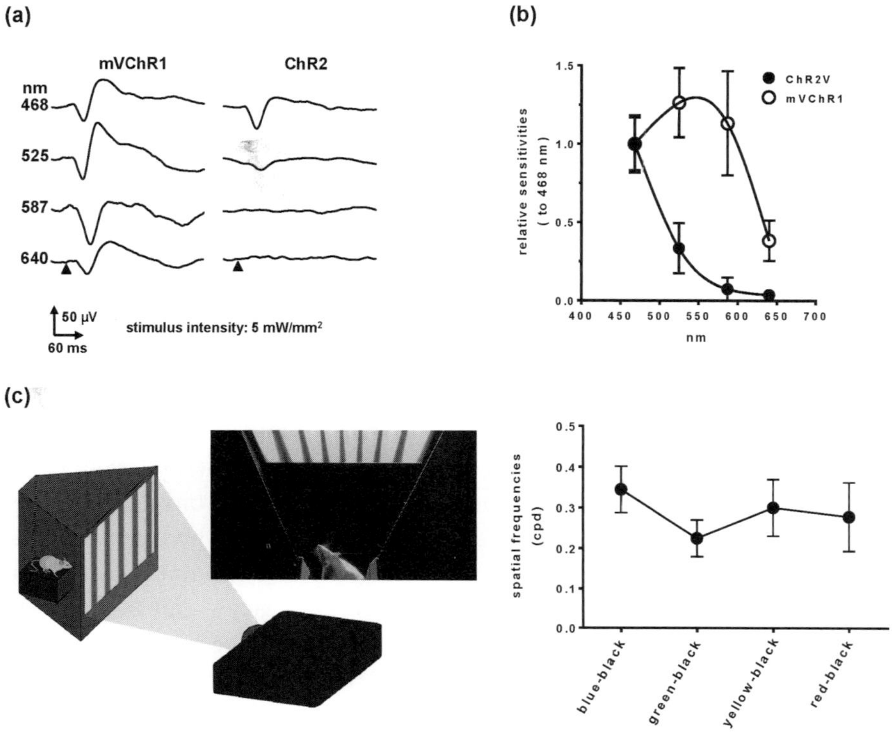

Figure 26.4 Visual function in rats subjected to *mVChR1* gene transduction. (A) VEPs for photic stimulation at various wavelengths. (B) Comparison of wavelength sensitivities in rates undergoing *mVChR1* and *ChR2* gene transduction. (C) Measurement of spatial frequency thresholds using a projector. This figure was modified from Tomita *et al.* (2014).

wavelength of 468–640 nm, and VEPs were recorded (Figure 26.4A and B) (Tomita *et al.*, 2014). These results suggested that while transducing a single gene cannot achieve color vision, it could produce vision of the entire spectrum of visible light.

We previously reported that rats with vision restored through *ChR2* gene transduction were able to recognize a rotating striped pattern. However, we also reported that the vision created with ChR2 had lower light sensitivity than did normal vision, necessitating high illumination in order to enable recognition of the striped pattern (Tomita *et al.*, 2010). Meanwhile, as mVChR1 exhibits broad wavelength sensitivity, behavioral analysis of visual function in RCS rats with transduced *mVChR1* requires a system that is capable of displaying striped patterns in a variety of colors at high illumination. Therefore, we constructed a system that uses a projector to project striped patterns onto a translucent screen (Figure 26.4C, top) in order to measure the visual acuity that is restored through *mVChR1* gene transduction. Rats with transduced *mVChR1* genes responded to all colors of stripes, and the spatial frequency threshold was in the range of 0.2–0.4 (blue–red) (Figure 26.4C, bottom). As tests cannot be conducted at the same light intensity on normal rats (the light would be so bright that they would not look at the screen), a neutral density filter was used to attenuate the light intensity by

10- to 20-fold in order to measure the spatial frequency threshold. The results indicated a spatial frequency threshold for normal rats in the range of 0.3–0.5 (blue–red), which tended to become lower as the wavelengths became longer (from blue to red).

As such, the results of behavioral analysis also indicated the possibility that, despite the large discrepancy with normal rats in terms of light sensitivity, *mVChR1* gene transduction could produce advanced vision.

26.5 Safety Studies

The clinical application of gene therapy requires the following: (1) a vector to be used for gene transduction; and (2) research on the side effects of the gene product to be transduced. As stated above, AAV vectors are currently considered the safest viral vectors. In gene therapy using viral vectors, possible serious side effects include inflammatory response to the vectors and cancerous transition of cells. Generally, the target gene is inserted into the host genome by using the viral vector. For this reason, insertion of the target gene into genomic regions that control cell proliferation, including tumor-suppressor genes, can inhibit their function. Wild-type AAV has a property that causes it to be incorporated into a specific site on chromosome 19 of the host cell (Giraud *et al.*, 1994; Linden *et al.*, 1996); however, the AAV that is used in gene therapy lacks this insertion ability. The likelihood of its insertion into the host genome is therefore extremely low, making the target gene transduced by the AAV an episome (Giraud *et al.*, 1995), with minimal likelihood of it provoking cancerous transition. There are no adverse reaction reports of cancer transition in clinical trials using AAV. Moreover, the side effects of transduced gene products are completely unknown. Until now, there have been no practical cases of gene therapy using genes that are not found in mammals, and the most important task in order to enable clinical application of this gene therapy will be investigation into the side effects of gene products.

Rats have been used to investigate immunological responses to gene transduction, including fluctuations in lymphocytes, inflammatory responses and the sustainability of effects, and it has been reported that no serious side effects were observed throughout the lifetime of the rats (Sugano *et al.*, 2016). Furthermore, these assessments are currently (as of November 2015) being conducted in higher animals (marmosets). We want to study the side effects carefully in order to implement "gene therapy to reconstruct visual function."

26.6 Perspective

We believe that the closest field of clinical application for optogenetics in treatment is its use as a method for restoring vision in cases of blindness due to photoreceptor cell degeneration. Vision regeneration research using optogenetics is not limited to the endowment of light sensitivity to ganglion cells (tertiary neurons) as presented here. Diverse treatment techniques are currently being studied, including: methods to activate only the ON-route in order to specifically express *ChR2* in ON-type bipolar cells (secondary neurons) (Lagali *et al.*, 2008); methods to transduce a light-activated chloride channel gene (*Natronomonas pharaonis* halorhodopsin) into photoreceptor cells (primary neurons) (Busskamp *et al.*, 2010); and methods of vision reconstruction by using light-sensitive compounds rather than gene therapy alone (Tochitsky *et al.*, 2014). The effectiveness of such methods is steadily being confirmed in each of these studies. In addition to being a useful experimental model of a part of the central nervous system, the eye is predicted to produce literally "visible" results regarding efficacy in clinical applications, and methods of vision reconstruction for the visually impaired using optogenetics technologies are highly anticipated. The greatest concern that has been raised regarding the application of such gene therapies in humans is that they use genes that are originally absent in humans, and the completion of safety studies focusing on the clinical setting will be an important future challenge. Given that optogenetics has been established as a powerful research tool in the field of neuroscience, its application in humans is expected to lead to further progress in optogenetics research.

ACKNOWLEDGMENTS

This work was partly supported by Grants-in-Aid for Scientific Research from the Ministry of Education, Culture, Sports, Science, and Technology of Japan (nos. 25462747, 24390393 and 26670750); the Program for the Promotion of Fundamental Studies in Health Sciences of the National Institute of Biomedical Innovation (no. 10-06, NIBIO); the Translational Research Network program (no. 3805140342); and the Japanese Retinitis Pigmentosa Society. We express our heartfelt appreciation to Ichiro Hagimori of Narita Animal Science Laboratory, whose enormous support and insightful comments were invaluable during the course of this study.

REFERENCES

Bainbridge, J.W., Mehat M.S., Sundaram V., *et al.* (2015). Long-term effect of gene therapy on Leber's congenital amaurosis, *New England Journal of Medicine*, **372**: 1887–1897.

Busskamp, V., Duebel J., Blaya, J., *et al.* (2010). Genetic reactivation of cone photoreceptors restores visual responses in retinitis pigmentosa. *Science*, **329**: 413–417.

Giraud, C., Winocour, E., and Berns, K.I. (1994). Site-specific integration by adeno-associated virus is directed by a cellular DNA sequence. *Proceedings of the National Academy of Sciences of the United States of America*, **91**: 10039–10043.

Giraud, C., Winocour, E., and Berns, K.I. (1995). Recombinant junctions formed by site-specific integration of adeno-associated virus into an episome. *Journal of Virology*, **69**: 6917–6924.

Higashimoto, T., Urbinati S., Perumbeti A., *et al.* (2007). The woodchuck hepatitis virus post-transcriptional regulatory element reduces readthrough transcription from retroviral vectors. *Gene Therapy*, **14**: 1298–1304.

Kantor, B., McCown, T., Leone, P., and Gray, S.J. (2014). Clinical applications involving CNS gene transfer. *Advances in Genetics*, **87**: 71–124.

Lagali, P.S., Balya D., Awatramani G.B., *et al.* (2008). Light-activated channels targeted to ON bipolar cells restore visual function in retinal degeneration. *Nature Neuroscience*, **11**: 667–675.

Linden, R.M., Ward, P., Giraud, C., Winocour, E., and Berns, K.I. (1996). Site-specific integration by adeno-associated virus. *Proceedings of the National Academy of Sciences of the United States of America*, **93**: 11288–11294.

Nagel, G., Szellas T., Huhn W., *et al.* (2003). Channelrhodopsin-2, a directly light-gated cation-selective membrane channel. *Proceedings of the National Academy of Sciences of the United States of America*, **100**: 13940–13945.

Niwa, H., Yamamura, K., and Miyazaki, J. (1991). Efficient selection for high-expression transfectants with a novel eukaryotic vector. *Gene*, **108**: 193–199.

Rizzo, S., Belting C., Cinelli L., *et al.* (2014). The Argus II Retinal Prosthesis: 12-month outcomes from a single-study center. *American Journal of Ophthalmology*, **157**: 1282–1290.

Santos, A., Humayun M.S., de Juan E. Jr., *et al.* (1997). Preservation of the inner retina in retinitis pigmentosa. A morphometric analysis. *Archives of Ophthalmology*, **115**: 511–515.

Schnepp, B.C., Clark, K.R., Klemanski, D.L., Pacak, C.A., and Johnson, P.R. (2003). Genetic fate of recombinant adeno-associated virus vector genomes in muscle. *Journal of Virology*, **77**: 3495–3504.

Strettoi, E., Pignatelli, V., Rossi, C., Porciatti, V., and Falsini, B. (2003). Remodeling of second-order neurons in the retina of *rd/rd* mutant mice. *Vision Research*, **43**: 867–877.

Sugano, E., Isago, H., Wang, Z., Murayama, N., Tamai, M., and Tomita, H. (2010). Immune responses to adeno-associated virus type 2 encoding channelrhodopsin-2 in a genetically blind rat model for gene therapy. *Gene Therapy*, **18**: 266–274.

Sugano, E., Tabata K., Takahashi M., *et al.* (2016). Local and systemic responses following intravitreous injection of AAV2-encoded modified Volvox channelrhodopsin-1 in a genetically blind rat model. *Gene Therapy*, **23**: 158–166.

Tochitsky, I., Polosukhina A., Degtyar V.E., *et al.* (2014). Restoring visual function to blind mice with a photoswitch that exploits electrophysiological remodeling of retinal ganglion cells. *Neuron*, **81**: 800–813.

Tomita, H., Sugano E., Isago H., *et al.* (2010). Channelrhodopsin-2 gene transduced into retinal ganglion cells restores functional vision in genetically blind rats. *Experimental Eye Research*, **90**: 429–436.

Tomita, H., Sugano E., Murayama M., *et al.* (2014). Restoration of the majority of the visual spectrum by using modified *Volvox* channelrhodopsin-1. *Molecular Therapy*, **22**: 1434–1440.

Zemelman, B.V., Lee, G.A., Ng, M., and Miesenbock, G. (2002). Selective photostimulation of genetically chARGed neurons. *Neuron*, **33**: 15–22.

27 Optogenetic Dissection of a Top-down Prefrontal to Hippocampus Memory Circuit

Priyamvada Rajasethupathy

27.1 Introduction

One of the great mysteries of the human mind is its ability to form memories. The ability to faithfully store and retrieve information is an important adaptive quality that allows humans to reflect on, engage with and respond to an ever-changing environment. Disturbances of memory significantly diminish quality of life and also contribute to a number of disorders, including Alzheimer's, fronto-temporal dementias, drug addictions and post-traumatic stress disorder. Pioneering studies to understand memory processes have been transformative in illuminating the molecular and physiological mechanisms that occur in individual neurons and individual synapses (Kandel *et al.*, 2014). There exists a relative void, however, in our understanding of how individual neurons contribute to a network representation of a memory. Furthermore, the contributions of diverse brain regions, together with the precise functional connections between these areas that may mediate aspects of memory storage and retrieval, are only partially understood. The major limitation for these more circuit-level investigations of memory processes has been the lack of tools to record and manipulate genetically defined ensembles of neurons in local as well as long-range circuits with millisecond precision and in real time in the behaving animal. The development of such tools to manipulate (optogenetics) and record (two-photon cellular resolution imaging) neural activity from large ensembles and especially at projection terminals has opened the door to truly investigating the brain-wide circuits underlying memory processing in the mammalian brain (Deisseroth and Schnitzer, 2013; Deisseroth, 2015). These tools are powerful when performed independently, but, when applied together, observations in neural activity can inform manipulations in real time, thereby effectively "closing the loop" on targeted interventions that can causally define cognitive behavior. In this chapter, I will provide an overview of what has been learned about memory circuits using optogenetics thus far, and then highlight and discuss in detail one recent study that uses an all-optical approach to the interrogation of neural circuits (one-photon manipulation with concomitant two-photon imaging) in order to reveal the mechanisms underlying top-down memory processes in the mammalian brain.

27.2 Optogenetic Dissection of Hippocampal Circuits in Memory Processes

One of the first applications of optogenetics in order to address memory circuits was in the real-time manipulation of hippocampal neurons during learning and retrieval of a contextual fear conditioning task. In this task, mice are trained to associate an aversive unconditioned stimulus (foot shock) with an otherwise neutral conditioned stimulus (a particular context), and upon being placed in that same context at least 24 hours later, they produce a conditioned fear response (freezing). Pharmacological, electrical and genetic experiments have provided significant insight into the roles of the hippocampus and prefrontal cortex (PFC) during memory acquisition, storage and retrieval (Frankland and Bontempi, 2005), but the relative real-time contributions of specific sub-regions and cell types within the hippocampus or PFC have not been extensively investigated. Optogenetic disruption of specified sub-regions of the hippocampus revealed that the CA1 subfield is necessary for both the acquisition and retrieval of fearful contextual memories (Goshen *et al.*, 2011; Lovett-Barron *et al.*, 2014), as expected from previous studies, but that the dentate gyrus (DG) subfield is necessary only for acquisition (Andrews-Zwilling *et al.*, 2012; Kheirbek *et al.*, 2013). This may be explained by the fact that direct cortical inputs to CA3/CA2/CA1 subfields can bypass the DG, thereby making the DG dispensable for memory retrieval.

Furthermore, while it has long been expected from pharmacologic and lesion studies that the CA1 stores memories only temporarily (<1 month), real-time optogenetic manipulation of CA1 revealed that it is not only necessary for immediate recall of a memory, but also for remote recall (even 12 weeks after the formation of a memory), and interestingly also for the entire duration of the retrieval period (Goshen *et al.*, 2011). These results provide a new framework for thinking about memory processes as engaging the hippocampus throughout the lifetime of that memory. The reason for the contradictory results that have been obtained using optogenetic studies versus pharmacological/genetic studies likely lies in the temporal dynamics of these techniques; given that pharmacological or genetic manipulations are orders of magnitude slower than neural activity, they likely enable compensatory mechanisms to drive the ensuing behavioral response, whereas optogenetic inhibition on the millisecond timescale reveals the underlying real-time role of the hippocampus during retrieval. Indeed, when optogenetic inhibition was delivered on a pharmacological timescale (30 minutes), compensatory mechanisms in the anterior cingulate (AC) were engaged and there was no impairment in recall, whereas precise illumination disrupted the recall (Goshen *et al.*, 2011).

27.3 Using Optogenetics to Understand Memory Allocation and Retrieval

Long-standing questions in memory research have related to the identity and functionality of the neurons participating in the memory trace: which neurons are chosen to participate, and are the same neurons that initially encode the memory later used to retrieve the memory? These questions have been addressed elegantly by a series of studies focused on the lateral amygdala, initially based on clever genetic and chemogenetic manipulations (Han *et al.*, 2007; Rejmers *et al.*, 2007; Han *et al.*, 2009; Yiu *et al.*, 2014), but more recently rigorously addressed through optogenetics. For instance, it was shown that during memory formation, only a fraction of eligible neurons are allocated for that memory (Han *et al.*, 2007; Rejmers *et al.*, 2007; Han *et al.*, 2009), and they are chosen based on their expression of the transcription factor CREB and on their current state of excitability (Yiu *et al.*, 2014). The initially allocated neurons were then preferentially recruited during retrieval of that memory (Rejmers *et al.*, 2007) and, furthermore, their activity was found to be necessary for memory retrieval, since genetic ablation or suppression of the CREB-overexpressing cells (and not an equivalently sized random population of neurons) was sufficient to erase the memory (Han *et al.*, 2009). In order to then bias the network so as to provide a synthetic and arbitrary pattern of allocation, optogenetics was used to make a subset of cells more excitable, which then intriguingly revealed (and confirmed) that fear memories were funneled into the more excitable neurons.

These experiments reveal that the initially allocated neurons are used and are necessary for future retrieval of that same memory. But are these neurons sufficient? Here, optogenetics has played an important role in answering this question. In a series of experiments designed to express channelrhodopsin-2 (ChR2) specifically in DG neurons that were activated during a contextual memory task (using cFos-tTA mice to drive ChR2 in a doxycycline-dependent manner), it was shown that optogenetic reactivation of the neurons that were used to initially encode the memory was sufficient for induction of memory retrieval (Liu *et al.*, 2012). A similar experiment was also conducted using pharmacologic activation of the initial ensemble (using DREADs), confirming the same result (Garner *et al.*, 2012). These experiments were then followed by another study demonstrating that an arbitrary set of neurons could be forced into participating in a contextual fear memory representation (by pairing optical activation of those neurons with application of shocks to the animal), and that these same neurons are then sufficient to elicit the retrieval of the fear response at a later time (Ramirez *et al.*, 2015). These results were then extended to allocate arbitrary sets of neurons so as to encode "false" or synthetic memories not only for aversive memories, but also for appetitive memories (Choi *et al.*, 2011; Redondo *et al.*, 2014; Root *et al.*, 2014). In summary, these studies point to the ability to elicit the behavioral output of a specific memory by directly activating the population of neurons that were used to encode the memory and, more intriguingly, it is also possible to impose an alternative, artificial memory on a chosen neural ensemble to then drive the behavior of interest. These results, especially the latter conclusion, would not have been possible without the use of optogenetic tools that enable the "playing-in" of artificial patterns of neural activity that are linked to animal behavior.

27.4 Entorhinal and Prefrontal Involvement in Memory Processes

Apart from hippocampal and amygdalar circuits, learning and memory processes have been optogenetically investigated in the upstream entorhinal cortex and PFC regions of the brain (Goshen, 2014). The last several decades of memory research has focused on dissecting the role of entorhinal inputs to the hippocampus in shaping memory representations and, dovetailing on this rich knowledge stream, optogenetic manipulation of entorhinal inputs to the hippocampus have confirmed much of what we knew before and have also revealed new insights into how hippocampal circuits receive inputs in order to reliably encode features of time and space (Zhang *et al.*, 2013; Buetfering *et al.*, 2014). In contrast to the extensively studied entorhinal inputs to the hippocampus (the canonical "bottom-up" memory circuits), there has been a relatively minimal focus on investigating the role of PFC inputs to the hippocampus (the proposed "top-down" memory circuits). Few studies have uncovered a role of the PFC (through hippocampal relay regions) in modulating the specificity versus generalizability of contextual memories (Xu and Sudhof, 2013) and refining these contextual memories based on expectation and goal-oriented behavior (Ito *et al.*, 2015). However, only recently, has there been a study that discovered the existence of a direct, top-down PFC to hippocampus circuit, and optogenetic manipulation of this circuit revealed its important roles in performing memory retrieval (Rajasethupathy *et al.*, 2015). I describe below the details of this study.

27.5 Discovery of a Novel Top-down (AC–CA) Memory Circuit

To identify potentially novel top-down inputs to the hippocampus in a global and unbiased fashion, Rajasethupathy *et al.* (2015) injected into the hippocampal CA1/CA3 region a retrograde tracer that is capable of transducing axons and cell bodies directly (but is not capable of moving beyond the directly transduced cells) (Wickersham *et al.*, 2007); this tracer was a rabies virus with its glycoprotein gene replaced by the tdTomato label gene, or RV-tdT. Supporting the validity of this approach, they observed robust tdT labeling that was largely limited to distinct brain regions with known inputs to the hippocampus, such as the medial septum, contralateral CA3 and entorhinal cortex. However, in addition to these well-known projections to the hippocampus, they also identified a previously uncharacterized input arising from the dorsal anterior cingulate (AC) cortex and adjacent frontal cortical association cortex (Figure 27.1A), both of which are reciprocally connected with the mediodorsal thalamic nucleus – a defining feature of the PFC in rodents. These neurons were found approximately 200–300 μm from the midline and displayed long apical dendrites extending toward the pia, suggesting preferential labeling of layer 2/3 principal neurons. To further validate the existence of this novel prefrontal to hippocampus projection, they researchers then injected an anterograde label (AAV5-CaMKIIα::EYFP, which neither effectively transduces axons nor moves across cells) targeting direct expression in excitatory

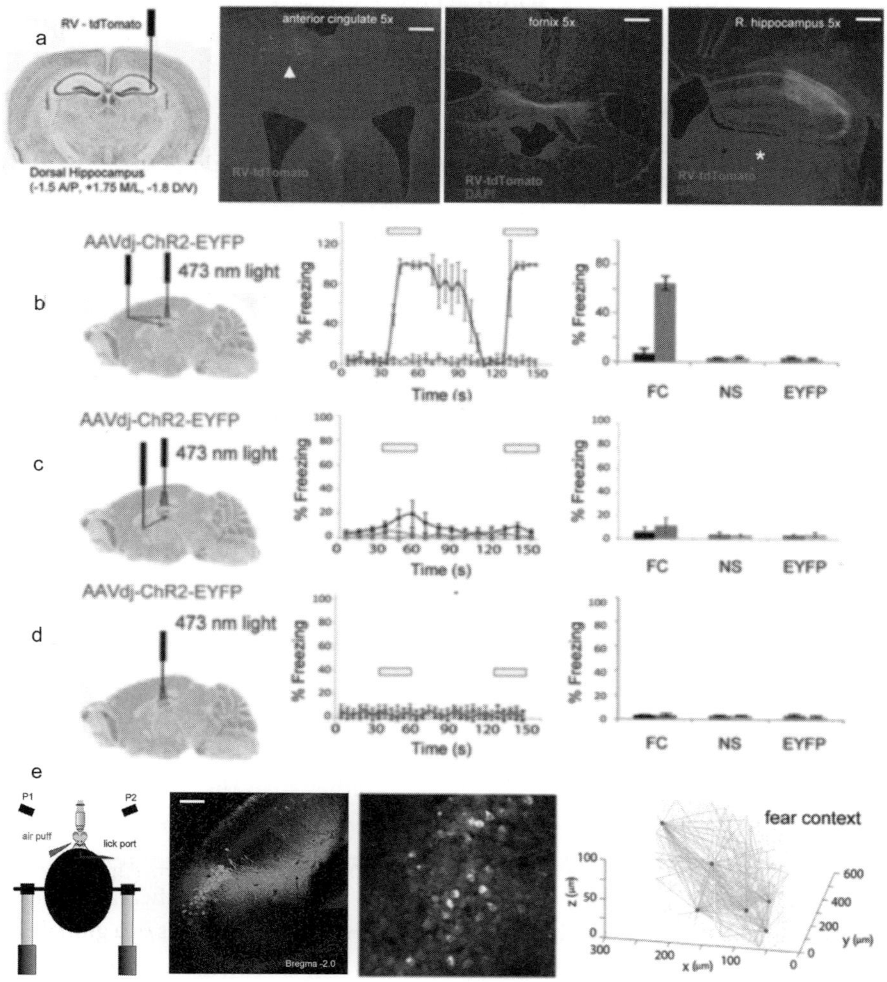

Figure 27.1 (A) Retrograde rabies tracing from the hippocampus reveals afferent fibers originating in the AC (arrowhead). There was no leak of rabies virus into the medial dorsal thalamus (asterisk), a known projection target of PFC. Scale bar: 300 μm. (B) Stimulating AC–CA projections resulted in significant light-induced memory retrieval (freezing) that was not observed in mice without opsin (EYFP) or mice that had not been fear conditioned (no shock). (C) Stimulating medial septal projections to the hippocampus demonstrated a strikingly smaller trend. (D) Directly stimulating the hippocampus demonstrated no light-dependent freezing. (E) Visualizing population dynamics in CA2/CA3 during contextual fear conditioning in a virtual environment: the emergence of hub neurons during learning. (A black-and-white version of this figure will appear in some formats. For the color version, please refer to the plate section.)

neurons of the dorsal AC. They detected the expected presence of yellow fluorescence-filled projection terminals bilaterally in the striatum and ipsilaterally in the medial dorsal nucleus of the thalamus, both areas that are known to receive strong projections from the PFC, but also bilaterally in the hippocampus. Inputs to CA3 and CA1 were observed in both the stratum oriens and stratum radiatum (interestingly, sparing CA2 and the DG).

To determine whether these prefrontal projections gave rise to direct the monosynaptic drive of hippocampal neurons, the researchers transduced principal neurons of the dorsomedial PFC, consisting primarily of AC, with an adeno-associated virus encoding a channelrhodopsin, and performed acute-slice, whole-cell patch clamp recordings of light-driven excitatory postsynaptic currents in CA1/CA3 pyramidal cell bodies. Cell responses in CA1 and CA3 were reliable, fast (<3.5 ms on average) and sustained (following 10-Hz stimulation trains); this is therefore consistent with the presence of an efficacious and direct monosynaptic connection from the AC to CA3 and CA1 (termed AC–CA).

27.6 The AC–CA Projection Is Necessary and Sufficient for Memory Retrieval

To probe the functional significance of this projection, Rajasethupathy *et al.* (2015) conducted a series of optogenetic experiments in order to manipulate this pathway before and after contextual fear conditioning, and also in the setting of memory extinction and re-instatement. They injected AAVdj-ChR2-EYFP (or AAVdj-EYFP in a parallel cohort) into the AC and targeted light stimulation to terminals in the hippocampus. Recruiting cells contributing to the AC–CA projection by this means, the researchers observed significant freezing in response to optical stimulation in the neutral context only in mice that had previously been fear conditioned (Figure 27.1B). To ensure that the observed light-driven behavioral response was indeed due to reactivation of a fear memory, rather than due to a direct, non-specific drive of fear behavior, the very same mice were subjected to several days of contingency degradation by exposure to the original, conditioned context without shock, after which stimulation with light failed to produce significant freezing. Interestingly, re-instatement of the fear memory once again reliably produced freezing. Importantly, they found that a simple, non-specific drive of the hippocampus (either through direct activation of hippocampal ensembles or of medial septal input to the hippocampus) is not sufficient to induce memory retrieval and that, instead, the AC–CA projection provides access to cells driving the expression of a recently acquired contextual memory (Figure 27.1C and D).

The aforementioned experiments demonstrated that activating the AC–CA projecting cells was sufficient to induce contextual memory retrieval, but Rajasethupathy *et al.* (2015) next asked: is it required? To do so, they targeted the eNpHR3.0 opsin to cells giving rise to the AC–CA projection, with light targeted focally and bilaterally to AC–CA cell bodies, and observed striking deficits in both the latency and strength of the fear response. They further showed that this effect was fully reversible and also specific to retrieving hippocampal-dependent memories.

Taken together, their anatomical, electrophysiological and behavioral data revealed the existence of a previously uncharacterized monosynaptic prefrontal to hippocampus projection. When this circuit is inhibited, fear-conditioned mice are unable to retrieve the fear memory with the same strength or speed as control counterparts, indicating its endogenous importance for memory retrieval. In contrast, activation of this circuit is sufficient for robust fear memory retrieval in recently conditioned mice, but not in naive unconditioned mice, mice in which the memory had been extinguished, mice not expressing opsin or mice receiving other types of direct or indirect drive of hippocampus.

27.7 Hub Neurons Emerge during Learning

If it were possible to observe in real time the influence of this projection on hippocampal cell population activity, clues might emerge regarding the cellular features of the network activity that mediates retrieval. Rajasethupathy *et al.* (2015) therefore adapted the fear-conditioning paradigm into a behaving virtual reality preparation, wherein head-fixed mice navigate on an axially fixed track ball under a two-photon microscope (Dombeck *et al.*, 2010) to enable imaging of hippocampal neural populations during learning and retrieval. In this paradigm, mildly water-deprived mice were habituated to the virtual environment before initiation of a 3-day protocol involving exposure to two contexts on day 1, training with aversive air puffs associated with one context (fear) but not the other (neutral) on day 2 and memory retrieval testing (fear memory assessed by amount of lick suppression) (Mahoney and Ayres, 1976; Bouton and Bolles, 1980; Lovett-Barron *et al.*, 2014) in both contexts on days 3 and 30. For imaging during this behavior, mice were injected with the genetically encoded Ca^{2+} indicator GCaMP6m (Chen *et al.*, 2013) driven by a CaMKIIα promoter fragment in order to be expressed chiefly in the pyramidal cells of the hippocampus. Mice were implanted with a cranial window over CA2/CA3 and imaged daily in both contexts during training and retrieval, using fast (30 Hz), volumetric (500 × 500 μm x/y, 100-μm depth), two-photon imaging, providing access to >500 neurons in any given field of view (Figure 27.1E).

To capitalize on this simultaneous broad scope and single-cell resolution during contextual memory testing, Rajasethupathy *et al.* (2015) first sought to identify the population features of the CA2/CA3 functional network that differed consistently across days or contexts. By binning the number of correlated partners that exist for each neuron, they found that there was a significant shift in the functional connectivity of the network from random organization on day 1 before learning to a more ordered and hierarchical organization on day 3 in the fear context after learning, with the emergence of a small population of "hub-like" neurons displaying extensive functional connectivity in the network (Figure 27.1E).

To better understand the functional impact of these hub neurons, they then focused their analysis on the activity of the entire network at times when the hub neurons were active. They found that the activity of hub neurons precedes population-wide synchronous events, which is consistent with these neurons playing a role in recruiting network activity. Interestingly, these broad synchronous events (which were significantly more prevalent in the memory context than a neutral context) consisted of cell assemblies that were essentially orthogonal and with no apparent anatomical organization, suggesting recruitment of multiple separable ensembles (that potentially recur over long enough timescales) during memory retrieval.

27.8 AC–CA Projections Preferentially Recruit Hub Neurons during Memory Retrieval

Taken together, these volumetric imaging studies of CA3 during memory retrieval demonstrated the emergence of a sparse set of neurons with hub-like properties, which could therefore serve as efficient and potent routes of access for a globally distant network during memory retrieval. If functionally important in this way, Rajasethupathy *et al.* (2015) hypothesized that indeed these hub neurons might be preferentially recruited by top-down projections during memory retrieval. To test this possibility, they sought to stimulate AC–CA projections to induce memory retrieval as they had done earlier, but also simultaneously to image the postsynaptic CA2/CA3 network in order to observe local dynamics. To do so, they first developed a new red-shifted opsin termed bReaChES (combining the strengths of existing opsins C1V1 [Yizhar *et al.*, 2011] and ReaChR [Lin *et al.*, 2013]) that could provide strong photocurrents, high spike fidelity and, most importantly, robust trafficking in long-range projections. To then enable simultaneous single-photon stimulation of the projection along with two-photon imaging of the target population, mice were injected with AAVdj-CaMKII::GCAMP6m in CA3 and AAV8-CaMKII::bReaChES in the AC, with cannula placement over CA2/CA3 for optical access. They then found that stimulation of the AC–CA projections led to the recruitment of a different population of neurons after training as compared with before training, revealing a functional target shift in the influence of the AC–CA projections that was associated with memory formation. Furthermore, while relatively few (~5%) of the hub-like neurons in either fear or neutral contexts were recruited by stimulation of AC–CA projections before training, a strikingly increased fraction (~20%) of the hub-like neurons in the fear context were recruited by the AC–CA projection after training; in contrast, recruitment of non-hub-like neurons in the fear context and recruitment of any neurons in the neutral context remained unchanged and low. These results further demonstrated swift reorganization of the functional impact of the AC–CA projections, with preferential recruitment of hub neurons representing the recently formed contextual fear memory. Together, these findings revealed a potentially general means by which top-down circuit influences could organize and engage with salient memory representations in order to enable efficient retrieval.

27.9 Conclusion

Optogenetics has been used to shed light on a wide variety of questions involving memory circuits, with much work already having focused on delineating the role of bottom-up circuits in memory allocation, processing and retrieval. Only recently has an exploration into the optogenetic dissection of top-down memory circuits begun. Specifically, the ability of the AC–CA top-down circuit to selectively recruit hub neurons (highly functional connectivity neurons) in the hippocampus during memory retrieval suggests that hub neurons may be serving as efficient routes of access during the retrieval of salient memory representations. The observed emergence of hub neurons and increasing functional sparseness of memory-associated networks during learning are emerging themes (Vinje and Gallant, 2000; Olshausen and Field, 2004; Yassin *et al.*, 2010; Komiyama *et al.*, 2010; Gdalyahu *et al.*, 2012). Furthermore, computational models of memory have suggested that hierarchical networks with few extensively connected hub nodes, as Rajasethupathy *et al.* (2015) described, are well suited for memory stability (since random degradation of the network will more likely affect non-hub nodes that are less consequential to the memory representation) and rapid recall (since access to hubs alone is likely to be sufficient to access the entire memory representation) (Buzsaki *et al.*, 2004; Bullmore and Sporns, 2009; Perin *et al.*, 2011). Interestingly, the ability of single neurons to drive behavior (Brecht *et al.*, 2004; Houweling and Brecht, 2008; Li *et al.*, 2009) has suggested the existence of hub neurons beyond memory processes alone, which is consistent with the fact that hierarchical network motifs have been identified broadly in neural systems (e.g. in functional magnetic resonance imaging studies of the primate brain [Achard *et al.*, 2006; He *et al.*, 2007; Hagmann *et al.*, 2008; Meunier *et al.*, 2010] and in *in vitro* studies of mammalian brain tissue [Song *et al.*, 2005; Yu *et al.*, 2008; Bonifazi *et al.*, 2009; Li *et al.*, 2010]). Specifically in the memory context, Rajasethupathy *et al.* (2015) suggest that the adaptive emergence of contextual memory-specific hub neurons may facilitate efficient access to the engram, which is suitable for memories that are strongly encoded via repetition, reward or saliency. The authors do note, however, that such top-down control via hubs may not be suitable for all memories; while hierarchical networks create stability and robustness, trade-offs may include susceptibility to overfitting of representations at the expense of adaptability and failure to generalize (Hermundstad *et al.*, 2011).

The identification of a direct top-down input to the hippocampus further opens the door to a range of questions regarding the mechanisms that the PFC uses in order to gate and/or integrate information flow to the hippocampus during top-down retrieval. It will also be important to determine whether and how top-down circuits interface (temporally and anatomically) with bottom-up circuits during memory retrieval. The increase in genetic tools (Cre driver lines, smaller and more efficient promoters driving viral expression and anterograde and retrograde tracers) to target specified cell populations based on cell type and connectivity, together with improvements in optical tools (for single-photon or two-photon imaging of large intact volumes), offer the promise of enabling the dissection of these and other circuit features contributing to the complex cognitive processing underlying memory processes in the mammalian brain.

REFERENCES

Achard, S., Salvador, R., Whitcher, B. *et al.* (2006). A resilient, low-frequency, small-world human brain functional network with highly connected association cortical hubs. *J. Neurosci.*, **26**, 63–72.

Andrews-Zwilling, Y., Gillespie, A.K., Kravitz, A.V. *et al.* (2012). Hilar GABAergic interneuron activity controls spatial learning, memory retrieval. *PLoS ONE*, **7**, e40555.

Bonifazi, P., Goldin, M., Picardo, M.A. *et al.* (2009). GABAergic hub neurons orchestrate synchrony in developing hippocampal networks. *Science*, **326**, 1419–24.

Bouton, M.E. and Bolles, R.C. (1980). Conditioned fear assessed by freezing and by the suppression of three different baselines. *Anim. Learn. Behav.* **8**, 429–434.

Brecht, M., Schneider, M. and Sakmann, B. (2004). Whisker movements evoked by stimulation of single pyramidal cells in rat motor cortex. *Nature*, **427**, 704–710.

Buetfering, C., Allen, K. and Monyer, H. (2014). Parvalbumin interneurons provide grid cell-driven recurrent inhibition in the medial entorhinal cortex. *Nat. Neurosci.*, **17**, 710–718.

Bullmore, E. and Sporns, O. (2009). Complex brain networks: graph theoretical analysis of structural and functional systems. *Nat. Rev. Neurosci.*, **10**, 186–198.

Buzsáki, G., Geisler, C., Henze, D.A. *et al.* (2004). Interneuron diversity series: circuit complexity and axon wiring economy of cortical interneurons. *Trends Neurosci.*, **27**, 186–193.

Chen, T.W., Wardill, T.J., Sun, Y. *et al.* (2013). Ultrasensitive fluorescent proteins for imaging neuronal activity. *Nature*, **499**, 295–300.

Choi, G.B., Stettler, D.D., Kallman, B.R. *et al.* (2011). Driving opposing behaviors with ensembles of piriform neurons. *Cell*, **146**, 1004–1015.

Deisseroth, K. (2015). Optogenetics: 10 years of microbial opsins in neuroscience. *Nat. Neurosci.*, **18**, 2013–2025.

Deisseroth, K. and Schnitzer, M.J. (2013). Engineering approaches to illuminating brain structure and dynamics. *Neuron*, **80**, 568–577.

Dombeck, D.A., Harvey, C.D., Tian, L. *et al.* (2010). Functional imaging of hippocampal place cells at cellular resolution during virtual navigation. *Nat. Neurosci.*, **13**, 1433–1440.

Frankland, P.W. and Bontempi, B. (2005). The organization of recent and remote memories. *Nat. Rev. Neurosci.*, **6**, 119–130.

Garner, A.R., Rowland, D.C., Hwang, S.Y. *et al.* (2012). Generation of a synthetic memory trace. *Science*, **335**, 1513–1516.

Gdalyahu, A., Tring, E., Polack, P. *et al.* (2012). Associative fear learning enhances sparse network coding in primary sensory cortex. *Neuron*, **75**, 121–132.

Gore, F., Schwartz, E.C., Brangers, B.C. *et al.* (2015). Neural representations of unconditioned stimuli in basolateral amygdala mediate innate and learned responses. *Cell*, **162**, 134–145.

Goshen, I. (2014). The optogenetic revolution in memory research. *Trends Neurosci.*, **37**, 511–522.

Goshen, I., Brodsky, M., Prakash, R. *et al.* (2011). Dynamics of retrieval strategies for remote memories. *Cell*, **147**, 678–689.

Hagmann, P., Cammoun, L., Gigandet, X., *et al.* (2008). Mapping the structural core of human cerebral cortex. *PLoS Biol.*, **6**, e159.

Han, J.H., Kushner, S.A., Yiu, A.P. *et al.* (2007). Neuronal competition and selection during memory formation. *Science*, **316**, 457–460.

Han, J.H., Kushner, S.A., Yiu, A.P. *et al.* (2009). Selective erasure of a fear memory. *Science*, **323**, 1492–1496.

He, Y., Chen, Z.J. and Evans, A. C. (2007). Small-world anatomical networks in the human brain revealed by cortical thickness from MRI. *Cereb. Cortex*, **17**, 2407–2419.

Hermundstad, A.M., Brown, K.S., Bassett, D.S. *et al.* (2011). Learning, memory, and the role of neural network architecture. *PLoS Comput. Biol.*, **7**, e1002063.

Houweling, A.R. and Brecht, M. (2008). Behavioural report of single neuron stimulation in somatosensory cortex. *Nature*, **451**, 65–68.

Ito, H.T., Zhang, S.J., Witter, M.P. *et al.* (2015). A prefrontal-thalamo-hippocampal circuit for goal-directed spatial navigation. *Nature*, **522**, 50–55.

Kandel, E.R., Dudai, Y. and Mayford, M.R. (2014). The molecular and systems biology of memory. *Cell*, **157**, 163–186.

Kheirbek, M.A., Drew, L.J., Burghardt, N.S. *et al.* (2013). Differential control of learning and anxiety along the dorsoventral axis of the dentate gyrus. *Neuron*, **77**, 955–968.

Komiyama, T., Sato, T.R., O'Connor, D.H. *et al.* (2010). Learning-related fine-scale specificity imaged in motor cortex circuits of behaving mice. *Nature*, **464**, 1182–1186.

Li, C.Y., Poo, M-M. and Dan, Y. (2009). Burst spiking of a single cortical neuron modifies global brain state. *Science*, **324**, 643–646.

Li, X., Ouyang, G., Usami, A. *et al.* (2010). Scale-free topology of the CA3 hippocampal network: a novel method to analyze functional neuronal assemblies. *Biophys. J.*, **98**, 1733–1741.

Lin, J.Y., Knutsen, P.M., Muller, A. *et al.* (2013). ReaChR: a red-shifted variant of channelrhodopsin enables deep transcranial optogenetic excitation. *Nat. Neurosci.*, **16**, 1499–1508.

Liu, X., Ramirez, S., Pang, P.T. *et al.* (2012). Optogenetic stimulation of a hippocampal engram activates fear memory recall. *Nature*, **484**, 381–385.

Lovett-Barron, M., Kaifosh, P., Kheirbek, M.A. *et al.* (2014). Dendritic inhibition in the hippocampus supports fear learning. *Science*, **343**, 857–863.

Mahoney, W.J. and Ayres, J.J.B. (1976). One-trial simultaneous and backward fear conditioning as reflected in conditioned suppression of licking in rats. *Anim. Learn. Behav.* **4**, 357–362.

Meunier, D., Lambiotte, R. and Bullmore, E.T. (2010). Modular and hierarchically modular organization of brain networks. *Front. Neurosci.*, **4**, 200.

Olshausen, B.A. and Field, D.J. (2004). Sparse coding of sensory inputs. *Curr. Opin. Neurobiol.*, **14**, 481–487.

Perin, R., Berger, T.K. and Markram, H. (2011). A synaptic organizing principle for cortical neuronal groups. *PNAS*, **108**, 5419–5424.

Ramirez, S., Liu, X., Lin, P.A. *et al.* (2013). Creating a false memory in the hippocampus. *Science*, **341**, 387–391.

Rajasethupathy, P., Sankaran, S., Marshel, J. *et al.* (2015). Projections from neocortex mediate top-down control of memory retrieval. *Nature*, **526**, 653–659.

Redondo, R.L., Kim, J., Arons, A.L. *et al.* (2014). Bidirectional switch of the valence associated with a hippocampal contextual memory engram. *Nature*, **513**, 426–430.

Reijmers, L.G., Perkins, B.L., Matsuo, N. *et al.* (2007). Localization of a stable neural correlate of associative memory. *Science*, **317**, 1230–1233.

Root, C.M., Denny, C.A., Hen, R. *et al.* (2014). The participation of cortical amygdala in innate, odour-driven behaviour. *Nature*, **515**, 269–273.

Song, S., Sjöström, P.J., Reigl, M. *et al.* (2005). Highly nonrandom features of synaptic connectivity in local cortical circuits. *PLoS Biol.*, **3**, e68.

Vinje, W.E. and Gallant, J.L. (2000). Sparse coding and decorrelation in primary visual cortex during natural vision. *Science*, **287**, 1273–1276.

Wickersham, I.R., Finke, S., Conzelmann, K.K. *et al.* (2007). Retrograde neuronal tracing with a deletion-mutant rabies virus. *Nat. Methods*, **4**, 47–49.

Xu, W. and Südhof, T.C. (2013). A neural circuit for memory specificity and generalization. *Science*, **339**, 1290–1295.

Yassin, L., Benedetti, B.L., Jouhanneau, J.S. *et al.* (2010). An embedded subnetwork of highly active neurons in the neocortex. *Neuron*, **68**, 1043–1050.

Yiu, A.P., Mercaldo, V., Yan, C. *et al.* (2014). Neurons are recruited to a memory trace based on relative neuronal excitability immediately before training. *Neuron*, **83**, 722–735.

Yizhar, O., Fenno, L.E., Prigge, M. *et al.* (2011). Neocortical excitation/inhibition balance in information processing and social dysfunction. *Nature*, **477**, 171–178.

Yu, S., Huang, D., Singer, W. *et al.* (2008). A small world of neuronal synchrony. *Cereb. Cortex*, **18**, 2891-2901.

Zhang, S.-J., Ye, J., Miao, C. *et al.* (2013). Optogenetic dissection of entorhinal–hippocampal functional connectivity. *Science*, **340**, 1232627.

Part VI

Optogenetics in Sleep, Prosthetics, and Epigenetics of Neurodegenerative Diseases

28 Optogenetic Dissection of Sleep–Wake Control: Evidence for a Thalamic Control of Sleep Architecture

Charles-Francois Vincent Latchoumane and Hee-Sup Shin

28.1 Introduction

Sleep is a periodic, global state that is remarkably conserved across all mammals. Wakefulness is characterized by cortical desynchronization and a wide range of brain oscillations in correlation with the exploration of, and stimulation from, an individual's environment, as well as the cognitive demand and the arousal level of the moment. Sleep onset, on the other hand, is driven cyclically following a circadian rhythm and homeostatic demands. The ultradian regulation of sleep is sub-differentiated into a synchronized sleep state and a desynchronized wake-like state with a low-muscle tone state termed non-rapid-eye movement (NREM) and rapid-eye movement (REM) sleep, respectively (Whalley, 2015). The cyclic alteration of sleep and wake states is thought to be hard-coded into the genetic and metabolic oscillations of the brain that occur in correlation with environmental drivers such as light–dark cycles (Pace-Schott and Hobson, 2002). Sleep stages are generally observed through the resulting oscillatory signatures of electroencephalograms (EEGs) in conjunction with muscle tone through electromyograms (EMGs), but are also understood through systemic changes in hormones, blood pressure, breathing rate, movement per minute and eye movement. Sleep is therefore the behavioral reflection of a complex and fine regulation of the metabolic and cognitive capacity of the body.

Although sleep is undeniably recognized as one of the most naturally beneficial forms of rest for corporal and mental health, our modern lifestyle has entrained a forceful habit of maintenance of wakefulness for both professional and recreational reasons and has led to the increased propensity for sleep-related disorders and its associated socio-economical disadvantages (Venkatraman *et al.*, 2011). Prolonged wakefulness and sleep deprivation are known, for example, to result in impaired cognition (Williamson and Feyer, 2000a; Diekelmann and Born, 2010), a drunk-like state (Williamson and Feyer, 2000b), hallucination (Babkoff *et al.*, 1989), hormonal deregulation (Haus and Smolensky, 2013) and immune system deregulation, and could cascade into an irreversible loss of overall homeostasis, eventually leading to death (Rechtschaffen *et al.*, 1983). The essential nature of sleep is evident, and although the past decade has helped elucidate part of its mysteries, many aspects of sleep–wake regulation and control remain to be understood.

Thus far, sleep studies have mostly focused on sleep disorders and on lesion and pharmacological manipulations of sleep centers and their associated neurotransmitters and neuropeptides. Optogenetics, a method for the genetically targeted manipulation of neuronal activity using light, was developed a decade ago (Zemelman *et al.*, 2002; Lima and Miesenböck, 2005; Boyden *et al.*, 2005) and offers new approaches for the confirmation and probing of neuronal circuits, such as the one involved in sleep–wake regulation (Adamantidis *et al.*, 2009). Optogenetics offers millisecond-precision stimulation/inhibition of neurons compared to pharmacological and lesion approaches. In addition, a high spatial resolution can also be achieved using optogenetics through viral delivery coupled with promoter-based targeting and the cre-loxP recombination-like system (e.g. Deisseroth, 2012). In comparison, electrical stimulations exercise a blunt excitation of all cells and passing fibers contained in an arbitrarily sized region of the brain, affecting synaptic plasticity and other metabolic reservoirs all together. Optogenetics offers a variety of solutions for the detailed study of functional connectivity and neural correlates of behavior and has seen a surge in use for the elucidation of sleep–wake control.

In this review, we focus on the most recent findings in relation to optogenetic probing of the neural correlates of sleep–wake regulation. We first revisit the main advantages and limitations of the optogenetic approach for the study of neuronal circuits. We then review the overall sleep–wake regulation network and the recent optogenetic studies associated with it. It has long been established that the cortex, thalamus and subcortical regions are mostly part of the end route of sleep–wake regulation. However, recent findings have pointed to a stronger possible modulation of consciousness and sleep–wake states through the thalamocortical (TC) pathway (Steriade *et al.*, 1993b; McCormick and Bal, 1997; Kim *et al.*, 2012; Taylor and Crunelli, 2015). We thus propose a particular perspective on the role of the TC network as a possible strong modulator of sleep architecture and arousal.

28.2 The Main Advantages and Disadvantages of Optogenetic Control of Neuronal Circuits

The contradictory findings from previous lesion and pharmacological studies might be understood through their mere technical limitations. Firstly, it is rather difficult to find a sleep center, or a region in the brain in general, with a clearly homogeneous cell-type population. Therefore, lesions permanently disabling a brain region will likely affect all cells equally, leading to a dominant visible effect that does not reflect the true nature of the circuitry involved. Secondly, the nature and target of each neuron within the same region, even among the same cell type, might differ greatly, leading to differential activation of centers with opposite and, therefore, canceling effects. Thirdly, gene manipulation and permanent lesions often result in a long-term compensatory effect that might mask the possible natural role of a specific region and might lead to the conclusion that a region is not necessary for a particular cognitive function, while the sufficiency of induction/maintenance for the same cognitive function might have been possible. Optogenetics clearly solves these three issues through: (1) the possible selective targeting of cell type and region via a cre-loxP recombination-like system; (2) the possibility to control axon terminals from a source region at its target region; and (3) the precise temporal modulation of the excitation and inhibition of neurons with control of the baseline and recovery for possible compensatory effects. Optogenetic stimulations, however, do not yet offer a proper way to control the contributions of neuropeptides that act on a wide spatial scale and a slower and longer timeframe (Arrigoni and Saper, 2014). Nonetheless, optogenetic stimulations powerfully modulate states such as wakefulness and sleep. Techniques involving chemogenetic, light-sensing G-proteins and light control of the expression of synaptic receptors/channels are rapidly being developed (Dong *et al.*, 2010; Aston-Jones and Deisseroth, 2013) and might allow their own detailed approached to the sleep–wake regulation at all scales of space and time.

Other limitations of optogenetics arise from the artificial nature of this technique. Optogenetic techniques necessitate the illumination of the brain tissue, requiring high coherent light (i.e. laser) intensity or high-frequency stimulation and could result in a heating effect that might influence the activity of neurons (Tye and Deisseroth, 2012). However, the impact of these effects could be estimated and optimized using newly available modeling tools (Stujenske *et al.*, 2015) or more sensitive probes that would require lower input intensities. Finally, the activation of light-sensitive channels is known to alter the internal balance of the cell, mostly through changes in cytoplasmic ion and proton concentrations. The long-term stimulation or strong activation following the high expression of a channel might cause toxicity and malfunctioning of the cell, as observed with halorhodopsin (Fenno *et al.*, 2011). Most research using optogenetics-based approaches now attempts to include control experiments for those limitations in order to provide more reliable conclusions.

28.3 Overview of the Sleep–Wake Control Centers through Optogenetic Probing

Sleep–wake regulation depends on the complex interaction of multiple, distributed centers throughout the brain. Those centers have been characterized based on their dominant activity or maximal/minimal firing rate in correlation with one or two of the three states associated with sleep: wakefulness, REM sleep and NREM sleep. The organization of each center has been extensively studied (Pace-Schott and Hobson, 2002; Adamantidis *et al.*, 2009; Whalley, 2015), and we present here a short summary of those findings with a focus on optogenetics-based approaches.

28.3.1 The Central Pacemakers

The suprachiasmatic nucleus (SCN) has long been considered to be the major circadian pacemaker in the brain, receiving photonic inputs from the retina and other arousal centers. The SCN affects other sleep–wake-promoting centers, mostly indirectly, through the intermediary modulation of the medial pre-optic area, sub-paraventricular zone, dorsomedial hypothalamic nucleus (Deurveilher and Semba, 2005) and basal forebrain. In addition, recent studies using optogenetics and chemogenetics point to a tight link between synaptic terminals of vaso-intestinal protein-expressing neurons, calcium concentration and gene expression for the regulation of SCN circadian rhythmicity (Brancaccio *et al.*, 2013; Brancaccio *et al.*, 2014).

28.3.2 Transmitters of Pacemaking

The ventrolateral and median preoptic areas of the anterior hypothalamus are two of the few sleep active centers found in the brain. They act on the wake-promoting center mainly through inhibitory GABAergic projections (Sherin *et al.*, 1996; Saito *et al.*, 2013).

The basal forebrain is another central transmitter of the circadian rhythm (Sarter and Bruno, 1999). The cell type-specific microcircuit probing of the basal forebrain revealed that glutamatergic neurons are the top effectors of the wake-promoting effect of basal forebrain optogenetic stimulations, probably through the modulation of the cholinergic neurons that promote cortical desynchronization (Sarter and Bruno, 1999; Irmak and de Lecea, 2013; Yang *et al.*, 2014) and the parvalbumin (PV) neurons that promote cortical γ-rhythms (Alitto and Dan, 2013). Most cells in the basal forebrain were found to be wake/REM active; however, somatostatin (SST)-expressing neurons were found to be NREM active, and optic stimulation of SST neurons increased the probability for NREM sleep induction (Xu *et al.*, 2015). This result contrasts with a recent demonstration of chemogenetic activation of GABAergic stimulation of basal forebrain neurons that showed a dominant wake-promoting role (Anaclet *et al.*, 2015). The anterior thalamus and basal forebrain are also sensitive to the accumulation of stressor molecules and hormones following prolonged wakefulness, such as adenosine and cortisol, which are indicative of sleep pressure, and these two structures could in turn be down-modulated in favor of the homeostatic balance of sleep (Blanco-Centurion *et al.*, 2006; Yang *et al.*, 2013).

28.3.3 Central Modulator of Arousal, NREM Sleep and REM Sleep

Following the advent of optogenetics, two specific groups of cell have gained much attention for their crucial role in sleep–wake control: the hypocretin/orexin neurons (Bonnavion and de Lecea, 2010) and the melanin-concentrating hormone (MCH) neurons (Luppi *et al.*, 2013). Disorders of the lateral hypothalamus and lesion studies involving hypocretin cells have demonstrated their critical role in the induction and maintenance of wakefulness. Hypocretin neurons are active during the wake state and REM-to-wake transitions, but are completely silent in other states, including quiet wake. The optogenetic stimulation of hypocretin neurons at higher than 5-Hz frequencies could induce and increase the probability of waking from NREM slow-wave sleep (SWS) and REM sleep (Adamantidis *et al.*, 2007), whereas silencing of these neurons through archaerhodopsin induced

SWS (Tsunematsu *et al.*, 2013). Since the projection pattern of hypocretin neurons is complex and manifold, these neurons play various roles in cognition and behavior (Peyron *et al.*, 1998); however, their control over sleep–wake induction/maintenance is thought to take place through the glutamatergic stimulation of the basal forebrain (Eggermann *et al.*, 2001; España *et al.*, 2005), the posterior hypothalamus (Nishino *et al.*, 2001) and the brain stem ascending arousal centers (Carter *et al.*, 2010). The lateral dorsal hypothalamic nucleus is highly populated by MCH neurons, which are mostly REM-active neurons. Optic stimulation of these neurons increases the inhibitory release of GABA neurotransmitters, which promote both NREM and REM sleep and, in particular, can stabilize REM sleep duration if excited or slow the REM θ frequency if inhibited (Jego *et al.*, 2013). For their role in sleep–wake regulation, hypocretin and MCH neurons seem to display overlapping innervations of arousal centers, but may bidirectionally modulate them through excitation and inhibition, respectively.

28.3.4 Major Effectors of Wake-promotion and Ultradian NREM–REM Rhythms

The brain stem is home to some of the most powerful arousal centers that innervate a large portion of the brain through direct axonal projections or via the diffusion of neuro-modulators. Arousal and wake-promoting centers include the pedunculopontine tegmental nucleus (PPT)/laterodorsal tegmentum (LDT: cholinergic), the dorsal raphe nucleus (DRN: serotonergic), the tuberomammillary nucleus (histaminergic) and the locus coeruleus (LC: noradrenergic) (Pace-Schott and Hobson, 2002; Carter *et al.*, 2010). Optogenetic excitation and inhibition of the LC increase wake and reduce NREM sleep (Carter *et al.*, 2010), respectively. The strong and long-term induction/maintenance of wakefulness is thought to require a combined activation of wake-promoting centers in order to help sustain a cortical depolarization, leading to activation, desynchronization and arousal. The ultradian rhythms (i.e. alternation of NREM–REM sleep) involve the complex interaction of multiple REM-off/REM-on cell types of the mesopontine tegmental region and other pontine centers, such as the periaqueductal central gray, which are usually understood and integrated in the flip-flop model (Lu *et al.*, 2006) and the reciprocal interaction model (McCarley and Massaquoi, 1992; Pace-Schott and Hobson, 2002). For example, the optogenetic stimulation of the cholinergic neurons of the PPT or LDT could promote wakefulness, but more efficiently promote REM sleep for minute-long stimulations (Dort *et al.*, 2015).

28.3.5 Downstream Modulators of Sleep and Wakefulness

The anterior hypothalamus, basal forebrain and mesopontine tegmental centers all project to the cortico-thalamo-limbic regions through an ascending arousal pathway (Sarter and Bruno, 1999). The limbic regions are another type of relay that may modulate cortex depolarization through cholinergic and adrenergic drives and promote attention and arousal. In particular, the thalamus and the cortex are thought to orchestrate and provide a readout of the final brain state, and their oscillations and firings are the main indicators that are used in sleep staging in mammalians (Steriade *et al.*, 1993b). In the next section, we will detail the role of the TC network in sleep and its possible role in sleep–wake generation.

28.4 The TC System and the Signature Unconsciousness during Sleep

Sleep is the result of long-term adaptation to the Earth's diurnal cycles. While many genetic and physiological indicators of sleep exist (Pace-Schott and Hobson, 2002), the global changes within the brain are some of the main signs of sleep in mammalians. The long history of sleep staging has shown that the cortical EEG and muscle tone (EMG) are some of the easiest, most reliable indices for the characterization of the sleep state. In addition, one major characteristic of the sleep state is the loss of consciousness that accompanies it (Lindsley, 1960; Hobson and Pace-Schott, 2002), which is often tested and quantified on the basis of the loss of responsiveness to commands or arousing stimuli such as verbal responses, finger tapping, righting reflexes, loud sounds and painful stimulations. Since most conscious and unconscious expressions of cognition must transit through the cortex and thalamus, the TC loop (composed of corticothalamic [CT], TC and thalamic reticular neurons) has been thought to play a crucial role in the expression of unconsciousness (Van der Werf *et al.*, 2002; Schiff, 2008; Alkire *et al.*, 2009; Gummadavelli *et al.*, 2015) that arises during sleep, but also during states such as anesthesia and seizures.

28.4.1 Thalamus to Cortex

The TC loop has been extensively studied (Steriade and Llinás, 1988; Steriade *et al.*, 1993b). The thalamus is often separated into lower/higher order, or specific/non-specific, based on its input/output profile. Specific thalamic nuclei, such as the ventrobasal, lateral geniculate and medial geniculate thalamic nuclei, often play the role of sensory relay stations where sensory signals from the brain stem transit to reach their corresponding cortical processor, usually ending in layer IV of cortical columns (Llinás *et al.*, 1998). Non-specific thalamic nuclei are not directly associated with the relay of sensory information, instead relaying internal signals from other brain structures and participating in different types of cognition, ranging from emotion to memory to attention. The thalamus is mostly composed of excitatory neurons, except for the thalamic reticular nucleus (TRN), which is almost exclusively GABAergic (Houser *et al.*, 1980; Steriade, 2001) and has one of the strongest expressions of low-threshold t-type calcium channels (Cav3.1, Cav3.2 and Cav3.3) in the brain (Talley *et al.*, 1999). Thalamic neurons are wake/REM active and show correlation with arousal as well, but are not silent during NREM sleep (Steriade *et al.*, 1993b). Rather, the conjunction of t-type calcium channel de-inactivation following the intense hyperpolarizing GABAergic/cholinergic drive of the NREM state induces a shift in the firing pattern of thalamic cells from tonic to phasic/bursting (Steriade *et al.*, 1993b). The bursting properties of thalamic neurons may have multiple functions, including sensory gating (McCormick and Bal, 1994), top-down differentiation of stimulus saliency (Montemurro *et al.*, 2007; Mease *et al.*, 2014), dendritic integration (Cueni *et al.*, 2008) or downstream short-term/long-term plasticity changes (Linden *et al.*, 2009).

The oscillations (<1 Hz) and δwaves (1–4 Hz) are synchronized oscillations of UP- and DOWN-state characteristics of NREM sleep and might be generated and sustained by the thalamus (Crunelli *et al.*, 2015). Optogenetic stimulation of TC neurons was shown to induce the cortical UP state (Poulet *et al.*, 2012) and TC/TRN neurons can sustain an UP and DOWN state in isolated conditions (Crunelli and Hughes, 2010). Due to its central position and its distributed projection pattern, the TC loop seems essential in helping to synchronize long-range cortical regions (Stroh *et al.*, 2013; Mesbah-Oskui *et al.*, 2014;). During sleep, slow, synchronized activity of the cortex can lead to the interruption of inter-regional communication and might result in the loss of global information transfer and, eventually, loss of consciousness. The closed-loop optogenetic inhibition of thalamic neurons was able to interrupt seizures through the disabling of thalamic synchronization (Paz *et al.*, 2013). Similar closed-loop optogenetic activation (5–200-ms light pulses) of ventrobasal thalamic neurons during spontaneous absence seizures could interrupt the behavior arrest and could also initiate such behaviors during a state of low arousal (Taylor and Crunelli, 2015). The role of the thalamus on the cortex is, therefore, to relay the preprocessed sensory–motor stimuli (active/quiet wake) and implement positive feedback onto cortical activity, enabling synchronization and gating during slow/δ-wave/sleep spindle oscillations (Steriade *et al.*, 1993a; De Gennaro and Ferrara, 2003; Crunelli *et al.*, 2015).

28.4.2 Top-down Cortical Control of Thalamic Activity

The cortical pyramidal neurons of layer V/VI are known to project to the TC neurons through excitatory glutamatergic projections. Cortical layer VI can recruit thalamic neurons and alter the response to sensory afferent inputs by switching their response mode from burst to tonic through depolarization using single short-pulse optogenetic stimulation (Mease *et al.*, 2014). The mechanism involved in the cortical top-down control of TC neurons is thought to involve metabotropic glutamatergic receptors (mGluR) and their coupling with phospholipase C β4, as well as with t-type calcium channels (McCormick and von Krosigk, 1992; Cheong *et al.*, 2008; Cheong *et al.*, 2009). The CT projections can therefore alter the excitability and firing mode of thalamic neurons and, as a result, modulate the frequency, intensity and synchronization of thalamic activity (Contreras *et al.*, 1996; Blumenfeld and McCormick, 2000). Between the excitatory crosstalk of TC and CT neurons stands the major inhibitory input and "guardian" of thalamic gating, the TRN.

28.4.3 The TRN

The TRN is primarily composed of a majority of GABAergic neurons with a large overlap of PV- and SST-expressing cells (Pinault, 2004). TRN cells only project to TC neurons, but receive input from a large number of major brain centers, including arousal centers. The TRN is considered to be a major pacemaker and generator of thalamic oscillation. Optic stimulation of TRN cells was shown to induce burst (Halassa *et al.*, 2011) and to control the initiation and termination of fast waxing and waning sleep spindle oscillations *in vivo* (Barthó *et al.*, 2014). TRN cells are also able to generate and interrupt slower oscillations including the slow wave-and-spike and wave discharges associated with absence seizures (Pinault, 2004; Paz *et al.*, 2013; Lee *et al.*, 2014; Roux *et al.*, 2014).

28.5 Optogenetic Probing of the TC Pathway for the Control of Sleep

Although heavily involved in the generation of NREM sleep oscillations, the TC loop has been mostly considered in an arousal-promoting role and less as a sleep promoter. However, the removal of the thalamus in a cat was shown to cause important reductions in NREM and REM sleep (Villablanca and Salinas-Zeballos, 1972). In particular, recent optogenetic stimulation of the TRN has shown that it is possible to induce burst in this section of the TC loop and produce spindles (Halassa *et al.*, 2011; Kim *et al.*, 2012). TRN-generated spindles have been found to strongly correlate with sleep quality and arousal thresholds (De Gennaro and Ferrara, 2003; Astori *et al.*, 2013). Kim *et al.* demonstrated for the first time a possible direct role of the thalamic spindles in influencing sleep architecture (Kim *et al.*, 2012). The down-modulation of arousal was later confirmed by continuous pulse stimulation of the TRN (Lewis *et al.*, 2015). Optically induced spindles increased the probability of NREM to REM transitions, which in turn increased the total time spent in NREM sleep (Kim *et al.*, 2012). The more sustained burst and spindle oscillations in SK2-overexpressing mice also result in an increase in arousal threshold in the mice, increasing the number of duration of bouts of NREM sleep (Wimmer *et al.*, 2012). Therefore, the thalamus might be able to indirectly delay or gate arousal through spindles and TRN bursting; however, the exact mechanism that links the spindles, spindle density, NREM–REM transition and sleep remains to be discovered.

28.6 Future Directions and Discussion

The optogenetic dissection of sleep–wake regulation has brought even more interest to an already popular field. Although most of the current discoveries are solid confirmations of already-known pathways and mechanisms, a more detail dissection of the microcircuit *in vivo* allows us to observe the natural responses of a complex system by manipulating its processor units: the neuron, its multiple cell types and their properties.

However, the ascending arousal system is highly redundant in its role of promoting the depolarization of cortical and subcortical neurons and facilitating desynchronization through different types of monoaminergic and neurotransmitter types, confusing the timing of slow diffusion and direct excitation. Arousal centers such as the LC can reduce wakefulness and induce a wake-to-NREM transition following photo-inhibition and photo-excitation, respectively, indicating that they can both initiate and maintain an aroused state (Carter *et al.*, 2010). Hypocretin neurons of the lateral hypothalamus project to and excite arousal centers such as the LC and DRN, orchestrating wakefulness. Interestingly, excitation of hypocretin cells increases the probability of transition from NREM sleep to wake (Adamantidis *et al.*, 2007), but the silencing of these neurons can cause all mice to enter SWS within 40 s in the light period of the circadian rhythm, but not the dark period (Tsunematsu *et al.*, 2013). Therefore, the strength of a center in sleep–wake promotion might vary greatly based on the circadian rhythm itself and the combined activation of similar centers. A recipe for the full control of the induction and maintenance of the wake, NREM sleep and REM sleep states remains to be found. Closed-loop monitoring of population firing and optogenetic manipulation of several arousal centers could investigate this complex, multicenter control of the brain's global state in a circadian-dependent manner. Therapeutic approaches to inhibiting the proper centers at night might help to restore a regular circadian rhythm in disorders of sleep such as hypo-/hyper-somnia and neuropathic pain (Ito *et al.*, 2013).

The mechanism underlying the TC control of sleep and consciousness remains widely unexplored as well. Is it possible that synchronized oscillations in the cortex and thalamus could modulate ascending arousal signaling? Where does the transition from light sleep to deep sleep originate from? Are there questions that could be dissected using combined opto- and chemo-genetics for the specific targeting of axon terminals and long-range aminergic and neuropeptidic modulation?

Sleep architecture is known to dynamically change over the course of a mammalian life. From newborn to adulthood, a polyphasic, REM-heavy sleep architecture is gradually replaced by a dominant, nocturnal NREM sleep (Feinberg, 1974; Welsh *et al.*, 1986). Recent work using chemogenetic stimulation could functionally differentiate two populations of the pontine tegmentum originating from the same embryonic cell linage. Medial and lateral pontine glutamatergic neurons could increase NREM sleep and increase wake, respectively, at the expense of REM sleep (Hayashi *et al.*, 2015). These authors also found that the mechanism for the regulation of REM sleep by these two populations of the pontine tegmentum was through the excitation of GABAergic neurons of the deep mesencephalic nucleus that, in turn, inhibited the REM-promoting neurons of the sub-laterodorsal nucleus. This work raises the following questions for the future understanding of sleep and, in particular, REM sleep: (1) why is REM sleep dominant in the neonate or even needed at all? (2) How does cell migration and differentiation from a common cell linage affect the propensity for REM/NREM sleep in neonate mammalians? (3) How could cells of similar embryonic linage and promoters regulate sleep in species with no obvious REM and NREM sleep, such as reptiles?

Most importantly, consciousness as a self-reflecting, stable and reproducible state naturally rises and fades during sleep–wake cycles. Several centers are thought to contribute to and control the emergence of consciousness (Van der Werf *et al.*, 2002; Schiff, 2008; Mathur, 2014; Leeman-Markowski *et al.*, 2015);

however, optogenetics-based approaches might help us to understand how consciousness is modulated in relation to arousal, wake–NREM–REM cycles and brain oscillations. Optogenetics, without being a magic wand, might play a crucial role in the discovery and dissection of the complex networks involved in one of the great mysteries of neuroscience: the control of sleep.

ACKNOWLEDGMENTS

This review is performed under the support of the grant IBS-R001-D1 from the Institute for Basic Science, Korea.

REFERENCES

Adamantidis, A., Carter, M.C., De Lecea, L., 2009. Optogenetic deconstruction of sleep–wake circuitry in the brain. *Front. Mol. Neurosci.* **2**, 31.

Adamantidis, A.R., Zhang, F., Aravanis, A.M., Deisseroth, K., de Lecea, L., 2007. Neural substrates of awakening probed with optogenetic control of hypocretin neurons. *Nature* **450**, 420–424.

Alitto, H.J., Dan, Y., 2013. Cell-type-specific modulation of neocortical activity by basal forebrain input. *Front. Syst. Neurosci.* **6**, 79.

Alkire, M.T., Asher, C.D., Franciscus, A.M., Hahn, E.L., 2009. Thalamic microinfusion of antibody to a voltage-gated potassium channel restores consciousness during anesthesia. *Anesthesiology* **110**, 766–773.

Anaclet, C., Pedersen, N.P., Ferrari, L.L., Venner, A., Bass, C.E., Arrigoni, E., Fuller, P.M., 2015. Basal forebrain control of wakefulness and cortical rhythms. *Nat. Commun.* **6**, 8744.

Arrigoni, E., Saper, C.B., 2014. What optogenetic stimulation is telling us (and failing to tell us) about fast neurotransmitters and neuromodulators in brain circuits for wake–sleep regulation. *Curr. Opin. Neurobiol.* **29**, 165–171.

Aston-Jones, G., Deisseroth, K., 2013. Recent advances in optogenetics and pharmacogenetics. *Brain Res.* **1511**, 1–5.

Astori, S., Wimmer, R.D., Lüthi, A., 2013. Manipulating sleep spindles – expanding views on sleep, memory, and disease. *Trends Neurosci.* **36**, 738–748.

Babkoff, H., Sing, H.C., Thorne, D.R., Genser, S.G., Hegge, F.W., 1989. Perceptual distortions and hallucinations reported during the course of sleep deprivation. *Percept. Mot. Skills* **68**, 787–798.

Barthó, P., Slézia, A., Mátyás, F., Faradzs-Zade, L., Ulbert, I., Harris, K.D., Acsády, L., 2014. Ongoing network state controls the length of sleep spindles via inhibitory activity. *Neuron* **82**, 1367–1379.

Blanco-Centurion, C., Xu, M., Murillo-Rodriguez, E., Gerashchenko, D., Shiromani, A.M., Salin-Pascual, R.J., Hof, P.R., Shiromani, P.J., 2006. Adenosine and sleep homeostasis in the basal forebrain. *J. Neurosci.* **26**, 8092–8100.

Blumenfeld, H., McCormick, D.A., 2000. Corticothalamic inputs control the pattern of activity generated in thalamocortical networks. *J. Neurosci.* **20**, 5153–5162.

Bonnavion, P., de Lecea, L., 2010. Hypocretins in the control of sleep and wakefulness. *Curr. Neurol. Neurosci. Rep.* **10**, 174–179.

Boyden, E.S., Zhang, F., Bamberg, E., Nagel, G., Deisseroth, K., 2005. Millisecond-timescale, genetically targeted optical control of neural activity. *Nat. Neurosci.* **8**, 1263–1268.

Brancaccio, M., Enoki, R., Mazuski, C.N., Jones, J., Evans, J.A., Azzi, A., 2014. Network-mediated encoding of circadian time: the suprachiasmatic nucleus (SCN) from genes to neurons to circuits, and back. *J. Neurosci.* **34**, 15192–15199.

Brancaccio, M., Maywood, E.S., Chesham, J.E., Loudon, A.S.I., Hastings, M.H., 2013. A Gq-Ca^{2+} axis controls circuit-level encoding of circadian time in the suprachiasmatic nucleus. *Neuron* **78**, 714–728.

Carter, M.E., Yizhar, O., Chikahisa, S., Nguyen, H., Adamantidis, A., Nishino, S., Deisseroth, K., de Lecea, L., 2010. Tuning arousal with optogenetic modulation of locus coeruleus neurons. *Nat. Neurosci.* **13**, 1526–1533.

Cheong, E., Lee, S., Choi, B.J., Sun, M., Lee, C.J., Shin, H.-S., 2008. Tuning thalamic firing modes via simultaneous modulation of T- and L-type Ca^{2+} channels controls pain sensory gating in the thalamus. *J. Neurosci.* **28**, 13331–13340.

Cheong, E., Zheng, Y., Lee, K., Lee, J., Kim, S., Sanati, M., Lee, S., Kim, Y.-S., Shin, H.-S., 2009. Deletion of phospholipase C β4 in thalamocortical relay nucleus leads to absence seizures. *Proc. Natl. Acad. Sci.* **106**, 21912–21917.

Contreras, D., Destexhe, A., Sejnowski, T.J., Steriade, M., 1996. Control of spatiotemporal coherence of a thalamic oscillation by corticothalamic feedback. *Science* **274**, 771–774.

Crunelli, V., David, F., Lőrincz, M.L., Hughes, S.W., 2015. The thalamocortical network as a single slow wave-generating unit. *Curr. Opin. Neurobiol.* **31**, 72–80.

Crunelli, V., Hughes, S.W., 2010. The slow (<1 Hz) rhythm of non-REM sleep: a dialogue between three cardinal oscillators. *Nat. Neurosci.* **13**, 9–17.

Cueni, L., Canepari, M., Luján, R., Emmenegger, Y., Watanabe, M., Bond, C.T., Franken, P., Adelman, J.P., Lüthi, A., 2008. T-type Ca^{2+} channels, SK2 channels and SERCAs gate sleep-related oscillations in thalamic dendrites. *Nat. Neurosci.* **11**, 683–692.

De Gennaro, L., Ferrara, M., 2003. Sleep spindles: an overview. *Sleep Med. Rev.* **7**, 423–440.

Deisseroth, K., 2012. Optogenetics and psychiatry: applications, challenges, and opportunities. *Biol. Psychiatry* **71**, 1030–1032.

Deurveilher, S., Semba, K., 2005. Indirect projections from the suprachiasmatic nucleus to major arousal-promoting cell groups in rat: implications for the circadian control of behavioural state. *Neuroscience* **130**, 165–183.

Diekelmann, S., Born, J., 2010. The memory function of sleep. *Nat. Rev. Neurosci.* **11**, 114–126.

Dong, S., Allen, J.A., Farrell, M., Roth, B.L., 2010. A chemical-genetic approach for precise spatio-temporal control of cellular signaling. *Mol. Biosyst.* **6**, 1376–1380.

Dort, C.J.V., Zachs, D.P., Kenny, J.D., Zheng, S., Goldblum, R.R., Gelwan, N.A., Ramos, D.M., Nolan, M.A., Wang, K., Weng, F.-J., Lin, Y., Wilson, M.A., Brown, E.N., 2015. Optogenetic activation of cholinergic neurons in the PPT or LDT induces REM sleep. *Proc. Natl. Acad. Sci.* **112**, 584–589.

Eggermann, E., Serafin, M., Bayer, L., Machard, D., Saint-Mleux, B., Jones, B.E., Mühlethaler, M., 2001. Orexins/hypocretins excite basal forebrain cholinergic neurones. *Neuroscience* **108**, 177–181.

España, R.A., Reis, K.M., Valentino, R.J., Berridge, C.W., 2005. Organization of hypocretin/ orexin efferents to locus coeruleus and basal forebrain arousal-related structures. *J. Comp. Neurol.* **481**, 160–178.

Feinberg, I., 1974. Changes in sleep cycle patterns with age. *J. Psychiatr. Res.* **10**, 283–306.

Fenno, L., Yizhar, O., Deisseroth, K., 2011. The development and application of optogenetics. *Annu. Rev. Neurosci.* **34**, 389–412.

Gummadavelli, A., Motelow, J.E., Smith, N., Zhan, Q., Schiff, N.D., Blumenfeld, H., 2015. Thalamic stimulation to improve level of consciousness after seizures: Evaluation of electrophysiology and behavior. *Epilepsia* **56**, 114–124.

Halassa, M.M., Siegle, J.H., Ritt, J.T., Ting, J.T., Feng, G., Moore, C.I., 2011. Selective optical drive of thalamic reticular nucleus generates thalamic bursts and cortical spindles. *Nat. Neurosci.* **14**, 1118–1120.

Haus, E.L., Smolensky, M.H., 2013. Shift work and cancer risk: potential mechanistic roles of circadian disruption, light at night, and sleep deprivation. *Sleep Med. Rev.* **17**, 273–284.

Hayashi, Y., Kashiwagi, M., Yasuda, K., Ando, R., Kanuka, M., Sakai, K., Itohara, S., 2015. Cells of a common developmental origin regulate REM/non-REM sleep and wakefulness in mice. *Science* **350**, 957–961.

Hobson, J.A., Pace-Schott, E.F., 2002. The cognitive neuroscience of sleep: neuronal systems, consciousness and learning. *Nat. Rev. Neurosci.* **3**, 679–693.

Houser, C.R., Vaughn, J.E., Barber, R.P., Roberts, E., 1980. GABA neurons are the major cell type of the nucleus reticularis thalami. *Brain Res.* **200**, 341–354.

Irmak, S.O., de Lecea, L., 2013. Basal forebrain cholinergic modulation of sleep transitions. *Sleep* **37**, 1941–1951.

Ito, H., Yanase, M., Yamashita, A., Kitabatake, C., Hamada, A., Suhara, Y., Narita, M., Ikegami, D., Sakai, H., Yamazaki, M., Narita, M., 2013. Analysis of sleep disorders under pain using an optogenetic tool: possible involvement of the activation of dorsal raphe nucleus-serotonergic neurons. *Mol. Brain* **6**, 59.

Jego, S., Glasgow, S.D., Herrera, C.G., Ekstrand, M., Reed, S.J., Boyce, R., Friedman, J., Burdakov, D., Adamantidis, A.R., 2013. Optogenetic identification of a rapid eye movement sleep modulatory circuit in the hypothalamus. *Nat. Neurosci.* **16**, 1637–1643.

Kim, A., Latchoumane, C., Lee, S., Kim, G.B., Cheong, E., Augustine, G.J., Shin, H.-S., 2012. Optogenetically induced sleep spindle rhythms alter sleep architectures in mice. *Proc. Natl. Acad. Sci.* **109**, 20673–20678.

Lee, S.E., Lee, J., Latchoumane, C., Lee, B., Oh, S.-J., Saud, Z.A., Park, C., Sun, N., Cheong, E., Chen, C.-C. *et al.*, 2014. Rebound burst firing in the reticular thalamus is not essential for pharmacological absence seizures in mice. *Proc. Natl. Acad. Sci.* **111**, 11828–11833.

Leeman-Markowski, B.A., Smart, O.L., Faught, R.E., Gross, R.E., Meador, K.J., 2015. Cessation of gamma activity in the dorsomedial nucleus associated with loss of consciousness during focal seizures. *Epilepsy Behav.* **51**, 215–220.

Lewis, L.D., Voigts, J., Flores, F.J., Schmitt, L.I., Wilson, M.A., Halassa, M.M., Brown, E.N., 2015. Thalamic reticular nucleus induces fast and local modulation of arousal state. *Elife* **4**, e08760.

Lima, S.Q., Miesenböck, G., 2005. Remote control of behavior through genetically targeted photostimulation of neurons. *Cell* **121**, 141–152.

Linden, M.L., Heynen, A.J., Haslinger, R.H., Bear, M.F., 2009. Thalamic activity that drives visual cortical plasticity. *Nat. Neurosci.* **12**, 390–392.

Lindsley, D.B., 1960. Attention, consciousness, sleep, and wakefulness, in: Magoun, H.W., Hall, V. (Eds.), *Handbook of Physiology*. American Physiological Society.

Llinás, R., Ribary, U., Contreras, D., Pedroarena, C., 1998. The neuronal basis for consciousness. *Philos. Trans. R. Soc. B Biol. Sci.* **353**, 1841–1849.

Lu, J., Sherman, D., Devor, M., Saper, C.B., 2006. A putative flip–flop switch for control of REM sleep. *Nature* **441**, 589–594.

Luppi, P.H., Peyron, C., Fort, P., 2013. Role of MCH neurons in paradoxical (REM) sleep control. *Sleep* **36**, 1775–1776.

Mathur, B.N., 2014. The claustrum in review. *Front. Syst. Neurosci.* **8**, 48.

McCarley, R.W., Massaquoi, S.G., 1992. Neurobiological structure of the revised limit cycle reciprocal interaction model of REM cycle control. *J. Sleep Res.* **1**, 132–137.

McCormick, D.A., Bal, T., 1994. Sensory gating mechanisms of the thalamus. *Curr. Opin. Neurobiol.* **4**, 550–556.

McCormick, D.A., Bal, T., 1997. Sleep and arousal: thalamocortical mechanisms. *Annu. Rev. Neurosci.* **20**, 185–215.

McCormick, D.A., von Krosigk, M., 1992. Corticothalamic activation modulates thalamic firing through glutamate "metabotropic" receptors. *Proc. Natl. Acad. Sci.* **89**, 2774–2778.

Mease, R.A., Krieger, P., Groh, A., 2014. Cortical control of adaptation and sensory relay mode in the thalamus. *Proc. Natl. Acad. Sci.* **111**, 6798–6803.

Mesbah-Oskui, L., Orser, B.A., Horner, R.L., 2014. Thalamic δ-subunit containing GABAA receptors promote electrocortical signatures of deep non-REM sleep but do not mediate the effects of etomidate at the thalamus *in vivo*. *J. Neurosci.* **34**, 12253–12266.

Montemurro, M.A., Panzeri, S., Maravall, M., Alenda, A., Bale, M.R., Brambilla, M., Petersen, R.S., 2007. Role of precise spike timing in coding of dynamic vibrissa stimuli in somatosensory thalamus. *J. Neurophysiol.* **98**, 1871–1882.

Nishino, S., Fujiki, N., Ripley, B., Sakurai, E., Kato, M., Watanabe, T., Mignot, E., Yanai, K., 2001. Decreased brain histamine content in hypocretin/orexin receptor-2 mutated narcoleptic dogs. *Neurosci. Lett.* **313**, 125–128.

Pace-Schott, E.F., Hobson, J.A., 2002. The neurobiology of sleep: genetics, cellular physiology and subcortical networks. *Nat. Rev. Neurosci.* **3**, 591–605.

Paz, J.T., Davidson, T.J., Frechette, E.S., Delord, B., Parada, I., Peng, K., Deisseroth, K., Huguenard, J.R., 2013. Closed-loop optogenetic control of thalamus as a tool for interrupting seizures after cortical injury. *Nat. Neurosci.* **16**, 64–70.

Peyron, C., Tighe, D.K., van den Pol, A.N., Lecea, L. de, Heller, H.C., Sutcliffe, J.G., Kilduff, T.S., 1998. Neurons containing hypocretin (Orexin) project to multiple neuronal systems. *J. Neurosci.* **18**, 9996–10015.

Pinault, D., 2004. The thalamic reticular nucleus: structure, function and concept. *Brain Res. Rev.* **46**, 1–31.

Poulet, J.F., Fernandez, L.M., Crochet, S., Petersen, C.C., 2012. Thalamic control of cortical states. *Nat. Neurosci.* **15**, 370–372.

Rechtschaffen, A., Gilliland, M.A., Bergmann, B.M., Winter, J.B., 1983. Physiological correlates of prolonged sleep deprivation in rats. *Science* **221**, 182–184.

Roux, L., Stark, E., Sjulson, L., Buzsáki, G., 2014. *In vivo* optogenetic identification and manipulation of GABAergic interneuron subtypes. *Curr. Opin. Neurobiol.* **26**, 88–95.

Saito, Y.C., Tsujino, N., Hasegawa, E., Akashi, K., Abe, M., Mieda, M., Sakimura, K., Sakurai, T., 2013. GABAergic neurons in the preoptic area send direct inhibitory projections to orexin neurons. *Front. Neural Circuits* **7**, 192.

Sarter, M., Bruno, J.P., 1999. Cortical cholinergic inputs mediating arousal, attentional processing and dreaming: differential afferent regulation of the basal forebrain by telencephalic and brainstem afferents. *Neuroscience* **95**, 933–952.

Schiff, N.D., 2008. Central thalamic contributions to arousal regulation and neurological disorders of consciousness. *Ann. N. Y. Acad. Sci.* **1129**, 105–118.

Sherin, J.E., Shiromani, P.J., McCarley, R.W., Saper, C.B., 1996. Activation of ventrolateral preoptic neurons during sleep. *Science* **271**, 216–219.

Steriade, M., 2001. The GABAergic reticular nucleus: a preferential target of corticothalamic projections. *Proc. Natl. Acad. Sci.* **98**, 3625–3627.

Steriade, M., Contreras, D., Dossi, R.C., Nunez, A., 1993a. The slow (< 1 Hz) oscillation in reticular thalamic and thalamocortical neurons: scenario of sleep rhythm generation in interacting thalamic and neocortical networks. *J. Neurosci.* **13**, 3284–3299.

Steriade, M., Llinás, R.R., 1988. The functional states of the thalamus and the associated neuronal interplay. *Physiol. Rev.* **68**, 649–742.

Steriade, M., McCormick, D.A., Sejnowski, T.J., 1993b. Thalamocortical oscillations in the sleeping and aroused brain. *Science* **262**, 679–685.

Stroh, A., Adelsberger, H., Groh, A., Rühlmann, C., Fischer, S., Schierloh, A., Deisseroth, K., Konnerth, A., 2013. Making waves: initiation and propagation of corticothalamic Ca^{2+} waves *in vivo*. *Neuron* **77**, 1136–1150.

Stujenske, J.M., Spellman, T., Gordon, J.A., 2015. Modeling the spatiotemporal dynamics of light and heat propagation for *in vivo* optogenetics. *Cell Rep.* **12**, 525–534.

Talley, E.M., Cribbs, L.L., Lee, J.-H., Daud, A., Perez-Reyes, E., Bayliss, D.A., 1999. Differential distribution of three members of a gene family encoding low voltage-activated (t-type) calcium channels. *J. Neurosci.* **19**, 1895–1911.

Taylor, H.L., Crunelli, V., 2015. Optogenetic drive of thalamocortical neurons can block and induce experimental absence seizures in freely moving animals. *Proc. Physiol. Soc.* 34.

Tsunematsu, T., Tabuchi, S., Tanaka, K.F., Boyden, E.S., Tominaga, M., Yamanaka, A., 2013. Long-lasting silencing of orexin/hypocretin neurons using archaerhodopsin induces slow-wave sleep in mice. *Behav. Brain Res.* **255**, 64–74.

Tye, K.M., Deisseroth, K., 2012. Optogenetic investigation of neural circuits underlying brain disease in animal models. *Nat. Rev. Neurosci.* **13**, 251–266.

Van der Werf, Y.D., Witter, M.P., Groenewegen, H.J., 2002. The intralaminar and midline nuclei of the thalamus. Anatomical and functional evidence for participation in processes of arousal and awareness. *Brain Res. Rev.* **39**, 107–140.

Venkatraman, V., Huettel, S.A., Chuah, L.Y.M., Payne, J.W., Chee, M.W.L., 2011. Sleep deprivation biases the neural mechanisms underlying economic preferences. *J. Neurosci.* **31**, 3712–3718.

Villablanca, J., Salinas-Zeballos, M.E., 1972. Sleep–wakefulness, EEG and behavioral studies of chronic cats without the thalamus: the "athalamic" cat. *Arch. Ital. Biol.* **110**, 383–411.

Welsh, D.K., Richardson, G.S., Dement, W.C., 1986. Effect of age on the circadian pattern of sleep and wakefulness in the mouse. *J. Gerontol.* **41**, 579–586.

Whalley, K., 2015. Sleep: dissecting sleep circuits. *Nat. Rev. Neurosci.* **16**, 704.

Williamson, A., Feyer, A., 2000a. Moderate sleep deprivation produces impairments in cognitive and motor performance equivalent to legally prescribed levels of alcohol intoxication. *Occup. Environ. Med.* **57**, 649–655.

Williamson, A.M., Feyer, A.-M., 2000b. Moderate sleep deprivation produces impairments in cognitive and motor performance equivalent to legally prescribed levels of alcohol intoxication. *Occup. Environ. Med.* **57**, 649–655.

Wimmer, R.D., Astori, S., Bond, C.T., Rovó, Z., Chatton, J.-Y., Adelman, J.P., Franken, P., Lüthi, A., 2012. Sustaining sleep spindles through enhanced SK2-channel activity consolidates sleep and elevates arousal threshold. *J. Neurosci.* **32**, 13917–13928.

Xu, M., Chung, S., Zhang, S., Zhong, P., Ma, C., Chang, W.-C., Weissbourd, B., Sakai, N., Luo, L., Nishino, S., Dan, Y., 2015. Basal forebrain circuit for sleep-wake control. *Nat. Neurosci.* **18**, 1641–1647.

Yang, C., Franciosi, S., Brown, R.E., 2013. Adenosine inhibits the excitatory synaptic inputs to basal forebrain cholinergic, GABAergic, and parvalbumin neurons in mice. *Front. Neurol.* **4**, 77.

Yang, C., McKenna, J.T., Zant, J.C., Winston, S., Basheer, R., Brown, R.E., 2014. Cholinergic neurons excite cortically projecting basal forebrain GABAergic neurons. *J. Neurosci.* **34**, 2832–2844.

Zemelman, B.V., Lee, G.A., Ng, M., Miesenböck, G., 2002. Selective photostimulation of genetically ChARGed neurons. *Neuron* **33**, 15–22.

29 **Optogenetics and Auditory Implants**

Jenny X. Chen, Elliott Kozin, M. Christian Brown, and Daniel J. Lee

29.1 Introduction

The World Health Organization estimates that 360 million people, representing nearly 5% of the world's population, suffer from disabling hearing loss with functional, social and economic consequences. The majority of these patients have mild to moderate sensorineural hearing loss, and there are no medical or surgical therapies to regenerate the inner ear as of 2016. Hearing aids remain the only option in these cases. In contrast, cochlear implants (CIs) and auditory brainstem implants (ABIs) have become standards of care for deaf children and adults for providing partial restoration of hearing (O'Donoghue, 2013). Auditory implants convert sound into electrical pulses that are used to stimulate neurons of the auditory pathways via a multichannel electrode array. Although meaningful percepts are generated, auditory implant users struggle with background noise, sound localization and musical appreciation.

Due to the limitations of existing CI and ABI technology, several auditory research groups have used optogenetics to explore the possibility of using light in place of electricity for neural stimulation. The rationale is that, unlike electricity, light can be focused, theoretically resulting in greater numbers of independent channels of frequency information. Harnessing this greater number of channels might improve performance. Focused stimulation would also reduce side effects from off-target activation of non-auditory pathways, a problem of great concern in ABIs. Overall, optogenetics has the potential to provide improved outcomes for patients who use an auditory implant.

This chapter begins by examining the history of auditory implant development with a focus on ABIs because their limitations are much greater than those of CIs. We then discuss how recent basic science investigations into optogenetics by our group and others offer glimpses into the future of hearing implants based on light. We conclude by identifying obstacles to the translation of optogenetic tools in order to improve clinical auditory implants, pinpointing how future clinical and scientific studies might overcome these challenges.

29.2 The History of Auditory Implants

The first attempt to electrically stimulate the ear dates back to the eighteenth century, shortly after Alessandro Volta invented the first electrolytic cell. During one experiment, Volta connected a battery to two metal rods and discovered that when he inserted the rods into his ear canals he felt a jolt and heard a crackling noise (Eshraghi *et al.*, 2012). In 1957, Djourno and Eyries electrically stimulated the auditory nerve of a bilaterally deaf patient during surgery, causing the patient to report sound sensations (Eisen, 2003). Subsequent experiments in the 1960s spurred the development of the first CI in 1972, wherein single electrodes were placed into the cochleae of patients with profound hearing loss, including children. This first CI was rudimentary, but users benefited from sound awareness that enhanced lip-reading abilities (Fretz and Fravel, 1985).

Modern CI technology provides meaningful sound and speech perception without the need for lip-reading in the majority of pediatric and adult users. In the normal-hearing cochlea, hair cells and their respective afferent nerve fibers are arranged along a tonotopic gradient such that the basal cochlea is maximally stimulated by high-frequency stimuli, while the apical cochlea is stimulated by low-frequency stimuli. Flexible, multichannel CI arrays are surgically placed into the scala tympani of the spiral-shaped cochlea via the round window to approximate this frequency-specific or tonotopic organization of the inner ear (Figure 29.1B).

A small subset of children and adults with profound hearing loss are not candidates for CIs due to anatomic constraints, such as absent/hypoplastic or scarred cochleae or cochlear nerves. One group of patients who develop progressive bilateral deafness and do not benefit from CIs have a rare, autosomal dominant syndrome called neurofibromatosis type 2 (NF2), which is found in 1 in 33,000 newborns (Evans, 1993). This syndrome is associated with mutations of the tumor-suppressor gene *NF2* (chromosome 22q) and results in the development of multiple brain, skull base and spinal cord tumors. The hallmark of NF2 is bilateral vestibular schwannomas, benign tumors that arise on the vestibular nerve in both ears. Growth and/or surgical resection of these tumors results in profound hearing loss. Because the auditory nerve is damaged or severed, cochlear implantation is not a hearing-rehabilitative option for deaf NF2 patients, except in rare cases in which the anatomic integrity of the nerve is preserved.

ABIs bypass the cochlea and cochlear nerve to directly stimulate the cochlear nucleus, and were originally developed at the House Ear Institute in order to provide hearing sensations to NF2 patients (Eisenberg, 2015). The first ABI was placed in 1979 in a woman with NF2 and included a percutaneous connector that coupled the externally worn microphone and speech processor to the surgically implanted receiver stimulator. Two ball electrodes were placed on the surface of the patient's cochlear nucleus in the brainstem. Transition to a wireless transcutaneous connector and multichannel array culminated in a 21-electrode ABI that was eventually approved for NF2 patients by the Food and Drug Administration (FDA) in 2000. The majority of ABI users with NF2 do not have full restoration of speech comprehension, but they experience sound

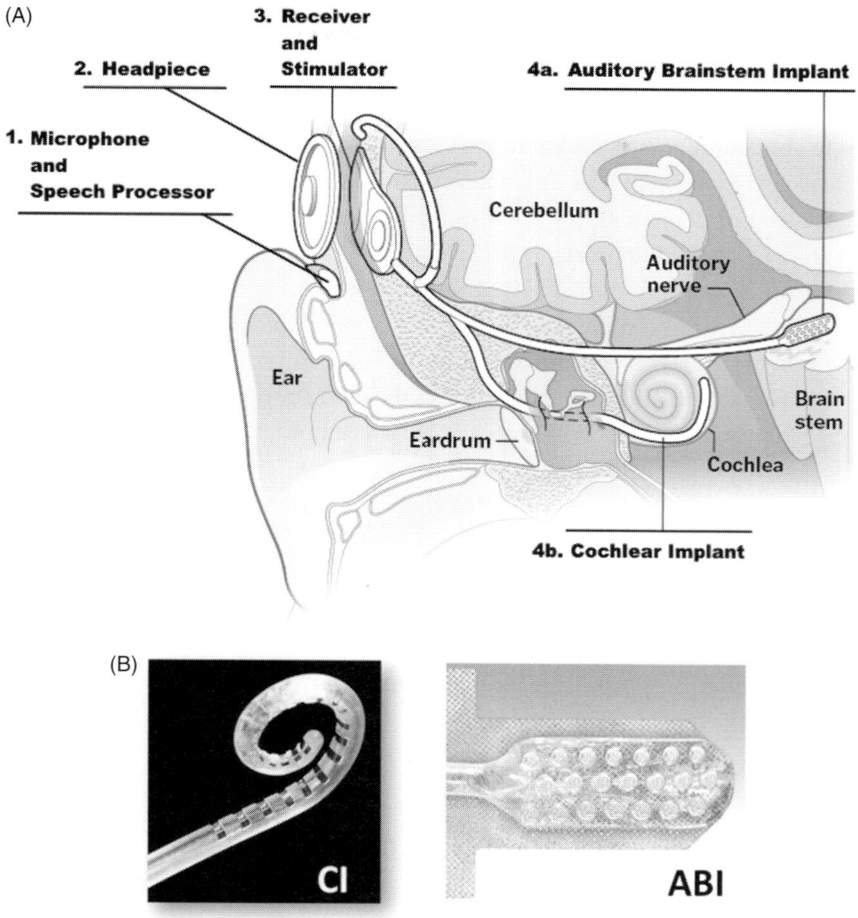

Figure 29.1 Contemporary auditory implants stimulate peripheral (CI) or central (ABI) neurons using multichannel electrode arrays. (A) Sound from the external environment is collected by (1) the microphone and transformed into an electrical code in the speech processor that also contains the battery. This information is sent from (2) an external transmitter, called the headpiece, to (3) a surgically implanted receiver/simulator that is powered through induction from the speech processor. The external and internal components are brought into close proximity across the skin using magnets of opposite polarity in the headpiece and the receiver/stimulator. The receiver/stimulator is connected by wires leading to an electrode array. The CI array, receiver/stimulator and wires are placed during a 2hour out-patient procedure called a mastoidectomy. The electrode array is placed into the scala tympani of the cochlea (4a). In contrast, the ABI is placed during a more extensive surgical approach called a craniotomy, which exposes the region of the auditory nerve termination in the brainstem center called the cochlear nucleus. The ABI uses the same speech processor as the CI and activates (4b) an array of electrodes that stimulate cochlear nucleus neurons. Image adapted from (Neff, 2014). (B) The CI is a multichannel array in a Silastic sheath and approximates the first one to two turns of the spiral-shaped cochlea. The CI electrically stimulates the first-order auditory neurons, called spiral ganglion cells, in a pattern that follows the tonotopic organization of the cochlea, where the most basal electrodes are used to encode the highest acoustic frequencies and more apical electrodes encode lower frequencies. In contrast, the ABI electrode is a multichannel array and is placed on the surface of the cochlear nucleus.
As of July 2016, there are no ABI devices that are FDA approved and the electrode that is shown is from a device that is now being employed under compassionate use indications and with the oversight of the FDA. (A black-and-white version of this figure will appear in some formats. For the color version, please refer to the plate section.)

sensations that enhance lip-reading abilities (Ebinger *et al.*, 2000). Penetrating electrodes were also developed to improve the device; unfortunately, this design did not improve speech perception (Otto *et al.*, 2008).

New indications for ABIs under clinical investigation in the USA and abroad include congenitally deaf children with severe malformations of the cochlea or cochlear nerve hypoplasia or aplasia, or post-lingually deaf children and adults with scarring or damage of the cochlea from meningitis, otosclerosis or trauma (Colletti *et al.*, 2014; Puram *et al.*, 2016; Sennaroğlu *et al.*, 2016;).

The modern auditory implant has five functional components (Figure 29.1A). External to the patient, there is a microphone, a speech processor and a headpiece with a transmitting coil. The microphone and speech processor is worn around the ear, collecting sound information from the environment and delivering it to the speech processor. The contemporary ABI speech processor is identical to that of CIs and is worn behind and around the pinna. The speech processor is a computer that separates sounds into their composite frequency spectra and organizes the information into an electrical code. This code is then conveyed to a transmitter that delivers the signal to the internal components of the implant: the receiver and the stimulating electrode array. The transmitter sends signals wirelessly to a receiver that is the surgically implanted under the skin at a position above and behind the pinna. The receiver in turn conveys the coded signals to the electrode array. In CIs, a one-dimensional, linear electrode array is surgically placed into the scala tympani of the cochlea (Figure 29.1B). Electrical stimulation of first-order auditory neurons called spiral ganglion cells results in the propagation of an action potential along the cochlear nerve and into the cochlear nucleus of the brainstem. Eventually, the higher auditory pathways are activated, resulting in a sound percept. The ABI is a modified CI with a two-dimensional, paddle-type electrode array (Figure 29.1B). It is placed on the surface of the cochlear nucleus through a more invasive approach involving a craniotomy. There, it stimulates second-order cochlear nucleus neurons, which subsequently activate higher auditory pathways.

29.3 Modern Implant Performance

As of 2015, over 300 000 people have received CIs and more than 1000 patients have received ABIs (Kaplan *et al.*, 2015). Auditory implant surgery can be life changing: post-lingually deafened adults report substantial improvements in open set speech understanding by as soon as 3 months after CI surgery (Geier *et al.*, 1999; Waltzman *et al.*, 1999). Pediatric CI users benefit from dramatic gains in speech and language development. In one study following children for 10–14 years after implant surgery, 77% of patients produced speech that was intelligible to an average listener, and all patients were in school or engaged in their communities (Beadle *et al.*, 2005).

Mean auditory outcomes of ABI users are lower than those for CI users. They range from modest gains in sound appreciation and environmental sound detection (Ebinger *et al.*, 2000; Nevison *et al.*, 2002) to the acquisition of meaningful open set sentence recognition in a minority of users (Colletti *et al.*, 2012). ABIs also enhance lip-reading (Nevison *et al.*, 2002; Otto *et al.*, 2004; Otto *et al.*, 2008).

While initially developed for adult patients with NF2, the indications for ABIs have expanded. Presently, the indications include both adult and pediatric patients who are not candidates for CIs due to congenital or acquired conditions, including absent or deficient cochlea or cochlear nerves or scarring of the cochlea. Outcomes for these expanded populations suggest that ABIs can be effective tools for hearing rehabilitation in patients with non-tumor causes of hearing loss. Recent follow-up of tumor and non-tumor pediatric patients implanted with the ABI shows significant promise: in one study of 64 consecutive children, all patients demonstrated significant improvement in auditory perception (Colletti *et al.*, 2014). Seven children (11%) were able to converse via telephone within 3 years of implantation. All children showed continued improvements over time. A 2015 systematic review of pediatric ABI users concluded that all users had improved sound detection (as measured on a Categories of Auditory Performance scale), and children without disabilities were more likely to have better outcomes (Noij *et al.*, 2015). Mean outcomes of non-tumor adult ABI users are better than NF2 ABI users; although the reasons for this difference are not understood, they may include tumor-related injury of the brainstem (Brackmann *et al.*, 1993; Matthies *et al.*, 2000; Lim *et al.*, 2008; Colletti *et al.*, 2009).

The auditory midbrain implant (AMI) is an experimental device that stimulates at an even higher level of the auditory pathway: the inferior colliculus (IC). One idea supporting the use of the AMI is that it bypasses the cochlear nucleus, which may be compromised by tumor or the surgery to remove it. To date, three patients have been implanted in clinical trials, and all three patients demonstrated hearing responses (Lim *et al.*, 2008). The patient with the most optimally positioned AMI gained sound perception and significant improvements in consonant and number recognition, as well as enhancements in lip-reading accuracy (Lim *et al.*, 2008).

29.4 Limitations of Electrical Implants

Despite ongoing advances in the hardware and software components of auditory implants, specific challenges remain for patients in search of hearing restoration. One key limitation of all auditory prostheses (CIs, ABIs and AMIs) remains: the use of electricity results in a broad pattern of activation of neurons in the general vicinity of the activated electrode. This pattern is expected to be much broader than that produced by sound in an individual with normal hearing. In addition, overlapping current fields from different electrodes, or channel interaction, reduce the number of effective channels available for transmitting information (Boëx *et al.*, 2003; Baskent, 2006; Stickney *et al.*, 2006; Bingabr *et al.*, 2008). For example, while modern CIs employ one to two dozen electrodes, studies suggest that no more than four to eight distinct sites should be active at any one time because of the substantial overlap from nearby electrodes as they sit in the conductive fluid of the perilymph (Fishman *et al.*, 1997; Garnham *et al.*, 2002).

This makes poor use of the nearly 30 000 tonopically organized spiral ganglion neurons that are fine tuned to specific frequencies and may reduce speech comprehension, especially in noisy environments (22,23). Processing strategies such as continuous interleaved stimulation are designed to minimize electrode channel interaction (Wilson, 2013).

Similarly, studies of ABIs indicate that increasing the number of electrodes beyond five does not significantly improve perceptual outcomes (Kuchta *et al.*, 2004). The broad spread of current and overlapping fields of electrical stimulation may be even more important for ABIs as the central auditory pathways consist of complex networks of excitatory and inhibitory interneurons that may all be affected by non-specific electrical stimulation. Current spread often results in off-target effects, causing ABI users to experience side effects such as pain and dizziness due to the stimulation of adjacent cranial nerve centers in the brainstem, such as the trigeminal or vestibular nerves (Colletti *et al.*, 2010).

29.5 Optical Stimulation Techniques

Alternative strategies for auditory neuronal stimulation are being explored in order to improve spatial resolution. Radiant energy has been proposed as an alternative approach to an electric current, since light can be focused in order to activate smaller regions, thus reducing channel interaction (Wells *et al.*, 2005; Peron and Svoboda, 2011). Two optical strategies are under investigation for use in auditory implants: infrared neural stimulation (INS) and optogenetics.

29.5.1 Infrared Neural Stimulation

INS refers to the exposure of *unmodified* neural tissue to pulsed, mid-wavelength radiant energy delivered with an optical fiber in order to elicit an action potential. Infrared light has been shown to depolarize unmodified neurons with significantly reduced spread of neural activation (<300 μm) compared with electrical stimulation (Wells *et al.*, 2005). This technology has been applied *in vivo* to stimulate embryonic quail hearts (Jenkins *et al.*, 2010), non-primate visual cortex (Cayce *et al.*, 2014) and gerbil cochleae (Izzo *et al.*, 2007). Advantages of INS include non-contact stimulation and there being no need for genetic manipulation of the target tissue. However, there are several aspects of INS that may affect its translation into the clinic. Infrared light generates thermal gradients in order to depolarize neurons, which results in the accumulation of heat energy that could be harmful to tissues (Wells *et al.*, 2007a; Wells, *et al.*, 2007b; Richter and Tan, 2014). Moreover, the energy required for infrared stimulation exceeds what is practical for use in portable prostheses (Richter *et al.*, 2011). Recently, the use of INS has been further challenged by the finding that INS does not trigger auditory activity in the completely deafened auditory system, suggesting that neuronal responses could be an opto-acoustic artefact mediated by mechanical stimulation of residual hair cells (Verma *et al.*, 2014; Thompson *et al*, 2015). Finally, the long-term repercussions of radiant infrared energy on sensitive neural tissue have not been well characterized.

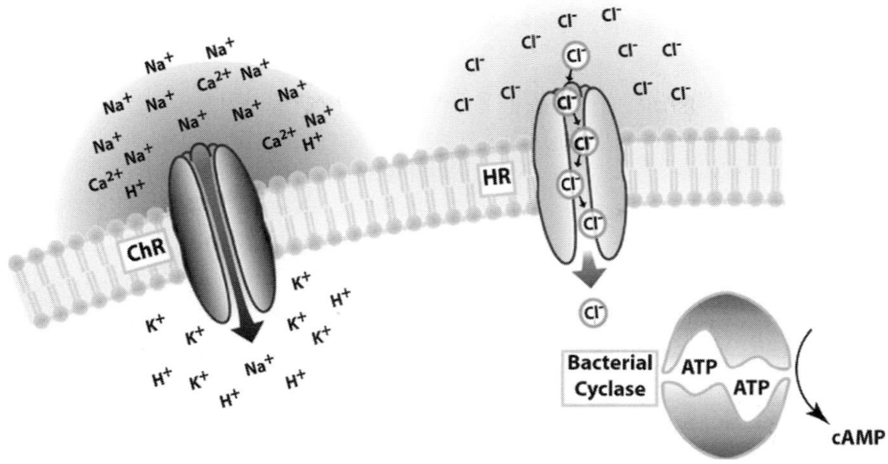

Figure 29.2 Fundamentals of optogenetics. Radiant exposure of opsins to specific wavelengths of light opens these channels to allow the influx of cations into the interior of the neuron. In the case of ChR2 (left), the influx of cations depolarizes the neuron, whereas in the case of halorhodopsin (HR; right), the influx of anions hyperpolarizes the neuron. Expression of opsins on peripheral or central auditory neurons requires transgenic, viral or stem cell-mediated approaches prior to stimulation with light. Figure adapted from Yizhar *et al.* (2011). (A black-and-white version of this figure will appear in some formats. For the color version, please refer to the plate section.)

29.5.2 Optogenetics

Optogenetics is a technique that is in widespread use in neuroscience research. This technique uses genetically modified neurons that express light-gated membrane channels called "opsins" (Figure 29.2). One of the first opsins used in neuroscience is called "channelrhodopsin-2" (ChR2). ChR2 was isolated from the green algae *Chlamydomonas reinhardtii* and is expressed on the surface of neurons using viral vector-mediated delivery (Boyden *et al.*, 2005; Han and Boyden, 2007). In the presence of blue light, channelrhodopsins are ion channels that open to allow inward flow of cations, which depolarizes neurons. ChR2-expressing neurons can be activated by focused beams of light with spatial selectivity on the scale of micrometers (Bernstein *et al.*, 2008). Preliminary evidence suggests that enhanced spatial resolution in the peripheral and central auditory system can be achieved with optogenetics. In transgenic ChR2-expressing mice, light stimulation of the cochlea using a 250-μm optical fiber was compared with monopolar electrical stimulation and 31-kHz tone bursts (Hernandez *et al.*, 2014). The spread of excitation using optogenetics (475 ± 65.5 μm) was significantly more focused than in electrical stimulation (828.6 ± 101.7 um) and comparable to acoustic responses (666.7 ± 95.7 μm). Narrowing the laser beam diameter using adjustable collimators can improve spatial selectivity in the auditory brainstem (Hight *et al.*, 2015a). Models of the dimensions of activated neural tissue suggest that the depth and diameter of a beam of light in neural tissue varies from a few micrometers to several hundred micrometers as a function of the beam power (Hirschbiegel, 2015).

Mutating isolated amino acid residues in ChR2 has resulted in an extended family of opsins (Figure 29.3), including opsins with red-shifted light sensitivity

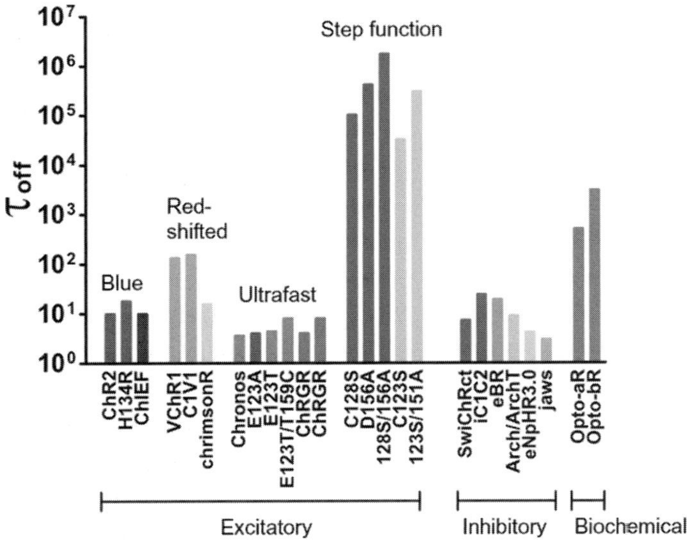

Figure 29.3 Classes of opsins. Chart showing the off-time constants (in milliseconds) of a variety of opsins. Since the discovery of ChR2, a range of opsins has been discovered or bioengineered from existing opsins, including red-shifted, ultrafast, on/off step function and inhibitory opsins. The ideal opsins for the auditory system are low threshold with ultrafast on/off kinetics and a brief refractory period for encoding the high pulse trains that are required for speech perception. Figure adapted from Guru *et al.* (2015). (A black-and-white version of this figure will appear in some formats. For the color version, please refer to the plate section.)

(such as VChR1, C1V1 and chrimsonR) (Lin *et al.*, 2013), opsins with faster kinetics (E123T, also known as ChETA) (Gunaydin *et al.*, 2010; Lin, 2011) and opsins that can be switched on for increased periods of time with a brief pulse of light, enabling a stable step in membrane potential (e.g. by modifying the C128 position of ChR2) (Berndt *et al.*, 2009). Other families of naturally occurring opsins that have been discovered include inhibitory opsins such as halorhodopsins (e.g. eNpHR3.0), which are ion channels that are selective for chloride ions and allow for the inactivation of specific neural circuits (Gradinaru *et al.*, 2010). The development of inhibitors is of particular importance as it allows us to conduct of loss-of-function experiments at the level of the neuron.

Optogenetic tools have been used in a variety of basic science and translational applications. Early experiments mapped neural circuits in brain slices and probed a range of animal models in order to determine the neural populations involved in memory formation, learning, vision, wakening and sensation (Boyden, 2011). Animal disease models have been created using optogenetics in order to study depression, schizophrenia, addition, Parkinson's disease and epilepsy, among other neuropsychiatric phenomena (Tye and Deisseroth, 2012). Outside the central nervous system, clinical fields that have demonstrated potential applications of optogenetics include ophthalmology, wherein optogenetics is under investigation for vision restoration (Barrett *et al.*, 2014), and cardiology, wherein optogenetics may be used for cardiac pacing (Nussinovitch and Gepstein, 2015). In 2009, ChR2 was shown to be safe for use in awake non-human primate brains, portending future clinical applications (Han *et al.*, 2009).

29.5.3 Optogenetics in the Auditory Nervous System

This section details the various considerations for applying optogenetic technologies to the auditory nervous system, exploring strategies for opsin delivery and tissue-specific expression, the temporal characteristics of optogenetic tools, their energy requirements and the electronic design of an optogenetic implant. We conclude by summarizing the literature on the *in vivo* and behavioral studies of optogenetic implants in animal models.

29.5.3.1 Opsin Delivery

One obstacle for the future application of optogenetics is the dependence on viral vectors for delivery of the opsin. A range of ethical and technical hurdles stands between current studies that manipulate virally delivered opsins in the laboratory and the future use of optogenetic technologies in human patients. The need for a viral vector to deliver opsins offers specific benefits and risks. Adeno-associated viruses (AAVs) do not integrate into the genome, but enable lasting expression of opsins in the central auditory system (for up to at least 18 months in murine neurons) without causing cellular toxicity (Shimano *et al.*, 2013). Notably, the use of live viruses in humans is being addressed across many fields that depend on the viral expression of novel genes as a component of therapy (Kay *et al.*, 2001). To date, AAV delivery has proven safe for use in the human eye for the treatment of Leber's congenital amaurosis, and it is poised for application in a number of other clinical areas (Asokan *et al.*, 2012). Genetic manipulation of peripheral sensory neurons is affected by many variables, including transduction method, host immunity, virus genus and strain, virus titer and virus design (with cell type-specific or -non-specific promoters). For example, studies of AAV serotypes for gene delivery into L4–L5 dorsal root ganglions in mice confirmed that specific strains offer significantly more efficient gene transduction and gene expression than others (Mason *et al.*, 2010). Optogenetic manipulation of the peripheral auditory nervous system is possible using ChR2-expressing transgenic mouse lines (Hernandez *et al.*, 2014), but this model is not applicable in the clinic. Transduction of spiral ganglion cells with opsin-expressing AAV2/1 viruses has been successfully documented *in vitro* and *in vivo* in neonatal mice (Meng *et al.*, 2014).

29.5.3.2 Tissue-specific Expression of Opsins

Given the limitations of CIs and ABIs, several groups have explored the utility of optogenetic tools to mimic patterns of natural auditory information. This is particularly crucial for central auditory prostheses, given that the large drop in performance of ABIs as compared with CIs may be attributable to the enormous complexity of the central nervous system as compared with peripheral nerve fibers. While modern ABIs electrically stimulate vast areas of interconnected neurons and supporting cells, optogenetic tools might one day enable stimulation of targeted auditory networks. The ability to deliver opsins to specific neural networks or neuron types promises exciting new ways to generate and propagate complex signals along auditory pathways.

To date, the level of cell selectivity of opsin expression has been limited to discriminating between neuronal and non-neuronal cell populations. For example, Thy1.2 and Synapsin1 promoters have been used to drive expression in auditory neural populations (Hernandez *et al.*, 2014; Guo *et al.*, 2015). Future experiments may attempt to achieve control of very narrow cell populations as more is learned through *in situ* hybridization and quantitative polymerase chain reaction experiments about differential gene expression patterns in subsets of neurons both in the cochlea and in the auditory brainstem. Such experiments have revealed that the CamKII promoter is preferentially expressed in pyramidal neurons in the cortex (Dittgen *et al.*, 2004), and that N-methyl-D-aspartate receptor subunit I is differentially expressed in cell types of the cochlear nucleus (Bilak *et al.*, 1996). However, thoughtful promoter selection is important in order to ensure optimal opsin delivery; for example, AAV is limited by its small 4.8-kb size and cannot efficiently package long promoter sequences (Wu *et al.*, 2010). Changing promoters may also impact light-driven responses in unexpected ways: preliminary reports from ChR2-expressing transgenic mice suggest that a VGLUT-2 promoter does not mediate light-driven responses in the cochlear nucleus, even though diffuse opsin expression was confirmed on fluorescence microscopy (Kozin *et al.*, 2015).

Concurrently, a palette of opsins that are responsive to non-overlapping wavelengths of light can used to drive more complex patterns of neuronal activity, which might be used to simulate more natural signals compared to the large areas of neuron activation caused by untargeted electrical stimulation. Investigations into highly red-shifted opsins such as Chrimson raise the possibility of introducing multiple classes of opsins simultaneously that are responsive to different frequencies of light (Klapoetke *et al.*, 2014). Multiple opsins can be introduced either via numerous viral constructs or a single construct expressing multiple channelrhodopsins that are stochastically expressed, such as in "Autobow" mice (Cai *et al.*, 2013). Combined with smaller and more dynamic optical arrays, these strategies can greatly improve the intricacy of the transmitted code in specific neural populations. With the use of multichannel micro-LED arrays, new strategies for speech processing will need to be developed in order to code incoming sound into complex patterns of optical stimulation. These future experiments will simultaneously explore a more nuanced understanding of the natural patterns of neuronal firing that transmit the most accurate information to higher levels of the brain, as read out by behavioral and neurophysiological experiments.

29.5.3.3 Temporal Characteristics of Opsin Response

For specific application in the auditory nervous system, a fundamental problem remains: most existing channelrhodopsins have inherently slow channel kinetics, which reduces their ability to deliver high-frequency neural stimulation above 40–50 Hz (Boyden *et al.*, 2005; Darrow *et al.*, 2015). One opsin that has been shown to stimulate responses at over 100 Hz is ChETA, but this opsin was demonstrated in later experiments to have significantly lower light sensitivity and strong desensitization (Lin, 2011). The ability to entrain high rates with significant temporal fidelity is important for application in the auditory system as central auditory neurons are known to fire at several hundred Hertz with exceptional precision in order to encode the rapidly changing characteristics of human speech (Miller *et al.*, 2006; Golding and Oertel, 2012).

Recently, a novel channelrhodopsin named "Chronos" was isolated from the algal species *Stigeoclonium helveticum*. Patch clamp recordings from cultured neurons and brain slices initially demonstrated that Chronos has the fastest channel kinetics of any channelrhodopsin described to date, combined with high light sensitivity (Figure 29.3) (Klapoetke *et al.*, 2014). The speed of this new opsin has been attributed to its narrow deactivation time of approximately 3 ms, compared with a ChR2 deactivation time of 20 ms. Neurophysiological comparisons of Chronos and standard ChR2 in the central auditory pathways *in vivo* show convincing evidence that Chronos-mediated responses cause entrainment to pulse rates of over 200 Hz (Figure 29.4) (Guo *et al.*, 2015; Hight *et al.*, 2015b). Interestingly, for both ChR2 and Chronos, there was still an unsynchronized response to even higher pulse rates.

29.5.3.4 Optoelectronics

Optical stimulation in auditory prostheses could be achieved either through the internal generation of light at the stimulated region or through guiding light into the organ from an external source. Both strategies are under active investigation in the fields of material sciences and bioengineering in order to create flexible

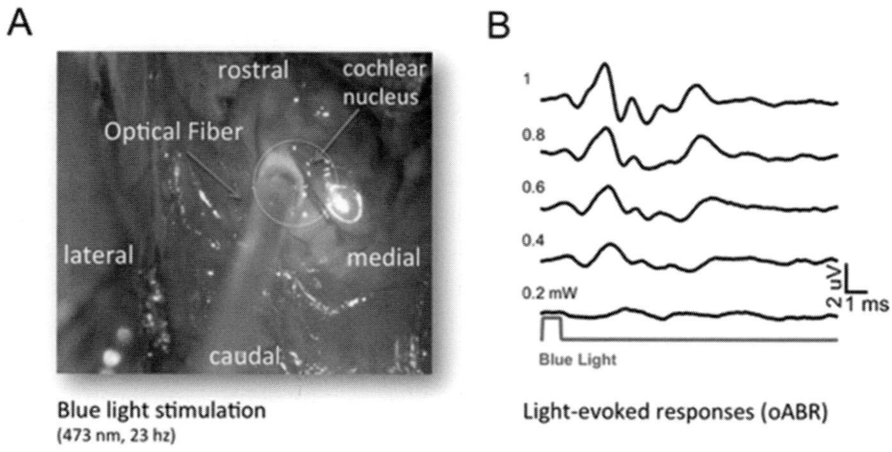

Figure 29.4 Optogenetic model of the ABI. (A) Experiment to stimulate the mouse cochlear nucleus with light conveyed by an optical fiber (400 μm diameter). Transduction of auditory brainstem neurons was achieved by direct inoculation of the cochlear nucleus via posterior craniotomy with an AAV vector with ChR2 or Chronos under the control of a ubiquitous promoter (CAG). Following a 1 month incubation period, the cochlear nucleus was re-exposed and stimulated with pulsed blue light (laser wavelength 475 nm for ChR2 or 500 nm for Chronos; 14–448 pulses s^{-1}, train durations 300–500 ms, light intensity 0–13 mW). (B) Resulting far-field responses are optically evoked auditory brainstem responses. These multi-peaked waveforms resembled sound-evoked auditory brainstem responses. No responses were observed from postmortem subjects or from control mice without cochlear nucleus opsin expression. Figure adapted from Hight et al. (2015b). (A black-and-white version of this figure will appear in some formats. For the color version, please refer to the plate section.)

arrays for cochlear application or stiff arrays for brainstem implantation (Kim *et al.*, 2013; Goßler, 2014). Internal light generation may be accomplished through thin-film LEDs; these arrays have been created in a variety of colors and can be miniaturized and even integrated in flexible substrates (Tian *et al.*, 2016) that are suitable for future use *in vivo*. Micro-LEDs have been developed that can elicit responses in ChR2-expressing neurons with micrometer and millisecond resolutions (Grossman *et al.*, 2010). Flexible micro-LED arrays with approximately 100 LEDs per square centimeter have been shown to be suitable for implantation into rat cochleae (Keppeler *et al.*, 2015). Inorganic LEDs have been miniaturized to a "cellular scale" (in one report, such LEDs were miniaturized to 100 μm thick with lateral dimensions of 1 mm^2), enabling spatially precise delivery of photons with reduced thermal spreading (Kim *et al.*, 2013). Recently, a soft, stretchable, wireless optoelectronic system was developed that was composed of a radiofrequency harvesting unit that amplifies signals from an external transmitter and multiplies the voltages in order to power LEDs (Park *et al.*, 2015).

If active solutions prove difficult to miniaturize and optimize for tissue safety, external solutions such as wave-guiding arrays might be favorable, because separation of the energy source from the tissues offers safety and device stability. Wave guides have been demonstrated in optogenetic animal models (Hernandez *et al.*, 2014), but light energy can be lost during transfer.

29.5.3.5 Power Requirements

The power requirements of an optogenetically based auditory implant for use in humans are currently unknown. In most studies, optoelectronics are relatively simple, such as intracochlear stimulation with 250-μm optical fibers. One study found that optical stimulation of spiral ganglion neurons expressing ChR2 could occur with an order of magnitude less energy than for infrared stimulation: optically evoked auditory brainstem responses (ABRs) were elicited with 2 μJ of blue light entering a 1-mm^2 cochleostomy, as compared with 16–150 μJ mm^{-2} for infrared light (Hernandez *et al.*, 2014). Energy requirements could be further reduced by using intracochlear emitters placed closer to targeted cells and by increasing the light sensitivity of advanced opsins such as Chronos (Klapoetke *et al.*, 2014).

29.5.3.6 *In Vivo* Studies

Several landmark *in vivo* studies have demonstrated the utility of optogenetics in designing auditory implants. Shimano *et al.* first demonstrated that ChR2 expression in the dorsal cochlear nucleus could activate auditory neurons in a rat model (Shimano *et al.*, 2013). Using AAV2 with a ubiquitous CAG promoter, ChR2 tagged with GFP was expressed in all layers of the dorsal cochlear nucleus and persisted for 18 months. The opsin expression did not impact acoustic responses as measured by ABR thresholds in response to sound. Blue light stimulation of the cochlear nucleus after surgical craniotomy resulted in sustained activity in the cochlear nucleus neurons. Darrow *et al.* further demonstrated that auditory far-field responses could be evoked (optical

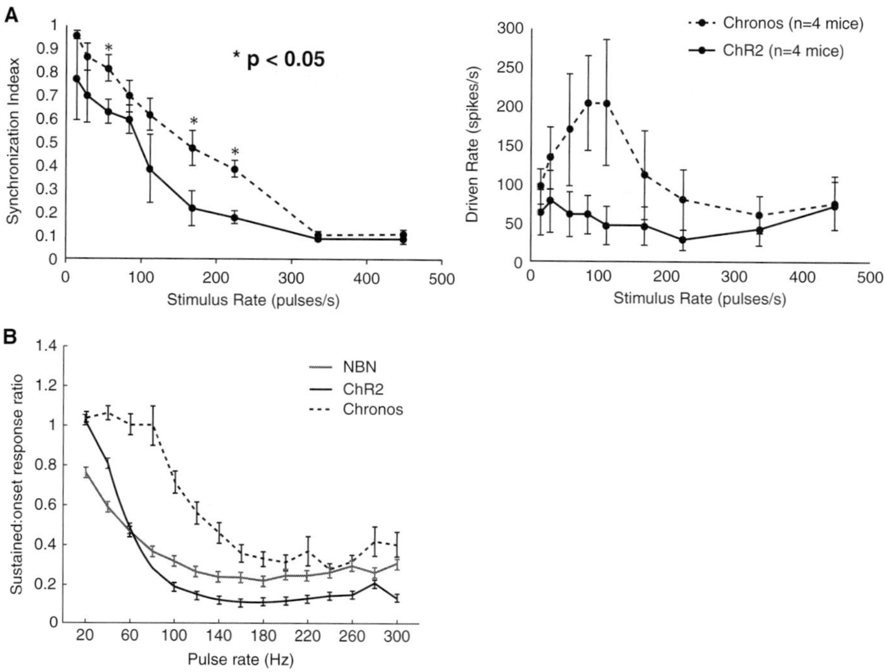

Figure 29.5 (A) Plot of synchronization index and (B) driven rate as a function of stimulus rate in an *in vivo* model of the ABI (figures from Hight *et al.*, 2015b). Pulsed blue light was delivered to the photosensitized cochlear nucleus (see Figure 29.5A). Responses comprised multi-unit activity recorded from the IC. Compared to ChR2, responses in Chronos animals showed higher synchronization rates (and driven rates). (C) Optogenetic responses in an *in vivo* AMI model (optrode placed in the IC for stimulation and recording). The plot illustrates firing rate adaptation as represented by the ratio of spikes evoked by the first and the following stimulus pulses during each train of optogenetic or acoustic stimulation. Chronos is able to support higher stimulus rates with more robust responses compared to ChR2 or sound stimulation using narrow band noise. Figure from Guo *et al.* (2015).

ABRs) by the optogenetic control of cochlear nucleus neurons using pulsed blue light delivered by an optical fiber, and that multiunit activity could be directly recorded during these same experiments from higher auditory centers (the IC and the auditory cortex) (Figure 29.5) (Darrow *et al.*, 2015). No optically induced responses were detected in sham-injected mice, ruling out an opto-phonic artifact. Similarly, in transgenic mice that express ChR2 in spiral ganglion cells, optical stimulation of the cochlea evokes ABRs and local field potentials upstream in the IC (Hernandez *et al.*, 2014). Other opsins, including the new ultrafast opsin Chronos, have also been tested *in vivo* (see Section 29.5.3.3).

29.5.3.7 Behavioral Studies

One study to date has examined the behavioral read out of an optogenetic auditory implant. Guo *et al.* created the first mouse model of a midbrain auditory implant, comparing ChR2 and Chronos opsins (Figure 29.6) (Guo *et al.*, 2015). After unilateral IC injections of saline or viral constructs, mice were trained to perform an auditory avoidance task. An optic fiber assembly was implanted into

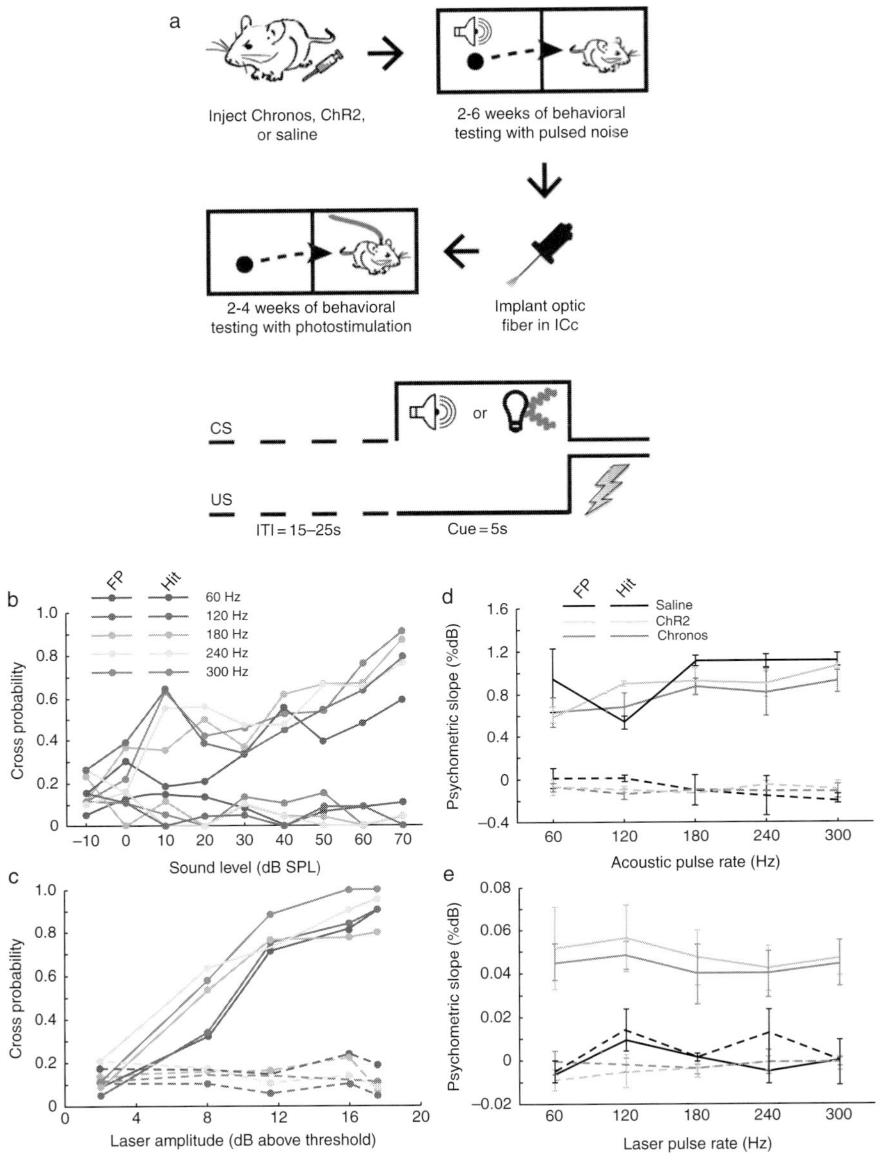

Figure 29.6 Behavioral detection of photostimulation mediated by ChR2 is similar to sound stimulation, but was not superior in Chronos-expressing mice. (A) Animals with Chronos, ChR2 or saline injected into the IC (midbrain) were trained to perform a behavioral detection task in response to pulsed noise. Animals cross the shuttle box to avoid a foot shock. After successful training, animals were implanted with chronic optical fibers at the site of prior injections and the detection task was repeated with photostimulation. Correct and false-positive crossing probabilities are calculated as a function of (B) sound level and (C) laser amplitude. The slope of the linear fit of psychometric curves is used as a proxy for pulse detection. The psychometric slopes for ChR2- and Chronos-expressing mice are not significantly different across changing rates of (D) acoustic or (E) laser stimulation. Figure adapted from Guo *et al.* (2015). (A black-and-white version of this figure will appear in some formats. For the color version, please refer to the plate section.)

the photosensitized midbrain and the detection task was repeated with photostimulation in place of acoustic input. Chronos- and ChR2-expressing mice exhibited similar detection slopes, while saline-injected animals performed at chance. In neurophysiological experiments, Chronos was better able to entrain high pulse rates than ChR2 (see Sections 29.5.3.3 and 29.5.3.6). However, these results did not translate into a perceptual advantage in behavioral experiments.

Responses to optogenetic stimulation of the auditory midbrain are poorly encoded and quickly suppressed at the level of the auditory cortex (Guo *et al.*, 2015). Similar suppressive responses have been shown in recordings of electrical stimulation. Further research into the perceptual outputs of optogenetic stimulation is necessary through more complicated discrimination tasks that force animals to report the saliency of light input at different frequencies when using newer opsins such as Chronos as compared with slower opsins.

29.5.4 Evaluating the Functional Outcomes of Optogenetic Implants

The final hurdle in using optogenetic technology in auditory implants depends on finding convincing evidence in preclinical and clinical trials that optogenetic implants will offer improved patient outcomes compared with existing electric-only devices. Optogenetic tools appear primed to do just this, given prior explanations of how light can be better targeted than non-specific electrical stimulation.

Ultimately, the success of auditory implants over the past four decades serves to emphasize that the auditory prosthesis is actually a testament to the remarkable ability of the human brain to adapt. It is surprising that the small number of electrodes in modern CIs can allow the majority of post-lingually deafened patients to perceive speech and even some elements of music (Kohlberg *et al.*, 2014). In patients with ABIs, results have been more varied for different surgical candidates, but have been particularly successful for younger non-NF2 patients, no doubt due to improved neural plasticity as compared with older individuals. A 12-year follow-up of deaf children with ABIs found that 41% developed open set speech recognition (Colletti *et al.*, 2014). Nevertheless, there is great opportunity for novel optogenetic auditory prosthetics to improve on current technology with better stimulation of the complex neural circuits that govern auditory perception.

29.6 Conclusion

Optogenetic technology has tremendous potential and opens new doors in the designing of better auditory prostheses. Recent studies have demonstrated that novel opsins such as Chronos can encode a wide range of temporal stimulation rates with better discriminability compared with older generations of opsins. Simultaneous advances in the safe application of viral vectors in clinical therapies and in the design of miniature optical arrays that are suitable for use in auditory implants may expedite the start of human trials. These projects underscore the complexity of the central auditory neural circuits that prostheses aim to manipulate and set a foundation for future experiments aiming at a superior optogenetic auditory implant.

REFERENCES

Asokan, Aravind, David V. Schaffer, and R. Jude Samulski. 2012. "The AAV Vector Toolkit: Poised at the Clinical Crossroads." *Molecular Therapy* **20** (4): 699–708.

Barrett, John, Roland Berlinguer-Palmini, and Patrick Degenaar. 2014. "Optogenetic Approaches to Retinal Prosthesis." *Visual Neuroscience* **31** (4–5): 345–54.

Baskent, Deniz. 2006. "Speech Recognition in Normal Hearing and Sensorineuronal Hearing Loss as a Function of the Number of Spectral Channels." *The Journal of the Acoustic Society of America* **120** (5): 2908–25.

Beadle, Elizabeth A. R., Dyan J. McKinley, Thomas P. Nikolopoulos, Jackie Brough, Gerard M. O'Donoghue, and Sue M. Archbold. 2005. "Long-Term Functional Outcomes and Academic-Occupational Status in Implanted Children after 10 to 14 Years of Cochlear Implant Use." *Otology & Neurotology: Official Publication of the American Otological Society, American Neurotology Society [and] European Academy of Otology and Neurotology* **26** (6): 1152–60.

Berndt, André, Ofer Yizhar, Lisa A. Gunaydin, Peter Hegemann, and Karl Deisseroth. 2009. "Bi-Stable Neural State Switches." *Nature Neuroscience* **12** (2): 229–34.

Bernstein, Jacob G., Xue Han, Michael A. Henninger, Emily Y. Ko, Xiaofeng Qian, Giovanni Talei Franzesi, Jackie P. McConnell, Patrick Stern, Robert Desimone, and Edward S. Boyden. 2008. "Prosthetic Systems for Therapeutic Optical Activation and Silencing of Genetically-Targeted Neurons." *Proceedings – Society of Photo-Optical Instrumentation Engineers* **6854**: 68540H.

Bilak, M. M., S. R. Bilak, and D. K. Morest. 1996. "Differential Expression of N-Methyl-D-Aspartate Receptor in the Cochlear Nucleus of the Mouse." *Neuroscience* **75** (4): 1075–97.

Bingabr, Mohamed, Blas Espinoza-Varas, and Philipos C. Loizou. 2008. "Simulating the Effect of Spread of Excitation in Cochlear Implants." *Hearing Research* **241** (1–2): 73–9.

Boëx, Colette, Chloé de Balthasar, Maria-Izabel Kós, and Marco Pelizzone. 2003. "Electrical Field Interactions in Different Cochlear Implant Systems." *The Journal of the Acoustical Society of America* **114** (4 Pt 1): 2049–57.

Boyden, Edward S. 2011. "A History of Optogenetics: The Development of Tools for Controlling Brain Circuits with Light." *F1000 Biology Reports* **3**: 11.

Boyden, Edward S., Feng Zhang, Ernst Bamberg, Georg Nagel, and Karl Deisseroth. 2005. "Millisecond-Timescale, Genetically Targeted Optical Control of Neural Activity." *Nature Neuroscience* **8** (9): 1263.

Brackmann, D. E., W. E. Hitselberger, R. A. Nelson, J. Moore, M. D. Waring, F. Portillo, R. V. Shannon, and F. F. Telischi. 1993. "Auditory Brainstem Implant: I. Issues in Surgical Implantation." *Otolaryngology – Head and Neck Surgery: Official Journal of American Academy of Otolaryngology – Head and Neck Surgery* **108** (6): 624–33.

Cai, Dawen, Kimberly B. Cohen, Tuanlian Luo, Jeff W. Lichtman, and Joshua R. Sanes. 2013. "New Tools for the Brainbow Toolbox." *Nature Methods* **10** (6): 540–47.

Cayce, Jonathan M., Robert M. Friedman, Gang Chen, E. Duco Jansen, Anita Mahadevan-Jansen, and Anna W. Roe. 2014. "Infrared Neural Stimulation of Primary Visual Cortex in Non-Human Primates." *NeuroImage* **84**: 181–90.

Christian Goßler and Colin Bierbrauer. 2014. "GaN-Based Micro-LED Arrays on Flexible Substrates for Optical Cochlear Implants." *Journal of Physics D: Applied Physics* **47** (20): 205401.

Colletti, Liliana, Robert Shannon, and Vittorio Colletti. 2012. "Auditory Brainstem Implants for Neurofibromatosis Type 2." *Current Opinion in Otolaryngology & Head and Neck Surgery* **20** (5): 353–57.

Colletti, Liliana, Robert V. Shannon, and Vittorio Colletti. 2014. "The Development of Auditory Perception in Children Following Auditory Brainstem Implantation." *Audiology & Neuro-Otology* **19** (6): 386–94.

Colletti, Vittorio, Robert Shannon, Marco Carner, Sheila Veronese, and Liliana Colletti. 2009. "Outcomes in Nontumor Adults Fitted with the Auditory Brainstem Implant:

10 Years' Experience." *Otology & Neurotology: Official Publication of the American Otological Society, American Neurotology Society [and] European Academy of Otology and Neurotology* **30** (5): 614–18.

Colletti, Vittorio, Robert V. Shannon, Marco Carner, Sheila Veronese, and Liliana Colletti. 2010. "Complications in Auditory Brainstem Implant Surgery in Adults and Children." *Otology & Neurotology: Official Publication of the American Otological Society, American Neurotology Society [and] European Academy of Otology and Neurotology* **31** (4): 558–64.

Darrow, Keith N., Michaël C. C. Slama, Elliott D. Kozin, Maryanna Owoc, Kenneth Hancock, Judith Kempfle, Albert Edge, *et al.* 2015. "Optogenetic Stimulation of the Cochlear Nucleus Using Channelrhodopsin-2 Evokes Activity in the Central Auditory Pathways." *Brain Research* **1599**: 44–56.

Dittgen, Tanjew, Axel Nimmerjahn, Shoji Komai, Pawel Licznerski, Jack Waters, Troy W. Margrie, Fritjof Helmchen, Winfried Denk, Michael Brecht, and Pavel Osten. 2004. "Lentivirus-Based Genetic Manipulations of Cortical Neurons and Their Optical and Electrophysiological Monitoring *In Vivo*." *Proceedings of the National Academy of Sciences of the United States of America* **101** (52): 18206–11.

Ebinger, K., S. Otto, J. Arcaroli, S. Staller, and P. Arndt. 2000. "Multichannel Auditory Brainstem Implant: US Clinical Trial Results." *The Journal of Laryngology and Otology*. **27**: 50–3.

Eisen, Marc D. 2003. "Djourno, Eyries, and the First Implanted Electrical Neural Stimulator to Restore Hearing." *Otology & Neurotology: Official Publication of the American Otological Society, American Neurotology Society [and] European Academy of Otology and Neurotology* **24** (3): 500–6.

Eisenberg, Laurie S. 2015. "The Contributions of William F. House to the Field of Implantable Auditory Devices." *Hearing Research* **322**: 52–56.

Eshraghi, Adrien A., Ronen Nazarian, Fred F. Telischi, Suhrud M. Rajguru, Eric Truy, and Chhavi Gupta. 2012. "The Cochlear Implant: Historical Aspects and Future Prospects." *Anatomical Record (Hoboken, N.J.: 2007)* **295** (11): 1967–80.

Evans, D. Gareth. 1993. "Neurofibromatosis 2." In *GeneReviews(®)*, edited by Roberta A. Pagon, Margaret P. Adam, Holly H. Ardinger, Stephanie E. Wallace, Anne Amemiya, Lora JH Bean, Thomas D. Bird, *et al.*, Seattle (WA): University of Washington, Seattle. http://www.ncbi.nlm.nih.gov/books/NBK1201/.

Fishman, K. E., R. V. Shannon, and W. H. Slattery. 1997. "Speech Recognition as a Function of the Number of Electrodes Used in the SPEAK Cochlear Implant Speech Processor." *Journal of Speech, Language, and Hearing Research: JSLHR* **40** (5): 1201–15.

Fretz, R. J. and R. P. Fravel. 1985. "Design and Function: A Physical and Electrical Description of the 3M House Cochlear Implant System." *Ear and Hearing* **6** (3 Suppl.): 14S–19S.

Garnham, Carolyn, Martin O'Driscoll, Richard Ramsden, and Shakeel Saeed. 2002. "Speech Understanding in Noise with a Med-El COMBI 40+ Cochlear Implant Using Reduced Channel Sets." *Ear and Hearing* **23** (6): 540–52.

Geier, L., M. Barker, L. Fisher, and J. Opie. 1999. "The Effect of Long-Term Deafness on Speech Recognition in Postlingually Deafened Adult CLARION Cochlear Implant Users." *The Annals of Otology, Rhinology & Laryngology. Supplement* **177**: 80–3.

Golding, Nace L. and Donata Oertel. 2012. "Synaptic Integration in Dendrites: Exceptional Need for Speed." *The Journal of Physiology* **590** (Pt 22): 5563–69.

Gradinaru, Viviana, Feng Zhang, Charu Ramakrishnan, Joanna Mattis, Rohit Prakash, Ilka Diester, Inbal Goshen, Kimberly R. Thompson, and Karl Deisseroth. 2010. "Molecular and Cellular Approaches for Diversifying and Extending Optogenetics." *Cell* **141** (1): 154–65.

Grossman, Nir, Vincent Poher, Matthew S. Grubb, Gordon T. Kennedy, Konstantin Nikolic, Brian McGovern, Rolando Berlinguer Palmini, *et al.* 2010. "Multi-Site Optical Excitation Using ChR2 and Micro-LED Array." *Journal of Neural Engineering* **7** (1): 16004.

Gunaydin, Lisa A., Ofer Yizhar, André Berndt, Vikaas S. Sohal, Karl Deisseroth, and Peter Hegemann. 2010. "Ultrafast Optogenetic Control." *Nature Neuroscience* **13** (3): 387–92.

Guo, Wei, Ariel E. Hight, Jenny X. Chen, Nathan C. Klapoetke, Kenneth E. Hancock, Barbara G. Shinn-Cunningham, Edward S. Boyden, Daniel J. Lee, and Daniel B. Polley. 2015. "Hearing the Light: Neural and Perceptual Encoding of Optogenetic Stimulation in the Central Auditory Pathway." *Scientific Reports* **5**: 10319.

Guru, Akash, Ryan J. Post, Yi-Yun Ho, and Melissa R. Warden. 2015. "Making Sense of Optogenetics." *International Journal of Neuropsychopharmacology* **18** (11): pyv079.

Han, Xue and Edward S. Boyden. 2007. "Multiple-Color Optical Activation, Silencing, and Desynchronization of Neural Activity, with Single-Spike Temporal Resolution." *PLoS ONE* **2** (3): e299.

Han, Xue, Xiaofeng Qian, Jacob G. Bernstein, Hui-Hui Zhou, Giovanni Talei Franzesi, Patrick Stern, Roderick T. Bronson, Ann M. Graybiel, Robert Desimone, and Edward S. Boyden. 2009. "Millisecond-Timescale Optical Control of Neural Dynamics in the Nonhuman Primate Brain." *Neuron* **62** (2): 191–8.

Hernandez, Victor H., Anna Gehrt, Kirsten Reuter, Zhizi Jing, Marcus Jeschke, Alejandro Mendoza Schulz, Gerhard Hoch, *et al.* 2014. "Optogenetic Stimulation of the Auditory Pathway." *The Journal of Clinical Investigation* **124** (3): 1114–29.

Hight, Ariel E., Elliott D. Kozin, Keith Darrow, Ashton Lehmann, Edward Boyden, M. Christian Brown, and Daniel J. Lee. 2015b. "Superior Temporal Resolution of Chronos versus Channelrhodopsin-2 in an Optogenetic Model of the Auditory Brainstem Implant." *Hearing Research* **322**: 235–41.

Hight, Ariel E., Elliott Kozin, Xiankai Meng, Amélie Guex, Alyson Kaplan, Stephanie Lacour, Albert Edge, M. Christian Brown, and Daniel Lee. 2015a. "Optogenetic Control of Cochlear Nucleus Neurons Using Spatially Focused Beams from a Laser Collimator [Abstract]." *Association for Research in Otolaryngology Abstracts Midwinter Meeting 2015*, February.

Hirschbiegel, Constantin. 2015. "The Propagation of Optical Impulses in Optogenetic Auditory Brainstem Implants [Master's Thesis]."

Izzo, Agnella D., Joseph T. Walsh, E. Duco Jansen, Mark Bendett, Jim Webb, Heather Ralph, and Claus-Peter Richter. 2007. "Optical Parameter Variability in Laser Nerve Stimulation: A Study of Pulse Duration, Repetition Rate, and Wavelength." *IEEE Transactions on Bio-Medical Engineering* **54** (6 Pt 1): 1108–14.

Jenkins, M. W., A. R. Duke, S. Gu, H. J. Chiel, H. Fujioka, M. Watanabe, E. D. Jansen, and A. M. Rollins. 2010. "Optical Pacing of the Embryonic Heart." *Nature Photonics* **4**: 623–26.

Kang, Robert, Grace Liu Nimmons, Ward Drennan, Jeff Longnion, Chad Ruffin, Kaibao Nie, Jong Ho Won, Tina Worman, Bevan Yueh, and Jay Rubinstein. 2009. "Development and Validation of the University of Washington Clinical Assessment of Music Perception Test." *Ear and Hearing* **30** (4): 411–8.

Kaplan, Alyson B., Elliott D. Kozin, Sidharth V. Puram, Maryanna S. Owoc, Parth V. Shah, Ariel E. Hight, Rosh K.V. Sethi, Aaron K. Remenschneider, and Daniel J. Lee. 2015. "Auditory Brainstem Implant Candidacy in the United States in Children 0–17 Years Old." *International Journal of Pediatric Otorhinolaryngology* **79** (3): 310–5.

Kay, Mark A., Joseph C. Glorioso, and Luigi Naldini. 2001. "Viral Vectors for Gene Therapy: The Art of Turning Infectious Agents into Vehicles of Therapeutics." *Nature Medicine* **7** (1): 33–40.

Keppeler, Daniel, Victor Hernandez, Anna Gehrt, Marcus Jeschke, Christian Wrobel, Gerhard Hoch, Christian Goßler, *et al.* 2015. "Towards Optogenetic Stimulation of the Auditory Pathway: Implementing the Fast Channelrhodopsin Chronos and Multichannel microLED Arrays [Abstract]." *Association for Research in Otolaryngology Abstracts Midwinter Meeting 2015*, February.

Kim, Tae-Il, Jordan G. McCall, Yei Hwan Jung, Xian Huang, Edward R. Siuda, Yuhang Li, Jizhou Song, *et al.* 2013. "Injectable, Cellular-Scale Optoelectronics with Applications for Wireless Optogenetics." *Science (New York, N.Y.)* **340** (6129): 211–6.

Klapoetke, Nathan C., Yasunobu Murata, Sung Soo Kim, Stefan R. Pulver, Amanda Birdsey-Benson, Yong Ku Cho, Tania K. Morimoto, *et al.* 2014. "Independent Optical Excitation of Distinct Neural Populations." *Nature Methods* **11** (3): 338–46.

Kohlberg, Gavriel, Jaclyn B. Spitzer, Dean Mancuso, and Anil K. Lalwani. 2014. "Does Cochlear Implantation Restore Music Appreciation?" *The Laryngoscope* **124** (3): 587–8.

Kozin, Elliott, Ariel E. Hight, Xiankai Meng, Alyson Kaplan, Ashton Lehmann, Edward Boyden, Albert Edge, M. Christian Brown, and Daniel Lee. 2015. "Comparison of Virally-Mediated Transfer of ChR2 to Transgenic Expression of ChR2 in the Cochlear Nucleus: Implications for Development of an Optical Auditory Brainstem Implant [Abstract]." *Association for Research in Otolaryngology Abstracts Midwinter Meeting 2015*, February.

Kuchta, Johannes, Steven R. Otto, Robert V. Shannon, William E. Hitselberger, and Derald E. Brackmann. 2004. "The Multichannel Auditory Brainstem Implant: How Many Electrodes Make Sense?" *Journal of Neurosurgery* **100** (1): 16–23.

Lim, Hubert H., Thomas Lenarz, David J. Anderson, and Minoo Lenarz. 2008. "The Auditory Midbrain Implant: Effects of Electrode Location." *Hearing Research* **242** (1–2): 74–85.

Lin, John Y. 2011. "A User's Guide to Channelrhodopsin Variants: Features, Limitations and Future Developments." *Experimental Physiology* **96** (1): 19–25.

Lin, John Y., Per Magne Knutsen, Arnaud Muller, David Kleinfeld, and Roger Y. Tsien. 2013. "ReaChR: A Red-Shifted Variant of Channelrhodopsin Enables Deep Transcranial Optogenetic Excitation." *Nature Neuroscience* **16** (10): 1499–508.

Mason, Matthew R.J., Erich M.E. Ehlert, Ruben Eggers, Chris W. Pool, Stephan Hermening, Angelina Huseinovic, Eric Timmermans, Bas Blits, and Joost Verhaagen. 2010. "Comparison of AAV Serotypes for Gene Delivery to Dorsal Root Ganglion Neurons." *Molecular Therapy* **18** (4): 715–24.

Matthies, C., S. Thomas, M. Moshrefi, A. Lesinski-Schiedat, C. Frohne, R.D. Battmer, T. Lenarz, and M. Samii. 2000. "Auditory Brainstem Implants: Current Neurosurgical Experiences and Perspective." *The Journal of Laryngology and Otology* **27**: 32–6.

Meng, Xiankai, Elliott Kozin, Gang Q. Li, Ruth Anne Eatock, Daniel Lee, and Albert Edge. 2014. "Adeno-Associated Virus Vector Delivery of Channelrhodopsin-2 Into Spiral Ganglion Neurons [Abstract]." *Association for Research in Otolaryngology Abstracts Midwinter Meeting 2014*, February.

Miller, Charles A., Paul J. Abbas, Barbara K. Robinson, Kirill V. Nourski, Fawen Zhang, and Fuh-Cherng Jeng. 2006. "Electrical Excitation of the Acoustically Sensitive Auditory Nerve: Single-Fiber Responses to Electric Pulse Trains." *Journal of the Association for Research in Otolaryngology* **7** (3): 195–210.

Nevison, Barry, Roland Laszig, Wolf-Peter Sollmann, Thomas Lenarz, Olivier Sterkers, Richard Ramsden, Bernard Fraysse, *et al.*, 2002. "Results from a European Clinical Investigation of the Nucleus Multichannel Auditory Brainstem Implant." *Ear and Hearing* **23** (3): 170–83.

Noij, Kimberley S., Elliott D. Kozin, Rosh Sethi, Parth V. Shah, Alyson B. Kaplan, Barbara Herrmann, Aaron Remenschneider, and Daniel J. Lee. 2015. "Systematic Review of Nontumor Pediatric Auditory Brainstem Implant Outcomes." *Otolaryngology–Head and Neck Surgery: Official Journal of American Academy of Otolaryngology-Head and Neck Surgery* **153** (5): 739–50.

Nussinovitch, Udi and Lior Gepstein. 2015. "Optogenetics for *In Vivo* Cardiac Pacing and Resynchronization Therapies." *Nature Biotechnology* **33** (7): 750–4.

O'Donoghue, Gerard. 2013. "Cochlear Implants – Science, Serendipity, and Success." *New England Journal of Medicine* **369** (13): 1190–3.

Otto, Steven R., Derald E. Brackmann, and William Hitselberger. 2004. "Auditory Brainstem Implantation in 12- to 18-Year-Olds." *Archives of Otolaryngology – Head & Neck Surgery* **130** (5): 656–9.

Otto, Steven R., Robert V. Shannon, Eric P. Wilkinson, William E. Hitselberger, Douglas B. McCreery, Jean K. Moore, and Derald E. Brackmann. 2008. "Audiologic Outcomes with the Penetrating Electrode Auditory Brainstem Implant." *Otology & Neurotology:*

Official Publication of the American Otological Society, American Neurotology Society [and] European Academy of Otology and Neurotology **29** (8): 1147–54.

Park, Sung Il, Daniel S. Brenner, Gunchul Shin, Clinton D. Morgan, Bryan A. Copits, Ha Uk Chung, Melanie Y. Pullen, *et al.* 2015. "Soft, Stretchable, Fully Implantable Miniaturized Optoelectronic Systems for Wireless Optogenetics." *Nature Biotechnology* **33** (12): 1280–6.

Peron, Simon and Karel Svoboda. 2011. "From Cudgel to Scalpel: Toward Precise Neural Control with Optogenetics." *Nature Methods* **8** (1): 30.

Puram, Sidharth V., Samuel R. Barber, Elliott D. Kozin, Parth Shah, Aaron Remenschneider, Barbara S. Herrmann, Ann-Christine Duhaime, Fred G. Barker, and Daniel J. Lee. 2016. "Outcomes Following Pediatric Auditory Brainstem Implant Surgery: Early Experiences in a North American Center." *Otolaryngology – Head and Neck Surgery: Official Journal of American Academy of Otolaryngology – Head and Neck Surgery* **155** (1): 133–8.

Richter, Claus-Peter and Xiaodong Tan. 2014. "Photons and Neurons." *Hearing Research* **311**: 72–88.

Richter, C.-P., S. M. Rajguru, A. I. Matic, E. L. Moreno, A. J. Fishman, A. M. Robinson, E. Suh, and J. T. Walsh. 2011. "Spread of Cochlear Excitation during Stimulation with Pulsed Infrared Radiation: Inferior Colliculus Measurements." *Journal of Neural Engineering* **8** (5): 56006.

Sennaroğlu, Levent, Gonca Sennaroğlu, Esra Yücel, Burçak Bilginer, Gamze Atay, M. Demir Bajin, Burçe Özgen Mocan, *et al.*, 2016. "Long-Term Results of ABI in Children With Severe Inner Ear Malformations." *Otology & Neurotology: Official Publication of the American Otological Society, American Neurotology Society [and] European Academy of Otology and Neurotology* **37** (7): 865–72.

Shimano, T., B. Fyk-Kolodziej, N. Mirza, M. Asako, K. Tomoda, S. Bledsoe, Z.H. Pan, S. Molitor, and A.G. Holt. 2013. "Assessment of the AAV-Mediated Expression of Channelrhodopsin-2 and Halorhodopsin in Brainstem Neurons Mediating Auditory Signaling." *Brain Research* **1511**: 138–52.

Stickney, Ginger S., Philipos C. Loizou, Lakshmi N. Mishra, Peter F. Assmann, Robert V. Shannon, and Jane M. Opie. 2006. "Effects of Electrode Design and Configuration on Channel Interactions." *Hearing Research* **211** (1–2): 33–45.

Thompson, Alexander C., James B. Fallon, Andrew K. Wise, Scott A. Wade, Robert K. Shepherd, and Paul R. Stoddart. 2015. "Infrared Neural Stimulation Fails to Evoke Neural Activity in the Deaf Guinea Pig Cochlea." *Hearing Research* **324**: 46–53.

Tian, Pengfei, Jonathan J.D. McKendry, Erdan Gu, Zhizhong Chen, Yongjian Sun, Guoyi Zhang, Martin D. Dawson, and Ran Liu. 2016. "Fabrication, Characterization and Applications of Flexible Vertical InGaN Micro-Light Emitting Diode Arrays." *Optics Express* **24** (1): 699–707.

Tye, Kay M. and Karl Deisseroth. 2012. "Optogenetic Investigation of Neural Circuits Underlying Brain Disease in Animal Models." *Nature Reviews Neuroscience* **13** (4): 251–66.

Verma, Rohit U., Amélie A. Guex, Kenneth E. Hancock, Nedim Durakovic, Colette M. McKay, Michaël C.C. Slama, M. Christian Brown, and Daniel J. Lee. 2014. "Auditory Responses to Electric and Infrared Neural Stimulation of the Rat Cochlear Nucleus." *Hearing Research* **310**: 69–75.

Waltzman, S.B., N.L. Cohen, and J.T. Roland. 1999. "A Comparison of the Growth of Open-Set Speech Perception between the Nucleus 22 and Nucleus 24 Cochlear Implant Systems." *The American Journal of Otology* **20** (4): 435–41.

Wells, Jonathon, Chris Kao, Peter Konrad, Tom Milner, Jihoon Kim, Anita Mahadevan-Jansen, and E. Duco Jansen. 2007a. "Biophysical Mechanisms of Transient Optical Stimulation of Peripheral Nerve." *Biophysical Journal* **93** (7): 2567–80.

Wells, Jonathon, Chris Kao, Karthik Mariappan, Jeffrey Albea, E. Duco Jansen, Peter Konrad, and Anita Mahadevan-Jansen. 2005. "Optical Stimulation of Neural Tissue *In Vivo*." *Optics Letters* **30** (5): 504–6.

Wells, Jonathon, Sharon Thomsen, Peter Whitaker, E. Duco Jansen, Chris C. Kao, Peter E. Konrad, and Anita Mahadevan-Jansen. 2007b. "Optically Mediated Nerve Stimulation: Identification of Injury Thresholds." *Lasers in Surgery and Medicine* **39** (6): 513–26.

Wilson, Blake S. 2013. "Toward Better Representations of Sound with Cochlear Implants." *Nature Medicine* **19** (10): 1245–8.

Wu, Zhijian, Hongyan Yang, and Peter Colosi. 2010. "Effect of Genome Size on AAV Vector Packaging." *Molecular Therapy* **18** (1): 80–6.

Yizhar, Ofer, Lief E. Fenno, Thomas J. Davidson, Murtaza Mogri, and Karl Deisseroth. 2011. "Optogenetics in Neural Systems." *Neuron* **71** (1): 9–34.

Zeltner, Brie. 2014. "University Hospitals, Cleveland Clinic Offer Rare Surgery to Profoundly Deaf Patients; Auditory Brainstem Implant Technology May Have More Uses in the Future." *Cleveland.com*. March 24. http://www.cleveland.com/healthfit/index.ssf/2014/03/university_hospitals_cleveland_1.html.

30 Optogenetic Stimulation for Cochlear Prosthetics

Victor H. Hernandez Gonzalez and Tobias Moser

30.1 Introduction

In the hearing sense, sound is transduced by the hair cells of the organ of Corti in the inner ear and encoded by spiral ganglion neurons (SGNs). This transduction has an extraordinary temporal and spectral resolution and an amazing intensity sensibility.

Unfortunately, according to the World Health Organization, more than 360 million people in the world suffer from disabling hearing loss, representing over 5% of the world's population; 32 million of these people are children, and it is expected that the number of people with hearing loss will double by 2030 (WHO, 2015), making it the most common sensory impairment. Nevertheless, today it is possible to offer to these patients successful support with hearing aids and cochlear implants (CIs).

Alessandro Volta first stimulated the auditory system with electricity in the late eighteenth century by inserting two metal rods into his own ears. In the last 40 years, electrical hearing stimulation with the use of CIs has made great progress. CIs are designed to help patients with profound sensorineural deafness and are based on direct electrical stimulation on the acoustic nerve, bypassing the region of physiological loss. With more than 450 000 CIs having been implanted worldwide, CIs are the most successful neuroprosthesis available today, giving these patients open speech comprehension and almost normal integration into scholarship, labor and society.

Nevertheless, even though the great advantages of CIs are more than evident, there are still significant failures in their performance that are driving efforts to overcome these limitations. In this chapter, we will highlight one of those efforts: stimulating the auditory nerve not with an electrical current, but with light by employing optogenetics.

30.2 Need for Alternatives to Electrical Stimulation

As noted above, hearing disabilities are the most common sensory deficit. Approximately 20% of people with hearing loss would benefit from hearing aids and CIs, representing 56 million potential users of these devices (WHO, 2015). CIs are therefore an extraordinary therapeutic option for many of the patients who suffer from sensorineural hearing loss.

Modern CIs are the most successful neuroprosthesis available today, providing the majority of implanted patients with speech comprehension, and some of them are even able to make phone calls. This sensory prosthesis converts sounds into electrical signals that are then sent to electrodes organized in 8 to 24 specific sites inserted directly into the cochlea, which has more than 30 000 SGNs organized along the tonotopic axis of the organ. In spite of these great advantages, CIs suffer from poor frequency resolution. The very low number of electrodes dramatically limits the stimulation of this tonotopic arrangement. This is due to the widespread electrical field around an electrode contact (Kral *et al.*, 1998) leading to crosstalk and gross neural interactions (Shannon, 1983), reducing the number of independent frequency channels; therefore, it is clear that increasing the number of channels would not increase the resolution of the stimuli (Fishman *et al.*, 1997; Friesen *et al.*, 2001). Electrical coding of the intensity of sound is also limited, with an output dynamic range typically below 20 dB (Zeng *et al.*, 2002; Zeng *et al.*, 2008); this compression also contributes to reducing speech perception in noisy environments (Zeng *et al.*, 2002). It has also been proposed that poor pitch resolution is due to the limits of electrical stimulation of the cochlea. Zeng *et al.* (2014) compared electrical and acoustical pitch perception and found that the frequency-electrode function was twice as compressed in the frequency range, that was one to two octaves lower than the normal acoustic hearing range and that frequency discrimination was 24 times worse than in normal-hearing controls. Their interpretation was that these abnormalities could be due to a combination of the broad electric field, distant intracochlear electrode placement and non-uniform spiral ganglion cell distribution due to the variable number of surviving neurons after deafness and the implantation itself (Zeng *et al.*, 2014). These problems are not easily resolved as they are inherent to the implants and the characteristics of each organ to be implanted.

In an effort to overcome these limitations, some alternatives are to use bipolar, tripolar or partially tripolar electrical stimulation (Donaldson *et al.*, 2005; Srinivasan *et al.*, 2012; Srinivasan *et al.*, 2013; Wu and Luo, 2014); however, these approaches have the disadvantage of greater power consumption.

It is clear that all of these limitations make CIs insufficient for providing fine tonal perception to implanted patients (Shannon *et al.*, 2004), which is needed for speech comprehension in a noisy background, the use of the phone without visual cues, the appreciation of music or speech perception in tonal languages such as mandarin (Fetterman and Domico, 2002; Wu *et al.*, 2013; Kohlberg *et al.*, 2014; Sousa *et al.*, 2015). Thus, optical stimulation and particularly optogenetics has emerged as an attainable alternative for stimulating the auditory system and therefore as a therapeutic strategy, mainly for one of its most frequent and severe presentations, sensorineural hearing loss.

30.3 Optical Stimulation

Focusing on etiology, different strategies, ranging from pharmacology to genetics and molecular biology, are areas of intense research for the treatment or cure of deafness and other hearing impairments, but have been without success until now.

In recent years, optical stimulation has emerged to offer a new possibility of stimulating SGNs. Light activation of the auditory nerve is a hopeful prospect that has taken concrete shape, offering great improvements in frequency resolution, as light can be conveniently focused in order to activate a very limited number of

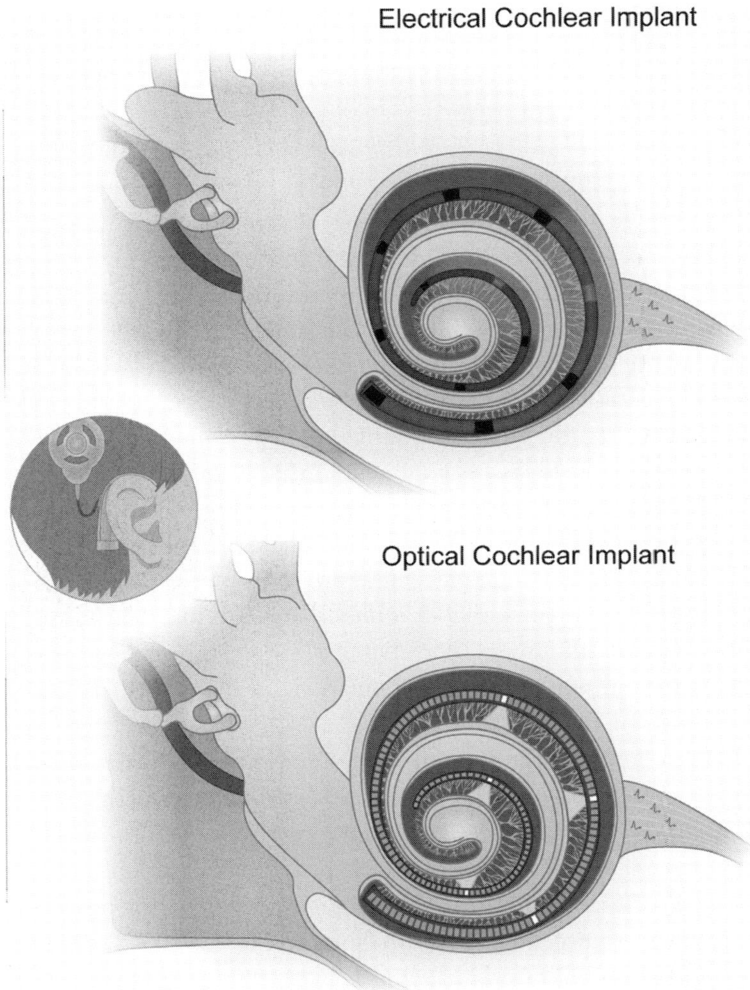

Figure 30.1 Electrical versus optical spiral ganglion (SGN) neuron stimulation. Top: electrical cochlear implants (blue) use between 8 and 24 electrode contacts (black and red) to stimulate SGNs (yellow). The very low number of electrodes and the current spread around an electrode contact (red halo) dramatically reduce the stimulation along the cochlear tonotopic axis, in this way limiting the frequency resolution of electrical coding. Bottom: focused optical stimulation promises spatially confined activation of SGNs, allowing for a higher number of independent stimulation channels and improving the frequency and intensity resolution of sound coding. (A black-and-white version of this figure will appear in some formats. For the color version, please refer to the plate section.)

SGNs. This confined area of stimulation may also be able to increase the dynamic range of coding, both varying the intensity of stimulation in individual channels and modulating the number of active neighboring channels in a more controlled way compared to electrical CIs (Figure 30.1).

One of the optical methods that has been proposed to evoke responses in the auditory nerve is infrared (IR) stimulation. The research in this field started and has been led by Richter and co-workers. They demonstrated that IR stimulation can evoke an auditory response (Izzo *et al.*, 2006) and characterized the minimal duration of the pulse, as well as its radiant exposure (Izzo *et al.*, 2007; Izzo *et al.*, 2008), the minimal rate (Izzo *et al.*, 2007; Littlefield *et al.*, 2010) and, very importantly, the spatial spread of IR stimulation, showing that it is comparable with that of an acoustic pure tone (Richter *et al.*, 2011). Notwithstanding all of these results, some questions regarding IR stimulation are still open. For example, the laser beam is able to produce optoacoustic effects (Teudt *et al.*, 2011; Schultz *et al.*, 2012), and this effect is also supported by the finding that after deafening gerbils with aminoglycosides, the optically evoked compound action potential thresholds either increased or even disappeared (Richter *et al.*, 2008; Verma *et al.*, 2014). The energy requirements of a laser beam are much higher than traditional electrical implants (Zierhofer *et al.*, 1995; Littlefield *et al.*, 2010; Richter *et al.*, 2011) and, most importantly, the way in which IR stimulation could evoke neural responses has not been thoroughly elucidated, although different explanations have been proposed (Izzo *et al.*, 2008; Richter *et al.*, 2011; Shapiro *et al.*, 2012). It is therefore crucial for us to understand the real mechanism underlying IR stimulation of the nervous system and, even more so, to progress toward clinical applicability and the development of IR optical implants.

As an alternative to IR stimulation, optogenetics achieves optical stimulation of the auditory system through a mechanism that has been clearly characterized. Moreover, optogenetics today offers a flexible approach, allowing modulation of the stimulus in multiple ways (i.e. light wavelength, duration, the possibility of using a combination of different optogenetic proteins for modulation of the response, etc.). A more detailed explanation about the general mechanisms of optogenetics will be provided later in this chapter.

30.4 Optogenetic Stimulation of the Auditory System and Optical Auditory Therapeutic Strategies

Optogenetics is a technique that uses light to control neuron activity. To achieve this control, light-sensitive proteins that are naturally found in different species of microbes are genetically introduced into the nervous system. In such a way, depending on the kinetics of the protein, it is possible to change the activity of the neural circuit of interest or even of a single neuron under light stimulation. In 2002, the group of Spudich was the first to identify that a couple of rhodopsins are responsible for phototaxis in the microscopic alga *Chlamydomonas reinhardtii* (Sineshchekov *et al.*, 2002). That same year, Nagel and collaborators characterized for the first time the electrical properties of this protein and called it channelrhodopsin1-A (Nagel *et al.*, 2002). One year later, they expressed channelrhodopsin2

(ChR2), a related protein of the same alga, in oocytes and mammalian cells, showing that it is a cation-selective channel, and they proposed its use in controlling neural activity (Nagel *et al.*, 2003). Soon after, the group of Karl Deisseroth exploited this idea for the first time, showing that ChR2 is safely and stably expressed in neurons and, when activated with light, it is possible to evoke action and synaptic potentials (Boyden *et al.*, 2005).

Since then, ChR2 has undergone different modifications at the molecular level to change its kinetics and light sensibility. Thanks to this and to different strategies of gene delivery, optogenetics today has great spatiotemporal accuracy and has been used to help us to understand a wide range of different neural processes, as well as to open up the possibility of therapy for neurological disorders.

Wild-type ChR2 has a peak of activation of 470 nm (Nagel *et al.*, 2003). This is of course an advantage over IR stimulation, as short wavelengths produce less heating and therefore the optoacoustic effect is also reduced. This limits the possibility of damage. Nevertheless, it also represents a disadvantage, as the scattering is larger and thus the ability to penetrate deeply into tissues is very low (Aravanis *et al.*, 2007). Another disadvantage of ChR2 is that it has a very slow time constant of closing (Nagel *et al.*, 2003), limiting the possibility of high-frequency stimulation rates. In an effort to improve the slow kinetics of the channel, to change the light sensitivity and to move to red shifted-sensitive proteins, many different mutations of the channel have been created (Yizhar *et al.*, 2011; Klapoetke *et al.*, 2014; Hososhima *et al.*, 2015). All of these new technological advances have also provided a boost for the development of new optoelectronic devices, even at the micro and nano levels (Deisseroth *et al.*, 2006; Aravanis *et al.*, 2007; Anikeeva *et al.*, 2012; Park *et al.*, 2015). Actually, some of these molecular variants have already been used to demonstrate neural activation with auditory implants (Hernandez *et al.*, 2014; Hight *et al.*, 2015). It is not the scope of this chapter to review their use in basic neuroscience, and therefore we will focus on therapeutic research, specifically hearing prosthesis development.

The initial question to address is how to express an exogenous protein in the hearing nervous system. One strategy is to develop transgenic animals that express ChR in a stable and widespread manner. This has been achieved using specific promoters or with viral vectors. For a detailed review of this field, the reader can refer to Knöpfel and Boyden (2012) and Ting and Feng (2013).

The first applications of optogenetics for driving the activity of SGNs used transgenic mice and rats expressing ChR2 in the nervous system, including the SGNs, under the Thy1.2 promoter (Wang *et al.*, 2007; Tomita *et al.*, 2009). In these animals, Hernandez and co-workers showed for the first time that it was possible to record auditory activity, which they called the optical auditory brainstem response (oABR), when micro-LEDs (μLEDs) or fiber-coupled lasers inserted through the round window or opening of a small cochleostomy illuminated a small region of the cochlea (Hernandez *et al.*, 2014). These responses differed from auditory ABRs (aABRs; evoked by clicks) in waveform, number of waves and amplitude, but were more comparable to electrical ABRs. This difference was attributed to the recruitment of more SGNs and a higher synchrony of firing induced by optogenetic versus acoustic stimulation. In the cochlea, the channel was expressed only in the soma and neurites of the SGNs, such that optogenetically evoked auditory activity was safely attributed to SGNs and not to hair cell stimulation. These responses were specific for the ChR2-mediated activation of SGNs, as they were absent in wild-type control animals or in ChR2 mice where lidocaine or tetrodotoxin were applied directly to the cochleostomy. When light was projected onto their intact cochleae, it was not possible to observe any response due to the high diffraction index of the bone, indicating that the light must directly illuminate the SGNs. The amplitude of oABR, measured as the N1 amplitude, increased with light intensity and pulse duration (up to ~2 ms at 4 mW mm^{-2}), but there was little variability during repeated stimulations. Nevertheless, as expected, oABR amplitudes declined when stimulus rates were raised above 20 Hz, but remained sizable (tens of μV) up to 70 Hz. oABR amplitudes decreased and latency increased during higher-frequency stimulation.

The activity evoked in the cochlea also activated deeper regions of the auditory pathway. Extracellular recordings from individual neurons in the auditory nerve and cochlear nucleus clearly revealed activation by cochlear optogenetics. The high reliability and temporal precision of the action potentials recorded in the auditory nerve are remarkable. This precision was slightly lost and jitter appeared when recording from the ipsilateral cochlear nucleus.

One of the main objectives of optical stimulation is to improve the frequency resolution shown by modern CIs. To do this, it is important to clearly show tonotopic optogenetic stimulation. This was tested by using two different strategies. In the first one, light stimulation was used to mask the cochlear response to a pure tone. Such "light-on-tone" masking provided a first estimate of the spread of light-evoked cochlear excitation. The result was that aABRs were masked by optogenetic stimulation of SGNs in a frequency-dependent manner: higher sound pressure levels were required to evoke detectable aABRs at 8 and 12 kHz, but not at 4 or 16–32 kHz. The frequency range at which light masking was observed was corroborated with the tonotopic position of the cochleostomy related to the tracing of the basilar membrane, as identified by high-resolution X-ray phase-contrast tomography (Bartels *et al.*, 2013). The frequency range embraced by a typical cochleostomy was estimated to be 8–12 kHz based on a previously established frequency map of the mouse cochlea (Müller *et al.*, 2005). These findings indicated that the transcochlear μLED illumination indeed affected sound coding by SGNs and that this effect was restricted to a tonotopic range of less than an octave. The second strategy to test cochlear spread of excitation involved using multi-electrode array recordings of local field potentials from the central nucleus of the IC, which provides access to the tonotopically ordered input of the auditory pathway. In this case, the response was evoked by the insertion of a 250-μm optical fiber through the round window into the scala tympani. This approach also permitted the recording of the activity in the nucleus evoked by monopolar electric stimulation. In this way, it was possible to compare the cochlear spread of excitation between auditory, electrical and optogenetic stimulation. Optogenetic responses showed spatial tuning that was at least as narrow as that for acoustic stimulation, as would be expected with stimulation of the high-frequency cochlear base; however, responses to electrical stimulation were spatially more extended.

In the same work, the ability of cochlear optogenetics to activate the auditory pathway in mouse models of human deafness was shown. One model of the early development of genetic human deafness is the line called the pachanga mouse, mouse carrying the D1767G point mutation in the Otoferlin ear protein, which codes for the human deafness locus DFNB9. In order to express ChR2 in their SGNs, $Otof^{Pga/Pga}$ mice were crossed with the Thy1.2-driven ChR2 transgenic strain. Transcochlear μLED stimulation also elicited oABRs in these $Otof^{Pga/Pga}$-ChR2 mice. Acquired deafness models were also used. Administration of furosemide in ChR2 mice collapsed the endocochlear potential, which was corroborated by the disappearing of aABRs 30 minutes after the injection. In this case, optogenetic stimulation was also able to evoke responses in the auditory pathway.

It is desirable to use cochlear optogenetics in other species and translated into clinical applications, so an alternative to transgenic animals must be proposed. As noted above, viral vectors are an alternative. Virus-mediated optogenetics offers flexibility for testing various ChR variants. Hernandez and co-workers focused on using adeno-associated virus (AAV) vectors in order to express a variant of ChR2 called CatCh (Hernandez *et al.*, 2014), which shows higher Ca^{2+} permeability and induces 70-fold greater neuron light sensitivity compared with wild-type ChR2 (Kleinlogel *et al.*, 2011). By transuterine injection of AAV2/6, which carried the transgene under the control of the human synapsin promoter, directly into the otocyst of embryos of 11.5 days post-coitus, it was possible to find postnatal expression in SGNs. This expression was functional, and oABRs could be elicited after stimulation of the basal cochlea via a 250μm plastic optical fiber coupled to a laser and inserted through the round window. Importantly, it was possible to evoke responses with light pulses of as short as 200 μs, and it was also possible to record action potentials in the auditory nerve, which, for strong stimuli, occurred with shorter latencies than those of SGNs in ChR2 mice.

These pioneering experiments clearly show that cochlear optogenetic stimulation is feasible and promises better frequency resolution than traditional electrical CIs. They open up the possibility to translate this technology to the bedside, offering an improvement in the hearing functionality of implanted patients and improving their quality of life.

Sometimes, however, it is not possible to implant a cochlear prosthesis. This is the case in trauma involving the inner ear or tumors, typically vestibular schwannomas. In these cases, auditory brainstem implants (ABIs) have been proposed (Brackmann *et al.*, 1993; Shannon *et al.*, 1993). These are based on direct electrical stimulation of the cochlear nuclei. ABIs can provide partial restoration of hearing, but often fail to enable open speech understanding (Colletti *et al.*, 2009; Herrmann *et al.*, 2015; Noij *et al.*, 2015). Recently, optogenetics has also achieved these treatment efforts. Direct *in vitro* photostimulation of the auditory brainstem neurons of transgenic rats carrying ChR2 elicited photocurrents in the stellate and bushy cells of the anteroventral cochlear nucleus, mimicking the tonic and phasic firing patterns, respectively, found with electrical current injection, while medial nucleus of the trapezoid body principal cells showed adaptive firing (Hernandez *et al.*, 2014). Optogenetic virus-mediated transfection of the cochlear nucleus has been successfully performed without apparent deleterious effects on hearing even 18 months after expression (Shimano *et al.*, 2013).

After the end of light stimulation, ChR2 closes with a very low time constant (in the order of milliseconds); therefore, sustained action potential rates are limited to below 50 Hz (Boyden *et al.*, 2005). One of the next objectives in auditory optogenetics is to improve the maximal firing rates. SGNs are able to follow acoustic and electrical stimulation up to around 300 Hz (Miller *et al.*, 2006). It is evident that most ChR proteins are not able to reach this rate of activity. Recently, a new member of the family has appeared, called Chronos, which has faster kinetics (having the fastest inactivation time constant described so far) and is more light sensitive (Klapoetke *et al.*, 2014). Therefore, it seems that Chronos would enable neurons to follow stimulation at hundreds of Hertz. Indeed, this new protein has already been tested in a murine ABI model in which Chronos has the ability to drive the auditory system at higher stimulation rates than ChR2 (Hight *et al.*, 2015). This puts Chronos at an advantage as a very attractive target for optogenetic hearing restoration with ABIs and CIs.

30.5 Outlook

It is clear now that optogenetics is a real possibility for the stimulation of the auditory pathway, marking the start line for the development of new otoprostheses, either in the cochlea or the brainstem. There is still much to do. Even though Chronos seems to be an excellent tool for reaching a higher rate of stimulation, red-shifted ChR will help to promote optogenetics as a valuable stimulation option in auditory research and prosthetics. Before jumping to human translation, optogenetics must be established for species other than murine models, including non-human primates. Thus, vectors and application techniques need to be tested before thinking about their applicability in preclinical trials.

Technological issues must also be solved. Multichannel optical devices, with hundreds of active sites, each one individually controlled and coded, will need to be developed. There is also the obstacle of the small space in which they should be implanted. The optical implants have to be tested in chronic implanted animal models and then assessed for their reliability and safety, in addition to their auditory perception and frequency and intensity resolution of coding. This will force the research to become a multi- and trans-disciplinary effort.

REFERENCES

Anikeeva, P., Andalman, A.S., Witten, I., Warden, M., Goshen, I., Grosenick, L., Gunaydin, L.A., Frank, L.M., Deisseroth, K., 2012. Optetrode: a multichannel readout for optogenetic control in freely moving mice. *Nat. Neurosci.* **15**, 163–170.

Aravanis, A.M., Wang, L.-P., Zhang, F., Meltzer, L.A., Mogri, M.Z., Schneider, M.B., Deisseroth, K., 2007. An optical neural interface: in vivo control of rodent motor cortex with integrated fiberoptic and optogenetic technology. *J. Neural Eng.* **4**, S143–S156.

Bartels, M., Hernandez, V.H., Krenkel, M., Moser, T., Salditt, T., 2013. Phase contrast tomography of the mouse cochlea at microfocus x-ray sources. *Appl. Phys. Lett.* **103**, 083703.

Boyden, E.S., Zhang, F., Bamberg, E., Nagel, G., Deisseroth, K., 2005. Millisecond-timescale, genetically targeted optical control of neural activity. *Nat. Neurosci.* **8**, 1263–1268.

Brackmann, D.E., Hitselberger, W.E., Nelson, R.A., Moore, J., Waring, M.D., Portillo, F., Shannon, R.V., Telischi, F.F., 1993. Auditory brainstem implant: I. Issues in surgical implantation. *Otolaryngol. Head Neck Surg.* **108**, 624–633.

Colletti, V., Shannon, R.V., Carner, M., Veronese, S., Colletti, L., 2009. Progress in restoration of hearing with the auditory brainstem implant. *Prog. Brain Res.* **175**, 333–345.

Deisseroth, K., Feng, G., Majewska, A.K., Miesenböck, G., Ting, A., Schnitzer, M.J., 2006. Next-generation optical technologies for illuminating genetically targeted brain circuits. *J. Neurosci.* **26**, 10380–10386.

Donaldson, G.S., Kreft, H.A., Litvak, L., 2005. Place-pitch discrimination of single- versus dual-electrode stimuli by cochlear implant users (L). *J. Acoust. Soc. Am.* **118**, 623–626.

Fetterman, B.L., Domico, E.H., 2002. Speech recognition in background noise of cochlear implant patients. *Otolaryngol. Head Neck Surg.* **126**, 257–263.

Fishman, K.E., Shannon, R.V., Slattery, W.H., 1997. Speech recognition as a function of the number of electrodes used in the SPEAK cochlear implant speech processor. *J. Speech Lang. Hear. Res.* **40**, 1201–1215.

Friesen, L.M., Shannon, R.V., Baskent, D., Wang, X., 2001. Speech recognition in noise as a function of the number of spectral channels: comparison of acoustic hearing and cochlear implants. *J. Acoust. Soc. Am.* **110**, 1150–1163.

Hernandez, V.H., Gehrt, A., Reuter, K., Jing, Z., Jeschke, M., Mendoza Schulz, A., Hoch, G., Bartels, M., Vogt, G., Garnham, C.W., Yawo, H., Fukazawa, Y., Augustine, G.J., Bamberg, E., Kügler, S., Salditt, T., de Hoz, L., Strenzke, N., Moser, T., 2014. Optogenetic stimulation of the auditory pathway. *J. Clin. Invest.* **124**, 1114–1129.

Herrmann, B.S., Brown, M.C., Eddington, D.K., Hancock, K.E., Lee, D.J., 2015. Auditory brainstem implant: electrophysiologic responses and subject perception. *Ear Hear.* **36**, 368–376.

Hight, A.E., Kozin, E.D., Darrow, K., Lehmann, A., Boyden, E., Brown, M.C., Lee, D.J., 2015. Superior temporal resolution of Chronos versus channelrhodopsin-2 in an optogenetic model of the auditory brainstem implant. *Hear. Res.* **322**, 235–241.

Hososhima, S., Yuasa, H., Ishizuka, T., Hoque, M.R., Yamashita, T., Yamanaka, A., Sugano, E., Tomita, H., Yawo, H., 2015. Near-infrared (NIR) up-conversion optogenetics. *Sci. Rep.* **5**, 16533.

Izzo, A.D., Richter, C.-P., Jansen, E.D., Walsh, J.T., 2006. Laser stimulation of the auditory nerve. *Lasers Surg. Med.* **38**, 745–753.

Izzo, A.D., Walsh, J.T., Jansen, E.D., Bendett, M., Webb, J., Ralph, H., Richter, C.-P., 2007. Optical parameter variability in laser nerve stimulation: a study of pulse duration, repetition rate, and wavelength. *IEEE Trans. Biomed. Eng.* **54**, 1108–1114.

Izzo, A.D., Walsh, J.T., Ralph, H., Webb, J., Bendett, M., Wells, J., Richter, C.-P., 2008. Laser stimulation of auditory neurons: effect of shorter pulse duration and penetration depth. *Biophys. J.* **94**, 3159–3166.

Klapoetke, N.C., Murata, Y., Kim, S.S., Pulver, S.R., Birdsey-Benson, A., Cho, Y.K., Morimoto, T.K., Chuong, A.S., Carpenter, E.J., Tian, Z., Wang, J., Xie, Y., Yan, Z., Zhang, Y., Chow, B.Y., Surek, B., Melkonian, M., Jayaraman, V., Constantine-Paton, M., Wong,

G.K.-S., Boyden, E.S., 2014. Independent optical excitation of distinct neural populations. *Nat. Methods* **11**, 338–346.

Kleinlogel, S., Feldbauer, K., Dempski, R.E., Fotis, H., Wood, P.G., Bamann, C., Bamberg, E., 2011. Ultra light-sensitive and fast neuronal activation with the Ca^{2+}-permeable channelrhodopsin CatCh. *Nat. Neurosci.* **14**, 513–518.

Knöpfel, Boyden, 2012. *Optogenetics: Tools for Controlling and Monitoring Neuronal Activity.* Elsevier.

Kohlberg, G., Spitzer, J.B., Mancuso, D., Lalwani, A.K., 2014. Does cochlear implantation restore music appreciation? *The Laryngoscope* **124**, 587–588.

Kral, A., Hartmann, R., Mortazavi, D., Klinke, R., 1998. Spatial resolution of cochlear implants: the electrical field and excitation of auditory afferents. *Hear. Res.* **121**, 11–28.

Littlefield, P.D., Vujanovic, I., Mundi, J., Matic, A.I., Richter, C.-P., 2010. Laser stimulation of single auditory nerve fibers. *The Laryngoscope* **120**, 2071–2082.

Miller, C.A., Abbas, P.J., Robinson, B.K., Nourski, K.V., Zhang, F., Jeng, F.-C., 2006. Electrical excitation of the acoustically sensitive auditory nerve: single-fiber responses to electric pulse trains. *J. Assoc. Res. Otolaryngol.* **7**, 195–210.

Müller, M., von Hünerbein, K., Hoidis, S., Smolders, J.W.T., 2005. A physiological place-frequency map of the cochlea in the CBA/J mouse. *Hear. Res.* **202**, 63–73.

Nagel, G., Ollig, D., Fuhrmann, M., Kateriya, S., Musti, A.M., Bamberg, E., Hegemann, P., 2002. Channelrhodopsin-1: a light-gated proton channel in green algae. *Science* **296**, 2395–2398.

Nagel, G., Szellas, T., Huhn, W., Kateriya, S., Adeishvili, N., Berthold, P., Ollig, D., Hegemann, P., Bamberg, E., 2003. Channelrhodopsin-2, a directly light-gated cation-selective membrane channel. *Proc. Natl. Acad. Sci. U. S. A.* **100**, 13940–13945.

Noij, K.S., Kozin, E.D., Sethi, R., Shah, P.V., Kaplan, A.B., Herrmann, B., Remenschneider, A., Lee, D.J., 2015. Systematic review of nontumor pediatric auditory brainstem implant outcomes. *Otolaryngol. Head Neck Surg.* **153**, 739–750.

Park, S.I., Brenner, D.S., Shin, G., Morgan, C.D., Copits, B.A., Chung, H.U., Pullen, M.Y., Noh, K.N., Davidson, S., Oh, S.J., Yoon, J., Jang, K.-I., Samineni, V.K., Norman, M., Grajales-Reyes, J.G., Vogt, S.K., Sundaram, S.S., Wilson, K.M., Ha, J.S., Xu, R., Pan, T., Kim, T.-I., Huang, Y., Montana, M.C., Golden, J.P., Bruchas, M.R., Gereau, R.W., Rogers, J.A., 2015. Soft, stretchable, fully implantable miniaturized optoelectronic systems for wireless optogenetics. *Nat. Biotechnol.* **33**, 1280–1286.

Richter, C.-P., Bayon, R., Izzo, A.D., Otting, M., Suh, E., Goyal, S., Hotaling, J., Walsh, J.T., 2008. Optical stimulation of auditory neurons: effects of acute and chronic deafening. *Hear. Res.* **242**, 42–51.

Richter, C.-P., Rajguru, S.M., Matic, A.I., Moreno, E.L., Fishman, A.J., Robinson, A.M., Suh, E., Walsh, J.T., 2011. Spread of cochlear excitation during stimulation with pulsed infrared radiation: inferior colliculus measurements. *J. Neural Eng.* **8**, 056006.

Schultz, M., Baumhoff, P., Maier, H., Teudt, I.U., Krüger, A., Lenarz, T., Kral, A., 2012. Nanosecond laser pulse stimulation of the inner ear-a wavelength study. *Biomed. Opt. Express* **3**, 3332–3345.

Shannon, R.V., 1983. Multichannel electrical stimulation of the auditory nerve in man. II. Channel interaction. *Hear. Res.* **12**, 1–16.

Shannon, R.V., Fayad, J., Moore, J., Lo, W.W., Otto, S., Nelson, R.A., O'Leary, M., 1993. Auditory brainstem implant: II. Postsurgical issues and performance. *Otolaryngol. Head Neck Surg.* **108**, 634–642.

Shannon, R.V., Fu, Q.-J., Galvin, J., 2004. The number of spectral channels required for speech recognition depends on the difficulty of the listening situation. *Acta Otolaryngol. Suppl.* **552**, 50–54.

Shapiro, M.G., Homma, K., Villarreal, S., Richter, C.-P., Bezanilla, F., 2012. Infrared light excites cells by changing their electrical capacitance. *Nat. Commun.* **3**, 736.

Shimano, T., Fyk-Kolodziej, B., Mirza, N., Asako, M., Tomoda, K., Bledsoe, S., Pan, Z.H., Molitor, S., Holt, A.G., 2013. Assessment of the AAV-mediated expression of channelrhodopsin-2 and halorhodopsin in brainstem neurons mediating auditory signaling. *Brain Res.* 1511, 138–152.

Sineshchekov, O.A., Jung, K.-H., Spudich, J.L., 2002. Two rhodopsins mediate phototaxis to low- and high-intensity light in *Chlamydomonas reinhardtii. Proc. Natl. Acad. Sci. U. S. A.* **99**, 8689–8694.

Sousa, A.F., Carvalho, A.C., Couto, M.I., Tsuji, R.K., Goffi-Gomez, M.V., Bento, R.F., Matas, C.G., Befi-Lopes, D.M., 2015. Telephone Usage and Cochlear Implant: Auditory Training Benefits. *Int. Arch. Otorhinolaryngol.* **19**, 269–272.

Srinivasan, A.G., Padilla, M., Shannon, R.V., Landsberger, D.M., 2013. Improving speech perception in noise with current focusing in cochlear implant users. *Hear. Res.* **299**, 29–36.

Srinivasan, A.G., Shannon, R.V., Landsberger, D.M., 2012. Improving virtual channel discrimination in a multi-channel context. *Hear. Res.* **286**, 19–29.

Teudt, I.U., Maier, H., Richter, C.-P., Kral, A., 2011. Acoustic events and "optophonic" cochlear responses induced by pulsed near-infrared laser. *IEEE Trans. Biomed. Eng.* **58**, 1648–1655.

Ting, J.T., Feng, G., 2013. Development of transgenic animals for optogenetic manipulation of mammalian nervous system function: progress and prospects for behavioral neuroscience. *Behav. Brain Res.* **255**, 3–18.

Tomita, H., Sugano, E., Fukazawa, Y., Isago, H., Sugiyama, Y., Hiroi, T., Ishizuka, T., Mushiake, H., Kato, M., Hirabayashi, M., Shigemoto, R., Yawo, H., Tamai, M., 2009. Visual properties of transgenic rats harboring the channelrhodopsin-2 gene regulated by the Thy-1.2 promoter. *PLoS One* **4**, e7679.

Verma, R.U., Guex, A.A., Hancock, K.E., Durakovic, N., McKay, C.M., Slama, M.C.C., Brown, M.C., Lee, D.J., 2014. Auditory responses to electric and infrared neural stimulation of the rat cochlear nucleus. *Hear. Res.* **310**, 69–75.

Wang, H., Peca, J., Matsuzaki, M., Matsuzaki, K., Noguchi, J., Qiu, L., Wang, D., Zhang, F., Boyden, E., Deisseroth, K., Kasai, H., Hall, W.C., Feng, G., Augustine, G.J., 2007. High-speed mapping of synaptic connectivity using photostimulation in Channelrhodopsin-2 transgenic mice. *Proc. Natl. Acad. Sci. U. S. A.* **104**, 8143–8148.

WHO, 2015. Deafness and hearing loss. [www document]. URL http://www.who.int/mediacentre/factsheets/fs300/en/.

Wu, C.-C., Luo, X., 2014. Electrode spanning with partial tripolar stimulation mode in cochlear implants. *J. Assoc. Res. Otolaryngol.* **15**, 1023–1036.

Wu, C.-M., Liu, T.-C., Wang, N.-M., Chao, W.-C., 2013. Speech perception and communication ability over the telephone by Mandarin-speaking children with cochlear implants. *Int. J. Pediatr. Otorhinolaryngol.* **77**, 1295–1302.

Yizhar, O., Fenno, L.E., Davidson, T.J., Mogri, M., Deisseroth, K., 2011. Optogenetics in neural systems. *Neuron* **71**, 9–34.

Zeng, F.-G., Grant, G., Niparko, J., Galvin, J., Shannon, R., Opie, J., Segel, P., 2002. Speech dynamic range and its effect on cochlear implant performance. *J. Acoust. Soc. Am.* **111**, 377–386.

Zeng, F.-G., Rebscher, S., Harrison, W., Sun, X., Feng, H., 2008. Cochlear implants: system design, integration, and evaluation. *IEEE Rev. Biomed. Eng.* **1**, 115–142.

Zeng, F.-G., Tang, Q., Lu, T., 2014. Abnormal pitch perception produced by cochlear implant stimulation. *PLoS One* **9**, e88662.

Zierhofer, C.M., Hochmair-Desoyer, I.J., Hochmair, E.S., 1995. Electronic design of a cochlear implant for multichannel high-rate pulsatile stimulation strategies. *IEEE Trans. Rehabil. Eng.* **3**, 112–116.

31 The Role of Amino Acids in Neurodegenerative and Addictive Diseases

Joan Fallon

31.1 Introduction

Amino acids have long been described as the building blocks of proteins. Amino acids are classified as essential, non-essential or semi-essential. Essential amino acids are those that the body cannot synthesize or otherwise obtain endogenously in both children and adults. These essential amino acids must be obtained exogenously by eating a protein and breaking the protein down into its component amino acids through enzymatic hydrolysis. Semi-essential amino acids are those that young children must obtain exogenously since they cannot be synthesized or procured endogenously in the pediatric population. Semi-essential amino acids can be procured or synthesized endogenously in adults. Non-essential amino acids are those that the body can procure or synthesize endogenously in both adults and children (Matthews, 1972).

Amino acids are cleaved from the ingested protein as mono-, di-, tri-, tetra- and higher-peptide fragments. Amino acid pool depletion and so a lack of building blocks for new proteins represents one way in which diminished numbers of amino acids, especially essential amino acids, could impact neurodegeneration in a number of well-known conditions such as autism, Parkinson's disease and schizophrenia. Furthermore, recent literature points to the significant role the essential amino acid phenylalanine plays in the pathophysiology of addiction, especially cocaine addiction, quite possibly by the ΔFosB theory of brain change resulting in addiction (Nestler *et al.*, 2001).

Emerging evidence demonstrates that amino acids are not only involved in protein synthesis, but also play a significant role in other important neurological functions. For example, those essential amino acids that are cleaved by the digestive enzyme chymotrypsin – phenylalanine and tryptophan – are necessary components of the body's synthesis of dopamine and serotonin, respectively. There is a known serotonin deficiency seen in children with autism (Schain and Freedman, 1961; Matthew, 1972; Anderson *et al.*, 1987; Coutinho *et al.* 2004).

Many such correlations are beginning to emerge, and we will examine some of them in this chapter.

31.2 Physiology

Classical thinking with respect to amino acids identifies them as the building blocks of new proteins synthesized by the body. Essential amino acids need to be ingested in the form of protein and are cleaved from that protein. They cannot be obtained any other way or synthesized by the body. Furthermore, the administration of these essential amino acids directly into the gastrointestinal tract results in poor uptake and utilization (Matthews, 1972). The semi-essential amino acids act similarly to essential amino acids in small children and the non-essential amino acids can be synthesized by the body without any additional input from ingested exogenous proteins.

While this classical view has been held for decades, we now know that the demarcation between essential and non-essential is not as well defined as previously believed. For example, the endogenous synthesis of many non-essential amino acids is under the direct control of essential amino acids. So while in theory a non-essential amino acid can be synthesized, an essential amino acid may be necessary for that synthesis to occur.

Furthermore, the essential amino acid methionine, for example, serves as the initiation codon at the beginning of all protein synthesis in humans. This is also a very important amino acid with respect to RNA replication. Leucine is the only amino acid that is required to optimally stimulate muscle protein synthesis. Both methionine and leucine and the processes they enable are necessary in childhood for growth and repair (Fafournoux *et al.*, 2000; Norton and Layman, 2006).

In addition to their roles in neurotransmitters, regulators of protein degradation, nitrogen physiology and catabolism, amino acids are emerging as important regulators of gene expression. This role has been described in the literature over the last 15 years (Drabkin and Rajbhandary, 1998; Fafournoux *et al.*, 2000; Norton and Layman, 2006; Avruch *et al.*, 2009).

Starvation of certain amino acids can actually determine the up-regulation or down-regulation of specific genes. For example, Fafournoux *et al.* (2000) noted that gene expression is strongly induced in response to methionine starvation. The *CHOP* gene is of particular interest because of its role in controlling neural pruning in the developing brain through apoptosis. Autopsied brains of autistic individuals show an excess number of neurons and neuronal connections, suggesting that part of the cellular pathology in the disorder involves a failure of normal neural pruning mechanisms (Fafournoux *et al.*, 2000; Avruch *et al.*, 2009; Morimoto, 2012; Tang *et al.*, 2014).

There is a wide variety of mechanisms by which amino acids can affect the regulation of gene expression. Experiments conducted in yeast have shown that regulation can proceed through very specific forms of control, such as those that are reliant upon end products of enzymes associated with a particular biosynthetic pathway. For example, the synthesis of leucine is controlled by the transcriptional activator Leu3p in response to the availability of leucine (Tang *et al.*, 2014).

An additional example of regulation is found in the deprivation of certain amino acids or stress, such as that from heat shock. In these instances, a lack of specific amino acids causes the corresponding tRNAs to accumulate, which in turn leads to an increased production of a transcriptional regulator.

There have been well over 500 genes identified that can be affected by the presence of absence of specific amino acids. Through the outside availability of food sources or the ability of the body to utilize those sources, this relatively new finding in gene regulation represents an important source of epigenetic influence affecting the course of disease pathology.

We will see in Section 31.3 how new theories of cocaine addiction and the role of the brain transcription factor ΔFosB are interdependent with the essential amino acid phenylalanine. The changes in ΔFosB after cocaine use result in the elongation of the dendritic spines in the presence of a depleted source of phenylalanine.

The role of amino acids and especially amino acid pool depletion, particularly of essential amino acids, could impact autism and other neurodegenerative conditions. Clearly, the role of amino acids is no longer relegated to simply being the "building blocks of protein."

31.3 Pathophysiology

Disturbances in the serotonergic system in autism have been noted since Schain and Freedman's original report in 1961 of elevated blood serotonin levels in patients with autism. Further studies showed that when serotonin levels of platelet-poor fractions of blood from autistic individuals were studied, individuals with autism did not show elevated levels.

Thus, elevated levels of blood serotonin in autistic patients were associated with the platelet fraction. Furthermore, the lack of serotonin in the non-platelet fraction of blood suggested either hyperactivity of a transport mechanism into the platelets or a defect of transporters allowing the serotonin to move out of the platelets. Subsequent studies demonstrated the presence of a potent serotonin uptake transporter that functions to accumulate serotonin in the platelets, accounting for the altered compartmentalization observed in the blood of autistic patients (Schain *et al.*, 1961; Anderson *et al.*, 1987; Coutinho *et al.*, 2004; Naushad *et al.*, 2013).

These studies brought attention to the serotonin transporter gene *HTT* as a possible candidate gene in autism. Early studies using selective serotonin reuptake inhibitors (SSRIs) suggested some effect. However, the latest review of published studies does not support the effectiveness of SSRIs, with some evidence of harm. Studies surrounding the related gene locus *SLC6A4*, however, have also been inconsistent, and genetic linkage studies for several candidate genes continue (Naushad *et al.*, 2013; Williams *et al.*, 2013).

Deciphering the genetic code over the last three decades has revealed the presence of three separate stop codons and a single start codon in DNA. The start codon is so designated because this is the point at which the information for creating any particular protein begins in the genome. This information is initially stored in DNA and consists of a sequence of purine and pyrimidine bases. The code sequence is always the same and it specifically codes for the amino acid methionine. Thus, methionine is the first amino acid contained in all proteins created by the body. By virtue of its status as the initial amino acid for all protein synthesis, methionine is absolutely essential for life (Fafournoux *et al.*, 2000; Naushad *et al.*, 2013; Williams *et al.*, 2013).

Therefore, a lack of amino acids can affect multiple processes in the human body, including, but not limited to, the formation of neurotransmitters, epigenetic influences, protein catabolism and anabolism, multiple levels of transcription and the persistence of addictive behaviors. The fact that the lack or presence of amino acids can have an epigenetic effect and potentially turn on or off gene expression adds to the importance of the availability of a complete amino acid pool.

While the complex balance of amino acids is necessary for life, there are basic tenets of that balance that hold true for all human proteomics:

- An amino acid response system has been identified. This system signals and identifies deficiencies and imbalances in amino acids in the human body (Balasubramanian *et al.*, 2013).
- Dietary intake of amino acids is typically not balanced to exactly match the body's demands for various amino acids (Felig, 1975).
- Amino acids taken via the diet must be chemically modified and rearranged to provide adequate levels of all of the amino acids needed (Felig, 1975; Norton and Layman, 2006; Avruch *et al.*, 2009; Balasubramanian *et al.*, 2013).
- Essential amino acids can only be procured from ingested protein. They cannot be manufactured in the body (Felig, 1975; Avruch *et al.*, 2009; Balasubramanian *et al.*, 2013).
- Non-essential amino acids can be manufactured in the body; however, in some cases, they are under the direct control of essential amino acids (Felig, 1975; Fafourinoux *et al.*, 2000; Balasubramanian *et al.*, 2013).
- Amino acid absorption pathways in the gastrointestinal mucosa are configured to transport di- and tri-peptides, not single amino acids. Supplementation with single amino acids will not be sufficient for the replacement of amino acids due to a lack of protein ingestion (Fairclough *et al.*, 1980).
- In humans, unlike other mammals, levels of protein in the diet cannot induce digestive proteases to pathologic levels. Amylases and, to some degree, lipases can be induced by the level of carbohydrate and fat in the diet (Boivin, 1990).
- There is a significant number of pathways in the body for balancing the pool of amino acids, both for synthesis and for degradation. The number of enzymes and co-enzymes necessary for each pathway creates a great potential for disease when there is a deficiency in the pathway (Fairclough *et al.*, 1980; Boivin, 1990).

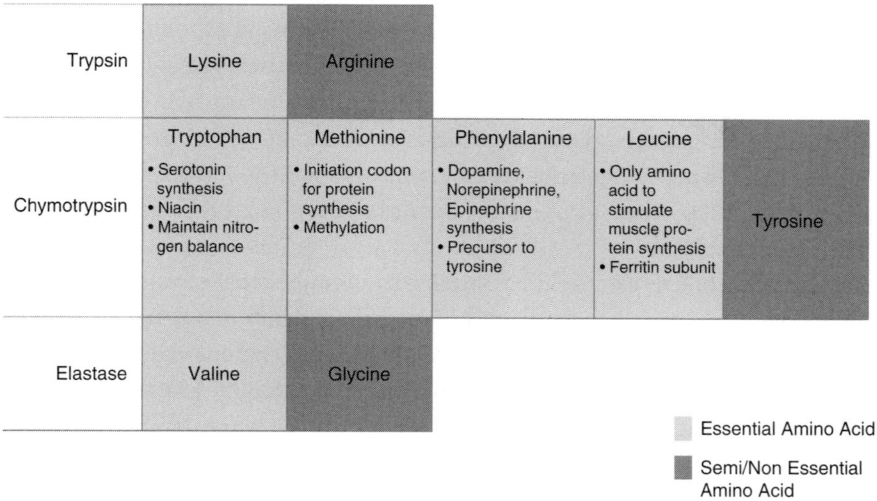

Figure 31.1 Major proteases found in pancreatic enzymes and their cleaved amino acids.

- The genetic influences of the amino acids may account for certain disease states (Felig, 1975; Fairclough *et al.*, 1980; Boivin, 1990).
- Disruption of the amino acid sequence, even by just one amino acid, can have profound consequences for growth and development.
  The consequences may be physiological, cognitive, psychological and behavioral. Some of these are inheritable and many are fatal (Boivin, 1990).

The essential and non-essential amino acids have the ability to control protein synthesis, gene regulation, nitrogen balance and neurotransmission, among others. The importance of the amino acid supply and the ability to synthesize non-essential amino acids and to obtain essential amino acids from the diet are paramount. The inability of the body to directly synthesize essential amino acids and the fact that the essential amino acids play roles in the synthesis of non-essential amino acids, as in the case of asparagine, which is under the control of methionine (an essential amino acid, as shown in Figure 31.1), underscores the importance of the enzyme's activity and its ability to aid in the breakdown of protein and to supply the body's amino acid pool.

Subsequent to the discovery of pathologically low levels of chymotrypsin (Heil *et al.*, 2014), multiple studies were independently undertaken in a large subgroup of children with autism by academic researchers in order to study amino acids and their effects on children with autism. These studies have demonstrated that there appear to be low amounts of circulating amino acids in children with autism (Arnold *et al.*, 2003; Evans *et al.*, 2008; Naushad *et al.*, 2013).

In 2003, Arnold *et al.* reported an amino acid deficit in children with autism spectrum disorder. While children with a restricted diet showed a greater deficiency in key essential amino acids, all autistic children had deficient amino acid profiles compared to controls. Additionally, Evans *et al.* (2008) described lower-than-expected excretion of amino acids from autistic patients, suggesting the presence of altered metabolic homeostasis. This finding is key in that the quality of food ingestion, as stated earlier, is not the underlying cause of the lack of amino acids demonstrated in children with autism (Schedl *et al.*, 1968; Norton and Layman, 2006; Evans *et al.*, 2008; Munasinghe *et al.*, 2010; Naushad *et al.*, 2013; Daly *et al.*, 2014).

Various academic researchers have studied the administration of amino acids in children with autism and were unable to successfully demonstrate improvements in autism behaviors. Studies with secretin, plant-based enzymes and other drugs have been similarly unsuccessful. Specifically, in a paper by Munasinghe *et al.* (2010) in which plant-based digestive enzymes were given to children in an office-based clinical trial, no efficacy was shown. Plant enzymes are not suitable for delivering sufficient proteases to children with autism, as evidenced by early Phase I studies (Chez *et al.*, 2000; Munasinghe *et al.*, 2010; Williams *et al.*, 2012; Heil *et al.*, 2014).

The utilization and absorption of amino acids and the preferential uptake in the gastrointestinal system (specifically in the duodenum) is accomplished by the absorption of di- and tri-peptides through the mucosa. There appears to be a preferential absorption (in volume and rate) of di- and tri-peptides over single amino acids, as well as tetrapeptides and higher peptides. This is an important aspect of enzyme replacement utilization for those who lack chymotrypsin. For children with autism presenting with low/pathological levels of chymotrypsin, the need to deliver di- and tri-peptides to the small intestinal mucosa for absorption is paramount (Fairclough *et al.*, 1980).

In a paper in *Gut*, Fairclough *et al.* (1980) describe the ways in which the amino acid channels are set up in the small intestines. Figure 31.1 is derived from this paper. These researchers describe not only the specific breakdown of protein into di- and tri-peptides, but also the absorption of the amino acids in those forms that are preferential and more expeditious, as evidenced by the two- to three-times faster absorption when proteins are in the form of di- and tri-peptides.

The absorption channels that allow for the maximized absorption of di- and tri-peptides in the mucosal wall of the small intestine do not support the administration of single amino acids, which are present in many dietary supplements. These single amino acids are mainly destroyed by the gastric acid of the stomach. Furthermore, these absorption channels are the preferential way in which these di- and tri-peptides are absorbed.

In work undertaken in the late 1990s, the author demonstrated that a large subgroup of children with autism exhibit low levels of fecal chymotrypsin, thus signaling the potential for a lack of protein digestion, as chymotrypsin is one of the main protease enzymes in the human body. The inability to induce pathological levels pointed to the fact that this subgroup may actually have a deficit in protein digestion. As chymotrypsin cleaves phenylalanine, tryptophan, methionine and leucine, the neurological implications of the potential lack of amino acids in the generalized amino acid pool was profound.

The lack of this enzyme and the connection to neurological function cannot be ignored. It has been reported in the literature that children with autism have low levels of tryptophan as well as low amounts of serotonin. Serotonin is a neurotransmitter that is found in the gastrointestinal tract (80%), brain and platelets in humans and is thought to be involved in the brain's control over behavior. Functions such as appetite, mood, aggression, sleep, memory, cognition, motor function and sexual function have been cited as potential targets of serotonin's influence. Furthermore, serotonin may play a role in neuroendocrine signaling (Shain and Freedman, 1961; Coutinho *et al.*, 2004; Guesnet and Alessandri, 2011).

Deficiencies in serotonin have been linked to depression, increased appetite and other disorders. Regarding such deficiencies having been linked to depression, there are a large number of antidepressants that target serotonin metabolism. One of these classes of drugs, the selective serotonin reuptake inhibitors, prevents the body from reabsorbing serotonin so that it can remain in the brain and so be available for the synapses (Anderson *et al.*, 1987; Coutinho *et al.*, 2004; Guesnet and Alessandri, 2011).

The emergence of evidence with respect to cocaine addiction and the role of ΔFosB that builds up in the brains of cocaine addicts is implicated in the persistence of cocaine addiction (McClung and Nestler, 2003). The presence of cocaine in the body will lead to increased utilization of dopamine at the sites of dopamine transporters. The pleasure/reward mechanism, loss of control and compulsive behaviors are controlled by the limbic system, which contains cells that are highly responsive to dopamine. These responsive cells are designed to trigger pleasurable responses that make us both feel good and want to repeat the experience. In some instances, the drive for sexual pleasure and gratification can promote a desire to mate and result in species survival. On the other hand, the same mechanism can help to keep us repeating pleasurable yet destructive behaviors, such as those seen in addiction (Nestler, 2001).

Scientists have searched for the biological basis of this seemingly constructive and destructive mechanism, and current theories revolve around the transcription factor ΔFosB. ΔFosB is a unique transcription factor that is derived from the *fosB* gene via alternative splicing, and therefore is known as a genetic transcription factor (McClung *et al.*, 2004).

One potentially key type of cocaine-related change that appears to last for many months after the last cocaine exposure, and perhaps longer, is an alteration in the physical structure of nerve cells in the nucleus accumbens (NAc). Chronic cocaine exposure causes these cells to extend and sprout new offshoots on their dendrites (Nestler, 2001; Robinson and Berridge, 2001). Dendrites are the branch-like fibers that grow out from nerve cell bodies and collect incoming signals from other nerve cells. Just as a bigger antenna picks up more radio waves, more dendrite branches in the NAc will theoretically collect a greater volume of nerve signals coming from other regions (e.g. the hippocampus, amygdala and frontal cortex). Because cocaine addiction decreases stores of phenylalanine, elongation of the dendritic spines occurs in the presence of a depleted source of phenylalanine. The converse is true in the case of phenyl ketone urea, where there is a build-up of phenylalanine, causing the dendrites to become stubs.

Furthermore, Nestler (2001) hypothesized that the formation of ΔFosB from the FosB in the NAc is through a "leucine zipper" mechanism, in which leucine is found at every seventh position. He further argued that the elongation of the dendritic spines found in the presence of the ΔFosB is regulated by the presence or absence of phenylalanine. This addition thereby results in an altered ratio of leucine to phenylalanine.

ΔFosB is naturally present in small quantities in the cells of the NAc in the brain, and chronic cocaine exposure causes it to accumulate at high levels (Nestler *et al.*, 2001). It has been postulated by some that ΔFosB might play a role as a molecular "switch" in the transition from drug abuse to addiction, mainly for the reasons given in the remainder of this section.

Once created, a molecule of ΔFosB lasts for 6–8 weeks before breaking apart chemically (Nestler *et al.*, 2001). Therefore, each new episode of cocaine abuse exacerbates the build-up of ΔFosB that has accumulated from all previous episodes over approximately 2 months. If someone is abusing cocaine daily, the levels of ΔFosB will be extremely elevated all of the time.

ΔFosB causes the elongation and formation of dendritic spines, which are postulated to induce and maintain cocaine cravings. In the presence of phenylalanine, the elongation recedes to the normal state.

Mice with elevated ΔFosB exhibit a set of behaviors that correspond to human addictive behaviors, while mice with normal levels of ΔFosB do not. Conversely, blocking the build-up of ΔFosB in mice during a regimen of cocaine exposure reduces these behaviors (Nestler *et al.*, 2001).

Chronic administration of cocaine has recently been shown to increase ΔFosB in several brain regions, most specifically the NAc. It also builds up in such areas as the frontal cortex and amygdala (McClung *et al.*, 2004). The accumulations of ΔFosB in the frontal cortex and the amygdala are much smaller than those that cocaine causes in the NAc (McClung and Nestler, 2003).

ΔFosB apparently causes more than 25% of all chronic cocaine-induced changes in gene expression in the NAc, a finding that highlights the dominant role of this transcription factor in mediating cocaine's genetic effects in the brain. One of the genes stimulated by ΔFosB is an enzyme called cyclin-dependent kinase-5 (CDK5), which promotes nerve cell growth. This finding has shed new light on mechanisms underlying cocaine's very long-lasting effects on the brain (Nestler, 2001). Because ΔFosB is induced in the brain specifically and remains in these brain regions for long periods of time, it has been proposed that ΔFosB acts as a sustained "molecular switch" that first initiates and then maintains some of the long-term adaptations of the brain in response to chronic perturbations (Nestler *et al.*, 2001; McClung *et al.*, 2004).

31.4 Conclusion

The role of amino acids and especially essential amino acids has begun to emerge as an important aspect of neurological function. Essential amino acids as well as non-essential amino acids play important roles in neurological function by directly affecting gene expression. The role of amino acids in gene expression opens up new doors for the study of epigenetics and disease, especially neurodegenerative diseases.

REFERENCES

American Psychiatric Association (2013). *Diagnostic and Statistical Manual of Mental Disorders* (5th ed.) Washington, DC.

Anderson, G., Feibel, F., & Cohen, D. (1987). Determination of serotonin in whole blood, platelet-rich plasma, platelet-poor plasma and plasma ultrafiltrate. *Life Sciences*, **40** (11), 1063–1070.

Arnold, G., Hyman, S., Mooney, R., & Kirby, R. (2003). Plasma amino acids profiles in children with autism: potential risk of nutritional deficiencies. *Journal of Autism and Developmental Disorders*, **33**(4), 449–454.

Avruch, J., Long, X., Ortiz-Vega, S., Rapley, J., Papageorgiou, A., & Dai, N. (2009). Amino acid regulation of TOR complex 1. *AJP: Endocrinology and Metabolism*, **296**(4), E592–E602.

Balasubramanian, M., Butterworth, E., & Kilberg, M. (2013). Asparagine synthetase: regulation by cell stress and involvement in tumor biology. *American Journal of Physiology: Endocrinology and Metabolism*, **304**(8), E789–E799.

Chez, M., Buchanan, C., Bagan, B., Hammer, M., McCarthy, K., Ovrutskaya, I., *et al.* (2000). Secretin and autism: a two-part clinical investigation. *Journal of Autism and Developmental Disorders*, **30**(2), 87–94.

Coutinho, A., Oliveira, G., Morgadinho, T., Fesel, C., Macedo, T., Bento, C., *et al.* (2004). Variants of the serotonin transporter gene (*SLC6A4*) significantly contribute to hyperserotonemia in autism. *Molecular Psychiatry*, **9**(3), 264–271.

Daly, E., Ecker, C., Hallahan, B., Deeley, Q., Craig, M., Murphy, C., *et al.* (2014). Response inhibition and serotonin in autism: a functional MRI study using acute tryptophan depletion. *Brain*, **137**(Pt 9), 2600–2610.

Drabkin, H. & Rajbhandary, U. (1998). Initiation of protein synthesis in mammalian cells with codons other than AUG and amino acids other than methionine. *Molecular and Cellular Biology*, **18**(9), 5140–5147.

Evans, C., Dunstan, H., Rothkirch, T., Roberts, T., Reichelt, K., Cosford, R., *et al.* (2008). Altered amino acid excretion in children with autism. *Nutritional Neuroscience*, **11**(1), 9–17.

Fafournoux, P., Bruhat, A., & Jousse, C. (2000). Amino acid regulation of gene expression. *Biochemical Journal*, **351**(Pt 1), 1–12.

Fairclough, P., Hegarty, J., Silk, D., & Clark, M. (1980). Comparison of the absorption of two protein hydrolysates and their effects on water and electrolyte movements in the human jejunum. *Gut*, **21**(10), 829–834.

Felig, P. (1975). Amino acid metabolism in man. *Annual Review of Biochemistry*, **44**, 933–955.

First, M., Spitzer, R., Gibbon, M., & Williams, J. (2002). *Structured Clinical Interview for DSM-IV-TR Axis I Disorders, Research Version, Patient Edition*. New York: Biometrics Research, New York State Psychiatric Institute.

Guesnet, P. & Alessandri, J. (2011). Docosahexaenoic acid (DHA) and the developing central nervous system (CNS) – implications for dietary recommendations. *Biochimie*, **93**(1), 7–12.

Heil, M., Pearson, D., & Fallon, J. (2014). Low endogenous fecal chymotrypsin: a possible biomarker for autism. Poster presented at the annual IMFAR Conference on Autism, Atlanta, GA.

Matthews, D. (1972). Intestinal absorption of amino acids and peptides. *Proceedings of the Nutrition Society*, **31**(2), 171–177.

McClung, C., Ulery, P., Perrotti, L., Zachariou, V., Berton, O., & Nestler, E. (2004). ΔFosB: a molecular switch for long-term adaptation in the brain. *Molecular Brain Research*, **132**(2), 146–154.

McClung, C. & Nestler, E. (2003). Regulation of gene expression and cocaine reward by CREB and ΔFosB. *Nature Neuroscience*, **6**(11), 1208–1215.

Morimoto, R. (2012). The heat shock response: Systems biology of proteotoxic stress in aging and disease. *Cold Spring Harbor Symposia on Quantitative Biology*, **76**, 91–99.

Munasinghe, S., Oliff, C., Finn, J., & Wray, J. (2010). Digestive enzyme supplementation for autism spectrum disorders: A double-blind randomized controlled trial. *Journal of Autism and Developmental Disorders*, **40**(9), 1131–1138.

Naushad, S., Jain, J., Prasad, C., Naik, U., & Akella, R. (2013). Autistic children exhibit distinct plasma amino acid profile. *Indian Journal of Biochemistry and Biophysics*, **50** (5), 474–478.

Nestler, E. (2001). Molecular basis of long-term plasticity underlying addiction. *Nature Reviews Neuroscience*, **2**(2), 119–128.

Nestler, E., Barrot, M., & Self, D. (2001). ΔFosB: a sustained molecular switch for addiction. *Proceedings of the National Academy of Sciences*, **98**(20), 11042–11046.

Norton, L., & Layman, D. (2006). Leucine regulates translation initiation of protein synthesis in skeletal muscle after exercise. *The Journal of Nutrition*, **136**(2), 533S–537S.

Patton, J., Stanford, M., & Barratt, E. (1995). Factor structure of the Barratt impulsiveness scale. *Journal of Clinical Psychology*, **51**, 768–774.

Rivest, J., Bernier, J., & Pomar, C. (2000). A dynamic model of protein digestion in the small intestine of pigs. *Journal of Animal Science*, **78**, 328–240.

Robinson, T. & Berridge, K. (2001). Incentive-sensitization and addiction. *Addiction*, **96**(1), 103–114.

Schain, R., & Freedman, D. (1961). Studies on 5-hydroxyindole metabolism in autistic and other mentally retarded children. *The Journal of Pediatrics*, **58**, 315–320.

Schedl, H., Pierce, C., Rider, A., & Clifton, J. (1968). Absorption of L-methionine from the human small intestine. *Journal of Clinical Investigation*, **47**(2), 417–425.

Tang, G., Gudsnuk, K., Kuo, S., Cotrina, M., Rosoklija, G., Sosunov, A., *et al.* (2014). Loss of mTOR-dependent macroautophagy causes autistic-like synaptic pruning deficits. *Neuron*, **83**(5), 1482–1482.

Williams, K., Wheeler, D., Silove, N., & Hazell, P. (2013). Selective serotonin reuptake inhibitors (SSRIs) for autism spectrum disorders (ASD). *The Cochrane Database of Systematic Reviews*, **8**, CD004677.

32 Optogenetics in Deep Brain Stimulation: Ethical Considerations

Sabine Müller and Henrik Walter

32.1 Introduction

Although optogenetic deep brain stimulation (DBS) has not yet been tested in human beings, it can be considered as the next logical step in the development of DBS. DBS has been used to treat more than 100 000 patients suffering from advanced Parkinson's disease, and since 1999, it has been tested in several hundreds of psychiatric patients. It can be expected that optogenetics might improve DBS in two different ways: first, it might help to improve the selection of the DBS targets by illuminating disease mechanisms; second, optogenetic DBS might replace electrical DBS because it might be more efficient to activate or to selectively block those cell types whose malfunction underlies movement or psychiatric disorders.

Although this development is still in its infancy, a bioethical debate on optogenetics has already begun (Walter and Müller, 2013; Schleiermacher, 2013; Brewer and Nicolai, 2014; Gilbert *et al.*, 2014; Hess, 2014; Maslen and Savulescu, 2014; Müller and Walter, 2014; Canli, 2015). The bioethical discussion has started so early because many modern medical ethicists think ahead instead of simply reacting to technological developments. Thinking in advance in biomedical ethics should not be misunderstood as speculative ethics (although this also exists). Rather, ethical investigations of emerging technologies should be based on empirical facts and should investigate the expectable risks and benefits for patients and society. Ethical analyses of emerging technologies should not hinder medical research through speculation about hypothetical risks that are based mainly on science fiction or animosity toward technology (Müller and Walter, 2014). In the very beginning of a new technology, the ethical evaluation can only be preliminary, because the empirical basis of the evaluation is still very thin.

32.2 Optogenetic Research for Improving Electrical DBS

DBS of the subthalamic nucleus (STN) reduces Parkinsonian symptoms. However, because DBS affects neurons as well as afferent axons, it was unclear how DBS causes these effects. Optogenetics research has illuminated a potential mechanism of DBS: the therapeutic effect of DBS on Parkinsonism might be caused by stimulation-induced alterations of the activity of the pathway between the M1 motor cortical region and the STN. Furthermore, this optogenetic research points to other methods of therapeutic stimulation, such as direct cortical stimulation (LaLumiere, 2011).

When optogenetics research explains the mechanisms underlying DBS, it might contribute to improving DBS, as well as transcranial magnetic stimulation (TMS). In particular, it might help with exploring the parameters and sites of stimulation that are necessary for symptomatic relief (LaLumiere, 2011).

32.3 Optogenetic DBS

High-frequency DBS (usually ~130 Hz) has been considered to be a method that creates temporary lesions by inhibiting the targeted area through an electrical current. In DBS, all cells of a given region are subjected to the effects of the electrical current. Therefore, electrical DBS has two major drawbacks, namely the electrophysiological non-specificity and the biological non-specificity. High-frequency DBS yields a mostly unpredictable mixed pattern of inhibition of cell somata and activation of axons, which can yield opposite effects. Furthermore, within the target area, multiple neurons with different biochemical characteristics are addressed in the same way. Electrical DBS simultaneously inhibits activating glutamatergic projection neurons and inhibiting GABAergic interneurons, which counteract each other. Thus, the net effect of inhibition is reduced (Sturm, 2013). In contrast, optogenetic DBS is expected to allow for either inhibiting or activating specifically targeted neuron types by choosing light colors (i.e. by stimulating with different wavelengths) (Pastrana, 2011; Deisseroth, 2012; Sturm, 2013).

An optogenetic DBS system would first require the injection of a virus that carries the opsin gene into the cells of interest. Then, laser light has to be delivered into the brain through an optical fiber using a chronically implanted cannula that is affixed to the skull (Zhang *et al.*, 2010; Pama *et al.*, 2013).

Additionally, optogenetic DBS might allow an electrical recording of brain activity during stimulation, since the latter does not interfere with electrical recording (Pastrana, 2011). This might support the development of closed-loop DBS systems, which apply a current only when needed.

A potential further development of optogenetic DBS would be a real-time feedback system, in which the opsin genes turn themselves off once the membrane expression reaches an effective level or once an electrode detects adequate functioning in order to prevent crowding endogenous receptors. This might be supported by a built-in regulatory sequence on the vector that can be acted upon with a drug (Gradinaru, 2013).

An optogenetic DBS system that also collects brain activity data would allow for the storing of these data inside and/or outside the patient (e.g. on the physician's laptop or on the clinic's server) (Müller and Walter, 2014). This might support the fine tuning of the stimulation.

Optogenetic DBS might have several clinical advantages: first, it will be more efficient than electrical stimulation, because the opposite effects of inhibition and activation would not occur (Gradinaru, 2013; Sturm, 2013); second, optogenetic DBS might allow us to specifically address the pathogenesis of neurological or psychiatric disorders (Sturm, 2013); and third, the incidence of unintended (psychiatric) side effects might be reduced if the sorts of neurons that are involved in a certain neurological or psychiatric disorder are exclusively manipulated (LaLumiere, 2011; Müller and Walter, 2014).

To date, optogenetic DBS has not been tested in human beings. However, animal research looks promising: Kravitz *et al.* (2010) successfully treated Parkinsonian mice through optogenetic DBS by directly activating the basal ganglia circuitry. This completely rescued deficits in freezing, bradykinesia and locomotor initiation. Yoon *et al.* (2016) reduced levodopa-induced dyskinesias in Parkinsonian rats by optogenetic inhibition of the STN. Covington *et al.* (2010) performed optogenetic stimulation of the medial prefrontal cortex in mice, which effectively reduced social withdrawal and anhedonia. Chen *et al.* (2013) showed that optogenetic prelimbic cortex stimulation significantly prevented cocaine-addicted rats from compulsive cocaine seeking. Liu *et al.* (2012) reactivated a fearful memory and even implanted a false memory in mice by reactivating those hippocampus neurons that had encoded the fearful memory.

The first experiments in primates utilizing optogenetic stimulation in the sensorimotor domain showed clear excitatory and inhibitory effects as measured by electrical recordings, but they did not evoke movements (Diester *et al.*, 2011). Cavanaugh *et al.* (2012) successfully induced deficits in saccadic eye movement by optogenetic inactivation of superior colliculus neurons in monkeys.

32.4 Future Clinical Applications of Optogenetic DBS

Presently, we cannot say whether optogenetic DBS will ever be used in humans. Recently, *Nature Neuroscience* asked several optogenetics experts whether they felt that optogenetics would ever become a clinical tool for treating human diseases (Adamantidis *et al.*, 2015). The experts were generally more optimistic with regard to treating the retina than other parts of the brain. Ivan Soltesz argued that the brain regions located closest to the dura are expected to be the first candidates for applications, particularly for treating epilepsy. However, Anatol Kreitzer and Christian Lüscher do not expect applications of optogenetics for treating brain diseases in the next 10 years. There are many obstacles to overcome, such as cell type-specific targeting, stability of opsin expression, viral long-term toxicity (Christian Lüscher), the uniform delivery of light over broad areas deep in the human brain and the risk that optogenetic proteins cause inflammation or disruption of cellular processes (Anatol Kreitzer). Opsin expression is problematic because viral vectors typically do not induce very stable expression as the protein continues to accumulate; accumulation might jeopardize the natural cellular machine.

Christian Lüscher does not support optogenetic DBS, but rather wants to develop optogenetically inspired DBS protocols (Adamantidis *et al.*, 2015). In the same spirit, Antonio Bonci does not propose developing optogenetic DBS, but rather using repetitive TMS (rTMS) and DBS based on knowledge gained in optogenetics studies. His enthusiasm came from reactions to a paper of his group published in *Nature* in 2013 (Chen *et al.*, 2013), which described how compulsively cocaine-seeking rats were successfully treated through optogenetic prelimbic cortex stimulation. This animal study led directly to a clinical study with rTMS of the frontal cortex in drug-addicted patients (Adamantidis *et al.*, 2015).

Whether or not a combination of optogenetics and DBS will ever make it to the clinic remains to be seen, but, if it does, we expect that it will be used as a first step to improving electrical DBS. In a second step, optogenetic DBS might even replace electrical DBS, provided that its safety challenges can be adequately addressed.

32.5 Ethical Evaluation

In order to ethically evaluate a new medical technology, firstly its benefits, risks and adverse effects on patients have to be considered.

Benefits for patients can consist of improvements to the quality of life and/or the prolongation of (quality-adjusted) life expectancy. To be beneficial, a given therapy needs to have clinically relevant effectiveness, sustainability of effect and the non-existence of a less noxious therapy. The justification for developing optogenetic DBS must at least make plausible that a clear benefit compared to existing therapies can be expected.

Presently, the clinical efficacy of optogenetic DBS cannot be evaluated, since it has not yet been tested in humans. However, theoretical reasons speak to optogenetic DBS having both a greater efficacy and fewer adverse effects on mood, cognition, personality or behavior as compared to electrical DBS. This would be a strong ethical argument for its further development and for clinical trials.

The risks and adverse effects of a therapy can only be justified if the benefit would override the harm. Several risks of optogenetic DBS should be addressed before it is tested in humans. Additionally, its clinical benefit is still unclear.

Probably the greatest challenges are opsin delivery and opsin tolerability. Opsin delivery is associated with all of the challenges of gene therapy. Opsin tolerability depends firstly on whether opsins will trigger the immune system, and secondly on whether they compromise the integrity of the membrane in which they are expressed (Gradinaru, 2013). Additional safety issues relate to whether opsins at the required concentration have any toxic effects and whether activated opsins have any systemic adverse effects on the brain (Walter and Müller, 2013).

The important issue of immunoreactions has not yet been explored in the human brain. Research in rodents cannot answer this question, because their immune systems are different from those of humans. In order to investigate this issue, experiments in either non-human primates or in human volunteers will have to be performed. Both alternatives are ethically highly problematic. According to the content of the recent brain initiatives of Japan and China, which focus on non-human primate models of human disease (Okano *et al.*, 2015), it can be reasonably speculated that such research might be conducted in these Asian countries rather than in European countries or the USA.

Further challenges are related to the placement of the light device, light penetration and heating (Gradinaru, 2013). Furthermore, the risks of brain surgery and implantation of the device have to be considered, which are well known from electrical DBS.

A safety study in monkeys has assessed the safety of two viral vector systems, two human promotors and three opsins. Although histological work-up confirmed high and well-tolerated expression, there were hints of aggregations that might have negative effects on cell health. Furthermore, these researchers found cortical damage caused by the optrodes (Diester *et al.*, 2011).

Therefore, before optogenetic DBS can be tested in clinical trials, more research is needed in order to obtain a sufficient level of safety.

The ethical principle of respect for the patient's autonomy demands that physicians respect the patient's autonomous decisions about medical therapies and support their capacity for autonomous decision-making. Only patients with a sufficient capacity for autonomy can give valid informed consent. If optogenetic DBS would be tested as an alternative to electrical DBS in Parkinsonian and psychiatric patients, their capacity for informed consent is of crucial importance, because many neurological and mental disorders can reduce the capacity for autonomy. However, for these patients, the capacity to consent is not denied generally, but has to be assessed individually.

A particularly careful investigation of the capacity to consent is necessary in the case of new interventions whose efficacy, risks, rates of adverse effects and long-term consequences are not yet known. Furthermore, it is necessary to protect patients from the so-called therapeutic misconception (i.e. the subjects of a clinical study fail to adequately recognize the key differences between treatment and clinical research) (Lidz *et al.*, 2004). On the other hand, respect for autonomy implies respecting the autonomous decisions of patients who want to undergo new and risky treatments after being informed comprehensively.

Of particular ethical relevance are the effects of optogenetic DBS on the minds and personalities of patients. This concern stems from observations in the practice of electrical DBS in neurology and psychiatry, where unwanted personality changes occur in a small fraction of patients. Although DBS-induced personality changes are not generally ethically problematic, they clearly are problematic if they either cause suffering or occur against the patient's will. Furthermore, the possibility of external control of the mind raises the fundamental concern that it might allow for mind control. Indeed, under electrical DBS, some patients develop a feeling of self-alienation, since their mood, personality and behavior become the objects of the decisions of the physicians (Walter and Müller, 2013; Müller and Walter, 2014).

Finally, the possibility of storing brain activity data raises severe ethical issues with regard to data security, data safety and mental privacy. These problems should be considered before the first clinical trials with optogenetics-based DBS are started (Müller and Walter, 2014).

Since at present neither the risks nor the benefits of optogenetic DBS can be specified sufficiently, further basic research is required. We plead for a mandatory register of all clinical interventions and clinical trials in optogenetic DBS, since otherwise the same publication bias will inevitably occur as in studies with electrical DBS (Walter and Müller, 2013).

REFERENCES

Adamantidis, A., Arber, S., Bains, J. S., Bamberg, E., Bonci, A., Buzsáki, G., *et al.* 2015. Optogenetics: 10 years after ChR2 in neurons – views from the community. *Nature Neuroscience*, **18**(9): 1202–1212.

Brewer, C. D. & Nicolai, E., 2014. Reassessing the ethical importance of efficacy and autonomy in optogenetics trials. *AJOB Neuroscience*, **5**(3): 16–18.

Canli, T., 2015. Neurogenethics: an emerging discipline at the intersection of ethics, neuroscience, and genomics. *Applied & Translational Genomics*, **5**: 18–22.

Cavanaugh, J., Monosov, I. E., McAlonan, K., Berman, R., Smith, M. K., Cao, V., *et al.* 2012. Optogenetic inactivation modifies monkey visuomotor behavior. *Neuron*, **76**(5): 901–907.

Chen, B. T., Yau H.-J., Hatch, C., Kusumoto-Yoshida, I, Cho, S. L., Hopf, W., *et al.* 2013. Rescuing cocaine-induced prefrontal cortex hypoactivity prevents compulsive cocaine seeking. *Nature*, **496**: 359–364.

Covington, H. E., Lobo, M. K., Maze, I., Vialou, V., Hyman, J. M., Zaman, S., *et al.* 2010. Antidepressant effect of optogenetic stimulation of the medial prefrontal cortex. *Journal of Neuroscience*, 30(48): 16082–16090.

Deisseroth, K., 2012. Optogenetics and psychiatry: applications, challenges, and opportunities. *Biological Psychiatry*, **71**: 1030–1032.

Diester, I., Kaufman, M. T., Mogri, M., Pashaie, R., Goo, W., Yizhar, O., *et al.* 2011. An optogenetic tool box designed for primates. *Nature Neuroscience*, **14**(3): 387–397.

Gilbert, F., Harris A. R., & Kapsa, R. M. I. 2014. Controlling brain cells with light: ethical considerations for optogenetic clinical trials. *AJOB Neuroscience* **5**(3): 3–11.

Gradinaru, V. 2013. Optogenetics to benefit human health: opportunities and challenges. In: Hegemann, P. & Sigrist, S. (Eds.): *Optogenetics*. Berlin & Boston: De Gruyter: 127–131.

Hess, P., 2014. The ethical dilemmas of experimental invasive brain technology. *AJOB Neuroscience*, **5**(3): 18–20.

Kravitz, A. V., Freeze, B. S., Parker, P. R., Kay, K., Thwin, M. T., Deisseroth, K., *et al.* 2010. Regulation of parkinsonian motor behaviours by optogenetic control of basalganglia circuitry. *Nature*, **466**: 622–626.

LaLumiere, R. T., 2011. A new technique for controlling the brain: optogenetics and its potential for the use in research and the clinic. *Brain Stimulation*, **4**: 1–6.

Lidz, C. W., Appelbaum, P. S., Grisso, T., & Renaud, M. 2004. Therapeutic misconception and the appreciation of risks in clinical trials. *Social Science & Medicine*, **58**: 1689–1697.

Liu, X., Ramirez, S., Pang, P. T., Puryear, C. B., Govindarajan, A., Deisseroth, K., *et al.* 2012. Optogenetic stimulation of a hippocampal engram activates fear memory recall. *Nature*, **484**: 381–385.

Maslen, H. & Savulescu, J., 2014. First Phase 1 optogenetics trials should be conducted in people who are dying. *AJOB Neuroscience*, **5**(3): 16–18.

Müller, S. & Walter, H., 2014. Neither speculative nor narrow-minded ethics is needed for optogenetics-based DBS in psychiatry and neurology. *AJOB Neuroscience*, **5**(3): 12–20.

Okano, H., Miyawaki, A., & Kasai, K., 2015. Brain/MINDS: brain-mapping project in Japan. *Phil. Trans. R. Soc. B*, **370**: 20140310.

Pama, E. A. C., Colzato, L. S., & Hommel, B., 2013. Optogenetics as a neuromodulation tool in cognitive neuroscience. *Frontiers in Psychology*, **4**: 610.

Pastrana, E., 2011. Nature methods primer: optogenetics: controlling cell function with light. *Nature Methods*, **8**(1): 24–25.

Schleiermacher, S., 2013. History in the making: the ethics of optogenetics. In: Hegemann, P. & Sigrist, S. (Eds.): *Optogenetics*. Berlin & Boston: De Gruyter: 199–200.

Sturm, V., 2013. Potential of optogenetics in deep brain stimulation. In: Hegemann, P. & Sigrist, S. (Eds.): *Optogenetics*. Berlin & Boston: De Gruyter: 157–160.

Yoon, H. H., Min, J., Hwang, E., Lee, C. J., Suh, J.-K., Hwang, O., *et al.* 2016. Optogenetic inhibition of the subthalamic nucleus reduces levodopa-induced dyskinesias in a rat model of Parkinson's disease. *Stereotactic and Functional Neurosurgery*, **94**: 41–53.

Walter, H. & Müller, S., 2013. Optogenetics as a new therapeutic tool in medicine? A view from the principles of biomedical ethics. In: Hegemann, P. & Sigrist, S. (Eds.): *Optogenetics*. Berlin & Boston: De Gruyter: 201–211.

Zhang, F., Gradinaru, V., Adamantidis, A. R., Durand, R., Airan, R. D., de Lecea, L., *et al.* 2010. Optogenetic interrogation of neural circuits: technology for probing mammalian brain structures. *Nature Protocols*, **5**: 439–456.

Index